T0342350

Impulsive and Hybrid Dynamical Systems

PRINCETON SERIES IN APPLIED MATHEMATICS

Edited by

Ingrid Daubechies, *Princeton University*
Weinan E, *Princeton University*
Jan Karel Lenstra, *Eindhoven University*
Endre Süli, *University of Oxford*

The Princeton Series in Applied Mathematics publishes high quality advanced texts and monographs in all areas of applied mathematics. Books include those of a theoretical and general nature as well as those dealing with the mathematics of specific applications areas and real-world situations.

Impulsive and Hybrid Dynamical Systems

Stability, Dissipativity, and Control

Wassim M. Haddad

VijaySekhar Chellaboina

Sergey G. Nersesov

PRINCETON UNIVERSITY PRESS

PRINCETON AND OXFORD

Published by Princeton University Press, 41 William Street, Princeton, New Jersey 08540

In the United Kingdom: Princeton University Press, 3 Market Place, Woodstock, Oxfordshire OX20 1SY

Library of Congress Cataloging-in-Publication Data

Haddad, Wassim M., 1961–
    Impulsive and hybrid dynamical systems : stability, dissipativity, and control.
/ Wassim M. Haddad, VijaySekhar Chellaboina, and Sergey G. Nersesov.
        p. cm. (Princeton series in applied mathematics)
    Includes bibliographical references and index.
    ISBN-13: 978-691-12715-6 (cl : alk. paper)
    ISBN-10: 0-691-12715-8 (cl : alk. paper)
    1. Automatic control. 2. Control theory. 3. Dynamics. 4. Discrete-time systems. I. Chellaboina, VijaySekhar, 1970– II. Nersesov, Sergey G., 1976– III. Title. IV. Series.

TJ213.H23 2006
003'.85—dc22   2005056496

British Library Cataloging-in-Publication Data is available

This book has been composed in Times Roman in LaTeX

The publisher would like to acknowledge the authors of this volume for providing the camera-ready copy from which this book was printed.

Printed on acid-free paper. ∞

pup.princeton.edu

Printed in the United States of America

10 9 8 7 6 5 4 3 2 1

The highest form of pure thought is in mathematics.

—Plato

History shows that those heads of empires who have encouraged the cultivation of mathematics, the common source of all exact sciences, are also those whose reigns have been the most brilliant and whose glory is the most durable.

—Michel Chasles

A scientist worthy of the name, above all a mathematician, experiences in his work the same impression as an artist; his pleasure is as great and of the same nature.

—Henri Poincaré

Mathematics, rightly viewed, possesses not only truth, but supreme beauty—a beauty ... sublimely pure, capable of a stern perfection such as only the greatest art can show.

—Bertrand Russell

There is no branch of mathematics, however abstract, which may not some day be applied to phenomena of the real world.

—Nikolai Lobachevsky

From the intrinsic evidence of his creation, the Great Architect of the Universe now begins to appear as a pure mathematician.

—Sir James Jeans

This, therefore, is mathematics: she reminds you of the invisible forms of the soul; she gives life to her own discoveries; she awakens the mind and purifies the intellect; she brings light to our intrinsic ideas; and she abolishes oblivion and ignorance which are ours by birth.

—Proclus

Contents

Preface

Dynamical systems theory holds the supreme position among all mathematical disciplines as it provides the foundation for unlocking many of the mysteries in nature and the universe which involve the evolution of time. Dynamical systems theory is used to study ecological systems, geological systems, biological systems, economic systems, pharmacological systems, physiological systems, neural systems, cognitive systems, and physical systems (e.g., mechanics, quantum mechanics, thermodynamics, fluids, magnetic fields, galaxies, etc.), to cite but a few examples. Many of these systems involve an interacting mixture of continuous and discrete dynamics exhibiting discontinuous flows on appropriate manifolds, and hence, give rise to *hybrid* dynamics. The increasingly complex nature of engineering systems involving controller architectures with real-time embedded software also gives rise to hybrid systems, wherein the continuous mathematics of the system dynamics and control interact with the discrete mathematics of logic and computer science.

Modern complex engineering systems additionally involve multiple modes of operation placing stringent demands on controller design and implementation of increasing complexity. Such systems typically possess a multiechelon hierarchical hybrid control architecture characterized by continuous-time dynamics at the lower levels of the hierarchy and discrete-time dynamics at the higher levels of the hierarchy. The lower-level units directly interact with the dynamical system to be controlled while the higher-level units receive information from the lower-level units as inputs and provide (possibly discrete) output commands which serve to coordinate and reconcile the (sometimes competing) actions of the lower-level units. The hierarchical controller organization reduces processor cost and controller complexity by breaking up the processing task into relatively small pieces and decomposing the fast and slow control functions. Typically, the higher-level units perform logical checks that determine system mode operation, while the lower-level units execute continuous-variable commands for a given system mode of operation. The ability of developing an analysis and control design framework for hybrid

dynamical systems is imperative in light of the increasingly complex
nature of dynamical systems which have interacting continuous-time
dynamics as well as discrete-event dynamics, such as advanced high
performance tactical fighter aircraft, variable-cycle gas turbine engines, air and ground transportation systems, and swarms of air and
space vehicles.

Hybrid dynamical systems is an emerging discipline within dynamical systems theory and control, and hence, the term *hybrid system* has
many meanings to different researchers and practitioners. We define
a hybrid dynamical system as an *interacting* countable collection of
dynamical systems involving a mixture of continuous-time dynamics
and discrete events that includes impulsive dynamical systems, hierarchical systems, and switching systems as special cases. In this
monograph we develop a unified analysis and control design framework for impulsive and hybrid dynamical systems using a Lyapunov
and dissipative systems approach. The monograph is written from a
system-theoretic point of view and can be viewed as a contribution
to mathematical system theory and control system theory. The material in this book is thus intended to complement the monographs
on qualitative analysis, asymptotic analysis, and stability analysis of
impulsive dynamical systems [12–14, 93, 148].

After a brief introduction on impulsive and hybrid dynamical systems in Chapter 1, fundamental stability theory for nonlinear impulsive dynamical systems is developed in Chapter 2. In Chapter 3, we
extend classical dissipativity theory to impulsive dynamical systems.
Chapter 4 provides a treatment of nonnegative and compartmental
impulsive dynamical systems. A detailed treatment of vector dissipativity theory for large-scale impulsive dynamical systems is given
in Chapter 5, while stability results for feedback interconnections of
impulsive dynamical systems are given in Chapter 6. In Chapters 7
and 8 we develop energy-based hybrid controllers for Euler-Lagrange,
port-controlled Hamiltonian, and dissipative dynamical systems. A
detailed treatment of optimal hybrid control is given in Chapter 9,
while Chapters 10 and 11 provide extensions to hybrid disturbance
rejection control and hybrid robust control, respectively. Next, Chapter 12 develops a unified dynamical systems framework for a general
class of systems possessing left-continuous flows which include hybrid,
impulsive, and switching systems as special cases. Finally, in Chapter 13 we generalize Poincaré's theorem to left-continuous dynamical
systems for analyzing the stability of periodic orbits of impulsive and
hybrid dynamical systems.

The first author would like to thank V. Lakshmikantham for his valuable discussions on impulsive differential equations over the recent years. In addition, the authors thank V. Lakshmikantham, Anthony N. Michel, and Tomohisa Hayakawa for their constructive comments and feedback. In some parts of the monograph we have relied on work we have done jointly with Dennis S. Bernstein, Sanjay P. Bhat, Qing Hui, Nataša A. Kablar, and Alexander V. Roup; it is a pleasure to acknowledge their contributions.

The results reported in this monograph were obtained at the School of Aerospace Engineering, Georgia Institute of Technology, Atlanta, the Department of Mechanical and Aerospace Engineering, University of Missouri, Columbia, and the Department of Mechanical, Aerospace, and Biomedical Engineering of the University of Tennessee, Knoxville, between June 1999 and July 2005. The research support provided by the Air Force Office of Scientific Research and the National Science Foundation over the years has been instrumental in allowing us to explore basic research topics that have led to some of the material in this monograph. We are indebted to them for their support.

Atlanta, Georgia, USA, March 2006, *Wassim M. Haddad*
Knoxville, Tennessee, USA, March 2006, *VijaySekhar Chellaboina*
Villanova, Pennsylvania, USA, March 2006, *Sergey G. Nersesov*

Chapter One

Introduction

1.1 Impulsive and Hybrid Dynamical Systems

Modern complex engineering systems are highly interconnected and mutually interdependent, both physically and through a multitude of information and communication network constraints. The complexity of modern controlled dynamical systems is further exacerbated by the use of hierarchical embedded control subsystems within the feedback control system, that is, abstract decision-making units performing logical checks that identify system mode operation and specify the continuous-variable subcontroller to be activated. These multiechelon systems (see Figure 1.1) are classified as *hybrid* systems (see [6, 126, 161] and the numerous references therein) and involve an *interacting* countable collection of dynamical systems possessing a hierarchical structure characterized by continuous-time dynamics at the lower-level units and logical decision-making units at the higher level of the hierarchy. The lower-level units directly interact with the dynamical system to be controlled, while the logical decision-making, higher-level units receive information from the lower-level units as inputs and provide (possibly discrete) output commands, which serve to coordinate and reconcile the (sometimes competing) actions of the lower-level units.

The hierarchical controller organization reduces processor cost and controller complexity by breaking up the processing task into relatively small pieces and decomposing the fast and slow control functions. Typically, the higher-level units perform logical checks that determine system mode operation, while the lower-level units execute continuous-variable commands for a given system mode of operation. Due to their multiechelon hierarchical structure, hybrid dynamical systems are capable of simultaneously exhibiting continuous-time dynamics, discrete-time dynamics, logic commands, discrete events, and resetting events. Such systems include dynamical switching systems [29, 101, 140], nonsmooth impact systems [28, 32], biological systems [93], sampled-data systems [71], discrete-event systems [139], intelligent vehicle/highway systems [113], constrained mechanical sys-

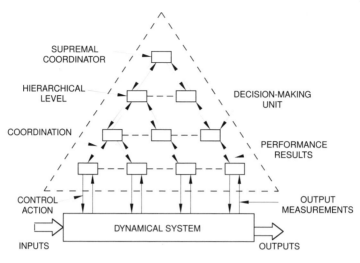

Figure 1.1 Multiechelon dynamical system [87].

tems [28], and flight control systems [158], to cite but a few examples.

The mathematical descriptions of many hybrid dynamical systems can be characterized by impulsive differential equations [12, 14, 79, 93, 148]. Impulsive dynamical systems can be viewed as a subclass of hybrid systems and consist of three elements—namely, a continuous-time differential equation, which governs the motion of the dynamical system between impulsive or resetting events; a difference equation, which governs the way the system states are instantaneously changed when a resetting event occurs; and a criterion for determining when the states of the system are to be reset. Since impulsive systems can involve impulses at variable times, they are in general time-varying systems, wherein the resetting events are both a function of time and the system's state. In the case where the resetting events are defined by a prescribed sequence of times which are independent of the system state, the equations are known as *time-dependent differential equations* [12, 14, 35, 61, 62, 93]. Alternatively, in the case where the resetting events are defined by a manifold in the state space that is independent of time, the equations are autonomous and are known as *state-dependent differential equations* [12, 14, 35, 61, 62, 93].

Hybrid and impulsive dynamical systems exhibit a very rich dynamical behavior. In particular, the trajectories of hybrid and impulsive dynamical systems can exhibit multiple complex phenomena such as *Zeno* solutions, *noncontinuability* of solutions or *deadlock*, *beating* or *livelock*, and *confluence* or merging of solutions. A Zeno solution involves a system trajectory with infinitely many resettings in finite time. Deadlock corresponds to a dynamical system state from which no continuation, continuous or discrete, is possible. A hybrid dynamical system experiences beating when the system trajectory encounters the same resetting surface a finite or infinite number of times in zero time. Finally, confluence involves system solutions that coincide after a certain point in time. These phenomena, along with the breakdown of many of the fundamental properties of classical dynamical system theory, such as continuity of solutions and continuous dependence of solutions on the system's initial conditions, make the analysis of hybrid and impulsive dynamical systems extremely challenging.

The range of applications of hybrid and impulsive dynamical systems is not limited to controlled dynamical systems. Their usage arises in several different fields of science, including computer science, mathematical programming, and modeling and simulation. In computer science, discrete program verification and logic is interwoven with a continuous environment giving rise to hybrid dynamical systems. Specifically, computer software systems interact with the physical system to admit feedback algorithms that improve system performance and system robustness. Alternatively, in mathematical linear and nonlinear optimization with inequality constraints, changes in continuous and discrete states can be computed by a switching dynamic framework. Modeling and simulating complex dynamical systems with multiple modes of operation involving multiple system transitions also give rise to hybrid dynamical systems. Among the earliest investigations of dynamical systems involving continuous dynamics and discrete switchings can be traced back to relay control systems and bang-bang optimal control.

Dynamical systems involving an interacting mixture of continuous and discrete dynamics abound in nature and are not limited to engineering systems with programmable logic controllers. Hybrid systems arise naturally in biology, physiology, pharmacology, economics, biocenology, demography, chemistry, neuroscience, impact mechanics, quantum mechanics, systems with shock effects, and cosmology, among numerous other fields of science. For example, mechanical systems subject to unilateral constraints on system positions give rise to hybrid dynamical systems. These systems involve discontinuous solutions, wherein discontinuities arise primarily from impacts (or collisions) when the system trajectories encounter the unilateral constraints. In physiological systems the blood pressure and blood flow to different tissues of the human body are controlled to provide sufficient oxygen to the cells of each organ. Certain organs such as the kidneys normally require higher blood flows than is necessary to satisfy basic oxygen needs. However, during stress (such as hemorrhage) when perfusion pressure falls, perfusion of certain regions (e.g., brain and heart) takes precedence over perfusion of other regions, and hierarchical controls (overriding controls) shut down flow to these other regions. This shutting down process can be modeled as a resetting event giving rise to a hybrid system. As another example, biomolecular genetic systems also combine discrete events, wherein a gene is turned on or off for transcription, with continuous dynamics involving concentrations of chemicals in a given cell. Even though many scientists and engineers recognize that a large number of life science and engineering systems are hybrid in nature, these systems have been traditionally modeled, analyzed, and designed as purely discrete or purely continuous systems. The reason for this is that only recently has the theory of impulsive and hybrid dynamical systems been sufficiently developed to fully capture the interaction between the continuous and discrete dynamics of these systems.

Even though impulsive dynamical systems were first formulated by Mil'man and Myshkis [123, 124],[1] the fundamental theory of impulsive differential equations is developed in the monographs by Bainov, Lakshmikantham, Perestyuk, Samoilenko, and Simeonov [12–14, 93, 148]. These monographs develop qualitative solution properties, existence of solutions, asymptotic properties of solutions, and stability theory of impulsive dynamical systems. In this monograph we build on the results of [12–14, 93, 148] to develop invariant set stability theorems, partial stability, Lagrange stability, boundedness and ultimate boundedness, dissipativity theory, vector dissipativity theory, energy-based hybrid control, optimal control, disturbance rejection control, and robust control for nonlinear impulsive and hybrid dynamical systems.

1.2 A Brief Outline of the Monograph

The main objective of this monograph is to develop a general analysis and control design framework for nonlinear impulsive and hybrid

---

[1]Mil'man and Myshkis were the first to develop qualitative analysis results for impulsive dynamical systems. However, work on impact and hybrid systems can be traced back to ancient Greek scientists and mathematicians such as Aristotle, Archimedes, and Heron. Problems related to Heron's work on hybrid automata (Περί αὐτοματοποιητικῆς) as well as problems on impact dynamics attracted the interest of numerous physicists and mathematicians who followed with relevant contributions made in the last three centuries. Notable contributions include the work of Leibniz, Newton, (Jacob) Bernoulli, d'Alembert, Poisson, Huygens, Coriolis, Darboux, Routh, Appell, and Lyapunov.

dynamical systems. The main contents of the monograph are as follows. In Chapter 2, we establish notation and definitions, and develop stability theory for nonlinear impulsive dynamical systems. Specifically, Lyapunov stability theorems are developed for time-dependent and state-dependent impulsive dynamical systems. Furthermore, we state and prove a fundamental result on positive limit sets for state-dependent impulsive dynamical systems. Using this result, we generalize the Krasovskii-LaSalle invariant set theorem to impulsive dynamical systems. In addition, partial stability, Lagrange stability, boundedness, ultimate boundedness, and stability theorems via vector Lyapunov functions are also established.

In Chapter 3, we extend the notion of dissipative dynamical systems [165,166] to develop the concept of dissipativity for impulsive dynamical systems. Specifically, the classical concepts of system storage functions and system supply rates are extended to impulsive dynamical systems. In addition, we develop extended Kalman-Yakubovitch-Popov conditions in terms of the hybrid system dynamics for characterizing dissipativeness via system storage functions and hybrid supply rates for impulsive dynamical systems. Furthermore, a generalized hybrid energy balance interpretation involving the system's stored or accumulated energy, dissipated energy over the continuous-time dynamics, and dissipated energy at the resetting instants is given. Specialization of these results to passive and nonexpansive impulsive systems is also provided. In Chapter 4, we extend the results of Chapters 2 and 3 to develop stability and dissipativity results for impulsive nonnegative and compartmental dynamical systems.

In Chapter 5, we develop vector dissipativity notions for large-scale nonlinear impulsive dynamical systems. In particular, we introduce a generalized definition of dissipativity for large-scale nonlinear impulsive dynamical systems in terms of a hybrid vector inequality, a vector hybrid supply rate, and a vector storage function. Dissipativity properties of the large-scale impulsive system are shown to be determined from the dissipativity properties of the individual impulsive subsystems making up the large-scale system and the nature of the system interconnections. Using the concepts of dissipativity and vector dissipativity, in Chapter 6 we develop feedback interconnection stability results for impulsive nonlinear dynamical systems. General stability criteria are given for Lyapunov, asymptotic, and exponential stability of feedback impulsive dynamical systems. In the case of quadratic hybrid supply rates corresponding to net system power and weighted input-output energy, these results generalize the positivity and small gain theorems to the case of nonlinear impulsive dynamical systems.

In Chapter 7, we develop a hybrid control framework for impulsive port-controlled Hamiltonian systems. In particular, we obtain constructive sufficient conditions for hybrid feedback stabilization that provide a shaped energy function for the closed-loop system while preserving a hybrid Hamiltonian structure at the closed-loop level. A novel class of energy-based hybrid controllers is proposed in Chapter 8 as a means for achieving enhanced energy dissipation in Euler-Lagrange, port-controlled Hamiltonian, and dissipative dynamical systems. These controllers combine a logical switching architecture with continuous dynamics to guarantee that the system plant energy is strictly decreasing across resetting events. The general framework leads to closed-loop systems described by impulsive differential equations. In addition, we construct hybrid controllers that guarantee that the closed-loop system is consistent with basic thermodynamic principles. In particular, the existence of an entropy function for the closed-loop system is established that satisfies a hybrid Clausius-type inequality. Extensions to hybrid Euler-Lagrange systems and impulsive dynamical systems are also developed.

In Chapter 9, a unified framework for hybrid feedback optimal and inverse optimal control involving a hybrid nonlinear nonquadratic performance functional is developed. It is shown that the hybrid cost functional can be evaluated in closed form as long as the cost functional considered is related in a specific way to an underlying Lyapunov function that guarantees asymptotic stability of the nonlinear closed-loop impulsive system. Furthermore, the Lyapunov function is shown to be a solution of a steady-state, hybrid Hamilton-Jacobi-Bellman equation. Extensions of the hybrid feedback optimal control framework to disturbance rejection control and robust control are addressed in Chapters 10 and 11, respectively.

In Chapter 12, we develop a unified dynamical systems framework for a general class of systems possessing left-continuous flows, that is, left-continuous dynamical systems. These systems are shown to generalize virtually all existing notions of dynamical systems and include hybrid, impulsive, and switching dynamical systems as special cases. Furthermore, we generalize dissipativity, passivity, and nonexpansivity theory to left-continuous dynamical systems. Specifically, the classical concepts of system storage functions and supply rates are extended to left-continuous dynamical systems providing a generalized hybrid system energy interpretation in terms of stored energy, dissipated energy over the continuous-time dynamics, and dissipated energy over the resetting events. Finally, the generalized dissipativity notions are used to develop general stability criteria for feedback interconnections of left-continuous dynamical systems. These results generalize the positivity and small gain theorems to the case of left-continuous and hybrid dynamical systems.

Finally, in Chapter 13 we generalize Poincaré's theorem to dynamical systems possessing left-continuous flows to address the stability of limit cycles and periodic orbits of left-continuous, hybrid, and impulsive dynamical systems. It is shown that the resetting manifold provides a natural hyperplane for defining a Poincaré return map. In the special case of impulsive dynamical systems, we show that the Poincaré map replaces an nth-order impulsive dynamical system by an $(n-1)$th-order discrete-time system for analyzing the stability of periodic orbits.

Chapter Two

Stability Theory for Nonlinear Impulsive Dynamical Systems

2.1 Introduction

One of the most basic issues in system theory is stability of dynamical systems. System stability is characterized by analyzing the response of a dynamical system to small perturbations in the system states. Specifically, an equilibrium point of a dynamical system is said to be *stable* if, for small values of initial disturbances, the perturbed motion remains in an arbitrarily prescribed small region of the state space. More precisely, stability is equivalent to continuity of solutions as a function of the system initial conditions over a neighborhood of the equilibrium point uniformly in time. If, in addition, all solutions of the dynamical system approach the equilibrium point for large values of time, then the equilibrium point is said to be *asymptotically stable*.

The most complete contribution to the stability analysis of nonlinear dynamical systems was introduced in the late nineteenth century by the Russian mathematician Alexandr Mikhailovich Lyapunov in his seminal work entitled *The General Problem of the Stability of Motion* [110–112]. Lyapunov's results, which include the direct and indirect methods, along with the Krasovskii-LaSalle invariance principle [15, 91, 98, 99], provide a powerful framework for analyzing the stability of nonlinear dynamical systems as well as designing feedback controllers which guarantee closed-loop system stability. Lyapunov's *direct method* states that if a positive-definite function of the states of a given dynamical system can be constructed for which its time rate of change due to perturbations in a neighborhood of the system's equilibrium is always negative or zero, then the system's equilibrium point is stable or, equivalently, *Lyapunov stable*. Alternatively, if the time rate of change of the positive definite function is strictly negative, then the system's equilibrium point is asymptotically stable.

In light of the increasingly complex nature of dynamical system analysis and design, such as nonsmooth impact systems [28, 32], biological systems [93], demographic systems [106], hybrid systems [30,

169], sampled-data systems [71], discrete-event systems [139], systems with shock effects, and feedback systems with impulsive or resetting controls [35, 61, 62], dynamical systems exhibiting discontinuous flows on appropriate manifolds arise naturally. The mathematical descriptions of such systems can be characterized by impulsive differential equations [12, 14, 79, 93, 148]. To analyze the stability of dynamical systems with impulsive effects, Lyapunov stability results have been presented in the literature [12, 92–95, 105, 148, 153, 154, 170]. In particular, local and global asymptotic stability conclusions of an equilibrium point of a given impulsive dynamical system are provided if a smooth (at least continuously differentiable) positive-definite function of the nonlinear system states (Lyapunov function) can be constructed for which its time rate of change over the continuous-time dynamics is strictly negative and its difference across the resetting times is negative. However, unlike dynamical systems possessing continuous flows, Krasovskii-LaSalle-type invariant set stability theorems [15, 91, 98, 99] have not been addressed for impulsive dynamical systems. This is in spite of the fact that systems theory with impulsive effects has dominated the Russian and Eastern European literature [12, 14, 79, 92–95, 148, 153, 154]. There appear to be (at least) two reasons for this state of affairs, namely, solutions of impulsive dynamical systems are *not* continuous in time and are *not* continuous functions of the system's initial conditions, which are two key properties needed to establish invariance of positive limit sets, and hence an invariance principle.

In this chapter, we develop Lyapunov and invariant set stability theorems for nonlinear impulsive dynamical systems. In particular, invariant set theorems are derived, wherein system trajectories converge to the largest invariant set of Lyapunov level surfaces of the impulsive dynamical system. For state-dependent impulsive dynamical systems with continuously differentiable Lyapunov functions defined on a compact positively invariant set (with respect to the nonlinear impulsive system), the largest invariant set is contained in a hybrid level surface composed of a union involving vanishing Lyapunov derivatives and differences of the system dynamics over the continuous-time trajectories and the resetting instants, respectively. In addition, if the Lyapunov derivative along the continuous-time system trajectories is negative semidefinite and no system trajectories can stay indefinitely at points where the function's derivative or difference identically vanishes, then the system's equilibrium is asymptotically stable. These results provide less conservative conditions for examining the stability of state-dependent impulsive dynamical systems as compared to the classical results presented in [12,93,148,153,170]. In addition, partial stability, Lagrange stability, boundedness, ultimate boundedness, stability of time-dependent impulsive dynamical systems, and stability theorems via vector Lyapunov functions are also established.

2.2 Nonlinear Impulsive Dynamical Systems

In this section, we develop notation and introduce some basic properties of impulsive dynamical systems [12,14,79,92–95,105,148,153,154]. The notation used in this monograph is fairly standard. Specifically, $\mathbb{R}$ denotes the set of real numbers, $\overline{\mathbb{Z}}_+$ denotes the set of nonnegative integers, $\mathbb{Z}_+$ denotes the set of positive integers, $\mathbb{R}^n$ denotes the set of $n \times 1$ column vectors, $\mathbb{R}^{n \times m}$ denotes the set of $n \times m$ real matrices, $\mathbb{S}^n$ denotes the set of $n \times n$ symmetric matrices, $\mathbb{N}^n$ denotes the set of $n \times n$ nonnegative-definite matrices, $\mathbb{P}^n$ denotes the set of $n \times n$ positive-definite matrices, $(\cdot)^{\mathrm{T}}$ denotes transpose, $(\cdot)^{\#}$ denotes group generalized inverse, and I_n or I denotes the $n \times n$ identity matrix. Furthermore, $\mathcal{L}_2$ denotes the space of square-integrable Lebesgue measurable functions on $[0, \infty)$ and ℓ_2 denotes the space of square-summable sequences on $\overline{\mathbb{Z}}_+$. In addition, we denote the boundary, the interior, and the closure of the set $\mathcal{S}$ by $\partial \mathcal{S}$, $\overset{\circ}{\mathcal{S}}$, and $\overline{\mathcal{S}}$, respectively.

We write $\| \cdot \|$ for the Euclidean vector norm, $\mathcal{R}(M)$ and $\mathcal{N}(M)$ for the range space and the null space of a matrix M, $\mathrm{spec}(M)$ for the spectrum of the square matrix M, $\mathrm{ind}(M)$ for the index of M (that is, the size of the largest Jordan block of M associated with $\lambda = 0$, where $\lambda \in \mathrm{spec}(M)$), $\otimes$ for the Kronecker product, and $\oplus$ for the Kronecker sum. Furthermore, we write $V'(x)$ for the Fréchet derivative of V at x, $\mathcal{B}_\varepsilon(\alpha)$, $\alpha \in \mathbb{R}^n$, $\varepsilon > 0$, for the open ball centered at α with radius ε, $M \geq 0$ (respectively, $M > 0$) to denote the fact that the Hermitian matrix M is nonnegative (respectively, positive) definite, inf to denote infimum (that is, the greatest lower bound), sup to denote supremum (that is, the least upper bound), and $x(t) \to \mathcal{M}$ as $t \to \infty$ to denote that $x(t)$ approaches the set $\mathcal{M}$ (that is, for each $\varepsilon > 0$ there exists $T > 0$ such that $\mathrm{dist}(x(t), \mathcal{M}) < \varepsilon$ for all $t > T$, where $\mathrm{dist}(p, \mathcal{M}) \triangleq \inf_{x \in \mathcal{M}} \|p - x\|$). Finally, the notions of openness, convergence, continuity, and compactness that we use throughout the monograph refer to the topology generated on $\mathbb{R}^n$ by the norm $\| \cdot \|$.

As discussed in Chapter 1, an impulsive dynamical system consists of three elements:

i) a continuous-time dynamical equation, which governs the motion of the system between resetting events;

   ii) a difference equation, which governs the way the states are instantaneously changed when a resetting event occurs; and

   iii) a criterion for determining when the states of the system are to be reset.

Thus, an impulsive dynamical system has the form

$$\dot{x}(t) = f_{\mathrm{c}}(x(t)), \quad x(0) = x_0, \quad (t, x(t)) \notin \mathcal{S}, \tag{2.1}$$

$$\Delta x(t) = f_{\mathrm{d}}(x(t)), \quad (t, x(t)) \in \mathcal{S}, \tag{2.2}$$

where $t \geq 0$, $x(t) \in \mathcal{D} \subseteq \mathbb{R}^n$, $\mathcal{D}$ is an open set with $0 \in \mathcal{D}$, $\Delta x(t) \triangleq x(t^+) - x(t)$, where $x(t^+) \triangleq x(t) + f_{\mathrm{d}}(x(t)) = \lim_{\varepsilon \to 0} x(t+\varepsilon)$, $x(t) \in \mathcal{Z}$, $f_{\mathrm{c}} : \mathcal{D} \to \mathbb{R}^n$ is continuous, $f_{\mathrm{d}} : \mathcal{S} \to \mathbb{R}^n$ is continuous, and $\mathcal{S} \subset [0, \infty) \times \mathcal{D}$ is the *resetting set*. A function $x : \mathcal{I}_{x_0} \to \mathcal{D}$ is a *solution* to the impulsive dynamical system (2.1) and (2.2) on the interval $\mathcal{I}_{x_0} \subseteq \mathbb{R}$ with initial condition $x(0) = x_0$, if $x(\cdot)$ is left-continuous and $x(t)$ satisfies (2.1) and (2.2) for all $t \in \mathcal{I}_{x_0}$.

We assume that the continuous-time dynamics $f_{\mathrm{c}}(\cdot)$ are such that the solution to (2.1) is jointly continuous in t and x_0 between resetting events. A sufficient condition ensuring this is Lipschitz continuity of $f_{\mathrm{c}}(\cdot)$. Alternatively, uniqueness of solutions in forward time along with the continuity of $f_{\mathrm{c}}(\cdot)$ ensure that solutions to (2.1) between resetting events are continuous functions of the initial conditions $x_0 \in \mathcal{D}$ even when $f_{\mathrm{c}}(\cdot)$ is not Lipschitz continuous on $\mathcal{D}$ (see [41, Theorem 4.3, p. 59]). More generally, $f_{\mathrm{c}}(\cdot)$ need not be continuous. In particular, if $f_{\mathrm{c}}(\cdot)$ is discontinuous but bounded and $x(\cdot)$ is the unique solution to (2.1) between resetting events in the sense of Filippov [44], then continuous dependence of solutions between resetting events with respect to the initial conditions hold [44]. We refer to the differential equation (2.1) as the *continuous-time dynamics*, and we refer to the difference equation (2.2) as the *resetting law*. In addition, we use the notation $s(t, \tau, x_0)$ to denote the solution $x(t)$ of (2.1) and (2.2) at time $t \geq \tau$ with initial condition $x(\tau) = x_0$. Finally, a point $x_{\mathrm{e}} \in \mathcal{D}$ is an *equilibrium point* of (2.1) and (2.2) if and only if $s(t, \tau, x_{\mathrm{e}}) = x_{\mathrm{e}}$ for all $\tau \geq 0$ and $t \geq \tau$. Note that $x_{\mathrm{e}} \in \mathcal{D}$ is an equilibrium point of (2.1) and (2.2) if and only if $f_{\mathrm{c}}(x_{\mathrm{e}}) = 0$ and $f_{\mathrm{d}}(x_{\mathrm{e}}) = 0$.

For a particular trajectory $x(t)$, we let t_k denote the kth instant of time at which $(t, x(t))$ intersects $\mathcal{S}$, and we call the times t_k the *resetting times*. Thus, the trajectory of the system (2.1) and (2.2) from the initial condition $x(0) = x_0$ is given by $\psi(t, 0, x_0)$ for $0 < t \leq t_1$, where $\psi(t, 0, x_0)$ denotes the solution to continuous-time dynamics (2.1). If and when the trajectory reaches a state $x_1 \triangleq x(t_1)$ satisfying $(t_1, x_1) \in \mathcal{S}$, then the state is instantaneously transferred to $x_1^+ \triangleq x_1 + f_{\mathrm{d}}(x_1)$, according to the resetting law (2.2). The trajectory $x(t)$, $t_1 < t \leq t_2$, is then given by $\psi(t, t_1, x_1^+)$, and so on. Note that the solution $x(t)$ of (2.1) and (2.2) is left-continuous, that is, it is continuous everywhere except at the resetting times t_k, and

$$x_k \triangleq x(t_k) = \lim_{\varepsilon \to 0^+} x(t_k - \varepsilon), \tag{2.3}$$

$$x_k^+ \triangleq x(t_k) + f_{\mathrm{d}}(x(t_k)) = \lim_{\varepsilon \to 0^+} x(t_k + \varepsilon), \tag{2.4}$$

for $k = 1, 2, \ldots$.

To ensure the well-posedness of the resetting times we make the following additional assumptions:

A1. If $(t, x(t)) \in \overline{\mathcal{S}} \backslash \mathcal{S}$, then there exists $\varepsilon > 0$ such that, for all $0 < \delta < \varepsilon$,

$$\psi(t + \delta, t, x(t)) \notin \mathcal{S}.$$

A2. If $(t_k, x(t_k)) \in \partial \mathcal{S} \cap \mathcal{S}$, then there exists $\varepsilon > 0$ such that, for all $0 \leq \delta < \varepsilon$,

$$\psi(t_k + \delta, t_k, x(t_k) + f_{\mathrm{d}}(x(t_k))) \notin \mathcal{S}.$$

Assumption A1 ensures that if a trajectory reaches the closure of $\mathcal{S}$ at a point that does not belong to $\mathcal{S}$, then the trajectory must be directed away from $\mathcal{S}$, that is, a trajectory cannot enter $\mathcal{S}$ through a point that belongs to the closure of $\mathcal{S}$ but not to $\mathcal{S}$. Furthermore, A2 ensures that when a trajectory intersects the resetting set $\mathcal{S}$, it instantaneously exits $\mathcal{S}$. Finally, we note that if $x_0 \in \mathcal{S}$ then the system initially resets to $x_0^+ = x_0 + f_{\mathrm{d}}(x_0) \notin \mathcal{S}$, which serves as the initial condition for continuous-time dynamics (2.1).

It follows from A2 that resetting removes the pair (t_k, x_k) from the resetting set $\mathcal{S}$. Thus, immediately after resetting occurs, the continuous-time dynamics (2.1), and not the resetting law (2.2), becomes the active element of the impulsive dynamical system. Furthermore, it follows from A1 and A2 that no trajectory can intersect the interior of $\mathcal{S}$. Specifically, it follows from A1 that a trajectory can only reach $\mathcal{S}$ through a point belonging to both $\mathcal{S}$ and its boundary. And, from A2, it follows that if a trajectory reaches a point in $\mathcal{S}$ that is on the boundary of $\mathcal{S}$, then the trajectory is instantaneously removed from $\mathcal{S}$. Since a continuous trajectory starting outside of $\mathcal{S}$ and intersecting the interior of $\mathcal{S}$ must first intersect the boundary of $\mathcal{S}$, it follows that no trajectory can reach the interior of $\mathcal{S}$.

To show that the resetting times t_k are well defined and distinct, assume $T = \inf\{t \in \overline{\mathbb{R}}_+ : \psi(t, 0, x_0) \in \mathcal{S}\} < \infty$. Now, *ad absurdum*, suppose t_1 is not well defined, that is, $\min\{t \in \overline{\mathbb{R}}_+ : \psi(t, 0, x_0) \in \mathcal{S}\}$ does not exist. Since $\psi(\cdot, 0, x_0)$ is continuous, it follows that $\psi(T, 0, x_0) \in \partial\mathcal{S}$ and since, by assumption, $\min\{t \in \overline{\mathbb{R}}_+ : \psi(t, 0, x_0) \in \mathcal{S}\}$ does not exist it follows that $\psi(T, 0, x_0) \in \overline{\mathcal{S}} \backslash \mathcal{S}$. Note that $\psi(t, 0, x_0) = s(t, 0, x_0)$, for every t such that $\psi(\tau, 0, x) \notin \mathcal{S}$ for all $0 \le \tau \le t$. Now, it follows from A1 that there exists $\varepsilon > 0$ such that $s(T + \delta, 0, x_0) = \psi(T + \delta, 0, x_0)$, $\delta \in (0, \varepsilon)$, which implies that $\inf\{t \in \overline{\mathbb{R}}_+ : \psi(t, 0, x_0) \in \mathcal{S}\} > T$, which is a contradiction. Hence, $\psi(T, 0, x_0) \in \partial\mathcal{S} \cap \mathcal{S}$ and $\inf\{t \in \overline{\mathbb{R}}_+ : \psi(t, 0, x_0) \in \mathcal{S}\} = \min\{t \in \overline{\mathbb{R}}_+ : \psi(t, 0, x_0) \in \mathcal{D}\}$, which implies that the first resetting time t_1 is well defined for all initial conditions $x_0 \in \mathcal{D}$. Next, it follows from A2 that t_2 is also well defined and $t_2 \ne t_1$. Repeating the above arguments it follows that the resetting times t_k are well defined and distinct.

Since the resetting times are well defined and distinct, and since the solution to (2.1) exists and is unique, it follows that the solution of the impulsive dynamical system (2.1) and (2.2) also exists and is unique over a forward time interval. However, it is important to note that the analysis of impulsive dynamical systems can be quite involved. In particular, such systems can exhibit Zenoness and beating, as well as confluence, wherein solutions exhibit infinitely many resettings in a finite time, encounter the same resetting surface a finite or infinite number of times in zero time, and coincide after a certain point in time. In this monograph we allow for the possibility of confluence and Zeno solutions; however, A2 precludes the possibility of beating. Furthermore, since *not* every bounded solution of an impulsive dynamical system over a forward time interval can be extended to infinity due to Zeno solutions, we assume that existence and uniqueness of solutions are satisfied in forward time. For details see [12, 14, 93]. The following two examples demonstrate some of this rich behavior inherent in impulsive dynamical systems.

Example 2.1   Consider the scalar impulsive dynamical system originally studied in [12] given by

$$\dot{x}(t) = 0, \quad x(0) = x_0, \quad (t, x(t)) \notin \mathcal{S}, \tag{2.5}$$
$$\Delta x(t) = x^2(t) \operatorname{sgn} x(t) - x(t), \quad (t, x(t)) \in \mathcal{S}, \tag{2.6}$$

where $\operatorname{sgn} x \triangleq x/|x|$, $x \ne 0$, $\operatorname{sgn}(0) \triangleq 0$, and $\mathcal{S} = \{(t, x) \in \overline{\mathbb{R}}_+ \times \mathbb{R} :$

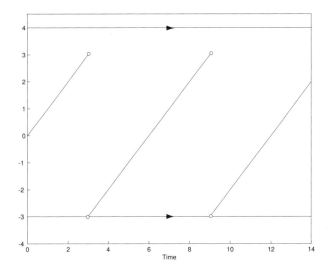

Figure 2.1 System trajectories for $x_0 = -3$ and $x_0 = 4$.

$|x| < 3$ and $t = x + 6k, \ k \in \overline{\mathbb{Z}}_+\}$. The trajectory $s(t, 0, x_0)$ of (2.5) and (2.6) with $|x_0| \geq 3$ does not intersect $\mathcal{S}$, and hence, is continuous (see Figure 2.1). Alternatively, if $1 < |x_0| < 3$, then the trajectory $s(t, 0, x_0)$ reaches $\mathcal{S}$ a finite number of times. In particular, the trajectory $s(t, 0, 2^{1/4})$ reaches $\mathcal{S}$ three times with $t_3 = 2$ (see Figure 2.2). If $0 < x_0 < 1$, then $s(t, 0, x_0)$ reaches $\mathcal{S}$ infinitely many times and $\lim_{k \to \infty} t_k = \infty$ and $\lim_{k \to \infty} x(t_k) = 0$ (see Figure 2.3). If, alternatively, $-1 < x_0 < 0$, then the trajectory $s(t, 0, x_0)$ reaches $\mathcal{S}$ infinitely many times in a finite time. In this case, $\lim_{k \to \infty} t_k = 6$ and $\lim_{k \to \infty} x(t_k) = 0$ (see Figure 2.4). Finally, (2.5) and (2.6) exhibits confluence. In particular, the trajectories $s(t, 0, 2^{1/4})$ and $s(t, 0, 4)$ coincide after $t > 2$ (see Figures 2.1 and 2.2). $\triangle$

Example 2.2   In this example we consider an impulsive system with a nonconvergent Zeno solution. Specifically, consider the impulsive dynamical system with continuous-time dynamics given by

$$\dot{x}(t) = \begin{bmatrix} 0 \\ 0 \\ -|r(x(t)) - 1|\, \mathrm{sgn}(x_3(t)) \end{bmatrix}, \quad x(0) = x_0, \quad x(t) \notin \mathcal{Z},$$

$$(2.7)$$

where $x = [x_1, x_2, x_3]^{\mathrm{T}}$ and $r(x) \triangleq \sqrt{x_1^2 + x_2^2}$, and discrete-time dy-

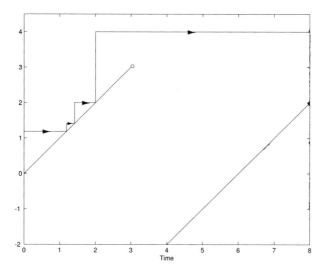

Figure 2.2 System trajectory for $x_0 = 2^{1/4}$.

namics given by

$$\Delta x(t) = \begin{bmatrix} \left(\frac{2r(x(t))-1}{r(x(t))}\right) \cos[\theta(x(t)) + \ln r(x(t))] \\ \left(\frac{2r(x(t))-1}{r(x(t))}\right) \sin[\theta(x(t)) + \ln r(x(t))] \\ \frac{(r(x(t))-1)^3}{r(x(t))} \end{bmatrix} - x(t),$$

$$x(t) \in \mathcal{Z}, \quad (2.8)$$

where $\theta(x) \triangleq \tan^{-1}\left(\frac{x_2}{x_1}\right)$, $\tau_k(x_0) \triangleq t_k$, and resetting set $\mathcal{Z} \triangleq \{x \in \mathbb{R}^3 : x_3 = 0\}$. Note that if $r(x_0) > 1$ then

$$r(x(t_k)) = \frac{(k+1)r(x_0) - k}{k\, r(x_0) - (k-1)}, \quad (2.9)$$

which implies that $\lim_{k\to\infty} r(x(t_k)) = 1$. Furthermore,

$$\theta(x(t_k)) = \theta(x_0) + \sum_{i=1}^{k} \ln r(x(t_i)) = \theta(x_0) + \ln\left[\frac{(r(x_0) - 1)k + r(x_0)}{r(x_0)}\right],$$

$$(2.10)$$

and hence, $\theta(x(t_k)) \to \infty$ as $k \to \infty$. Finally, $x_3(t_k) = \frac{(r(t_k)-1)^3}{r(t_k)} \to 0$ as $k \to \infty$. Hence, the sequence of impact points is bounded but does not converge (see Figure 2.5).

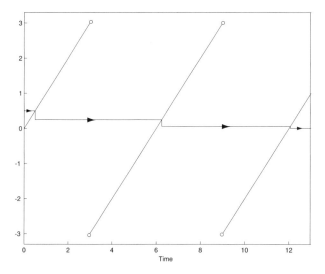

Figure 2.3 System trajectory for $x_0 = 1/2$.

Finally, to show that the solution of this system is Zeno note that

$$\tau_1(x_0) = \frac{|x_3(0)|}{|r(x_0) - 1|}, \tag{2.11}$$

$$\tau_2(x_0) - \tau_1(x_0) = \frac{|x_3(\tau_1^+(x_0))|}{|r(\tau_1^+(x_0)) - 1|} = (r(x_0) - 1)^2, \tag{2.12}$$

and

$$\tau_{k+1}(x_0) - \tau_k(x_0) = \frac{|x_3(\tau_k^+(x_0))|}{|r(\tau_k^+(x_0)) - 1|} = (r(\tau_k(x_0)) - 1)^2. \tag{2.13}$$

Now, using the fact that $r(\tau_k(x_0)) - 1 = \frac{r(x_0)-1}{k\,r(x_0)-(k-1)}$, it follows that

$$\tau_{k+1}(x_0) - \tau_k(x_0) = \left(\frac{r(x_0) - 1}{k\,r(x_0) - (k - 1)}\right)^2 = \frac{1}{\left(k + \frac{1}{r(x_0)-1}\right)^2}. \tag{2.14}$$

Since the series $\sum_{k=1}^{\infty} \frac{1}{(k+a)^2}$ converges for all $a \in \mathbb{R}$, $\lim_{k \to \infty} \tau_k(x_0) = \tau_1(x_0) + \sum_{k=1}^{\infty}[\tau_{k+1}(x_0) - \tau_k(x_0)]$ exists, and hence, the trajectory of the system (2.7) and (2.8) is Zeno.   △

Clearly, Example 2.2 shows that not every bounded Zeno solution is extendable. In fact, the impulsive dynamical system (2.7) and (2.8) is not even left-continuous at $t = \tau(x_0)$ for if it were left-continuous, then necessarily $\lim_{k \to \infty} s(\tau_k(x_0), 0, x_0) = s(\tau(x_0), 0, x_0)$, where $\tau(x_0)$

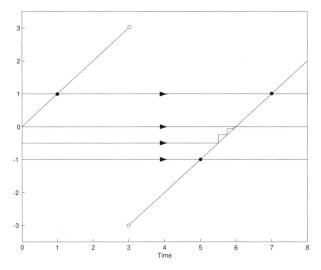

Figure 2.4 System trajectories for $x_0 = -1/2$, $x_0 = 1$, $x_0 = -1$, and $x_0 = 0$.

denotes the Zeno time (accumulation time). However, if a Zeno solution is convergent and the continuous and discrete parts of the states converge to a unique value at the Zeno time $\tau(x_0)$, then by reinitializing the impulsive dynamical system at $\tau(x_0)$ using $s(\tau(x_0), 0, x_0)$ as the system initial condition, a Zeno solution can be extended.

In [12, 79, 92–95, 105, 153, 154], the resetting set $\mathcal{S}$ is defined in terms of a countable number of functions $\tau_k : \mathcal{D} \to (0, \infty)$, and is given by

$$\mathcal{S} = \bigcup_k \{(\tau_k(x), x) : x \in \mathcal{D}\}. \tag{2.15}$$

The analysis of impulsive dynamical systems with a resetting set of the form (2.15) can be quite involved. Furthermore, since impulsive dynamical systems of the form (2.1) and (2.2) involve impulses at variable times they are time-varying systems. In this monograph, we will consider impulsive dynamical systems involving two distinct forms of the resetting set $\mathcal{S}$. In the first case, the resetting set is defined by a prescribed sequence of times which are independent of the state x. These equations are thus called *time-dependent impulsive dynamical systems*. In the second case, the resetting set is defined by a region in the state space that is independent of time. These equations are called *state-dependent impulsive dynamical systems*.

Time-dependent impulsive dynamical systems can be written as (2.1) and (2.2) with $\mathcal{S}$ defined as

$$\mathcal{S} \triangleq \mathcal{T} \times \mathcal{D}, \tag{2.16}$$

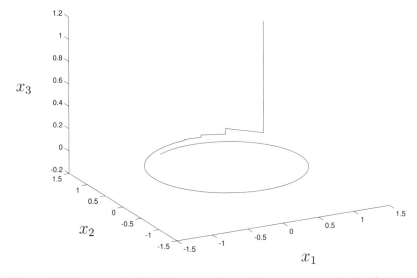

Figure 2.5 Phase portrait of Zeno system with nonconverging trajectory.

where

$$\mathcal{T} \triangleq \{t_1, t_2, \dots\} \tag{2.17}$$

and $0 \le t_1 < t_2 < \cdots$ are prescribed resetting times. Now, (2.1) and (2.2) can be rewritten in the form of the *time-dependent impulsive dynamical system*

$$\dot{x}(t) = f_{\rm c}(x(t)), \quad x(0) = x_0, \quad t \ne t_k, \tag{2.18}$$
$$\Delta x(t) = f_{\rm d}(x(t)), \quad t = t_k. \tag{2.19}$$

Since $0 \notin \mathcal{T}$ and $t_k < t_{k+1}$, it follows that Assumptions A1 and A2 are satisfied. Since time-dependent impulsive dynamical systems involve impulses at a fixed sequence of times, they are time-varying systems.

Example 2.3 To show that time-dependent impulsive dynamical systems are time-varying systems consider the scalar time-dependent impulsive dynamical system

$$\dot{x}(t) = 0, \quad x(t_0) = 0, \quad t \ne 2, \tag{2.20}$$
$$\Delta x(t) = 1, \quad t = 2. \tag{2.21}$$

Since

$$s(t, 1, 0) = \begin{cases} 0, & 1 < t \le 2, \\ 1, & t > 2, \end{cases} \tag{2.22}$$

and

$$s(t-1,0,0) = \begin{cases} 0, & 1 < t \le 3, \\ 1, & t > 3, \end{cases} \tag{2.23}$$

it follows that $s(t,1,0) \ne s(t-1,0,0)$, $2 < t \le 3$, and hence, (2.20) and (2.21) is time varying. △

State-dependent impulsive dynamical systems can be written as (2.1) and (2.2) with $\mathcal{S}$ defined as

$$\mathcal{S} \triangleq [0,\infty) \times \mathcal{Z}, \tag{2.24}$$

where $\mathcal{Z} \subset \mathcal{D}$. Therefore, (2.1) and (2.2) can be rewritten in the form of the *state-dependent impulsive dynamical system*

$$\dot{x}(t) = f_c(x(t)), \quad x(0) = x_0, \quad x(t) \notin \mathcal{Z}, \tag{2.25}$$
$$\Delta x(t) = f_d(x(t)), \quad x(t) \in \mathcal{Z}. \tag{2.26}$$

We assume that if $x \in \mathcal{Z}$, then $x + f_d(x) \notin \mathcal{Z}$. In addition, we assume that if at time t the trajectory $x(t) \in \overline{\mathcal{Z}} \backslash \mathcal{Z}$, then there exists $\varepsilon > 0$ such that for all $0 < \delta < \varepsilon$, $x(t + \delta) \notin \mathcal{Z}$. These assumptions represent the specialization of A1 and A2 for the particular resetting set (2.24). It follows from these assumptions that for a particular initial condition, the resetting times $\tau_k(x_0) \triangleq t_k$ are distinct and well defined. Since the resetting set $\mathcal{Z}$ is a subset of the state space and is independent of time, state-dependent impulsive dynamical systems are time-invariant systems.

Finally, note that if $x^* \in \mathcal{D}$ satisfies $f_d(x^*) = 0$, then $x^* \notin \mathcal{Z}$. To see this, suppose $x^* \in \mathcal{Z}$. Then $x^* + f_d(x^*) = x^* \in \mathcal{Z}$, contradicting A2. Thus, if $x = x_e$ is an equilibrium point of (2.25) and (2.26), then $x_e \notin \mathcal{Z}$, and hence, $x_e \in \mathcal{D}$ is an equilibrium point of (2.25) and (2.26) if and only if $f_c(x_e) = 0$. In addition, note that it follows from the definition of $\tau_k(\cdot)$ that $\tau_1(x) > 0$, $x \notin \mathcal{Z}$, and $\tau_1(x) = 0$, $x \in \mathcal{Z}$. Finally, since for every $x \in \mathcal{Z}$, $x + f_d(x) \notin \mathcal{Z}$, it follows that $\tau_2(x) = \tau_1(x) + \tau_1(x + f_d(x)) > 0$.

2.3 Stability Theory of Impulsive Dynamical Systems

In this section, we present Lyapunov, asymptotic, and exponential stability theorems for nonlinear state-dependent impulsive dynamical systems. The following definition introduces several types of stability corresponding to the zero solution $x(t) \equiv 0$ of (2.25) and (2.26).

Definition 2.1 *i) The zero solution $x(t) \equiv 0$ to (2.25) and (2.26) is Lyapunov stable if, for all $\varepsilon > 0$, there exists $\delta = \delta(\varepsilon) > 0$ such that if $\|x(0)\| < \delta$, then $\|x(t)\| < \varepsilon$, $t \geq 0$.*

ii) The zero solution $x(t) \equiv 0$ to (2.25) and (2.26) is asymptotically *stable if it is Lyapunov stable and there exists $\delta > 0$ such that if $\|x(0)\| < \delta$, then $\lim_{t \to \infty} x(t) = 0$.*

iii) The zero solution $x(t) \equiv 0$ to (2.25) and (2.26) is exponentially *stable if there exist positive constants α, β, and δ such that if $\|x(0)\| < \delta$, then $\|x(t)\| \leq \alpha \|x(0)\| e^{-\beta t}$, $t \geq 0$.*

iv) The zero solution $x(t) \equiv 0$ to (2.25) and (2.26) is globally asymptotically *stable if it is Lyapunov stable and for all $x(0) \in \mathbb{R}^n$, $\lim_{t \to \infty} x(t) = 0$.*

v) The zero solution $x(t) \equiv 0$ to (2.25) and (2.26) is globally exponentially *stable if there exist positive constants α and β such that $\|x(t)\| \leq \alpha \|x(0)\| e^{-\beta t}$, $t \geq 0$, for all $x(0) \in \mathbb{R}^n$.*

vi) Finally, the zero solution $x(t) \equiv 0$ to (2.25) and (2.26) is unstable *if it is not Lyapunov stable.*

. **Theorem 2.1** *Consider the nonlinear impulsive dynamical system $\mathcal{G}$ given by (2.25) and (2.26). Suppose there exists a continuously differentiable function $V : \mathcal{D} \to [0, \infty)$ satisfying $V(0) = 0$, $V(x) > 0$, $x \in \mathcal{D}$, $x \neq 0$, and*

$$V'(x) f_{\mathrm{c}}(x) \leq 0, \quad x \notin \mathcal{Z}, \tag{2.27}$$
$$V(x + f_{\mathrm{d}}(x)) \leq V(x), \quad x \in \mathcal{Z}. \tag{2.28}$$

Then the zero solution $x(t) \equiv 0$ to (2.25) and (2.26) is Lyapunov stable. Furthermore, if the inequality (2.27) is strict for all $x \neq 0$, then the zero solution $x(t) \equiv 0$ to (2.25) and (2.26) is asymptotically stable. Alternatively, if there exist scalars $\alpha, \beta, \varepsilon > 0$, and $p \geq 1$ such that

$$\alpha \|x\|^p \leq V(x) \leq \beta \|x\|^p, \quad x \in \mathcal{D}, \tag{2.29}$$
$$V'(x) f_{\mathrm{c}}(x) \leq -\varepsilon V(x), \quad x \notin \mathcal{Z}, \tag{2.30}$$

and (2.28) holds, then the zero solution $x(t) \equiv 0$ to (2.25) and (2.26) is exponentially stable. Finally, if, in addition, $\mathcal{D} = \mathbb{R}^n$ and

$$V(x) \to \infty \text{ as } \|x\| \to \infty, \tag{2.31}$$

then the above asymptotic and exponential stability results are global.

Proof. Let $\varepsilon > 0$ be such that $\mathcal{B}_\varepsilon(0) \subseteq \mathcal{D}$. Since $\overline{\mathcal{B}}_\varepsilon(0)$ is compact and $f_d(x)$, $x \in \overline{\mathcal{Z}}$, is continuous, it follows that

$$\eta \triangleq \max \left\{ \varepsilon, \ \max_{\overline{\mathcal{B}}_\varepsilon(0) \cap \overline{\mathcal{Z}}} \|x + f_d(x)\| \right\} \tag{2.32}$$

exists. Next, let $\alpha \triangleq \min_{x \in \mathcal{D}: \ \varepsilon \leq \|x\| \leq \eta} V(x)$. Note $\alpha > 0$ since $0 \notin \partial \mathcal{B}_\varepsilon(0)$ and $V(x) > 0$, $x \in \mathcal{D}$, $x \neq 0$. Next, let $\beta \in (0, \alpha)$ and define $\mathcal{D}_\beta \triangleq \{x \in \mathcal{B}_\varepsilon(0) : \ V(x) \leq \beta\}$. Now, let $x_0 \in \mathcal{D}_\beta$ and note that for $\mathcal{S} = [0, \infty) \times \mathcal{Z}$ it follows from Assumptions A1 and A2 that the resetting times $\tau_k(x_0)$ are well defined and distinct for every trajectory of (2.25) and (2.26).

Prior to the first resetting time, we can determine the value of $V(x(t))$ as

$$V(x(t)) = V(x(0)) + \int_0^t V'(x(\tau)) f_c(x(\tau)) d\tau, \quad t \in [0, \tau_1(x_0)]. \tag{2.33}$$

Between consecutive resetting times $\tau_k(x_0)$ and $\tau_{k+1}(x_0)$, we can determine the value of $V(x(t))$ as its initial value plus the integral of its rate of change along the trajectory $x(t)$, that is,

$$V(x(t)) = V(x(\tau_k(x_0)) + f_d(x(\tau_k(x_0)))) + \int_{\tau_k(x_0)}^t V'(x(\tau)) f_c(x(\tau)) d\tau,$$
$$t \in (\tau_k(x_0), \tau_{k+1}(x_0)], \tag{2.34}$$

for $k = 1, 2, \ldots$. Adding and subtracting $V(x(\tau_k(x_0)))$ to and from the right hand side of (2.34) yields

$$V(x(t)) = V(x(\tau_k(x_0))) + [V(x(\tau_k(x_0)) + f_d(x(\tau_k(x_0))))$$
$$- V(x(\tau_k(x_0)))] + \int_{\tau_k(x_0)}^t V'(x(\tau)) f_c(x(\tau)) d\tau,$$
$$t \in (\tau_k(x_0), \tau_{k+1}(x_0)], \tag{2.35}$$

and in particular, at time $\tau_{k+1}(x_0)$,

$$V(x(\tau_{k+1}(x_0))) = V(x(\tau_k(x_0))) + [V(x(\tau_k(x_0)) + f_d(x(\tau_k(x_0))))$$
$$- V(x(\tau_k(x_0)))] + \int_{\tau_k(x_0)}^{\tau_{k+1}(x_0)} V'(x(\tau)) f_c(x(\tau)) d\tau.$$
$$\tag{2.36}$$

By recursively substituting (2.36) into (2.35) and ultimately into (2.33), we obtain

$$V(x(t)) = V(x(0)) + \int_0^t V'(x(\tau))f_{\mathrm{c}}(x(\tau))\mathrm{d}\tau$$
$$+ \sum_{i=1}^k [V(x(\tau_i(x_0)) + f_{\mathrm{d}}(x(\tau_i(x_0)))) - V(x(\tau_i(x_0)))],$$
$$t \in (\tau_k(x_0), \tau_{k+1}(x_0)]. \quad (2.37)$$

If we allow $t_0 \triangleq 0$ and $\sum_{i=1}^0 \triangleq 0$, then (2.37) is valid for $k \in \overline{\mathbb{Z}}_+$. From (2.37) and (2.28) we obtain

$$V(x(t)) \le V(x(0)) + \int_0^t V'(x(\tau))f_{\mathrm{c}}(x(\tau))\mathrm{d}\tau, \quad t \ge 0. \quad (2.38)$$

Furthermore, it follows from (2.27) that

$$V(x(t)) \le V(x(0)) \le \beta, \quad x(0) \in \mathcal{D}_\beta, \quad t \ge 0. \quad (2.39)$$

Next, suppose, *ad absurdum*, there exists $T > 0$ such that $\|x(T)\| \ge \varepsilon$. Hence, since $\|x_0\| < \varepsilon$, there exists $t_1 \in (0, T]$ such that either $\|x(t_1)\| = \varepsilon$ or $x(t_1) \in \mathcal{Z}$, $\|x(t_1)\| < \varepsilon$, and $\|x(t_1^+)\| \ge \varepsilon$. If $\|x(t_1)\| = \varepsilon$ then $V(x(t_1)) \ge \alpha > \beta$, which is a contradiction. Alternatively, if $x(t_1) \in \mathcal{Z}$, $\|x(t_1)\| < \varepsilon$, and $\|x(t_1^+)\| \ge \varepsilon$, then $\|x(t_1^+)\| = \|x(t_1) + f_{\mathrm{d}}(x(t_1))\| \le \eta$, which implies that $V(x(t_1^+)) \ge \alpha > \beta$ contradicting (2.39). Hence, $V(x(t)) \le \beta$ and $\|x(t)\| < \varepsilon$, $t \ge 0$, which implies that $\mathcal{D}_\beta$ is a positive invariant set (see Definition 2.3) with respect to (2.25) and (2.26). Next, since $V(\cdot)$ is continuous and $V(0) = 0$, there exists $\delta = \delta(\varepsilon) \in (0, \varepsilon)$ such that $V(x) < \beta$, $x \in \mathcal{B}_\delta(0)$. Since $\mathcal{B}_\delta(0) \subset \mathcal{D}_\beta \subset \mathcal{B}_\varepsilon(0) \subseteq \mathcal{D}$ and $\mathcal{D}_\beta$ is a positive invariant set with respect to (2.25) and (2.26), it follows that for all $x_0 \in \mathcal{B}_\delta(0)$, $x(t) \in \mathcal{B}_\varepsilon(0)$, $t \ge 0$, which establishes Lyapunov stability.

To prove asymptotic stability, suppose (2.28) and (2.37) hold, and let $x_0 \in \mathcal{B}_\delta(0)$. Then it follows that $x(t) \in \mathcal{B}_\varepsilon(0)$, $t \ge 0$. However, $V(x(t))$, $t \ge 0$, is monotonically decreasing and bounded from below by zero. Next, it follows from (2.28) and (2.37) that

$$V(x(t)) - V(x(s)) \le \int_s^t V'(x(\tau))f_{\mathrm{c}}(x(\tau))\mathrm{d}\tau, \quad t > s, \quad (2.40)$$

and, assuming strict inequality in (2.27), we obtain

$$V(x(t)) < V(x(s)), \quad t > s, \quad (2.41)$$

provided $x(s) \neq 0$. Now, suppose, *ad absurdum*, $x(t)$, $t \geq 0$, does not converge to zero. This implies that $V(x(t))$, $t \geq 0$, is lower bounded by a positive number, that is, there exists $L > 0$ such that $V(x(t)) \geq L > 0$, $t \geq 0$. Hence, by continuity of $V(\cdot)$ there exists $\delta' > 0$ such that $V(x) < L$, $x \in \mathcal{B}_{\delta'}(0)$, which further implies that $x(t) \notin \mathcal{B}_{\delta'}(0)$, $t \geq 0$. Next, let $L_1 \triangleq \min_{\delta' \leq \|x\| \leq \varepsilon} -V'(x)f_c(x)$ which implies that $-V'(x)f_c(x) \geq L_1$, $\delta' \leq \|x\| \leq \varepsilon$, and hence, it follows from (2.37) that

$$V(x(t)) - V(x_0) \leq \int_0^t V'(x(\tau))f_c(x(\tau))\mathrm{d}\tau \leq -L_1 t, \qquad (2.42)$$

and hence, for all $x_0 \in \mathcal{B}_\delta(0)$,

$$V(x(t)) \leq V(x_0) - L_1 t.$$

Letting $t > \frac{V(x_0) - L}{L_1}$, it follows that $V(x(t)) < L$, which is a contradiction. Hence, $x(t) \to 0$ as $t \to \infty$, establishing asymptotic stability.

To show exponential stability, note that it follows from (2.29), (2.30), and (2.28) that the zero solution $x(t) \equiv 0$ to (2.25) and (2.26) is asymptotically stable. Hence, there exists $\delta > 0$ such that for all $x_0 \in \mathcal{B}_\delta(0)$, $x(t) \to 0$ as $t \to \infty$. Next, let $x_0 \in \mathcal{B}_\delta(0)$ and note that it follows from (2.30) that prior to the first resetting time

$$\dot{V}(x(t)) \leq -\varepsilon V(x(t)), \quad 0 \leq t \leq \tau_1(x_0), \quad x_0 \in \mathcal{B}_\delta(0), \quad (2.43)$$

which implies that

$$V(x(t)) \leq V(x_0)e^{-\varepsilon t}, \quad 0 \leq t \leq \tau_1(x_0), \quad x_0 \in \mathcal{B}_\delta(0). \quad (2.44)$$

Similarly, between the first and second resetting times

$$\dot{V}(x(t)) \leq -\varepsilon V(x(t)), \quad \tau_1(x_0) < t \leq \tau_2(x_0), \quad x_0 \in \mathcal{B}_\delta(0), \quad (2.45)$$

which, using (2.28) and (2.44), yields

$$\begin{aligned} V(x(t)) &\leq V(x(\tau_1(x_0)) + f_\mathrm{d}(x(\tau_1(x_0))))e^{-\varepsilon(t - \tau_1(x_0))} \\ &\leq V(x(\tau_1(x_0)))e^{-\varepsilon(t - \tau_1(x_0))} \\ &\leq V(x_0)e^{-\varepsilon\tau_1(x_0)}e^{-\varepsilon(t - \tau_1(x_0))} \\ &= V(x_0)e^{-\varepsilon t}, \quad \tau_1(x_0) < t \leq \tau_2(x_0), \quad x_0 \in \mathcal{B}_\delta(0). \quad (2.46) \end{aligned}$$

Recursively repeating the above arguments for $t \in (\tau_k(x_0), \tau_{k+1}(x_0)]$, $k = 3, 4, \ldots$, it follows that

$$V(x(t)) \leq V(x_0)e^{-\varepsilon t}, \quad t \geq 0, \quad x_0 \in \mathcal{B}_\delta(0). \quad (2.47)$$

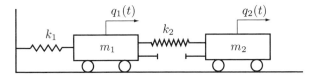

Figure 2.6 Two-mass system with constraint buffers.

Now, it follows from (2.29) and (2.47) that for all $t \geq 0$,

$$\alpha\|x(t)\|^p \leq V(x(t)) \leq V(x_0)e^{-\varepsilon t} \leq \beta\|x_0\|^p e^{-\varepsilon t}, \quad x_0 \in \mathcal{B}_\delta(0), \quad (2.48)$$

and hence,

$$\|x(t)\| \leq \left(\frac{\beta}{\alpha}\right)^{1/p} \|x_0\| e^{-\frac{\varepsilon}{p}t}, \quad t \geq 0, \quad x_0 \in \mathcal{B}_\delta(0), \quad (2.49)$$

establishing exponential stability of the zero solution $x(t) \equiv 0$ to (2.25) and (2.26).

Finally, global asymptotic and exponential stability follow from standard arguments. Specifically, let $x(0) \in \mathbb{R}^n$, and let $\beta \triangleq V(x(0))$. It follows from (2.31) that there exists $\varepsilon > 0$ such that $V(x) > \beta$ for all $x \in \mathbb{R}^n$ such that $\|x\| > \varepsilon$. Hence, $\|x(0)\| \leq \varepsilon$, and, since $V(x(t))$ is strictly decreasing, it follows that $\|x(t)\| < \varepsilon$, $t > 0$. The remainder of the proof is identical to the proof of asymptotic (respectively, exponential) stability. □

In the proof of Theorem 2.1, we note that assuming strict inequality in (2.27), the inequality (2.41) is obtained *provided* $x(s) \neq 0$. This proviso is necessary since it may be possible to reset the states to the origin, in which case $x(s) = 0$ for a finite value of s. In this case, for $t > s$, we have $V(x(t)) = V(x(s)) = V(0) = 0$. This situation does not present a problem, however, since reaching the origin in finite time is a stronger condition than reaching the origin as $t \to \infty$.

Example 2.4   Consider the two-mass, two-spring system with buffer constraints of length $\frac{L}{2}$ shown in Figure 2.6. Between collisions the system dynamics, with state variables defined in Figure 2.6, are given by

$$m_1\ddot{q}_1(t) + (k_1 + k_2)q_1(t) - k_2 q_2(t) = 0, \quad q_1(0) = q_{01}, \quad \dot{q}_1(0) = \dot{q}_{01},$$
$$t \geq 0, \quad (2.50)$$
$$m_2\ddot{q}_2(t) - k_2 q_1(t) + k_2 q_2(t) = 0, \quad q_2(0) = q_{02}, \quad \dot{q}_2(0) = \dot{q}_{02}.$$
$$(2.51)$$

At the instant of a collision, the velocities of the masses change according to the law of conservation of linear momentum and the loss of kinetic energy due to a collision so that

$$m_1 \dot{q}_1(t_k^+) + m_2 \dot{q}_2(t_k^+) = m_1 \dot{q}_1(t_k) + m_2 \dot{q}_2(t_k), \qquad (2.52)$$

$$\dot{q}_1(t_k^+) - \dot{q}_2(t_k^+) = -e(\dot{q}_1(t_k) - \dot{q}_2(t_k)), \qquad (2.53)$$

where $e \in [0, 1)$ is the coefficient of restitution. Solving (2.52) and (2.53) for $\dot{q}_1(t_k^+)$ and $\dot{q}_2(t_k^+)$, the resetting dynamics are given by

$$\Delta \dot{q}_1(t_k) = \dot{q}_1(t_k^+) - \dot{q}_1(t_k) = -\frac{(1+e)m_2}{m_1 + m_2}(\dot{q}_1(t_k) - \dot{q}_2(t_k)), \quad (2.54)$$

$$\Delta \dot{q}_2(t_k) = \dot{q}_2(t_k^+) - \dot{q}_2(t_k) = \frac{(1+e)m_1}{m_1 + m_2}(\dot{q}_1(t_k) - \dot{q}_2(t_k)). \quad (2.55)$$

Defining $x_1 \triangleq q_1$, $x_2 \triangleq \dot{q}_1$, $x_3 \triangleq q_2$, and $x_4 \triangleq \dot{q}_2$, we can rewrite (2.50), (2.51), (2.54), and (2.55) in state space form (2.25) and (2.26) with $x \triangleq [x_1, x_2, x_3, x_4]^{\mathrm{T}}$,

$$f_c(x) = \begin{bmatrix} x_2 \\ -\frac{(k_1 + k_2)}{m_1} x_1 + \frac{k_2}{m_1} x_3 \\ x_4 \\ \frac{k_2}{m_2} x_1 - \frac{k_2}{m_2} x_3 \end{bmatrix}, \quad f_d(x) = \begin{bmatrix} 0 \\ -\frac{(1+e)m_2}{m_1 + m_2}(x_2 - x_4) \\ 0 \\ \frac{(1+e)m_1}{m_1 + m_2}(x_2 - x_4) \end{bmatrix}, \qquad (2.56)$$

$\mathcal{D} = \mathbb{R}^4$, and $\mathcal{Z} = \{x \in \mathbb{R}^4 : x_1 - x_3 = L, \ x_2 > x_4\}$. Note that $x_e = 0$ is an equilibrium point of the system.

To analyze the stability of the zero solution $x(t) \equiv 0$ consider the Lyapunov function candidate

$$V(x) = \frac{1}{2}\left[m_1 x_2^2 + m_2 x_4^2 + k_1 x_1^2 + k_2(x_3 - x_1)^2 \right], \quad x \in \mathcal{D}. \quad (2.57)$$

Now, it follows that $\dot{V}(x) = 0$, $x \in \mathcal{D}$, $x \notin \mathcal{Z}$, and

$$\Delta V(x) = \frac{(e^2 - 1)m_1 m_2 (x_2 - x_4)^2}{2(m_1 + m_2)} \le 0, \quad x \in \mathcal{Z}, \qquad (2.58)$$

and hence, by Theorem 2.1 the zero solution $x(t) \equiv 0$ to (2.50), (2.51), (2.54), and (2.55) is Lyapunov stable. $\triangle$

To examine the stability of linear[1] state-dependent impulsive systems set $f_c(x) = A_c x$ and $f_d(x) = (A_d - I_n)x$ in Theorem 2.1. Considering the quadratic Lyapunov function candidate $V(x) = x^{\mathrm{T}} P x$,

---

[1]Impulsive dynamical systems with $f_c(x) = Ax$ and $f_d(x) = (A_d - I)x$ are *not* linear. However, this minor abuse in terminology provides a natural way of

where $P > 0$, it follows from Theorem 2.1 that the conditions

$$x^{\mathrm{T}}(A_{\mathrm{c}}^{\mathrm{T}}P + PA_{\mathrm{c}})x < 0, \quad x \notin \mathcal{Z}, \tag{2.59}$$

$$x^{\mathrm{T}}(A_{\mathrm{d}}^{\mathrm{T}}PA_{\mathrm{d}} - P)x \leq 0, \quad x \in \mathcal{Z}, \tag{2.60}$$

establish asymptotic stability for linear state-dependent impulsive systems. These conditions are implied by $P > 0$, $A_{\mathrm{c}}^{\mathrm{T}}P + PA_{\mathrm{c}} < 0$, and $A_{\mathrm{d}}^{\mathrm{T}}PA_{\mathrm{d}} - P \leq 0$, which can be solved using a Linear Matrix Inequality (LMI) feasibility problem [27].

2.4 An Invariance Principle for State-Dependent Impulsive Dynamical Systems

In this section, we develop an invariance principle for state-dependent impulsive dynamical systems. The following key assumption and several definitions are needed for the statement of the next fundamental result on positive limit sets for impulsive dynamical systems. Recall that a state-dependent impulsive dynamical system is time invariant, and hence, $s(t + \tau, \tau, x_0) = s(t, 0, x_0)$ for all $x_0 \in \mathcal{D}$, $t, \tau \in [0, \infty)$. For simplicity of exposition, in the remainder of this section we denote the trajectory $s(t, 0, x_0)$ by $s(t, x_0)$. Furthermore, let $\mathcal{T}_{x_0} \triangleq [0, \infty) \setminus \{\tau_1(x_0), \tau_2(x_0), \ldots\}$.

Assumption 2.1 *Consider the impulsive dynamical system $\mathcal{G}$ given by (2.25) and (2.26), and let $s(t, x_0)$, $t \geq 0$, denote the solution to (2.25) and (2.26) with initial condition x_0. Then, for every $x_0 \in \mathcal{D}$ and for every $\varepsilon > 0$ and $t \in \mathcal{T}_{x_0}$, there exists $\delta(\varepsilon, x_0, t) > 0$ such that if $\|x_0 - y\| < \delta(\varepsilon, x_0, t)$, $y \in \mathcal{D}$, then $\|s(t, x_0) - s(t, y)\| < \varepsilon$.*

Assumption 2.1 is a generalization of the standard continuous dependence property for dynamical systems with continuous flows to dynamical systems with left-continuous flows. Specifically, by letting $\mathcal{T}_{x_0} = \overline{\mathcal{T}}_{x_0} = [0, \infty)$, where $\overline{\mathcal{T}}_{x_0}$ denotes the closure of the set $\mathcal{T}_{x_0}$, the *quasi-continuous dependence property* (i.e., Assumption 2.1) specializes to the classical continuous dependence of solutions of a given dynamical system with respect to the system's initial conditions $x_0 \in \mathcal{D}$ [162]. If, in addition, $x_0 = 0$, $s(t, 0) = 0$, $t \geq 0$, and $\delta(\varepsilon, 0, t)$ can be chosen independent of t, then continuous dependence implies the classical Lyapunov stability of the zero trajectory

---

differentiating between impulsive dynamical systems with nonlinear vector fields versus impulsive dynamical systems with linear vector fields, and considerably simplifies the presentation.

$s(t, 0) = 0$, $t \geq 0$. Hence, Lyapunov stability of motion can be interpreted as continuous dependence of solutions uniformly in t for all $t \geq 0$. Conversely, continuous dependence of solutions can be interpreted as Lyapunov stability of motion for every fixed time t [162]. Analogously, Lyapunov stability of motion of an impulsive dynamical system as defined in [93] can be interpreted as *quasi-continuous dependence of solutions* (i.e., Assumption 2.1) uniformly in t for all $t \in \mathcal{T}_{x_0}$.

Recall that for $x \in \mathcal{D}$, the map $s(\cdot, x) : \mathbb{R} \to \mathcal{D}$ defines the *solution curve* or *trajectory* of (2.25) and (2.26) through the point x in $\mathcal{D}$. Identifying $s(\cdot, x)$ with its graph, the trajectory or *orbit* of a point $x_0 \in \mathcal{D}$ is defined as the motion along the curve

$$\mathcal{O}_{x_0} \triangleq \{x \in \mathcal{D} : \ x = s(t, x_0), \ t \in \mathbb{R}\}. \tag{2.61}$$

For $t \geq 0$, we define the *positive orbit* through the point $x_0 \in \mathcal{D}$ as the motion along the curve

$$\mathcal{O}_{x_0}^+ \triangleq \{x \in \mathcal{D} : \ x = s(t, x_0), \ t \geq 0\}. \tag{2.62}$$

Similarly, the *negative orbit* through the point $x_0 \in \mathcal{D}$ is defined as

$$\mathcal{O}_{x_0}^- \triangleq \{x \in \mathcal{D} : \ x = s(t, x_0), \ t \leq 0\}. \tag{2.63}$$

Hence, the orbit $\mathcal{O}_x$ of a point $x \in \mathcal{D}$ is given by $\mathcal{O}_x^+ \cup \mathcal{O}_x^- = \{s(t, x) : t \geq 0\} \cup \{s(t, x) : \ t \leq 0\}$.

Definition 2.2 *A point $p \in \mathcal{D}$ is a* positive limit point *of the trajectory $s(\cdot, x)$ of (2.25) and (2.26) if there exists a monotonic sequence $\{t_n\}_{n=0}^\infty$ of positive numbers, with $t_n \to \infty$ as $n \to \infty$, such that $s(t_n, x) \to p$ as $n \to \infty$. A point $q \in \mathcal{D}$ is a* negative limit point *of the trajectory $s(\cdot, x)$ of (2.25) and (2.26) if there exists a monotonic sequence $\{t_n\}_{n=0}^\infty$ of negative numbers, with $t_n \to -\infty$ as $n \to \infty$, such that $s(t_n, x) \to q$ as $n \to \infty$. The set of all positive limit points of $s(t, x)$, $t \geq 0$, is the* positive limit set *$\omega(x)$ of $s(\cdot, x)$ of (2.25) and (2.26). The set of all negative limit points of $s(t, x)$, $t \leq 0$, is the* negative limit set *$\alpha(x)$ of $s(\cdot, x)$ of (2.25) and (2.26).*

In the literature, the positive limit set is often referred to as the ω-*limit set* while the negative limit set is referred to as the α-*limit set*. Note that $p \in \mathcal{D}$ is a positive limit point of the trajectory $s(t, x_0)$, $t \geq 0$, if and only if there exists a monotonic sequence $\{t_n\}_{n=0}^\infty \subset \mathcal{T}_{x_0}$, with $t_n \to \infty$ as $n \to \infty$, such that $s(t_n, x_0) \to p$ as $n \to \infty$. To see this, let $p \in \omega(x_0)$ and recall that $\mathcal{T}_{x_0}$ is a dense subset of the semi-infinite interval $[0, \infty)$. In this case, it follows that there exists an unbounded sequence $\{t_n\}_{n=0}^{\infty}$, with $t_n \to \infty$ as $n \to \infty$, such that $\lim_{n \to \infty} s(t_n, x_0) = p$. Hence, for every $\varepsilon > 0$, there exists $n > 0$ such that $\|s(t_n, x_0) - p\| < \varepsilon/2$. Furthermore, since $s(\cdot, x_0)$ is left-continuous and $\mathcal{T}_{x_0}$ is a dense subset of $[0, \infty)$, there exists $\hat{t}_n \in \mathcal{T}_{x_0}$, $\hat{t}_n \le t_n$, such that $\|s(\hat{t}_n, x_0) - s(t_n, x_0)\| < \varepsilon/2$, and hence, $\|s(\hat{t}_n, x_0) - p\| \le \|s(t_n, x_0) - p\| + \|s(\hat{t}_n, x_0) - s(t_n, x_0)\| < \varepsilon$. Using this procedure, with $\varepsilon = 1, 1/2, 1/3, \ldots$, we can construct an unbounded sequence $\{\hat{t}_k\}_{k=1}^{\infty} \subset \mathcal{T}_{x_0}$ such that $\lim_{k \to \infty} s(\hat{t}_k, x_0) = p$. Hence, $p \in \omega(x_0)$ if and only if there exists a monotonic sequence $\{t_n\}_{n=0}^{\infty} \subset \mathcal{T}_{x_0}$, with $t_n \to \infty$ as $n \to \infty$, such that $s(t_n, x_0) \to p$ as $n \to \infty$. For the next definition, $s_t(x)$ denotes the *flow* $s(t, \cdot) : \mathcal{D} \to \mathbb{R}^n$ of (2.25) and (2.26) for a given $t \in \mathbb{R}$.

Definition 2.3 *A set* $\mathcal{M} \subset \mathcal{D} \subseteq \mathbb{R}^n$ *is a* positively invariant set *with respect to the nonlinear dynamical system (2.25) and (2.26) if* $s_t(\mathcal{M}) \subseteq \mathcal{M}$ *for all* $t \ge 0$, *where* $s_t(\mathcal{M}) \triangleq \{s_t(x) : x \in \mathcal{M}\}$. *A set* $\mathcal{M} \subset \mathcal{D} \subseteq \mathbb{R}^n$ *is a* negatively invariant set *with respect to the nonlinear dynamical system (2.25) and (2.26) if* $s_t(\mathcal{M}) \subseteq \mathcal{M}$ *for all* $t \le 0$. *A set* $\mathcal{M} \subseteq \mathcal{D}$ *is an* invariant set *with respect to the dynamical system (2.25) and (2.26) if* $s_t(\mathcal{M}) = \mathcal{M}$ *for all* $t \in \mathbb{R}$.

In the case where $t \ge 0$ in (2.25) and (2.26), note that a set $\mathcal{M} \subseteq \mathcal{D}$ is a negatively invariant set with respect to the nonlinear dynamical system (2.25) and (2.26) if, for every $y \in \mathcal{M}$ and every $t \ge 0$, there exists $x \in \mathcal{M}$ such that $s(t, x) = y$ and $s(\tau, x) \in \mathcal{M}$ for all $\tau \in [0, t]$. Hence, if $\mathcal{M}$ is negatively invariant, then $\mathcal{M} \subseteq s_t(\mathcal{M})$ for all $t \ge 0$; the converse, however, is not generally true. Furthermore, a set $\mathcal{M} \subset \mathcal{D}$ is an invariant set with respect to (2.25) and (2.26) (defined over $t \ge 0$) if $s_t(\mathcal{M}) = \mathcal{M}$ for all $t \ge 0$. Note that a set $\mathcal{M}$ is invariant if and only if $\mathcal{M}$ is positively and negatively invariant.

Definition 2.4 *The trajectory* $s(\cdot, x)$ *of (2.25) and (2.26) is* bounded *if there exists* $\gamma > 0$ *such that* $\|s(t, x)\| < \gamma$, $t \in \mathbb{R}$.

Next, we state and prove a fundamental result on positive limit sets for impulsive dynamical systems. This result generalizes the classical results on positive limit sets to systems with left-continuous flows. An analogous result holds for negative limit sets and is left as an exercise for the reader. Furthermore, we use the notation $x(t) \to \mathcal{M} \subseteq \mathcal{D}$ as $t \to \infty$ to denote that $x(t)$ approaches $\mathcal{M}$, that is, for each $\varepsilon > 0$, there exists $T > 0$ such that $\text{dist}(x(t), \mathcal{M}) < \varepsilon$ for all $t > T$.

Theorem 2.2 *Consider the impulsive dynamical system $\mathcal{G}$ given by (2.25) and (2.26), assume Assumption 2.1 holds, and suppose that for $x_0 \in \mathcal{D}$ the trajectory $s(t, x_0)$ of $\mathcal{G}$ is bounded for all $t \geq 0$. Then the positive limit set $\omega(x_0)$ of $s(t, x_0)$, $t \geq 0$, is a nonempty, compact invariant set. Furthermore, $s(t, x_0) \rightarrow \omega(x_0)$ as $t \rightarrow \infty$.*

Proof. Let $s(t, x_0)$, $t \geq 0$, denote the solution to $\mathcal{G}$ with initial condition $x_0 \in \mathcal{D}$. Since $s(t, x_0)$ is bounded for all $t \geq 0$, it follows from the Bolzano-Weierstrass theorem [146] that every sequence in the positive orbit $\mathcal{O}_{x_0}^+ \triangleq \{s(t, x_0) : t \in [0, \infty)\}$ has at least one accumulation point $y \in \mathcal{D}$ as $t \rightarrow \infty$, and hence, $\omega(x_0)$ is nonempty. Furthermore, since $s(t, x_0)$, $t \geq 0$, is bounded it follows that $\omega(x_0)$ is bounded. To show that $\omega(x_0)$ is closed let $\{y_i\}_{i=0}^{\infty}$ be a sequence contained in $\omega(x_0)$ such that $\lim_{i \rightarrow \infty} y_i = y \in \mathcal{D}$. Now, since $y_i \rightarrow y$ as $i \rightarrow \infty$ it follows that for every $\varepsilon > 0$, there exists i such that $\|y - y_i\| < \varepsilon/2$. Next, since $y_i \in \omega(x_0)$ it follows that for every $T > 0$, there exists $t \geq T$ such that $\|s(t, x_0) - y_i\| < \varepsilon/2$. Hence, it follows that for every $\varepsilon > 0$ and $T > 0$, there exists $t \geq T$ such that $\|s(t, x_0) - y\| \leq \|s(t, x_0) - y_i\| + \|y_i - y\| < \varepsilon$, which implies that $y \in \omega(x_0)$, and hence, $\omega(x_0)$ is closed. Thus, since $\omega(x_0)$ is closed and bounded, $\omega(x_0)$ is compact.

Next, to show positive invariance of $\omega(x_0)$ let $y \in \omega(x_0)$ so that there exists an increasing unbounded sequence $\{t_n\}_{n=0}^{\infty} \subset \mathcal{T}_{x_0}$ such that $s(t_n, x_0) \rightarrow y$ as $n \rightarrow \infty$. Now, it follows from Assumption 2.1 that for every $\varepsilon > 0$ and $t \in \mathcal{T}_y$, there exists $\delta(\varepsilon, y, t) > 0$ such that $\|y - z\| < \delta(\varepsilon, y, t)$, $z \in \mathcal{D}$, implies $\|s(t, y) - s(t, z)\| < \varepsilon$ or, equivalently, for every sequence $\{y_i\}_{i=1}^{\infty}$ converging to y and $t \in \mathcal{T}_y$, $\lim_{i \rightarrow \infty} s(t, y_i) = s(t, y)$. Now, since by assumption there exists a unique solution to $\mathcal{G}$, it follows that the semi-group property $s(\tau, s(t, x_0)) = s(t + \tau, x_0)$ for all $x_0 \in \mathcal{D}$ and $t, \tau \in [0, \infty)$ holds. Furthermore, since $s(t_n, x_0) \rightarrow y$ as $n \rightarrow \infty$, it follows from the semi-group property that $s(t, y) = s(t, \lim_{n \rightarrow \infty} s(t_n, x_0)) = \lim_{n \rightarrow \infty} s(t + t_n, x_0) \in \omega(x_0)$ for all $t \in \mathcal{T}_y$. Hence, $s(t, y) \in \omega(x_0)$ for all $t \in \mathcal{T}_y$. Next, let $t \in [0, \infty) \backslash \mathcal{T}_y$ and note that, since $\mathcal{T}_y$ is dense in $[0, \infty)$, there exists a sequence $\{\tau_n\}_{n=0}^{\infty}$ such that $\tau_n \leq t$, $\tau_n \in \mathcal{T}_y$, and $\lim_{n \rightarrow \infty} \tau_n = t$. Now, since $s(\cdot, y)$ is left-continuous it follows that $\lim_{n \rightarrow \infty} s(\tau_n, y) = s(t, y)$. Finally, since $\omega(x_0)$ is closed and $s(\tau_n, y) \in \omega(x_0)$, $n = 1, 2, \ldots$, it follows that $s(t, y) = \lim_{n \rightarrow \infty} s(\tau_n, y) \in \omega(x_0)$. Hence, $s_t(\omega(x_0)) \subseteq \omega(x_0)$, $t \geq 0$, establishing positive invariance of $\omega(x_0)$.

Now, to show invariance of $\omega(x_0)$ let $y \in \omega(x_0)$ so that there exists an increasing unbounded sequence $\{t_n\}_{n=0}^{\infty}$ such that $s(t_n, x_0) \rightarrow y$ as $n \rightarrow \infty$. Next, let $t \in \mathcal{T}_{x_0}$ and note that there exists $N \in \mathbb{Z}_+$

such that $t_n > t$, $n \geq N$. Hence, it follows from the semi-group property that $s(t, s(t_n - t, x_0)) = s(t_n, x_0) \to y$ as $n \to \infty$. Now, it follows from the Bolzano-Weierstass theorem [146] that there exists a subsequence z_{n_k} of the sequence $z_n = s(t_n - t, x_0)$, $n = N, N + 1, \ldots$, such that $z_{n_k} \to z \in \mathcal{D}$ and, by definition, $z \in \omega(x_0)$. Next, it follows from Assumption 2.1 that $\lim_{k \to \infty} s(t, z_{n_k}) = s(t, \lim_{k \to \infty} z_{n_k})$, and hence, $y = s(t, z)$, which implies that $\omega(x_0) \subseteq s_t(\omega(x_0))$, $t \in \mathcal{T}_{x_0}$. Next, let $t \in [0, \infty) \backslash \mathcal{T}_{x_0}$, let $\hat{t} \in \mathcal{T}_{x_0}$ be such that $\hat{t} > t$, and consider $y \in \omega(x_0)$. Now, there exists $\hat{z} \in \omega(x_0)$ such that $y = s(\hat{t}, \hat{z})$, and it follows from the positive invariance of $\omega(x_0)$ that $z = s(\hat{t} - t, \hat{z}) \in \omega(x_0)$. Furthermore, it follows from the semi-group property of $\mathcal{G}$ (i.e., $s(\tau, s(t, x_0)) = s(t + \tau, x_0)$ for all $x_0 \in \mathcal{D}$ and $t, \tau \in [0, \infty)$) that $s(t, z) = s(t, s(\hat{t} - t, \hat{z})) = s(\hat{t}, \hat{z}) = y$, which implies that for all $t \in [0, \infty) \backslash \mathcal{T}_{x_0}$ and for every $y \in \omega(x_0)$, there exists $z \in \omega(x_0)$ such that $y = s(t, z)$. Hence, $\omega(x_0) \subseteq s_t(\omega(x_0))$, $t \geq 0$. Now, using positive invariance of $\omega(x_0)$ it follows that $s_t(\omega(x_0)) = \omega(x_0)$, $t \geq 0$, establishing invariance of the positive limit set $\omega(x_0)$.

Finally, to show $s(t, x_0) \to \omega(x_0)$ as $t \to \infty$, suppose, *ad absurdum*, $s(t, x_0) \not\to \omega(x_0)$ as $t \to \infty$. In this case, there exists an $\varepsilon > 0$ and a sequence $\{t_n\}_{n=0}^{\infty}$, with $t_n \to \infty$ as $n \to \infty$, such that

$$\inf_{p \in \omega(x_0)} \|s(t_n, x_0) - p\| \geq \varepsilon, \quad n \geq 0.$$

However, since $s(t, x_0)$, $t \geq 0$, is bounded, the bounded sequence $\{s(t_n, x_0)\}_{n=0}^{\infty}$ contains a convergent subsequence $\{s(t_n^*, x_0)\}_{n=0}^{\infty}$ such that $s(t_n^*, x_0) \to p^* \in \omega(x_0)$ as $n \to \infty$, which contradicts the original supposition. Hence, $s(t, x_0) \to \omega(x_0)$ as $t \to \infty$. $\square$

Note that the compactness of the positive limit set $\omega(x_0)$ depends only on the boundedness of the trajectory $s(t, x_0)$, $t \geq 0$, whereas left-continuity and Assumption 2.1 are key in proving invariance of the positive limit set $\omega(x_0)$. In classical dynamical systems, where the trajectory $s(\cdot, \cdot)$ is assumed to be continuous in both its arguments, both the left-continuity and the quasi-continuous dependence properties are trivially satisfied. Finally, we note that unlike dynamical systems with continuous flows, the positive limit set of an impulsive dynamical system may not be connected.

Example 2.5 To demonstrate the importance of the quasi-continuous dependence property for the invariance of positive limit sets, consider the state-dependent impulsive dynamical system

$$\dot{x}(t) = -x(t), \quad x(t) \neq 0, \tag{2.64}$$

$$\Delta x(t) = 1, \quad x(t) = 0, \tag{2.65}$$

where $t \geq 0$, $x(t) \in \mathbb{R}$, and $x(0) = x_0$. In this case, the trajectory $s(\cdot, \cdot)$ is given by $s(0, x_0) = x_0$, $x_0 \in \mathbb{R}$, and for all $t > 0$,

$$s(t, x_0) = \begin{cases} e^{-t} x_0, & x_0 \neq 0, \\ e^{-t}, & x_0 = 0, \end{cases} \tag{2.66}$$

which shows that for every $x_0 \in \mathbb{R}$, the trajectory $s(t, x_0)$ is left-continuous in t and approaches the positive limit set containing only the origin. However, note that the dynamical system does not satisfy the quasi-continuous dependence property and the origin is *not* an invariant set. $\triangle$

2.5 Necessary and Sufficient Conditions for Quasi-Continuous Dependence

In this section, we develop necessary and sufficient conditions for quasi-continuous dependence of solutions for state-dependent impulsive dynamical systems. The following result provides sufficient conditions that guarantee that the impulsive dynamical system $\mathcal{G}$ given by (2.25) and (2.26) satisfies Assumption 2.1. For this result, the following definition of stability with respect to a compact positively invariant set is needed.

Definition 2.5 *Let $\mathcal{M} \subset \mathcal{D}$ be a compact positively invariant set for the nonlinear impulsive dynamical system (2.25) and (2.26). $\mathcal{M}$ is Lyapunov stable if for every open neighborhood $\mathcal{O}_1 \subseteq \mathcal{D}$ of $\mathcal{M}$, there exists an open neighborhood $\mathcal{O}_2 \subseteq \mathcal{O}_1$ of $\mathcal{M}$ such that $x(t) \in \mathcal{O}_1$ for all $x_0 \in \mathcal{O}_2$ and $t \geq 0$.*

Proposition 2.1 *Consider the nonlinear impulsive dynamical system $\mathcal{G}$ given by (2.25) and (2.26). Assume A1 and A2 hold, and assume that either of the following statements holds:*

i) *For all $x_0 \in \mathcal{D}$, $0 \leq \tau_1(x_0) < \infty$, $\tau_1(\cdot)$ is continuous, and $\lim_{k \to \infty} \tau_k(x_0) \to \infty$.*

ii) *For all $x_0 \notin \overline{\mathcal{Z}} \backslash \mathcal{Z}$, $0 \leq \tau_1(x_0) < \infty$, $\tau_1(\cdot)$ is continuous, $\lim_{k \to \infty} \tau_k(x_0) \to \infty$, and $\overline{\mathcal{Z}} \backslash \mathcal{Z}$ is a Lyapunov stable, compact positively invariant set with respect to the dynamical system $\mathcal{G}$.*

Then $\mathcal{G}$ satisfies Assumption 2.1.

Proof. To show that $i)$ implies Assumption 2.1, assume that for all $x_0 \notin \mathcal{Z}, 0 < \tau_1(x_0) < \infty, \tau_1(\cdot)$ is continuous, and $\lim_{k \to \infty} \tau_k(x_0) \to \infty$. In this case, it follows from the definition of $\tau_k(x_0)$ that for every $x_0 \in \mathcal{D}$ and $k \in \{1, 2, \ldots, \}$,

$$\tau_k(x_0) = \tau_{k-j}(x_0) + \tau_j[s(\tau_{k-j}(x_0), x_0) + f_\mathrm{d}(s(\tau_{k-j}(x_0), x_0))],$$
$$j = 1, \ldots, k, \quad (2.67)$$

where $\tau_0(x_0) \triangleq 0$. Since $f_\mathrm{c}(\cdot)$ is such that the solutions to (2.25) are continuous with respect to the initial conditions between resetting events, it follows that for every $k = 0, 1, \ldots$, and $t \in (\tau_k(x_0), \tau_{k+1}(x_0)]$, $\psi(\cdot, \cdot)$ is continuous in both its arguments. Specifically, note that since $\tau_1(x_0)$ is continuous it follows that $\eta_1(x_0) \triangleq s(\tau_1(x_0), x_0) = \psi(\tau_1(x_0), x_0)$ is continuous on $\mathcal{D}$. Hence, it follows from (2.67) and the continuity of $f_\mathrm{d}(\cdot)$ that $\tau_2(x_0) = \tau_1(x_0) + \tau_1[s(\tau_1(x_0), x_0) + f_\mathrm{d}(s(\tau_1(x_0), x_0))]$ is also continuous which implies that $\eta_2(x_0) \triangleq s(\tau_2(x_0), x_0) = \psi(\tau_2(x_0) - \tau_1(x_0), \eta_1(x_0) + f_\mathrm{d}(\eta_1(x_0))$ is continuous on $\mathcal{D}$. By recursively repeating this procedure for $k = 3, 4, \ldots$, it follows that $\tau_k(x_0)$ and $\eta_k(x_0) \triangleq s(\tau_k(x_0), x_0)$ are continuous on $\mathcal{D}$.

Next, let $t \in \mathcal{I}_{x_0}$ be such that $\tau_k(x_0) < t < \tau_{k+1}(x_0)$. Now, noting that $s(t, x_0) = \psi(t - \tau_k(x_0), s(\tau_k(x_0), x_0) + f_\mathrm{d}(s(\tau_k(x_0), x_0)))$, it follows from the continuity of $f_\mathrm{d}(\cdot)$ and $\tau_k(\cdot)$ that $s(t, x_0)$ is a continuous function of x_0 for all $t \in \mathcal{I}_{x_0}$ such that $\tau_k(x_0) < t < \tau_{k+1}(x_0)$ for some k. Hence, since $\lim_{k \to \infty} \tau_k(x_0) \to \infty$, $\mathcal{G}$ satisfies Assumption 2.1. Next, consider the case in which $x_0 \in \mathcal{Z}$. Note that in this case $\tau_1(x_0) = 0$ and $\tau_2(x_0) = \tau_1(x_0 + f_\mathrm{d}(x_0))$. Since $x_0 \in \mathcal{Z}$, it follows that $x_0 + f_\mathrm{d}(x_0) \notin \mathcal{Z}$, and since $\tau_1(\cdot)$ is continuous on $\mathcal{D}$ and $f_\mathrm{d}(\cdot)$ is continuous on $\mathcal{Z}$, it follows that $\tau_2(\cdot)$ is continuous on $\mathcal{Z}$. Now, Assumption 2.1 for all $x_0 \in \mathcal{Z}$ can be shown as above.

Alternatively, if $ii)$ is satisfied then as in the proof of $i)$ it can be shown that for all $x_0 \in \mathcal{D}, x_0 \notin \overline{\mathcal{Z}} \backslash \mathcal{Z}, s(t, x_0)$ is a continuous function of x_0 for all $t \in \mathcal{I}_{x_0}$. Next, if $\overline{\mathcal{Z}} \backslash \mathcal{Z}$ is a Lyapunov stable, compact positively invariant set with respect to $\mathcal{G}$, then for all $x_0 \in \overline{\mathcal{Z}} \backslash \mathcal{Z}, \mathcal{I}_{x_0} = [0, \infty)$. Now, the continuity of $s(t, x_0)$ for all $t \in [0, \infty)$ follows from the Lyapunov stability of $\overline{\mathcal{Z}} \backslash \mathcal{Z}$, and hence, $\mathcal{G}$ satisfies Assumption 2.1. $\square$

If, for every $x_0 \in \mathcal{D}$, the solution $s(t, x_0)$ to (2.25) and (2.26) is a Zeno solution, that is, $\lim_{k \to \infty} \tau_k(x_0) \to \tau(x_0) < \infty$, and the resetting sequence $\{\tau_k(x_0)\}_{k=0}^{\infty}$ is uniformly convergent in x_0, then condition $ii)$ of Proposition 2.1 implies that $\mathcal{G}$ satisfies Assumption 2.1. To see this, note that since $\{\tau_k(\cdot)\}_{k=1}^{\infty}$ is a uniformly convergent sequence of continuous functions, it follows that $\tau(\cdot)$ is a continuous function. Now, noting that for all $t > \tau(x_0)$, $t \in \mathcal{T}_{x_0}$, $s(t, x_0) = \psi(t - \tau(x_0), s(\tau^+(x_0), x_0))$, it follows that $s(t, x_0)$ is a continuous function of x_0 for all $t \in \mathcal{T}_{x_0}$, which proves that $\mathcal{G}$ satisfies Assumption 2.1.

Proposition 2.1 requires that the first resetting time $\tau_1(\cdot)$ be continuous at $x_0 \in \mathcal{D}$. The following result provides sufficient conditions for establishing the continuity of $\tau_1(\cdot)$ at $x_0 \in \mathcal{D}$.

Proposition 2.2 *Consider the nonlinear impulsive dynamical system $\mathcal{G}$ given by (2.25) and (2.26). Assume there exists a continuously differentiable function $\mathcal{X} : \mathcal{D} \to \mathbb{R}$ such that the resetting set of $\mathcal{G}$ is given by $\mathcal{Z} = \{x \in \mathcal{D} : \mathcal{X}(x) = 0\}$ and $\mathcal{X}'(x) f_c(x) \neq 0$, $x \in \mathcal{Z}$. Then $\tau_1(\cdot)$ is continuous at $x_0 \in \mathcal{D}$, where $0 < \tau_1(x_0) < \infty$.*

Proof. Let $x_0 \notin \overline{\mathcal{Z}}$ be such that $0 < \tau_1(x_0) < \infty$. It follows from the definition of $\tau_1(\cdot)$ that $s(t, x_0) = \psi(t, x_0)$, $t \in [0, \tau_1(x_0)]$, $\mathcal{X}(s(t, x_0)) \neq 0$, $t \in (0, \tau_1(x_0))$, and $\mathcal{X}(s(\tau_1(x_0), x_0)) = 0$. Without loss of generality, let $\mathcal{X}(s(t, x_0)) > 0$, $t \in (0, \tau_1(x_0))$. Since $\hat{x} \triangleq \psi(\tau_1(x_0), x_0) \in \overline{\mathcal{Z}}$, it follows by assumption that $\mathcal{X}'(\hat{x}) f_c(\hat{x}) \neq 0$, and hence, there exists $\theta > 0$ such that $\mathcal{X}(\psi(t, \hat{x})) > 0$, $t \in [-\theta, 0)$, and $\mathcal{X}(\psi(t, \hat{x})) < 0$, $t \in (0, \theta]$. (This fact can be easily shown by expanding $\mathcal{X}(\psi(t, x))$ via a Taylor series expansion about $\hat{x}$ and using the fact that $\mathcal{X}'(\hat{x}) f_c(\hat{x}) \neq 0$.) Hence, $\mathcal{X}(\psi(t, x_0)) > 0$, $t \in [\hat{t}_1, \tau_1(x_0))$, and $\mathcal{X}(\psi(t, x_0)) < 0$, $t \in (\tau_1(x_0), \hat{t}_2]$, where $\hat{t}_1 \triangleq \tau_1(x_0) - \theta$ and $\hat{t}_2 \triangleq \tau_1(x_0) + \theta$.

Next, let $\varepsilon \triangleq \min\{\mathcal{X}(\psi(\hat{t}_1, x_0)), \mathcal{X}(\psi(\hat{t}_2, x_0))\}$. Now, since $\mathcal{X}(\cdot)$ and $\psi(\cdot, \cdot)$ are jointly continuous, it follows that there exists $\delta > 0$ such that

$$\sup_{0 \leq t \leq \hat{t}_2} |\mathcal{X}(\psi(t, x)) - \mathcal{X}(\psi(t, x_0))| < \varepsilon, \quad x \in \mathcal{B}_\delta(x_0), \quad (2.68)$$

which implies that $\mathcal{X}(\psi(\hat{t}_1, x)) > 0$ and $\mathcal{X}(\psi(\hat{t}_2, x)) < 0$, $x \in \mathcal{B}_\delta(x_0)$. Hence, it follows that $\hat{t}_1 < \tau_1(x) < \hat{t}_2$, $x \in \mathcal{B}_\delta(x_0)$. The continuity of $\tau_1(\cdot)$ at x_0 now follows immediately by noting that θ can be chosen arbitrarily small. $\qquad \square$

The first assumption in Proposition 2.2 implies that the resetting set $\mathcal{Z}$ is an embedded submanifold [80], while the second assumption assures that the solution of $\mathcal{G}$ is not tangent to the resetting set $\mathcal{Z}$. The next result provides a partial converse to Proposition 2.1. For this result, we introduce the following assumption in place of A1 and A2.

A3. $f_c(\cdot)$ is locally Lipschitz continuous on $\mathcal{D}$, $\mathcal{Z}$ is closed, and $f_d(x) \neq 0$, $x \in \mathcal{Z} \backslash \partial \mathcal{Z}$. If $x \in \partial \mathcal{Z}$ such that $f_d(x) = 0$, then $f_c(x) = 0$. If $x \in \mathcal{Z}$ such that $f_d(x) \neq 0$, then $x + f_d(x) \notin \mathcal{Z}$.

The following definitions are needed for the statement of the next result.

Definition 2.6 *Let* $\mathcal{D} \subseteq \mathbb{R}^n$, $f : \mathcal{D} \to \mathbb{R}$, *and* $x \in \mathcal{D}$. f *is lower-semicontinuous at* $x \in \mathcal{D}$ *if for every sequence* $\{x_n\}_{n=0}^{\infty} \subset \mathcal{D}$ *such that* $\lim_{n \to \infty} x_n = x$, $f(x) \leq \lim \inf_{n \to \infty} f(x_n)$.

Note that a function $f : \mathcal{D} \to \mathbb{R}$ is lower-semicontinuous at $x \in \mathcal{D}$ if and only if for each $\alpha \in \mathbb{R}$ the set $\{x \in \mathcal{D} : f(x) > \alpha\}$ is open. Equivalently, a bounded function $f : \mathcal{D} \to \mathbb{R}$ is lower-semicontinuous at $x \in \mathcal{D}$ if and only if for each $\varepsilon > 0$, there exists $\delta > 0$ such that $\|x - y\| < \delta$, $y \in \mathcal{D}$, implies $f(x) - f(y) \leq \varepsilon$.

Definition 2.7 *Let* $\mathcal{D} \subseteq \mathbb{R}^n$, $f : \mathcal{D} \to \mathbb{R}$, *and* $x \in \mathcal{D}$. f *is upper-semicontinuous at* $x \in \mathcal{D}$ *if for every sequence* $\{x_n\}_{n=0}^{\infty} \subset \mathcal{D}$ *such that* $\lim_{n \to \infty} x_n = x$, $f(x) \geq \lim \sup_{n \to \infty} f(x_n)$, *or, equivalently, for each* $\alpha \in \mathbb{R}$ *the set* $\{x \in \mathcal{D} : f(x) < \alpha\}$ *is open.*

As in the case of continuous functions, a function f is said to be lower- (respectively, upper-) semicontinuous on $\mathcal{D}$ if f is lower- (respectively, upper-) semicontinuous at every point $x \in \mathcal{D}$. Clearly, if f is both lower- and upper-semicontinuous, then f is continuous.

Proposition 2.3 *Consider the nonlinear impulsive dynamical system $\mathcal{G}$ given by (2.25) and (2.26), and assume A3 holds. If $\mathcal{G}$ satisfies Assumption 2.1, then $\tau_1(\cdot)$ is lower-semicontinuous at every $x \notin \mathcal{Z}$. Furthermore, for every $x \notin \mathcal{Z}$ such that $\tau_1(x) < \infty$, $\tau_1(\cdot)$ is continuous at x. Finally, for every $x \notin \mathcal{Z}$ such that $\tau_1(x) = \infty$, $\tau_1(x_n) \to \infty$ for every sequence $\{x_n\}_{n=1}^{\infty}$ such that $x_n \to x$.*

Proof. Assume $\mathcal{G}$ satisfies Assumption 2.1. Let $x \notin \mathcal{Z}$ and let $\{x_n\}_{n=1}^{\infty} \notin \mathcal{Z}$ be such that $x_n \to x$ and $\tau_1(x_1) \geq \tau_1(x_2) \geq \cdots \geq \tau_- \overset{\triangle}{=} \lim_{n \to \infty} \tau_1(x_n)$. First, assume $\tau_1(x_1) < \infty$ so that $\tau_-, \tau_1(x_2), \ldots < \infty$. Since $f_c(\cdot)$ is locally Lipschitz continuous on $\mathcal{D}$ it follows that $\psi(\cdot, \cdot)$ is jointly continuous, and hence, it follows that $\psi(\tau_1(x_n), x_n) \to \psi(\tau_-, x)$ as $n \to \infty$. Next, since $\mathcal{Z}$ is closed and $\psi(\tau_1(x_n), x_n) \in \mathcal{Z}$ for every $n = 1, 2, \ldots$, it follows that $\psi(\tau_-, x) \in \mathcal{Z}$ which implies that $\tau_- \geq \tau_1(x) = \inf\{t \in \overline{\mathbb{R}}_+ : \psi(t, x) \in \mathcal{Z}\}$, establishing the lower semicontinuity of $\tau_1(\cdot)$ at x. Alternatively, if $\tau_- = \infty$ so that

$\tau_1(x_1) = \tau_1(x_2) = \cdots = \infty$, lower semicontinuity follows trivially since $\tau_1(x) \leq \tau_- = \infty$.

Next, note that since $f_c(\cdot)$ is locally Lipschitz continuous on $\mathcal{D}$ it follows that $\psi(t,x)$, $t \geq 0$, cannot converge to any equilibrium in a finite time, and hence, $f_c(\psi(\tau_1(x), x)) \neq 0$, which implies that $f_d(\psi(\tau_1(x), x)) \neq 0$. Let $\{x_n\}_{n=1}^{\infty} \notin \mathcal{Z}$ be such that $x_n \to x$ as $n \to \infty$ and $\tau_1(x_1) \leq \tau_1(x_2) \leq \cdots \leq \tau_+ \triangleq \lim_{n\to\infty} \tau_1(x_n)$. Suppose, *ad absurdum*, $\tau_+ > \tau_1(x)$, let $\varepsilon > 0$ be such that $\tau_1(x) < \tau_+ - \varepsilon < \tau_2(x)$, and let $M \in \mathbb{Z}_+$ be such that $\tau_+ - \varepsilon < \tau_1(x_n)$, $n > M$. Now, since $\mathcal{G}$ satisfies Assumption 2.1, it follows that $s(\tau_+ - \varepsilon, x_n) \to s(\tau_+ - \varepsilon, x)$, and for every $n \geq M$, $s(\tau_+ - \varepsilon, x_n) = \psi(\tau_+ - \varepsilon, x_n)$. Furthermore, $\lim_{n\to\infty} s(\tau_+ - \varepsilon, x_n) = \lim_{n\to\infty} \psi(\tau_+ - \varepsilon, x_n) = \psi(\tau_+ - \varepsilon, x)$. Hence,

$$\psi(\tau_+ - \varepsilon - \tau_1(x), \psi(\tau_1(x), x) + f_d(\psi(\tau_1(x), x)))$$
$$= s(\tau_+ - \varepsilon, x)$$
$$= \lim_{n\to\infty} s(\tau_+ - \varepsilon, x_n)$$
$$= \psi(\tau_+ - \varepsilon, x)$$
$$= \psi(\tau_+ - \varepsilon - \tau_1(x), \psi(\tau_1(x), x)). \tag{2.69}$$

Now, since $f_c(\cdot)$ is locally Lipschitz continuous on $\mathcal{D}$ it follows that the solution $\psi(t,x)$, $t \in \mathbb{R}$, is unique both forward and backward in time, and hence, it follows that $\psi(\tau_1(x), x) = \psi(\tau_1(x), x) + f_d(\psi(\tau_1(x), x))$, or, equivalently, $f_d(\psi(\tau_1(x), x)) = 0$, which is a contradiction. Hence, $\tau_+ \leq \tau_1(x)$, and thus, $\tau_1(\cdot)$ is upper-semicontinuous at x. Hence, $\tau_1(\cdot)$ is continuous at x.

Finally, let $x \notin \mathcal{Z}$ be such that $\tau_1(x) = \infty$ and let $\{x_n\}_{n=1}^{\infty} \in \mathcal{Z}$ be such that $x_n \to x$. Suppose, *ad absurdum*, that $\{\tau_1(x_n)\}_{n=1}^{\infty}$ has a bounded subsequence $\{\tau_1(x_{n_j})\}_{j=1}^{\infty}$. Let $\tau \triangleq \lim_{j\to\infty} \tau_1(x_{n_j}) < \infty$. Now, since $\psi(\cdot, \cdot)$ is jointly continuous it follows that $\lim_{j\to\infty} \psi(\tau_1(x_{n_j}), x_{n_j}) = \psi(\tau, x)$. Next, since $\mathcal{Z}$ is closed and $\psi(\tau_1(x_{n_j}), x_{n_j}) \in \mathcal{Z}$, $j = 1, 2, \ldots$, it follows that $\psi(\tau, x) \in \mathcal{Z}$ which implies that $\tau_1(x) = \inf\{t \in \overline{\mathbb{R}}_+ : \psi(t,x) \notin \mathcal{Z}\} \leq \tau < \infty$, which is a contradiction. Hence, $\lim_{n\to\infty} \tau_1(x_n) = \infty$. $\square$

The following result shows that all convergent Zeno solutions to (2.25) and (2.26) converge to $\overline{\mathcal{Z}} \setminus \mathcal{Z}$ if A1 and A2 hold, while all convergent Zeno solutions converge to an equilibrium point if A3 holds.

Proposition 2.4 *Consider the nonlinear impulsive dynamical system $\mathcal{G}$ given by (2.25) and (2.26). If the trajectory $s(t, x_0)$, $t \geq 0$, to (2.25) and (2.26) is convergent, bounded, and Zeno, that is, there exists $\tau(x_0) < \infty$ such that $\tau_k(x_0) \to \tau(x_0)$ as $k \to \infty$ and*

$\lim_{k\to\infty} s(\tau_k(x_0), x_0) = s(\tau(x_0), x_0)$, *then the following statements hold:*

   i) If A1 and A2 hold, and $\tau_2(\cdot)$ is continuous on $\mathcal{Z}$, then $s(\tau(x_0), x_0)$ $\in \overline{\mathcal{Z}} \backslash \mathcal{Z}$.

   ii) If A3 holds, then $s(\tau(x_0), x_0)$ is an equilibrium point.

Proof. If the trajectory $s(t, x_0)$, $t \geq 0$, is Zeno, then there exists $\tau(x_0) < \infty$ such that $\tau_k(x_0) \to \tau(x_0)$ as $k \to \infty$ and, since $\tau_1(x_0) < \tau_2(x_0) < \cdots < \tau(x_0)$, it follows that $\tau_1(x_0) < \infty$. Next, note that there exists $y_1 \in \mathcal{Z}$ such that $s(\tau_1(x_0), x_0) = y_1$, and hence, it follows that $\tau_2(x_0) = \tau_1(x_0) + \tau_1(y_1 + f_d(y_1)) = \tau_1(x_0) + \tau_2(y_1)$. By recursively repeating this procedure for $k = 3, 4, \ldots$, it follows that

$$\tau_k(x_0) = \tau_1(x_0) + \sum_{i=1}^{k-1} \tau_2(y_i),$$

where $y_i \triangleq s(\tau_i(x_0), x_0)$, $i = 1, 2, \ldots$. Now, since $\tau(x_0) = \lim_{k\to\infty} \tau_k$ (x_0) it follows that $\tau(x_0) = \tau_1(x_0) + \sum_{k=1}^{\infty} \tau_2(y_k)$. Hence, it follows that $\tau_2(y_k) \to 0$ as $k \to \infty$. Now, if the trajectory $s(t, x_0)$, $t \geq 0$, is bounded, then the sequence $\{y_k\}_{k=0}^{\infty}$ is also bounded and it follows from the Bolzano-Weierstrass theorem [146] that there exists a convergent subsequence $\{y_{k_i}\}_{i=1}^{\infty}$ such that $\lim_{i\to\infty} y_{k_i} = y \in \overline{\mathcal{Z}}$. Hence, since $s(\cdot, x_0)$ is left-continuous, it follows that $y = \lim_{i\to\infty} y_{k_i} = \lim_{i\to\infty} s(\tau_{k_i}(x_0), x_0) = s(\lim_{i\to\infty} \tau_{k_i}(x_0), x_0) = s(\tau(x_0), x_0)$.

   i) Assume A1 and A2 hold, and assume $\tau_2(\cdot)$ is continuous on $\mathcal{Z}$. Next, *ad absurdum*, suppose $y \in \mathcal{Z}$. Since $\tau_2(\cdot)$ is continuous on $\mathcal{Z}$ it follows that $\tau_2(y) = \tau_2(\lim_{i\to\infty} y_{k_i}) = \lim_{i\to\infty} \tau_2(y_{k_i}) = 0$, which contradicts the fact that $\tau_2(x) > 0$, $x \in \mathcal{Z}$. Thus, $y \in \overline{\mathcal{Z}} \backslash \mathcal{Z}$ or, equivalently, $s(\tau(x_0), x_0) \in \overline{\mathcal{Z}} \backslash \mathcal{Z}$.

   ii) Finally, assume A3 holds. Furthermore, note that $y_k = \psi(\tau_2(y_{k-1}), y_{k-1} + f_d(y_{k-1}))$, $k = 2, 3, \ldots$, and since $\psi(\cdot, \cdot)$ is jointly continuous and $\tau_2(y_k) \to 0$ as $k \to \infty$, it follows that

$$\begin{aligned}
y &= \lim_{k\to\infty} y_k \\
&= \psi(\lim_{k\to\infty} \tau_2(y_k), \lim_{k\to\infty} (y_k + f_d(y_k))) \\
&= \psi(0, y + f_d(y)) \\
&= y + f_d(y),
\end{aligned}$$

which implies that $f_d(y) = 0$. Now, since $\mathcal{Z}$ is closed it follows that $y \in \mathcal{Z}$, and since $f_d(y) = 0$, it follows from A3 that $f_c(y) = 0$, which proves the result. $\qquad\qquad\square$

2.6 Invariant Set Theorems for State-Dependent Impulsive Dynamical Systems

In this section, we generalize the Krasovskii-LaSalle invariance principle to state-dependent impulsive dynamical systems. This result characterizes impulsive dynamical system limit sets in terms of continuously differentiable functions. In particular, we show that the system trajectories converge to an invariant set contained in a union of level surfaces characterized by the continuous-time dynamics and the resetting system dynamics. Henceforth, we assume that $f_c(\cdot)$, $f_d(\cdot)$, and $\mathcal{Z}$ are such that the dynamical system $\mathcal{G}$ given by (2.25) and (2.26) satisfies Assumption 2.1. For the next result $V^{-1}(\gamma)$ denotes the γ-*level set* of $V(\cdot)$, that is, $V^{-1}(\gamma) \triangleq \{x \in \mathcal{D}_c : V(x) = \gamma\}$, where $\gamma \in \mathbb{R}$, $\mathcal{D}_c \subseteq \mathcal{D}$, and $V : \mathcal{D}_c \to \mathbb{R}$ is a continuously differentiable function, and let $\mathcal{M}_\gamma$ denote the largest invariant set (with respect to $\mathcal{G}$) contained in $V^{-1}(\gamma)$.

Theorem 2.3 *Consider the impulsive dynamical system $\mathcal{G}$ given by (2.25) and (2.26), assume $\mathcal{D}_c \subset \mathcal{D}$ is a compact positively invariant set with respect to (2.25) and (2.26), and assume that there exists a continuously differentiable function $V : \mathcal{D}_c \to \mathbb{R}$ such that*

$$V'(x)f_c(x) \leq 0, \quad x \in \mathcal{D}_c, \quad x \notin \mathcal{Z}, \tag{2.70}$$
$$V(x + f_d(x)) \leq V(x), \quad x \in \mathcal{D}_c, \quad x \in \mathcal{Z}. \tag{2.71}$$

Let $\mathcal{R} \triangleq \{x \in \mathcal{D}_c : x \notin \mathcal{Z}, V'(x)f_c(x) = 0\} \cup \{x \in \mathcal{D}_c : x \in \mathcal{Z}, V(x + f_d(x)) = V(x)\}$ and let $\mathcal{M}$ denote the largest invariant set contained in $\mathcal{R}$. If $x_0 \in \mathcal{D}_c$, then $x(t) \to \mathcal{M}$ as $t \to \infty$.

Proof. Using identical arguments as in the proof of Theorem 2.1 it follows that for all $t \in (\tau_k(x_0), \tau_{k+1}(x_0)]$,

$$V(x(t)) - V(x(0)) = \int_0^t V'(x(\tau))f_c(x(\tau))\mathrm{d}\tau$$
$$+ \sum_{i=1}^k [V(x(\tau_i(x_0)) + f_d(x(\tau_i(x_0)))) - V(x(\tau_i(x_0)))].$$

Hence, it follows from (2.70) and (2.71) that $V(x(t)) \leq V(x(0))$, $t \geq 0$. Using a similar argument it follows that $V(x(t)) \leq V(x(\tau))$, $t \geq \tau$, which implies that $V(x(t))$ is a nonincreasing function of time. Since $V(\cdot)$ is continuous on a compact set $\mathcal{D}_c$ there exists $\beta \in \mathbb{R}$ such that $V(x) \geq \beta$, $x \in \mathcal{D}_c$. Furthermore, since $V(x(t))$, $t \geq 0$, is nonincreasing, $\gamma_{x_0} \triangleq \lim_{t \to \infty} V(x(t))$, $x_0 \in \mathcal{D}_c$, exists. Now, for all $y \in \omega(x_0)$ there exists an increasing unbounded sequence $\{t_n\}_{n=0}^{\infty}$ such that $x(t_n) \to y$ as $n \to \infty$, and, since $V(\cdot)$ is continuous, it follows that $V(y) = V(\lim_{n \to \infty} x(t_n)) = \lim_{n \to \infty} V(x(t_n)) = \gamma_{x_0}$. Hence, $y \in V^{-1}(\gamma_{x_0})$ for all $y \in \omega(x_0)$, or, equivalently, $\omega(x_0) \subseteq V^{-1}(\gamma_{x_0})$.

Now, since $\mathcal{D}_c$ is compact and positively invariant, it follows that $x(t)$, $t \geq 0$, is bounded for all $x_0 \in \mathcal{D}_c$, and hence, it follows from Theorem 2.2 that $\omega(x_0)$ is a nonempty, compact invariant set. Thus, $\omega(x_0)$ is a subset of the largest invariant set contained in $V^{-1}(\gamma_{x_0})$, that is, $\omega(x_0) \subseteq \mathcal{M}_{\gamma_{x_0}}$. Hence, for every $x_0 \in \mathcal{D}_c$, there exists $\gamma_{x_0} \in \mathbb{R}$ such that $\omega(x_0) \subseteq \mathcal{M}_{\gamma_{x_0}}$, where $\mathcal{M}_{\gamma_{x_0}}$ is the largest invariant set contained in $V^{-1}(\gamma_{x_0})$, which implies that $V(x) = \gamma_{x_0}$, $x \in \omega(x_0)$. Now, since $\mathcal{M}_{\gamma_{x_0}}$ is a invariant set, it follows that for all $x(0) \in \mathcal{M}_{\gamma_{x_0}}$, $x(t) \in \mathcal{M}_{\gamma_{x_0}}$, $t \geq 0$, and hence, $\dot{V}(x(t)) \triangleq \frac{\mathrm{d}V(x(t))}{\mathrm{d}t} = V'(x(t))f_c(x(t)) = 0$, for all $x(t) \notin \mathcal{Z}$, and $V(x(t) + f_d(x(t))) = V(x(t))$, for all $x(t) \in \mathcal{Z}$. Thus, $\mathcal{M}_{\gamma_{x_0}}$ is contained in $\mathcal{M}$ which is the largest invariant set contained in $\mathcal{R}$. Hence, $x(t) \to \mathcal{M}$ as $t \to \infty$. $\square$

Example 2.6 Consider the nonlinear state-dependent impulsive dynamical system

$$
\begin{bmatrix} \dot{x}_1(t) \\ \dot{x}_2(t) \\ \dot{x}_3(t) \\ \dot{x}_4(t) \end{bmatrix} = \begin{bmatrix} x_3(t) \\ x_4(t) \\ x_1(t) - 2x_4(t) - \dfrac{x_1(t)}{\sqrt{x_1^2(t) + x_2^2(t)}} \\ x_2(t) + 2x_3(t) - \dfrac{x_2(t)}{\sqrt{x_1^2(t) + x_2^2(t)}} \end{bmatrix},
$$

$$
\begin{bmatrix} x_1(0) \\ x_2(0) \\ x_3(0) \\ x_4(0) \end{bmatrix} = \begin{bmatrix} x_{10} \\ x_{20} \\ x_{30} \\ x_{40} \end{bmatrix}, \quad x(t) \notin \overline{\mathcal{Z}}, \quad (2.72)
$$

$$
\begin{bmatrix} \dot{x}_1(t) \\ \dot{x}_2(t) \\ \dot{x}_3(t) \\ \dot{x}_4(t) \end{bmatrix} = \begin{bmatrix} -x_2(t) \\ x_1(t) \\ -x_4(t) \\ x_3(t) \end{bmatrix}, \quad x(t) \in \overline{\mathcal{Z}} \backslash \mathcal{Z}, \quad (2.73)
$$

$$
\begin{bmatrix} \Delta x_1(t) \\ \Delta x_2(t) \\ \Delta x_3(t) \\ \Delta x_4(t) \end{bmatrix} = \begin{bmatrix} 0 \\ 0 \\ -(1+e)(x_2(t) + x_3(t)) \\ (1+e)(x_1(t) - x_4(t)) \end{bmatrix}, \quad x(t) \in \mathcal{Z}, \quad (2.74)
$$

where $t \geq 0$, $x_1(t), x_2(t), x_3(t), x_4(t) \in \mathbb{R}$, $e \in (0,1)$, $\mathcal{Z} = \{x \in \mathcal{D}_c : x_1^2 + x_2^2 = 1, x_1x_3 + x_2x_4 < 0\}$, $x \triangleq [x_1 \ x_2 \ x_3 \ x_4]^{\mathrm{T}}$, and

$\mathcal{D} = \mathcal{D}_c = \{x \in \mathbb{R}^4 : x_1^2 + x_2^2 \geq 1, \ x_1 x_4 - x_2 x_3 = x_1^2 + x_2^2\}$. First, note that $\overline{\mathcal{Z}} = \{x \in \mathcal{D}_c : x_1^2 + x_2^2 = 1, \ x_1 x_3 + x_2 x_4 \leq 0\}$, and hence, $\overline{\mathcal{Z}} \backslash \mathcal{Z} = \{x \in \mathcal{D}_c : x_1^2 + x_2^2 = 1, \ x_1 x_3 + x_2 x_4 = 0\}$ which can be shown to be a compact invariant set with respect to the dynamical system $\mathcal{G}$ given by (2.72)–(2.74). Furthermore, note that $\mathcal{D}_c$ is an invariant set with respect to the impulsive dynamical system (2.72)–(2.74). To see this, consider the function $\phi(x) \triangleq x_1 x_4 - x_2 x_3 - x_1^2 - x_2^2$ and note that $\dot{\phi}(x)$ is identically zero along the solutions of (2.72), and $\phi(x + \Delta x) - \phi(x) = 0$ for all $x \in \mathcal{D}_c$.

Next, we use Proposition 2.1 to show that the dynamical system (2.72)–(2.74) satisfies Assumption 2.1. To see this, note that it can be shown that

$$\tau_1(x) = \begin{cases} \dfrac{x_1 x_3 + x_2 x_4 + \sqrt{2(x_1^2 + x_2^2)(\sqrt{x_1^2 + x_2^2} - 1) + (x_1 x_3 + x_2 x_4)^2}}{\sqrt{x_1^2 + x_2^2}}, & x_1^2 + x_2^2 > 1, \\ 2(x_1 x_3 + x_2 x_4), & x_1^2 + x_2^2 = 1, \\ & x_1 x_3 + x_2 x_4 > 0, \end{cases}$$
(2.75)

which shows that $\tau_1(x)$ is continuous for all $x \notin \overline{\mathcal{Z}}$. Furthermore, it follows from (2.75) that $\tau_1(x) \to 0$ as $x \to \partial \mathcal{Z}$ which implies that $\tau_1(\cdot)$ is continuous on $\mathcal{D}$. Finally, it can be shown that for all $x \in \mathcal{D}$, the sequence $\{\tau_k(x)\}_{k=1}^{\infty}$ is a uniformly convergent sequence. Now, it follows from $ii)$ of Proposition 2.1 that the dynamical system $\mathcal{G}$ given by (2.72)–(2.74) satisfies Assumption 2.1.

Next, (2.72)–(2.74) can be written in the form of (2.25) and (2.26) with

$$f_c(x) = \begin{bmatrix} x_3 \\ x_4 \\ x_1 - 2x_4 - \dfrac{x_1}{\sqrt{x_1^2 + x_2^2}} \\ x_2 + 2x_3 - \dfrac{x_2}{\sqrt{x_1^2 + x_2^2}} \end{bmatrix}, \quad x \notin \overline{\mathcal{Z}},$$

$$f_c(x) = \begin{bmatrix} -x_2 \\ x_1 \\ -x_4 \\ x_3 \end{bmatrix}, \quad x \in \overline{\mathcal{Z}} \backslash \mathcal{Z},$$

and

$$f_d(x) = \begin{bmatrix} 0 \\ 0 \\ (1+e)(x_2 - x_3) \\ (1+e)(x_1 - x_4) \end{bmatrix}, \quad x \in \mathcal{Z}.$$

Now, consider the function $V : \mathcal{D}_c \to \mathbb{R}$ given by

$$V(x) = (x_1^2 + x_2^2)^{1/2} + \tfrac{1}{2}\frac{(x_1 x_3 + x_2 x_4)^2}{x_1^2 + x_2^2}, \qquad (2.76)$$

and note that $V'(x)f_c(x) = 0$ for all $x \notin \mathcal{Z}$, which implies that $\overline{\mathcal{Z}}\backslash\mathcal{Z}$ is Lyapunov stable. Furthermore, since $e \in (0,1)$ note that $V(x + f_d(x)) = V(x)$ if and only if $x_1 x_3 + x_2 x_4 = 0$. Hence, the set $\{x \in \mathcal{Z} : V(x + f_d(x)) = V(x)\} = \emptyset$ and the set $\mathcal{R} = \mathcal{D}_c\backslash\mathcal{Z}$. Now, note that the largest invariant set $\mathcal{M}$ contained in $\mathcal{R} = \mathcal{D}_c\backslash\mathcal{Z}$ is $\{x \in \mathcal{D}_c : x_1^2 + x_2^2 = 1,\ x_1 x_3 + x_2 x_4 = 0\}$, and hence, it follows from Theorem 2.3 that the solution $x(t)$, $t \geq 0$, to (2.72)–(2.74) approaches the invariant set $\{x \in \mathcal{D}_c : x_1^2 + x_2^2 = 1,\ x_1 x_3 + x_2 x_4 = 0\}$ as $t \to \infty$ for all initial conditions contained in $\mathcal{D}_c$.

Finally, Figure 2.7 shows the phase portrait of the states x_1 versus x_2 for the initial condition $[x_1(0)\ x_2(0)\ x_3(0)\ x_4(0)]^{\mathrm{T}} = [2\ 0\ 0\ 2]^{\mathrm{T}} \in \mathcal{D}_c$. Alternatively, this can also be shown using Proposition 2.3. Specifically, it follows from Proposition 2.3 that $(x_1(t), x_2(t)) \to \overline{\mathcal{Z}}\backslash\mathcal{Z}$ as $t \to \tau(x_0)$ and since $\overline{\mathcal{Z}}\backslash\mathcal{Z}$ is an invariant set, it follows from Theorem 2.3 that $(x_1(t), x_2(t)) \to \{x \in \mathcal{D}_c : x_1^2 + x_2^2 = 1,\ x_1 x_3 + x_2 x_4 = 0\}$ as $t \to \infty$. $\triangle$

The following corollaries to Theorem 2.3 present sufficient conditions that guarantee local asymptotic stability of the nonlinear impulsive dynamical system (2.25) and (2.26). For these results, recall that if the zero solution $x(t) \equiv 0$ to (2.25) and (2.26) is asymptotically stable, then the *domain of attraction* $\mathcal{D}_A \subseteq \mathcal{D}$ of (2.25) and (2.26) is given by

$$\mathcal{D}_A \triangleq \{x_0 \in \mathcal{D} : \text{if } x(t_0) = x_0,\ \text{then } \lim_{t\to\infty} x(t) = 0\}. \qquad (2.77)$$

Corollary 2.1 *Consider the nonlinear impulsive dynamical system (2.25) and (2.26), assume $\mathcal{D}_c \subset \mathcal{D}$ is a compact positively invariant set with respect to (2.25) and (2.26) such that $0 \in \overset{\circ}{\mathcal{D}}_c$, and assume there exists a continuously differentiable function $V : \mathcal{D}_c \to \mathbb{R}$ such that $V(0) = 0$, $V(x) > 0$, $x \neq 0$, and (2.70) and (2.71) are satisfied. Furthermore, assume that the set $\mathcal{R} \triangleq \{x \in \mathcal{D}_c : x \notin \mathcal{Z},\ V'(x)f_c(x) = 0\} \cup \{x \in \mathcal{D}_c : x \in \mathcal{Z},\ V(x + f_d(x)) = V(x)\}$ contains no invariant set other than the set $\{0\}$. Then the zero solution $x(t) \equiv 0$ to (2.25) and (2.26) is asymptotically stable and $\mathcal{D}_c$ is a subset of the domain of attraction of (2.25) and (2.26).*

Proof. Lyapunov stability of the zero solution $x(t) \equiv 0$ to (2.25) and (2.26) follows from Theorem 2.1. Next, it follows from Theorem

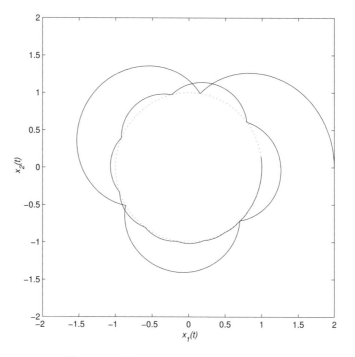

Figure 2.7 Phase portrait of x_1 versus x_2.

2.3 that if $x_0 \in \mathcal{D}_c$, then $\omega(x_0) \subseteq \mathcal{M}$, where $\mathcal{M}$ denotes the largest invariant set contained in $\mathcal{R}$, which implies that $\mathcal{M} = \{0\}$. Hence, $x(t) \to \mathcal{M} = \{0\}$ as $t \to \infty$, establishing asymptotic stability of the zero solution $x(t) \equiv 0$ to (2.25) and (2.26). $\square$

Setting $\mathcal{D} = \mathbb{R}^n$ and requiring $V(x) \to \infty$ as $\|x\| \to \infty$ in Corollary 2.1, it follows that the zero solution $x(t) \equiv 0$ to (2.25) and (2.26) is globally asymptotically stable. Similar remarks hold for Corollaries 2.2 and 2.3 below.

Example 2.7   Consider a bouncing ball, with coefficient of restitution $e \in (0,1)$, on a horizontal surface under a normalized gravitational field. Modeling the surface collisions as instantaneous, it follows from Newton's equations of motion that the bouncing ball dynamics are characterized by the state-dependent impulsive differential equations

$$\left[\begin{array}{c} \dot{x}_1(t) \\ \dot{x}_2(t) \end{array} \right] = \left[\begin{array}{c} x_2(t) \\ -\mathrm{sgn}(x_1) \end{array} \right], \quad \left[\begin{array}{c} x_1(0) \\ x_2(0) \end{array} \right] = \left[\begin{array}{c} x_{10} \\ x_{20} \end{array} \right],$$
$$(x_1(t), x_2(t)) \notin \mathcal{Z}, \quad (2.78)$$

$$\begin{bmatrix} \Delta x_1(t) \\ \Delta x_2(t) \end{bmatrix} = \begin{bmatrix} 0 \\ -(1+e)x_2(t) \end{bmatrix}, \qquad (x_1(t), x_2(t)) \in \mathcal{Z}, \quad (2.79)$$

where $t \geq 0$, $x_1(t), x_2(t) \in \mathbb{R}$, $x_1(t) \geq 0$, $\mathrm{sgn}(x_1) \triangleq x_1/|x_1|$, $x_1 \neq 0$, $\mathrm{sgn}(0) \triangleq 0$, $\mathcal{Z} = \{(x_1, x_2) \in \mathcal{D} : x_1 = 0, \ x_2 < 0\}$, and $\mathcal{D} = \{(x_1, x_2) \in \mathbb{R}^2 : x_1 \geq 0\}$.

First, we use Proposition 2.1 to show that the impulsive dynamical system (2.78) and (2.79) satisfies Assumption 2.1. Note that $\overline{\mathcal{Z}} = \{(x_1, x_2) \in \mathbb{R}^2 : x_1 = 0, \ x_2 \leq 0\}$, and hence, $\overline{\mathcal{Z}}\backslash\mathcal{Z} = \{(0,0)\}$ is a compact invariant set with respect to the dynamical system $\mathcal{G}$ given by (2.78) and (2.79). Next, it can be shown that

$$\tau_1(x_1, x_2) = \begin{cases} x_2 + \sqrt{x_2^2 + 2x_1}, & x_1 > 0, \\ 2x_2, & x_1 = 0, \quad x_2 > 0, \end{cases} \qquad (2.80)$$

which shows that $\tau_1(x_1, x_2)$ is continuous for all $(x_1, x_2) \notin \overline{\mathcal{Z}}$. Furthermore, it follows from (2.80) that $\tau_1(x) \to 0$ as $x \to \partial\mathcal{Z}$ which implies that $\tau_1(\cdot)$ is continuous on $\mathcal{D}$. Finally, it can be shown that for all $(x_1, x_2) \in \mathcal{D}$, the sequence $\{\tau_k(x_1, x_2)\}_{k=1}^{\infty}$ is a uniformly convergent sequence. Now, it follows from $ii)$ of Proposition 2.1 that the dynamical system $\mathcal{G}$ given by (2.78) and (2.79) satisfies Assumption 2.1.

Next, (2.78) and (2.79) can be written in the form of (2.25) and (2.26) with $x \triangleq [x_1, \ x_2]^{\mathrm{T}}$, $f_\mathrm{c}(x) = [x_2, \ -\mathrm{sgn}(x_1)]^{\mathrm{T}}$, and $f_\mathrm{d}(x) = [0, \ -(1+e)x_2]^{\mathrm{T}}$. Now, consider the function $V : \mathbb{R}^2 \to \mathbb{R}$ given by $V(x) = x_1 + \frac{1}{2}x_2^2$ and note that $V'(x)f_\mathrm{c}(x) = 0$ for all $x \notin \mathcal{Z}$, which implies that $\overline{\mathcal{Z}}\backslash\mathcal{Z} = \{(0,0)\}$ is Lyapunov stable. Furthermore, since $e \in (0,1)$, note that $V(x + f_\mathrm{d}(x)) = V(x)$ if and only if $x_2 = 0$. Hence, the set $\{(x_1, x_2) \in \mathcal{Z} : V(x + f_\mathrm{d}(x)) = V(x)\} = \emptyset$ and the set $\mathcal{R} = \{(x_1, x_2) \in \mathbb{R}^2 : x_1 \geq 0\}\backslash\mathcal{Z}$. Now, note that the largest invariant set $\mathcal{M}$ contained in $\mathcal{R} = \{(x_1, x_2) \in \mathbb{R}^2 : x_1 \geq 0\}\backslash\mathcal{Z}$ is $\{(0,0)\}$, and hence, since $V(x)$ is radially unbounded, it follows from Theorem 2.3 that $(x_1(t), x_2(t)) \to (0,0)$ as $t \to \infty$. Alternatively, this can also be shown using Proposition 2.3. Specifically, it follows from Proposition 2.3 that $(x_1(t), x_2(t)) \to \overline{\mathcal{Z}}\backslash\mathcal{Z} = \{(0,0)\}$ as $t \to \tau(x_0)$ and since $\overline{\mathcal{Z}}\backslash\mathcal{Z} = \{(0,0)\}$ is an invariant set it follows from Corollary 2.1 that $(x_1(t), x_2(t)) \to \{(0,0)\}$ as $t \to \infty$. $\triangle$

Corollary 2.2 *Consider the nonlinear impulsive dynamical system (2.25) and (2.26), assume $\mathcal{D}_\mathrm{c} \subset \mathcal{D}$ is a compact positively invariant set with respect to (2.25) and (2.26) such that $0 \in \overset{\circ}{\mathcal{D}}_\mathrm{c}$, and assume* *there exists a continuously differentiable function $V : \mathcal{D}_c \to \mathbb{R}$ such that $V(0) = 0$, $V(x) > 0$, $x \neq 0$,*

$$V'(x)f_c(x) < 0, \quad x \in \mathcal{D}_c, \quad x \notin \mathcal{Z}, \quad x \neq 0, \qquad (2.81)$$

and (2.71) is satisfied. Then the zero solution $x(t) \equiv 0$ to (2.25) and (2.26) is asymptotically stable and $\mathcal{D}_c$ is a subset of the domain of attraction of (2.25) and (2.26).

Proof. It follows from (2.81) that $V'(x)f_c(x) = 0$ for all $x \in \mathcal{D}_c \backslash \mathcal{Z}$ if and only if $x = 0$. Hence, $\mathcal{R} = \{0\} \cup \{x \in \mathcal{D}_c : x \in \mathcal{Z}, V(x + f_d(x)) = V(x)\}$, which contains no invariant set other than $\{0\}$. Now, the result follows as a direct consequence of Corollary 2.1. $\qquad \square$

Corollary 2.3 *Consider the nonlinear impulsive dynamical system (2.25) and (2.26), assume $\mathcal{D}_c \subset \mathcal{D}$ is a compact positively invariant set with respect to (2.25) and (2.26) such that $0 \in \overset{\circ}{\mathcal{D}}_c$, and assume that for all $x_0 \in \mathcal{D}_c$, $x_0 \neq 0$, there exists $\tau \geq 0$ such that $x(\tau) \in \mathcal{Z}$, where $x(t)$, $t \geq 0$, denotes the solution to (2.25) and (2.26) with the initial condition x_0. Furthermore, assume there exists a continuously differentiable function $V : \mathcal{D}_c \to \mathbb{R}$ such that $V(0) = 0$, $V(x) > 0$, $x \neq 0$,*

$$V(x + f_d(x)) - V(x) < 0, \quad x \in \mathcal{D}_c, \quad x \in \mathcal{Z}, \qquad (2.82)$$

and (2.70) is satisfied. Then the zero solution $x(t) \equiv 0$ to (2.25) and (2.26) is asymptotically stable and $\mathcal{D}_c$ is a subset of the domain of attraction of (2.25) and (2.26).

Proof. It follows from (2.82) that $\mathcal{R} = \{x \in \mathcal{D}_c : x \notin \mathcal{Z}, V'(x)f_c(x) = 0\}$. Since, for all $x_0 \in \mathcal{D}_c$, $x_0 \neq 0$, there exists $\tau \geq 0$ such that $x(\tau) \in \mathcal{Z}$, it follows that the largest invariant set contained in $\mathcal{R}$ is $\{0\}$. Now, the result is as a direct consequence of Corollary 2.1. $\qquad \square$

2.7 Partial Stability of State-Dependent Impulsive Dynamical Systems

In many engineering applications, *partial stability*, that is, stability with respect to part of the system's states, is often necessary. In particular, partial stability arises in the study of electromagnetics [173], inertial navigation systems [155], spacecraft stabilization via gimballed gyroscopes and/or flywheels [163], combustion systems [9], vibrations in rotating machinery [108], and biocenology [144], to cite but a few examples. For example, in the field of biocenology involving Lotka-Volterra predator-prey models of population dynamics with age structure, if some of the species preyed upon are left alone, then the corresponding population increases without bound while a subset of the prey species remains stable [144, pp. 260–269]. The need to consider partial stability in the aforementioned systems arises from the fact that stability notions involve equilibrium coordinates as well as a hyperplane of coordinates that is closed but *not* compact. Hence, partial stability involves motion lying in a subspace instead of an equilibrium point.

Additionally, *partial stabilization*, that is, closed-loop stability with respect to part of the closed-loop system's state, also arises in many engineering applications [108,163]. Specifically, in spacecraft stabilization via gimballed gyroscopes asymptotic stability of an equilibrium position of the spacecraft is sought, while requiring Lyapunov stability of the axis of the gyroscope relative to the spacecraft [163]. Alternatively, in the control of rotating machinery with mass imbalance, spin stabilization about a nonprincipal axis of inertia requires motion stabilization with respect to a subspace instead of the origin [108]. Perhaps the most common application where partial stabilization is necessary is adaptive control, wherein asymptotic stability of the closed-loop plant states is guaranteed without necessarily achieving parameter error convergence.

In this section, we introduce the notion of partial stability for nonlinear state-dependent impulsive dynamical systems. Specifically, consider the nonlinear state-dependent impulsive dynamical system

$$\dot{x}_1(t) = f_{1c}(x_1(t), x_2(t)), \quad x_1(0) = x_{10}, \quad (x_1(t), x_2(t)) \notin \mathcal{Z}, \quad (2.83)$$
$$\dot{x}_2(t) = f_{2c}(x_1(t), x_2(t)), \quad x_2(0) = x_{20}, \quad (x_1(t), x_2(t)) \notin \mathcal{Z}, \quad (2.84)$$
$$\Delta x_1(t) = f_{1d}(x_1(t), x_2(t)), \quad (x_1(t), x_2(t)) \in \mathcal{Z}, \quad (2.85)$$
$$\Delta x_2(t) = f_{2d}(x_1(t), x_2(t)), \quad (x_1(t), x_2(t)) \in \mathcal{Z}, \quad (2.86)$$

where $t \geq 0$, $x_1 \in \mathcal{D} \subseteq \mathbb{R}^{n_1}$, $\mathcal{D}$ is an open set such that $0 \in \mathcal{D}$, $x_2 \in \mathbb{R}^{n_2}$, $\Delta x_1(t) = x_1(t^+) - x_1(t)$, $\Delta x_2(t) = x_2(t^+) - x_2(t)$, $f_{1c} : \mathcal{D} \times \mathbb{R}^{n_2} \to \mathbb{R}^{n_1}$ is such that for every $x_2 \in \mathbb{R}^{n_2}$, $f_{1c}(0, x_2) = 0$ and $f_{1c}(\cdot, x_2)$ is locally Lipschitz in x_1, $f_{2c} : \mathcal{D} \times \mathbb{R}^{n_2} \to \mathbb{R}^{n_2}$ is such that for every $x_1 \in \mathcal{D}$, $f_{2c}(x_1, \cdot)$ is locally Lipschitz continuous on $\mathcal{D}$ in x_2, $f_{1d} : \mathcal{D} \times \mathbb{R}^{n_2} \to \mathbb{R}^{n_1}$ is continuous and $f_{1d}(0, x_2) = 0$ for all $x_2 \in \mathbb{R}^{n_2}$, $f_{2d} : \mathcal{D} \times \mathbb{R}^{n_2} \to \mathbb{R}^{n_2}$ is continuous, and $\mathcal{Z} \subset \mathcal{D} \times \mathbb{R}^{n_2}$.

For a particular trajectory $x(t) = (x_1(t), x_2(t))$, $t \geq 0$, we let t_k (=

$\tau_k(x_{10}, x_{20}))$ denote the kth instant of time at which $x(t)$ intersects $\mathcal{Z}$. Furthermore, we make the following assumptions:

A1'. If $x(t) \in \overline{\mathcal{Z}} \backslash \mathcal{Z}$, then there exists $\varepsilon > 0$ such that, for all $0 < \delta < \varepsilon$, $x(t + \delta) \notin \mathcal{Z}$.

A2'. If $x(t_k) = [x_1^{\mathrm{T}}(t_k), x_2^{\mathrm{T}}(t_k)]^{\mathrm{T}} \in \partial \mathcal{Z} \cap \mathcal{Z}$, then the system states reset to $x^+(t_k) \triangleq [x_1^{\mathrm{T}}(t_k^+), x_2^{\mathrm{T}}(t_k^+)]^{\mathrm{T}} = x(t_k) + f_{\mathrm{d}}(x_1(t_k), x_2(t_k))$, according to the resetting law (2.85) and (2.86), which serves as the initial condition for the continuous-time dynamics (2.83) and (2.84).

Assumption A1' is a specialization of A1 for the particular resetting set (2.24). Furthermore, A2' is a specialization of A2 to the partial stability problem. The following definition introduces several types of partial stability of the nonlinear state-dependent impulsive dynamical system (2.83)–(2.86).

Definition 2.8 *i) The nonlinear impulsive dynamical system (2.83)–(2.86) is Lyapunov stable with respect to x_1 if, for every $\varepsilon > 0$ and $x_{20} \in \mathbb{R}^{n_2}$, there exists $\delta = \delta(\varepsilon, x_{20}) > 0$ such that $\|x_{10}\| < \delta$ implies that $\|x_1(t)\| < \varepsilon$ for all $t \geq 0$ (see Figure 2.8(a)).*

ii) The nonlinear impulsive dynamical system (2.83)–(2.86) is Lyapunov stable with respect to x_1 uniformly in x_{20} if, for every $\varepsilon > 0$, there exists $\delta = \delta(\varepsilon) > 0$ such that $\|x_{10}\| < \delta$ implies that $\|x_1(t)\| < \varepsilon$ for all $t \geq 0$ and for all $x_{20} \in \mathbb{R}^{n_2}$.

iii) The nonlinear impulsive dynamical system (2.83)–(2.86) is asymptotically stable with respect to x_1 if it is Lyapunov stable with respect to x_1 and, for every $x_{20} \in \mathbb{R}^{n_2}$, there exists $\delta = \delta(x_{20}) > 0$ such that $\|x_{10}\| < \delta$ implies that $\lim_{t \to \infty} x_1(t) = 0$ (see Figure 2.8(b)).

iv) The nonlinear impulsive dynamical system (2.83)–(2.86) is asymptotically stable with respect to x_1 uniformly in x_{20} if it is Lyapunov stable with respect to x_1 uniformly in x_{20} and there exists $\delta > 0$ such that $\|x_{10}\| < \delta$ implies that $\lim_{t \to \infty} x_1(t) = 0$ uniformly in x_{10} and x_{20} for all $x_{20} \in \mathbb{R}^{n_2}$.

v) The nonlinear impulsive dynamical system (2.83)–(2.86) is globally asymptotically stable with respect to x_1 if it is Lyapunov stable with respect to x_1 and $\lim_{t \to \infty} x_1(t) = 0$ for all $x_{10} \in \mathbb{R}^{n_1}$ and $x_{20} \in \mathbb{R}^{n_2}$.

vi) The nonlinear impulsive dynamical system (2.83)–(2.86) is globally asymptotically stable with respect to x_1 uniformly in x_{20} if it is Lyapunov stable with respect to x_1 uniformly in x_{20} and $\lim_{t \to \infty} x_1(t) = 0$ uniformly in x_{10} and x_{20} for all $x_{10} \in \mathbb{R}^{n_1}$ and $x_{20} \in \mathbb{R}^{n_2}$.

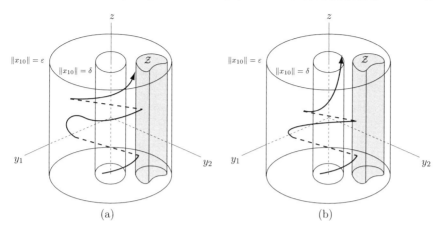

Figure 2.8 (a) Partial Lyapunov stability with respect to x_1. (b) Partial asymptotic stability with respect to x_1. $x_1 = [y_1, y_2]^{\mathrm{T}}$, $x_2 = z$, and $x = [x_1^{\mathrm{T}}, x_2^{\mathrm{T}}]^{\mathrm{T}}$.

vii) *The nonlinear impulsive dynamical system (2.83)–(2.86) is* exponentially stable with respect to x_1 *uniformly in* x_{20} *if there exist scalars* $\alpha, \beta, \delta > 0$ *such that* $\|x_{10}\| < \delta$ *implies that* $\|x_1(t)\| \le \alpha\|x_{10}\|e^{-\beta t}$, $t \ge 0$, *for all* $x_{20} \in \mathbb{R}^{n_2}$.

viii) *The nonlinear impulsive dynamical system (2.83)–(2.86) is* globally exponentially stable with respect to x_1 *uniformly in* x_{20} *if there exist scalars* $\alpha, \beta > 0$ *such that* $\|x_1(t)\| \le \alpha\|x_{10}\|e^{-\beta t}$, $t \ge 0$, *for all* $x_{10} \in \mathbb{R}^{n_1}$ *and* $x_{20} \in \mathbb{R}^{n_2}$.

Next, we present sufficient conditions for partial stability of the nonlinear state-dependent impulsive dynamical system (2.83)–(2.86). For notational convenience define $f_c(x_1, x_2) \triangleq [f_{1c}^{\mathrm{T}}(x_1, x_2), f_{2c}^{\mathrm{T}}(x_1, x_2)]^{\mathrm{T}}$ and $f_d(x_1, x_2) \triangleq [f_{1d}^{\mathrm{T}}(x_1, x_2), f_{2d}^{\mathrm{T}}(x_1, x_2)]^{\mathrm{T}}$. Furthermore, define

$$\dot{V}(x_1, x_2) \triangleq V'(x_1, x_2)f_c(x_1, x_2), \quad (x_1, x_2) \notin \mathcal{Z}, \tag{2.87}$$
$$\Delta V(x_1, x_2) \triangleq V(x_1 + f_{1d}(x_1, x_2), x_2 + f_{2d}(x_1, x_2)) - V(x_1, x_2),$$
$$(x_1, x_2) \in \mathcal{Z}, \tag{2.88}$$

for a given continuously differentiable function $V : \mathcal{D} \times \mathbb{R}^{n_2} \to \mathbb{R}$. Finally, we assume that the solution $(x_1(t), x_2(t))$ to (2.83)–(2.86) exists and is unique for all $t \ge 0$. It is important to note that unlike standard theory, the existence of a Lyapunov function $V(x_1, x_2)$ satisfying the conditions in Theorem 2.4 below is *not* sufficient to ensure that all solutions of (2.83)–(2.86) starting in $\mathcal{D} \times \mathbb{R}^{n_2}$ can be extended to infinity since none of the states of (2.83)–(2.86) serve as an independent variable. We do note, however, that continuous differentiability of $f_{1c}(\cdot,\cdot)$ and $f_{2c}(\cdot,\cdot)$ and continuity of $f_{1d}(\cdot,\cdot)$ and $f_{2d}(\cdot,\cdot)$ provide a sufficient condition for the existence and uniqueness of solutions to (2.83)–(2.86) for a forward time interval.

For the next result we assume, without loss of generality, that $x_0 = [x_{10}^T, x_{20}^T]^T \notin \mathcal{Z}$ so that the continuous-time dynamics (2.83) and (2.84) are active until the first resetting time. If $x_0 \in \mathcal{Z}$, then it follows from assumption A2′ that the system initially resets to x_0^+, which serves as the initial condition for the continuous-time dynamics (2.83) and (2.84). The following definition of class $\mathcal{K}$ and class $\mathcal{K}_\infty$ functions is needed.

Definition 2.9 *A continuous function* $\gamma : [0,a) \to [0,\infty)$, *where* $a \in (0,\infty]$, *is of* class $\mathcal{K}$ *if it is strictly increasing and* $\gamma(0) = 0$. *A continuous function* $\gamma : [0,\infty) \to [0,\infty)$ *is of* class $\mathcal{K}_\infty$ *if it is strictly increasing,* $\gamma(0) = 0$, *and* $\gamma(s) \to \infty$ *as* $s \to \infty$.

Theorem 2.4 *Consider the nonlinear state-dependent impulsive dynamical system (2.83)–(2.86). Then the following statements hold:*

i) *If there exist a continuously differentiable function* $V : \mathcal{D} \times \mathbb{R}^{n_2} \to \mathbb{R}$ *and a class* $\mathcal{K}$ *function* $\alpha(\cdot)$ *such that*

$$V(0, x_2) = 0, \quad x_2 \in \mathbb{R}^{n_2}, \tag{2.89}$$
$$\alpha(\|x_1\|) \le V(x_1, x_2), \quad (x_1, x_2) \in \mathcal{D} \times \mathbb{R}^{n_2}, \tag{2.90}$$
$$\dot{V}(x_1, x_2) \le 0, \quad (x_1, x_2) \in \mathcal{D} \times \mathbb{R}^{n_2}, \quad (x_1, x_2) \notin \mathcal{Z}, \tag{2.91}$$
$$\Delta V(x_1, x_2) \le 0, \quad (x_1, x_2) \in \mathcal{Z}, \tag{2.92}$$

then the nonlinear state-dependent impulsive dynamical system given by (2.83)–(2.86) is Lyapunov stable with respect to x_1.

ii) *If there exist a continuously differentiable function* $V : \mathcal{D} \times \mathbb{R}^{n_2} \to \mathbb{R}$ *and class* $\mathcal{K}$ *functions* $\alpha(\cdot)$ *and* $\beta(\cdot)$ *satisfying (2.90)–(2.92) and*

$$V(x_1, x_2) \le \beta(\|x_1\|), \quad (x_1, x_2) \in \mathcal{D} \times \mathbb{R}^{n_2}, \tag{2.93}$$

then the nonlinear state-dependent impulsive dynamical system given by (2.83)–(2.86) is Lyapunov stable with respect to x_1 *uniformly in* x_{20}.

iii) *If there exist a continuously differentiable function* $V : \mathcal{D} \times \mathbb{R}^{n_2} \to \mathbb{R}$ *and class* $\mathcal{K}$ *functions* $\alpha(\cdot)$, $\beta(\cdot)$, *and* $\gamma(\cdot)$ *satisfying*

(2.90), (2.92), (2.93), and

$$\dot{V}(x_1, x_2) \leq -\gamma(\|x_1\|), \quad (x_1, x_2) \in \mathcal{D} \times \mathbb{R}^{n_2}, \quad (x_1, x_2) \notin \mathcal{Z},$$
$$(2.94)$$

then the nonlinear state-dependent impulsive dynamical system given by (2.83)–(2.86) is asymptotically stable with respect to x_1 uniformly in x_{20}.

iv) *If $\mathcal{D} = \mathbb{R}^{n_1}$ and there exist a continuously differentiable function $V : \mathbb{R}^{n_1} \times \mathbb{R}^{n_2} \to \mathbb{R}$, a class $\mathcal{K}$ function $\gamma(\cdot)$, and class $\mathcal{K}_\infty$ functions $\alpha(\cdot)$ and $\beta(\cdot)$ satisfying (2.90) and (2.92)–(2.94), then the nonlinear state-dependent impulsive dynamical system given by (2.83)–(2.86) is globally asymptotically stable with respect to x_1 uniformly in x_{20}.*

v) *If there exist a continuously differentiable function $V : \mathcal{D} \times \mathbb{R}^{n_2} \to \mathbb{R}$ and positive constants α, β, γ, and $p \geq 1$ satisfying (2.92) and*

$$\alpha \|x_1\|^p \leq V(x_1, x_2) \leq \beta \|x_1\|^p, \quad (x_1, x_2) \in \mathcal{D} \times \mathbb{R}^{n_2}, \quad (2.95)$$
$$\dot{V}(x_1, x_2) \leq -\gamma \|x_1\|^p, \quad (x_1, x_2) \in \mathcal{D} \times \mathbb{R}^{n_2}, \quad (x_1, x_2) \notin \mathcal{Z},$$
$$(2.96)$$

then the nonlinear state-dependent impulsive dynamical system given by (2.83)–(2.86) is exponentially stable with respect to x_1 uniformly in x_{20}.

vi) *If $\mathcal{D} = \mathbb{R}^{n_1}$ and there exist a continuously differentiable function $V : \mathbb{R}^{n_1} \times \mathbb{R}^{n_2} \to \mathbb{R}$ and positive constants α, β, γ, and $p \geq 1$ satisfying (2.92), (2.95), and (2.96), then the nonlinear state-dependent impulsive dynamical system given by (2.83)–(2.86) is globally exponentially stable with respect to x_1 uniformly in x_{20}.*

Proof. *i)* Prior to the first resetting time $t_1 \triangleq \tau_1(x_0)$, $V(x_1(t), x_2(t))$ is given by

$$V(x_1(t), x_2(t)) = V(x_1(0), x_2(0))$$
$$+ \int_0^t V'(x_1(\tau), x_2(\tau)) f_c(x_1(\tau), x_2(\tau)) d\tau, \quad t \in [0, t_1].$$
$$(2.97)$$

Between consecutive resetting times $t_k \triangleq \tau_k(x_0)$ and $t_{k+1} \triangleq \tau_{k+1}(x_0)$, we can determine the value of $V(x_1(t), x_2(t))$ as its initial value plus the integral of its rate of change along the trajectory $x(t)$, that is,

$$
\begin{aligned}
V(x_1(t), x_2(t)) = {} & V(x_1(t_k) + f_{1d}(x_1(t_k), x_2(t_k)),\ x_2(t_k) \\
& + f_{2d}(x_1(t_k), x_2(t_k))) \\
& + \int_{t_k}^{t} V'(x_1(\tau), x_2(\tau)) f_c(x_1(\tau), x_2(\tau)) \mathrm{d}\tau, \\
& \hspace{4cm} t \in (t_k, t_{k+1}], \hspace{1cm} (2.98)
\end{aligned}
$$

for $k = 1, 2, \ldots$. Adding and subtracting $V(x_1(t_k), x_2(t_k))$ to and from the right hand side of (2.98) yields

$$
\begin{aligned}
V(x_1(t), x_2(t)) = {} & V(x_1(t_k), x_2(t_k)) \\
& + [V(x_1(t_k) + f_{1d}(x_1(t_k), x_2(t_k)),\ x_2(t_k) \\
& + f_{2d}(x_1(t_k), x_2(t_k))) - V(x_1(t_k), x_2(t_k))] \\
& + \int_{t_k}^{t} V'(x_1(\tau), x_2(\tau)) f_c(x_1(\tau), x_2(\tau)) \mathrm{d}\tau
\end{aligned}
$$

$$(2.99)$$

for all $t \in (t_k, t_{k+1}]$, and in particular, at time t_{k+1},

$$
\begin{aligned}
V(x_1(t_{k+1}), x_2(t_{k+1})) = {} & V(x_1(t_k), x_2(t_k)) \\
& + [V(x_1(t_k) + f_{1d}(x_1(t_k), x_2(t_k)),\ x_2(t_k) \\
& + f_{2d}(x_1(t_k), x_2(t_k))) - V(x_1(t_k), x_2(t_k))] \\
& + \int_{t_k}^{t_{k+1}} V'(x_1(\tau), x_2(\tau)) f_c(x_1(\tau), x_2(\tau)) \mathrm{d}\tau.
\end{aligned}
$$

$$(2.100)$$

By recursively substituting (2.100) into (2.99) and ultimately into (2.98), we obtain

$$
\begin{aligned}
V(x_1(t), x_2(t)) = {} & V(x_1(0), x_2(0)) \\
& + \int_{0}^{t} V'(x_1(\tau), x_2(\tau)) f_c(x_1(\tau), x_2(\tau)) \mathrm{d}\tau \\
& + \sum_{i \in \mathbb{Z}_{[0,t)}} [V(x_1(t_i) + f_{1d}(x_1(t_i), x_2(t_i)),\ x_2(t_i) \\
& + f_{2d}(x_1(t_i), x_2(t_i))) - V(x_1(t_i), x_2(t_i))], \\
& \hspace{4cm} t \geq 0, \hspace{0.5cm} (2.101)
\end{aligned}
$$

where $\mathbb{Z}_{[0,t)} \triangleq \{i \in \overline{\mathbb{Z}}_+ : 0 \leq t_i < t\}$. From (2.101) and (2.92) we obtain

$$
V(x_1(t), x_2(t)) \leq V(x_1(0), x_2(0))
$$

$$+ \int_0^t V'(x_1(\tau), x_2(\tau)) f_c(x_1(\tau), x_2(\tau)) \mathrm{d}\tau, \quad t \geq 0.$$

$$(2.102)$$

Furthermore, it follows from (2.102) that

$$V(x_1(t), x_2(t)) \leq V(x_1(s), x_2(s))$$
$$+ \int_s^t V'(x_1(\tau), x_2(\tau)) f_c(x_1(\tau), x_2(\tau)) \mathrm{d}\tau, \quad t > s,$$

$$(2.103)$$

which, using (2.91), implies that $V(x_1(t), x_2(t)), t \geq 0$, is a nonincreasing function of time.

Next, let $x_{20} \in \mathbb{R}^{n_2}$, let $\varepsilon > 0$ be such that $\mathcal{B}_\varepsilon(0) \triangleq \{x_1 \in \mathbb{R}^{n_1} : \|x_1\| < \varepsilon\} \subset \mathcal{D}$, define $\eta \triangleq \alpha(\varepsilon)$, and define $\mathcal{D}_\eta \triangleq \{x_1 \in \mathcal{B}_\varepsilon(0) : V(x_1, x_{20}) < \eta\}$. Since $V(\cdot, \cdot)$ is continuous and $V(0, x_{20}) = 0$ it follows that $\mathcal{D}_\eta$ is nonempty and there exists $\delta = \delta(\varepsilon, x_{20}) > 0$ such that $V(x_1, x_{20}) < \eta$, $x_1 \in \mathcal{B}_\delta(0)$. Hence, $\mathcal{B}_\delta(0) \subseteq \mathcal{D}_\eta$. Next, since $V(x_1(t), x_2(t))$ is a nonincreasing function of time it follows that for every $x_{10} \in \mathcal{B}_\delta(0) \subseteq \mathcal{D}_\eta$,

$$\alpha(\|x_1(t)\|) \leq V(x_1(t), x_2(t)) \leq V(x_{10}, x_{20}) < \eta = \alpha(\varepsilon).$$

Thus, for every $x_{10} \in \mathcal{B}_\delta(0) \subseteq \mathcal{D}_\eta$, $x_1(t) \in \mathcal{B}_\varepsilon(0)$, $t \geq 0$, establishing Lyapunov stability with respect to x_1.

ii) Let $\varepsilon > 0$ and let $\mathcal{B}_\varepsilon(0)$ and η be given as in the proof of *i)*. Now, let $\delta = \delta(\varepsilon) > 0$ be such that $\beta(\delta) = \alpha(\varepsilon)$. Hence, it follows from (2.93) that for all $(x_{10}, x_{20}) \in \mathcal{B}_\delta(0) \times \mathbb{R}^{n_2}$,

$$\alpha(\|x_1(t)\|) \leq V(x_1(t), x_2(t)) \leq V(x_{10}, x_{20}) < \beta(\delta) = \alpha(\varepsilon),$$

and hence, $x_1(t) \in \mathcal{B}_\varepsilon(0)$, $t \geq 0$.

iii) Lyapunov stability uniformly in x_{20} follows from *ii)*. Next, let $\varepsilon > 0$ and $\delta = \delta(\varepsilon) > 0$ be such that for every $x_{10} \in \mathcal{B}_\delta(0)$, $x_1(t) \in \mathcal{B}_\varepsilon(0)$, $t \geq 0$, (the existence of such a (δ, ε) pair follows from uniform Lyapunov stability) and assume that (2.94) holds. Since (2.94) implies (2.91) it follows that for every $x_{10} \in \mathcal{B}_\delta(0)$, $V(x_1(t), x_2(t))$ is a nonincreasing function of time and, since $V(\cdot, \cdot)$ is bounded from below, it follows from the Bolzano-Weierstass theorem [146] that there exists $L \geq 0$ such that $\lim_{t \to \infty} V(x_1(t), x_2(t)) = L$.

Now, suppose, *ad absurdum*, for some $x_{10} \in \mathcal{B}_\delta(0)$, $L > 0$. Since $V(\cdot, \cdot)$ is continuously differentiable and $V(0, x_2) = 0$ for all $x_2 \in \mathbb{R}^{n_2}$ it follows that $\mathcal{D}_L \triangleq \{x_1 \in \mathcal{B}_\varepsilon(0) : V(x_1, x_2) \leq L \text{ for all } x_2 \in \mathbb{R}^{n_2}\}$

is nonempty and $x_1(t) \notin \mathcal{D}_L$, $t \geq 0$. Thus, as in the proof of i), there exists $\hat{\delta} > 0$ such that $\mathcal{B}_{\hat{\delta}}(0) \subset \mathcal{D}_L$. Hence, it follows from (2.92) and (2.94) that for the given $x_{10} \in \mathcal{B}_{\delta}(0)$ and $t \geq 0$,

$$
\begin{aligned}
V(x_1(t), x_2(t)) = V(x_{10}, x_{20}) &+ \int_0^t \dot{V}(x_1(\tau), x_2(\tau)) \mathrm{d}\tau \\
&+ \sum_{i \in \mathbb{Z}_{[0,t)}} [V(x_1(t_i) + f_{1d}(x_1(t_i), x_2(t_i)), x_2(t_i) \\
&\quad + f_{2d}(x_1(t_i), x_2(t_i))) - V(x_1(t_i), x_2(t_i))] \\
\leq V(x_{10}, x_{20}) &- \int_0^t \gamma(\|x_1(s)\|) \mathrm{d}s \\
\leq V(x_{10}, x_{20}) &- \gamma(\hat{\delta}) t. \qquad (2.104)
\end{aligned}
$$

Letting $t \geq \frac{V(x_{10}, x_{20}) - L}{\gamma(\hat{\delta})}$, it follows that $V(x_1(t), x_2(t)) \leq L$, which is a contradiction. Hence, $L = 0$, and, since $x_{10} \in \mathcal{B}_{\delta}(0)$ was chosen arbitrarily, it follows that $V(x_1(t), x_2(t)) \to 0$ as $t \to \infty$ for all $x_{10} \in \mathcal{B}_{\delta}(0)$. Now, since $V(x_1(t), x_2(t)) \geq \alpha(\|x_1(t)\|) \geq 0$, $t \geq 0$, it follows that $\alpha(\|x_1(t)\|) \to 0$ or, equivalently, $x_1(t) \to 0$, $t \to \infty$, establishing asymptotic stability with respect to x_1 uniformly in x_{20}.

iv) Let $\delta > 0$ be such that $\|x_{10}\| < \delta$. Since $\alpha(\cdot)$ is a class $\mathcal{K}_{\infty}$ function, it follows that there exists $\varepsilon > 0$ such that $\beta(\delta) \leq \alpha(\varepsilon)$. Now, (2.92) and (2.94) imply that $V(x_1(t), x_2(t))$ is a nonincreasing function of time, and hence, it follows from (2.93) that $\alpha(\|x_1(t)\|) \leq V(x_1(t), x_2(t)) \leq V(x_{10}, x_{20}) < \beta(\delta) \leq \alpha(\varepsilon)$, $t \geq 0$. Hence, $x_1(t) \in \mathcal{B}_{\varepsilon}(0)$, $t \geq 0$. Now, the proof follows as in the proof of iii).

v) Let $\varepsilon > 0$ and $\mathcal{B}_{\varepsilon}(0)$ be given as in the proof of i), and let $\eta \triangleq \alpha \varepsilon$ and $\delta = \frac{\eta}{\beta}$. Now, (2.96) implies that $\dot{V}(x_1, x_2) \leq 0$, and hence, as in the proof of ii), it follows that for all $(x_{10}, x_{20}) \in \mathcal{B}_{\delta}(0) \times \mathbb{R}^{n_2}$, $x_1(t) \in \mathcal{B}_{\varepsilon}(0)$, $t \geq 0$. Furthermore, it follows from (2.95) and (2.96) that for all $t \geq 0$ and $(x_{10}, x_{20}) \in \mathcal{B}_{\delta}(0) \times \mathbb{R}^{n_2}$,

$$
\dot{V}(x_1(t), x_2(t)) \leq -\gamma \|x_1(t)\|^p \leq -\frac{\gamma}{\beta} V(x_1(t), x_2(t)), \quad 0 \leq t \leq t_1,
$$

which implies that

$$
V(x_1(t), x_2(t)) \leq V(x_{10}, x_{20}) e^{-\frac{\gamma}{\beta} t}, \quad 0 \leq t \leq t_1. \qquad (2.105)
$$

Similarly, between the first and second resetting times

$$
\dot{V}(x_1(t), x_2(t)) \leq -\gamma \|x_1(t)\|^p \leq -\frac{\gamma}{\beta} V(x_1(t), x_2(t)), \quad t_1 < t \leq t_2,
$$

which, using (2.92) and (2.105), yields

$$
\begin{aligned}
V(x_1(t), x_2(t)) &\leq V(x_1(t_1) + f_{1d}(x_1(t_1), x_2(t_1)), x_2(t_1) \\
&\quad + f_{2d}(x_1(t_1), x_2(t_1)))e^{-\frac{\gamma}{\beta}(t-t_1)} \\
&\leq V(x_1(t_1), x_2(t_1))e^{-\frac{\gamma}{\beta}(t-t_1)} \\
&\leq V(x_{10}, x_{20})e^{-\frac{\gamma}{\beta}t_1}e^{-\frac{\gamma}{\beta}(t-t_1)} \\
&= V(x_{10}, x_{20})e^{-\frac{\gamma}{\beta}t}, \quad t_1 < t \leq t_2.
\end{aligned} \tag{2.106}
$$

Recursively repeating the above arguments for $t_k < t \leq t_{k+1}$, $k = 3, 4, \ldots$, it follows that

$$
V(x_1(t), x_2(t)) \leq V(x_{10}, x_{20})e^{-\frac{\gamma}{\beta}t}, \quad t \geq 0. \tag{2.107}
$$

Now, it follows from (2.95) and (2.107) that, for all $t \geq 0$,

$$
\alpha \|x_1(t)\|^p \leq V(x_1(t), x_2(t)) \leq V(x_{10}, x_{20})e^{-\frac{\gamma}{\beta}t} \leq \beta \|x_{10}\|^p e^{-\frac{\gamma}{\beta}t},
$$

and hence,

$$
\|x_1(t)\| \leq \left(\frac{\beta}{\alpha}\right)^{1/p} \|x_{10}\| e^{-\frac{\gamma}{\beta p}t}, \quad t \geq 0,
$$

establishing exponential stability with respect to x_1 uniformly in x_{20}.
vi) The proof follows as in iv) and v). $\qquad\square$

By setting $n_1 = n$ and $n_2 = 0$, Theorem 2.4 specializes to the case of nonlinear state-dependent impulsive dynamical systems of the form

$$
\begin{aligned}
\dot{x}_1(t) &= f_{1c}(x_1(t)), & x_1(0) &= x_{10}, & x_1(t) &\notin \mathcal{Z}, & (2.108) \\
\Delta x_1(t) &= f_{1d}(x_1(t)), & x_1(t) &\in \mathcal{Z}. & & & (2.109)
\end{aligned}
$$

In this case, Lyapunov (respectively, asymptotic) stability with respect to x_1 and Lyapunov (respectively, asymptotic) stability with respect to x_1 uniformly in x_{20} are equivalent to Lyapunov (respectively, asymptotic) stability of nonlinear state-dependent impulsive dynamical systems. Furthermore, note that in this case there exists a continuously differentiable function $V : \mathcal{D} \to \mathbb{R}$ such that (2.90), (2.92), (2.93), and (2.94) hold if and only if $V(\cdot)$ is such that $V(0) = 0$, $V(x_1) > 0$, $x_1 \in \mathcal{D}$, $x_1 \neq 0$, $V'(x_1)f_{1c}(x_1) < 0$, $x_1 \notin \mathcal{Z}$, $x_1 \neq 0$, and $\Delta V(x_1) \leq 0$, $x_1 \in \mathcal{Z}$. In addition, if $\mathcal{D} = \mathbb{R}^{n_1}$, then there exist class $\mathcal{K}_\infty$ functions $\alpha(\cdot)$ and $\beta(\cdot)$, and a continuously differentiable function $V(\cdot)$ such that (2.90), (2.92), (2.93), and (2.94) hold if and only if $V(\cdot)$ is such that $V(0) = 0$, $V(x_1) > 0$, $x_1 \in \mathbb{R}^{n_1}$, $x_1 \neq 0$,

$V'(x_1)f_{1c}(x_1) < 0$, $x_1 \notin \mathcal{Z}$, $x_1 \neq 0$, $\Delta V(x_1) \leq 0$, $x_1 \in \mathcal{Z}$, and $V(x_1) \to \infty$ as $\|x_1\| \to \infty$. Hence, in this case, Theorem 2.4 collapses to the Lyapunov stability theorem for state-dependent impulsive dynamical systems given by Theorem 2.1.

In the case of state-dependent impulsive dynamical systems $\mathcal{G}$ satisfying the quasi-continuous dependence assumption (Assumption 2.1), Theorem 2.3 shows that bounded system trajectories of $\mathcal{G}$ approach the largest invariant set $\mathcal{M}$ characterized by the set of all points in a compact set $\mathcal{D}$ of the state space involving a union of level surfaces characterized by vanishing Lyapunov derivatives and differences of the impulsive system dynamics. In the case of partially stable systems, however, it is not generally clear on how to define the set $\mathcal{M}$ since $\dot{V}(x_1, x_2)$ and $\Delta V(x_1, x_2)$ are functions of both x_1 and x_2. However, if $\dot{V}(x_1, x_2) \leq -W(x_1) \leq 0$ and $\Delta V(x_1, x_2) \leq 0$, where $W : \mathcal{D} \subseteq \mathbb{R}^{n_1} \to \mathbb{R}$ is continuous and nonnegative definite, then a set $\mathcal{R} \supset \mathcal{M}$ can be defined as the set of points where $W(x_1)$ identically vanishes, that is, $\mathcal{R} = \{x_1 \in \mathcal{D}, (x_1, x_2) \notin \mathcal{Z} : W(x_1) = 0\}$. In this case, as shown in the next theorem, the partial system trajectories $x_1(t)$ approach $\mathcal{R}$ as t tends to infinity.

Theorem 2.5 *Consider the nonlinear impulsive dynamical system given by (2.83)–(2.86) and assume that there exists $\varepsilon > 0$ such that $\tau_{k+1}(x_0) - \tau_k(x_0) > \varepsilon$ for all $k \in \overline{\mathbb{Z}}_+$ and $x_0 \in \mathcal{D} \times \mathbb{R}^{n_2}$. Furthermore, assume there exist a continuously differentiable function $V : \mathcal{D} \times \mathbb{R}^{n_2} \to \mathbb{R}$ and continuous functions $W_1 : \mathcal{D} \to \mathbb{R}$, $W_2 : \mathcal{D} \to \mathbb{R}$, and $W : \mathcal{D} \to \mathbb{R}$, such that $W_1(\cdot)$ and $W_2(\cdot)$ are positive definite, $W(\cdot)$ is nonnegative definite, $\dot{W}(x_1(\cdot))$ is bounded, and*

$$W_1(x_1) \leq V(x_1, x_2) \leq W_2(x_1), \quad (x_1, x_2) \in \mathcal{D} \times \mathbb{R}^{n_2}, \quad (2.110)$$
$$\dot{V}(x_1, x_2) \leq -W(x_1), \quad (x_1, x_2) \in \mathcal{D} \times \mathbb{R}^{n_2}, \quad (x_1, x_2) \notin \mathcal{Z}, \quad (2.111)$$
$$\Delta V(x_1, x_2) \leq 0, \quad (x_1, x_2) \in \mathcal{D} \times \mathbb{R}^{n_2}, \quad (x_1, x_2) \in \mathcal{Z}. \quad (2.112)$$

Then the impulsive dynamical system (2.83)–(2.86) is Lyapunov stable with respect to x_1 uniformly in x_{20} and there exists $\mathcal{D}_0 \subset \mathcal{D}$ such that for all $(x_{10}, x_{20}) \in \mathcal{D}_0 \times \mathbb{R}^{n_2}$, $x_1(t) \to \mathcal{R}$ as $t \to \infty$, where $\mathcal{R} \triangleq \{x_1 \in \mathcal{D}, (x_1, x_2) \notin \mathcal{Z} : W(x_1) = 0\}$. If, in addition, $\mathcal{D} = \mathbb{R}^{n_1}$ and $W_1(\cdot)$ is radially unbounded, then $x_1(t) \to \mathcal{R} = \{x_1 \in \mathbb{R}^{n_1}, (x_1, x_2) \notin \mathcal{Z} : W(x_1) = 0\}$ as $t \to \infty$ for all $(x_{10}, x_{20}) \in \mathbb{R}^{n_1} \times \mathbb{R}^{n_2}$.

Proof. Lyapunov stability of the system (2.83)–(2.86) with respect to x_1 uniformly in x_{20} follows from Theorem 2.4 by noting that, since $W_1(\cdot)$ and $W_2(\cdot)$ are positive definite functions, there exist $r > 0$ and class $\mathcal{K}$ functions $\alpha, \beta : [0, r) \rightarrow [0, \infty)$ such that $\mathcal{B}_r(0) \subset \mathcal{D}$ and $\alpha(\|x_1\|) \leq W_1(x_1)$, $x_1 \in \mathcal{B}_r(0)$, and $W_2(x_1) \leq \beta(\|x_1\|)$, $x_1 \in \mathcal{B}_r(0)$. Next, it can be shown as in the proof of Theorem 2.4 that, for all $(x_{10}, x_{20}) \in \mathcal{D} \times \mathbb{R}^{n_2}$,

$$V(x_1(t), x_2(t)) \leq V(x_{10}, x_{20}) - \int_0^t W(x_1(s)) \mathrm{d}s, \quad t \geq 0. \quad (2.113)$$

Now, since $W(\cdot)$ is nonnegative it follows that $V(\cdot, \cdot)$ is a nonincreasing function of time. Next, let $\hat{\delta} > 0$ and choose $0 < \delta < \min\{\hat{\delta}, r\}$ such that $\mathcal{B}_\delta(0) \in \overset{\circ}{\mathcal{D}}$, let $\eta = \min_{\|x_1\| = \delta} W_1(x_1)$, and define $\mathcal{D}_0 \triangleq \{x_1 \in \mathcal{B}_\delta(0) : W_2(x_1) \leq \eta\}$. Since $V(\cdot, \cdot)$ is a nonincreasing function of time it follows that $x_1(t) \in \mathcal{B}_\delta(0)$ for all $(x_{10}, x_{20}) \in \mathcal{D}_0 \times \mathbb{R}^{n_2}$. Now, it follows from (2.113) that $V(x_1(t), x_2(t))$ is bounded for all $t \geq 0$. Thus, since $W(\cdot)$ is nonnegative it follows from (2.113) that $\lim_{t \to \infty} \int_0^t W(x_1(s)) \mathrm{d}s$ exists and is bounded. Next, note that

$$\lim_{t \to \infty} \int_0^t W(x_1(s)) \mathrm{d}s = \sum_{k=0}^{\infty} \int_{t_k}^{t_{k+1}} W(x_1(s)) \mathrm{d}s, \quad (2.114)$$

where t_k, $k \in \overline{\mathbb{Z}}_+$, are resetting times and $t_0 \triangleq 0$, and hence, $\lim_{k \to \infty} \int_{t_k}^{t_{k+1}} W(x_1(s)) \mathrm{d}s = 0$. Now, suppose, *ad absurdum*, that $\lim_{t \to \infty} W(x_1(t)) \neq 0$ for some $\hat{x}_0 \in \mathcal{D}_0 \times \mathbb{R}^{n_2}$. In this case, there exists a sequence $\{\hat{t}_i\}_{i=1}^{\infty}$ and a positive number $\alpha > 0$ such that $W(x_1(\hat{t}_i)) = \alpha$ for all $i = 1, 2, \ldots$. Furthermore, since $\dot{W}(x_1(\cdot))$ is bounded it follows that there exists $\lambda > 0$ such that $|\dot{W}(x_1(t))| < \lambda$ for all $t \geq 0$. Hence, if $\alpha > \varepsilon\lambda$, then $\int_{t_k}^{t_{k+1}} W(x_1(s)) \mathrm{d}s > \varepsilon(\alpha - \varepsilon\lambda)$ for all $k \in \overline{\mathbb{Z}}_+$, and if $\alpha \leq \varepsilon\lambda$, then $\int_{t_k}^{t_{k+1}} W(x_1(s)) \mathrm{d}s > \frac{\alpha^2}{2\lambda}$ for all $k \in \overline{\mathbb{Z}}_+$, which is a contradiction since $\lim_{k \to \infty} \int_{t_k}^{t_{k+1}} W(x_1(s)) \mathrm{d}s = 0$. Thus, for all $x_0 \in \mathcal{D}_0 \times \mathbb{R}^{n_2}$, $W(x_1(t)) \rightarrow 0$ as $t \rightarrow \infty$, which proves the result. Finally, if, in addition, $\mathcal{D} = \mathbb{R}^{n_1}$ and $W_1(\cdot)$ is radially unbounded, then, as in the proof of *iii*) of Theorem 2.4, for every $x_{10} \in \mathbb{R}^{n_1}$ there exists $\varepsilon, \delta > 0$ such that $x_{10} \in \mathcal{B}_\delta(0)$ and $x_1(t) \in \mathcal{B}_\varepsilon(0)$, $t \geq 0$. Now, the proof follows by repeating the above arguments. $\square$

Theorem 2.5 shows that the partial system trajectories $x_1(t)$ approach $\mathcal{R}$ as t tends to infinity. However, since the positive limit set of the partial trajectories $x_1(t)$ is a subset of $\mathcal{R}$, Theorem 2.5 is a weaker result than an invariance theorem wherein one would conclude that the partial trajectory $x_1(t)$ approaches the largest invariant set $\mathcal{M}$ contained in $\mathcal{R}$. This is not true in general for partially stable impulsive systems since the positive limit set of a partial trajectory $x_1(t)$, $t \geq 0$, is not an invariant set.

2.8 Stability of Time-Dependent Impulsive Dynamical Systems

In this section, we use the results of Section 2.7 to develop stability theorems for nonlinear time-dependent impulsive dynamical systems. Specifically, we consider the time-dependent impulsive dynamical system

$$\dot{x}(t) = f_c(t, x(t)), \quad x(t_0) = x_0, \quad t \neq t_k, \qquad (2.115)$$
$$\Delta x(t) = f_d(t, x(t)), \quad t = t_k, \qquad (2.116)$$

where $t \geq t_0$, $t_0 < t_1 < t_2 < \cdots$ are prescribed resetting times, $x(t) \in \mathcal{D}$, $t \geq t_0$, $\mathcal{D} \subseteq \mathbb{R}^n$ is an open set such that $0 \in \mathcal{D}$, $f_c : [t_0, \hat{t}) \times \mathcal{D} \to \mathbb{R}^n$ is such that $f_c(\cdot, \cdot)$ is jointly continuous in x and t, $f_c(t, 0) = 0$ for every $t \in [t_0, \hat{t})$, $f_c(t, \cdot)$ is locally Lipschitz continuous on $\mathcal{D}$ in x uniformly in t for all t in compact subsets of $[0, \infty)$, $f_d : [t_0, \hat{t}) \times \mathcal{D} \to \mathbb{R}^n$ is such that $f_d(\cdot, \cdot)$ is jointly continuous in x and t, and $f_d(t, 0) = 0$ for every $t \in [t_0, \hat{t})$.

Under the above assumptions the solution to the continuous-time dynamics (2.115) exists and is unique, which, due to continuity of the resetting dynamics (2.116), implies that the solution $x(t)$, $t \geq t_0$, to (2.115) and (2.116) exists and is unique over the interval $[t_0, \hat{t})$. Since time-dependent impulsive dynamical systems involve impulses at a fixed sequence of times, they are time-varying systems. The following definition provides eight types of stability for time-dependent impulsive dynamical systems.

Definition 2.10 *i) The nonlinear time-dependent impulsive dynamical system given by (2.115) and (2.116) is Lyapunov stable if, for every $\varepsilon > 0$ and $t_0 \in [0, \infty)$, there exists $\delta = \delta(\varepsilon, t_0) > 0$ such that $\|x_0\| < \delta$ implies that $\|x(t)\| < \varepsilon$ for all $t \geq t_0$.*

ii) The nonlinear time-dependent impulsive dynamical system given by (2.115) and (2.116) is uniformly Lyapunov stable if, for every $\varepsilon > 0$, there exists $\delta = \delta(\varepsilon) > 0$ such that $\|x_0\| < \delta$ implies that $\|x(t)\| < \varepsilon$ for all $t \geq t_0$ and for all $t_0 \in [0, \infty)$.

iii) The nonlinear time-dependent impulsive dynamical system given by (2.115) and (2.116) is asymptotically stable if it is Lyapunov stable and for every $t_0 \in [0, \infty)$, there exists $\delta = \delta(t_0) > 0$ such that $\|x_0\| < \delta$ implies that $\lim_{t \to \infty} x(t) = 0$.

iv) *The nonlinear time-dependent impulsive dynamical system given by (2.115) and (2.116) is* uniformly asymptotically stable *if it is uniformly Lyapunov stable and there exists* $\delta > 0$ *such that* $\|x_0\| < \delta$ *implies that* $\lim_{t \to \infty} x(t) = 0$ *uniformly in* t_0 *and* x_0 *for all* $t_0 \in [0, \infty)$.

v) *The nonlinear time-dependent impulsive dynamical system given by (2.115) and (2.116) is* globally asymptotically stable *if it is Lyapunov stable and* $\lim_{t \to \infty} x(t) = 0$ *for all* $x_0 \in \mathbb{R}^n$ *and* $t_0 \in [0, \infty)$.

vi) *The nonlinear time-dependent impulsive dynamical system given by (2.115) and (2.116) is* globally uniformly asymptotically stable *if it is uniformly Lyapunov stable and* $\lim_{t \to \infty} x(t) = 0$ *uniformly in* t_0 *and* x_0 *for all* $x_0 \in \mathbb{R}^n$ *and* $t_0 \in [0, \infty)$.

vii) *The nonlinear time-dependent impulsive dynamical system given by (2.115) and (2.116) is* (uniformly) exponentially stable *if there exist scalars* $\alpha, \beta, \delta > 0$ *such that* $\|x_0\| < \delta$ *implies that* $\|x(t)\| \leq \alpha\|x_0\|e^{-\beta t}$, $t \geq t_0$ *and* $t_0 \in [0, \infty)$.

viii) *The nonlinear time-dependent impulsive dynamical system given by (2.115) and (2.116) is* globally (uniformly) exponentially stable *if there exist scalars* $\alpha, \beta > 0$ *such that* $\|x(t)\| \leq \alpha\|x_0\|e^{-\beta t}$, $t \geq t_0$, *for all* $x_0 \in \mathbb{R}^n$ *and* $t_0 \in [0, \infty)$.

Next, using Theorem 2.4 we present sufficient conditions for stability of the nonlinear time-dependent impulsive dynamical system (2.115) and (2.116). For the following result define

$$\dot{V}(t, x) \triangleq \frac{\partial V}{\partial x}(t, x) f_{\mathrm{c}}(t, x) + \frac{\partial V}{\partial t}(t, x), \quad t \neq t_k, \quad (2.117)$$

$$\Delta V(t, x) \triangleq V(t, x + f_{\mathrm{d}}(t, x)) - V(t, x), \quad t = t_k, \quad (2.118)$$

for a given continuously differentiable function $V : [0, \infty) \times \mathcal{D} \to \mathbb{R}$.

Theorem 2.6 *Consider the nonlinear time-dependent impulsive dynamical system (2.115) and (2.116). Then the following statements hold:*

i) *If there exist a continuously differentiable function* $V : [0, \infty) \times \mathcal{D} \to \mathbb{R}$ *and a class* $\mathcal{K}$ *function* $\alpha(\cdot)$ *such that*

$$V(t, 0) = 0, \quad t \in [0, \infty), \quad (2.119)$$

$$\alpha(\|x\|) \leq V(t, x), \quad (t, x) \in [0, \infty) \times \mathcal{D}, \quad (2.120)$$

$$\dot{V}(t, x) \leq 0, \quad (t, x) \in [0, \infty) \times \mathcal{D}, \quad t \neq t_k, \quad (2.121)$$

$$\Delta V(t, x) \leq 0, \quad x \in \mathcal{D}, \quad t = t_k, \quad (2.122)$$

then the nonlinear time-dependent impulsive dynamical system (2.115) and (2.116) is Lyapunov stable.

ii) If there exist a continuously differentiable function $V : [0, \infty) \times \mathcal{D} \to \mathbb{R}$ and class $\mathcal{K}$ functions $\alpha(\cdot)$ and $\beta(\cdot)$ satisfying (2.120)–(2.122) and

$$V(t, x) \leq \beta(\|x\|), \quad (t, x) \in [0, \infty) \times \mathcal{D}, \qquad (2.123)$$

then the nonlinear time-dependent impulsive dynamical system (2.115) and (2.116) is uniformly Lyapunov stable.

iii) If there exist a continuously differentiable function $V : [0, \infty) \times \mathcal{D} \to \mathbb{R}$ and class $\mathcal{K}$ functions $\alpha(\cdot)$, $\beta(\cdot)$, and $\gamma(\cdot)$ satisfying (2.120), (2.122), (2.123), and

$$\dot{V}(t, x) \leq -\gamma(\|x\|), \quad (t, x) \in [0, \infty) \times \mathcal{D}, \quad t \neq t_k, \qquad (2.124)$$

then the nonlinear time-dependent impulsive dynamical system (2.115) and (2.116) is uniformly asymptotically stable.

iv) If $\mathcal{D} = \mathbb{R}^n$ and there exist a continuously differentiable function $V : [0, \infty) \times \mathbb{R}^n \to \mathbb{R}$, a class $\mathcal{K}$ function $\gamma(\cdot)$, and class $\mathcal{K}_\infty$ functions $\alpha(\cdot)$ and $\beta(\cdot)$ satisfying (2.120) and (2.122)–(2.124), then the nonlinear time-dependent impulsive dynamical system (2.115) and (2.116) is globally uniformly asymptotically stable.

v) If there exist a continuously differentiable function $V : [0, \infty) \times \mathcal{D} \to \mathbb{R}$ and positive constants $\alpha, \beta, \gamma, p \geq 1$ satisfying (2.122) and

$$\alpha\|x\|^p \leq V(t, x) \leq \beta\|x\|^p, \quad (t, x) \in [0, \infty) \times \mathcal{D}, \qquad (2.125)$$
$$\dot{V}(t, x) \leq -\gamma\|x\|^p, \quad (t, x) \in [0, \infty) \times \mathcal{D}, \quad t \neq t_k, \qquad (2.126)$$

then the nonlinear time-dependent impulsive dynamical system (2.115) and (2.116) is uniformly exponentially stable.

vi) If $\mathcal{D} = \mathbb{R}^n$ and there exist a continuously differentiable function $V : [0, \infty) \times \mathbb{R}^n \to \mathbb{R}$ and positive constants α, β, γ, and $p \geq 1$ satisfying (2.122), (2.125), and (2.126), then the nonlinear time-dependent impulsive dynamical system (2.115) and (2.116) is globally uniformly exponentially stable.

Proof. Let $n_1 = n$, $n_2 = 1$, $x_1(t - t_0) = x(t)$, $x_2(t - t_0) = t$, $f_{1c}(x_1, x_2) = f_c(t, x)$, $f_{2c}(x_1, x_2) = 1$, $f_{1d}(x_1, x_2) = f_d(t, x)$, $f_{2d}(x_1, x_2) = 0$, and $\mathcal{Z} = \mathcal{D} \times \mathcal{T}$, where $\mathcal{T} \triangleq \{t_1, t_2, \ldots\}$ denotes the set of prescribed resetting times. Now, note that with $\tau = t - t_0$, the solution

$x(t)$, $t \geq t_0$, to the nonlinear time-dependent impulsive dynamical system (2.115) and (2.116) is equivalently characterized by the solution $x_1(\tau)$, $\tau \geq 0$, to the nonlinear state-dependent impulsive dynamical system

$$\dot{x}_1(\tau) = f_{1c}(x_1(\tau), x_2(\tau)), \quad x_1(0) = x_0, \quad (x_1(\tau), x_2(\tau)) \notin \mathcal{Z},$$
$$(2.127)$$
$$\dot{x}_2(\tau) = 1, \quad x_2(0) = t_0, \quad (x_1(\tau), x_2(\tau)) \notin \mathcal{Z}, \quad (2.128)$$
$$\Delta x_1(\tau) = f_{1d}(x_1(\tau), x_2(\tau)), \quad (x_1(\tau), x_2(\tau)) \in \mathcal{Z}, \quad (2.129)$$
$$\Delta x_2(\tau) = 0, \quad (x_1(\tau), x_2(\tau)) \in \mathcal{Z}, \quad (2.130)$$

where $\tau \geq 0$ and $\dot{x}_1(\cdot)$ and $\dot{x}_2(\cdot)$ denote differentiation with respect to τ. Furthermore, note that since $f_c(t, 0) = 0$, $f_d(t, 0) = 0$, $t \geq t_0$, it follows that $f_{1c}(0, x_2) = 0$ and $f_{1d}(0, x_2) = 0$ for all $x_2 \in \mathbb{R}$, respectively.

Next, note that the resetting set $\mathcal{Z} = \mathcal{D} \times \mathcal{T}$ consists of hyperplanes in $\mathbb{R}^{n_1+1}$ parallel to $\mathbb{R}^{n_1}$ such that when the trajectory $(x_1(\tau), x_2(\tau))$, $\tau \geq 0$, intersects one of these hyperplanes the system resets according to the resetting law (2.129) and (2.130) to another point on the hyperplane. Hence, (2.127)–(2.130) satisfy assumptions A1$'$ and A2$'$. To see this, note that since $\overline{\mathcal{Z}} \backslash \mathcal{Z} = (\overline{\mathcal{D}} \times \overline{\mathcal{T}}) \backslash (\mathcal{D} \times \mathcal{T}) = \partial\mathcal{D} \times \emptyset$ and $(x_1(\tau), x_2(\tau)) \notin \partial\mathcal{D} \times \emptyset$, $\tau \geq 0$, then A1$'$ is satisfied. Furthermore, since $\partial\mathcal{Z} \cap \mathcal{Z} = (\partial\mathcal{D} \times \overline{\mathcal{T}}) \cap (\mathcal{D} \times \mathcal{T}) = \emptyset \times \mathcal{T}$, it follows that $(x_1(\tau), x_2(\tau)) \notin \partial\mathcal{Z} \cap \mathcal{Z}$, and hence, A2$'$ holds. Now, the result is a direct consequence of Theorem 2.4. $\qquad\square$

In light of Theorem 2.6 it follows that Theorem 2.4 can be trivially extended to address partial stability for *state/time-dependent impulsive dynamical systems*. Specifically, consider the state/time-dependent impulsive dynamical system given by

$$\dot{x}_1(t) = f_{1c}(t, x_1(t), x_2(t)), \quad x_1(t_0) = x_{10},$$
$$(t, x_1(t), x_2(t)) \notin ([0, \infty) \times \mathcal{Z}) \cup (\mathcal{T} \times \mathcal{D} \times \mathbb{R}^{n_2}), \quad (2.131)$$
$$\dot{x}_2(t) = f_{2c}(t, x_1(t), x_2(t)), \quad x_2(t_0) = x_{20},$$
$$(t, x_1(t), x_2(t)) \notin ([0, \infty) \times \mathcal{Z}) \cup (\mathcal{T} \times \mathcal{D} \times \mathbb{R}^{n_2}), \quad (2.132)$$
$$\Delta x_1(t) = f_{1d}(t, x_1(t), x_2(t)),$$
$$(t, x_1(t), x_2(t)) \in ([0, \infty) \times \mathcal{Z}) \cup (\mathcal{T} \times \mathcal{D} \times \mathbb{R}^{n_2}), \quad (2.133)$$
$$\Delta x_2(t) = f_{2d}(t, x_1(t), x_2(t)),$$
$$(t, x_1(t), x_2(t)) \in ([0, \infty) \times \mathcal{Z}) \cup (\mathcal{T} \times \mathcal{D} \times \mathbb{R}^{n_2}), \quad (2.134)$$

where $t \geq t_0$, $x_1 \in \mathcal{D}$, $\mathcal{D} \subseteq \mathbb{R}^{n_1}$ is an open set such that $0 \in \mathcal{D}$,

$x_2 \in \mathbb{R}^{n_2}$, $f_{1c} : [t_0, \hat{t}) \times \mathcal{D} \times \mathbb{R}^{n_2} \to \mathbb{R}^{n_1}$ is such that for all $x_2 \in \mathbb{R}^{n_2}$ and $t \in [t_0, \hat{t})$, $f_{1c}(t, 0, x_2) = 0$ and for every $x_2 \in \mathbb{R}^{n_2}$ and $t \in [t_0, \hat{t})$, $f_{1c}(t, \cdot, x_2)$ is locally Lipschitz in x_1, $f_{2c} : [t_0, \hat{t}) \times \mathcal{D} \times \mathbb{R}^{n_2} \to \mathbb{R}^{n_2}$ is such that for every $x_1 \in \mathcal{D}$, $f_{2c}(\cdot, x_1, \cdot)$ is locally Lipschitz in x_2, $f_{1d} : [t_0, \hat{t}) \times \mathcal{D} \times \mathbb{R}^{n_2} \to \mathbb{R}^{n_1}$ is continuous and $f_{1d}(t, 0, x_2) = 0$ for all $x_2 \in \mathbb{R}^{n_2}$ and $t \in [t_0, \hat{t})$, $f_{2d} : [t_0, \hat{t}) \times \mathcal{D} \times \mathbb{R}^{n_2} \to \mathbb{R}^{n_2}$ is continuous, $\mathcal{Z} \subset \mathcal{D} \times \mathbb{R}^{n_2}$, and $\mathcal{T} = \{t_1, t_2, \ldots\}$ is the set of prescribed resetting times.

Next, let $\hat{x}_1(t - t_0) = x_1(t)$, $\hat{x}_2(t - t_0) = [x_2^{\mathrm{T}}(t)\ t]^{\mathrm{T}}$, $\hat{f}_{1c}(\hat{x}_1, \hat{x}_2) = f_{1c}(t, x_1, x_2)$, $\hat{f}_{2c}(\hat{x}_1, \hat{x}_2) = [f_{2c}^{\mathrm{T}}(t, x_1, x_2)\ 1]^{\mathrm{T}}$, $\hat{f}_{1d}(\hat{x}_1, \hat{x}_2) = f_{1d}(t, x_1, x_2)$, $\hat{f}_{2d}(\hat{x}_1, \hat{x}_2) = [f_{2d}^{\mathrm{T}}(t, x_1, x_2)\ 0]^{\mathrm{T}}$, and $\hat{\mathcal{Z}} = (\mathcal{Z} \times [0, \infty)) \cup (\mathcal{D} \times \mathbb{R}^{n_2} \times \mathcal{T})$. Now, note that with $\tau = t - t_0$, the solution $(x_1(t), x_2(t))$, $t \geq t_0$, to the nonlinear state/time-dependent impulsive dynamical system (2.131)–(2.134) is equivalently characterized by the solution $(\hat{x}_1(\tau), \hat{x}_2(\tau))$, $\tau \geq 0$, to the nonlinear (autonomous) state-dependent impulsive dynamical system

$$\dot{\hat{x}}_1(\tau) = \hat{f}_{1c}(\hat{x}_1(\tau), \hat{x}_2(\tau)), \quad \hat{x}_1(0) = x_{10}, \quad (\hat{x}_1(\tau), \hat{x}_2(\tau)) \notin \hat{\mathcal{Z}},$$
$$(2.135)$$
$$\dot{\hat{x}}_2(\tau) = \hat{f}_{2c}(\hat{x}_1(\tau), \hat{x}_2(\tau)), \quad \hat{x}_2(0) = [x_{20}^{\mathrm{T}}\ t_0]^{\mathrm{T}}, \quad (\hat{x}_1(\tau), \hat{x}_2(\tau)) \notin \hat{\mathcal{Z}},$$
$$(2.136)$$
$$\Delta\hat{x}_1(\tau) = \hat{f}_{1d}(\hat{x}_1(\tau), \hat{x}_2(\tau)), \quad (\hat{x}_1(\tau), \hat{x}_2(\tau)) \in \hat{\mathcal{Z}}, \qquad (2.137)$$
$$\Delta\hat{x}_2(\tau) = \hat{f}_{2d}(\hat{x}_1(\tau), \hat{x}_2(\tau)), \quad (\hat{x}_1(\tau), \hat{x}_2(\tau)) \in \hat{\mathcal{Z}}, \qquad (2.138)$$

where $\dot{\hat{x}}_1(\cdot)$ and $\dot{\hat{x}}_2(\cdot)$ denote differentiation with respect to τ. Hence, Theorem 2.4 can be used to derive sufficient conditions for partial stability results for the nonlinear state/time-dependent impulsive dynamical systems of the form (2.131)–(2.134). In this case, it is important to note that partial stability may be uniform with respect to either or both of x_{20} and t_0.

Next, we consider the time-dependent impulsive dynamical system with periodic resettings given by

$$\dot{x}(t) = f_c(x(t)), \quad x(t_0) = x_0, \quad t \neq t_k, \qquad (2.139)$$
$$\Delta x(t) = f_d(x(t)), \quad t = t_k, \qquad (2.140)$$

where $x(t) \in \mathcal{D}$, $t \geq t_0$, $\mathcal{D} \subseteq \mathbb{R}^n$ is an open set such that $0 \in \mathcal{D}$, $t_k = kT$, $k = 1, 2, \ldots$, where $T > 0$, $f_c : \mathcal{D} \to \mathbb{R}^n$ is locally Lipschitz continuous on $\mathcal{D}$, $f_c(0) = 0$, and $f_d : \mathcal{D} \to \mathbb{R}^n$ is such that $f_d(\cdot)$ is continuous and $f_d(0) = 0$. In this case, since the vector fields of (2.139) and (2.140) are time-independent, the n-dimensional phase portrait of (2.139) and (2.140) is not affected by the periodic resettings of the time variable. That is, when $t = T$, then t is reset to zero. The time-dependent impulsive dynamical system (2.139) and (2.140) with periodic resettings can hence be equivalently characterized as a state-dependent impulsive dynamical system with an additional state representing time, that is,

$$\dot{\tilde{x}}(t) = \tilde{f}_{\mathrm{c}}(\tilde{x}(t)), \quad \tilde{x}(t_0) = \tilde{x}_0, \quad \tilde{x}(t) \notin \tilde{\mathcal{Z}}, \tag{2.141}$$
$$\Delta \tilde{x}(t) = \tilde{f}_{\mathrm{d}}(\tilde{x}(t)), \quad \tilde{x}(t) \in \tilde{\mathcal{Z}}, \tag{2.142}$$

where

$$\tilde{x}(t) = \begin{bmatrix} x(t) \\ \tau(t) \end{bmatrix}, \quad \tilde{x}(t_0) = \begin{bmatrix} x_0 \\ \tau_0 \end{bmatrix}, \tag{2.143}$$

$$\tilde{f}_{\mathrm{c}}(\tilde{x}) = \begin{bmatrix} f_{\mathrm{c}}(x) \\ 1 \end{bmatrix}, \quad \tilde{f}_{\mathrm{d}}(\tilde{x}) = \begin{bmatrix} f_{\mathrm{d}}(x) \\ -T \end{bmatrix}, \tag{2.144}$$

$\tilde{\mathcal{Z}} = \{\tilde{x} = [x^{\mathrm{T}}, \tau]^{\mathrm{T}} \in \mathcal{D} \times [0, T] : \tau = T\}$, $x_0 \in \mathcal{D}$, and $\tau_0 = t_0 \in [0, T)$. Note that the solution $x(t)$, $t \geq t_0$, to (2.139) and (2.140) is equivalently characterized by the partial solution $x(t)$, $t \geq t_0$, to (2.141) and (2.142).

Theorem 2.7 *Consider the time-dependent impulsive dynamical system (2.139) and (2.140) with $t_k = kT$, $k = 1, 2, \ldots$, and $T > 0$. Assume that $\mathcal{D}_{\mathrm{c}} \subset \mathcal{D}$ is a compact positively invariant set with respect to (2.139) and (2.140). Furthermore, assume there exists a continuously differentiable function $V : \mathcal{D}_{\mathrm{c}} \to \mathbb{R}$ such that $V(0) = 0$, $V(x) > 0$, $x \neq 0$, $x \in \mathcal{D}_{\mathrm{c}}$, and*

$$V'(x) f_{\mathrm{c}}(x) \leq 0, \quad x \in \mathcal{D}_{\mathrm{c}}, \tag{2.145}$$
$$V(x + f_{\mathrm{d}}(x)) - V(x) \leq 0, \quad x \in \mathcal{D}_{\mathrm{c}}. \tag{2.146}$$

Let $\mathcal{R}_\gamma \triangleq \{x \in \mathcal{D}_{\mathrm{c}} : V(x) = \gamma\}$, where $\gamma > 0$, and let $\mathcal{M}_\gamma$ denote the largest invariant set contained in $\mathcal{R}_\gamma$. If for each $\gamma > 0$, $\mathcal{M}_\gamma$ contains no system trajectory, then the zero solution $x(t) \equiv 0$ to (2.139) and (2.140) is uniformly asymptotically stable.

Proof. Uniform Lyapunov stability follows from $ii)$ of Theorem 2.6 since $V(x)$, $x \in \mathcal{D}_{\mathrm{c}}$, is a positive-definite function on a compact set $\mathcal{D}_{\mathrm{c}}$. To show asymptotic stability, consider the state-dependent representation (2.141) and (2.142) of the time-dependent impulsive dynamical system (2.139) and (2.140). Note that for the impulsive dynamical system given by (2.141) and (2.142) the state variable $\tau(t)$, $t \geq t_0$, is defined over the interval $[0, T]$, and hence, the set $\mathcal{D}_c \times [0, T]$ is a compact positively invariant set with respect to (2.141) and (2.142). Furthermore, defining $\tilde{\mathcal{X}}(\tilde{x}) \triangleq \tau - T$, $\tilde{x} \in \mathcal{D}_c \times [0, T]$, it follows that $\tilde{\mathcal{Z}} = \{\tilde{x} \in \mathcal{D}_c \times [0, T] : \tilde{\mathcal{X}}(\tilde{x}) = 0\}$. Note that $\tilde{\mathcal{X}}'(\tilde{x}) \tilde{f}_c(\tilde{x}) = 1 \neq 0$, $\tilde{x} \in \tilde{\mathcal{Z}}$, and hence, it follows from Proposition 2.2 that $\tau_1(\cdot)$ is continuous at $\tilde{x}_0 \in \mathcal{D}_c \times [0, T]$, where $0 < \tau_1(\tilde{x}_0) < \infty$. (Specifically, $\tau_1(\tilde{x}_0) = T - t_0$ which is continuous at $\tilde{x}_0 \in \mathcal{D}_c \times [0, T]$.)

Next, since (2.141) and (2.142) possesses an infinite number of resettings, it follows from Proposition 2.1 that (2.141) and (2.142) satisfy Assumption 2.1. Now, it follows from Theorem 2.2 that the positive limit set $\omega(\tilde{x}_0)$ of (2.141) and (2.142) with $\tilde{x}_0 \in \mathcal{D}_c \times [0, T]$ is a nonempty, compact invariant set, which further implies that the positive limit set $\omega(x_0)$ of (2.139) and (2.140) with $x_0 \in \mathcal{D}_c$ is a nonempty, compact invariant set. Next, it follows from (2.145) and (2.146) that $V(x(t))$ is nonincreasing for all $t \geq 0$, which implies that, since $V(\cdot)$ is continuous, $V(x(t)) \to \gamma \geq 0$ as $t \to \infty$. Hence, the positive limit set $\omega(x_0)$ is contained in $\mathcal{M}_\gamma$. Now, since by assumption $\mathcal{M}_\gamma$ contains no system trajectory for each $\gamma > 0$, it follows that $\gamma = 0$, establishing uniform asymptotic stability of the zero solution $x(t) \equiv 0$ to (2.139) and (2.140).   □

Finally, we analyze a time-varying *periodic* dynamical system as a special case of a state-dependent impulsive dynamical system. Consider the nonlinear periodic dynamical system given by

$$\dot{x}(t) = f(t, x(t)), \quad x(t_0) = x_0, \quad t \geq t_0, \tag{2.147}$$

where $x(t) \in \mathbb{R}^n$, $t \geq t_0$, $f : [t_0, \infty) \times \mathbb{R}^n \to \mathbb{R}^n$ is such that $f(t + T, x) = f(t, x)$, $t \geq 0$, $x \in \mathbb{R}^n$, where $T > 0$ is given. Furthermore, we assume $f(\cdot, \cdot)$ is such that for every $t_0 \in \mathbb{R}_+$ and $x_0 \in \mathbb{R}^n$, there exists a unique solution $x(t)$, $t \geq t_0$, to (2.147). Note that with $\tau = t - t_0$, the solution $x(t)$, $t \geq t_0$, to the nonlinear time-varying dynamical system (2.147) is equivalently characterized by the solution $x_1(\tau)$, $\tau \geq 0$, of the nonlinear autonomous dynamical system

$$\dot{x}_1(\tau) = f(x_2(\tau), x_1(\tau)), \quad x_1(0) = x_0, \quad \tau \geq 0, \tag{2.148}$$
$$\dot{x}_2(\tau) = 1, \quad x_2(0) = t_0, \tag{2.149}$$

where $\dot{x}_1(\cdot)$ and $\dot{x}_2(\cdot)$ denote differentiation with respect to τ.

Next, using the fact that $f(t + T, x) = f(t, x)$, where $t \geq t_0$ and $x \in \mathbb{R}^n$, the solution $x(t)$, $t \geq t_0$, of the nonlinear time-varying dynamical system (2.147) can be also characterized as a solution $x_1(\tau)$, $\tau \geq 0$, of nonlinear state-dependent impulsive dynamical system given by

$$\begin{bmatrix} \dot{x}_1(\tau) \\ \dot{x}_2(\tau) \end{bmatrix} = \begin{bmatrix} f(x_2(\tau), x_1(\tau)) \\ 1 \end{bmatrix}, \quad \begin{bmatrix} x_1(0) \\ x_2(0) \end{bmatrix} = \begin{bmatrix} x_0 \\ t_0 \end{bmatrix},$$
$$(x_1(\tau), x_2(\tau)) \notin \mathcal{Z}, \qquad (2.150)$$

$$\begin{bmatrix} \Delta x_1(\tau) \\ \Delta x_2(\tau) \end{bmatrix} = \begin{bmatrix} 0 \\ -T \end{bmatrix}, \quad (x_1(\tau), x_2(\tau)) \in \mathcal{Z}, \qquad (2.151)$$

where $\mathcal{Z} = \{(x_1, x_2) \in \mathbb{R}^n \times \mathbb{R} : x_2 = T\}$. Note that the solution to (2.150) and (2.151) is bounded if the solution $x(t)$, $t \geq t_0$, to (2.147) is bounded. Now, it follows from Propositions 2.2 and 2.1 that the nonlinear state-dependent dynamical system given by (2.150) and (2.151) satisfies Assumption 2.1. Hence, it follows from Theorem 2.2 that for every $(t_0, x_0) \in [0, \infty) \times \mathbb{R}^n$, the positive limit set $\omega(t_0, x_0)$ of the solution $(x_1(\tau), x_2(\tau))$, $\tau \geq 0$, to (2.150) and (2.151) is nonempty, bounded, and invariant or, equivalently, for every $(t_0, x_0) \in [0, \infty) \times \mathbb{R}^n$ the positive limit set $\omega(t_0, x_0)$ of the solution $x(t)$, $t \geq t_0$, to the nonlinear periodic dynamical system (2.147) is nonempty, bounded, and invariant.

Although both solutions $x_1(\tau)$, $\tau \geq 0$, to (2.148) and (2.149) and (2.150) and (2.151), are equivalent to the solution $x(t)$, $t \geq t_0$, to (2.147), note that the solution to (2.148) and (2.149) is always unbounded, whereas the solution of the impulsive dynamical system (2.150) and (2.151) is always bounded if the solution $x(t)$, $t \geq t_0$, to (2.147), is bounded. The assumption that $f(\cdot, \cdot)$ is periodic is critical in casting (2.147) as a state-dependent impulsive dynamical system (2.150) and (2.151). In light of the above, it follows that the positive limit set of (2.147) is nonempty, bounded, and invariant. This of course is a classical result for time-varying periodic dynamical systems [162, p. 153].

2.9 Lagrange Stability, Boundedness, and Ultimate Boundedness

In the previous sections we introduced the concepts of stability and partial stability for nonlinear impulsive dynamical systems. In certain engineering applications, however, it is more natural to ascertain whether for every system initial condition in a ball of radius δ the solution of the nonlinear impulsive dynamical system is bounded. This leads to the notions of *Lagrange stability*, *boundedness*, and *ultimate boundedness*. These notions are closely related to what is known in the literature as *practical stability*. In this section, we present Lyapunov-like theorems for boundedness and ultimate boundedness of nonlinear impulsive dynamical systems.

Definition 2.11 *i) The nonlinear state-dependent impulsive dynamical system given by (2.83)–(2.86) is* Lagrange stable *with respect to x_1 if, for every $x_{10} \in \mathcal{D}$ and $x_{20} \in \mathbb{R}^{n_2}$, there exists $\varepsilon = \varepsilon(x_{10}, x_{20}) > 0$ such that $\|x_1(t)\| < \varepsilon$, $t \geq 0$.*

ii) The nonlinear state-dependent impulsive dynamical system given by (2.83)–(2.86) is bounded *with respect to x_1 uniformly in x_2 if, for every $x_{20} \in \mathbb{R}^{n_2}$, there exists $\gamma > 0$ such that, for every $\delta \in (0, \gamma)$, there exists $\varepsilon = \varepsilon(\delta) > 0$ such that $\|x_{10}\| < \delta$ implies $\|x_1(t)\| < \varepsilon$, $t \geq 0$. The nonlinear state-dependent impulsive dynamical system (2.83)–(2.86) is* globally bounded *with respect to x_1 uniformly in x_2 if, for every $x_{20} \in \mathbb{R}^{n_2}$ and $\delta \in (0, \infty)$, there exists $\varepsilon = \varepsilon(\delta) > 0$ such that $\|x_{10}\| < \delta$ implies $\|x_1(t)\| < \varepsilon$, $t \geq 0$.*

iii) The nonlinear state-dependent impulsive dynamical system given by (2.83)–(2.86) is ultimately bounded *with respect to x_1 uniformly in x_2 with bound ε if, for every $x_{20} \in \mathbb{R}^{n_2}$, there exists $\gamma > 0$ such that, for every $\delta \in (0, \gamma)$, there exists $T = T(\delta, \varepsilon) > 0$ such that $\|x_{10}\| < \delta$ implies $\|x_1(t)\| < \varepsilon$, $t \geq T$. The nonlinear state-dependent impulsive dynamical system (2.83)–(2.86) is* globally ultimately bounded *with respect to x_1 uniformly in x_2 with bound ε if, for every $x_{20} \in \mathbb{R}^{n_2}$ and $\delta \in (0, \infty)$, there exists $T = T(\delta, \varepsilon) > 0$ such that $\|x_{10}\| < \delta$ implies $\|x_1(t)\| < \varepsilon$, $t \geq T$.*

Note that if a nonlinear state-dependent impulsive dynamical system is globally bounded with respect to x_1 uniformly in x_2, then it is Lagrange stable with respect to x_1. Alternatively, if a nonlinear state-dependent impulsive dynamical system is (globally) bounded with respect to x_1 uniformly in x_2, then there exists $\varepsilon > 0$ such that it is (globally) ultimately bounded with respect to x_1 uniformly in x_2 with a bound ε. Conversely, if a nonlinear state-dependent impulsive dynamical system is (globally) ultimately bounded with respect to x_1 uniformly in x_2 with a bound ε, then it is (globally) bounded with respect to x_1 uniformly in x_2. The following results present Lyapunov-like theorems for boundedness and ultimate boundedness. For these results recall that $\dot{V}(x_1, x_2) = V'(x_1, x_2) f_c(x_1, x_2)$, where $f_c(x_1, x_2) = [f_{1c}^{\mathrm{T}}(x_1, x_2) \ f_{2c}^{\mathrm{T}}(x_1, x_2)]^{\mathrm{T}}$, and $\Delta V(x_1, x_2) = V(x_1 + f_{1d}(x_1, x_2), x_2 + f_{2d}(x_1, x_2)) - V(x_1, x_2)$, for a given continuously differentiable function $V : \mathcal{D} \times \mathbb{R}^{n_2} \to \mathbb{R}$.

Theorem 2.8 *Consider the nonlinear state-dependent impulsive dynamical system (2.83)–(2.86). Assume there exist a continuously dif-*

ferentiable function $V : \mathcal{D} \times \mathbb{R}^{n_2} \to \mathbb{R}$ *and class* $\mathcal{K}$ *functions* $\alpha(\cdot)$ *and* $\beta(\cdot)$ *such that*

$$\alpha(\|x_1\|) \leq V(x_1, x_2) \leq \beta(\|x_1\|), \quad (x_1, x_2) \in \mathcal{D} \times \mathbb{R}^{n_2}, \quad (2.152)$$

$$\dot{V}(x_1, x_2) \leq 0, \quad x_1 \in \mathcal{D}, \quad \|x_1\| \geq \mu, \quad x_2 \in \mathbb{R}^{n_2}, \quad (x_1, x_2) \notin \mathcal{Z},$$
$$(2.153)$$

$$\Delta V(x_1, x_2) \leq 0, \quad x_1 \in \mathcal{D}, \quad \|x_1\| \geq \mu, \quad x_2 \in \mathbb{R}^{n_2}, \quad (x_1, x_2) \in \mathcal{Z},$$
$$(2.154)$$

where $\mu > 0$ *is such that* $\mathcal{B}_{\alpha^{-1}(\beta(\mu))}(0) \subset \mathcal{D}$. *Furthermore, assume* $\theta \triangleq \sup_{(x_1,x_2)\in\overline{\mathcal{B}}_\mu(0)\times\mathbb{R}^{n_2}\cap\mathcal{Z}} V(x_1 + f_{1\mathrm{d}}(x_1, x_2), x_2 + f_{2\mathrm{d}}(x_1, x_2))$ *ex-ists. Then the nonlinear state-dependent impulsive dynamical system (2.83)–(2.86) is bounded with respect to* x_1 *uniformly in* x_2. *Further-more, for every* $\delta \in (0, \gamma)$, $x_{10} \in \overline{\mathcal{B}}_\delta(0)$ *implies that* $\|x_1(t)\| \leq \varepsilon$, $t \geq 0$, *where*

$$\varepsilon = \varepsilon(\delta) \triangleq \alpha^{-1}(\max\{\eta, \beta(\delta)\}), \quad (2.155)$$

where $\eta \geq \max\{\beta(\mu), \theta\}$ *and* $\gamma \triangleq \sup\{r > 0 : \mathcal{B}_{\alpha^{-1}(\beta(r))}(0) \subset \mathcal{D}\}$. *If, in addition,* $\mathcal{D} = \mathbb{R}^{n_1}$ *and* $\alpha(\cdot)$ *is a class* $\mathcal{K}_\infty$ *function, then the nonlinear state-dependent impulsive dynamical system (2.83)–(2.86) is globally bounded with respect to* x_1 *uniformly in* x_2 *and for every* $x_{10} \in \mathbb{R}^{n_1}$, $\|x_1(t)\| \leq \varepsilon$, $t \geq 0$, *where* ε *is given by (2.155) with* $\delta = \|x_{10}\|$.

Proof. First, let $\delta \in (0, \mu]$ and assume $\|x_{10}\| \leq \delta$. If $\|x_1(t)\| \leq \mu$, $t \geq 0$, then it follows from (2.152) that $\|x_1(t)\| \leq \mu \leq \alpha^{-1}(\beta(\mu)) \leq \alpha^{-1}(\eta)$, $t \geq 0$. Alternatively, if there exists $T > 0$ such that $\|x_1(T)\| > \mu$, then it follows that there exists $\tau < T$ such that either $\|x_1(\tau)\| = \mu$, $(x_1(\tau), x_2(\tau)) \notin \mathcal{Z}$, and $\|x_1(t)\| > \mu$, $t \in (\tau, T]$, or $(x_1(\tau), x_2(\tau)) \in \mathcal{Z}$, $\|x_1(\tau)\| \leq \mu$, and $\|x_1(t)\| > \mu$, $t \in (\tau, T]$. Hence, it follows from (2.152)–(2.154) that

$$\alpha(\|x_1(T)\|) \leq V(x_1(T), x_2(T)) \leq V(x_1(\tau), x_2(\tau)) \leq \beta(\mu) \leq \eta,$$

if $\|x_1(\tau)\| = \mu$ and $(x_1(\tau), x_2(\tau)) \notin \mathcal{Z}$, or

$$\begin{aligned}
\alpha(\|x_1(T)\|) &\leq V(x_1(T), x_2(T)) \leq V(x_1(\tau^+), x_2(\tau^+)) \\
&= V(x_1(\tau) + f_{1\mathrm{d}}(x_1(\tau), x_2(\tau)), x_2(\tau) + f_{2\mathrm{d}}(x_1(\tau), x_2(\tau))) \\
&\leq \theta \\
&\leq \eta,
\end{aligned}$$

if $\|x_1(\tau)\| \leq \mu$ and $(x_1(\tau), x_2(\tau)) \in \mathcal{Z}$. In either case, it follows that $\|x_1(T)\| \leq \alpha^{-1}(\eta)$.

Next, let $\delta \in (\mu, \gamma)$ and assume $x_{10} \in \overline{\mathcal{B}}_\delta(0)$ and $\|x_{10}\| > \mu$. Now, for every $\hat{t} > 0$ such that $\|x_1(t)\| \geq \mu$, $t \in [0, \hat{t}]$, it follows from (2.152) and (2.153) that

$$\alpha(\|x_1(t)\|) \leq V(x_1(t), x_2(t)) \leq V(x_{10}, x_{20}) \leq \beta(\delta), \quad t \geq 0,$$

which implies that $\|x_1(t)\| \leq \alpha^{-1}(\beta(\delta))$, $t \in [0, \hat{t}]$. Next, if there exists $T > 0$ such that $\|x_1(T)\| \leq \mu$, then it follows as in the proof of the first case that $\|x_1(t)\| \leq \alpha^{-1}(\eta)$, $t \geq T$. Hence, if $x_{10} \in \mathcal{B}_\delta(0) \backslash \mathcal{B}_\mu(0)$, then $\|x_1(t)\| \leq \alpha^{-1}(\max\{\eta, \beta(\delta)\})$, $t \geq 0$. Finally, if $\mathcal{D} = \mathbb{R}^{n_1}$ and $\alpha(\cdot)$ is a class $\mathcal{K}_\infty$ function it follows that $\beta(\cdot)$ is a class $\mathcal{K}_\infty$ function, and hence $\gamma = \infty$. Hence, the nonlinear state-dependent impulsive dynamical system (2.83)–(2.86) is globally bounded with respect to x_1 uniformly in x_2. $\qquad\square$

Theorem 2.9 *Consider the nonlinear state-dependent impulsive dynamical system (2.83)–(2.86). Assume there exist a continuously differentiable function $V : \mathcal{D} \times \mathbb{R}^{n_2} \to \mathbb{R}$ and class $\mathcal{K}$ functions $\alpha(\cdot)$ and $\beta(\cdot)$ such that (2.152) and (2.154) hold. Furthermore, assume that there exists a continuous function $W : \mathcal{D} \to \mathbb{R}$ such that $W(x_1) > 0$, $\|x_1\| > \mu$, and*

$$\dot{V}(x_1, x_2) \leq -W(x_1), \quad x_1 \in \mathcal{D}, \quad \|x_1\| > \mu, \quad x_2 \in \mathbb{R}^{n_2},$$
$$(x_1, x_2) \notin \mathcal{Z}, \; (2.156)$$

where $\mu > 0$ is such that $\mathcal{B}_{\alpha^{-1}(\beta(\mu))}(0) \subset \mathcal{D}$. Finally, assume $\theta \triangleq \sup_{(x_1, x_2) \in \overline{\mathcal{B}}_\mu(0) \times \mathbb{R}^{n_2} \cap \mathcal{Z}} V(x_1 + f_{1d}(x_1, x_2), x_2 + f_{2d}(x_1, x_2))$ exists. Then the nonlinear state-dependent impulsive dynamical system given by (2.83)–(2.86) is ultimately bounded with respect to x_1 uniformly in x_2 with bound $\varepsilon \triangleq \alpha^{-1}(\eta)$, where $\eta > \max\{\beta(\mu), \theta\}$. Furthermore, $\limsup_{t \to \infty} \|x_1(t)\| \leq \alpha^{-1}(\beta(\mu))$. If, in addition, $\mathcal{D} = \mathbb{R}^{n_1}$ and $\alpha(\cdot)$ is a class $\mathcal{K}_\infty$ function, then the nonlinear state-dependent impulsive dynamical system (2.83)–(2.86) is globally ultimately bounded with respect to x_1 uniformly in x_2 with bound ε.

Proof. First, let $\delta \in (0, \mu]$ and assume $\|x_{10}\| \leq \delta$. As in the proof of Theorem 2.8, it follows that $\|x_1(t)\| \leq \alpha^{-1}(\eta) = \varepsilon$, $t \geq 0$. Next, let $\delta \in (\mu, \gamma)$, where $\gamma \triangleq \sup\{r > 0 : \mathcal{B}_{\alpha^{-1}(\beta(r))}(0) \subset \mathcal{D}\}$, and assume $x_{10} \in \mathcal{B}_\delta(0)$ and $\|x_{10}\| > \mu$. In this case, it follows from Theorem 2.8 that $\|x_1(t)\| \leq \alpha^{-1}(\max\{\eta, \beta(\delta)\})$, $t \geq 0$. Suppose, *ad absurdum*, $\|x_1(t)\| \geq \beta^{-1}(\eta)$, $t \geq 0$, or, equivalently, $x_1(t) \in \mathcal{O} \triangleq$

$\mathcal{B}_{\alpha^{-1}(\beta(\delta))}(0)\backslash\mathcal{B}_{\beta^{-1}(\eta)}(0)$, $t \geq 0$. Since $\overline{\mathcal{O}}$ is compact, $W(\cdot)$ is continuous, and $W(x_1) > 0$, $\|x_1\| \geq \beta^{-1}(\eta) > \mu$, it follows from Weierstrass' theorem [146] that $k \triangleq \min_{x_1 \in \overline{\mathcal{O}}} W(x_1) > 0$ exists. Hence, it follows from (2.154) and (2.156) that

$$V(x_1(t), x_2(t)) \leq V(x_{10}, x_{20}) - kt, \quad t \geq 0, \tag{2.157}$$

which implies that

$$\alpha(\|x_1(t)\|) \leq \beta(\|x_{10}\|) - kt \leq \beta(\delta) - kt, \quad t \geq 0. \tag{2.158}$$

Now, letting $t > \beta(\delta)/k$, it follows that $\alpha(\|x_1(t)\|) < 0$, which is a contradiction. Hence, there exists $T = T(\delta, \eta) > 0$ such that $\|x_1(T)\| < \beta^{-1}(\eta)$. Thus, it follows from Theorem 2.8 that $\|x_1(t)\| \leq \alpha^{-1}(\beta(\beta^{-1}(\eta))) = \alpha^{-1}(\eta)$, $t \geq T$, which proves that the nonlinear state-dependent impulsive dynamical system (2.83)–(2.86) is ultimately bounded with respect to x_1 uniformly in x_2 with bound $\varepsilon = \alpha^{-1}(\eta)$. Furthermore, $\limsup_{t \to \infty} \|x_1(t)\| \leq \alpha^{-1}(\beta(\mu))$.

Finally, if $\mathcal{D} = \mathbb{R}^{n_1}$ and $\alpha(\cdot)$ is a class $\mathcal{K}_\infty$ function it follows that $\beta(\cdot)$ is a class $\mathcal{K}_\infty$ function, and hence, $\gamma = \infty$. Hence, the nonlinear state-dependent impulsive dynamical system (2.83)–(2.86) is globally ultimately bounded with respect to x_1 uniformly in x_2 with bound ε. $\square$

Next, we specialize Theorems 2.8 and 2.9 to nonlinear time-dependent impulsive dynamical systems. The following definition is needed for these results.

Definition 2.12 *i) The nonlinear time-dependent impulsive dynamical system given by (2.115) and (2.116) is* Lagrange stable *if, for every $x_0 \in \mathbb{R}^n$ and $t_0 \in \mathbb{R}$, there exists $\varepsilon = \varepsilon(t_0, x_0) > 0$ such that $\|x(t)\| < \varepsilon$, $t \geq t_0$.*

ii) The nonlinear time-dependent impulsive dynamical system given by (2.115) and (2.116) is uniformly bounded *if there exists $\gamma > 0$ such that, for every $\delta \in (0, \gamma)$, there exists $\varepsilon = \varepsilon(\delta) > 0$ such that $\|x_0\| < \delta$ implies $\|x(t)\| < \varepsilon$, $t \geq t_0$. The nonlinear time-dependent impulsive dynamical system (2.115) and (2.116) is* globally uniformly bounded *if, for every $\delta \in (0, \infty)$, there exists $\varepsilon = \varepsilon(\delta) > 0$ such that $\|x_0\| < \delta$ implies $\|x(t)\| < \varepsilon$, $t \geq t_0$.*

iii) The nonlinear time-dependent impulsive dynamical system given by (2.115) and (2.116) is uniformly ultimately bounded with bound ε *if there exists $\gamma > 0$ such that, for every $\delta \in (0, \gamma)$, there exists $T = T(\delta, \varepsilon) > 0$ such that $\|x_0\| < \delta$ implies $\|x(t)\| < \varepsilon$, $t \geq t_0 + T$.*

The nonlinear time-dependent impulsive dynamical system (2.115) and (2.116) is globally uniformly ultimately bounded with bound ε *if, for every* $\delta \in (0, \infty)$, *there exists* $T = T(\delta, \varepsilon) > 0$ *such that* $\|x_0\| < \delta$ *implies* $\|x(t)\| < \varepsilon$, $t \geq t_0 + T$.

For the following result define

$$\dot{V}(t,x) \triangleq \frac{\partial V}{\partial t} + \frac{\partial V}{\partial x}(t,x)f_c(t,x)$$

and

$$\Delta V(t,x) \triangleq V(t, x + f_d(t,x)) - V(t,x),$$

where $V : \mathbb{R} \times \mathcal{D} \to \mathbb{R}$ is a given continuously differentiable function.

Corollary 2.4 *Consider the nonlinear time-dependent impulsive dynamical system (2.115) and (2.116). Assume there exist a continuously differentiable function* $V : \mathbb{R} \times \mathcal{D} \to \mathbb{R}$ *and class* $\mathcal{K}$ *functions* $\alpha(\cdot)$ *and* $\beta(\cdot)$ *such that*

$$\alpha(\|x\|) \leq V(t,x) \leq \beta(\|x\|), \quad x \in \mathcal{D}, \quad t \in \mathbb{R}, \qquad (2.159)$$
$$\dot{V}(t,x) \leq 0, \quad x \in \mathcal{D}, \quad \|x\| \geq \mu, \quad t \neq t_k, \qquad (2.160)$$
$$\Delta V(t,x) \leq 0, \quad x \in \mathcal{D}, \quad \|x\| \geq \mu, \quad t = t_k, \qquad (2.161)$$

where $\mu > 0$ *is such that* $\mathcal{B}_{\alpha^{-1}(\beta(\mu))}(0) \subset \mathcal{D}$. *Furthermore, assume* $\sup_{(t,x) \in \{t_1, t_2, \dots\} \times \mathcal{D}} V(t, x + f_d(t,x))$ *exists. Then the nonlinear time-dependent impulsive dynamical system (2.115) and (2.116) is uniformly bounded. If, in addition,* $\mathcal{D} = \mathbb{R}^n$ *and* $\alpha(\cdot)$ *is a class* $\mathcal{K}_\infty$ *function, then the nonlinear time-dependent impulsive dynamical system (2.115) and (2.116) is globally uniformly bounded.*

Proof. Let $n_1 = n$, $n_2 = 1$, $x_1(t - t_0) = x(t)$, $x_2(t - t_0) = t$, $f_{1c}(x_1, x_2) = f_c(t, x)$, $f_{2c}(x_1, x_2) = 1$, $f_{1d}(x_1, x_2) = f_d(t, x)$, $f_{2d}(x_1, x_2) = 0$, and $\mathcal{Z} = \mathcal{D} \times \mathcal{T}$, where $\mathcal{T} \triangleq \{t_1, t_2, \dots\}$ denotes the set of prescribed resetting times. Now, note that with $\tau = t - t_0$, the solution $x(t)$, $t \geq t_0$, to the nonlinear time-dependent impulsive dynamical system (2.115) and (2.116) is equivalently characterized by the solution $x_1(\tau)$, $\tau \geq 0$, to the nonlinear state-dependent impulsive dynamical system

$$\dot{x}_1(\tau) = f_{1c}(x_1(\tau), x_2(\tau)), \quad x_1(0) = x_0, \quad (x_1(\tau), x_2(\tau)) \notin \mathcal{Z},$$
$$\qquad (2.162)$$
$$\dot{x}_2(\tau) = 1, \quad x_2(0) = t_0, \quad (x_1(\tau), x_2(\tau)) \notin \mathcal{Z}, \qquad (2.163)$$
$$\Delta x_1(\tau) = f_{1d}(x_1(\tau), x_2(\tau)), \quad (x_1(\tau), x_2(\tau)) \in \mathcal{Z}, \qquad (2.164)$$
$$\Delta x_2(\tau) = 0, \quad (x_1(\tau), x_2(\tau)) \in \mathcal{Z}, \qquad (2.165)$$

where $\tau \geq 0$ and $\dot{x}_1(\cdot)$ and $\dot{x}_2(\cdot)$ denote differentiation with respect to τ. Furthermore, note that since $f_c(t, 0) = 0$, $f_d(t, 0) = 0$, $t \geq t_0$, it follows that $f_{1c}(0, x_2) = 0$ and $f_{1d}(0, x_2) = 0$ for all $x_2 \in \mathbb{R}$, respectively.

Next, note that the resetting set $\mathcal{Z} = \mathcal{D} \times \mathcal{T}$ consists of hyperplanes in $\mathbb{R}^{n_1+1}$ parallel to $\mathbb{R}^{n_1}$ such that when the trajectory $(x_1(\tau), x_2(\tau))$, $\tau \geq 0$, intersects one of these hyperplanes the system resets according to the resetting law (2.164) and (2.165) to another point on the hyperplane. Hence, (2.162)–(2.165) satisfy Assumptions A1$'$ and A2$'$. To see this, note that since $\overline{\mathcal{Z}} \backslash \mathcal{Z} = (\overline{\mathcal{D}} \times \overline{\mathcal{T}}) \backslash (\mathcal{D} \times \mathcal{T}) = \partial \mathcal{D} \times \varnothing$ and $(x_1(\tau), x_2(\tau)) \notin \partial \mathcal{D} \times \varnothing$, $\tau \geq 0$, then A1$'$ is satisfied. Furthermore, since $\partial \mathcal{Z} \cap \mathcal{Z} = (\partial \mathcal{D} \times \overline{\mathcal{T}}) \cap (\mathcal{D} \times \mathcal{T}) = \varnothing \times \mathcal{T}$, it follows that $(x_1(\tau), x_2(\tau)) \notin \partial \mathcal{Z} \cap \mathcal{Z}$, and hence, A2$'$ holds. Now, the result is a direct consequence of Theorem 2.8. $\square$

Corollary 2.5 *Consider the nonlinear time-dependent impulsive dynamical system (2.115) and (2.116). Assume there exist a continuously differentiable function $V : \mathbb{R} \times \mathcal{D} \to \mathbb{R}$ and class $\mathcal{K}$ functions $\alpha(\cdot)$ and $\beta(\cdot)$ such that (2.159) and (2.161) hold. Furthermore, assume that there exists a continuous function $W : \mathcal{D} \to \mathbb{R}$ such that $W(x) > 0$, $\|x\| > \mu$, and*

$$\dot{V}(t, x) \leq -W(x), \quad x \in \mathcal{D}, \quad \|x\| > \mu, \quad t \neq t_k, \tag{2.166}$$

where $\mu > 0$ is such that $\mathcal{B}_{\alpha^{-1}(\beta(\mu))}(0) \subset \mathcal{D}$. Finally, assume $\theta \triangleq \sup_{(t,x)\in\{t_1,t_2,\dots\}\times\mathcal{D}} V(t, x + f_d(t, x))$ exists. Then the nonlinear time-dependent impulsive dynamical system given by (2.115) and (2.116) is uniformly ultimately bounded with bound $\varepsilon \triangleq \alpha^{-1}(\eta)$, where $\eta \triangleq \max\{\beta(\mu), \theta\}$. Furthermore, $\limsup_{t\to\infty} \|x(t)\| \leq \alpha^{-1}(\beta(\mu))$. If, in addition, $\mathcal{D} = \mathbb{R}^n$ and $\alpha(\cdot)$ is a class $\mathcal{K}_\infty$ function, then the nonlinear time-dependent impulsive dynamical system (2.115) and (2.116) is globally uniformly ultimately bounded with bound ε.

Proof. The proof is an immediate consequence of Theorem 2.9 using similar arguments as in the proof of Corollary 2.4 and, hence, is omitted. $\square$

Finally, we specialize Corollaries 2.4 and 2.5 to nonlinear state-dependent impulsive dynamical systems. For these results we need the following specialization of Definition 2.12.

Definition 2.13 *i) The nonlinear state-dependent impulsive dynamical system given by (2.25) and (2.26) is* Lagrange stable *if, for every $x_0 \in \mathbb{R}^n$, there exists $\varepsilon = \varepsilon(x_0) > 0$ such that $\|x(t)\| < \varepsilon$, $t \geq 0$.*

ii) The nonlinear state-dependent impulsive dynamical system given by (2.25) and (2.26) is bounded *if there exists $\gamma > 0$ such that, for every $\delta \in (0, \gamma)$, there exists $\varepsilon = \varepsilon(\delta) > 0$ such that $\|x_0\| < \delta$ implies $\|x(t)\| < \varepsilon$, $t \geq 0$. The nonlinear state-dependent impulsive dynamical system (2.25) and (2.26) is* globally bounded *if, for every $\delta \in (0, \infty)$, there exists $\varepsilon = \varepsilon(\delta) > 0$ such that $\|x_0\| < \delta$ implies $\|x(t)\| < \varepsilon$, $t \geq 0$.*

iii) The nonlinear state-dependent impulsive dynamical system given by (2.25) and (2.26) is ultimately bounded with bound ε *if there exists $\gamma > 0$ such that, for every $\delta \in (0, \gamma)$, there exists $T = T(\delta, \varepsilon) > 0$ such that $\|x_0\| < \delta$ implies $\|x(t)\| < \varepsilon$, $t \geq T$. The nonlinear state-dependent impulsive dynamical system (2.25) and (2.26) is* globally ultimately bounded with bound ε *if, for every $\delta \in (0, \infty)$, there exists $T = T(\delta, \varepsilon) > 0$ such that $\|x_0\| < \delta$ implies $\|x(t)\| < \varepsilon$, $t \geq T$.*

Corollary 2.6 *Consider the nonlinear state-dependent impulsive dynamical system (2.25) and (2.26). Assume there exist a continuously differentiable function $V : \mathcal{D} \to \mathbb{R}$ and class $\mathcal{K}$ functions $\alpha(\cdot)$ and $\beta(\cdot)$ such that*

$$\alpha(\|x\|) \leq V(x) \leq \beta(\|x\|), \quad x \in \mathcal{D}, \tag{2.167}$$

$$V'(x)f_\mathrm{c}(x) \leq 0, \quad x \in \mathcal{D}, \quad x \notin \mathcal{Z}, \quad \|x\| \geq \mu, \tag{2.168}$$

$$V(x + f_\mathrm{d}(x)) - V(x) \leq 0, \quad x \in \mathcal{D}, \quad x \in \mathcal{Z}, \quad \|x\| \geq \mu, \tag{2.169}$$

where $\mu > 0$ is such that $\mathcal{B}_{\alpha^{-1}(\beta(\mu))}(0) \subset \mathcal{D}$. Furthermore, assume $\sup_{x \in \overline{\mathcal{B}}_\mu(0) \cap \mathcal{Z}} V(x + f_\mathrm{d}(x))$ exists. Then the nonlinear state-dependent impulsive dynamical system (2.25) and (2.26) is bounded. If, in addition, $\mathcal{D} = \mathbb{R}^n$ and $V(x) \to \infty$ as $\|x\| \to \infty$, then the nonlinear state-dependent impulsive dynamical system (2.25) and (2.26) is globally bounded.

Proof. The result is a direct consequence of Corollary 2.4. $\square$

Corollary 2.7 *Consider the nonlinear state-dependent impulsive dynamical system (2.25) and (2.26). Assume there exists a continuously differentiable function $V : \mathcal{D} \to \mathbb{R}$ and class $\mathcal{K}$ functions $\alpha(\cdot)$ and $\beta(\cdot)$ such that (2.167) and (2.169) hold, and*

$$V'(x)f_\mathrm{c}(x) < 0, \quad x \in \mathcal{D}, \quad x \notin \mathcal{Z}, \quad \|x\| > \mu, \tag{2.170}$$

where $\mu > 0$ is such that $\mathcal{B}_{\alpha^{-1}(\beta(\mu))}(0) \subset \mathcal{D}$ with $\eta > \beta(\mu)$. Furthermore, assume $\theta \triangleq \sup_{x \in \overline{\mathcal{B}}_\mu(0) \cap \mathcal{Z}} V(x + f_\mathrm{d}(x))$ exists. Then the nonlinear state-dependent impulsive dynamical system (2.25) and (2.26) is ultimately bounded with bound $\varepsilon \triangleq \alpha^{-1}(\eta)$, where $\eta \triangleq \max\{\beta(\mu), \theta\}$. Furthermore, $\limsup_{t \to \infty} \|x(t)\| \le \alpha^{-1}(\beta(\mu))$. If, in addition, $\mathcal{D} = \mathbb{R}^n$ and $V(x) \to \infty$ as $\|x\| \to \infty$, then the nonlinear state-dependent impulsive dynamical system (2.25) and (2.26) is globally ultimately bounded with bound ε.

Proof. The result is a direct consequence of Corollary 2.5.   □

2.10 Stability Theory via Vector Lyapunov Functions

In this section, we introduce the notion of vector Lyapunov functions for stability analysis of nonlinear impulsive dynamical systems. The use of vector Lyapunov functions in dynamical system theory offers a very flexible framework since each component of the vector Lyapunov function can satisfy less rigid requirements as compared to a single scalar Lyapunov function. Specifically, since for many nonlinear dynamical systems constructing a system Lyapunov function can be a difficult task, weakening the hypothesis on the Lyapunov function enlarges the class of Lyapunov functions that can be used for analyzing system stability. Moreover, in certain applications, such as the analysis of large-scale nonlinear dynamical systems, several Lyapunov functions arise naturally from the stability properties of each individual subsystem. To develop the theory of vector Lyapunov functions for nonlinear impulsive dynamical systems, we first introduce some results on vector differential inequalities and the *vector comparison principle*. The following definitions introduce the notions of class $\mathcal{W}$ and class $\mathcal{W}_\mathrm{d}$ functions involving *quasi-monotone increasing* and *nondecreasing* functions, respectively.

Definition 2.14 ([151]) *A function $w_\mathrm{c} = [w_{\mathrm{c}1}, \ldots, w_{\mathrm{c}q}]^\mathrm{T} : \mathbb{R}^q \to \mathbb{R}^q$ is of class $\mathcal{W}$ if $w_{\mathrm{c}i}(z') \le w_{\mathrm{c}i}(z'')$, $i = 1, \ldots, q$, for all $z', z'' \in \mathbb{R}^q$ such that $z'_j \le z''_j$, $z'_i = z''_i$, $j = 1, \ldots, q$, $i \ne j$, where z_i denotes the ith component of z.*

Definition 2.15 ([96]) *A function $w_\mathrm{d} = [w_{\mathrm{d}1}, \ldots, w_{\mathrm{d}q}]^\mathrm{T} : \mathbb{R}^q \to \mathbb{R}^q$ is of class $\mathcal{W}_\mathrm{d}$ if $w_{\mathrm{d}i}(z') \le w_{\mathrm{d}i}(z'')$, $i = 1, \ldots, q$, for all $z', z'' \in \mathbb{R}^q$ such that $z'_i \le z''_i$, $i = 1, \ldots, q$.*

Note that if $w_d(\cdot) \in \mathcal{W}_d$, then $w_d(\cdot) \in \mathcal{W}$. If $w_c(\cdot) \in \mathcal{W}$ we say that w_c satisfies the *Kamke condition* [88, 164]. Note that if $w_c(z) = W_c z$, where $W_c \in \mathbb{R}^{q \times q}$, then the function $w_c(\cdot)$ is of class $\mathcal{W}$ if and only if W_c is *essentially nonnegative*, that is, all the off-diagonal entries of the matrix W_c are nonnegative. Alternatively, if $w_d(z) = W_d z$, where $W_d \in \mathbb{R}^{q \times q}$, then the function $w_d(\cdot)$ is of class $\mathcal{W}_d$ if and only if W_d is *nonnegative*, that is, all the entries of the matrix W_d are nonnegative. Furthermore, note that it follows from Definition 2.14 that every scalar ($q = 1$) function $w_c(z)$ is of class $\mathcal{W}$. Next, we consider the nonlinear comparison system given by

$$\dot{z}(t) = w_c(z(t)), \quad z(t_0) = z_0, \quad t \in \mathcal{I}_{z_0}, \tag{2.171}$$

where $z(t) \in \mathcal{Q} \subseteq \mathbb{R}^q$, $t \in \mathcal{I}_{z_0}$, is the comparison system state vector, $\mathcal{I}_{z_0} \subseteq \mathcal{T} \subseteq [0, \infty)$ is the maximal interval of existence of a solution $z(t)$ of (2.171), $\mathcal{Q}$ is an open set, $0 \in \mathcal{Q}$, and $w_c : \mathcal{Q} \to \mathbb{R}^q$ is Lipschitz continuous on $\mathcal{Q}$. For the results of this section we write $x \geq\geq 0$ (respectively, $x >> 0$), $x \in \mathbb{R}^n$, to indicate that every component of x is nonnegative (respectively, positive). Furthermore, we denote the nonnegative and positive orthants of $\mathbb{R}^n$ by $\overline{\mathbb{R}}_+^n$ and $\mathbb{R}_+^n$, respectively. That is, if $x \in \mathbb{R}^n$, then $x \in \overline{\mathbb{R}}_+^n$ and $x \in \mathbb{R}_+^n$ are equivalent, respectively, to $x \geq\geq 0$ and $x >> 0$.

Proposition 2.5 *Consider the nonlinear comparison system (2.171). Assume that the function $w_c : \mathcal{Q} \to \mathbb{R}^q$ is continuous and $w_c(\cdot)$ is of class $\mathcal{W}$. If there exists a continuously differentiable vector function $V = [v_1, \ldots, v_q]^{\mathrm{T}} : \mathcal{I}_{z_0} \to \mathcal{Q}$ such that*

$$\dot{V}(t) << w_c(V(t)), \quad t \in \mathcal{I}_{z_0}, \tag{2.172}$$

then $V(t_0) << z_0$, $z_0 \in \mathcal{Q}$, implies

$$V(t) << z(t), \quad t \in \mathcal{I}_{z_0}, \tag{2.173}$$

where $z(t)$, $t \in \mathcal{I}_{z_0}$, is the solution to (2.171).

Proof. Since $V(t)$, $t \in \mathcal{I}_{z_0}$, is continuous it follows that for sufficiently small $\tau > 0$,

$$V(t) << z(t), \quad t \in [t_0, t_0 + \tau]. \tag{2.174}$$

Now, suppose, *ad absurdum*, inequality (2.173) does not hold on the entire interval $\mathcal{I}_{z_0}$. Then there exists $\hat{t} \in \mathcal{I}_{z_0}$ such that $V(t) << z(t)$, $t \in [t_0, \hat{t})$, and for at least one $i \in \{1, \ldots, q\}$,

$$v_i(\hat{t}) = z_i(\hat{t}) \tag{2.175}$$

and

$$v_j(\hat{t}) \leq z_j(\hat{t}), \quad j \neq i, \quad j = 1,\ldots,q. \tag{2.176}$$

Since $w_c(\cdot) \in \mathcal{W}$, it follows from (2.172), (2.175), and (2.176) that

$$\dot{v}_i(\hat{t}) < w_{ci}(V(\hat{t})) \leq w_{ci}(z(\hat{t})) = \dot{z}_i(\hat{t}), \tag{2.177}$$

which, along with (2.175), implies that for sufficiently small $\hat{\tau} > 0$, $v_i(t) > z_i(t)$, $t \in [\hat{t} - \hat{\tau}, \hat{t})$. This contradicts the fact that $V(t) <<$ $z(t)$, $t \in [t_0, \hat{t})$, and establishes (2.173). □

Next, we present a stronger version of Proposition 2.5 where the strict inequalities are replaced by soft inequalities.

Proposition 2.6 *Consider the nonlinear comparison system (2.171). Assume that the function $w_c : \mathcal{Q} \to \mathbb{R}^q$ is continuous and $w_c(\cdot)$ is of class $\mathcal{W}$. Let $z(t)$, $t \in \mathcal{I}_{z_0}$, be the solution to (2.171) and $[t_0, t_0+\tau] \subseteq \mathcal{I}_{z_0}$ be a compact interval. If there exists a continuously differentiable vector function $V : [t_0, t_0 + \tau] \to \mathcal{Q}$ such that*

$$\dot{V}(t) \leq\leq w_c(V(t)), \quad t \in [t_0, t_0 + \tau], \tag{2.178}$$

then

$$V(t_0) \leq\leq z_0, \quad z_0 \in \mathcal{Q}, \tag{2.179}$$

implies

$$V(t) \leq\leq z(t), \quad t \in [t_0, t_0 + \tau]. \tag{2.180}$$

Proof. Consider the family of comparison systems given by

$$\dot{z}(t) = w_c(z(t)) + \tfrac{\varepsilon}{n}\mathbf{e}, \quad z(t_0) = z_0 + \tfrac{\varepsilon}{n}\mathbf{e}, \tag{2.181}$$

where $\varepsilon > 0$, $n \in \overline{\mathbb{Z}}_+$, $\mathbf{e} \triangleq [1,\ldots,1]^{\mathrm{T}}$, and $t \in \mathcal{I}_{z_0 + \frac{\varepsilon}{n}\mathbf{e}}$, and let the solution to (2.181) be denoted by $s_{(n)}(t, z_0 + \tfrac{\varepsilon}{n}\mathbf{e})$, $t \in \mathcal{I}_{z_0 + \frac{\varepsilon}{n}\mathbf{e}}$. Now, it follows from Theorem 3 of [42, p. 17] that there exists a compact interval $[t_0, t_0+\tau] \subseteq \mathcal{I}_{z_0}$ such that $s_{(n)}(t, z_0+\tfrac{\varepsilon}{n}\mathbf{e})$, $t \in [t_0, t_0+\tau]$, is defined for all sufficiently large n. Moreover, it follows from Proposition 2.5 that

$$V(t) << s_{(n)}(t, z_0 + \tfrac{\varepsilon}{n}\mathbf{e}) << s_{(m)}(t, z_0 + \tfrac{\varepsilon}{m}\mathbf{e}), \quad n > m,$$
$$t \in [t_0, t_0 + \tau], \tag{2.182}$$

for all sufficiently large $m \in \overline{\mathbb{Z}}_+$. Since the functions $s_{(n)}(t, z_0 + \tfrac{\varepsilon}{n}\mathbf{e})$, $t \in [t_0, t_0 + \tau]$, $n \in \overline{\mathbb{Z}}_+$, are continuous in t, decreasing in n, and bounded from below, it follows that the sequence of functions $s_{(n)}(\cdot, z_0 + \frac{\varepsilon}{n}\mathbf{e})$ converges uniformly on the compact interval $[t_0, t_0 + \tau]$ as $n \to \infty$, that is, there exists a continuous function $\hat{z} : [t_0, t_0 + \tau] \to Q$ such that

$$s_{(n)}(t, z_0 + \tfrac{\varepsilon}{n}\mathbf{e}) \to \hat{z}(t), \quad n \to \infty, \tag{2.183}$$

uniformly on $[t_0, t_0 + \tau]$. Hence, it follows from (2.182) and (2.183) that

$$V(t) \leq\leq \hat{z}(t), \quad t \in [t_0, t_0 + \tau]. \tag{2.184}$$

Next, note that it follows from (2.181) that

$$s_{(n)}(t, z_0 + \tfrac{\varepsilon}{n}\mathbf{e}) = z_0 + \tfrac{\varepsilon}{n}\mathbf{e} + \int_{t_0}^{t} w_{\mathrm{c}}(s_{(n)}(\sigma, z_0 + \tfrac{\varepsilon}{n}\mathbf{e}))\mathrm{d}\sigma,$$
$$t \in [t_0, t_0 + \tau], \tag{2.185}$$

which implies that $\hat{z}(t_0) = z_0$ and, since $w_{\mathrm{c}}(\cdot)$ is a continuous function, $w_{\mathrm{c}}(s_{(n)}(t, z_0 + \frac{\varepsilon}{n}\mathbf{e})) \to w_{\mathrm{c}}(\hat{z}(t))$ as $n \to \infty$ uniformly on $[t_0, t_0 + \tau]$. Hence, taking the limit as $n \to \infty$ on both sides of (2.185) yields

$$\hat{z}(t) = z_0 + \int_{t_0}^{t} w_{\mathrm{c}}(\hat{z}(\sigma))\mathrm{d}\sigma, \quad t \in [t_0, t_0 + \tau], \tag{2.186}$$

which implies that $\hat{z}(t)$ is the solution to (2.171) on the interval $[t_0, t_0 + \tau]$. Hence, by uniqueness of solutions of (2.171) we obtain that $\hat{z}(t) = z(t)$, $[t_0, t_0 + \tau]$. This along with (2.184) proves the result. $\qquad\square$

Next, consider the nonlinear dynamical system given by

$$\dot{x}(t) = f(x(t)), \quad x(t_0) = x_0, \quad t \in \mathcal{I}_{x_0}, \tag{2.187}$$

where $x(t) \in \mathcal{D} \subseteq \mathbb{R}^n$, $t \in \mathcal{I}_{x_0}$, is the system state vector, $\mathcal{I}_{x_0}$ is the maximal interval of existence of a solution $x(t)$ of (2.187), $\mathcal{D}$ is an open set, $0 \in \mathcal{D}$, and $f(\cdot)$ is Lipschitz continuous on $\mathcal{D}$. The following result is a direct consequence of Proposition 2.6.

Corollary 2.8 *Consider the nonlinear dynamical system (2.187). Assume there exists a continuously differentiable vector function $V : \mathcal{D} \to \mathcal{Q} \subseteq \mathbb{R}^q$ such that*

$$V'(x)f(x) \leq\leq w_{\mathrm{c}}(V(x)), \quad x \in \mathcal{D}, \tag{2.188}$$

where $w_{\mathrm{c}} : \mathcal{Q} \to \mathbb{R}^q$ *is a continuous function,* $w_{\mathrm{c}}(\cdot) \in \mathcal{W}$, *and*

$$\dot{z}(t) = w_{\mathrm{c}}(z(t)), \quad z(t_0) = z_0, \quad t \in \mathcal{I}_{z_0}, \tag{2.189}$$

has a unique solution $z(t)$, $t \in \mathcal{I}_{z_0}$, *where* $x(t)$, $t \in \mathcal{I}_{x_0}$, *is a solution to* (2.187). *If* $[t_0, t_0 + \tau] \subseteq \mathcal{I}_{x_0} \cap \mathcal{I}_{z_0}$ *is a compact interval, then*

$$V(x_0) \leq\leq z_0, \quad z_0 \in \mathcal{Q}, \tag{2.190}$$

implies

$$V(x(t)) \leq\leq z(t), \quad t \in [t_0, t_0 + \tau]. \tag{2.191}$$

Proof. Define $\eta(t) \triangleq V(x(t))$, $t \in \mathcal{I}_{x_0}$, and note that (2.188) implies

$$\dot{\eta}(t) \leq\leq w_{\mathrm{c}}(\eta(t)), \quad t \in \mathcal{I}_{x_0}. \tag{2.192}$$

Moreover, if $[t_0, t_0 + \tau] \subseteq \mathcal{I}_{x_0} \cap \mathcal{I}_{z_0}$ is a compact interval, then it follows from Proposition 2.6, with $V(x_0) = \eta(t_0) \leq\leq z_0$, that

$$V(x(t)) = \eta(t) \leq\leq z(t), \quad t \in [t_0, t_0 + \tau], \tag{2.193}$$

which establishes the result. $\qquad\qquad\qquad\qquad\qquad\qquad\square$

If in (2.187) $f : \mathbb{R}^n \to \mathbb{R}^n$ is globally Lipschitz continuous, then (2.187) has a unique solution $x(t)$ for all $t \geq t_0$. An alternative sufficient condition for global existence and uniqueness of solutions to (2.187) is continuous differentiability of $f : \mathbb{R}^n \to \mathbb{R}^n$ and uniform boundedness of $f'(x)$ on $\mathbb{R}^n$. Note that if the solutions to (2.187) and (2.189) are globally defined for all $x_0 \in \mathcal{D}$ and $z_0 \in \mathcal{Q}$, then the result of Corollary 2.8 holds for any arbitrarily large but compact interval $[t_0, t_0 + \tau] \subset \overline{\mathbb{R}}_+$. For the remainder of this section we assume that the solutions to the systems (2.187) and (2.189) are defined for all $t \geq t_0$. Continuous differentiability of $f(\cdot)$ and $w_{\mathrm{c}}(\cdot)$ provides a sufficient condition for the existence and uniqueness of solutions to (2.187) and (2.189) for all $t \geq t_0$.

Next, we consider the state-dependent impulsive dynamical system given by

$$\dot{x}(t) = f_{\mathrm{c}}(x(t)), \quad x(t_0) = x_0, \quad x(t) \notin \mathcal{Z}, \quad t \in \mathcal{I}_{x_0}, \tag{2.194}$$

$$\Delta x(t) = f_{\mathrm{d}}(x(t)), \quad x(t) \in \mathcal{Z}, \tag{2.195}$$

where $x(t) \in \mathcal{D} \subseteq \mathbb{R}^n$, $t \in \mathcal{I}_{x_0}$, is the system state vector, $\mathcal{I}_{x_0}$ is the maximal interval of existence of a solution $x(t)$ to (2.194) and (2.195), $\mathcal{D}$ is an open set, $0 \in \mathcal{D}$, $f_{\mathrm{c}} : \mathcal{D} \to \mathbb{R}^n$ is Lipschitz continuous and satisfies $f_c(0) = 0$, $f_d : \mathcal{D} \to \mathbb{R}^n$ is continuous, $\Delta x(t) \triangleq x(t^+) - x(t)$, and $\mathcal{Z} \subset \mathcal{D} \subseteq \mathbb{R}^n$ is the resetting set. We assume that A1 and A2 hold so that the required properties for the existence and uniqueness of solutions to (2.194) and (2.195) are satisfied.

Theorem 2.10 *Consider the impulsive dynamical system (2.194) and (2.195). Assume there exists a continuously differentiable vector function $V : \mathcal{D} \to \mathcal{Q} \subseteq \mathbb{R}^q$ such that*

$$V'(x)f_c(x) \leq\leq w_c(V(x)), \quad x \notin \mathcal{Z}, \tag{2.196}$$
$$V(x + f_d(x)) \leq\leq V(x) + w_d(V(x)), \quad x \in \mathcal{Z}, \tag{2.197}$$

where $w_c : \mathcal{Q} \to \mathbb{R}^q$ and $w_d : \mathcal{Q} \to \mathbb{R}^q$ are continuous functions, $w_c(\cdot) \in \mathcal{W}$, $w_d(\cdot) \in \mathcal{W}_d$, and the comparison impulsive dynamical system

$$\dot{z}(t) = w_c(z(t)), \quad z(t_0) = z_0, \quad x(t) \notin \mathcal{Z}, \quad t \in \mathcal{I}_{z_0}, \tag{2.198}$$
$$\Delta z(t) = w_d(z(t)), \quad x(t) \in \mathcal{Z}, \tag{2.199}$$

has a unique solution $z(t)$, $t \in \mathcal{I}_{z_0}$. If $[t_0, t_0 + \tau] \subseteq \mathcal{I}_{x_0} \cap \mathcal{I}_{z_0}$ is a compact interval, then

$$V(x_0) \leq\leq z_0, \quad z_0 \in \mathcal{Q}, \tag{2.200}$$

implies

$$V(x(t)) \leq\leq z(t), \quad t \in [t_0, t_0 + \tau], \tag{2.201}$$

where $x(t)$, $t \in \mathcal{I}_{x_0}$, is the solution to (2.194) and (2.195) with initial condition $x_0 \in \mathcal{D}$.

Proof. Without loss of generality, let $x_0 \notin \mathcal{Z}$, $x_0 \in \mathcal{D}$. If $x_0 \in \mathcal{Z}$, then by Assumption A2, $x_0 + f_d(x_0) \notin \mathcal{Z}$ serves as the initial condition for the continuous-time dynamics. If for $x_0 \notin \mathcal{Z}$ the solution $x(t) \notin \mathcal{Z}$ for all $t \in [t_0, t_0 + \tau]$, then the result follows from Corollary 2.8. Next, suppose the interval $[t_0, t_0 + \tau]$ contains the resetting times $\tau_k(x_0) < \tau_{k+1}(x_0)$, $k \in \{1, 2, \ldots, m\}$. Consider the compact interval $[t_0, \tau_1(x_0)]$ and let $V(x_0) \leq\leq z_0$. Then it follows from (2.196) and Corollary 2.8 that

$$V(x(t)) \leq\leq z(t), \quad t \in [t_0, \tau_1(x_0)], \tag{2.202}$$

where $z(t)$, $t \in \mathcal{I}_{z_0}$, is the solution to (2.198). Now, since $w_d(\cdot) \in \mathcal{W}_d$ it follows from (2.197) and (2.202) that

$$\begin{aligned} V(x(\tau_1^+(x_0))) &\leq\leq V(x(\tau_1(x_0))) + w_d(V(x(\tau_1(x_0)))) \\ &\leq\leq z(\tau_1(x_0)) + w_d(z(\tau_1(x_0))) \\ &= z(\tau_1^+(x_0)). \end{aligned} \tag{2.203}$$

Consider the compact interval $[\tau_1^+(x_0), \tau_2(x_0)]$. Since $V(x(\tau_1^+(x_0))) \leq\leq z(\tau_1(x_0))$, it follows from (2.196) that

$$V(x(t)) \leq\leq z(t), \quad t \in [\tau_1^+(x_0), \tau_2(x_0)]. \qquad (2.204)$$

Repeating the above arguments for $t \in [\tau_k^+(x_0), \tau_{k+1}(x_0)]$, $k = 3, \ldots, m$, yields (2.201). Finally, in the case of infinitely many resettings over the time interval $[t_0, t_0 + \tau]$, let $\lim_{k \to \infty} \tau_k(x_0) = \tau_\infty(x_0) \in (t_0, t_0 + \tau]$. In this case, $[t_0, \tau_\infty(x_0)] = [t_0, \tau_1(x_0)] \cup \left[\bigcup_{k=1}^{\infty} [\tau_k(x_0), \tau_{k+1}(x_0)] \right]$. Repeating the above arguments, the result can be shown for the interval $[t_0, \tau_\infty(x_0)]$. $\qquad \square$

If the solutions to (2.194), (2.195), and (2.198), (2.199) are globally defined for all $x_0 \in \mathcal{D}$ and $z_0 \in \mathcal{Q}$, then the result of Theorem 2.10 holds for any arbitrarily large but compact interval $[t_0, t_0 + \tau] \subset \overline{\mathbb{R}}_+$. For the remainder of this section we assume that the solutions to the systems (2.194), (2.195), and (2.198), (2.199) are defined for all $t \geq t_0$.

Theorem 2.11 *Consider the impulsive dynamical system (2.194) and (2.195). Assume that there exist a continuously differentiable vector function $V : \mathcal{D} \to \mathcal{Q} \cap \overline{\mathbb{R}}_+^q$ and a positive vector $p \in \mathbb{R}_+^q$ such that $V(0) = 0$, the scalar function $v : \mathcal{D} \to \mathbb{R}_+$ defined by $v(x) \triangleq p^T V(x)$, $x \in \mathcal{D}$, is such that $v(x) > 0$, $x \neq 0$, and*

$$V'(x) f_c(x) \leq\leq w_c(V(x)), \quad x \notin \mathcal{Z}, \qquad (2.205)$$
$$V(x + f_d(x)) \leq\leq V(x) + w_d(V(x)), \quad x \in \mathcal{Z}, \qquad (2.206)$$

where $w_c : \mathcal{Q} \to \mathbb{R}^q$ and $w_d : \mathcal{Q} \to \mathbb{R}^q$ are continuous, $w_c(\cdot) \in \mathcal{W}$, $w_d(\cdot) \in \mathcal{W}_d$, and $w_c(0) = 0$. Then the following statements hold:

i) If the zero solution $z(t) \equiv 0$ to

$$\dot{z}(t) = w_c(z(t)), \quad z(t_0) = z_0, \quad x(t) \notin \mathcal{Z}, \quad t \geq t_0, \quad (2.207)$$
$$\Delta z(t) = w_d(z(t)), \quad x(t) \in \mathcal{Z}, \qquad (2.208)$$

is Lyapunov stable, then the zero solution $x(t) \equiv 0$ to (2.194) and (2.195) is Lyapunov stable.

ii) If the zero solution $z(t) \equiv 0$ to (2.207) and (2.208) is asymptotically stable, then the zero solution $x(t) \equiv 0$ to (2.194) and (2.195) is asymptotically stable.

iii) If $\mathcal{D} = \mathbb{R}^n$, $\mathcal{Q} = \mathbb{R}^q$, $v : \mathbb{R}^n \to \overline{\mathbb{R}}_+$ is radially unbounded, and the zero solution $z(t) \equiv 0$ to (2.207) and (2.208) is globally asymptotically stable, then the zero solution $x(t) \equiv 0$ to (2.194) and (2.195) is globally asymptotically stable.

iv) If there exist constants $\nu \geq 1$, $\alpha > 0$, and $\beta > 0$ such that $v : \mathcal{D} \to \overline{\mathbb{R}}_+$ satisfies

$$\alpha \|x\|^{\nu} \leq v(x) \leq \beta \|x\|^{\nu}, \quad x \in \mathcal{D}, \qquad (2.209)$$

and the zero solution $z(t) \equiv 0$ to (2.207) and (2.208) is exponentially stable, then the zero solution $x(t) \equiv 0$ to (2.194) and (2.195) is exponentially stable.

v) If $\mathcal{D} = \mathbb{R}^n$, $\mathcal{Q} = \mathbb{R}^q$, there exist constants $\nu \geq 1$, $\alpha > 0$, and $\beta > 0$ such that $v : \mathbb{R}^n \to \overline{\mathbb{R}}_+$ satisfies (2.209), and the zero solution $z(t) \equiv 0$ to (2.207) and (2.208) is globally exponentially stable, then the zero solution $x(t) \equiv 0$ to (2.194) and (2.195) is globally exponentially stable.

Proof. Assume there exist a continuously differentiable vector function $V : \mathcal{D} \to \mathcal{Q} \cap \mathbb{R}^q_+$ and a positive vector $p \in \mathbb{R}^q_+$ such that $v(x) = p^{\mathrm{T}}V(x)$, $x \in \mathcal{D}$, is positive definite, that is, $v(0) = 0$ and $v(x) > 0$, $x \neq 0$. Since $v(x) = p^{\mathrm{T}}V(x) \leq \max_{i=1,\ldots,q}\{p_i\}\mathbf{e}^{\mathrm{T}}V(x)$, $x \in \mathcal{D}$, where $\mathbf{e} \triangleq [1,\ldots,1]^{\mathrm{T}}$, the function $\mathbf{e}^{\mathrm{T}}V(x)$, $x \in \mathcal{D}$, is also positive definite. Thus, there exist $r > 0$ and class $\mathcal{K}$ functions $\alpha, \beta : [0, r] \to \overline{\mathbb{R}}_+$ such that $\mathcal{B}_r(0) \subset \mathcal{D}$ and

$$\alpha(\|x\|) \leq \mathbf{e}^{\mathrm{T}}V(x) \leq \beta(\|x\|), \quad x \in \mathcal{B}_r(0). \qquad (2.210)$$

i) Let $\varepsilon > 0$ and choose $0 < \hat{\varepsilon} < \min\{\varepsilon, r\}$. It follows from Lyapunov stability of (2.207) and (2.208) that there exists $\mu = \mu(\hat{\varepsilon}) = \mu(\varepsilon) > 0$ such that if $\|z_0\|_1 < \mu$, where $\|z\|_1 \triangleq \sum_{i=1}^q |z_i|$ and z_i is the ith component of z, then $\|z(t)\|_1 < \alpha(\hat{\varepsilon})$, $t \geq t_0$. Now, choose $z_0 = V(x_0) \geq\geq 0$, $x_0 \in \mathcal{D}$. Since $V(x)$, $x \in \mathcal{D}$, is continuous, the function $\mathbf{e}^{\mathrm{T}}V(x)$, $x \in \mathcal{D}$, is also continuous. Hence, for $\mu = \mu(\hat{\varepsilon}) > 0$ there exists $\delta = \delta(\mu(\hat{\varepsilon})) = \delta(\varepsilon) > 0$ such that $\delta < \hat{\varepsilon}$ and if $\|x_0\| < \delta$, then $\mathbf{e}^{\mathrm{T}}V(x_0) = \mathbf{e}^{\mathrm{T}}z_0 = \|z_0\|_1 < \mu$, which implies that $\|z(t)\|_1 < \alpha(\hat{\varepsilon})$, $t \geq t_0$. In addition, it follows from (2.205) and (2.206), and Theorem 2.10 that $0 \leq\leq V(x(t)) \leq\leq z(t)$ on any compact interval $[t_0, t_0 + \tau]$, and hence, $\mathbf{e}^{\mathrm{T}}z(t) = \|z(t)\|_1$, $[t_0, t_0 + \tau]$. Let $\tau > t_0$ be such that $x(t) \in \mathcal{B}_r(0)$, $t \in [t_0, t_0 + \tau]$. Thus, using (2.210), it follows that for $\|x_0\| < \delta$,

$$\alpha(\|x(t)\|) \leq \mathbf{e}^{\mathrm{T}}V(x(t)) \leq \mathbf{e}^{\mathrm{T}}z(t) < \alpha(\hat{\varepsilon}), \quad t \in [t_0, t_0 + \tau], \quad (2.211)$$

which implies $\|x(t)\| < \hat{\varepsilon} < \varepsilon$, $t \in [t_0, t_0 + \tau]$. Now, suppose, *ad absurdum*, that for some $x_0 \in \mathcal{B}_\delta(0)$ there exists $\hat{t} > t_0 + \tau$ such that $\|x(\hat{t})\| \geq \hat{\varepsilon}$. Then, for $z_0 = V(x_0)$ and the compact interval $[t_0, \hat{t}]$

it follows from Theorem 2.10 that $V(x(\hat{t})) \leq\leq z(\hat{t})$, which implies that $\alpha(\hat{\varepsilon}) \leq \alpha(\|x(\hat{t})\|) \leq \mathbf{e}^{\mathrm{T}} V(x(\hat{t})) \leq \mathbf{e}^{\mathrm{T}} z(\hat{t}) < \alpha(\hat{\varepsilon})$. This is a contradiction, and hence, for a given $\varepsilon > 0$ there exists $\delta = \delta(\varepsilon) > 0$ such that for all $x_0 \in \mathcal{B}_\delta(0)$, $\|x(t)\| < \varepsilon$, $t \geq t_0$, which implies Lyapunov stability of the zero solution $x(t) \equiv 0$ to (2.194) and (2.195).

$ii)$ It follows from $i)$ and the asymptotic stability of (2.207) and (2.208) that the zero solution $x(t) \equiv 0$ to (2.194) and (2.195) is Lyapunov stable, and there exists $\mu > 0$ such that if $\|z_0\|_1 < \mu$, then $\lim_{t\to\infty} z(t) = 0$. As in $i)$, choose $z_0 = V(x_0) \geq\geq 0$, $x_0 \in \mathcal{D}$. It follows from Lyapunov stability of the zero solution $x(t) \equiv 0$ to (2.194) and (2.195), and the continuity of $V : \mathcal{D} \to \mathcal{Q} \cap \overline{\mathbb{R}}_+^q$ that there exists $\delta = \delta(\mu) > 0$ such that if $\|x_0\| < \delta$, then $\|x(t)\| < r$, $t \geq t_0$, and $\mathbf{e}^{\mathrm{T}} V(x_0) = \mathbf{e}^{\mathrm{T}} z_0 = \|z_0\|_1 < \mu$. Thus, by asymptotic stability of (2.207) and (2.208), for any arbitrary $\varepsilon > 0$ there exists $T = T(\varepsilon) > t_0$ such that $\|z(t)\|_1 < \alpha(\varepsilon)$, $t \geq T$. Thus, it follows from (2.205) and (2.206) and Theorem 2.10 that $0 \leq\leq V(x(t)) \leq\leq z(t)$ on any compact interval $[T, T + \tau]$, and hence, $\mathbf{e}^{\mathrm{T}} z(t) = \|z(t)\|_1$, $t \in [T, T + \tau]$, and, by (2.210),

$$\alpha(\|x(t)\|) \leq \mathbf{e}^{\mathrm{T}} V(x(t)) \leq \mathbf{e}^{\mathrm{T}} z(t) < \alpha(\varepsilon), \quad t \in [T, T + \tau]. \quad (2.212)$$

Now, suppose, $ad\ absurdum$, that for some $x_0 \in \mathcal{B}_\delta(0)$, $\lim_{t\to\infty} x(t) \neq 0$, that is, there exists a sequence $\{t_n\}_{n=1}^\infty$, with $t_n \to \infty$ as $n \to \infty$, such that $\|x(t_n)\| \geq \hat{\varepsilon}$, $n \in \mathbb{Z}_+$, for some $0 < \hat{\varepsilon} < r$. Choose $\varepsilon = \hat{\varepsilon}$ and $\hat{t} > T + \tau$ such that at least one $t_n \in [T, \hat{t}]$. Then it follows from (2.212) that $\alpha(\varepsilon) \leq \alpha(\|x(t_n)\|) < \alpha(\varepsilon)$, which is a contradiction. Hence, there exists $\delta > 0$ such that for all $x_0 \in \mathcal{B}_\delta(0)$, $\lim_{t\to\infty} x(t) = 0$ which along with Lyapunov stability implies asymptotic stability of the zero solution $x(t) \equiv 0$ to (2.194) and (2.195).

$iii)$ Suppose $\mathcal{D} = \mathbb{R}^n$, $\mathcal{Q} = \mathbb{R}^q$, $v : \mathbb{R}^n \to \overline{\mathbb{R}}_+$ is radially unbounded, and the zero solution $z(t) \equiv 0$ to (2.207) and (2.208) is globally asymptotically stable. In this case, $V : \mathbb{R}^n \to \overline{\mathbb{R}}_+^q$ satisfies (2.210) for all $x \in \mathbb{R}^n$, where the functions $\alpha, \beta : \mathbb{R}_+ \to \mathbb{R}_+$ are of class $\mathcal{K}_\infty$. Furthermore, Lyapunov stability of the zero solution $x(t) \equiv 0$ to (2.194) and (2.195) follows from $i)$. Next, for any $x_0 \in \mathbb{R}^n$ and $z_0 = V(x_0) \in \overline{\mathbb{R}}_+^q$, identical arguments as in $ii)$ can be used to show that $\lim_{t\to\infty} x(t) = 0$, which proves global asymptotic stability of the zero solution $x(t) \equiv 0$ to (2.194) and (2.195).

$iv)$ Suppose (2.209) holds. Since $p \in \mathbb{R}_+^q$, then

$$\hat{\alpha}\|x\|^\nu \leq \mathbf{e}^{\mathrm{T}} V(x) \leq \hat{\beta}\|x\|^\nu, \quad x \in \mathcal{D}, \quad (2.213)$$

where $\hat{\alpha} \triangleq \alpha / \max_{i=1,\dots,q}\{p_i\}$ and $\hat{\beta} \triangleq \beta / \min_{i=1,\dots,q}\{p_i\}$. It follows from the exponential stability of (2.207) and (2.208) that there exist positive constants γ, μ, and η such that if $\|z_0\|_1 < \mu$, then

$$\|z(t)\|_1 \le \gamma \|z_0\|_1 e^{-\eta(t-t_0)}, \quad t \ge t_0. \qquad (2.214)$$

Choose $z_0 = V(x_0) \ge\ge 0$, $x_0 \in \mathcal{D}$. By continuity of $V : \mathcal{D} \to \mathcal{Q} \cap \overline{\mathbb{R}}_+^q$, there exists $\delta = \delta(\mu) > 0$ such that for all $x_0 \in \mathcal{B}_\delta(0)$, $\mathbf{e}^{\mathrm{T}} V(x_0) = \mathbf{e}^{\mathrm{T}} z_0 = \|z_0\|_1 < \mu$. Furthermore, it follows from (2.213), (2.214), and Theorem 2.10 that for all $x_0 \in \mathcal{B}_\delta(0)$ the inequality

$$\hat{\alpha}\|x(t)\|^\nu \le \mathbf{e}^{\mathrm{T}} V(x(t)) \le \mathbf{e}^{\mathrm{T}} z(t) \le \gamma \|z_0\|_1 e^{-\eta(t-t_0)} \le \gamma \hat{\beta} \|x_0\|^\nu e^{-\eta(t-t_0)} \qquad (2.215)$$

holds on any compact interval $[t_0, t_0 + \tau]$. This in turn implies that for any $x_0 \in \mathcal{B}_\delta(0)$,

$$\|x(t)\| \le \left(\frac{\gamma\hat{\beta}}{\hat{\alpha}}\right)^{\frac{1}{\nu}} \|x_0\| e^{-\frac{\eta}{\nu}(t-t_0)}, \quad t \in [t_0, t_0 + \tau]. \qquad (2.216)$$

Now, suppose, *ad absurdum*, that for some $x_0 \in \mathcal{B}_\delta(0)$ there exists $\hat{t} > t_0 + \tau$ such that

$$\|x(\hat{t})\| > \left(\frac{\gamma\hat{\beta}}{\hat{\alpha}}\right)^{\frac{1}{\nu}} \|x_0\| e^{-\frac{\eta}{\nu}(\hat{t}-t_0)}. \qquad (2.217)$$

Then for the compact interval $[t_0, \hat{t}]$, it follows from (2.216) that $\|x(\hat{t})\| \le \left(\frac{\gamma\hat{\beta}}{\hat{\alpha}}\right)^{\frac{1}{\nu}} \|x_0\| e^{-\frac{\eta}{\nu}(\hat{t}-t_0)}$, which is a contradiction. Thus, inequality (2.216) holds for all $t \ge t_0$ establishing exponential stability of the zero solution $x(t) \equiv 0$ to (2.194) and (2.195).

v) The proof is identical to the proof of iv). $\qquad \qquad \square$

Note that for stability analysis each component of a vector Lyapunov function need not be positive definite, nor does it need to have a negative definite time derivative along the trajectories of (2.194) and (2.195). This provides more flexibility in searching for a vector Lyapunov function as compared to a scalar Lyapunov function for addressing the stability of impulsive dynamical systems. Finally, note that in the case where $w_{\mathrm{d}}(z) \equiv 0$, (2.207) and (2.208) specialize to a continuous-time dynamical system, and hence, standard stability methods can be used to examine the stability of (2.207).

Chapter Three

Dissipativity Theory for Nonlinear Impulsive Dynamical Systems

3.1 Introduction

In control engineering, dissipativity theory provides a fundamental framework for the analysis and control design of dynamical systems using an input-output system description based on system-energy-related considerations. The notion of energy here refers to abstract energy notions for which a physical system energy interpretation is not necessary. The dissipation hypothesis on dynamical systems results in a fundamental constraint on their dynamic behavior, wherein a dissipative dynamical system can deliver only a fraction of its energy to its surroundings and can store only a fraction of the work done to it. Many of the great landmarks of feedback control theory are associated with dissipativity theory. In particular, dissipativity theory provides the foundation for absolute stability theory; which in turn forms the basis of the Luré problem, as well as the circle and Popov criteria, which are extensively developed in the classical monographs by Aizerman and Gantmacher [1], Lefschetz [100], and Popov [142]. Since absolute stability theory concerns the stability of a dynamical system for classes of feedback nonlinearities which, as noted in [53,54], can readily be interpreted as an uncertainty model, it is not surprising that absolute stability theory (and hence dissipativity theory) also forms the basis of modern-day robust stability analysis and synthesis [53, 55, 66].

The key foundation in developing dissipativity theory for general nonlinear dynamical systems was presented by J. C. Willems [165,166] in his seminal two-part paper on dissipative dynamical systems. In particular, Willems [165] introduced the definition of dissipativity for general dynamical systems in terms of an inequality involving a generalized system power input, or *supply rate*, and a generalized energy function, or *storage function*. The storage function is bounded from below by the available system storage and bounded from above by the required supply. The available storage is the amount of internal generalized stored energy which can be extracted from the dynamical system, and the required supply is the amount of generalized energy that can be delivered to the dynamical system to transfer it from a state of minimum potential to a given state. Hence, as noted above, a dissipative dynamical system can deliver only a fraction of its stored generalized energy to its surroundings and can store only a fraction of generalized work done to it.

Dissipativity theory exploits the notion that numerous physical dynamical systems have certain input-output system properties related to conservation, dissipation, and transport of mass and energy. Such conservation laws are prevalent in dynamical systems such as mechanical, fluid, electromechanical, electrical, combustion, structural, biological, physiological, biomedical, ecological, and economic systems, as well as feedback control systems.

To see this, consider the single-degree-of-freedom spring-mass-damper mechanical system given by

$$M\ddot{x}(t) + C\dot{x}(t) + Kx(t) = u(t), \quad x(0) = x_0, \quad \dot{x}(0) = \dot{x}_0, \quad t \geq 0, \tag{3.1}$$

where $M > 0$ is the system mass, $C \geq 0$ is the system damping constant, $K \geq 0$ is the system stiffness, $x(t)$, $t \geq 0$, is the position of the mass M, and $u(t)$, $t \geq 0$, is an external force acting on the mass M. The energy of this system is given by

$$V_s(x, \dot{x}) = \tfrac{1}{2}M\dot{x}^2 + \tfrac{1}{2}Kx^2. \tag{3.2}$$

Now, assuming that the measured output of this system is the system velocity, that is, $y(t) = \dot{x}(t)$, it follows that the time rate of change of the system energy along the system trajectories is given by

$$\dot{V}_s(x, \dot{x}) = M\ddot{x}\dot{x} + Kx\dot{x} = uy - C\dot{x}^2. \tag{3.3}$$

Integrating (3.3) over the time interval $[0, T]$, it follows that

$$V_s(x(T), \dot{x}(T)) = V_s(x(0), \dot{x}(0)) + \int_0^T u(t)y(t)\mathrm{d}t - \int_0^T C\dot{x}^2(t)\mathrm{d}t, \tag{3.4}$$

which shows that the system energy at time $t = T$ is equal to the initial energy stored in the system plus the energy supplied to the system via the external force u minus the energy dissipated by the system damper. Equivalently, it follows from (3.3) that the rate of change in the system energy, or system power, is equal to the external supplied system power through the input port u minus the internal system power dissipated by the viscous damper. Note that in the case where the external input force u is zero and $C = 0$, that is, no system supply or dissipation is present, (3.3), or, equivalently, (3.4), shows that the system energy is constant. Furthermore, note that since $C \geq 0$ and $V(x(T), \dot{x}(T)) \geq 0$, $T \geq 0$, it follows from (3.4) that

$$\int_0^T u(t)y(t)\mathrm{d}t \geq -V_\mathrm{s}(x_0, \dot{x}_0), \qquad (3.5)$$

or, equivalently,

$$-\int_0^T u(t)y(t)\mathrm{d}t \leq V_\mathrm{s}(x_0, \dot{x}_0). \qquad (3.6)$$

Equation (3.6) shows that the energy that can be *extracted* from the system through its input-output ports is less than or equal to the initial energy stored in the system. This is precisely the notion of dissipativity.

Since Lyapunov functions can be viewed as generalizations of energy functions for nonlinear dynamical systems, the notion of dissipativity, with appropriate storage functions and supply rates, can be used to construct Lyapunov functions for nonlinear feedback systems by appropriately combining storage functions for each subsystem. Even though the original work on dissipative dynamical systems was formulated in the state space setting, describing the system dynamics in terms of continuous flows on appropriate manifolds, an input-output formulation for dissipative dynamical systems extending the notions of passivity [171], nonexpansivity [172], and conicity [147, 171] was presented in [73, 75, 127].

In this chapter we develop dissipativity theory for nonlinear impulsive dynamical systems. Specifically, we extend the notions of classical dissipativity theory using generalized storage functions and hybrid supply rates for impulsive dynamical systems. The overall approach provides an interpretation of a generalized hybrid energy balance for an impulsive dynamical system in terms of the stored or accumulated generalized energy, dissipated energy over the continuous-time dynamics, and dissipated energy at the resetting instants. Furthermore, as in the case of dynamical systems possessing continuous flows, we show that the set of all possible storage functions of an impulsive dynamical system forms a convex set, and is bounded from below by the system's available stored generalized energy which can be recovered from the system, and bounded from above by the system's required generalized energy supply needed to transfer the system from an initial state of minimum generalized energy to a given state. In addition, for time-dependent and state-dependent impulsive dynamical systems, we develop extended Kalman-Yakubovich-Popov algebraic conditions in terms of the system dynamics for characterizing dissipativeness via system storage functions for impulsive dynamical systems.

3.2 Dissipative Impulsive Dynamical Systems: Input-Output and State Properties

In this section, we extend dissipativity theory to nonlinear impulsive dynamical systems. Specifically, we consider controlled impulsive dynamical systems having the form

$$\dot{x}(t) = f_{\mathrm{c}}(x(t)) + G_{\mathrm{c}}(x(t))u_{\mathrm{c}}(t), \quad x(0) = x_0, \quad (t, x(t), u_{\mathrm{c}}(t)) \notin \mathcal{S}, \tag{3.7}$$

$$\Delta x(t) = f_{\mathrm{d}}(x(t)) + G_{\mathrm{d}}(x(t))u_{\mathrm{d}}(t), \quad (t, x(t), u_{\mathrm{c}}(t)) \in \mathcal{S}, \tag{3.8}$$

$$y_{\mathrm{c}}(t) = h_{\mathrm{c}}(x(t)) + J_{\mathrm{c}}(x(t))u_{\mathrm{c}}(t), \quad (t, x(t), u_{\mathrm{c}}(t)) \notin \mathcal{S}, \tag{3.9}$$

$$y_{\mathrm{d}}(t) = h_{\mathrm{d}}(x(t)) + J_{\mathrm{d}}(x(t))u_{\mathrm{d}}(t), \quad (t, x(t), u_{\mathrm{c}}(t)) \in \mathcal{S}, \tag{3.10}$$

where $t \geq 0$, $x(t) \in \mathcal{D} \subseteq \mathbb{R}^n$, $\mathcal{D}$ is an open set with $0 \in \mathcal{D}$, $\Delta x(t) = x(t^+) - x(t)$, $u_{\mathrm{c}}(t) \in U_{\mathrm{c}} \subseteq \mathbb{R}^{m_{\mathrm{c}}}$, $u_{\mathrm{d}}(t_k) \in U_{\mathrm{d}} \subseteq \mathbb{R}^{m_{\mathrm{d}}}$, t_k denotes the kth instant of time at which $(t, x(t), u_{\mathrm{c}}(t))$ intersects $\mathcal{S}$ for a particular trajectory $x(t)$ and input $u_{\mathrm{c}}(t)$, $y_{\mathrm{c}}(t) \in Y_{\mathrm{c}} \subseteq \mathbb{R}^{l_{\mathrm{c}}}$, $y_{\mathrm{d}}(t_k) \in Y_{\mathrm{d}} \subseteq \mathbb{R}^{l_{\mathrm{d}}}$, $f_{\mathrm{c}} : \mathcal{D} \to \mathbb{R}^n$ is Lipschitz continuous on $\mathcal{D}$ and satisfies $f_{\mathrm{c}}(0) = 0$, $G_{\mathrm{c}} : \mathcal{D} \to \mathbb{R}^{n \times m_{\mathrm{c}}}$, $f_{\mathrm{d}} : \mathcal{D} \to \mathcal{D}$ is continuous on $\mathcal{D}$ and satisfies $f_{\mathrm{d}}(0) = 0$, $G_{\mathrm{d}} : \mathcal{D} \to \mathbb{R}^{n \times m_{\mathrm{d}}}$, $h_{\mathrm{c}} : \mathcal{D} \to \mathbb{R}^{l_{\mathrm{c}}}$ and satisfies $h_{\mathrm{c}}(0) = 0$, $J_{\mathrm{c}} : \mathcal{D} \to \mathbb{R}^{l_{\mathrm{c}} \times m_{\mathrm{c}}}$, $h_{\mathrm{d}} : \mathcal{D} \to \mathbb{R}^{l_{\mathrm{d}}}$ and satisfies $h_{\mathrm{d}}(0) = 0$, $J_{\mathrm{d}} : \mathcal{D} \to \mathbb{R}^{l_{\mathrm{d}} \times m_{\mathrm{d}}}$, and $\mathcal{S} \subset [0, \infty) \times \mathcal{D} \times U_{\mathrm{c}}$ is the resetting set. Here, we assume that $u_{\mathrm{c}}(\cdot)$ and $u_{\mathrm{d}}(\cdot)$ are restricted to the class of *admissible* inputs consisting of measurable functions such that $(u_{\mathrm{c}}(t), u_{\mathrm{d}}(t_k)) \in U_{\mathrm{c}} \times U_{\mathrm{d}}$ for all $t \geq 0$ and $k \in \mathbb{Z}_{[0,t)} \triangleq \{k : 0 \leq t_k < t\}$, where the constraint set $U_{\mathrm{c}} \times U_{\mathrm{d}}$ is given with $(0, 0) \in U_{\mathrm{c}} \times U_{\mathrm{d}}$.

More precisely, for the impulsive dynamical system $\mathcal{G}$ given by (3.7)–(3.10) defined on the state space $\mathcal{D} \subseteq \mathbb{R}^n$, $\mathcal{U} \triangleq \mathcal{U}_{\mathrm{c}} \times \mathcal{U}_{\mathrm{d}}$ and $\mathcal{Y} \triangleq \mathcal{Y}_{\mathrm{c}} \times \mathcal{Y}_{\mathrm{d}}$ define an input and output space, respectively, consisting of left-continuous bounded U-valued and Y-valued functions on the semi-infinite interval $[0, \infty)$. The set $U \triangleq U_{\mathrm{c}} \times U_{\mathrm{d}}$, where $U_{\mathrm{c}} \subseteq \mathbb{R}^{m_{\mathrm{c}}}$ and $U_{\mathrm{d}} \subseteq \mathbb{R}^{m_{\mathrm{d}}}$, contains the set of input values, that is, for every $u = (u_{\mathrm{c}}, u_{\mathrm{d}}) \in \mathcal{U}$ and $t \in [0, \infty)$, $u(t) \in U$, $u_{\mathrm{c}}(t) \in U_{\mathrm{c}}$, and $u_{\mathrm{d}}(t_k) \in U_{\mathrm{d}}$. The set $Y \triangleq Y_{\mathrm{c}} \times Y_{\mathrm{d}}$, where $Y_{\mathrm{c}} \subseteq \mathbb{R}^{l_{\mathrm{c}}}$ and $Y_{\mathrm{d}} \subseteq \mathbb{R}^{l_{\mathrm{d}}}$, contains the set of output values, that is, for every $y = (y_{\mathrm{c}}, y_{\mathrm{d}}) \in \mathcal{Y}$

and $t \in [0, \infty)$, $y(t) \in Y$, $y_c(t) \in Y_c$, and $y_d(t_k) \in Y_d$. The spaces $\mathcal{U}$ and $\mathcal{Y}$ are assumed to be closed under the shift operator, that is, if $u(\cdot) \in \mathcal{U}$ (respectively, $y(\cdot) \in \mathcal{Y}$), then the function u_T (respectively, y_T) defined by $u_T \triangleq u(t+T)$ (respectively, $y_T \triangleq y(t+T)$) is contained in $\mathcal{U}$ (respectively, $\mathcal{Y}$) for all $T \geq 0$.

For convenience, we use the notation $s(t, \tau, x_0, u)$ to denote the solution $x(t)$ of (3.7) and (3.8) at time $t \geq \tau$ with initial condition $x(\tau) = x_0$, where $u = (u_c, u_d) : \mathbb{R} \times \mathcal{T} \to U_c \times U_d$ and $\mathcal{T} \triangleq \{t_1, t_2, \ldots\}$. Thus, the trajectory of the system (3.7) and (3.8) from the initial condition $x(0) = x_0$ is given by $\psi(t, 0, x_0, u)$ for $0 < t \leq t_1$. If and when the trajectory reaches a state $x_1 \triangleq x(t_1)$ satisfying $(t_1, x_1, u_1) \in \mathcal{S}$, where $u_1 \triangleq u_c(t_1)$, then the state is instantaneously transferred to $x_1^+ \triangleq x_1 + f_d(x_1) + G_d(x_1)u_d$, where $u_d \in U_d$ is a given input, according to the resetting law (3.8). The trajectory $x(t)$, $t_1 < t \leq t_2$, is then given by $\psi(t, t_1, x_1^+, u)$, and so on. As in the uncontrolled case, the solution $x(t)$ of (3.7) and (3.8) is left-continuous, that is, it is continuous everywhere except at the resetting times t_k, and

$$x_k \triangleq x(t_k) = \lim_{\varepsilon \to 0^+} x(t_k - \varepsilon), \tag{3.11}$$

$$x_k^+ \triangleq x(t_k) + f_d(x(t_k)) + G_d(x(t_k))u_d(t_k)$$
$$= \lim_{\varepsilon \to 0^+} x(t_k + \varepsilon), \quad u_d(t_k) \in U_d, \tag{3.12}$$

for $k = 1, 2, \ldots$. Furthermore, the analogs to Assumptions A1 and A2 become:

A1. If $(t, x(t), u_c(t)) \in \overline{\mathcal{S}} \backslash \mathcal{S}$, then there exists $\varepsilon > 0$ such that, for all $0 < \delta < \varepsilon$,

$$\psi(t + \delta, t, x(t), u_c(t + \delta)) \notin \mathcal{S}.$$

A2. If $(t_k, x(t_k), u_c(t_k)) \in \partial \mathcal{S} \cap \mathcal{S}$, then there exists $\varepsilon > 0$ such that, for all $0 \leq \delta < \varepsilon$ and $u_d(t_k) \in U_d$,

$$\psi(t_k + \delta, t_k, x(t_k) + f_d(x(t_k)) + G_d(x(t_k))u_d(t_k), u_c(t_k + \delta)) \notin \mathcal{S}.$$

Time-dependent impulsive dynamical systems can be written as (3.7)–(3.10) with $\mathcal{S}$ defined as

$$\mathcal{S} \triangleq \mathcal{T} \times \mathcal{D} \times U_c. \tag{3.13}$$

Now (3.7)–(3.10) can be rewritten in the form of the time-dependent impulsive dynamical system

$$\dot{x}(t) = f_{\mathrm{c}}(x(t)) + G_{\mathrm{c}}(x(t))u_{\mathrm{c}}(t), \quad x(0) = x_0, \quad t \neq t_k, \quad (3.14)$$
$$\Delta x(t) = f_{\mathrm{d}}(x(t)) + G_{\mathrm{d}}(x(t))u_{\mathrm{d}}(t), \quad t = t_k, \quad (3.15)$$
$$y_{\mathrm{c}}(t) = h_{\mathrm{c}}(x(t)) + J_{\mathrm{c}}(x(t))u_{\mathrm{c}}(t), \quad t \neq t_k, \quad (3.16)$$
$$y_{\mathrm{d}}(t) = h_{\mathrm{d}}(x(t)) + J_{\mathrm{d}}(x(t))u_{\mathrm{d}}(t), \quad t = t_k. \quad (3.17)$$

Since $0 \notin \mathcal{T}$ and $t_k < t_{k+1}$, it follows that Assumptions A1 and A2 are satisfied.

Standard continuous-time and discrete-time dynamical systems as well as sampled-data systems can be treated as special cases of impulsive dynamical systems. In particular, setting $f_{\mathrm{d}}(x) = 0$, $G_{\mathrm{d}}(x) = 0$, $h_{\mathrm{d}}(x) = 0$, and $J_{\mathrm{d}}(x) = 0$, it follows that (3.14)–(3.17) has an identical state trajectory as the nonlinear continuous-time system

$$\dot{x}(t) = f_{\mathrm{c}}(x(t)) + G_{\mathrm{c}}(x(t))u_{\mathrm{c}}(t), \quad x(0) = x_0, \quad t \geq 0, \quad (3.18)$$
$$y_{\mathrm{c}}(t) = h_{\mathrm{c}}(x(t)) + J_{\mathrm{c}}(x(t))u_{\mathrm{c}}(t). \quad (3.19)$$

Alternatively, setting $f_{\mathrm{c}}(x) = 0$, $G_{\mathrm{c}}(x) = 0$, $h_{\mathrm{c}}(x) = 0$, $J_{\mathrm{c}}(x) = 0$, $t_k = kT$, and $T = 1$, it follows that (3.14)–(3.17) has an identical state trajectory as the nonlinear discrete-time system

$$x(k+1) = f_{\mathrm{d}}(x(k)) + G_{\mathrm{d}}(x(k))u_{\mathrm{d}}(k), \quad x(0) = x_0, \quad k \in \overline{\mathbb{Z}}_+, \quad (3.20)$$
$$y_{\mathrm{d}}(k) = h_{\mathrm{d}}(x(k)) + J_{\mathrm{d}}(x(k))u_{\mathrm{d}}(k). \quad (3.21)$$

Finally, to show that (3.14)–(3.17) can be used to represent sampled-data systems, consider the continuous-time nonlinear system (3.18) and (3.19) with piecewise constant input $u_{\mathrm{c}}(t) = u_{\mathrm{d}}(t_k)$, $t \in (t_k, t_{k+1}]$, and sampled measurements $y_{\mathrm{d}}(t_k) = h_{\mathrm{d}}(x(t_k)) + J_{\mathrm{d}}(x(t_k))u_{\mathrm{d}}(t_k)$. Defining $\hat{x} = [x^{\mathrm{T}}, u_{\mathrm{c}}^{\mathrm{T}}]^{\mathrm{T}}$, it follows that the sampled-data system can be represented as

$$\dot{\hat{x}} = \hat{f}(\hat{x}(t)), \quad t \neq t_k, \quad (3.22)$$
$$\Delta \hat{x}(t) = \begin{bmatrix} 0 & 0 \\ 0 & -I \end{bmatrix} \hat{x}(t) + \begin{bmatrix} 0 \\ I \end{bmatrix} u_{\mathrm{d}}(t), \quad t = t_k, \quad (3.23)$$
$$y(t) = \hat{h}(\hat{x}(t)), \quad t \neq t_k, \quad (3.24)$$
$$y_{\mathrm{d}}(t) = \hat{h}_{\mathrm{d}}(\hat{x}(t)) + \hat{J}_{\mathrm{d}}(\hat{x}(t))u_{\mathrm{d}}(t), \quad t = t_k, \quad (3.25)$$

where

$$\hat{f}(\hat{x}) = \begin{bmatrix} f_{\mathrm{c}}(x) + G_{\mathrm{c}}(x)u_{\mathrm{c}} \\ 0 \end{bmatrix}, \quad \hat{h}(\hat{x}) = h_{\mathrm{c}}(x) + J_{\mathrm{c}}(x)u_{\mathrm{c}},$$
$$\hat{h}_{\mathrm{d}}(\hat{x}) = h_{\mathrm{d}}(x), \quad \hat{J}_{\mathrm{d}}(\hat{x}) = J_{\mathrm{d}}(x).$$

State-dependent impulsive dynamical systems can be written as (3.7)–(3.10) with $\mathcal{S}$ defined as

$$\mathcal{S} \triangleq [0, \infty) \times \mathcal{Z}, \qquad (3.26)$$

where $\mathcal{Z} \triangleq \mathcal{Z}_x \times U_c$ and $\mathcal{Z}_x \subset \mathcal{D}$. Therefore, (3.7)–(3.10) can be rewritten in the form of the state-dependent impulsive dynamical system

$$\dot{x}(t) = f_c(x(t)) + G_c(x(t))u_c(t), \quad x(0) = x_0, \quad (x(t), u_c(t)) \notin \mathcal{Z},$$
$$(3.27)$$

$$\Delta x(t) = f_d(x(t)) + G_d(x(t))u_d(t), \quad (x(t), u_c(t)) \in \mathcal{Z}, \qquad (3.28)$$
$$y_c(t) = h_c(x(t)) + J_c(x(t))u_c(t), \quad (x(t), u_c(t)) \notin \mathcal{Z}, \qquad (3.29)$$
$$y_d(t) = h_d(x(t)) + J_d(x(t))u_d(t), \quad (x(t), u_c(t)) \in \mathcal{Z}. \qquad (3.30)$$

We assume that if $(x, u_c) \in \mathcal{Z}$, then $(x + f_d(x) + G_d(x)u_d, u_c) \notin \mathcal{Z}$, $u_d \in U_d$. In addition, we assume that if at time t the trajectory $(x(t), u_c(t)) \in \overline{\mathcal{Z}} \backslash \mathcal{Z}$, then there exists $\varepsilon > 0$ such that for $0 < \delta < \varepsilon$, $(x(t + \delta), u_c(t + \delta)) \notin \mathcal{Z}$. These assumptions represent the specialization of A1 and A2 for the particular resetting set (3.26). Finally, in the case where $\mathcal{S} \triangleq [0, \infty) \times \mathcal{D} \times \mathcal{Z}_{u_c}$, where $\mathcal{Z}_{u_c} \subset U_c$, we refer to (3.27)–(3.30) as an *input-dependent impulsive dynamical system*, while in the case where $\mathcal{S} \triangleq ([0, \infty) \times \mathcal{Z}_x \times U_c) \cup ([0, \infty) \times \mathcal{D} \times \mathcal{Z}_{u_c})$ we refer to (3.27)–(3.30) as an *input/state-dependent impulsive dynamical system*.

Next, we develop dissipativity theory for nonlinear impulsive dynamical systems. Specifically, we consider nonlinear impulsive dynamical systems $\mathcal{G}$ of the form given by (3.7)–(3.10) with $t \in \mathbb{R}$, $(t, x(t), u_c(t)) \notin \mathcal{S}$, and $(t, x(t), u_c(t)) \in \mathcal{S}$ replaced by $\mathcal{X}(t, x(t), u_c(t)) \neq 0$ and $\mathcal{X}(t, x(t), u_c(t)) = 0$, respectively, where $\mathcal{X} : \mathbb{R} \times \mathcal{D} \times U_c \to \mathbb{R}$. Note that setting $\mathcal{X}(t, x(t), u_c(t)) = (t - t_1)(t - t_2) \cdots$, where $t_k \to \infty$ as $k \to \infty$, (3.7)–(3.10) reduce to (3.14)–(3.17), while setting $\mathcal{X}(t, x(t), u_c(t)) = \mathcal{X}(x(t), u_c(t))$, where $\mathcal{X} : \mathcal{D} \times U_c \to \mathbb{R}^n$ is a support function characterizing the manifold $\mathcal{Z}$, (3.7)–(3.10) reduce to (3.27)–(3.30). Furthermore, we assume that the system functions $f_c(\cdot)$, $f_d(\cdot)$, $G_c(\cdot)$, $G_d(\cdot)$, $h_c(\cdot)$, $h_d(\cdot)$, $J_c(\cdot)$, and $J_d(\cdot)$ are continuous mappings. In addition, for the nonlinear dynamical system (3.7) we assume that the required properties for the existence and uniqueness of solutions are satisfied such that (3.7) has a unique solution for all $t \in \mathbb{R}$ [14, 93].

For the impulsive dynamical system $\mathcal{G}$ given by (3.7)–(3.10) a function $(s_c(u_c, y_c), s_d(u_d, y_d))$, where $s_c : U_c \times Y_c \to \mathbb{R}$ and $s_d : U_d \times Y_d \to \mathbb{R}$ are such that $s_c(0, 0) = 0$ and $s_d(0, 0) = 0$, is called a *hybrid supply rate* if $s_c(u_c, y_c)$ is locally integrable for all input-output pairs satisfying (3.7)–(3.10), that is, for all input-output pairs $u_c(t) \in \mathcal{U}_c$ and $y_c(t) \in Y_c$ satisfying (3.7)–(3.10), $s_c(\cdot, \cdot)$ satisfies $\int_t^{\hat{t}} |s_c(u_c(s), y_c(s))| \, ds < \infty$ for all $t, \hat{t} \geq 0$. Note that since all input-output pairs $u_d(t_k) \in \mathcal{U}_d$ and $y_d(t_k) \in Y_d$ are defined for discrete instants, $s_d(\cdot, \cdot)$ satisfies $\sum_{k \in \mathbb{Z}_{[t,\hat{t})}} |s_d(u_d(t_k), y_d(t_k))| < \infty$, where $k \in \mathbb{Z}_{[t,\hat{t})} \triangleq \{k : t \leq t_k < \hat{t}\}$.

Definition 3.1 *An impulsive dynamical system $\mathcal{G}$ of the form (3.7)–(3.10) is dissipative with respect to the hybrid supply rate (s_c, s_d) if the dissipation inequality*

$$0 \leq \int_{t_0}^T s_c(u_c(t), y_c(t)) dt + \sum_{k \in \mathbb{Z}_{[t_0,T)}} s_d(u_d(t_k), y_d(t_k)), \quad T \geq t_0,$$

(3.31)

is satisfied for all $T \geq t_0$ and all $(u_c(\cdot), u_d(\cdot)) \in \mathcal{U}_c \times \mathcal{U}_d$ with $x(t_0) = 0$. An impulsive dynamical system $\mathcal{G}$ of the form (3.7)–(3.10) is exponentially dissipative with respect to the hybrid supply rate (s_c, s_d) if there exists a constant $\varepsilon > 0$, such that the dissipation inequality (3.31) is satisfied with $s_c(u_c(t), y_c(t))$ replaced by $e^{\varepsilon t} s_c(u_c(t), y_c(t))$ and $s_d(u_d(t_k), y_d(t_k))$ replaced by $e^{\varepsilon t_k} s_d(u_d(t_k), y_d(t_k))$, for all $T \geq t_0$ and all $(u_c(\cdot), u_d(\cdot)) \in \mathcal{U}_c \times \mathcal{U}_d$ with $x(t_0) = 0$. An impulsive dynamical system is lossless with respect to the hybrid supply rate (s_c, s_d) if $\mathcal{G}$ is dissipative with respect to the supply rate (s_c, s_d) and the dissipation inequality (3.31) is satisfied as an equality for all $T \geq t_0$ and all $(u_c(\cdot), u_d(\cdot)) \in \mathcal{U}_c \times \mathcal{U}_d$ with $x(t_0) = x(T) = 0$.

Next, define the *available storage* $V_a(t_0, x_0)$ of the impulsive dynamical system $\mathcal{G}$ by

$$V_a(t_0, x_0) \triangleq - \inf_{(u_c(\cdot), u_d(\cdot)), T \geq t_0} \left[\int_{t_0}^T s_c(u_c(t), y_c(t)) dt \right.$$

$$\left. + \sum_{k \in \mathbb{Z}_{[t_0,T)}} s_d(u_d(t_k), y_d(t_k)) \right], \quad (3.32)$$

where $x(t)$, $t \geq t_0$, is the solution to (3.7)–(3.10) with admissible inputs $(u_c(\cdot), u_d(\cdot)) \in \mathcal{U}_c \times \mathcal{U}_d$ and $x(t_0) = x_0$. Note that $V_a(t_0, x_0) \geq 0$ for all $(t, x) \in \mathbb{R} \times \mathcal{D}$ since $V_a(t_0, x_0)$ is the supremum over a set of numbers containing the zero element ($T = t_0$). It follows from (3.32) that the available storage of a nonlinear impulsive dynamical system $\mathcal{G}$ is the maximum amount of generalized stored energy which can be extracted from $\mathcal{G}$ at any time T. Furthermore, define the *available exponential storage* of the impulsive dynamical system $\mathcal{G}$ by

$$
V_{\mathrm{a}}(t_0, x_0) \triangleq - \inf_{(u_{\mathrm{c}}(\cdot), u_{\mathrm{d}}(\cdot)), T \geq t_0} \left[\int_{t_0}^{T} e^{\varepsilon t} s_{\mathrm{c}}(u_{\mathrm{c}}(t), y_{\mathrm{c}}(t)) \mathrm{d}t \right.
$$

$$
\left. + \sum_{k \in \mathbb{Z}_{[t_0, T)}} e^{\varepsilon t_k} s_{\mathrm{d}}(u_{\mathrm{d}}(t_k), y_{\mathrm{d}}(t_k)) \right],
$$

$$(3.33)$$

where $\varepsilon > 0$ and $x(t)$, $t \geq t_0$, is the solution of (3.7)–(3.10) with admissible inputs $(u_{\mathrm{c}}(\cdot), u_{\mathrm{d}}(\cdot)) \in \mathcal{U}_{\mathrm{c}} \times \mathcal{U}_{\mathrm{d}}$ and $x(t_0) = x_0$.

Note that in the case of (time-invariant) state-dependent impulsive dynamical systems, the available storage is time invariant, that is, $V_{\mathrm{a}}(t_0, x_0) = V_{\mathrm{a}}(x_0)$. Furthermore, the available exponential storage satisfies

$$
V_{\mathrm{a}}(t_0, x_0) = - \inf_{(u_{\mathrm{c}}(\cdot), u_{\mathrm{d}}(\cdot)), T \geq t_0} \left[\int_{t_0}^{T} e^{\varepsilon t} s_{\mathrm{c}}(u_{\mathrm{c}}(t), y_{\mathrm{c}}(t)) \mathrm{d}t \right.
$$

$$
\left. + \sum_{k \in \mathbb{Z}_{[t_0, T)}} e^{\varepsilon t_k} s_{\mathrm{d}}(u_{\mathrm{d}}(t_k), y_{\mathrm{d}}(t_k)) \right]
$$

$$
= -e^{\varepsilon t_0} \inf_{(u_{\mathrm{c}}(\cdot), u_{\mathrm{d}}(\cdot)), T \geq 0} \left[\int_{0}^{T} e^{\varepsilon t} s_{\mathrm{c}}(u_{\mathrm{c}}(t), y_{\mathrm{c}}(t)) \mathrm{d}t \right.
$$

$$
\left. + \sum_{k \in \mathbb{Z}_{[0, T)}} e^{\varepsilon t_k} s_{\mathrm{d}}(u_{\mathrm{d}}(t_k), y_{\mathrm{d}}(t_k)) \right]
$$

$$
= e^{\varepsilon t_0} \hat{V}_{\mathrm{a}}(x_0), \qquad (3.34)
$$

where

$$
\hat{V}_{\mathrm{a}}(x_0) \triangleq - \inf_{(u_{\mathrm{c}}(\cdot), u_{\mathrm{d}}(\cdot)), T \geq 0} \left[\int_{0}^{T} e^{\varepsilon t} s_{\mathrm{c}}(u_{\mathrm{c}}(t), y_{\mathrm{c}}(t)) \mathrm{d}t \right.
$$

$$
\left. + \sum_{k \in \mathbb{Z}_{[0, T)}} e^{\varepsilon t_k} s_{\mathrm{d}}(u_{\mathrm{d}}(t_k), y_{\mathrm{d}}(t_k)) \right]. \qquad (3.35)
$$

Next, we show that the available storage (respectively, available exponential storage) is finite if and only if $\mathcal{G}$ is dissipative (respectively, exponentially dissipative). In order to state this result we require two additional definitions.

Definition 3.2 *Consider the impulsive dynamical system $\mathcal{G}$ given by (3.7)–(3.10) with hybrid supply rate (s_c, s_d). A continuous nonnegative definite function $V_s : \mathbb{R} \times \mathcal{D} \to \mathbb{R}$ satisfying $V_s(t, 0) = 0$, $t \in \mathbb{R}$, and*

$$V_s(T, x(T)) \leq V_s(t_0, x(t_0)) + \int_{t_0}^T s_c(u_c(t), y_c(t))dt$$

$$+ \sum_{k \in \mathbb{Z}_{[t_0, T)}} s_d(u_d(t_k), y_d(t_k)), \tag{3.36}$$

where $x(t)$, $t \geq t_0$, is a solution to (3.7)–(3.10) with $(u_c(t), u_d(t_k)) \in U_c \times U_d$ and $x(t_0) = x_0$, is called a storage function *for $\mathcal{G}$. A continuous nonnegative-definite function $V_s : \mathbb{R} \times \mathcal{D} \to \mathbb{R}$ satisfying $V_s(t, 0) = 0$, $t \in \mathbb{R}$, and*

$$e^{\varepsilon T} V_s(T, x(T)) \leq e^{\varepsilon t_0} V_s(t_0, x(t_0)) + \int_{t_0}^T e^{\varepsilon t} s_c(u_c(t), y_c(t))dt$$

$$+ \sum_{k \in \mathbb{Z}_{[t_0, T)}} e^{\varepsilon t_k} s_d(u_d(t_k), y_d(t_k)), \tag{3.37}$$

where $\varepsilon > 0$, is called an exponential storage function *for $\mathcal{G}$.*

Note that $V_s(t, x(t))$ is left-continuous on $[t_0, \infty)$ and is continuous everywhere on $[t_0, \infty)$ except on an unbounded closed discrete set $\mathcal{T} = \{t_1, t_2, \ldots\}$, where $\mathcal{T}$ is the set of times when the jumps occur for $x(t)$, $t \geq t_0$.

Definition 3.3 *An impulsive dynamical system $\mathcal{G}$ is* completely reachable *if for all $(t_0, x_0) \in \mathbb{R} \times \mathcal{D}$, there exist a finite time $t_i \leq t_0$, square integrable input $u_c(t)$ defined on $[t_i, t_0]$, and input $u_d(t_k)$ defined on $k \in \mathbb{Z}_{[t_i, t_0)}$, such that the state $x(t)$, $t \geq t_i$, can be driven from $x(t_i) = 0$ to $x(t_0) = x_0$.*

Theorem 3.1 *Consider the impulsive dynamical system $\mathcal{G}$ given by (3.7)–(3.10) and assume that $\mathcal{G}$ is completely reachable. Then $\mathcal{G}$ is dissipative (respectively, exponentially dissipative) with respect to the hybrid supply rate (s_c, s_d) if and only if the available system storage $V_a(t_0, x_0)$ given by (3.32) (respectively, the available exponential system storage $V_a(t_0, x_0)$ given by (3.33)) is finite for all $t_0 \in \mathbb{R}$ and $x_0 \in \mathcal{D}$, and $V_a(t, 0) = 0$, $t \in \mathbb{R}$. Moreover, if $V_a(t, 0) = 0$, $t \in \mathbb{R}$, and $V_a(t_0, x_0)$ is finite for all $t_0 \in \mathbb{R}$ and $x_0 \in \mathcal{D}$, then $V_a(t, x)$, $(t, x) \in \mathbb{R} \times \mathcal{D}$, is a storage function (respectively, exponential storage* *function) for* $\mathcal{G}$. *Finally, all storage functions (respectively, exponential storage functions)* $V_s(t,x)$, $(t,x) \in \mathbb{R} \times \mathcal{D}$, *for* $\mathcal{G}$ *satisfy*

$$0 \leq V_a(t,x) \leq V_s(t,x), \quad (t,x) \in \mathbb{R} \times \mathcal{D}. \tag{3.38}$$

Proof. Suppose $V_a(t,0) = 0$, $t \in \mathbb{R}$, and $V_a(t,x)$, $(t,x) \in \mathbb{R} \times \mathcal{D}$, is finite. Now, it follows from (3.32) (with $T = t_0$) that $V_a(t,x) \geq 0$, $(t,x) \in \mathbb{R} \times \mathcal{D}$. Next, let $x(t)$, $t \geq t_0$, satisfy (3.7)–(3.10) with admissible inputs $(u_c(t), u_d(t_k))$, $t \geq t_0$, $k \in \mathbb{Z}_{[t_0,t)}$, and $x(t_0) = x_0$. Since $-V_a(t,x)$, $(t,x) \in \mathbb{R} \times \mathcal{D}$, is given by the infimum over all admissible inputs $(u_c(\cdot), u_d(\cdot)) \in \mathcal{U}_c \times \mathcal{U}_d$ and $T \geq t_0$ in (3.32), it follows that for all admissible inputs $(u_c(\cdot), u_d(\cdot))$ and $t \in [t_0, T]$,

$$-V_a(t_0, x_0) \leq \int_{t_0}^{T} s_c(u_c(t), y_c(t))\mathrm{d}t + \sum_{k \in \mathbb{Z}_{[t_0,T)}} s_d(u_d(t_k), y_d(t_k))$$

$$= \int_{t_0}^{t} s_c(u_c(s), y_c(s))\mathrm{d}s + \sum_{k \in \mathbb{Z}_{[t_0,t)}} s_d(u_d(t_k), y_d(t_k))$$

$$+ \int_{t}^{T} s_c(u_c(s), y_c(s))\mathrm{d}s + \sum_{k \in \mathbb{Z}_{[t,T)}} s_d(u_d(t_k), y_d(t_k)),$$

which implies

$$-V_a(t_0, x_0) - \int_{t_0}^{t} s_c(u_c(t), y_c(t))\mathrm{d}t - \sum_{k \in \mathbb{Z}_{[t_0,t)}} s_d(u_d(t_k), y_d(t_k))$$

$$\leq \int_{t}^{T} s_c(u_c(s), y_c(s))\mathrm{d}s + \sum_{k \in \mathbb{Z}_{[t,T)}} s_d(u_d(t_k), y_d(t_k)).$$

Hence,

$$V_a(t_0, x_0) + \int_{t_0}^{t} s_c(u_c(t), y_c(t))\mathrm{d}t + \sum_{k \in \mathbb{Z}_{[t_0,t)}} s_d(u_d(t_k), y_d(t_k))$$

$$\geq - \inf_{(u_c(\cdot), u_d(\cdot)), T \geq t} \left[\int_{t}^{T} s_c(u_c(s), y_c(s))\mathrm{d}s \right.$$

$$\left. + \sum_{k \in \mathbb{Z}_{[t,T)}} s_d(u_d(t_k), y_d(t_k)) \right]$$

$$= V_a(t, x(t))$$

$$\geq 0, \tag{3.39}$$

which implies that

$$\int_{t_0}^{t} s_c(u_c(t), y_c(t)) dt + \sum_{k \in \mathbb{Z}_{[t_0,t)}} s_d(u_d(t_k), y_d(t_k)) \geq -V_a(t_0, x_0). \quad (3.40)$$

Hence, since by assumption $V_a(t_0, 0) = 0$, $t_0 \in \mathbb{R}$, $\mathcal{G}$ is dissipative with respect to the hybrid supply rate (s_c, s_d). Furthermore, $V_a(t, x)$, $(t, x) \in \mathbb{R} \times \mathcal{D}$, is a storage function for $\mathcal{G}$.

Conversely, suppose $\mathcal{G}$ is dissipative with respect to the hybrid supply rate (s_c, s_d) and let $t_0 \in \mathbb{R}$ and $x_0 \in \mathcal{D}$. Since $\mathcal{G}$ is completely reachable it follows that there exists $\hat{t} \leq t < t_0$, $u_c(t)$, $t \geq \hat{t}$, and $u_d(t_k)$, $k \in \mathbb{Z}_{[\hat{t},t_0)}$, such that $x(\hat{t}) = 0$ and $x(t_0) = x_0$. Hence, since $\mathcal{G}$ is dissipative with respect to the hybrid supply rate (s_c, s_d) it follows that, for all $T \geq t_0$,

$$0 \leq \int_{\hat{t}}^{T} s_c(u_c(t), y_c(t)) dt + \sum_{k \in \mathbb{Z}_{[\hat{t},T)}} s_d(u_d(t_k), y_d(t_k))$$

$$= \int_{\hat{t}}^{t_0} s_c(u_c(t), y_c(t)) dt + \sum_{k \in \mathbb{Z}_{[\hat{t},t_0)}} s_d(u_d(t_k), y_d(t_k))$$

$$+ \int_{t_0}^{T} s_c(u_c(t), y_c(t)) dt + \sum_{k \in \mathbb{Z}_{[t_0,T)}} s_d(u_d(t_k), y_d(t_k)),$$

and hence, there exists $W : \mathbb{R} \times \mathcal{D} \to \mathbb{R}$ such that

$$-\infty < W(t_0, x_0) \leq \int_{t_0}^{T} s_c(u_c(t), y_c(t)) dt + \sum_{k \in \mathbb{Z}_{[t_0,T)}} s_d(u_d(t_k), y_d(t_k)).$$

$$(3.41)$$

Now, it follows from (3.41) that, for all $(t_0, x_0) \in \mathbb{R} \times \mathcal{D}$,

$$V_a(t_0, x_0) = - \inf_{(u_c(\cdot), u_d(\cdot)), T \geq t_0} \left[\int_{t_0}^{T} s_c(u_c(t), y_c(t)) dt \right.$$

$$\left. + \sum_{k \in \mathbb{Z}_{[t_0,T)}} s_d(u_d(t_k), y_d(t_k)) \right]$$

$$\leq -W(t_0, x_0), \quad (3.42)$$

and hence, the available storage $V_a(t, x)$, $(t, x) \in \mathbb{R} \times \mathcal{D}$, is finite. Furthermore, with $x(t_0) = 0$, it follows that for all admissible $u_c(t)$,

$t \geq t_0$, and $u_{\mathrm{d}}(t_k)$, $k \in \mathbb{Z}_{[t_0,\infty)}$,

$$\int_{t_0}^{T} s_{\mathrm{c}}(u_{\mathrm{c}}(t), y_{\mathrm{c}}(t))\mathrm{d}t + \sum_{k \in \mathbb{Z}_{[t_0,T)}} s_{\mathrm{d}}(u_{\mathrm{d}}(t_k), y_{\mathrm{d}}(t_k)) \geq 0, \quad T \geq t_0,$$

$$(3.43)$$

which implies that

$$\sup_{(u_{\mathrm{c}}(\cdot),u_{\mathrm{d}}(\cdot)),\, T \geq t_0} \left[-\int_{t_0}^{T} s_{\mathrm{c}}(u_{\mathrm{c}}(t), y_{\mathrm{c}}(t))\mathrm{d}t - \sum_{k \in \mathbb{Z}_{[t_0,T)}} s_{\mathrm{d}}(u_{\mathrm{d}}(t_k), y_{\mathrm{d}}(t_k)) \right]$$

$$\leq 0, \quad (3.44)$$

or, equivalently, $V_{\mathrm{a}}(t_0, x(t_0)) = V_{\mathrm{a}}(t_0, 0) \leq 0$. However, since $V_{\mathrm{a}}(t, x) \geq 0$, $(t, x) \in \mathbb{R} \times \mathcal{D}$, it follows that $V_{\mathrm{a}}(t_0, 0) = 0$, $t_0 \in \mathbb{R}$.

Moreover, if $V_{\mathrm{s}}(t, x)$, $(t, x) \in \mathbb{R} \times \mathcal{D}$, is a storage function then it follows that, for all $T \geq t_0$ and $x_0 \in \mathcal{D}$,

$$V_{\mathrm{s}}(t_0, x_0) \geq V_{\mathrm{s}}(T, x(T)) - \int_{t_0}^{T} s_{\mathrm{c}}(u_{\mathrm{c}}(t), y_{\mathrm{c}}(t))\mathrm{d}t$$

$$- \sum_{k \in \mathbb{Z}_{[t_0,T)}} s_{\mathrm{d}}(u_{\mathrm{d}}(t_k), y_{\mathrm{d}}(t_k))$$

$$\geq - \left[\int_{t_0}^{T} s_{\mathrm{c}}(u_{\mathrm{c}}(t), y_{\mathrm{c}}(t))\mathrm{d}t + \sum_{k \in \mathbb{Z}_{[t_0,T)}} s_{\mathrm{d}}(u_{\mathrm{d}}(t_k), y_{\mathrm{d}}(t_k)) \right],$$

which implies

$$V_{\mathrm{s}}(t_0, x_0) \geq - \inf_{(u_{\mathrm{c}}(\cdot),u_{\mathrm{d}}(\cdot)),\, T \geq t_0} \left[\int_{t_0}^{T} s_{\mathrm{c}}(u_{\mathrm{c}}(t), y_{\mathrm{c}}(t))\mathrm{d}t \right.$$

$$\left. + \sum_{k \in \mathbb{Z}_{[t_0,T)}} s_{\mathrm{d}}(u_{\mathrm{d}}(t_k), y_{\mathrm{d}}(t_k)) \right]$$

$$= V_{\mathrm{a}}(t_0, x_0).$$

Finally, the proof for the exponentially dissipative case follows a similar construction and, hence, is omitted. ☐

The following corollary is immediate from Theorem 3.1.

Corollary 3.1 *Consider the impulsive dynamical system $\mathcal{G}$ given by (3.7)–(3.10) and assume that $\mathcal{G}$ is completely reachable. Then $\mathcal{G}$ is dissipative (respectively, exponentially dissipative) with respect to the* *hybrid supply rate* (s_c, s_d) *if and only if there exists a continuous storage function (respectively, exponential storage function)* $V_s(t, x)$, $(t, x) \in \mathbb{R} \times \mathcal{D}$, *satisfying (3.36) (respectively, (3.37))*.

Proof. The result follows from Theorem 3.1 with $V_s(t, x) = V_a(t, x)$, $(t, x) \in \mathbb{R} \times \mathcal{D}$. $\quad\square$

The next result gives necessary and sufficient conditions for dissipativity and exponential dissipativity over an interval $t \in (t_k, t_{k+1}]$ involving the consecutive resetting times t_k and t_{k+1}.

Theorem 3.2 *Assume* $\mathcal{G}$ *is completely reachable.* $\mathcal{G}$ *is dissipative with respect to the hybrid supply rate* (s_c, s_d) *if and only if there exists a continuous, nonnegative-definite function* $V_s : \mathbb{R} \times \mathcal{D} \rightarrow \mathbb{R}$ *such that, for all* $k \in \overline{\mathbb{Z}}_+$,

$$V_s(\hat{t}, x(\hat{t})) - V_s(t, x(t)) \leq \int_t^{\hat{t}} s_c(u_c(s), y_c(s)) \mathrm{d}s, \quad t_k < t \leq \hat{t} \leq t_{k+1},$$
$$(3.45)$$

$$V_s(t_k, x(t_k) + f_d(x(t_k)) + G_d(x(t_k))u_d(t_k)) - V_s(t_k, x(t_k))$$
$$\leq s_d(u_d(t_k), y_d(t_k)). \quad (3.46)$$

Furthermore, $\mathcal{G}$ *is exponentially dissipative with respect to the hybrid supply rate* (s_c, s_d) *if and only if there exist a continuous, nonnegative-definite function* $V_s : \mathbb{R} \times \mathcal{D} \rightarrow \mathbb{R}$ *and a scalar* $\varepsilon > 0$ *such that*

$$e^{\varepsilon\hat{t}}V_s(\hat{t}, x(\hat{t})) - e^{\varepsilon t}V_s(t, x(t)) \leq \int_t^{\hat{t}} e^{\varepsilon s} s_c(u_c(s), y_c(s)) \mathrm{d}s,$$

$$t_k < t \leq \hat{t} \leq t_{k+1}, \quad (3.47)$$
$$V_s(t_k, x(t_k) + f_d(x(t_k)) + G_d(x(t_k))u_d(t_k)) - V_s(t_k, x(t_k))$$
$$\leq s_d(u_d(t_k), y_d(t_k)). \quad (3.48)$$

Finally, $\mathcal{G}$ *is lossless with respect to the hybrid supply rate* (s_c, s_d) *if and only if there exists a continuous, nonnegative-definite function* $V_s : \mathbb{R} \times \mathcal{D} \rightarrow \mathbb{R}$ *such that (3.45) and (3.46) are satisfied as equalities.*

Proof. Let $k \in \overline{\mathbb{Z}}_+$ and suppose $\mathcal{G}$ is dissipative with respect to the hybrid supply rate (s_c, s_d). Then, there exists a continuous nonnegative-definite function $V_s : \mathbb{R} \times \mathcal{D} \rightarrow \mathbb{R}$ such that (3.36) holds. Now, since for $t_k < t \leq \hat{t} \leq t_{k+1}$, $\mathbb{Z}_{[t,\hat{t})} = \emptyset$, (3.45) is immediate.

Next, note that

$$V_{\rm s}(t_k^+, x(t_k^+)) - V_{\rm s}(t_k, x(t_k)) \le \int_{t_k}^{t_k^+} s_{\rm c}(u_{\rm c}(s), y_{\rm c}(s)){\rm d}s$$
$$+ s_{\rm d}(u_{\rm d}(t_k), y_{\rm d}(t_k)), \qquad (3.49)$$

which, since $\mathbb{Z}_{[t_k, t_k^+)} = \{k\}$, implies (3.46).

Conversely, suppose (3.45) and (3.46) hold, let $\hat{t} \ge t \ge 0$, and let $\mathbb{Z}_{[t, \hat{t})} = \{i, i+1, \dots, j\}$. (Note that if $\mathbb{Z}_{[t, \hat{t})} = \emptyset$ the converse is a direct consequence of (3.36).) In this case, it follows from (3.45) and (3.46) that

$$V_{\rm s}(\hat{t}, x(\hat{t})) - V_{\rm s}(t, x(t)) = V_{\rm s}(\hat{t}, x(\hat{t})) - V_{\rm s}(t_j^+, x(t_j^+)) + V_{\rm s}(t_j^+, x(t_j^+))$$
$$- V_{\rm s}(t_{j-1}^+, x(t_{j-1}^+))$$
$$+ V_{\rm s}(t_{j-1}^+, x(t_{j-1}^+)) - \cdots - V_{\rm s}(t_i^+, x(t_i^+))$$
$$+ V_{\rm s}(t_i^+, x(t_i^+)) - V_{\rm s}(t, x(t))$$
$$= V_{\rm s}(\hat{t}, x(\hat{t})) - V_{\rm s}(t_j^+, x(t_j^+))$$
$$+ V_{\rm s}(t_j, x(t_j) + f_{\rm d}(x(t_j)) + G_{\rm d}(x(t_j))u_{\rm d}(t_j))$$
$$- V_{\rm s}(t_j, x(t_j)) + V_{\rm s}(t_j, x(t_j))$$
$$- V_{\rm s}(t_{j-1}^+, x(t_{j-1}^+)) + \cdots$$
$$+ V_{\rm s}(t_i, x(t_i) + f_{\rm d}(x(t_i)) + G_{\rm d}(x(t_i))u_{\rm d}(t_i))$$
$$- V_{\rm s}(t_i, x(t_i)) + V_{\rm s}(t_i, x(t_i)) - V_{\rm s}(t, x(t))$$
$$\le \int_{t_j^+}^{\hat{t}} s_{\rm c}(u_{\rm c}(s), y_{\rm c}(s)){\rm d}s + s_{\rm d}(u_{\rm d}(t_j), y_{\rm d}(t_j))$$
$$+ \int_{t_{j-1}^+}^{t_j} s_{\rm c}(u_{\rm c}(s), y_{\rm c}(s)){\rm d}s + \cdots$$
$$+ s_{\rm d}(u_{\rm d}(t_i), y_{\rm d}(t_i))$$
$$+ \int_t^{t_i} s_{\rm c}(u_{\rm c}(s), y_{\rm c}(s)){\rm d}s$$
$$= \int_t^{\hat{t}} s_{\rm c}(u_{\rm c}(s), y_{\rm c}(s)){\rm d}s$$
$$+ \sum_{k \in \mathbb{Z}_{[t, \hat{t})}} s_{\rm d}(u_{\rm d}(t_k), y_{\rm d}(t_k)),$$

which implies that $\mathcal{G}$ is dissipative with respect to the hybrid supply rate $(s_{\rm c}, s_{\rm d})$.

Finally, similar constructions show that $\mathcal{G}$ is exponentially dissipative (respectively, lossless) with respect to the hybrid supply rate (s_c, s_d) if and only if (3.47) and (3.48) are satisfied (respectively, (3.45) and (3.46) are satisfied as equalities). □

If in Theorem 3.2 $V_s(\cdot, x(\cdot))$ is continuously differentiable almost everywhere on $[t_0, \infty)$ except on an unbounded closed discrete set $\mathcal{T} = \{t_1, t_2, \ldots\}$, where $\mathcal{T}$ is the set of times when jumps occur for $x(t)$, then an equivalent statement for dissipativeness of the impulsive dynamical system $\mathcal{G}$ with respect to the hybrid supply rate (s_c, s_d) is

$$\dot{V}_s(t, x(t)) \leq s_c(u_c(t), y_c(t)), \quad t_k < t \leq t_{k+1}, \qquad (3.50)$$
$$\Delta V_s(t_k, x(t_k)) \leq s_d(u_d(t_k), y_d(t_k)), \quad k \in \overline{\mathbb{Z}}_+, \qquad (3.51)$$

where $\dot{V}_s(\cdot, \cdot)$ denotes the total derivative of $V_s(t, x(t))$ along the state trajectories $x(t)$, $t \in (t_k, t_{k+1}]$, of the impulsive dynamical system (3.7)–(3.10) and $\Delta V_s(t_k, x(t_k)) \triangleq V_s(t_k^+, x(t_k^+)) - V_s(t_k, x(t_k)) = V_s(t_k, x(t_k) + f_d(x(t_k)) + G_d(x(t_k))u_d(t_k)) - V_s(t_k, x(t_k))$, $k \in \overline{\mathbb{Z}}_+$, denotes the difference of the storage function $V_s(t, x)$ at the resetting times t_k, $k \in \overline{\mathbb{Z}}_+$, of the impulsive dynamical system (3.7)–(3.10). Furthermore, an equivalent statement for exponential dissipativeness of the impulsive dynamical system $\mathcal{G}$ with respect to the hybrid supply rate (s_c, s_d) is given by

$$\dot{V}_s(t, x(t)) + \varepsilon V_s(t, x(t)) \leq s_c(u_c(t), y_c(t)), \quad t_k < t \leq t_{k+1}, \quad (3.52)$$

and (3.51).

The following theorem provides sufficient conditions for guaranteeing that all storage functions (respectively, exponential storage functions) of a given dissipative (respectively, exponentially dissipative) impulsive dynamical system are positive definite. For this result we need the following definition.

Definition 3.4 *An impulsive dynamical system $\mathcal{G}$ given by (3.7)–(3.10) is* zero-state observable *if $(u_c(t), u_d(t_k)) \equiv (0,0)$ and $(y_c(t), y_d(t_k)) \equiv (0,0)$ implies $x(t) \equiv 0$. An impulsive dynamical system $\mathcal{G}$ given by (3.7)–(3.10) is* strongly zero-state observable *if $u_c(t) \equiv 0$ and $y_c(t) \equiv 0$ implies $x(t) \equiv 0$. Finally, an impulsive system $\mathcal{G}$ is* minimal *if it is zero-state observable and completely reachable.*

Note that strong zero-state observability is a stronger condition than zero-state observability. In particular, strong zero-state observability implies zero-state observability; however, the converse is not necessarily true.

Theorem 3.3 *Consider the nonlinear impulsive dynamical system $\mathcal{G}$ given by (3.7)–(3.10) and assume that $\mathcal{G}$ is completely reachable and zero-state observable. Furthermore, assume that $\mathcal{G}$ is dissipative (respectively, exponentially dissipative) with respect to the hybrid supply rate (s_c, s_d) and there exist functions $\kappa_c : Y_c \to U_c$ and $\kappa_d : Y_d \to U_d$ such that $\kappa_c(0) = 0$, $\kappa_d(0) = 0$, $s_c(\kappa_c(y_c), y_c) < 0$, $y_c \neq 0$, and $s_d(\kappa_d(y_d), y_d) < 0$, $y_d \neq 0$. Then all the storage functions (respectively, exponential storage functions) $V_s(t, x)$, $(t, x) \in \mathbb{R} \times \mathcal{D}$, for $\mathcal{G}$ are positive definite, that is, $V_s(\cdot, 0) = 0$ and $V_s(t, x) > 0$, $(t, x) \in \mathbb{R} \times \mathcal{D}$, $x \neq 0$.*

Proof. It follows from Theorem 3.1 that the available storage $V_a(t, x)$, $(t, x) \in \mathbb{R} \times \mathcal{D}$, is a storage function for $\mathcal{G}$. Next, suppose, *ad absurdum*, there exists $(t_0, x_0) \in \mathbb{R} \times \mathcal{D}$ such that $V_a(t_0, x_0) = 0$, $x_0 \neq 0$, or, equivalently,

$$
\inf_{(u_c(\cdot), u_d(\cdot)), T \geq t_0} \left[\int_{t_0}^{T} s_c(u_c(t), y_c(t)) dt + \sum_{k \in \mathbb{Z}_{[t_0, T)}} s_d(u_d(t_k), y_d(t_k)) \right]
$$

$$
= 0. \quad (3.53)
$$

Furthermore, suppose there exists $[t_s, t_f) \subset \mathbb{R}$ such that $y_c(t) \neq 0$, $t \in [t_s, t_f)$, or $y_d(t_k) \neq 0$, for some $k \in \mathbb{Z}_+$. Now, since there exists $\kappa_c : Y_c \to U_c$ and $\kappa_d : Y_d \to U_d$ such that $\kappa_c(0) = 0$, $\kappa_d(0) = 0$, $s_c(\kappa_c(y_c), y_c) < 0$, $y_c \neq 0$, and $s_d(\kappa_d(y_d), y_d) < 0$, $y_d \neq 0$, the infimum in (3.53) occurs at a negative value, which is a contradiction. Hence, $y_c(t) = 0$ for almost all $t \in \mathbb{R}$, and $y_d(t_k) = 0$ for all $k \in \mathbb{Z}_+$. Next, since $\mathcal{G}$ is zero-state observable it follows that $x = 0$, and hence, $V_a(t, x) = 0$ if and only if $x = 0$. The result now follows from (3.38). Finally, the proof for the exponentially dissipative case is similar and, hence, is omitted. $\square$

Next, we introduce the concept of a required supply for a nonlinear impulsive dynamical system given by (3.7)–(3.10). Specifically, define the *required supply* $V_r(t_0, x_0)$ of the nonlinear impulsive dynamical system $\mathcal{G}$ by

$$
V_r(t_0, x_0) \triangleq \inf_{(u_c(\cdot), u_d(\cdot)), T \leq t_0} \left[\int_{T}^{t_0} s_c(u_c(t), y_c(t)) dt \right.
$$

$$
\left. + \sum_{k \in \mathbb{Z}_{[T, t_0)}} s_d(u_d(t_k), y_d(t_k)) \right], \quad (3.54)
$$

where $x(t)$, $t \geq T$, is the solution of (3.7)–(3.10) with $x(T) = 0$ and $x(t_0) = x_0$. It follows from (3.54) that the required supply of a nonlinear impulsive dynamical system is the minimum amount of generalized energy which can be delivered to the impulsive dynamical system in order to transfer it from an initial state $x(T) = 0$ to a given state $x(t_0) = x_0$. Similarly, define the *required exponential supply* of the nonlinear impulsive dynamical system $\mathcal{G}$ by

$$V_{\mathrm{r}}(t_0, x_0) \triangleq \inf_{(u_{\mathrm{c}}(\cdot), u_{\mathrm{d}}(\cdot)), T \leq t_0} \left[\int_T^{t_0} e^{\varepsilon t} s_{\mathrm{c}}(u_{\mathrm{c}}(t), y_{\mathrm{c}}(t)) \mathrm{d}t \right.$$

$$\left. + \sum_{k \in \mathbb{Z}_{[T, t_0)}} e^{\varepsilon t_k} s_{\mathrm{d}}(u_{\mathrm{d}}(t_k), y_{\mathrm{d}}(t_k)) \right], \quad (3.55)$$

where $\varepsilon > 0$ and $x(t)$, $t \geq T$, is the solution of (3.7)–(3.10) with $x(T) = 0$ and $x(t_0) = x_0$. Note that since, with $x(t_0) = 0$, the infimum in (3.54) is zero it follows that $V_{\mathrm{r}}(t_0, 0) = 0$, $t_0 \in \mathbb{R}$.

Next, using the notion of the required supply, we show that all storage functions are bounded from above by the required supply and bounded from below by the available storage. Hence, as in the case of dynamical systems with continuous flows [166], a dissipative impulsive dynamical system can deliver to its surroundings only a fraction of its stored generalized energy and can store only a fraction of the generalized work done to it.

Theorem 3.4 *Consider the nonlinear impulsive dynamical system $\mathcal{G}$ given by (3.7)–(3.10) and assume that $\mathcal{G}$ is completely reachable. Then $\mathcal{G}$ is dissipative (respectively, exponentially dissipative) with respect to the hybrid supply rate $(s_{\mathrm{c}}, s_{\mathrm{d}})$ if and only if $0 \leq V_{\mathrm{r}}(t, x) < \infty$, $t \in \mathbb{R}$, $x \in \mathcal{D}$. Moreover, if $V_{\mathrm{r}}(t, x)$ is finite and nonnegative for all $(t_0, x_0) \in \mathbb{R} \times \mathcal{D}$, then $V_{\mathrm{r}}(t, x)$, $(t, x) \in \mathbb{R} \times \mathcal{D}$, is a storage function (respectively, exponential storage function) for $\mathcal{G}$. Finally, all storage functions (respectively, exponential storage functions) $V_{\mathrm{s}}(t, x)$, $(t, x) \in \mathbb{R} \times \mathcal{D}$, for $\mathcal{G}$ satisfy*

$$0 \leq V_{\mathrm{a}}(t, x) \leq V_{\mathrm{s}}(t, x) \leq V_{\mathrm{r}}(t, x) < \infty, \quad (t, x) \in \mathbb{R} \times \mathcal{D}. \quad (3.56)$$

Proof. Suppose $0 \leq V_{\mathrm{r}}(t, x) < \infty$, $(t, x) \in \mathbb{R} \times \mathcal{D}$. Next, let $x(t)$, $t \in \mathbb{R}$, satisfy (3.7)–(3.10) with admissible inputs $(u_{\mathrm{c}}(t), u_{\mathrm{d}}(t_k))$, $t \in \mathbb{R}$, $k \in \mathbb{Z}_{[t_0, t)}$, and $x(t_0) = x_0$. Since $V_{\mathrm{r}}(t, x)$, $(t, x) \in \mathbb{R} \times \mathcal{D}$, is given by the infimum over all admissible inputs $(u_{\mathrm{c}}(\cdot), u_{\mathrm{d}}(\cdot)) \in \mathcal{U}_{\mathrm{c}} \times \mathcal{U}_{\mathrm{d}}$ and $T \leq t_0$ in (3.54), it follows that for all admissible inputs $(u_{\mathrm{c}}(\cdot), u_{\mathrm{d}}(\cdot))$

and $T \leq t \leq t_0$,

$$V_r(t_0, x_0) \leq \int_T^{t_0} s_c(u_c(t), y_c(t)) dt + \sum_{k \in \mathbb{Z}_{[T,t_0)}} s_d(u_d(t_k), y_d(t_k))$$

$$= \int_T^t s_c(u_c(s), y_c(s)) ds + \sum_{k \in \mathbb{Z}_{[T,t)}} s_d(u_d(t_k), y_d(t_k))$$

$$+ \int_t^{t_0} s_c(u_c(s), y_c(s)) ds + \sum_{k \in \mathbb{Z}_{[t,t_0)}} s_d(u_d(t_k), y_d(t_k)),$$

and hence,

$$V_r(t_0, x_0) \leq \inf_{(u_c(\cdot), u_d(\cdot)), \, T \leq t} \left[\int_T^t s_c(u_c(s), y_c(s)) ds \right.$$

$$\left. + \sum_{k \in \mathbb{Z}_{[T,t)}} s_d(u_d(t_k), y_d(t_k)) \right]$$

$$+ \int_t^{t_0} s_c(u_c(s), y_c(s)) ds + \sum_{k \in \mathbb{Z}_{[t,t_0)}} s_d(u_d(t_k), y_d(t_k))$$

$$= V_r(t, x(t)) + \int_t^{t_0} s_c(u_c(s), y_c(s)) ds$$

$$+ \sum_{k \in \mathbb{Z}_{[t,t_0)}} s_d(u_d(t_k), y_d(t_k)), \tag{3.57}$$

which shows that $V_r(t, x)$, $(t, x) \in \mathbb{R} \times \mathcal{D}$, is a storage function for $\mathcal{G}$, and hence, $\mathcal{G}$ is dissipative.

Conversely, suppose $\mathcal{G}$ is dissipative with respect to the hybrid supply rate (s_c, s_d) and let $t_0 \in \mathbb{R}$ and $x_0 \in \mathcal{D}$. Since $\mathcal{G}$ is completely reachable it follows that there exist $T < t_0$, $u_c(t)$, $T \leq t < t_0$, and $u_d(t_k)$, $k \in \mathbb{Z}_{[T,\infty)}$, such that $x(T) = 0$ and $x(t_0) = x_0$. Hence, since $\mathcal{G}$ is dissipative with respect to the hybrid supply rate (s_c, s_d) it follows that, for all $T \leq t_0$,

$$0 \leq \int_T^{t_0} s_c(u_c(t), y_c(t)) dt + \sum_{k \in \mathbb{Z}_{[T,t_0)}} s_d(u_d(t_k), y_d(t_k)), \tag{3.58}$$

and hence,

$$0 \leq \inf_{(u_c(\cdot), u_d(\cdot)), \, T \leq t_0} \left[\int_T^{t_0} s_c(u_c(s), y_c(s)) ds \right.$$

$$+ \sum_{k \in \mathbb{Z}_{[T,t_0)}} s_{\mathrm{d}}(u_{\mathrm{d}}(t_k), y_{\mathrm{d}}(t_k)) \Bigg], \quad (3.59)$$

which implies that

$$0 \le V_{\mathrm{r}}(t_0, x_0) < \infty, \quad (t_0, x_0) \in \mathbb{R} \times \mathcal{D}. \quad (3.60)$$

Next, if $V_{\mathrm{s}}(\cdot, \cdot)$ is a storage function for $\mathcal{G}$, then it follows from Theorem 3.1 that

$$0 \le V_{\mathrm{a}}(t, x) \le V_{\mathrm{s}}(t, x), \quad (t, x) \in \mathbb{R} \times \mathcal{D}. \quad (3.61)$$

Furthermore, for all $T \in \mathbb{R}$ such that $x(T) = 0$ it follows that

$$V_{\mathrm{s}}(t_0, x_0) \le V_{\mathrm{s}}(T, 0) + \int_T^{t_0} s_{\mathrm{c}}(u_{\mathrm{c}}(t), y_{\mathrm{c}}(t)) \mathrm{d}t + \sum_{k \in \mathbb{Z}_{[T,t_0)}} s_{\mathrm{d}}(u_{\mathrm{d}}(t_k), y_{\mathrm{d}}(t_k)),$$

$$(3.62)$$

and hence,

$$V_{\mathrm{s}}(t_0, x_0) \le \inf_{(u_{\mathrm{c}}(\cdot), u_{\mathrm{d}}(\cdot)), \, T \le t_0} \left[\int_T^{t_0} s_{\mathrm{c}}(u_{\mathrm{c}}(t), y_{\mathrm{c}}(t)) \mathrm{d}t \right.$$

$$\left. + \sum_{k \in \mathbb{Z}_{[T,t_0)}} s_{\mathrm{d}}(u_{\mathrm{d}}(t_k), y_{\mathrm{d}}(t_k)) \right]$$

$$= V_{\mathrm{r}}(t_0, x_0)$$

$$< \infty, \quad (3.63)$$

which implies (3.56). Finally, the proof for the exponentially dissipative case follows a similar construction and, hence, is omitted. $\square$

In light of Theorems 3.1 and 3.4 the following result on lossless impulsive dynamical systems is immediate.

Theorem 3.5 *Consider the nonlinear impulsive dynamical system $\mathcal{G}$ given by (3.7)–(3.10) and assume $\mathcal{G}$ is completely reachable to and from the origin. Then $\mathcal{G}$ is lossless with respect to the hybrid supply rate $(s_{\mathrm{c}}, s_{\mathrm{d}})$ if and only if there exists a continuous storage function $V_{\mathrm{s}}(t, x)$, $(t, x) \in \mathbb{R} \times \mathcal{D}$, satisfying (3.36) as an equality. Furthermore, if $\mathcal{G}$ is lossless with respect to the hybrid supply rate $(s_{\mathrm{c}}, s_{\mathrm{d}})$, then $V_{\mathrm{a}}(t, x) = V_{\mathrm{r}}(t, x)$, and hence the storage function $V_{\mathrm{s}}(t, x)$, $(t, x) \in \mathbb{R} \times \mathcal{D}$, is unique and is given by*

$$V_{\mathrm{s}}(t_0, x_0) = - \int_{t_0}^{T_+} s_{\mathrm{c}}(u_{\mathrm{c}}(t), y_{\mathrm{c}}(t)) \mathrm{d}t - \sum_{\mathbb{Z}_{[t_0, T_+)}} s_{\mathrm{d}}(u_{\mathrm{d}}(t_k), y_{\mathrm{d}}(t_k))$$

$$= \int_{-T_-}^{t_0} s_c(u_c(t), y_c(t))dt + \sum_{\mathbb{Z}_{[-T_-,t_0)}} s_d(u_d(t_k), y_d(t_k)), \quad (3.64)$$

where $x(t)$, $t \geq t_0$, is the solution to (3.7) and (3.8) with $(u_c(\cdot), u_d(\cdot)) \in \mathcal{U}_c \times \mathcal{U}_d$ and $x(t_0) = x_0$, $x_0 \in \mathcal{D}$, for any $T_+ > t_0$ and $T_- > -t_0$ such that $x(-T_-) = 0$ and $x(T_+) = 0$.

Proof. Suppose $\mathcal{G}$ is lossless with respect to the hybrid supply rate (s_c, s_d). Since $\mathcal{G}$ is completely reachable to and from the origin it follows that, for every $x_0 \in \mathcal{D}$, there exist $T_-, T_+ > 0$, $u_c(t) \in \mathcal{U}_c$, $t \in [-T_-, T_+]$, $u_d(t_k) \in \mathcal{U}_d$, $k \in \mathbb{Z}_{[-T_-,T_+)}$, such that $x(-T_-) = 0$, $x(T_+) = 0$, and $x(0) = x_0$. Now, it follows that

$$0 = \int_{-T_-}^{T_+} s_c(u_c(t), y_c(t))dt + \sum_{\mathbb{Z}_{[-T_-,T_+)}} s_d(u_d(t_k), y_d(t_k))$$

$$= \int_{-T_-}^{t_0} s_c(u_c(t), y_c(t))dt + \sum_{\mathbb{Z}_{[-T_-,t_0)}} s_d(u_d(t_k), y_d(t_k))$$

$$+ \int_{t_0}^{T_+} s_c(u_c(t), y_c(t))dt + \sum_{\mathbb{Z}_{[t_0,T_+)}} s_d(u_d(t_k), y_d(t_k))$$

$$\geq \inf_{(u_c(\cdot),u_d(\cdot)),\, T \leq t_0} \int_{T}^{t_0} s_c(u_c(t), y_c(t))dt + \sum_{\mathbb{Z}_{[T,t_0)}} s_d(u_d(t_k), y_d(t_k))$$

$$+ \inf_{(u_c(\cdot),u_d(\cdot)),\, T \geq t_0} \int_{t_0}^{T} s_c(u_c(t), y_c(t))dt + \sum_{\mathbb{Z}_{[t_0,T)}} s_d(u_d(t_k), y_d(t_k))$$

$$= V_r(t_0, x_0) - V_a(t_0, x_0), \quad (3.65)$$

which implies that $V_r(t_0, x_0) \leq V_a(t_0, x_0)$, $(t_0, x_0) \in \mathbb{R} \times \mathcal{D}$. However, since by definition $\mathcal{G}$ is dissipative with respect to the hybrid supply rate (s_c, s_d) it follows from Theorem 3.4 that $V_a(t_0, x_0) \leq V_r(t_0, x_0)$, $(t_0, x_0) \in \mathbb{R} \times \mathcal{D}$, and hence, every storage function $V_s(t_0, x_0)$, $(t_0, x_0) \in \mathbb{R} \times \mathcal{D}$, satisfies $V_a(t_0, x_0) = V_s(t_0, x_0) = V_r(t_0, x_0)$. Furthermore, it follows that the inequality in (3.65) is indeed an equality, which implies (3.64).

Next, let $t_0, t, T \geq 0$ be such that $t_0 < t < T$, $x(T) = 0$. Hence, it follows from (3.64) that

$$0 = V_s(t_0, x(t_0)) + \int_{t_0}^{T} s_c(u_c(t), y_c(t))dt + \sum_{\mathbb{Z}_{[t_0,T)}} s_d(u_d(t_k), y_d(t_k))$$

$$= V_s(t_0, x(t_0)) + \int_{t_0}^{t} s_c(u_c(t), y_c(t))\mathrm{d}t + \sum_{\mathbb{Z}_{[t_0,t)}} s_d(u_d(t_k), y_d(t_k))$$

$$+ \int_{t}^{T} s_c(u_c(t), y_c(t))\mathrm{d}t + \sum_{\mathbb{Z}_{[t,T)}} s_d(u_d(t_k), y_d(t_k))$$

$$= V_s(t_0, x(t_0)) + \int_{t_0}^{t} s_c(u_c(t), y_c(t))\mathrm{d}t + \sum_{\mathbb{Z}_{[t_0,t)}} s_d(u_d(t_k), y_d(t_k))$$

$$- V_s(t, x(t)),$$

which implies that (3.36) is satisfied as an equality.

Conversely, if there exists a storage function $V_s(t, x)$, $(t, x) \in \mathbb{R} \times \mathcal{D}$, satisfying (3.36) as an equality it follows from Corollary 3.1 that $\mathcal{G}$ is dissipative with respect to the hybrid supply rate (s_c, s_d). Furthermore, for every $u_c(t) \in \mathcal{U}_c$, $t \geq t_0$, $u_d(t_k) \in \mathcal{U}_d$, $k \in \mathbb{Z}_{[t_0,t)}$ and $x(t_0) = x(t) = 0$, it follows from (3.36) (satisfied as an equality) that

$$\int_{t_0}^{t} s_c(u_c(t), y_c(t))\mathrm{d}t + \sum_{\mathbb{Z}_{[t_0,t)}} s_d(u_d(t_k), y_d(t_k)) = 0,$$

which implies that $\mathcal{G}$ is lossless with respect to the hybrid supply rate (s_c, s_d). $\square$

Finally, as a direct consequence of Theorems 3.1 and 3.4, we show that the set of all possible storage functions of an impulsive dynamical system forms a convex set. An identical result holds for exponential storage functions.

Proposition 3.1 *Consider the nonlinear impulsive dynamical system $\mathcal{G}$ given by (3.7)–(3.10) with available storage $V_a(t, x)$, $(t, x) \in \mathbb{R} \times \mathcal{D}$, and required supply $V_r(t, x)$, $(t, x) \in \mathbb{R} \times \mathcal{D}$, and assume that $\mathcal{G}$ is completely reachable. Then*

$$V_s(t, x) \triangleq \alpha V_a(t, x) + (1 - \alpha)V_r(t, x), \quad \alpha \in [0, 1], \qquad (3.66)$$

is a storage function for $\mathcal{G}$.

Proof. The result is a direct consequence of the complete reachability of $\mathcal{G}$ along with the dissipation inequality (3.36) by noting that if $V_a(t, x)$ and $V_r(t, x)$ satisfy (3.36), then $V_s(t, x)$ satisfies (3.36). $\square$

3.3 Extended Kalman-Yakubovich-Popov Conditions for Impulsive Dynamical Systems

In this section, we show that dissipativeness of an impulsive dynamical system can be characterized in terms of the system functions $f_c(\cdot)$, $G_c(\cdot)$, $h_c(\cdot)$, $J_c(\cdot)$, $f_d(\cdot)$, $G_d(\cdot)$, $h_d(\cdot)$, and $J_d(\cdot)$. First, we concentrate on the theory for dissipative time-dependent impulsive dynamical systems. Since in the case of dissipative state-dependent impulsive dynamical systems it follows from Assumptions A1 and A2 that, for $\mathcal{S} = [0, \infty) \times \mathcal{Z}$, the resetting times are well defined and distinct for every trajectory of (3.27) and (3.28), the theory of dissipative state-dependent impulsive dynamical systems closely parallels that of dissipative time-dependent impulsive dynamical systems, and hence, many of the results are similar. In the cases where the results for dissipative state-dependent impulsive dynamical systems deviate markedly from their time-dependent counterparts, we present a thorough treatment of these results.

For the results in this section we consider the special case of dissipative impulsive systems with quadratic hybrid supply rates and set $U_c = \mathbb{R}^{m_c}$ and $U_d = \mathbb{R}^{m_d}$. Specifically, let $Q_c \in \mathbb{S}^{l_c}$, $S_c \in \mathbb{R}^{l_c \times m_c}$, $R_c \in \mathbb{S}^{m_c}$, $Q_d \in \mathbb{S}^{l_d}$, $S_d \in \mathbb{R}^{l_d \times m_d}$, and $R_d \in \mathbb{S}^{m_d}$ be given and assume $s_c(u_c, y_c) = y_c^{\mathrm{T}} Q_c y_c + 2 y_c^{\mathrm{T}} S_c u_c + u_c^{\mathrm{T}} R_c u_c$ and $s_d(u_d, y_d) = y_d^{\mathrm{T}} Q_d y_d + 2 y_d^{\mathrm{T}} S_d u_d + u_d^{\mathrm{T}} R_d u_d$. For simplicity of exposition, in the remainder of the chapter we assume that for time-dependent impulsive dynamical systems the storage functions do not depend explicitly on time. This corresponds to the case in which $\mathcal{G}$ is time varying but the energy storage mechanism does not reflect this. However, this is not to say that system energy dissipation does not have a time-varying character. Furthermore, we assume that there exist functions $\kappa_c : \mathbb{R}^{l_c} \to \mathbb{R}^{m_c}$ and $\kappa_d : \mathbb{R}^{l_d} \to \mathbb{R}^{m_d}$ such that $\kappa_c(0) = 0$, $\kappa_d(0) = 0$, $s_c(\kappa_c(y_c), y_c) < 0$, $y_c \neq 0$, and $s_d(\kappa_d(y_d), y_d) < 0$, $y_d \neq 0$, so that the storage function $V_s(x)$, $x \in \mathbb{R}^n$, is positive definite, and we assume that $V_s(\cdot)$ is continuously differentiable.

Theorem 3.6 *Let* $Q_c \in \mathbb{S}^{l_c}$, $S_c \in \mathbb{R}^{l_c \times m_c}$, $R_c \in \mathbb{S}^{m_c}$, $Q_d \in \mathbb{S}^{l_d}$, $S_d \in \mathbb{R}^{l_d \times m_d}$, *and* $R_d \in \mathbb{S}^{m_d}$. *If there exist functions* $V_s : \mathbb{R}^n \to \mathbb{R}$, $L_c : \mathbb{R}^n \to \mathbb{R}^{p_c}$, $L_d : \mathbb{R}^n \to \mathbb{R}^{p_d}$, $W_c : \mathbb{R}^n \to \mathbb{R}^{p_c \times m_c}$, $W_d : \mathbb{R}^n \to \mathbb{R}^{p_d \times m_d}$, $P_{1u_d} : \mathbb{R}^n \to \mathbb{R}^{1 \times m_d}$, *and* $P_{2u_d} : \mathbb{R}^n \to \mathbb{N}^{m_d}$ *such that* $V_s(\cdot)$ *is continuously differentiable and positive definite,* $V_s(0) = 0$,

$$V_s(x + f_d(x) + G_d(x)u_d) = V_s(x + f_d(x)) + P_{1u_d}(x)u_d + u_d^{\mathrm{T}} P_{2u_d}(x)u_d,$$
$$x \in \mathbb{R}^n, \quad u_d \in \mathbb{R}^{m_d}, \qquad (3.67)$$

and, for all $x \in \mathbb{R}^n$,

$$0 = V_s'(x)f_c(x) - h_c^T(x)Q_ch_c(x) + L_c^T(x)L_c(x), \tag{3.68}$$

$$0 = \tfrac{1}{2}V_s'(x)G_c(x) - h_c^T(x)(Q_cJ_c(x) + S_c) + L_c^T(x)W_c(x), \tag{3.69}$$

$$0 = R_c + S_c^T J_c(x) + J_c^T(x)S_c + J_c^T(x)Q_cJ_c(x) - W_c^T(x)W_c(x), \tag{3.70}$$

$$0 = V_s(x + f_d(x)) - V_s(x) - h_d^T(x)Q_dh_d(x) + L_d^T(x)L_d(x), \tag{3.71}$$

$$0 = \tfrac{1}{2}P_{1u_d}(x) - h_d^T(x)(Q_dJ_d(x) + S_d) + L_d^T(x)W_d(x), \tag{3.72}$$

$$0 = R_d + S_d^T J_d(x) + J_d^T(x)S_d + J_d^T(x)Q_dJ_d(x) - P_{2u_d}(x)$$
$$- W_d^T(x)W_d(x), \tag{3.73}$$

then the nonlinear impulsive system $\mathcal{G}$ given by (3.14)–(3.17) is dissipative with respect to the quadratic hybrid supply rate $(s_c(u_c, y_c), s_d(u_d, y_d)) = (y_c^T Q_c y_c + 2y_c^T S_c u_c + u_c^T R_c u_c, y_d^T Q_d y_d + 2y_d^T S_d u_d + u_d^T R_d u_d)$. If, alternatively,

$$\mathcal{Z}_c(x) \triangleq R_c + S_c^T J_c(x) + J_c^T(x)S_c + J_c^T(x)Q_cJ_c(x) > 0, \quad x \in \mathbb{R}^n, \tag{3.74}$$

and there exist a continuously differentiable function $V_s : \mathbb{R}^n \to \mathbb{R}$ and matrix functions $P_{1u_d} : \mathbb{R}^n \to \mathbb{R}^{1 \times m_d}$ and $P_{2u_d} : \mathbb{R}^n \to \mathbb{N}^{m_d}$ such that $V_s(\cdot)$ is positive definite, $V_s(0) = 0$, (3.67) holds, and for all $x \in \mathbb{R}^n$,

$$\mathcal{Z}_d(x) \triangleq R_d + S_d^T J_d(x) + J_d^T(x)S_d + J_d^T(x)Q_dJ_d(x) - P_{2u_d}(x) > 0, \tag{3.75}$$

$$0 \geq V_s'(x)f_c(x) - h_c^T(x)Q_ch_c(x)$$
$$+ [\tfrac{1}{2}V_s'(x)G_c(x) - h_c^T(x)(Q_cJ_c(x) + S_c)]$$
$$\cdot \mathcal{Z}_c^{-1}(x)[\tfrac{1}{2}V_s'(x)G_c(x) - h_c^T(x)(Q_cJ_c(x) + S_c)]^T, \tag{3.76}$$

$$0 \geq V_s(x + f_d(x)) - V_s(x) - h_d^T(x)Q_dh_d(x)$$
$$+ [\tfrac{1}{2}P_{1u_d}(x) - h_d^T(x)(Q_dJ_d(x) + S_d)]$$
$$\cdot \mathcal{Z}_d^{-1}(x)[\tfrac{1}{2}P_{1u_d}(x) - h_d^T(x)(Q_dJ_d(x) + S_d)]^T, \tag{3.77}$$

then $\mathcal{G}$ is dissipative with respect to the quadratic hybrid supply rate $(s_c(u_c, y_c), s_d(u_d, y_d)) = (y_c^T Q_c y_c + 2y_c^T S_c u_c + u_c^T R_c u_c, y_d^T Q_d y_d + 2y_d^T S_d u_d + u_d^T R_d u_d)$.

Proof. For any admissible input $u_c(t)$, $t, \hat{t} \in \mathbb{R}$, $t_k < t \leq \hat{t} \leq t_{k+1}$, and $k \in \overline{\mathbb{Z}}_+$, it follows from (3.68)–(3.70) that

$$V_s(x(\hat{t})) - V_s(x(t)) = \int_t^{\hat{t}} \dot{V}_s(x(s)) ds$$

$$\leq \int_t^{\hat{t}} \Big[\dot{V}_s(x(s)) + [L_c(x(s)) + W_c(x(s))u_c(s)]^T$$

$$\cdot [L_c(x(s)) + W_c(x(s))u_c(s)]\Big] ds$$

$$= \int_t^{\hat{t}} [V_s'(x(s))(f_c(x(s)) + G_c(x(s))u_c(s))$$

$$+ L_c^T(x(s))L_c(x(s)) + 2L_c^T(x(s))W_c(x(s))u_c(s)$$

$$+ u_c^T(s)W_c^T(x(s))W_c(x(s))u_c(s)]ds$$

$$= \int_t^{\hat{t}} [h_c^T(x(s))Q_c h_c(x(s)) + 2h_c^T(x(s))(S_c$$

$$+ Q_c J_c(x(s)))u_c(s)$$

$$+ u_c^T(s)(J_c^T(x(s))Q_c J_c(x(s)) + S_c^T J_c(x(s))$$

$$+ J_c^T(x(s))S_c + R_c)u_c(s)]ds$$

$$= \int_t^{\hat{t}} [y_c^T(s)Q_c y_c(s) + 2y_c^T(s)S_c u_c(s)$$

$$+ u_c^T(s)R_c u_c(s)]ds$$

$$= \int_t^{\hat{t}} s_c(u_c(s), y_c(s))ds, \tag{3.78}$$

where $x(t)$, $t \in (t_k, t_{k+1}]$, satisfies (3.14) and $\dot{V}_s(\cdot)$ denotes the total derivative of the storage function along the trajectories $x(t)$, $t \in (t_k, t_{k+1}]$, of (3.14).

Next, for any admissible input $u_d(t_k)$, $t_k \in \mathbb{R}$, and $k \in \overline{\mathbb{Z}}_+$, it follows that

$$\Delta V_s(x(t_k)) = V_s(x(t_k) + f_d(x(t_k)) + G_d(x(t_k))u_d(t_k)) - V_s(x(t_k)), \tag{3.79}$$

where $\Delta V_s(\cdot)$ denotes the difference of the storage function at the resetting times t_k, $k \in \overline{\mathbb{Z}}_+$, of (3.15). Hence, it follows from (3.71)–(3.73), the structural storage function constraint (3.67), and (3.79), that for all $x \in \mathbb{R}^n$ and $u_d \in \mathbb{R}^{m_d}$,

$$\Delta V_s(x) = V_s(x + f_d(x) + G_d(x)u_d) - V_s(x)$$

$$= V_s(x + f_d(x)) - V_s(x) + P_{1u_d}(x)u_d + u_d^T P_{2u_d}(x)u_d$$

$$= h_d^T(x)Q_d h_d(x) - L_d^T(x)L_d(x)$$

$$+ 2[h_d^T(x)(Q_d J_d(x) + S_d) - L_d^T(x)W_d(x)]u_d$$

$$+ u_d^T[R_d + S_d^T J_d(x) + J_d^T(x)S_d + J_d^T(x)Q_d J_d(x)]$$

$$-\mathcal{W}_d^T(x)\mathcal{W}_d(x)]u_d$$
$$= s_d(u_d, y_d) - [L_d(x) + \mathcal{W}_d(x)u_d]^T[L_d(x) + \mathcal{W}_d(x)u_d]$$
$$\le s_d(u_d, y_d). \tag{3.80}$$

Now, using (3.78) and (3.80) the result is immediate from Theorem 3.2.

To show that (3.76) and (3.77) imply that $\mathcal{G}$ is dissipative with respect to the quadratic hybrid supply rate (s_c, s_d), note that (3.68)–(3.73) can be equivalently written as

$$\begin{bmatrix} \mathcal{A}_c(x) & \mathcal{B}_c(x) \\ \mathcal{B}_c^T(x) & \mathcal{C}_c(x) \end{bmatrix} = -\begin{bmatrix} L_c^T(x) \\ \mathcal{W}_c^T(x) \end{bmatrix}[\, L_c(x) \quad \mathcal{W}_c(x)\,] \le 0, \quad x \in \mathbb{R}^n,$$
$$\tag{3.81}$$

$$\begin{bmatrix} \mathcal{A}_d(x) & \mathcal{B}_d(x) \\ \mathcal{B}_d^T(x) & \mathcal{C}_d(x) \end{bmatrix} = -\begin{bmatrix} L_d^T(x) \\ \mathcal{W}_d^T(x) \end{bmatrix}[\, L_d(x) \quad \mathcal{W}_d(x)\,] \le 0, \quad x \in \mathbb{R}^n,$$
$$\tag{3.82}$$

where $\mathcal{A}_c(x) \triangleq V_s'(x)f_c(x) - h_c^T(x)Q_c h_c(x)$, $\mathcal{B}_c(x) \triangleq \frac{1}{2}V_s'(x)G_c(x) - h_c^T(x)(Q_c J_c(x) + S_c)$, $\mathcal{C}_c(x) \triangleq -(R_c + S_c^T J_c(x) + J_c^T(x)S_c + J_c^T(x)Q_c \cdot J_c(x))$, $\mathcal{A}_d(x) \triangleq V_s(x + f_d(x)) - V_s(x) - h_d^T(x)Q_d h_d(x)$, $\mathcal{B}_d(x) \triangleq \frac{1}{2}P_{1u_d}(x) - h_d^T(x)(Q_d J_d(x) + S_d)$, and $\mathcal{C}_d(x) \triangleq -(R_d + S_d^T J_d(x) + J_d^T(x)S_d + J_d^T(x)Q_d J_d(x) - P_{2u_d}(x))$. Now, for all invertible $\mathcal{T}_c \in \mathbb{R}^{(m_c+1)\times(m_c+1)}$ and $\mathcal{T}_d \in \mathbb{R}^{(m_d+1)\times(m_d+1)}$ (3.81) and (3.82) hold if and only if $\mathcal{T}_c^T(3.81)\mathcal{T}_c$ and $\mathcal{T}_d^T(3.82)\mathcal{T}_d$ hold. Hence, the equivalence of (3.68)–(3.73) to (3.76) and (3.77) in the case when (3.74) and (3.75) hold follows from the $(1,1)$ blocks of $\mathcal{T}_c^T(3.81)\mathcal{T}_c$ and $\mathcal{T}_d^T(3.82)\mathcal{T}_d$, where

$$\mathcal{T}_c \triangleq \begin{bmatrix} 1 & 0 \\ -\mathcal{C}_c^{-1}(x)\mathcal{B}_c^T(x) & I_{m_c} \end{bmatrix}, \quad \mathcal{T}_d \triangleq \begin{bmatrix} 1 & 0 \\ -\mathcal{C}_d^{-1}(x)\mathcal{B}_d^T(x) & I_{m_d} \end{bmatrix}.$$

□

The structural constraint (3.67) on the system storage function is similar to the structural constraint invoked in standard discrete-time nonlinear passivity theory [37, 38, 40, 103, 104]. This of course is not surprising since impulsive dynamical systems involve a hybrid formulation of continuous-time and discrete-time dynamics. In the case where $u_d = 0$, or $\mathcal{G}$ is lossless with respect to a quadratic supply rate, or $\mathcal{G}$ is dissipative with respect to a quadratic supply rate of the form $(s_c, 0)$, condition (3.67) is necessary and sufficient (see Theorems 3.7 and 3.9), and hence, is automatically satisfied. Similarly, in the case where $\mathcal{G}$ is linear and dissipative with respect to a quadratic supply rate, condition (3.67) is also necessary and sufficient (see Theorem 3.11). In general, however, it is extremely difficult, if not impossible, to obtain (algebraic) sufficient conditions for dissipativity with respect to quadratic hybrid supply rates for impulsive dynamical systems without the structural constraint (3.67). Similar remarks hold for discrete-time nonlinear systems.

Note that it follows from (3.52) that if the conditions in Theorem 3.6 are satisfied with (3.68) replaced by

$$0 = V_s'(x)f_c(x) + \varepsilon V_s(x) - h_c^T(x)Q_c h_c(x) + L_c^T(x)L_c(x), \quad x \in \mathbb{R}^n, \tag{3.83}$$

where $\varepsilon > 0$, then the nonlinear impulsive dynamical system $\mathcal{G}$ is exponentially dissipative. Similar remarks hold for Corollaries 3.2 and 3.3 below.

Using (3.68)–(3.73) it follows that, for $\hat{t} \geq t \geq 0$ and $k \in \mathbb{Z}_{[t,\hat{t})}$,

$$\int_t^{\hat{t}} s_c(u_c(s), y_c(s))ds + \sum_{k \in \mathbb{Z}_{[t,\hat{t})}} s_d(u_d(t_k), y_d(t_k))$$
$$= V_s(x(\hat{t})) - V_s(x(t))$$
$$+ \int_t^{\hat{t}} [L_c(x(s)) + \mathcal{W}_c(x(s))u_c(s)]^T$$
$$\cdot [L_c(x(s)) + \mathcal{W}_c(x(s))u_c(s)]ds$$
$$+ \sum_{k \in \mathbb{Z}_{[t,\hat{t})}} [L_d(x(t_k)) + \mathcal{W}_d(x(t_k))u_d(t_k)]^T$$
$$\cdot [L_d(x(t_k)) + \mathcal{W}_d(x(t_k))u_d(t_k)], \tag{3.84}$$

which can be interpreted as a *generalized energy* balance equation, where $V_s(x(\hat{t})) - V_s(x(t))$ is the stored or accumulated generalized energy of the impulsive dynamical system; the second path-dependent term on the right corresponds to the dissipated energy of the impulsive dynamical system over the continuous-time dynamics; and the third discrete term on the right corresponds to the dissipated energy at the resetting instants.

Equivalently, it follows from Theorem 3.2 that (3.84) can be rewritten as

$$\dot{V}_s(x(t)) = s_c(u_c(t), y_c(t)) - [L_c(x(t)) + \mathcal{W}_c(x(t))u_c(t)]^T$$
$$\cdot [L_c(x(t)) + \mathcal{W}_c(x(t))u_c(t)], \quad t_k < t \leq t_{k+1}, \tag{3.85}$$

$$\Delta V_{\mathrm{s}}(x(t_k)) = s_{\mathrm{d}}(u_{\mathrm{d}}(t_k), y_{\mathrm{d}}(t_k)) - [L_{\mathrm{d}}(x(t_k)) + \mathcal{W}_{\mathrm{d}}(x(t_k))u_{\mathrm{d}}(t_k)]^{\mathrm{T}}$$
$$\cdot [L_{\mathrm{d}}(x(t_k)) + \mathcal{W}_{\mathrm{d}}(x(t_k))u_{\mathrm{d}}(t_k)], \quad k \in \overline{\mathbb{Z}}_+, \tag{3.86}$$

which yields a set of generalized energy conservation equations. Specifically, (3.85) shows that the rate of change in generalized energy, or generalized power, over the time interval $t \in (t_k, t_{k+1}]$ is equal to the generalized system power input minus the internal generalized system power dissipated; (3.86) shows that the change of energy at the resetting times t_k, $k \in \overline{\mathbb{Z}}_+$, is equal to the external generalized system energy at the resetting times minus the generalized dissipated energy at the resetting times.

If $\mathcal{G}$, with $(u_{\mathrm{c}}(t), u_{\mathrm{d}}(t_k)) \equiv (0,0)$ and a continuously differentiable positive-definite, radially unbounded storage function, is dissipative with respect to a quadratic hybrid supply rate, and $Q_{\mathrm{c}} \leq 0$ and $Q_{\mathrm{d}} \leq 0$, then it follows that $\dot{V}_{\mathrm{s}}(x(t)) \leq y_{\mathrm{c}}^{\mathrm{T}}(t)Q_{\mathrm{c}}y_{\mathrm{c}}(t) \leq 0$, $t \geq 0$, and $\Delta V_{\mathrm{s}}(x(t_k)) \leq y_{\mathrm{d}}^{\mathrm{T}}(t_k)Q_{\mathrm{d}}y_{\mathrm{d}}(t_k) \leq 0$, $k \in \overline{\mathbb{Z}}_+$. Hence, the undisturbed $((u_{\mathrm{c}}(t), u_{\mathrm{d}}(t_k)) \equiv (0,0))$ nonlinear impulsive dynamical system (3.14)–(3.17) is Lyapunov stable. Alternatively, if $\mathcal{G}$, with $(u_{\mathrm{c}}(t), u_{\mathrm{d}}(t_k)) \equiv (0,0)$ and a continuously differentiable positive-definite, radially unbounded storage function, is exponentially dissipative with respect to a quadratic hybrid supply rate, and $Q_{\mathrm{c}} \leq 0$ and $Q_{\mathrm{d}} \leq 0$, then it follows that $\dot{V}_{\mathrm{s}}(x(t)) \leq -\varepsilon V_{\mathrm{s}}(x(t)) + y_{\mathrm{c}}^{\mathrm{T}}(t)Q_{\mathrm{c}}y_{\mathrm{c}}(t) \leq -\varepsilon V_{\mathrm{s}}(x(t))$, $t \geq 0$, and $\Delta V_{\mathrm{s}}(x(t_k)) \leq y_{\mathrm{d}}^{\mathrm{T}}(t_k)Q_{\mathrm{d}}y_{\mathrm{d}}(t_k) \leq 0$, $k \in \overline{\mathbb{Z}}_+$. Hence, the undisturbed nonlinear impulsive dynamical system (3.14)–(3.17) is asymptotically stable. If, in addition, there exist constants $\alpha, \beta > 0$ and $p \geq 1$ such that $\alpha \|x\|^p \leq V_{\mathrm{s}}(x) \leq \beta \|x\|^p$, $x \in \mathbb{R}^n$, then the undisturbed nonlinear impulsive dynamical system (3.14)–(3.17) is exponentially stable.

Next, we provide necessary and sufficient conditions for the case where $\mathcal{G}$ given by (3.14)–(3.17) is lossless with respect to a quadratic hybrid supply rate $(s_{\mathrm{c}}, s_{\mathrm{d}})$.

Theorem 3.7 *Let $Q_{\mathrm{c}} \in \mathbb{S}^{l_c}$, $S_{\mathrm{c}} \in \mathbb{R}^{l_c \times m_c}$, $R_{\mathrm{c}} \in \mathbb{S}^{m_c}$, $Q_{\mathrm{d}} \in \mathbb{S}^{l_d}$, $S_{\mathrm{d}} \in \mathbb{R}^{l_d \times m_d}$, and $R_{\mathrm{d}} \in \mathbb{S}^{m_d}$. Then the nonlinear impulsive system $\mathcal{G}$ given by (3.14)–(3.17) is lossless with respect to the quadratic hybrid supply rate $(s_{\mathrm{c}}(u_{\mathrm{c}}, y_{\mathrm{c}}), s_{\mathrm{d}}(u_{\mathrm{d}}, y_{\mathrm{d}})) = (y_{\mathrm{c}}^{\mathrm{T}}Q_{\mathrm{c}}y_{\mathrm{c}} + 2y_{\mathrm{c}}^{\mathrm{T}}S_{\mathrm{c}}u_{\mathrm{c}} + u_{\mathrm{c}}^{\mathrm{T}}R_{\mathrm{c}}u_{\mathrm{c}}, y_{\mathrm{d}}^{\mathrm{T}}Q_{\mathrm{d}}y_{\mathrm{d}} + 2y_{\mathrm{d}}^{\mathrm{T}}S_{\mathrm{d}}u_{\mathrm{d}} + u_{\mathrm{d}}^{\mathrm{T}}R_{\mathrm{d}}u_{\mathrm{d}})$ if and only if there exist functions $V_{\mathrm{s}} : \mathbb{R}^n \to \mathbb{R}$, $P_{1u_{\mathrm{d}}} : \mathbb{R}^n \to \mathbb{R}^{1 \times m_d}$, and $P_{2u_{\mathrm{d}}} : \mathbb{R}^n \to \mathbb{N}^{m_d}$ such that $V_{\mathrm{s}}(\cdot)$ is continuously differentiable and positive definite, $V_{\mathrm{s}}(0) = 0$, and, for all $x \in \mathbb{R}^n$, (3.67) holds and*

$$0 = V_{\mathrm{s}}'(x)f_{\mathrm{c}}(x) - h_{\mathrm{c}}^{\mathrm{T}}(x)Q_{\mathrm{c}}h_{\mathrm{c}}(x), \tag{3.87}$$
$$0 = \tfrac{1}{2}V_{\mathrm{s}}'(x)G_{\mathrm{c}}(x) - h_{\mathrm{c}}^{\mathrm{T}}(x)(Q_{\mathrm{c}}J_{\mathrm{c}}(x) + S_{\mathrm{c}}), \tag{3.88}$$

$$0 = R_c + S_c^T J_c(x) + J_c^T(x) S_c + J_c^T(x) Q_c J_c(x), \tag{3.89}$$
$$0 = V_s(x + f_d(x)) - V_s(x) - h_d^T(x) Q_d h_d(x), \tag{3.90}$$
$$0 = \tfrac{1}{2} P_{1u_d}(x) - h_d^T(x)(Q_d J_d(x) + S_d), \tag{3.91}$$
$$0 = R_d + S_d^T J_d(x) + J_d^T(x) S_d + J_d^T(x) Q_d J_d(x) - P_{2u_d}(x). \tag{3.92}$$

If, in addition, $V_s(\cdot)$ is two-times continuously differentiable, then

$$P_{1u_d}(x) = V_s'(x + f_d(x)) G_d(x),$$
$$P_{2u_d}(x) = \tfrac{1}{2} G_d^T(x) V_s''(x + f_d(x)) G_d(x).$$

Proof. Sufficiency follows as in the proof of Theorem 3.6. To show necessity, suppose that the nonlinear impulsive dynamical system $\mathcal{G}$ is lossless with respect to the quadratic hybrid supply rate (s_c, s_d). Then, it follows from Theorem 3.2 that for all $k \in \overline{\mathbb{Z}}_+$,

$$V_s(x(\hat{t})) - V_s(x(t)) = \int_t^{\hat{t}} s_c(u_c(s), y_c(s)) ds, \quad t_k < t \le \hat{t} \le t_{k+1}, \tag{3.93}$$

and

$$V_s(x(t_k) + f_d(x(t_k)) + G_d(x(t_k)) u_d(t_k)) = V_s(x(t_k)) + s_d(u_d(t_k), y_d(t_k)). \tag{3.94}$$

Now, dividing (3.93) by $\hat{t} - t^+$ and letting $\hat{t} \to t^+$, (3.93) is equivalent to

$$\dot{V}_s(x(t)) = V_s'(x(t))[f_c(x(t)) + G_c(x(t)) u_c(t)] = s_c(u_c(t), y_c(t)),$$
$$t_k < t \le t_{k+1}. \tag{3.95}$$

Next, with $t = 0$, it follows from (3.95) that

$$V_s'(x_0)[f_c(x_0) + G_c(x_0) u_c(0)] = s_c(u_c(0), y_c(0)), \quad x_0 \in \mathbb{R}^n,$$
$$u_c(0) \in \mathbb{R}^{m_c}. \tag{3.96}$$

Since $x_0 \in \mathbb{R}^n$ is arbitrary, it follows that

$$\begin{aligned}
V_s'(x)[f_c(x) + G_c(x) u_c] &= y_c^T Q_c y_c + 2 y_c^T S_c u_c + u_c^T R_c u_c \\
&= h_c^T(x) Q_c h_c(x) + 2 h_c^T(x)(Q_c J_c(x) + S_c) u_c \\
&\quad + u_c^T(R_c + S_c^T J_c(x) + J_c^T(x) S_c \\
&\quad + J_c^T(x) Q_c J_c(x)) u_c, \quad x \in \mathbb{R}^n, \quad u_c \in \mathbb{R}^{m_c}.
\end{aligned}$$

Now, equating coefficients of equal powers yields (3.87)–(3.89).

Next, it follows from (3.94) with $k = 1$ that

$$V_s(x(t_1) + f_d(x(t_1)) + G_d(x(t_1))u_d(t_1)) = V_s(x(t_1))$$
$$+ s_d(u_d(t_1), y_d(t_1)). \quad (3.97)$$

Now, since the continuous-time dynamics (3.14) are Lipschitz continuous on $\mathcal{D}$, it follows that for arbitrary $x \in \mathbb{R}^n$ there exists $x_0 \in \mathbb{R}^n$ such that $x(t_1) = x$. Hence, it follows from (3.97) that

$$V_s(x + f_d(x) + G_d(x)u_d) = V_s(x) + y_d^T Q_d y_d + 2y_d^T S_d u_d + u_d^T R_d u_d$$
$$= V_s(x) + h_d^T(x)Q_d h_d(x)$$
$$+ 2h_d^T(x)(Q_d J_d(x) + S_d)u_d$$
$$+ u_d^T(R_d + S_d^T J_d(x) + J_d^T(x)S_d$$
$$+ J_d^T(x)Q_d J_d(x))u_d, \quad x \in \mathbb{R}^n, \ u_d \in \mathbb{R}^{m_d}.$$
$$(3.98)$$

Since the right-hand side of (3.98) is quadratic in u_d it follows that $V_s(x + f_d(x) + G_d(x)u_d)$ is quadratic in u_d, and hence, there exists $P_{1u_d} : \mathbb{R}^n \to \mathbb{R}^{1 \times m_d}$ and $P_{2u_d} : \mathbb{R}^n \to \mathbb{N}^{m_d}$ such that

$$V_s(x + f_d(x) + G_d(x)u_d) = V_s(x + f_d(x)) + P_{1u_d}(x)u_d$$
$$+ u_d^T P_{2u_d}(x)u_d. \quad (3.99)$$

Now, using (3.99) and equating coefficients of equal powers in (3.98) yields (3.90)–(3.92).

Finally, if $V_s(\cdot)$ is two-times continuously differentiable, applying a Taylor series expansion on (3.99) about $u_d = 0$ yields

$$P_{1u_d}(x) = \left. \frac{\partial V_s(x + f_d(x) + G_d(x)u_d)}{\partial u_d} \right|_{u_d=0} = V_s'(x + f_d(x))G_d(x),$$
$$(3.100)$$

$$P_{2u_d}(x) = \left. \frac{\partial^2 V_s(x + f_d(x) + G_d(x)u_d)}{\partial u_d^2} \right|_{u_d=0} = \tfrac{1}{2}G_d^T(x)V_s''(x$$
$$+ f_d(x))G_d(x), \quad (3.101)$$

which proves the result. $\qquad \square$

The following result presents the state-dependent analog of Theorem 3.6.

Theorem 3.8 Let $Q_c \in \mathbb{S}^{l_c}$, $S_c \in \mathbb{R}^{l_c \times m_c}$, $R_c \in \mathbb{S}^{m_c}$, $Q_d \in \mathbb{S}^{l_d}$, $S_d \in \mathbb{R}^{l_d \times m_d}$, and $R_d \in \mathbb{S}^{m_d}$. If there exist functions $V_s : \mathbb{R}^n \to \mathbb{R}$,

$L_c : \mathbb{R}^n \to \mathbb{R}^{p_c}$, $L_d : \mathbb{R}^n \to \mathbb{R}^{p_d}$, $\mathcal{W}_c : \mathbb{R}^n \to \mathbb{R}^{p_c \times m_c}$, $\mathcal{W}_d : \mathbb{R}^n \to \mathbb{R}^{p_d \times m_d}$, $P_{1u_d} : \mathbb{R}^n \to \mathbb{R}^{1 \times m_d}$, and $P_{2u_d} : \mathbb{R}^n \to \mathbb{N}^{m_d}$ such that $V_s(\cdot)$ is continuously differentiable and positive definite, $V_s(0) = 0$,

$$V_s(x + f_d(x) + G_d(x)u_d) = V_s(x + f_d(x)) + P_{1u_d}(x)u_d + u_d^T P_{2u_d}(x)u_d,$$
$$x \in \mathcal{Z}_x, \ u_d \in \mathbb{R}^{m_d}, (3.102)$$

and

$$0 = V_s'(x)f_c(x) - h_c^T(x)Q_c h_c(x) + L_c^T(x)L_c(x), \quad x \notin \mathcal{Z}_x, \qquad (3.103)$$
$$0 = \tfrac{1}{2}V_s'(x)G_c(x) - h_c^T(x)(Q_c J_c(x) + S_c) + L_c^T(x)\mathcal{W}_c(x), \quad x \notin \mathcal{Z}_x,$$
$$(3.104)$$
$$0 = R_c + S_c^T J_c(x) + J_c^T(x)S_c + J_c^T(x)Q_c J_c(x) - \mathcal{W}_c^T(x)\mathcal{W}_c(x),$$
$$x \notin \mathcal{Z}_x, \qquad (3.105)$$
$$0 = V_s(x + f_d(x)) - V_s(x) - h_d^T(x)Q_d h_d(x) + L_d^T(x)L_d(x), \quad x \in \mathcal{Z}_x,$$
$$(3.106)$$
$$0 = \tfrac{1}{2}P_{1u_d}(x) - h_d^T(x)(Q_d J_d(x) + S_d) + L_d^T(x)\mathcal{W}_d(x), \quad x \in \mathcal{Z}_x,$$
$$(3.107)$$
$$0 = R_d + S_d^T J_d(x) + J_d^T(x)S_d + J_d^T(x)Q_d J_d(x) - P_{2u_d}(x)$$
$$- \mathcal{W}_d^T(x)\mathcal{W}_d(x), \quad x \in \mathcal{Z}_x, \qquad (3.108)$$

then the nonlinear impulsive system $\mathcal{G}$ given by (3.27)–(3.30) is dissipative with respect to the quadratic hybrid supply rate $(s_c(u_c, y_c), s_d(u_d, y_d)) = (y_c^T Q_c y_c + 2y_c^T S_c u_c + u_c^T R_c u_c, \ y_d^T Q_d y_d + 2y_d^T S_d u_d + u_d^T R_d u_d)$. If, alternatively,

$$\mathcal{Z}_c(x) \triangleq R_c + S_c^T J_c(x) + J_c^T(x)S_c + J_c^T(x)Q_c J_c(x) > 0, \ x \notin \mathcal{Z}_x,$$
$$(3.109)$$

and there exist a continuously differentiable function $V_s : \mathbb{R}^n \to \mathbb{R}$ and matrix functions $P_{1u_d} : \mathbb{R}^n \to \mathbb{R}^{1 \times m_d}$ and $P_{2u_d} : \mathbb{R}^n \to \mathbb{N}^{m_d}$ such that $V_s(\cdot)$ is positive definite, $V_s(0) = 0$, (3.102) holds, and

$$\mathcal{Z}_d(x) \triangleq R_d + S_d^T J_d(x) + J_d^T(x)S_d + J_d^T(x)Q_d J_d(x) - P_{2u_d}(x) > 0,$$
$$x \in \mathcal{Z}_x, \qquad (3.110)$$

$$0 \geq V_s'(x)f_c(x) - h_c^T(x)Q_c h_c(x)$$
$$+ [\tfrac{1}{2}V_s'(x)G_c(x) - h_c^T(x)(Q_c J_c(x) + S_c)]$$
$$\cdot \mathcal{Z}_c^{-1}(x)[\tfrac{1}{2}V_s'(x)G_c(x) - h_c^T(x)(Q_c J_c(x) + S_c)]^T, \quad x \notin \mathcal{Z}_x,$$
$$(3.111)$$

$$0 \geq V_s(x + f_d(x)) - V_s(x) - h_d^T(x)Q_d h_d(x)$$

$$+[\tfrac{1}{2}P_{1u_{\mathrm d}}(x) - h_{\mathrm d}^{\mathrm T}(x)(Q_{\mathrm d}J_{\mathrm d}(x) + S_{\mathrm d})]$$
$$\cdot\mathcal{Z}_{\mathrm d}^{-1}(x)[\tfrac{1}{2}P_{1u_{\mathrm d}}(x) - h_{\mathrm d}^{\mathrm T}(x)(Q_{\mathrm d}J_{\mathrm d}(x) + S_{\mathrm d})]^{\mathrm T}, \quad x \in \mathcal{Z}_x,$$

$$(3.112)$$

then $\mathcal{G}$ is dissipative with respect to the quadratic hybrid supply rate $(s_{\mathrm c}(u_{\mathrm c}, y_{\mathrm c}), s_{\mathrm d}(u_{\mathrm d}, y_{\mathrm d})) = (y_{\mathrm c}^{\mathrm T}Q_{\mathrm c}y_{\mathrm c} + 2y_{\mathrm c}^{\mathrm T}S_{\mathrm c}u_{\mathrm c} + u_{\mathrm c}^{\mathrm T}R_{\mathrm c}u_{\mathrm c}, y_{\mathrm d}^{\mathrm T}Q_{\mathrm d}y_{\mathrm d} + 2y_{\mathrm d}^{\mathrm T}S_{\mathrm d}u_{\mathrm d} + u_{\mathrm d}^{\mathrm T}R_{\mathrm d}u_{\mathrm d})$.

Proof. The proof is similar to the proof of Theorem 3.6. $\square$

Next, we provide two definitions of nonlinear impulsive dynamical systems which are dissipative (respectively, exponentially dissipative) with respect to hybrid supply rates of a specific form.

Definition 3.5 *An impulsive dynamical system $\mathcal{G}$ of the form (3.7)–(3.10) with $m_{\mathrm c} = l_{\mathrm c}$ and $m_{\mathrm d} = l_{\mathrm d}$ is* passive *(respectively,* exponentially passive*) if $\mathcal{G}$ is dissipative (respectively, exponentially dissipative) with respect to the hybrid supply rate $(s_{\mathrm c}(u_{\mathrm c}, y_{\mathrm c}), s_{\mathrm d}(u_{\mathrm d}, y_{\mathrm d})) = (2u_{\mathrm c}^{\mathrm T}y_{\mathrm c}, 2u_{\mathrm d}^{\mathrm T}y_{\mathrm d})$.*

Definition 3.6 *An impulsive dynamical system $\mathcal{G}$ of the form (3.7)–(3.10) is* nonexpansive *(respectively,* exponentially nonexpansive*) if $\mathcal{G}$ is dissipative (respectively, exponentially dissipative) with respect to the hybrid supply rate $(s_{\mathrm c}(u_{\mathrm c}, y_{\mathrm c}), s_{\mathrm d}(u_{\mathrm d}, y_{\mathrm d})) = (\gamma_{\mathrm c}^2 u_{\mathrm c}^{\mathrm T}u_{\mathrm c} - y_{\mathrm c}^{\mathrm T}y_{\mathrm c}, \gamma_{\mathrm d}^2 u_{\mathrm d}^{\mathrm T}u_{\mathrm d} - y_{\mathrm d}^{\mathrm T}y_{\mathrm d})$, where $\gamma_{\mathrm c}, \gamma_{\mathrm d} > 0$ are given.*

A mixed passive-nonexpansive formulation of $\mathcal{G}$ can also be considered. Specifically, one can consider impulsive dynamical systems $\mathcal{G}$ which are dissipative with respect to hybrid supply rates of the form $(s_{\mathrm c}(u_{\mathrm c}, y_{\mathrm c}), s_{\mathrm d}(u_{\mathrm d}, y_{\mathrm d})) = (2u_{\mathrm c}^{\mathrm T}y_{\mathrm c}, \gamma_{\mathrm d}^2 u_{\mathrm d}^{\mathrm T}u_{\mathrm d} - y_{\mathrm d}^{\mathrm T}y_{\mathrm d})$, where $\gamma_{\mathrm d} > 0$, and vice versa. Furthermore, supply rates for input strict passivity, output strict passivity, and input-output strict passivity can also be considered [74]. However, for simplicity of exposition we do not do so here.

The following results present the nonlinear versions of the Kalman-Yakubovich-Popov positive real lemma and the bounded real lemma for nonlinear impulsive systems $\mathcal{G}$ of the form (3.14)–(3.17).

Corollary 3.2 *Consider the nonlinear impulsive system $\mathcal{G}$ given by (3.14)–(3.17). If there exist functions $V_{\mathrm s} : \mathbb{R}^n \to \mathbb{R}$, $L_{\mathrm c} : \mathbb{R}^n \to \mathbb{R}^{p_{\mathrm c}}$, $L_{\mathrm d} : \mathbb{R}^n \to \mathbb{R}^{p_{\mathrm d}}$, $\mathcal{W}_{\mathrm c} : \mathbb{R}^n \to \mathbb{R}^{p_{\mathrm c} \times m_{\mathrm c}}$, $\mathcal{W}_{\mathrm d} : \mathbb{R}^n \to \mathbb{R}^{p_{\mathrm d} \times m_{\mathrm d}}$, $P_{1u_{\mathrm d}}$:*

$\mathbb{R}^n \to \mathbb{R}^{1 \times m_d}$, and $P_{2u_d} : \mathbb{R}^n \to \mathbb{N}^{m_d}$ such that $V_s(\cdot)$ is continuously differentiable and positive definite, $V_s(0) = 0$,

$$V_s(x + f_d(x) + G_d(x)u_d) = V_s(x + f_d(x)) + P_{1u_d}(x)u_d + u_d^T P_{2u_d}(x)u_d,$$
$$x \in \mathbb{R}^n, \quad u_d \in \mathbb{R}^{m_d}, \tag{3.113}$$

and, for all $x \in \mathbb{R}^n$,

$$0 = V_s'(x)f_c(x) + L_c^T(x)L_c(x), \tag{3.114}$$
$$0 = \tfrac{1}{2}V_s'(x)G_c(x) - h_c^T(x) + L_c^T(x)W_c(x), \tag{3.115}$$
$$0 = J_c(x) + J_c^T(x) - W_c^T(x)W_c(x), \tag{3.116}$$
$$0 = V_s(x + f_d(x)) - V_s(x) + L_d^T(x)L_d(x), \tag{3.117}$$
$$0 = \tfrac{1}{2}P_{1u_d}(x) - h_d^T(x) + L_d^T(x)W_d(x), \tag{3.118}$$
$$0 = J_d(x) + J_d^T(x) - P_{2u_d}(x) - W_d^T(x)W_d(x), \tag{3.119}$$

then $\mathcal{G}$ is passive. If, alternatively, $J_c(x) + J_c^T(x) > 0$, $x \in \mathbb{R}^n$, and there exist a continuously differentiable function $V_s : \mathbb{R}^n \to \mathbb{R}$ and matrix functions $P_{1u_d} : \mathbb{R}^n \to \mathbb{R}^{1 \times m_d}$ and $P_{2u_d} : \mathbb{R}^n \to \mathbb{N}^{m_d}$ such that $V_s(\cdot)$ is positive definite, $V_s(0) = 0$, (3.113) holds, and for all $x \in \mathbb{R}^n$,

$$0 < J_d(x) + J_d^T(x) - P_{2u_d}(x), \tag{3.120}$$
$$0 \geq V_s'(x)f_c(x) + [\tfrac{1}{2}V_s'(x)G_c(x) - h_c^T(x)]$$
$$\cdot [J_c(x) + J_c^T(x)]^{-1}[\tfrac{1}{2}V_s'(x)G_c(x) - h_c^T(x)]^T, \tag{3.121}$$
$$0 \geq V_s(x + f_d(x)) - V_s(x) + [\tfrac{1}{2}P_{1u_d}(x) - h_d^T(x)]$$
$$\cdot [J_d(x) + J_d^T(x) - P_{2u_d}(x)]^{-1}[\tfrac{1}{2}P_{1u_d}(x) - h_d^T(x)]^T, \tag{3.122}$$

then $\mathcal{G}$ is passive.

Proof. The result is a direct consequence of Theorem 3.6 with $l_c = m_c$, $l_d = m_d$, $Q_c = 0$, $Q_d = 0$, $S_c = I_{m_c}$, $S_d = I_{m_d}$, $R_c = 0$, and $R_d = 0$. Specifically, with $\kappa_c(y_c) = -y_c$ and $\kappa_d(y_d) = -y_d$, it follows that $s_c(\kappa_c(y_c), y_c) = -2y_c^T y_c < 0$, $y_c \neq 0$, and $s_d(\kappa_d(y_d), y_d) = -2y_d^T y_d < 0$, $y_d \neq 0$, so that all of the assumptions of Theorem 3.6 are satisfied.   □

Corollary 3.3 *Consider the nonlinear impulsive system $\mathcal{G}$ given by (3.14)–(3.17). If there exist functions $V_s : \mathbb{R}^n \to \mathbb{R}$, $L_c : \mathbb{R}^n \to \mathbb{R}^{p_c}$, $L_d : \mathbb{R}^n \to \mathbb{R}^{p_d}$, $W_c : \mathbb{R}^n \to \mathbb{R}^{p_c \times m_c}$, $W_d : \mathbb{R}^n \to \mathbb{R}^{p_d \times m_d}$, P_{1u_d} :*

$\mathbb{R}^n \to \mathbb{R}^{1 \times m_d}$, and $P_{2u_d} : \mathbb{R}^n \to \mathbb{N}^{m_d}$ such that $V_s(\cdot)$ is continuously differentiable and positive definite, $V_s(0) = 0$,

$$
\begin{aligned}
V_s(x + f_d(x) + G_d(x)u_d) &= V_s(x + f_d(x)) + P_{1u_d}(x)u_d \\
&\quad + u_d^T P_{2u_d}(x)u_d, \quad x \in \mathbb{R}^n, \quad u_d \in \mathbb{R}^{m_d},
\end{aligned}
\tag{3.123}
$$

and, for all $x \in \mathbb{R}^n$,

$$
\begin{aligned}
0 &= V_s'(x)f_c(x) + h_c^T(x)h_c(x) + L_c^T(x)L_c(x), & (3.124) \\
0 &= \tfrac{1}{2}V_s'(x)G_c(x) + h_c^T(x)J_c(x) + L_c^T(x)W_c(x), & (3.125) \\
0 &= \gamma_c^2 I_{m_c} - J_c^T(x)J_c(x) - W_c^T(x)W_c(x), & (3.126) \\
0 &= V_s(x + f_d(x)) - V_s(x) + h_d^T(x)h_d(x) + L_d^T(x)L_d(x), & (3.127) \\
0 &= \tfrac{1}{2}P_{1u_d}(x) + h_d^T(x)J_d(x) + L_d^T(x)W_d(x), & (3.128) \\
0 &= \gamma_d^2 I_{m_d} - J_d^T(x)J_d(x) - P_{2u_d}(x) - W_d^T(x)W_d(x), & (3.129)
\end{aligned}
$$

then $\mathcal{G}$ is nonexpansive. If, alternatively, $\gamma_c^2 I_{m_c} - J_c^T(x)J_c(x) > 0$, $x \in \mathbb{R}^n$, and there exist a continuously differentiable function $V_s : \mathbb{R}^n \to \mathbb{R}$ and matrix functions $P_{1u_d} : \mathbb{R}^n \to \mathbb{R}^{1 \times m_d}$ and $P_{2u_d} : \mathbb{R}^n \to \mathbb{N}^{m_d}$ such that $V_s(\cdot)$ is positive definite, $V_s(0) = 0$, (3.123) holds, and for all $x \in \mathbb{R}^n$,

$$
0 < \gamma_d^2 I_{m_d} - J_d^T(x)J_d(x) - P_{2u_d}(x),
\tag{3.130}
$$

$$
\begin{aligned}
0 &\geq V_s'(x)f_c(x) + h_c^T(x)h_c(x) + [\tfrac{1}{2}V_s'(x)G_c(x) + h_c^T(x)J_c(x)] \\
&\quad \cdot [\gamma_c^2 I_{m_c} - J_c^T(x)J_c(x)]^{-1}[\tfrac{1}{2}V_s'(x)G_c(x) + h_c^T(x)J_c(x)]^T, & (3.131)
\end{aligned}
$$

$$
\begin{aligned}
0 &\geq V_s(x + f_d(x)) - V_s(x) + h_d^T(x)h_d(x) + [\tfrac{1}{2}P_{1u_d}(x) + h_d^T(x)J_d(x)] \\
&\quad \cdot [\gamma_d^2 I_{m_d} - J_d^T(x)J_d(x) - P_{2u_d}(x)]^{-1}[\tfrac{1}{2}P_{1u_d}(x) + h_d^T(x)J_d(x)]^T,
\end{aligned}
\tag{3.132}
$$

then $\mathcal{G}$ is nonexpansive.

Proof. The result is a direct consequence of Theorem 3.6 with $Q_c = -I_{l_c}$, $Q_d = -I_{l_d}$, $S_c = 0$, $S_d = 0$, $R_c = \gamma_c^2 I_{m_c}$, and $R_d = \gamma_d^2 I_{m_d}$. Specifically, with $\kappa_c(y_c) = -\frac{1}{2\gamma_c} y_c$ and $\kappa_d(y_d) = -\frac{1}{2\gamma_d} y_d$, it follows that $s_c(\kappa_c(y_c), y_c) = -\frac{3}{4} y_c^T y_c < 0$, $y_c \neq 0$, and $s_d(\kappa_d(y_d), y_d) = -\frac{3}{4} y_d^T y_d < 0$, $y_d \neq 0$, so that all of the assumptions of Theorem 3.6 are satisfied. $\square$

Corollaries 3.2 and 3.3 also hold for dissipative state-dependent impulsive dynamical systems. In this case, however, $x \in \mathbb{R}^n$ is replaced with $x \notin \mathcal{Z}_x$ for (3.114)–(3.116) (respectively, (3.124)–(3.126)) and $x \in \mathcal{Z}_x$ for (3.113) (respectively, (3.123)) and (3.117)–(3.119) (respectively, (3.127)–(3.129)). Next, we provide necessary *and* sufficient conditions for dissipativity of a nonlinear impulsive dynamical system $\mathcal{G}$ of the form (3.14)–(3.17) in the case where $s_d(u_d, y_d) \equiv 0$ and $G_d(x) \equiv 0$.

Theorem 3.9 *Let $Q_c \in \mathbb{S}^{l_c}$, $S_c \in \mathbb{R}^{l_c \times m_c}$, and $R_c \in \mathbb{S}^{m_c}$. Then the nonlinear impulsive system $\mathcal{G}$ given by (3.14)–(3.17) with $G_d(x) \equiv 0$ is dissipative with respect to the supply rate $(s_c(u_c, y_c), s_d(u_d, y_d)) = (y_c^{\mathrm{T}} Q_c y_c + 2 y_c^{\mathrm{T}} S_c u_c + u_c^{\mathrm{T}} R_c u_c, 0)$ if and only if there exist functions $V_s :$ $\mathbb{R}^n \to \mathbb{R}$, $L_c : \mathbb{R}^n \to \mathbb{R}^{p_c}$, $L_d : \mathbb{R}^n \to \mathbb{R}^{p_d}$, and $\mathcal{W}_c : \mathbb{R}^n \to \mathbb{R}^{p_c \times m_c}$ such that $V_s(\cdot)$ is continuously differentiable and positive definite, $V_s(0) = 0$, and for all $x \in \mathbb{R}^n$,*

$$0 = V_s'(x) f_c(x) - h_c^{\mathrm{T}}(x) Q_c h_c(x) + L_c^{\mathrm{T}}(x) L_c(x), \tag{3.133}$$

$$0 = \tfrac{1}{2} V_s'(x) G_c(x) - h_c^{\mathrm{T}}(x)(Q_c J_c(x) + S_c) + L_c^{\mathrm{T}}(x) \mathcal{W}_c(x), \tag{3.134}$$

$$0 = R_c + S_c^{\mathrm{T}} J_c(x) + J_c^{\mathrm{T}}(x) S_c + J_c^{\mathrm{T}}(x) Q_c J_c(x) - \mathcal{W}_c^{\mathrm{T}}(x) \mathcal{W}_c(x), \tag{3.135}$$

$$0 = V_s(x + f_d(x)) - V_s(x) + L_d^{\mathrm{T}}(x) L_d(x). \tag{3.136}$$

Proof. Sufficiency follows from Theorem 3.6 with $Q_d = 0$, $S_d = 0$, $R_d = 0$, $G_d(x) = 0$, $P_{1u_d}(x) = 0$, and $P_{2u_d}(x) = 0$. Necessity follows from Theorem 3.2 using a similar construction as in the proof of Theorem 3.7. □

Note that in the case where $s_d(u_d, y_d) \equiv 0$ and $G_d(x) \equiv 0$, it follows from Theorem 3.9 that the nonlinear impulsive system $\mathcal{G}$ given by (3.14)–(3.17) is passive (respectively, nonexpansive) if and only if there exist functions $V_s : \mathbb{R}^n \to \mathbb{R}$, $L_c : \mathbb{R}^n \to \mathbb{R}^{p_c}$, $L_d : \mathbb{R}^n \to \mathbb{R}^{p_d}$, and $\mathcal{W}_c : \mathbb{R}^n \to \mathbb{R}^{p_c \times m_c}$ such that $V_s(\cdot)$ is continuously differentiable and positive definite, $V_s(0) = 0$, and (3.114)–(3.116) and (3.136) (respectively, (3.124)–(3.126) and (3.136)) are satisfied.

Finally, we present a key result on linearization of impulsive dynamical systems. For this result, we assume that there exist functions $\kappa_c : \mathbb{R}^{l_c} \to \mathbb{R}^{m_c}$ and $\kappa_d : \mathbb{R}^{l_d} \to \mathbb{R}^{m_d}$ such that $\kappa_c(0) = 0$, $\kappa_d(0) = 0$, $s_c(\kappa_c(y_c), y_c) < 0$, $y_c \neq 0$, $s_d(\kappa_d(y_d), y_d) < 0$, $y_d \neq 0$, and the available storage $V_a(x)$, $x \in \mathbb{R}^n$, is a three-times continuously differentiable function.

Theorem 3.10 *Let $Q_c \in \mathbb{S}^{l_c}$, $S_c \in \mathbb{R}^{l_c \times m_c}$, $R_c \in \mathbb{S}^{m_c}$, $Q_d \in \mathbb{S}^{l_d}$, $S_d \in \mathbb{R}^{l_d \times m_d}$, and $R_d \in \mathbb{S}^{m_d}$, and suppose that the nonlinear impulsive system $\mathcal{G}$ given by (3.14)–(3.17) is dissipative with respect to* *the quadratic hybrid supply rate* $(s_c(u_c, y_c), s_d(u_d, y_d)) = (y_c^T Q_c y_c + 2y_c^T S_c u_c + u_c^T R_c u_c, y_d^T Q_d y_d + 2y_d^T S_d u_d + u_d^T R_d u_d)$. *Then there exist matrices* $P \in \mathbb{R}^{n \times n}$, $L_c \in \mathbb{R}^{p_c \times n}$, $W_c \in \mathbb{R}^{p_c \times m_c}$, $L_d \in \mathbb{R}^{p_d \times n}$, *and* $W_d \in \mathbb{R}^{p_d \times m_d}$, *with* P *nonnegative definite, such that*

$$0 = A_c^T P + P A_c - C_c^T Q_c C_c + L_c^T L_c, \tag{3.137}$$
$$0 = P B_c - C_c^T (Q_c D_c + S_c) + L_c^T W_c, \tag{3.138}$$
$$0 = R_c + S_c^T D_c + D_c^T S_c + D_c^T Q_c D_c - W_c^T W_c, \tag{3.139}$$
$$0 = A_d^T P A_d - P - C_d^T Q_d C_d + L_d^T L_d, \tag{3.140}$$
$$0 = A_d^T P B_d - C_d^T (Q_d D_d + S_d) + L_d^T W_d, \tag{3.141}$$
$$0 = R_d + S_d^T D_d + D_d^T S_d + D_d^T Q_d D_d - B_d^T P B_d - W_d^T W_d, \tag{3.142}$$

where

$$A_c = \left.\frac{\partial f_c}{\partial x}\right|_{x=0}, \quad B_c = G_c(0), \quad C_c = \left.\frac{\partial h_c}{\partial x}\right|_{x=0}, \quad D_c = J_c(0), \tag{3.143}$$

$$A_d = \left.\frac{\partial f_d}{\partial x}\right|_{x=0} + I_n, \quad B_d = G_d(0), \quad C_d = \left.\frac{\partial h_d}{\partial x}\right|_{x=0}, \quad D_d = J_d(0). \tag{3.144}$$

If, in addition, (A_c, C_c) *and* (A_d, C_d) *are observable, then* $P > 0$.

Proof. First note that since $\mathcal{G}$ is dissipative with respect to the hybrid supply rate (s_c, s_d) it follows from Theorem 3.2 that there exists a storage function $V_s : \mathbb{R}^n \to \mathbb{R}$ such that, for all $k \in \overline{\mathbb{Z}}_+$,

$$V_s(x(\hat{t})) - V_s(x(t)) \leq \int_t^{\hat{t}} s_c(u_c(s), y_c(s)) ds, \quad t_k < t \leq \hat{t} \leq t_{k+1}, \tag{3.145}$$

and

$$V_s(x(t_k)) + f_d(x(t_k)) + G_d(x(t_k)) u_d(t_k)) \leq V_s(x(t_k)) + s_d(u_d(t_k), y_d(t_k)). \tag{3.146}$$

Now, dividing (3.145) by $\hat{t} - t^+$ and letting $\hat{t} \to t^+$, (3.145) is equivalent to

$$\dot{V}_s(x(t)) = V_s'(x(t))[f_c(x(t)) + G_c(x(t)) u_c(t)] \leq s_c(u_c(t), y_c(t)), \quad t_k < t \leq t_{k+1}. \tag{3.147}$$

Next, with $t = 0$, it follows that

$$V_s'(x_0)[f_c(x_0) + G_c(x_0) u_c(0)] \leq s_c(u_c(0), y_c(0)), \quad x_0 \in \mathbb{R}^n, \quad u_c(0) \in \mathbb{R}^{m_c}. \tag{3.148}$$

Since $x_0 \in \mathbb{R}^n$ is arbitrary, it follows that

$$V_s'(x)[f_c(x) + G_c(x)u_c] \le s_c(u_c, h_c(x) + J_c(x)u_c), \ x \in \mathbb{R}^n, \ u_c \in \mathbb{R}^{m_c}. \tag{3.149}$$

Furthermore, it follows from (3.146) with $k = 1$ that

$$V_s(x(t_1) + f_d(x(t_1)) + G_d(x(t_1))u_d(t_1)) \le V_s(x(t_1)) + s_d(u_d(t_1), y_d(t_1)). \tag{3.150}$$

Now, since the continuous-time dynamics (3.14) are Lipschitz, it follows that for arbitrary $x \in \mathbb{R}^n$ there exists $x_0 \in \mathbb{R}^n$ such that $x(t_1) = x$. Hence, it follows from (3.150) that

$$V_s(x + f_d(x) + G_d(x)u_d) \le V_s(x) + s_d(u_d, h_d(x) + J_d(x)u_d),$$
$$x \in \mathbb{R}^n, \quad u_d \in \mathbb{R}^{m_d}. \tag{3.151}$$

Next, it follows from (3.149) and (3.151) that there exist smooth functions $d_c : \mathbb{R}^n \times \mathbb{R}^{m_c} \to \mathbb{R}$ and $d_d : \mathbb{R}^n \times \mathbb{R}^{m_d} \to \mathbb{R}$ such that $d_c(x, u_c) \ge 0$, $d_c(0,0) = 0$, $d_d(x, u_d) \ge 0$, $d_d(0,0) = 0$, and

$$0 = V_s'(x)[f_c(x) + G_c(x)u_c] - s_c(u_c, h_c(x) + J_c(x)u_c) + d_c(x, u_c),$$
$$x \in \mathbb{R}^n, \quad u_c \in \mathbb{R}^{m_c}, \tag{3.152}$$
$$0 = V_s(x + f_d(x) + G_d(x)u_d) - V_s(x) - s_d(u_d, h_d(x) + J_d(x)u_d)$$
$$+ d_d(x, u_d), \quad x \in \mathbb{R}^n, \quad u_d \in \mathbb{R}^{m_d}. \tag{3.153}$$

Now, expanding $V_s(\cdot)$, $d_c(\cdot, \cdot)$, and $d_d(\cdot, \cdot)$ via a Taylor series expansion about $x = 0$, $u_c = 0$, and $u_d = 0$, and using the fact that $V_s(\cdot)$, $d_c(\cdot, \cdot)$, and $d_d(\cdot, \cdot)$ are nonnegative definite and $V_s(0) = 0$, $d_c(0,0) = 0$, and $d_d(0,0) = 0$, it follows that there exist matrices $P \in \mathbb{R}^{n \times n}$, $L_c \in \mathbb{R}^{p_c \times n}$, $W_c \in \mathbb{R}^{p_c \times m_c}$, $L_d \in \mathbb{R}^{p_d \times n}$, and $W_d \in \mathbb{R}^{p_d \times m_d}$, with P nonnegative definite, such that

$$V_s(x) = x^{\mathrm{T}} P x + V_r(x), \tag{3.154}$$
$$d_c(x, u_c) = (L_c x + W_c u_c)^{\mathrm{T}} (L_c x + W_c u_c) + d_{cr}(x, u_c), \tag{3.155}$$
$$d_d(x, u_d) = (L_d x + W_d u_d)^{\mathrm{T}} (L_d x + W_d u_d) + d_{dr}(x, u_d), \tag{3.156}$$

where $V_r : \mathbb{R}^n \to \mathbb{R}$, $d_{cr} : \mathbb{R}^n \times \mathbb{R}^{m_c} \to \mathbb{R}$, and $d_{dr} : \mathbb{R}^n \times \mathbb{R}^{m_d} \to \mathbb{R}$ contain the higher-order terms of $V_s(\cdot)$, $d_c(\cdot, \cdot)$, and $d_d(\cdot, \cdot)$, respectively.

Next, let $f_c(x) = A_c x + f_{cr}(x)$, $h_c(x) = C_c x + h_{cr}(x)$, $f_d(x) = (A_d - I_n)x + f_{dr}(x)$, and $h_d(x) = C_d x + h_{dr}(x)$, where $f_{cr}(\cdot)$, $h_{cr}(\cdot)$, $f_{dr}(\cdot)$, and $h_{dr}(\cdot)$ contain the nonlinear terms of $f_c(x)$, $h_c(x)$, $f_d(x)$, and $h_d(x)$, respectively, and let $G_c(x) = B_c + G_{cr}(x)$, $J_c(x) = D_c + J_{cr}(x)$, $G_d(x) = B_d + G_{dr}(x)$, $J_d(x) = D_d + J_{dr}(x)$, where $G_{cr}(x)$, $J_{cr}(x)$, $G_{dr}(x)$, and $J_{dr}(x)$ contain the nonconstant terms of $G_c(x)$, $J_c(x)$,

$G_\mathrm{d}(x)$, and $J_\mathrm{d}(x)$, respectively. Using the above expressions, (3.152) and (3.153) can be written as

$$
\begin{aligned}
0 = {}& 2x^\mathrm{T} P(A_\mathrm{c} x + B_\mathrm{c} u_\mathrm{c}) - (x^\mathrm{T} C_\mathrm{c}^\mathrm{T} Q_\mathrm{c} C_\mathrm{c} x + 2x^\mathrm{T} C_\mathrm{c}^\mathrm{T} Q_\mathrm{c} D_\mathrm{c} u_\mathrm{c} \\
& + u_\mathrm{c}^\mathrm{T} D_\mathrm{c}^\mathrm{T} Q_\mathrm{c} D_\mathrm{c} u_\mathrm{c} + 2x^\mathrm{T} C_\mathrm{c}^\mathrm{T} S_\mathrm{c} u_\mathrm{c} + 2u_\mathrm{c}^\mathrm{T} D_\mathrm{c}^\mathrm{T} S_\mathrm{c} u_\mathrm{c} + u_\mathrm{c}^\mathrm{T} R_\mathrm{c} u_\mathrm{c}) \\
& + (L_\mathrm{c} x + W_\mathrm{c} u_\mathrm{c})^\mathrm{T}(L_\mathrm{c} x + W_\mathrm{c} u_\mathrm{c}) + \delta_\mathrm{c}(x, u_\mathrm{c}), \qquad (3.157) \\
0 = {}& (A_\mathrm{d} x + B_\mathrm{d} u_\mathrm{d})^\mathrm{T} P(A_\mathrm{d} x + B_\mathrm{d} u_\mathrm{d}) - x^\mathrm{T} P x - (x^\mathrm{T} C_\mathrm{d}^\mathrm{T} Q_\mathrm{d} C_\mathrm{d} x \\
& + 2x^\mathrm{T} C_\mathrm{d}^\mathrm{T} Q_\mathrm{d} D_\mathrm{d} u_\mathrm{d} + u_\mathrm{d}^\mathrm{T} D_\mathrm{d}^\mathrm{T} Q_\mathrm{d} D_\mathrm{d} u_\mathrm{d} + 2x^\mathrm{T} C_\mathrm{d}^\mathrm{T} S_\mathrm{d} u_\mathrm{d} + 2u_\mathrm{d}^\mathrm{T} D_\mathrm{d}^\mathrm{T} S_\mathrm{d} u_\mathrm{d} \\
& + u_\mathrm{d}^\mathrm{T} R_\mathrm{d} u_\mathrm{d}) + (L_\mathrm{d} x + W_\mathrm{d} u_\mathrm{d})^\mathrm{T}(L_\mathrm{d} x + W_\mathrm{d} u_\mathrm{d}) + \delta_\mathrm{d}(x, u_\mathrm{d}), \quad (3.158)
\end{aligned}
$$

where $\delta_\mathrm{c}(x, u_\mathrm{c})$ and $\delta_\mathrm{d}(x, u_\mathrm{d})$ are such that

$$
\lim_{\|x\|^2 + \|u_\mathrm{c}\|^2 \to 0} \frac{|\,\delta_\mathrm{c}(x, u_\mathrm{c})\,|}{\|x\|^2 + \|u_\mathrm{c}\|^2} = 0, \qquad \lim_{\|x\|^2 + \|u_\mathrm{d}\|^2 \to 0} \frac{|\,\delta_\mathrm{d}(x, u_\mathrm{d})\,|}{\|x\|^2 + \|u_\mathrm{d}\|^2} = 0.
$$

Now, viewing (3.157) and (3.158) as the Taylor series expansion of (3.152) and (3.153), respectively, about $x = 0$, $u_\mathrm{c} = 0$, and $u_\mathrm{d} = 0$, it follows that

$$
\begin{aligned}
0 = {}& x^\mathrm{T}(A_\mathrm{c}^\mathrm{T} P + P A_\mathrm{c} - C_\mathrm{c}^\mathrm{T} Q_\mathrm{c} C_\mathrm{c} + L_\mathrm{c}^\mathrm{T} L_\mathrm{c}) x + 2x^\mathrm{T}(P B_\mathrm{c} - C_\mathrm{c}^\mathrm{T} S_\mathrm{c} \\
& - C_\mathrm{c}^\mathrm{T} Q_\mathrm{c} D_\mathrm{c} + L_\mathrm{c}^\mathrm{T} W_\mathrm{c}) u_\mathrm{c} + u_\mathrm{c}^\mathrm{T}(W_\mathrm{c}^\mathrm{T} W_\mathrm{c} - D_\mathrm{c}^\mathrm{T} Q_\mathrm{c} D_\mathrm{c} - D_\mathrm{c}^\mathrm{T} S_\mathrm{c} \\
& - S_\mathrm{c}^\mathrm{T} D_\mathrm{c} - R_\mathrm{c}) u_\mathrm{c}, \quad x \in \mathbb{R}^n, \quad u_\mathrm{c} \in \mathbb{R}^{m_\mathrm{c}}, \qquad (3.159) \\
0 = {}& x^\mathrm{T}(A_\mathrm{d}^\mathrm{T} P A_\mathrm{d} - P - C_\mathrm{d}^\mathrm{T} Q_\mathrm{d} C_\mathrm{d} + L_\mathrm{d}^\mathrm{T} L_\mathrm{d}) x + 2x^\mathrm{T}(A_\mathrm{d}^\mathrm{T} P B_\mathrm{d} - C_\mathrm{d}^\mathrm{T} S_\mathrm{d} \\
& - C_\mathrm{d}^\mathrm{T} Q_\mathrm{d} D_\mathrm{d} + L_\mathrm{d}^\mathrm{T} W_\mathrm{d}) u_\mathrm{d} + u_\mathrm{d}^\mathrm{T}(W_\mathrm{d}^\mathrm{T} W_\mathrm{d} - D_\mathrm{d}^\mathrm{T} Q_\mathrm{d} D_\mathrm{d} - D_\mathrm{d}^\mathrm{T} S_\mathrm{d} \\
& - S_\mathrm{d}^\mathrm{T} D_\mathrm{d} - R_\mathrm{d} + B_\mathrm{d}^\mathrm{T} P B_\mathrm{d}) u_\mathrm{d}, \quad x \in \mathbb{R}^n, \quad u_\mathrm{d} \in \mathbb{R}^{m_\mathrm{d}}. \qquad (3.160)
\end{aligned}
$$

Next, equating coefficients of equal powers in (3.159) and (3.160) yields (3.137)–(3.142).

Finally, to show that $P > 0$ in the case where $(A_\mathrm{c}, C_\mathrm{c})$ and $(A_\mathrm{d}, C_\mathrm{d})$ are observable, note that it follows from Theorem 3.6 and (3.137)–(3.142) that the linearized impulsive dynamical system $\mathcal{G}$ with storage function $V_\mathrm{s}(x) = x^\mathrm{T} P x$ is dissipative with respect to the quadratic hybrid supply rate $(s_\mathrm{c}(u_\mathrm{c}, y_\mathrm{c}), s_\mathrm{d}(u_\mathrm{d}, y_\mathrm{d}))$. Now, the positive definiteness of P follows from Theorem 3.3. $\square$

It is important to note that Theorem 3.10 does *not* hold for state-dependent impulsive dynamical systems. To see this, it need only be noted that (3.137)–(3.142) follow from (3.159) and (3.160) if and only if $x \in \mathbb{R}^n$. Finally, we note that linearization results for exponentially dissipative time-dependent impulsive dynamical systems can be derived in an analogous manner.

3.4 Specialization to Linear Impulsive Dynamical Systems

In this section, we specialize the results of Section 3.3 to the case of linear impulsive dynamical systems. Specifically, setting $f_c(x) = A_c x$, $G_c(x) = B_c$, $h_c(x) = C_c x$, $J_c(x) = D_c$, $f_d(x) = (A_d - I_n)x$, $G_d(x) = B_d$, $h_d(x) = C_d x$, and $J_d(x) = D_d$, the nonlinear time-dependent impulsive dynamical system given by (3.14)–(3.17) specializes to

$$\dot{x}(t) = A_c x(t) + B_c u_c(t), \quad x(0) = x_0, \quad t \neq t_k, \tag{3.161}$$

$$\Delta x(t) = (A_d - I_n)x(t) + B_d u_d(t), \quad t = t_k, \tag{3.162}$$

$$y_c(t) = C_c x(t) + D_c u_c(t), \quad t \neq t_k, \tag{3.163}$$

$$y_d(t) = C_d x(t) + D_d u_d(t), \quad t = t_k, \tag{3.164}$$

where $A_c \in \mathbb{R}^{n \times n}$, $B_c \in \mathbb{R}^{n \times m_c}$, $C_c \in \mathbb{R}^{l_c \times n}$, $D_c \in \mathbb{R}^{l_c \times m_c}$, $A_d \in \mathbb{R}^{n \times n}$, $B_d \in \mathbb{R}^{n \times m_d}$, $C_d \in \mathbb{R}^{l_d \times n}$, and $D_d \in \mathbb{R}^{l_d \times m_d}$.

Theorem 3.11 Let $Q_c \in \mathbb{S}^{l_c}$, $S_c \in \mathbb{R}^{l_c \times m_c}$, $R_c \in \mathbb{S}^{m_c}$, $Q_d \in \mathbb{S}^{l_d}$, $S_d \in \mathbb{R}^{l_d \times m_d}$, and $R_d \in \mathbb{S}^{m_d}$, consider the linear impulsive dynamical system $\mathcal{G}$ given by (3.161)–(3.164), and assume $\mathcal{G}$ is minimal. Then the following statements are equivalent:

i) $\mathcal{G}$ is dissipative with respect to the quadratic hybrid supply rate $(s_c(u_c, y_c), s_d(u_d, y_d)) = (y_c^T Q_c y_c + 2 y_c^T S_c u_c + u_c^T R_c u_c, \; y_d^T Q_d y_d + 2 y_d^T S_d u_d + u_d^T R_d u_d)$.

ii) There exist matrices $P \in \mathbb{R}^{n \times n}$, $L_c \in \mathbb{R}^{p_c \times n}$, $W_c \in \mathbb{R}^{p_c \times m_c}$, $L_d \in \mathbb{R}^{p_d \times n}$, and $W_d \in \mathbb{R}^{p_d \times m_d}$, with P positive definite, such that (3.137)–(3.142) are satisfied.

If, alternatively, $R_c + S_c^T D_c + D_c^T S_c + D_c^T Q_c D_c > 0$, then $\mathcal{G}$ is dissipative with respect to the quadratic hybrid supply rate $(s_c(u_c, y_c), s_d(u_d, y_d)) = (y_c^T Q_c y_c + 2 y_c^T S_c u_c + u_c^T R_c u_c, \; y_d^T Q_d y_d + 2 y_d^T S_d u_d + u_d^T R_d u_d)$ if and only if there exists an $n \times n$ positive-definite matrix P such that

$$0 < R_d + S_d^T D_d + D_d^T S_d + D_d^T Q_d D_d - B_d^T P B_d, \tag{3.165}$$

$$0 \geq A_c^T P + P A_c - C_c^T Q_c C_c + [P B_c - C_c^T (Q_c D_c + S_c)]$$
$$\cdot [R_c + S_c^T D_c + D_c^T S_c + D_c^T Q_c D_c]^{-1} [P B_c - C_c^T (Q_c D_c + S_c)]^T, \tag{3.166}$$

$$0 \geq A_d^T P A_d - P - C_d^T Q_d C_d + [A_d^T P B_d - C_d^T (Q_d D_d + S_d)][R_d$$
$$+ S_d^T D_d + D_d^T S_d + D_d^T Q_d D_d - B_d^T P B_d]^{-1}$$
$$\cdot [A_d^T P B_d - C_d^T (Q_d D_d + S_d)]^T. \tag{3.167}$$

Proof. The fact that $ii)$ implies $i)$ follows from Theorem 3.6 with $f_c(x) = A_c x$, $G_c(x) = B_c$, $h_c(x) = C_c x$, $J_c(x) = D_c$, $f_d(x) = (A_d - I_n)x$, $G_d(x) = B_d$, $h_d(x) = C_d x$, $J_d(x) = D_d$, $V_s(x) = x^T P x$, $L_c(x) = L_c x$, $L_d(x) = L_d x$, $W_c(x) = W_c$, and $W_d(x) = W_d$.

To show that $i)$ implies $ii)$, note that if the linear impulsive dynamical system given by (3.161)–(3.164) is dissipative, then it follows from Theorem 3.10 with $f_c(x) = A_c x$, $G_c(x) = B_c$, $h_c(x) = C_c x$, $J_c(x) = D_c$, $f_d(x) = (A_d - I_n)x$, $G_d(x) = B_d$, $h_d(x) = C_d x$, and $J_d(x) = D_d$ that there exist matrices $P \in \mathbb{R}^{n \times n}$, $L_c \in \mathbb{R}^{p_c \times n}$, $W_c \in \mathbb{R}^{p_c \times m_c}$, $L_d \in \mathbb{R}^{p_d \times n}$, and $W_d \in \mathbb{R}^{p_d \times m_d}$, with P positive definite, such that (3.137)–(3.142) are satisfied. Finally, (3.165)–(3.167) follow from (3.75)–(3.77) and Theorem 3.10 with the linearization given above. $\square$

The proof of Theorem 3.4 relies on Theorem 3.10 which *a priori* assumes that the storage function $V_s(x)$, $x \in \mathbb{R}^n$, is three-times continuously differentiable. Unlike linear, time-invariant dissipative dynamical systems with continuous flows, there does not always exists a smooth (i.e., infinitely differentiable) storage function $V_s(x)$, $x \in \mathbb{R}^n$, for linear dissipative impulsive dynamical systems.

Note that (3.137)–(3.142) are equivalent to

$$\begin{bmatrix} \mathcal{A}_c & \mathcal{B}_c \\ \mathcal{B}_c^T & \mathcal{D}_c \end{bmatrix} = \begin{bmatrix} L_c^T \\ W_c^T \end{bmatrix} \begin{bmatrix} L_c & W_c \end{bmatrix} \geq 0, \qquad (3.168)$$

$$\begin{bmatrix} \mathcal{A}_d & \mathcal{B}_d \\ \mathcal{B}_d^T & \mathcal{D}_d \end{bmatrix} = \begin{bmatrix} L_d^T \\ W_d^T \end{bmatrix} \begin{bmatrix} L_d & W_d \end{bmatrix} \geq 0, \qquad (3.169)$$

where

$$\mathcal{A}_c = -A_c^T P - P A_c + C_c^T Q_c C_c, \quad \mathcal{B}_c = -P B_c + C_c^T (Q_c D_c + S_c),$$
$$\mathcal{A}_d = P - A_d^T P A_d + C_d^T Q_d C_d, \quad \mathcal{B}_d = -A_d^T P B_d + C_d^T (Q_d D_d + S_d),$$
$$\mathcal{D}_c = R_c + S_c^T D_c + D_c^T S_c + D_c^T Q_c D_c,$$

and

$$\mathcal{D}_d = R_d + S_d^T D_d + D_d^T S_d + D_d^T Q_d D_d - B_d^T P B_d.$$

Hence, dissipativity of linear impulsive dynamical systems with respect to quadratic hybrid supply rates can be characterized via Linear Matrix Inequalities (LMIs) [27]. Similar remarks hold for the passivity and nonexpansivity results given in Corollaries 3.4 and 3.5, respectively.

The following results present generalizations of the positive real lemma [3] and the bounded real lemma [3] for linear impulsive systems, respectively.

Corollary 3.4 *Consider the linear impulsive dynamical system $\mathcal{G}$ given by (3.161)–(3.164) with $m_c = l_c$ and $m_d = l_d$, and assume $\mathcal{G}$ is minimal. Then the following statements are equivalent:*

i) $\mathcal{G}$ *is passive.*

ii) *There exist matrices $P \in \mathbb{R}^{n \times n}$, $L_c \in \mathbb{R}^{p_c \times n}$, $W_c \in \mathbb{R}^{p_c \times m_c}$, $L_d \in \mathbb{R}^{p_d \times n}$, and $W_d \in \mathbb{R}^{p_d \times m_d}$, with P positive definite, such that*

$$0 = A_c^T P + P A_c + L_c^T L_c, \tag{3.170}$$
$$0 = P B_c - C_c^T + L_c^T W_c, \tag{3.171}$$
$$0 = D_c + D_c^T - W_c^T W_c, \tag{3.172}$$
$$0 = A_d^T P A_d - P + L_d^T L_d, \tag{3.173}$$
$$0 = A_d^T P B_d - C_d^T + L_d^T W_d, \tag{3.174}$$
$$0 = D_d + D_d^T - B_d^T P B_d - W_d^T W_d. \tag{3.175}$$

If, alternatively, $D_c + D_c^T > 0$, then $\mathcal{G}$ is passive if and only if there exists an $n \times n$ positive-definite matrix P such that

$$0 < D_d + D_d^T - B_d^T P B_d, \tag{3.176}$$
$$0 \geq A_c^T P + P A_c + (P B_c - C_c^T)(D_c + D_c^T)^{-1}(P B_c - C_c^T)^T, \tag{3.177}$$
$$0 \geq A_d^T P A_d - P + (A_d^T P B_d - C_d^T)(D_d + D_d^T - B_d^T P B_d)^{-1}$$
$$\cdot (A_d^T P B_d - C_d^T)^T. \tag{3.178}$$

Proof. The result is a direct consequence of Theorem 3.11 with $m_c = l_c$, $m_d = l_d$, $Q_c = 0$, $S_c = I_{m_c}$, $R_c = 0$, $Q_d = 0$, $S_d = I_{m_d}$, and $R_d = 0$. $\qquad\square$

Equations (3.170)–(3.172) are identical in form to the equations appearing in the continuous-time positive real lemma [2] used to characterize positive realness for continuous-time linear systems in the state space, while (3.173)–(3.175) are identical in form to the equations appearing in the discrete-time positive real lemma [76]. This is not surprising since, as noted in Section 3.3, impulsive dynamical systems involve a hybrid formulation of continuous-time and discrete-time dynamics. A key difference, however, is the fact that in the impulsive case a *single* positive-definite matrix P is required to satisfy all six equations. Similar remarks hold for Corollary 3.5 below.

Corollary 3.5 *Consider the linear impulsive dynamical system $\mathcal{G}$ given by (3.161)–(3.164), and assume $\mathcal{G}$ is minimal. Then the following statements are equivalent:*

i) $\mathcal{G}$ is nonexpansive.

ii) There exist matrices $P \in \mathbb{R}^{n \times n}$, $L_c \in \mathbb{R}^{p_c \times n}$, $W_c \in \mathbb{R}^{p_c \times m_c}$, $L_d \in \mathbb{R}^{p_d \times n}$, and $W_d \in \mathbb{R}^{p_d \times m_d}$, with P positive definite, such that

$$0 = A_c^{\mathrm{T}} P + P A_c + C_c^{\mathrm{T}} C_c + L_c^{\mathrm{T}} L_c, \tag{3.179}$$

$$0 = P B_c + C_c^{\mathrm{T}} D_c + L_c^{\mathrm{T}} W_c, \tag{3.180}$$

$$0 = \gamma_c^2 I_{m_c} - D_c^{\mathrm{T}} D_c - W_c^{\mathrm{T}} W_c, \tag{3.181}$$

$$0 = A_d^{\mathrm{T}} P A_d - P + C_d^{\mathrm{T}} C_d + L_d^{\mathrm{T}} L_d, \tag{3.182}$$

$$0 = A_d^{\mathrm{T}} P B_d + C_d^{\mathrm{T}} D_d + L_d^{\mathrm{T}} W_d, \tag{3.183}$$

$$0 = \gamma_d^2 I_d - D_d^{\mathrm{T}} D_d - B_d^{\mathrm{T}} P B_d - W_d^{\mathrm{T}} W_d. \tag{3.184}$$

If, in addition, $\gamma_c^2 I_{m_c} - D_c^{\mathrm{T}} D_c > 0$, then $\mathcal{G}$ is nonexpansive if and only if there exists an $n \times n$ positive-definite matrix P such that

$$0 < \gamma_d^2 I_{m_d} - D_d^{\mathrm{T}} D_d - B_d^{\mathrm{T}} P B_d, \tag{3.185}$$

$$0 \geq A_c^{\mathrm{T}} P + P A_c + (P B_c + C_c^{\mathrm{T}} D_c)(\gamma_c^2 I_{m_c} - D_c^{\mathrm{T}} D_c)^{-1}(P B_c + C_c^{\mathrm{T}} D_c)^{\mathrm{T}}$$
$$+ C_c^{\mathrm{T}} C_c, \tag{3.186}$$

$$0 \geq A_d^{\mathrm{T}} P A_d - P + (A_d^{\mathrm{T}} P B_d + C_d^{\mathrm{T}} D_d)(\gamma_d^2 I_{m_d} - D_d^{\mathrm{T}} D_d - B_d^{\mathrm{T}} P B_d)^{-1}$$
$$\cdot (A_d^{\mathrm{T}} P B_d + C_d^{\mathrm{T}} D_d)^{\mathrm{T}} + C_d^{\mathrm{T}} C_d. \tag{3.187}$$

Proof. The result is a direct consequence of Theorem 3.11 with $Q_c = -I_{l_c}$, $S_c = 0$, $R_c = \gamma_c^2 I_{m_c}$, $Q_d = -I_{l_d}$, $S_d = 0$, and $R_d = \gamma_d^2 I_{m_d}$. $\qquad\square$

It follows from (3.83) that if (3.170) and (3.179) are replaced, respectively, by

$$0 = A_c^{\mathrm{T}} P + P A_c + \varepsilon P + L_c^{\mathrm{T}} L_c, \tag{3.188}$$

$$0 = A_c^{\mathrm{T}} P + P A_c + \varepsilon P + C_c^{\mathrm{T}} C_c + L_c^{\mathrm{T}} L_c, \tag{3.189}$$

where $\varepsilon > 0$, then (3.188) and (3.171)–(3.175) provide necessary and sufficient conditions for exponential passivity, while (3.189) and (3.180)–(3.184) provide necessary and sufficient conditions for exponential nonexpansivity. These conditions present generalizations of the strict positive real lemma and the strict bounded real lemma for linear impulsive systems, respectively.

It is important to note that the equivalence between (3.137)–(3.142) and dissipativity of a linear *state-dependent* impulsive dynamical system does *not* hold. In particular, for linear state-dependent impulsive dynamical systems, (3.137)–(3.142) are only sufficient conditions for dissipativity. The next result provides less conservative sufficient conditions for linear state-dependent dynamical systems. For this result we consider the linear state-dependent dynamical system

$$\dot{x}(t) = A_{\mathrm{c}}x(t) + B_{\mathrm{c}}u_{\mathrm{c}}(t), \quad x \notin \mathcal{Z}_x, \qquad (3.190)$$
$$\Delta x(t) = (A_{\mathrm{d}} - I_n)x(t) + B_{\mathrm{d}}u_{\mathrm{d}}(t), \quad x \in \mathcal{Z}_x, \qquad (3.191)$$
$$y_{\mathrm{c}}(t) = C_{\mathrm{c}}x(t) + D_{\mathrm{c}}u_{\mathrm{c}}(t), \quad x \notin \mathcal{Z}_x, \qquad (3.192)$$
$$y_{\mathrm{d}}(t) = C_{\mathrm{d}}x(t) + D_{\mathrm{d}}u_{\mathrm{d}}(t), \quad x \in \mathcal{Z}_x, \qquad (3.193)$$

where $A_{\mathrm{c}} \in \mathbb{R}^{n \times n}$, $B_{\mathrm{c}} \in \mathbb{R}^{n \times m_{\mathrm{c}}}$, $C_{\mathrm{c}} \in \mathbb{R}^{l_{\mathrm{c}} \times n}$, $D_{\mathrm{c}} \in \mathbb{R}^{l_{\mathrm{c}} \times m_{\mathrm{c}}}$, $A_{\mathrm{d}} \in \mathbb{R}^{n \times n}$, $B_{\mathrm{d}} \in \mathbb{R}^{n \times m_{\mathrm{d}}}$, $C_{\mathrm{d}} \in \mathbb{R}^{l_{\mathrm{d}} \times n}$, and $D_{\mathrm{d}} \in \mathbb{R}^{l_{\mathrm{d}} \times m_{\mathrm{d}}}$.

Theorem 3.12 Let $Q_{\mathrm{c}} \in \mathbb{S}^{l_{\mathrm{c}}}$, $S_{\mathrm{c}} \in \mathbb{R}^{l_{\mathrm{c}} \times m_{\mathrm{c}}}$, $R_{\mathrm{c}} \in \mathbb{S}^{m_{\mathrm{c}}}$, $Q_{\mathrm{d}} \in \mathbb{S}^{l_{\mathrm{d}}}$, $S_{\mathrm{d}} \in \mathbb{R}^{l_{\mathrm{d}} \times m_{\mathrm{d}}}$, and $R_{\mathrm{d}} \in \mathbb{S}^{m_{\mathrm{d}}}$, and consider the linear impulsive system $\mathcal{G}$ given by (3.190)–(3.193). If there exist matrices $P \in \mathbb{R}^{n \times n}$, $L_{\mathrm{c}} \in \mathbb{R}^{p_{\mathrm{c}} \times n}$, $W_{\mathrm{c}} \in \mathbb{R}^{p_{\mathrm{c}} \times m_{\mathrm{c}}}$, $L_{\mathrm{d}} \in \mathbb{R}^{p_{\mathrm{d}} \times n}$, and $W_{\mathrm{d}} \in \mathbb{R}^{p_{\mathrm{d}} \times m_{\mathrm{d}}}$, with P nonnegative definite, such that

$$0 = x^{\mathrm{T}}(A_{\mathrm{c}}^{\mathrm{T}}P + PA_{\mathrm{c}} - C_{\mathrm{c}}^{\mathrm{T}}Q_{\mathrm{c}}C_{\mathrm{c}} + L_{\mathrm{c}}^{\mathrm{T}}L_{\mathrm{c}})x, \quad x \notin \mathcal{Z}_x, \qquad (3.194)$$
$$0 = x^{\mathrm{T}}(PB_{\mathrm{c}} - C_{\mathrm{c}}^{\mathrm{T}}(Q_{\mathrm{c}}D_{\mathrm{c}} + S_{\mathrm{c}}) + L_{\mathrm{c}}^{\mathrm{T}}W_{\mathrm{c}}), \quad x \notin \mathcal{Z}_x, \qquad (3.195)$$
$$0 = R_{\mathrm{c}} + S_{\mathrm{c}}^{\mathrm{T}}D_{\mathrm{c}} + D_{\mathrm{c}}^{\mathrm{T}}S_{\mathrm{c}} + D_{\mathrm{c}}^{\mathrm{T}}Q_{\mathrm{c}}D_{\mathrm{c}} - W_{\mathrm{c}}^{\mathrm{T}}W_{\mathrm{c}}, \qquad (3.196)$$
$$0 = x^{\mathrm{T}}(A_{\mathrm{d}}^{\mathrm{T}}PA_{\mathrm{d}} - P - C_{\mathrm{d}}^{\mathrm{T}}Q_{\mathrm{d}}C_{\mathrm{d}} + L_{\mathrm{d}}^{\mathrm{T}}L_{\mathrm{d}})x, \quad x \in \mathcal{Z}_x, \qquad (3.197)$$
$$0 = x^{\mathrm{T}}(A_{\mathrm{d}}^{\mathrm{T}}PB_{\mathrm{d}} - C_{\mathrm{d}}^{\mathrm{T}}(Q_{\mathrm{d}}D_{\mathrm{d}} + S_{\mathrm{d}}) + L_{\mathrm{d}}^{\mathrm{T}}W_{\mathrm{d}}), \quad x \in \mathcal{Z}_x, \qquad (3.198)$$
$$0 = R_{\mathrm{d}} + S_{\mathrm{d}}^{\mathrm{T}}D_{\mathrm{d}} + D_{\mathrm{d}}^{\mathrm{T}}S_{\mathrm{d}} + D_{\mathrm{d}}^{\mathrm{T}}Q_{\mathrm{d}}D_{\mathrm{d}} - B_{\mathrm{d}}^{\mathrm{T}}PB_{\mathrm{d}} - W_{\mathrm{d}}^{\mathrm{T}}W_{\mathrm{d}}, \qquad (3.199)$$

then the linear impulsive dynamical system $\mathcal{G}$ is dissipative with respect to the quadratic hybrid supply rate $(s_{\mathrm{c}}(u_{\mathrm{c}}, y_{\mathrm{c}}), s_{\mathrm{d}}(u_{\mathrm{d}}, y_{\mathrm{d}})) = (y_{\mathrm{c}}^{\mathrm{T}}Q_{\mathrm{c}}y_{\mathrm{c}} + 2y_{\mathrm{c}}^{\mathrm{T}}S_{\mathrm{c}}u_{\mathrm{c}} + u_{\mathrm{c}}^{\mathrm{T}}R_{\mathrm{c}}u_{\mathrm{c}}, y_{\mathrm{d}}^{\mathrm{T}}Q_{\mathrm{d}}y_{\mathrm{d}} + 2y_{\mathrm{d}}^{\mathrm{T}}S_{\mathrm{d}}u_{\mathrm{d}} + u_{\mathrm{d}}^{\mathrm{T}}R_{\mathrm{d}}u_{\mathrm{d}})$.

Proof. The proof follows from Theorem 3.8 with $f_{\mathrm{c}}(x) = A_{\mathrm{c}}x$, $G_{\mathrm{c}}(x) = B_{\mathrm{c}}$, $h_{\mathrm{c}}(x) = C_{\mathrm{c}}x$, $J_{\mathrm{c}}(x) = D_{\mathrm{c}}$, $f_{\mathrm{d}}(x) = (A_{\mathrm{d}} - I_n)x$, $G_{\mathrm{d}}(x) = B_{\mathrm{d}}$, $h_{\mathrm{d}}(x) = C_{\mathrm{d}}x$, $J_{\mathrm{d}}(x) = D_{\mathrm{d}}$, $V_{\mathrm{s}}(x) = x^{\mathrm{T}}Px$, $L_{\mathrm{c}}(x) = L_{\mathrm{c}}x$, $L_{\mathrm{d}}(x) = L_{\mathrm{d}}x$, $W_{\mathrm{c}}(x) = W_{\mathrm{c}}$, and $W_{\mathrm{d}}(x) = W_{\mathrm{d}}$. $\qquad \square$

Similar results hold for passivity and nonexpansivity of linear state-dependent impulsive dynamical systems.

Chapter Four

Impulsive Nonnegative and Compartmental
Dynamical Systems

4.1 Introduction

Nonnegative systems [20, 57, 58, 132] are essential in capturing the phenomenological features of a wide range of dynamical systems involving dynamic states whose values are nonnegative. A subclass of nonnegative dynamical systems are compartmental systems [4, 25, 47, 50, 83, 84, 114, 115, 125, 149]. These systems are derived from mass and energy balance considerations and are comprised of homogeneous interconnected macroscopic subsystems or compartments which exchange variable quantities of material via intercompartmental flow laws. Since biological and physiological systems have numerous input-output properties related to conservation, dissipation, and transport of mass and energy, nonnegative and compartmental systems are remarkably effective in describing the essential features of these dynamical systems. The range of applications of nonnegative and compartmental systems is not limited to biological and medical systems. Their usage includes demographic, epidemic [83, 85], ecological [129], economic [21], telecommunications [45], transportation, power, and large-scale systems [151, 152].

In this chapter, we develop several basic mathematical results on stability and dissipativity of impulsive nonnegative and compartmental dynamical systems. Specifically, using *linear* Lyapunov functions we develop sufficient conditions for Lyapunov stability and asymptotic stability for impulsive nonnegative dynamical systems. The consideration of a linear Lyapunov function leads to a *new* set of Lyapunov-like equations for examining the stability of linear impulsive nonnegative systems. The motivation for using a linear Lyapunov function follows from the fact that the state of a nonnegative dynamical system is nonnegative, and hence, a linear Lyapunov function is a valid Lyapunov function candidate.

Next, using *linear* and *nonlinear* storage functions with *linear* hybrid supply rates we develop *new* notions of classical dissipativity theory and exponential dissipativity theory for impulsive nonnegative dynamical systems. The overall approach provides a new interpretation of a mass balance for impulsive nonnegative systems with linear hybrid supply rates and linear and nonlinear storage functions. Specifically, we show that dissipativity of an impulsive nonnegative dynamical system involving a linear storage function and a linear hybrid supply rate implies that the system mass transport (respectively, change in system mass) is equal to the supplied system flux (respectively, mass) over the continuous-time dynamics (respectively, the resetting instants) minus the expelled system flux (respectively, mass) over the continuous-time dynamics (respectively, the resetting instants). In addition, we develop *new* Kalman-Yakubovich-Popov equations for impulsive nonnegative systems for characterizing dissipativity with linear and nonlinear storage functions and linear hybrid supply rates.

4.2 Stability Theory for Nonlinear Impulsive Nonnegative Dynamical Systems

In this section, we provide sufficient conditions for stability of state-dependent impulsive nonnegative dynamical systems, that is, state-dependent impulsive dynamical systems whose solutions remain in the nonnegative orthant for nonnegative initial conditions. First, however, we establish notation and definitions that are necessary for developing the main results of this chapter. For $x \in \mathbb{R}^n$ we write $x \geq\geq 0$ (respectively, $x >> 0$) to indicate that every component of x is nonnegative (respectively, positive). In this case, we say that x is *nonnegative* or *positive*, respectively. Likewise $A \in \mathbb{R}^{n \times m}$ is *nonnegative* or *positive*[1] if every entry of A is nonnegative or positive, respectively, which is written as $A \geq\geq 0$ or $A >> 0$, respectively. Let $\overline{\mathbb{R}}_+^n$ and $\mathbb{R}_+^n$ denote the nonnegative and positive orthants of $\mathbb{R}^n$, that is, if $x \in \mathbb{R}^n$, then $x \in \overline{\mathbb{R}}_+^n$ and $x \in \mathbb{R}_+^n$ are equivalent, respectively, to $x \geq\geq 0$ and $x >> 0$. The following definition introduces the notion of Z-, M-, and essentially nonnegative matrices.

Definition 4.1 ([21]) *Let $A \in \mathbb{R}^{n \times n}$. A is a Z-matrix if $A_{(i,j)} \leq 0$, $i, j = 1, \ldots, n$, $i \neq j$. A is an M-matrix (respectively, a nonsingular M-matrix) if A is a Z-matrix and all the principal minors of A are*

---

[1]In this chapter it is important to distinguish between a square nonnegative (respectively, positive) matrix and a nonnegative-definite (respectively, positive-definite) matrix.

nonnegative (respectively, positive). A is essentially nonnegative if $-A$ *is a Z-matrix, that is,* $A_{(i,j)} \geq 0$, $i, j = 1, \ldots, n$, $i \neq j$.

The following definitions introduce the notions of essentially nonnegative and nonnegative vector fields.

Definition 4.2 *Let* $f_c = [f_{c1}, \ldots, f_{cn}]^T : \mathcal{D} \to \mathbb{R}^n$, *where* $\mathcal{D}$ *is an open subset of* $\mathbb{R}^n$ *that contains* $\overline{\mathbb{R}}_+^n$. *Then* f_c *is essentially nonnegative if* $f_{ci}(x) \geq 0$, *for all* $i = 1, \ldots, n$, *and* $x \in \overline{\mathbb{R}}_+^n$ *such that* $x_i = 0$, *where* x_i *denotes the ith entry of* x.

Definition 4.3 *Let* $f_d = [f_{d1}, \ldots, f_{dn}]^T : \mathcal{D} \to \mathbb{R}^n$, *where* $\mathcal{D}$ *is an open subset of* $\mathbb{R}^n$ *that contains* $\overline{\mathbb{R}}_+^n$. *Then* f_d *is nonnegative if* $f_{di}(x) \geq 0$, *for all* $i = 1, \ldots, n$, *and* $x \in \overline{\mathbb{R}}_+^n$.

Note that if $f_c(x) = A_c x$, where $A_c \in \mathbb{R}^{n \times n}$, then f_c is essentially nonnegative if and only if A_c is essentially nonnegative. Similarly, if $f_d(x) = (A_d - I_n)x$, where $A_d \in \mathbb{R}^{n \times n}$, then $x + f_d(x)$ is nonnegative for all $x \in \overline{\mathbb{R}}_+^n$ if and only if A_d is nonnegative.

Consider the nonlinear state-dependent impulsive dynamical system of the form

$$\dot{x}(t) = f_c(x(t)), \quad x(0) = x_0, \quad x(t) \notin \mathcal{Z}_x, \tag{4.1}$$
$$\Delta x(t) = f_d(x(t)), \quad x(t) \in \mathcal{Z}_x, \tag{4.2}$$

where $t \geq 0$, $x(t) \in \mathcal{D} \subseteq \mathbb{R}^n$, $\mathcal{D}$ is an open subset of $\mathbb{R}^n$ that contains $\overline{\mathbb{R}}_+^n$ with $0 \in \mathcal{D}$, $\Delta x(t) \triangleq x(t^+) - x(t)$, $f_c : \mathcal{D} \to \mathbb{R}^n$ is Lipschitz continuous and satisfies $f_c(0) = 0$, $f_d : \mathcal{D} \to \mathbb{R}^n$ is continuous, and $\mathcal{Z}_x \subset \mathcal{D}$ is the resetting set. Here, we assume that $f_c(\cdot)$ and $f_d(\cdot)$ are such that Assumptions A1, A2, and 2.1 hold. The next result shows that $\overline{\mathbb{R}}_+^n$ is an invariant set for (4.1) and (4.2) if $f_c : \mathcal{D} \to \mathbb{R}^n$ is essentially nonnegative and $f_d : \mathcal{D} \to \mathbb{R}^n$ is such that $x + f_d(x)$ is nonnegative for all $x \in \overline{\mathbb{R}}_+^n$.

Proposition 4.1 *Suppose* $\overline{\mathbb{R}}_+^n \subset \mathcal{D}$. *If* $f_c : \mathcal{D} \to \mathbb{R}^n$ *is essentially nonnegative and* $f_d : \mathcal{Z}_x \to \mathbb{R}^n$ *is such that* $x + f_d(x)$ *is nonnegative, then* $\overline{\mathbb{R}}_+^n$ *is an invariant set with respect to (4.1) and (4.2).*

Proof. Consider the continuous-time dynamical system given by

$$\dot{x}_c(t) = f_c(x_c(t)), \quad x_c(0) = x_{c0}, \quad t \geq 0. \tag{4.3}$$

Now, it follows from Theorem 3.1 of [23] (see also Proposition 6.1 of [57]) that since $f_c : \mathcal{D} \to \mathbb{R}^n$ is essentially nonnegative, $\overline{\mathbb{R}}_+^n$ is an invariant set with respect to (4.3), that is, if $x_{c0} \in \overline{\mathbb{R}}_+^n$, then $x_c(t) \in \overline{\mathbb{R}}_+^n$, $t \geq 0$. Now, since with $x_{c0} = x_0$, $x(t) = x_c(t)$, $0 \leq t \leq \tau_1(x_0)$, it follows that $x(t) \in \overline{\mathbb{R}}_+^n$, $0 \leq t \leq \tau_1(x_0)$. Next, since $f_d : \mathcal{Z}_x \to \mathbb{R}^n$ is such that $x + f_d(x)$ is nonnegative it follows that $x_1^+ = x(\tau_1(x_0)) + f_d(x(\tau_1(x_0))) \in \overline{\mathbb{R}}_+^n$. Now, since $s(t, x_0) = s(t - \tau_1(x_0), x_1^+)$, $\tau_1(x_0) < t \leq \tau_2(x_0)$, with $x_{c0} = x_1^+$, it follows that $x(t) = x_c(t - \tau_1(x_0)) \in \overline{\mathbb{R}}_+^n$, $\tau_1(x_0) < t \leq \tau_2(x_0)$, and hence, $x_2^+ = x(\tau_2(x_0)) + f_d(x(\tau_2(x_0))) \in \overline{\mathbb{R}}_+^n$. Repeating this procedure for $\tau_i(x_0)$, $i = 3, 4, \ldots$, it follows that $\overline{\mathbb{R}}_+^n$ is an invariant set with respect to (4.1) and (4.2). $\qquad\square$

It is important to note that unlike continuous-time nonnegative systems [57] and discrete-time nonnegative systems [58], Proposition 4.1 provides only sufficient conditions for assuring that $\overline{\mathbb{R}}_+^n$ is an invariant set with respect to (4.1) and (4.2). To see this, let $\mathcal{Z}_x = \partial \overline{\mathbb{R}}_+^n$ and assume $x + f_d(x)$, $x \in \mathcal{Z}_x$, is nonnegative. Then, $\overline{\mathbb{R}}_+^n$ remains invariant with respect to (4.1) and (4.2) irrespective of whether $f_c(\cdot)$ is essentially nonnegative or not.

Next, we specialize Proposition 4.1 to linear state-dependent impulsive dynamical systems of the form

$$\dot{x}(t) = A_c x(t), \quad x(0) = x_0, \quad x(t) \notin \mathcal{Z}_x, \qquad (4.4)$$
$$\Delta x(t) = (A_d - I_n)x(t), \quad x(t) \in \mathcal{Z}_x, \qquad (4.5)$$

where $t \geq 0$, $x(t) \in \overline{\mathbb{R}}_+^n$, $A_c \in \mathbb{R}^{n \times n}$ is essentially nonnegative, $A_d \in \mathbb{R}^{n \times n}$ is nonnegative, and $\mathcal{Z}_x \subset \overline{\mathbb{R}}_+^n$. Note that in this case Assumption A2 implies that if $x \in \mathcal{Z}_x$, then $A_d x \notin \mathcal{Z}_x$.

Proposition 4.2 *Let $A_c \in \mathbb{R}^{n \times n}$ and $A_d \in \mathbb{R}^{n \times n}$. If A_c is essentially nonnegative and A_d is nonnegative, then $\overline{\mathbb{R}}_+^n$ is an invariant set with respect to (4.4) and (4.5).*

Proof. The proof is a direct consequence of Proposition 4.1 with $f_c(x) = A_c x$ and $f_d(x) = (A_d - I_n)x$. $\qquad\square$

The following definition introduces several types of stability corresponding to the equilibrium solution $x(t) \equiv x_e$ of (4.1) and (4.2) whose solutions remain in the nonnegative orthant $\overline{\mathbb{R}}_+^n$.

Definition 4.4 *Let $\overline{\mathbb{R}}_+^n$ be invariant with respect to (4.1) and (4.2) and let $x_e \in \overline{\mathbb{R}}_+^n$. Then, the equilibrium solution $x(t) \equiv x_e$ of the impulsive nonnegative dynamical system (4.1) and (4.2) is Lyapunov stable if, for every $\varepsilon > 0$, there exists $\delta = \delta(\varepsilon) > 0$ such that if $x_0 \in$*

$\mathcal{B}_\delta(x_e) \cap \overline{\mathbb{R}}^n_+$, then $x(t) \in \mathcal{B}_\varepsilon(x_e) \cap \overline{\mathbb{R}}^n_+$, $t \geq 0$. *The equilibrium solution* $x(t) \equiv x_e$ *of the impulsive nonnegative dynamical system (4.1) and (4.2) is* asymptotically stable *if it is Lyapunov stable and there exists* $\delta > 0$ *such that if* $x_0 \in \mathcal{B}_\delta(x_e) \cap \overline{\mathbb{R}}^n_+$, *then* $\lim_{t \to \infty} x(t) = x_e$ *Finally, the equilibrium solution* $x(t) \equiv x_e$ *of the impulsive nonnegative dynamical system (4.1) and (4.2) is* globally asymptotically stable *if the previous statement holds for all* $x_0 \in \overline{\mathbb{R}}^n_+$.

Next, we present several key results on stability of nonlinear impulsive nonnegative dynamical systems. We note that the standard Lyapunov stability theorems and invariant set theorems developed in Chapter 2 for nonlinear impulsive dynamical systems can be used directly with the required sufficient conditions verified on $\overline{\mathbb{R}}^n_+$.

Theorem 4.1 *Suppose there exists a continuously differentiable function* $V : \overline{\mathbb{R}}^n_+ \to [0, \infty)$ *satisfying* $V(x_e) = 0$, $V(x) > 0$, $x \neq x_e$, *and*

$$V'(x) f_c(x) \leq 0, \quad x \notin \mathcal{Z}_x, \tag{4.6}$$
$$V(x + f_d(x)) \leq V(x), \quad x \in \mathcal{Z}_x. \tag{4.7}$$

Then the equilibrium solution $x(t) \equiv x_e$ *of the impulsive nonnegative dynamical system (4.1) and (4.2) is Lyapunov stable. Furthermore, if the inequality (4.6) is strict for all* $x \neq x_e$, *then the equilibrium solution* $x(t) \equiv x_e$ *of the impulsive nonnegative dynamical system (4.1) and (4.2) is asymptotically stable. Finally, if, in addition,* $V(x) \to \infty$ *as* $\|x\| \to \infty$, *then the above asymptotic stability result is global.*

Proof. The proof is identical to the proof of Theorem 2.1 with $\mathcal{D} = \overline{\mathbb{R}}^n_+$ and $\mathcal{Z}_x \subset \overline{\mathbb{R}}^n_+$. $\square$

Next, we present a generalized Krasovskii-LaSalle invariant set stability theorem for impulsive nonnegative dynamical systems. For this result we assume that $f_c(\cdot)$, $f_d(\cdot)$, and $\mathcal{Z}_x$ are such that Assumption 2.1 holds.

Theorem 4.2 *Consider the impulsive nonnegative dynamical system* $\mathcal{G}$ *given by (4.1) and (4.2), assume* $\mathcal{D}_c \subset \overline{\mathbb{R}}^n_+$ *is a compact positively invariant set with respect to (4.1) and (4.2), and assume that there exists a continuously differentiable function* $V : \mathcal{D}_c \to \mathbb{R}$ *such that*

$$V'(x) f_c(x) \leq 0, \quad x \in \mathcal{D}_c, \quad x \notin \mathcal{Z}_x, \tag{4.8}$$
$$V(x + f_d(x)) \leq V(x), \quad x \in \mathcal{D}_c, \quad x \in \mathcal{Z}_x. \tag{4.9}$$

Let $\mathcal{R} \triangleq \{x \in \mathcal{D}_c : x \notin \mathcal{Z}_x, V'(x) f_c(x) = 0\} \cup \{x \in \mathcal{D}_c : x \in \mathcal{Z}_x, V(x + f_d(x)) = V(x)\}$ *and let* $\mathcal{M}$ *denote the largest invariant set contained in* $\mathcal{R}$. *If* $x_0 \in \mathcal{D}_c$, *then* $x(t) \to \mathcal{M}$ *as* $t \to \infty$.

Proof. The proof is identical to the proof of Theorem 2.3 with $\mathcal{D} = \overline{\mathbb{R}}_+^n$. □

Finally, we give sufficient conditions for Lyapunov stability and asymptotic stability for linear impulsive nonnegative dynamical systems using *linear* Lyapunov functions.

Theorem 4.3 *Consider the linear impulsive dynamical system given by (4.4) and (4.5) where $A_c \in \mathbb{R}^{n \times n}$ is essentially nonnegative and $A_d \in \mathbb{R}^{n \times n}$ is nonnegative. Then the following statements hold:*

i) *If there exist vectors $p, r_c, r_d \in \mathbb{R}^n$ such that $p >> 0$, $r_c \geq\geq 0$, and $r_d \geq\geq 0$ satisfy*

$$0 = x^T(A_c^T p + r_c), \quad x \notin \mathcal{Z}_x, \tag{4.10}$$
$$0 = x^T(A_d^T p - p + r_d), \quad x \in \mathcal{Z}_x, \tag{4.11}$$

then the zero solution $x(t) \equiv 0$ to (4.4) and (4.5) is Lyapunov stable.

ii) *If there exist vectors $p, r_c, r_d \in \mathbb{R}^n$ such that $p >> 0$, $r_c >> 0$, and $r_d \geq\geq 0$ satisfy (4.10) and (4.11), then the zero solution $x(t) \equiv 0$ to (4.4) and (4.5) is asymptotically stable.*

Proof. The result is a direct consequence of Theorem 4.1 with $V(x) = p^T x$, $f_c(x) = A_c x$, and $f_d(x) = (A_d - I_n)x$. Specifically, in this case, $V'(x)f_c(x) = p^T A_c x = -r_c^T x \leq 0$, $x \notin \mathcal{Z}_x$, and $V(x + f_d(x)) - V(x) = p^T A_d x - p^T x = -r_d^T x \leq 0$, $x \in \mathcal{Z}_x$, so that all the conditions of Theorem 4.1 are satisfied, which proves Lyapunov stability. In the case where $r_c >> 0$ it follows that $V'(x)f_c(x) = p^T A_c x = -r_c^T x < 0$, $x \notin \mathcal{Z}_x$, which proves asymptotic stability. □

For asymptotic stability, conditions (4.10) and (4.11) are implied by $p >> 0$, $A_c^T p << 0$, and $(A_d - I_n)^T p \leq\leq 0$ which can be solved using a linear matrix inequality (LMI) feasibility problem [27]. Specifically, for a given $r_c \in \mathbb{R}^n$ and $r_d \in \mathbb{R}^n$, note that there exists $p \in \mathbb{R}^n$ such that

$$0 = A_c^T p + r_c, \tag{4.12}$$
$$0 = A_d^T p - p + r_d, \tag{4.13}$$

if and only if $\text{rank}[A \;\; r] = \text{rank}\, A$, where

$$A \triangleq \begin{bmatrix} A_c^T \\ (A_d - I_n)^T \end{bmatrix}, \qquad r \triangleq \begin{bmatrix} r_c \\ r_d \end{bmatrix}. \tag{4.14}$$

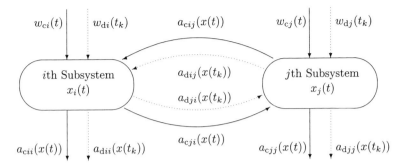

Figure 4.1 Nonlinear impulsive compartmental interconnected subsystem model.

Now, there exist $p >> 0$, $r_{\rm c} >> 0$, and $r_{\rm d} \geq\geq 0$ such that (4.12) and (4.13) are satisfied if and only if $p >> 0$ and $-Ap \geq\geq 0$.

4.3 Impulsive Compartmental Dynamical Systems

In this section, we specialize the results of Section 4.2 to impulsive compartmental dynamical systems. Specifically, we show that nonlinear impulsive compartmental dynamical systems are a special case of impulsive nonnegative dynamical systems. To see this, let $x_i(t)$, $i = 1,\ldots,n$, denote the mass (and hence a nonnegative quantity) of the ith subsystem of the impulsive compartmental system shown in Figure 4.1, let $a_{cii}(x) \geq 0$, $x \notin \mathcal{Z}_x$, denote the rate of flow of mass loss of the ith continuous-time subsystem, let $w_{ci}(t) \geq 0$, $t \geq 0$, $i = 1,\ldots,n$, denote the rate of mass inflow supplied to the ith continuous-time subsystem, and let $\phi_{cij}(x(t))$, $t \geq 0$, $i \neq j$, $i,j = 1,\ldots,n$, denote the net mass flow (or flux) from the jth continuous-time subsystem to the ith continuous-time subsystem given by $\phi_{cij}(x(t)) = a_{cij}(x(t)) - a_{cji}(x(t))$, where the rates of mass flows are such that $a_{cij}(x) \geq 0$, $x \notin \mathcal{Z}_x$, $i \neq j$, $i,j = 1,\ldots,n$.

Similarly, for the resetting dynamics, let $a_{dii}(x) \geq 0$, $x \in \mathcal{Z}_x$, denote the mass loss of the ith discrete-time subsystem, let $w_{di}(t_k) \geq 0$, $i = 1,\ldots,n$, denote the mass inflow supplied to the ith discrete-time subsystem, and let $\phi_{dij}(t_k)$, $i \neq j$, $i,j = 1,\ldots,n$, denote the net mass exchange from the jth discrete-time subsystem to the ith discrete-time subsystem given by $\phi_{dij}(x(t_k)) = a_{dij}(x(t_k)) - a_{dji}(x(t_k))$, where $t_k = \tau_k(x_0)$ and the mass flows are such that $a_{dij}(x) \geq 0$, $x \in \mathcal{Z}_x$, $i \neq j$, $i,j = 1,\ldots,n$.

A mass balance for the whole impulsive compartmental system

yields

$$\dot{x}_i(t) = -a_{cii}(x(t)) + \sum_{j=1, i \neq j}^{n} \phi_{cij}(x(t)) + w_{ci}(t), \quad x(t) \notin \mathcal{Z}_x,$$

$$i = 1, \ldots, n, \qquad (4.15)$$

$$\Delta x_i(t) = -a_{dii}(x(t)) + \sum_{j=1, i \neq j}^{n} \phi_{dij}(x(t)) + w_{di}(t), \quad x(t) \in \mathcal{Z}_x,$$

$$i = 1, \ldots, n, \qquad (4.16)$$

or, equivalently,

$$\dot{x}(t) = f_c(x(t)) + w_c(t), \quad x(0) = x_0, \quad x(t) \notin \mathcal{Z}_x, \qquad (4.17)$$
$$\Delta x(t) = f_d(x(t)) + w_d(t), \quad x(t) \in \mathcal{Z}_x, \qquad (4.18)$$

where $x(t) \triangleq [x_1(t), \ldots, x_n(t)]^T$, $w_c(t) \triangleq [w_{c1}(t), \ldots, w_{cn}(t)]^T$, $w_d(t) \triangleq [w_{d1}(t), \ldots, w_{dn}(t)]^T$, and for $i, j = 1, \ldots, n$,

$$f_{ci}(x) = -a_{cii}(x) + \sum_{j=1, i \neq j}^{n} [a_{cij}(x) - a_{cji}(x)], \qquad (4.19)$$

$$f_{di}(x) = -a_{dii}(x) + \sum_{j=1, i \neq j}^{n} [a_{dij}(x) - a_{dji}(x)]. \qquad (4.20)$$

Since all mass flows as well as compartment sizes are nonnegative, it follows that for all $i = 1, \ldots, n$, $f_{ci}(x) \geq 0$ for all $x \notin \mathcal{Z}_x$ whenever $x_i = 0$ and regardless of the values of x_j, $j \neq i$. Furthermore, $x_i + f_{di}(x) \geq 0$ for all $x \in \mathcal{Z}_x$. The above physical constraints are implied by $a_{cij}(x) \geq 0$, $a_{cii}(x) \geq 0$, $x \notin \mathcal{Z}_x$, $a_{dij}(x) \geq 0$, $a_{dii}(x) \geq 0$, $x \in \mathcal{Z}_x$, $w_{ci} \geq 0$, $w_{di} \geq 0$, for all $i, j = 1, \ldots, n$, and if $x_i = 0$, then $a_{cii}(x) = 0$ and $a_{cji}(x) = 0$ for all $i, j = 1, \ldots, n$, so that $\dot{x}_i \geq 0$. In this case, $f_c(x)$, $x \notin \mathcal{Z}_x$, is essentially nonnegative and $x + f_d(x)$, $x \in \mathcal{Z}_x$, is nonnegative, and hence, the impulsive compartmental model given by (4.15) and (4.16) is an impulsive nonnegative dynamical system.

Taking the total mass of the compartmental system $V(x) = e^T x = \sum_{i=1}^{n} x_i$, where $e^T \triangleq [1, \ldots, 1]$, as a Lyapunov function for the undisturbed (i.e., $w_c(t) \equiv 0$ and $w_d(t_k) \equiv 0$) system (4.15) and (4.16), and assuming $a_{cij}(0) = 0$, $i, j = 1, \ldots, n$, it follows that

$$\dot{V}(x) = \sum_{i=1}^{n} \dot{x}_i \qquad (4.21)$$

$$= -\sum_{i=1}^{n} a_{cii}(x) + \sum_{i=1}^{n}\sum_{j=1,i\neq j}^{n} [a_{cij}(x) - a_{cji}(x)] \quad (4.22)$$

$$= -\sum_{i=1}^{n} a_{cii}(x) \quad (4.23)$$

$$\leq 0, \quad x \notin \mathcal{Z}_x, \quad (4.24)$$

and

$$\Delta V(x) = \sum_{i=1}^{n} \Delta x_i \quad (4.25)$$

$$= -\sum_{i=1}^{n} a_{dii}(x) + \sum_{i=1}^{n}\sum_{j=1,i\neq j}^{n} [a_{dij}(x) - a_{dji}(x)] \quad (4.26)$$

$$= -\sum_{i=1}^{n} a_{dii}(x) \quad (4.27)$$

$$\leq 0, \quad x \in \mathcal{Z}_x, \quad (4.28)$$

which, by Theorem 4.1, shows that the zero solution $x(t) \equiv 0$ of the nonlinear impulsive compartmental system given by (4.15) and (4.16) is Lyapunov stable. If (4.15) and (4.16), with $w_c(t) \equiv 0$ and $w_d(t_k) \equiv 0$, has losses (outflows) from all compartments over the continuous-time dynamics, then $a_{cii}(x) > 0$, $x \notin \mathcal{Z}_x$, $x \neq 0$, and by Theorem 4.1 the zero solution $x(t) \equiv 0$ to (4.15) and (4.16) is asymptotically stable.

It is interesting to note that in the linear case $a_{cii}(x) = a_{cii}x_i$, $\phi_{cij}(x) = a_{cij}x_j - a_{cji}x_i$, $a_{dii}(x) = a_{dii}x_i$, and $\phi_{dij}(x) = a_{dij}x_j - a_{dji}x_i$, where $a_{cij} \geq 0$ and $a_{dij} \geq 0$, $i,j = 1,\dots,n$, so that (4.17) and (4.18) become

$$\dot{x}(t) = A_c x(t) + w_c(t), \quad x(0) = x_0, \quad x(t) \notin \mathcal{Z}_x, \quad (4.29)$$
$$\Delta x(t) = (A_d - I_n)x(t) + w_d(t), \quad x(t) \in \mathcal{Z}_x, \quad (4.30)$$

where for $i,j = 1,\dots,n$,

$$A_{c(i,j)} = \begin{cases} -\sum_{l=1}^{n} a_{cli}, & i = j, \\ a_{cij}, & i \neq j, \end{cases} \quad (4.31)$$

$$A_{d(i,j)} = \begin{cases} 1 - \sum_{l=1}^{n} a_{dli}, & i = j, \\ a_{dij}, & i \neq j. \end{cases} \quad (4.32)$$

Note that since at any given instant of time compartmental mass can only be transported, stored, or discharged but not created, and the maximum amount of mass that can be transported and/or discharged cannot exceed the mass in a compartment, it follows that $1 \geq \sum_{l=1}^{n} a_{\mathrm{d}li}$. Thus, A_{c} is an essentially nonnegative matrix and A_{d} is a nonnegative matrix, and hence, the impulsive compartmental model given by (4.29) and (4.30) is an impulsive nonnegative dynamical system.

The impulsive compartmental system (4.15) and (4.16) with no inflows, that is, $w_{\mathrm{c}i}(t) \equiv 0$ and $w_{\mathrm{d}i}(t_k) \equiv 0$, $i = 1, \ldots, n$, is said to be *inflow-closed*. Alternatively, if (4.15) and (4.16) possesses no losses (outflows) it is said to be *outflow-closed*. An impulsive compartmental system is said to be *closed* if it is inflow-closed and outflow-closed. Note that for a closed system $\dot{V}(x) = 0$, $x \notin \mathcal{Z}_x$, and $\Delta V(x) = 0$, $x \in \mathcal{Z}_x$, which shows that the total mass inside a closed system is conserved. Alternatively, in the case where $a_{\mathrm{c}ii}(x) \neq 0$, $x \notin \mathcal{Z}_x$, $a_{\mathrm{d}ii}(x) \neq 0$, $x \in \mathcal{Z}_x$, $w_{\mathrm{c}i}(t) \neq 0$, and $w_{\mathrm{d}i}(t_k) \neq 0$, $i = 1, \ldots, n$, it follows that (4.15) and (4.16) can be equivalently written as

$$\dot{x}(t) = [J_{\mathrm{c}n}(x(t)) - D_{\mathrm{c}}(x(t))] \left(\frac{\partial V}{\partial x}(x(t)) \right)^{\mathrm{T}} + w_{\mathrm{c}}(t), \quad x(t) \notin \mathcal{Z}_x,$$

$$(4.33)$$

$$\Delta x(t) = [J_{\mathrm{d}n}(x(t)) - D_{\mathrm{d}}(x(t))] \left(\frac{\partial V}{\partial x}(x(t)) \right)^{\mathrm{T}} + w_{\mathrm{d}}(t), \quad x(t) \in \mathcal{Z}_x,$$

$$(4.34)$$

where $J_{\mathrm{c}n}(x)$ and $J_{\mathrm{d}n}(x)$ are skew-symmetric matrix functions with $J_{\mathrm{c}n(i,i)}(x) = 0$, $J_{\mathrm{d}n(i,i)}(x) = 0$, $J_{\mathrm{c}n(i,j)}(x) = a_{\mathrm{c}ij}(x) - a_{\mathrm{c}ji}(x)$, and $J_{\mathrm{d}n(i,j)}(x) = a_{\mathrm{d}ij}(x) - a_{\mathrm{d}ji}(x)$, $i \neq j$, $D_{\mathrm{c}}(x) = \mathrm{diag}[a_{\mathrm{c}11}(x), a_{\mathrm{c}22}(x), \ldots, a_{\mathrm{c}nn}(x)] \geq\geq 0$, and $D_{\mathrm{d}}(x) = \mathrm{diag}[a_{\mathrm{d}11}(x), a_{\mathrm{d}22}(x), \ldots, a_{\mathrm{d}nn}(x)] \geq\geq 0$, $x \in \overline{\mathbb{R}}_{+}^{n}$.

In light of the above, an impulsive compartmental system is an *impulsive port-controlled Hamiltonian system* [70] with a Hamiltonian $\mathcal{H}(x) = V(x) = \mathbf{e}^{\mathrm{T}} x$ representing the total mass in the system, $D_{\mathrm{c}}(x)$ representing the outflow dissipation over the continuous-time dynamics, $D_{\mathrm{d}}(x)$ representing the outflow dissipation at the resetting instants, $w_{\mathrm{c}}(t)$ representing the supplied flux to the system over the continuous-time dynamics, and $w_{\mathrm{d}}(t_k)$ representing the supplied mass to the system at the resetting instants. This observation shows that impulsive compartmental systems are conservative systems. This will be further elaborated on in the following sections.

4.4 Dissipativity Theory for Impulsive Nonnegative Dynamical Systems

In this section, we extend the notion of dissipativity to nonlinear impulsive nonnegative dynamical systems. Specifically, we consider nonlinear impulsive dynamical systems $\mathcal{G}$ of the form

$$\dot{x}(t) = f_c(x(t)) + G_c(x(t))u_c(t), \quad x(0) = x_0, \quad (x(t), u_c(t)) \notin \mathcal{Z},$$
$$(4.35)$$
$$\Delta x(t) = f_d(x(t)) + G_d(x(t))u_d(t), \quad (x(t), u_c(t)) \in \mathcal{Z}, \quad (4.36)$$
$$y_c(t) = h_c(x(t)) + J_c(x(t))u_c(t), \quad (x(t), u_c(t)) \notin \mathcal{Z}, \quad (4.37)$$
$$y_d(t) = h_d(x(t)) + J_d(x(t))u_d(t), \quad (x(t), u_c(t)) \in \mathcal{Z}, \quad (4.38)$$

where $t \geq 0$, $x(t) \in \mathcal{D} \subseteq \mathbb{R}^n$, $\mathcal{D}$ is an open set with $0 \in \mathcal{D}$, $\Delta x(t) \triangleq x(t^+) - x(t)$, $u_c(t) \in U_c \subseteq \mathbb{R}^{m_c}$, $u_d(t_k) \in U_d \subseteq \mathbb{R}^{m_d}$, t_k denotes the kth instant of time at which $(x(t), u_c(t))$ intersects $\mathcal{Z} \subset \mathcal{D} \times U_c$ for a particular trajectory $x(t)$ and input $u_c(t)$, $y_c(t) \in Y_c \subseteq \mathbb{R}^{l_c}$, $y_d(t_k) \in Y_d \subseteq \mathbb{R}^{l_d}$, $f_c : \mathcal{D} \to \mathbb{R}^n$ is Lipschitz continuous and satisfies $f_c(0) = 0$, $G_c : \mathcal{D} \to \mathbb{R}^{n \times m_c}$, $f_d : \mathcal{D} \to \mathbb{R}^n$ is continuous, $G_d : \mathcal{D} \to \mathbb{R}^{n \times m_d}$, $h_c : \mathcal{D} \to \mathbb{R}^{l_c}$ and satisfies $h_c(0) = 0$, $J_c : \mathcal{D} \to \mathbb{R}^{l_c \times m_c}$, $h_d : \mathcal{D} \to \mathbb{R}^{l_d}$, $J_d : \mathcal{D} \to \mathbb{R}^{l_d \times m_d}$, and $\mathcal{Z} \subset \mathcal{D} \times U_c$.

Here, we assume that $u_c(\cdot)$ and $u_d(\cdot)$ are restricted to the class of *admissible* inputs consisting of measurable functions such that $(u_c(t), u_d(t_k)) \in U_c \times U_d$ for all $t \geq 0$ and $k \in \mathbb{Z}_{[0,t)} \triangleq \{k : 0 \leq t_k < t\}$, where the constraint set $U_c \times U_d$ is given with $(0,0) \in U_c \times U_d$. Furthermore, we assume that the set $\mathcal{Z} \triangleq \{(x, u_c) : \mathcal{X}(x, u_c) = 0\}$, where $\mathcal{X} : \mathcal{D} \times U_c \to \mathbb{R}$. In addition, we assume that the system functions $f_c(\cdot)$, $f_d(\cdot)$, $G_c(\cdot)$, $G_d(\cdot)$, $h_c(\cdot)$, $h_d(\cdot)$, $J_c(\cdot)$, and $J_d(\cdot)$ are continuous mappings. Finally, for the nonlinear dynamical system (4.35) we assume that the required properties for the existence and uniqueness of solutions are satisfied such that (4.35) has a unique solution for all $t \in \mathbb{R}$ [14, 93].

Next, we provide definitions and several results concerning dynamical systems of the form (4.35)–(4.38) with nonnegative inputs and nonnegative outputs.

Definition 4.5 *The nonlinear impulsive dynamical system $\mathcal{G}$ given by (4.35)–(4.38) with $x(0) = 0$ is* input-output[2] nonnegative *if the*

---

[2] The outputs here refer to measured outputs or observations and may have nothing to do with material outflows of the nonnegative compartmental system.

hybrid output $(y_c(t), y_d(t_k))$, $t \geq 0$, $k \in \overline{\mathbb{Z}}_+$, is nonnegative for every nonnegative hybrid input $(u_c(t), u_d(t_k))$, $t \geq 0$, $k \in \overline{\mathbb{Z}}_+$.

Definition 4.6 *The nonlinear impulsive dynamical system $\mathcal{G}$ given by (4.35)–(4.38) is* nonnegative *if for every $x(0) \in \overline{\mathbb{R}}_+^n$ and nonnegative hybrid input $(u_c(t), u_d(t_k))$, $t \geq 0$, $k \in \overline{\mathbb{Z}}_+$, the solution $x(t)$, $t \geq 0$, to (4.35) and (4.36) and the hybrid output $(y_c(t), y_d(t_k))$, $t \geq 0$, $k \in \overline{\mathbb{Z}}_+$, are nonnegative.*

Proposition 4.3 *Consider the nonlinear dynamical system $\mathcal{G}$ given by (4.35)–(4.38). If $f_c : \mathcal{D} \to \mathbb{R}^n$ is essentially nonnegative, $f_d : \mathcal{D} \to \mathbb{R}^n$ is such that $x + f_d(x)$ is nonnegative for all $x \in \overline{\mathbb{R}}_+^n$, $G_c(x) \geq\geq 0$, $G_d(x) \geq\geq 0$, $h_c(x) \geq\geq 0$, $h_d(x) \geq\geq 0$, $J_c(x) \geq\geq 0$, and $J_d(x) \geq\geq 0$, $x \in \overline{\mathbb{R}}_+^n$, then $\mathcal{G}$ is nonnegative.*

Proof. The proof is similar to the proof of Proposition 4.1 and, hence, is omitted. □

For the impulsive dynamical system $\mathcal{G}$ given by (4.35)–(4.38) let the function $(s_c(u_c, y_c), s_d(u_d, y_d))$, where $s_c : U_c \times Y_c \to \mathbb{R}$ and $s_d : U_d \times Y_d \to \mathbb{R}$ are such that $s_c(0, 0) = 0$ and $s_d(0, 0) = 0$, be a hybrid supply rate. The following definition introduces the notion of dissipativity and exponential dissipativity for a nonlinear impulsive nonnegative dynamical system. For this definition, define $\mathbb{Z}_{[t,\hat{t})} \triangleq \{k : t \leq t_k < \hat{t}\}$, $k \in \overline{\mathbb{Z}}_+$.

Definition 4.7 *The impulsive dynamical system $\mathcal{G}$ given by (4.35)–(4.38) is* exponentially dissipative *(respectively,* dissipative*) with respect to the hybrid supply rate (s_c, s_d) if there exists a continuous, nonnegative-definite function $V_s : \overline{\mathbb{R}}_+^n \to \overline{\mathbb{R}}_+$ called a* storage function *and a scalar $\varepsilon > 0$ (respectively, $\varepsilon = 0$) such that $V_s(0) = 0$ and the dissipation inequality*

$$e^{\varepsilon T} V_s(x(T)) \leq e^{\varepsilon t_0} V_s(x(t_0)) + \int_{t_0}^{T} e^{\varepsilon t} s_c(u_c(t), y_c(t)) dt$$

$$+ \sum_{k \in \mathbb{Z}_{[t_0, T)}} e^{\varepsilon t_k} s_d(u_d(t_k), y_d(t_k)), \quad T \geq t_0, \quad (4.39)$$

is satisfied for all $T \geq t_0$, where $x(t)$, $t \geq t_0$, is a solution of (4.35)–(4.38) with $(u_c(t), u_d(t_k)) \in U_c \times U_d$. The impulsive dynamical system given by (4.35)–(4.38) is lossless *with respect to the hybrid supply rate (s_c, s_d) if the dissipation inequality (4.39) is satisfied as an equality with $\varepsilon = 0$ for all $T \geq t_0$.*

The following result gives necessary and sufficient conditions for dissipativity over an interval $t \in (t_k, t_{k+1}]$ involving the consecutive resetting times t_k and t_{k+1}.

Theorem 4.4 *Assume $\mathcal{G}$ given by (4.35)–(4.38) is completely reachable. Then $\mathcal{G}$ is dissipative with respect to the hybrid supply rate (s_c, s_d) if and only if there exists a continuous, nonnegative-definite function $V_s : \overline{\mathbb{R}}_+^n \to \overline{\mathbb{R}}_+$ such that, for all $k \in \overline{\mathbb{Z}}_+$,*

$$V_s(x(\hat{t})) - V_s(x(t)) \leq \int_t^{\hat{t}} s_c(u_c(s), y_c(s)) \mathrm{d}s, \quad t_k < t \leq \hat{t} \leq t_{k+1},$$

$$(4.40)$$

$$V_s(x(t_k) + f_d(x(t_k)) + G_d(x(t_k)) u_d(t_k)) - V_s(x(t_k))$$
$$\leq s_d(u_d(t_k), y_d(t_k)). \quad (4.41)$$

Furthermore, $\mathcal{G}$ is exponentially dissipative with respect to the hybrid supply rate (s_c, s_d) if and only if there exist a continuous, nonnegative-definite function $V_s : \overline{\mathbb{R}}_+^n \to \overline{\mathbb{R}}_+$ and a scalar $\varepsilon > 0$ such that

$$e^{\varepsilon \hat{t}} V_s(x(\hat{t})) - e^{\varepsilon t} V_s(x(t)) \leq \int_t^{\hat{t}} e^{\varepsilon s} s_c(u_c(s), y_c(s)) \mathrm{d}s,$$

$$t_k < t \leq \hat{t} \leq t_{k+1}, \quad (4.42)$$
$$V_s(x(t_k) + f_d(x(t_k)) + G_d(x(t_k)) u_d(t_k)) - V_s(x(t_k))$$
$$\leq s_d(u_d(t_k), y_d(t_k)). \quad (4.43)$$

Finally, $\mathcal{G}$ is lossless with respect to the hybrid supply rate (s_c, s_d) if and only if there exists a continuous, nonnegative-definite function $V_s : \overline{\mathbb{R}}_+^n \to \overline{\mathbb{R}}_+$ such that (4.40) and (4.41) are satisfied as equalities.

Proof. The proof is identical to the proof of Theorem 3.2.   □

If $V_s(\cdot)$ is continuously differentiable, then an equivalent statement for dissipativeness of the impulsive dynamical system $\mathcal{G}$ with respect to the hybrid supply rate (s_c, s_d) is

$$\dot{V}_s(x(t)) \leq s_c(u_c(t), y_c(t)), \quad t_k < t \leq t_{k+1}, \quad (4.44)$$
$$\Delta V_s(x(t_k)) \leq s_d(u_d(t_k), y_d(t_k)), \quad k \in \overline{\mathbb{Z}}_+, \quad (4.45)$$

where $\dot{V}_s(\cdot)$ denotes the total derivative of $V_s(x(t))$ along the state trajectories $x(t)$, $t \in (t_k, t_{k+1}]$, of the impulsive dynamical system (4.35)–(4.38) and $\Delta V_s(x(t_k)) \triangleq V_s(x(t_k^+)) - V_s(x(t_k)) = V_s(x(t_k) +$

$f_d(x(t_k)) + G_d(x(t_k))u_d(t_k)) - V_s(x(t_k))$, $k \in \overline{\mathbb{Z}}_+$, denotes the difference of the storage function $V_s(x)$ at the resetting times t_k, $k \in \overline{\mathbb{Z}}_+$,
of the impulsive dynamical system (4.35)–(4.38). Furthermore, an
equivalent statement for exponential dissipativeness of the impulsive
dynamical system $\mathcal{G}$ with respect to the hybrid supply rate (s_c, s_d) is
given by

$$\dot{V}_s(x(t)) + \varepsilon V_s(x(t)) \leq s_c(u_c(t), y_c(t)), \quad t_k < t \leq t_{k+1}, \qquad (4.46)$$

and (4.45).

The following result presents Kalman-Yakubovich-Popov conditions
for impulsive nonnegative dynamical systems with linear hybrid sup-
ply rates of the form $(s_c(u_c, y_c), s_d(u_d, y_d)) = (q_c^T y_c + r_c^T u_c, q_d^T y_d + r_d^T u_d)$, where $q_c \in \mathbb{R}^{l_c}$, $q_c \neq 0$, $r_c \in \mathbb{R}^{m_c}$, $r_c \neq 0$, $q_d \in \mathbb{R}^{l_d}$, $q_d \neq 0$,
$r_d \in \mathbb{R}^{m_d}$, and $r_d \neq 0$. For the remainder of the section we assume
that $U_c = \overline{\mathbb{R}}_+^{m_c}$, $U_d = \overline{\mathbb{R}}_+^{m_d}$, and $\mathcal{Z} = \mathcal{Z}_x \times \overline{\mathbb{R}}^{m_c}$ so that resetting occurs
only when $x(t)$ intersects $\mathcal{Z}_x$.

Theorem 4.5 *Let $q_c \in \mathbb{R}^{l_c}$, $r_c \in \mathbb{R}^{m_c}$, $q_d \in \mathbb{R}^{l_d}$, and $r_d \in \mathbb{R}^{m_d}$.
Consider the nonlinear hybrid dynamical system $\mathcal{G}$ given by (4.35)–
(4.38) where $f_c : \mathcal{D} \to \mathbb{R}^n$ is essentially nonnegative, $f_d : \mathcal{Z}_x \to
\mathbb{R}^n$ is such that $x + f_d(x)$ is nonnegative, $G_c(x) \geq\geq 0$, $G_d(x) \geq\geq
0$, $h_c(x) \geq\geq 0$, $h_d(x) \geq\geq 0$, $J_c(x) \geq\geq 0$, and $J_d(x) \geq\geq 0$, $x \in
\overline{\mathbb{R}}_+^n$. If there exist functions $V_s : \overline{\mathbb{R}}_+^n \to \overline{\mathbb{R}}_+$, $\ell_c : \overline{\mathbb{R}}_+^n \to \overline{\mathbb{R}}_+$, $\ell_d :
\overline{\mathbb{R}}_+^n \to \overline{\mathbb{R}}_+$, $\mathcal{W}_c : \overline{\mathbb{R}}_+^n \to \overline{\mathbb{R}}_+^{m_c}$, $\mathcal{W}_d : \overline{\mathbb{R}}_+^n \to \overline{\mathbb{R}}_+^{m_d}$, and a scalar $\varepsilon >
0$ (respectively, $\varepsilon = 0$) such that $V_s(\cdot)$ is continuously differentiable,
nonnegative definite, $V_s(0) = 0$,*

$$V_s(x + f_d(x) + G_d(x)u_d) = V_s(x + f_d(x)) + V_s'(x + f_d(x))G_d(x)u_d,$$
$$x \in \mathcal{Z}_x, \quad u_d \in \overline{\mathbb{R}}_+^{m_d}, \qquad (4.47)$$

and

$$0 = V_s'(x)f_c(x) + \varepsilon V_s(x) - q_c^T h_c(x) + \ell_c(x), \quad x \notin \mathcal{Z}_x, \qquad (4.48)$$
$$0 = V_s'(x)G_c(x) - q_c^T J_c(x) - r_c^T + \mathcal{W}_c^T(x), \quad x \notin \mathcal{Z}_x, \qquad (4.49)$$
$$0 = V_s(x + f_d(x)) - V_s(x) - q_d^T h_d(x) + \ell_d(x), \quad x \in \mathcal{Z}_x, \qquad (4.50)$$
$$0 = V_s'(x + f_d(x))G_d(x) - q_d^T J_d(x) - r_d^T + \mathcal{W}_d^T(x), \quad x \in \mathcal{Z}_x, \qquad (4.51)$$

*then the nonlinear impulsive system $\mathcal{G}$ given by (4.35)–(4.38) is expo-
nentially dissipative (respectively, dissipative) with respect to the linear
hybrid supply rate $(s_c(u_c, y_c), s_d(u_d, y_d)) = (q_c^T y_c + r_c^T u_c, q_d^T y_d + r_d^T u_d)$.*

Proof. For any admissible input $u_c(t)$, $t, \hat{t} \in \mathbb{R}$, $t_k < t \leq \hat{t} \leq t_{k+1}$,
and $k \in \overline{\mathbb{Z}}_+$, it follows from (4.48) and (4.49) that for all $x \notin \mathcal{Z}_x$ and

$u_c \in \overline{\mathbb{R}}_+^{m_c}$,

$$
\begin{aligned}
\dot{V}_s(x) + \varepsilon V_s(x) &= V_s'(x)(f_c(x) + G_c(x)u_c) + \varepsilon V_s(x) \\
&= q_c^T h_c(x) - \ell_c(x) + q_c^T J_c(x)u_c + r_c^T u_c - \mathcal{W}_c^T(x)u_c \\
&= q_c^T y_c + r_c^T u_c - \ell_c(x) - \mathcal{W}_c^T(x)u_c \\
&\leq q_c^T y_c + r_c^T u_c \\
&= s_c(u_c, y_c).
\end{aligned}
\tag{4.52}
$$

Next, it follows from (4.50) and (4.51), and the structural storage function constraint (4.47) that for all $x \in \mathcal{Z}_x$ and $u_d \in \overline{\mathbb{R}}_+^{m_d}$,

$$
\begin{aligned}
\Delta V_s(x) &= V_s(x + f_d(x) + G_d(x)u_d) - V_s(x) \\
&= V_s(x + f_d(x)) - V_s(x) + V_s'(x + f_d(x))G_d(x)u_d \\
&= q_d^T h_d(x) - \ell_d(x) + q_d^T J_d(x)u_d + r_d^T u_d - \mathcal{W}_d^T(x)u_d \\
&= s_d(u_d, y_d) - \ell_d(x) - \mathcal{W}_d^T(x)u_d \\
&\leq s_d(u_d, y_d).
\end{aligned}
\tag{4.53}
$$

Now, using (4.52) and (4.53) the result follows from Theorem 4.4. $\square$

The structural constraint (4.47) on the system storage function is similar to the structural constraint invoked in standard nonlinear discrete-time dissipativity theory [38, 40] and the impulsive dissipativity theory presented in Chapter 3. However, since $V_s : \mathbb{R}_+^n \to \mathbb{R}_+$, we can take a first-order Taylor expansion in (4.47) as opposed to the second-order Taylor expansion as in Chapter 3 and [38, 40, 61].

As in standard dissipativity theory with quadratic supply rates [73], the concepts of linear supply rates and linear storage functions provide a generalized mass balance interpretation. Specifically, using (4.48)–(4.51), it follows that, for $\hat{t} \geq t \geq 0$ and $k \in \mathbb{Z}_{[t,\hat{t})}$,

$$
\int_t^{\hat{t}} [q_c^T y_c(s) + r_c^T u_c(s)] ds + \sum_{k \in \mathbb{Z}_{[t,\hat{t})}} [q_d^T y_d(t_k) + r_d^T u_d(t_k)]
$$

$$
= V_s(x(\hat{t})) - V_s(x(t)) + \int_t^{\hat{t}} [\ell_c^T(x(s))x(s) + \mathcal{W}_c^T(x(s))u_c(s)] ds
$$

$$
+ \sum_{k \in \mathbb{Z}_{[t,\hat{t})}} [\ell_d^T(x(t_k))x(t_k) + \mathcal{W}_d^T(x(t_k))u_d(t_k)],
\tag{4.54}
$$

which can be interpreted as a generalized mass balance equation where $V_s(x(\hat{t})) - V_s(x(t))$ is the stored mass of the nonlinear impulsive dynamical system; the second path-dependent term on the right corresponds to the expelled mass of the nonnegative system over the continuous-time dynamics; and the third discrete term on the right corresponds to the expelled mass at the resetting instants.

Equivalently, it follows from Theorem 4.4 that (4.54) can be rewritten as

$$\dot{V}_s(x(t)) = q_c^T y_c(t) + r_c^T u_c(t) - [\ell_c^T(x(t))x(t) + \mathcal{W}_c^T(x(t))u_c(t)],$$
$$t_k < t \leq t_{k+1}, \quad (4.55)$$

$$\Delta V_s(x(t_k)) = q_d^T y_d(t_k) + r_d^T u_d(t_k)$$
$$- [\ell_d^T(x(t_k))x(t_k) + \mathcal{W}_d^T(x(t_k))u_d(t_k)], \quad k \in \overline{\mathbb{Z}}_+, (4.56)$$

which yields a set of generalized mass conservation equations. Specifically, (4.55) and (4.56) show that the system mass transport (respectively, change in system mass) over the interval $t \in (t_k, t_{k+1}]$ (respectively, the resetting instants) is equal to the supplied system flux (respectively, mass) minus the expelled system flux (respectively, mass).

Note that if an impulsive nonnegative dynamical system $\mathcal{G}$, with a continuously differentiable, positive-definite storage function, is dissipative with respect to the linear hybrid supply rate $(q_c^T y_c + r_c^T u_c, q_d^T y_d + r_d^T u_d)$, and $q_c \leq\leq 0$, $q_d \leq\leq 0$, and $(u_c(t), u_d(t_k)) \equiv (0,0)$, then it follows that $\dot{V}_s(x(t)) \leq q_c^T y_c(t) \leq 0, t \geq 0$, and $\Delta V_s(x(t_k)) \leq q_d^T y_d(t_k) \leq 0, k \in \overline{\mathbb{Z}}_+$. Hence, the undisturbed $((u_c(t), u_d(t_k)) \equiv (0,0))$ system $\mathcal{G}$ is Lyapunov stable. Furthermore, if a nonnegative dynamical system $\mathcal{G}$, with a continuously differentiable, positive-definite storage function, is exponentially dissipative with respect to the linear hybrid supply rate $(q_c^T y_c + r_c^T u_c, q_d^T y_d + r_d^T u_d)$, and $q_c \leq\leq 0$, $q_d \leq\leq 0$, and $(u_c(t), u_d(t_k)) \equiv (0,0)$, then it follows that $\dot{V}_s(x(t)) \leq -\varepsilon V_s(x(t)) + q_c^T y_c(t) < 0$, $x(t) \neq 0, t \geq 0$, where $\varepsilon > 0$, and $\Delta V_s(x(t_k)) \leq q_d^T y_d(t_k) \leq 0, k \in \overline{\mathbb{Z}}_+$. Hence, the undisturbed (i.e., $(u_c(t), u_d(t_k)) \equiv (0,0)$) system $\mathcal{G}$ is asymptotically stable.

Next, we provide necessary and sufficient conditions for the case where $\mathcal{G}$ given by (4.35)–(4.38) is lossless with respect to the linear hybrid supply rate of the form $(s_c(u_c, y_c), s_d(u_d, y_d)) = (q_c^T y_c + r_c^T u_c, q_d^T y_d + r_d^T u_d)$.

Theorem 4.6 Let $q_c \in \mathbb{R}^{l_c}$, $r_c \in \mathbb{R}^{m_c}$, $q_d \in \mathbb{R}^{l_d}$, and $r_d \in \mathbb{R}^{m_d}$. Consider the nonlinear hybrid dynamical system $\mathcal{G}$ given by (4.35)–(4.38) where $f_c : \mathcal{D} \to \mathbb{R}^n$ is essentially nonnegative, $f_d : \mathcal{Z}_x \to \mathbb{R}^n$ is such that $x + f_d(x)$ is nonnegative, $G_c(x) \geq\geq 0$, $G_d(x) \geq\geq 0$, $h_c(x) \geq\geq 0$, $h_d(x) \geq\geq 0$, $J_c(x) \geq\geq 0$, and $J_d(x) \geq\geq 0$, $x \in \overline{\mathbb{R}}_+^n$. Then $\mathcal{G}$ is lossless with respect to the linear hybrid supply rate $(s_c(u_c, y_c), s_d(u_d, y_d)) = (q_c^T y_c + r_c^T u_c, q_d^T y_d + r_d^T u_d)$ if and only if *there exists a function* $V_s : \overline{\mathbb{R}}^n_+ \to \overline{\mathbb{R}}_+$ *such that* $V_s(\cdot)$ *is continuously differentiable, nonnegative definite,* $V_s(0) = 0$, *and for all* $x \in \mathcal{Z}_x$, $u_d \in \overline{\mathbb{R}}^{m_d}_+$, *(4.47) holds, and*

$$0 = V_s'(x)f_c(x) - q_c^T h_c(x), \quad x \notin \mathcal{Z}_x, \tag{4.57}$$

$$0 = V_s'(x)G_c(x) - q_c^T J_c(x) - r_c^T, \quad x \notin \mathcal{Z}_x, \tag{4.58}$$

$$0 = V_s(x + f_d(x)) - V_s(x) - q_d^T h_d(x), \quad x \in \mathcal{Z}_x, \tag{4.59}$$

$$0 = V_s'(x + f_d(x))G_d(x) - q_d^T J_d(x) - r_d^T, \quad x \in \mathcal{Z}_x. \tag{4.60}$$

Proof. Sufficiency follows as in the proof of Theorem 4.5. To show necessity, suppose that the nonlinear impulsive dynamical system $\mathcal{G}$ is lossless with respect to the linear hybrid supply rate (s_c, s_d). Then, it follows that for all $k \in \overline{\mathbb{Z}}_+$,

$$V_s(x(\hat{t})) - V_s(x(t)) = \int_t^{\hat{t}} s_c(u_c(s), y_c(s))ds, \quad t_k < t \le \hat{t} \le t_{k+1},$$
$$\tag{4.61}$$

and

$$V_s(x(t_k)) + f_d(x(t_k)) + G_d(x(t_k))u_d(t_k)) = V_s(x(t_k)) + s_d(u_d(t_k), y_d(t_k)).$$
$$\tag{4.62}$$

Now, dividing (4.61) by $\hat{t} - t^+$ and letting $\hat{t} \to t^+$, (4.61) is equivalent to

$$\dot{V}_s(x(t)) = V_s'(x(t))[f_c(x(t)) + G_c(x(t))u_c(t)] = s_c(u_c(t), y_c(t)),$$
$$t_k < t \le t_{k+1}. \tag{4.63}$$

Next, with $t = 0$, it follows from (4.63) that

$$V_s'(x_0)[f_c(x_0) + G_c(x_0)u_c(0)] = s_c(u_c(0), y_c(0)), \quad x_0 \notin \mathcal{Z}_x,$$
$$u_c(0) \in \overline{\mathbb{R}}^{m_c}_+. \tag{4.64}$$

Since $x_0 \notin \mathcal{Z}_x$ is arbitrary, it follows that

$$V_s'(x)[f_c(x) + G_c(x)u_c] = q_c^T y_c + r_c^T u_c = q_c^T h_c(x) + (r_c^T + q_c^T J_c(x))u_c,$$
$$x \notin \mathcal{Z}_x, \quad u_c \in \overline{\mathbb{R}}^{m_c}_+. \tag{4.65}$$

Now, setting $u_c = 0$ yields (4.57) which further yields (4.58). Next, it follows from (4.62) with $k = 1$ that

$$V_s(x(t_1) + f_d(x(t_1)) + G_d(x(t_1))u_d(t_1)) = V_s(x(t_1))$$
$$+ s_d(u_d(t_1), y_d(t_1)). \tag{4.66}$$

Now, since the continuous-time dynamics (4.35) are Lipschitz, it follows that for arbitrary $x \in \mathcal{Z}_x$ there exists $x_0 \notin \mathcal{Z}_x$ such that $x(t_1) = x$. Hence, it follows from (4.66) that

$$
\begin{aligned}
V_{\mathrm{s}}(x + f_{\mathrm{d}}(x) + G_{\mathrm{d}}(x)u_{\mathrm{d}}) &= V_{\mathrm{s}}(x) + q_{\mathrm{d}}^{\mathrm{T}} y_{\mathrm{d}} + r_{\mathrm{d}}^{\mathrm{T}} u_{\mathrm{d}} \\
&= V_{\mathrm{s}}(x) + q_{\mathrm{d}}^{\mathrm{T}} h_{\mathrm{d}}(x) + (r_{\mathrm{d}}^{\mathrm{T}} + q_{\mathrm{d}}^{\mathrm{T}} J_{\mathrm{d}}(x))u_{\mathrm{d}}, \\
& \qquad x \in \mathcal{Z}_x, \quad u_{\mathrm{d}} \in \overline{\mathbb{R}}_{+}^{m_{\mathrm{d}}}.
\end{aligned} \tag{4.67}
$$

Since the right-hand side of (4.67) is linear in u_{d} it follows that $V_{\mathrm{s}}(x + f_{\mathrm{d}}(x) + G_{\mathrm{d}}(x)u_{\mathrm{d}})$ is linear in u_{d}, and hence, there exists $P_{1u_{\mathrm{d}}} : \overline{\mathbb{R}}_{+}^{n} \to \mathbb{R}^{1 \times m_{\mathrm{d}}}$ such that

$$
V_{\mathrm{s}}(x + f_{\mathrm{d}}(x) + G_{\mathrm{d}}(x)u_{\mathrm{d}}) = V_{\mathrm{s}}(x + f_{\mathrm{d}}(x)) + P_{1u_{\mathrm{d}}}(x)u_{\mathrm{d}}. \tag{4.68}
$$

Since $V_{\mathrm{s}}(\cdot)$ is continuously differentiable, applying a Taylor series expansion on (4.68) about $u_{\mathrm{d}} = 0$ yields

$$
P_{1u_{\mathrm{d}}}(x) = \left. \frac{\partial V_{\mathrm{s}}(x + f_{\mathrm{d}}(x) + G_{\mathrm{d}}(x)u_{\mathrm{d}})}{\partial u_{\mathrm{d}}} \right|_{u_{\mathrm{d}}=0} = V_{\mathrm{s}}'(x + f_{\mathrm{d}}(x))G_{\mathrm{d}}(x).
$$
$$\tag{4.69}$$

Now, using (4.68) and equating coefficients of equal powers in (4.67) yields (4.59) and (4.60).   □

Next, we provide a key definition for impulsive nonnegative dynamical systems which are dissipative with respect to a special supply rate.

Definition 4.8 *An impulsive nonnegative dynamical system $\mathcal{G}$ given by (4.35)–(4.38) is* nonaccumulative *(respectively,* exponentially nonaccumulative*) if $\mathcal{G}$ is dissipative (respectively, exponentially dissipative) with respect to the hybrid supply rate $(s_{\mathrm{c}}(u_{\mathrm{c}}, y_{\mathrm{c}}), s_{\mathrm{d}}(u_{\mathrm{d}}, y_{\mathrm{d}})) = (\mathbf{e}^{\mathrm{T}} u_{\mathrm{c}} - \mathbf{e}^{\mathrm{T}} y_{\mathrm{c}}, \mathbf{e}^{\mathrm{T}} u_{\mathrm{d}} - \mathbf{e}^{\mathrm{T}} y_{\mathrm{d}})$.*

If $\mathcal{G}$ is nonaccumulative, then it follows that

$$
\dot{V}_{\mathrm{s}}(x(t)) \le \mathbf{e}^{\mathrm{T}} u_{\mathrm{c}}(t) - \mathbf{e}^{\mathrm{T}} y_{\mathrm{c}}(t), \quad t_k < t \le t_{k+1}, \tag{4.70}
$$
$$
\Delta V_{\mathrm{s}}(x(t_k)) \le \mathbf{e}^{\mathrm{T}} u_{\mathrm{d}}(t_k) - \mathbf{e}^{\mathrm{T}} y_{\mathrm{d}}(t_k), \quad k \in \overline{\mathbb{Z}}_{+}. \tag{4.71}
$$

If the components $u_{\mathrm{c}i}(\cdot)$, $i = 1, \ldots, m_{\mathrm{c}}$, and $u_{\mathrm{d}i}(\cdot)$, $i = 1, \ldots, m_{\mathrm{d}}$, of $u_{\mathrm{c}}(\cdot)$ and $u_{\mathrm{d}}(\cdot)$, respectively, denote flux and mass inputs of the impulsive system $\mathcal{G}$, and the components $y_{\mathrm{c}i}(\cdot)$, $i = 1, \ldots, l_{\mathrm{c}}$, and $y_{\mathrm{d}i}(\cdot)$, $i = 1, \ldots, l_{\mathrm{d}}$, of $y_{\mathrm{c}}(\cdot)$ and $y_{\mathrm{d}}(\cdot)$, respectively, denote flux and mass outputs of the hybrid system $\mathcal{G}$, then nonaccumulativity implies that the system mass transport (respectively, change in system mass) is always less than or equal to the difference between the system flux (respectively, mass) input and system flux (respectively, mass) output.

Next, we show that all impulsive compartmental systems with measured outputs corresponding to material outflows are nonaccumulative. Specifically, consider (4.33) and (4.34) with storage function $V_s(x) = e^T x$ and hybrid outputs

$$y_c = D_c(x) \left(\frac{\partial V}{\partial x}(x) \right)^T = [a_{c11}(x), a_{c22}(x), \ldots, a_{cnn}(x)]^T, \quad (4.72)$$

$$y_d = D_d(x) \left(\frac{\partial V}{\partial x}(x) \right)^T = [a_{d11}(x), a_{d22}(x), \ldots, a_{dnn}(x)]^T. \quad (4.73)$$

Now, it follows that

$$
\begin{aligned}
\dot{V}_s(x) &= e^T \left[[J_{cn}(x) - D_c(x)] \left(\frac{\partial V}{\partial x}(x) \right)^T + w_c \right] \\
&= e^T w_c - e^T y_c + e^T J_{cn}(x) e \\
&= e^T w_c - e^T y_c, \quad x \notin \mathcal{Z}_x,
\end{aligned}
\quad (4.74)
$$

and

$$
\begin{aligned}
\Delta V_s(x) &= e^T \left[[J_{dn}(x) - D_d(x)] \left(\frac{\partial V}{\partial x}(x) \right)^T + w_d \right] \\
&= e^T w_d - e^T y_d + e^T J_{dn}(x) e \\
&= e^T w_d - e^T y_d, \quad x \in \mathcal{Z}_x,
\end{aligned}
\quad (4.75)
$$

which shows that all impulsive compartmental systems are lossless with respect to the linear hybrid supply rate $(s_c, s_d) = (e^T w_c - e^T y_c, e^T w_d - e^T y_d)$. Alternatively, if the hybrid outputs y_c and y_d correspond to a partial observation of the material outflows, then it can easily be shown that the nonlinear impulsive compartmental system is dissipative with respect to the linear hybrid supply rate $(s_c, s_d) = (e^T w_c - e^T y_c, e^T w_d - e^T y_d)$.

4.5 Specialization to Linear Impulsive Dynamical Systems

In this section, we specialize the results of Section 4.4 to the case of linear impulsive dynamical systems. Specifically, setting $f_c(x) = A_c x$, $G_c(x) = B_c$, $h_c(x) = C_c x$, $J_c(x) = D_c$, $f_d(x) = (A_d - I_n)x$, $G_d(x) = B_d$, $h_d(x) = C_d x$, and $J_d(x) = D_d$, the nonnegative state-dependent impulsive dynamical system given by (4.35)–(4.38) specializes to

$$\dot{x}(t) = A_c x(t) + B_c u_c(t), \quad x(t) \notin \mathcal{Z}_x, \tag{4.76}$$

$$\Delta x(t) = (A_d - I_n)x(t) + B_d u_d(t), \quad x(t) \in \mathcal{Z}_x, \tag{4.77}$$

$$y_c(t) = C_c x(t) + D_c u_c(t), \quad x(t) \notin \mathcal{Z}_x, \tag{4.78}$$

$$y_d(t) = C_d x(t) + D_d u_d(t), \quad x(t) \in \mathcal{Z}_x, \tag{4.79}$$

where $A_c \in \mathbb{R}^{n \times n}$ is essentially nonnegative, and $B_c \in \mathbb{R}^{n \times m_c}$, $C_c \in \mathbb{R}^{l_c \times n}$, $D_c \in \mathbb{R}^{l_c \times m_c}$, $A_d \in \mathbb{R}^{n \times n}$, $B_d \in \mathbb{R}^{n \times m_d}$, $C_d \in \mathbb{R}^{l_d \times n}$, and $D_d \in \mathbb{R}^{l_d \times m_d}$ are nonnegative.

Theorem 4.7 *Let $q_c \in \mathbb{R}^{l_c}$, $r_c \in \mathbb{R}^{m_c}$, $q_d \in \mathbb{R}^{l_d}$, and $r_d \in \mathbb{R}^{m_d}$. Consider the linear impulsive dynamical system $\mathcal{G}$ given by (4.76)–(4.79) and assume that A_c is essentially nonnegative, A_d is nonnegative, $B_c \geq\geq 0$, $B_d \geq\geq 0$, $C_c \geq\geq 0$, $C_d \geq\geq 0$, $D_c \geq\geq 0$, and $D_d \geq\geq 0$. If there exist vectors $p \in \overline{\mathbb{R}}_+^n$, $l_c \in \overline{\mathbb{R}}_+^n$, $l_d \in \overline{\mathbb{R}}_+^n$, $w_c \in \overline{\mathbb{R}}_+^{m_c}$, $w_d \in \overline{\mathbb{R}}_+^{m_d}$, and a scalar $\varepsilon > 0$ (respectively, $\varepsilon = 0$) such that*

$$0 = x^T(A_c^T p + \varepsilon p - C_c^T q_c + l_c), \quad x \notin \mathcal{Z}_x, \tag{4.80}$$

$$0 = B_c^T p - D_c^T q_c - r_c + w_c, \tag{4.81}$$

$$0 = x^T(A_d^T p - p - C_d^T q_d + l_d), \quad x \in \mathcal{Z}_x, \tag{4.82}$$

$$0 = B_d^T p - D_d^T q_d - r_d + w_d, \tag{4.83}$$

then the linear impulsive dynamical system $\mathcal{G}$ given by (4.76)–(4.79) is exponentially dissipative (respectively, dissipative) with respect to the linear supply rate $(s_c(u_c, y_c), s_d(u_d, y_d)) = (q_c^T y_c + r_c^T u_c, q_d^T y_d + r_d^T u_d)$.

Proof. The proof follows from Theorem 4.5 with $f_c(x) = A_c x$, $G_c(x) = B_c$, $h_c(x) = C_c x$, $J_c(x) = D_c$, $f_d(x) = (A_d - I_n)x$, $G_d(x) = B_d$, $h_d(x) = C_d x$, $J_d(x) = D_d$, $V_s(x) = p^T x$, $\ell_c(x) = l_c^T x$, $\ell_d(x) = l_d^T x$, $\mathcal{W}_c(x) = w_c$, $\mathcal{W}_d(x) = w_d$. $\square$

For a given $l_c \in \mathbb{R}^n$, $w_c \in \mathbb{R}^{m_c}$, $l_d \in \mathbb{R}^n$, and $w_d \in \mathbb{R}^{m_d}$, note that if $\operatorname{rank}[M \; y] = \operatorname{rank} M$, where

$$M \triangleq \begin{bmatrix} A_c^T + \varepsilon I \\ B_c \\ A_d^T - I \\ B_d^T \end{bmatrix}, \quad y \triangleq \begin{bmatrix} C_c^T q_c - l_c \\ D_c^T q_c + r_c - w_c \\ C_d^T q_d - l_d \\ D_d^T q_d + r_d - w_d \end{bmatrix}, \tag{4.84}$$

then there exists $p \in \mathbb{R}^n$ such that (4.80)–(4.83) are satisfied. Now, if there exists $p \in \mathbb{R}^n$ such that inequalities

$$p \geq\geq 0, \tag{4.85}$$

$$z - Mp \geq\geq 0, \tag{4.86}$$

where

$$z \triangleq \begin{bmatrix} C_c^T q_c \\ D_c^T q_c + r_c \\ C_d^T q_d \\ D_d^T q_d + r_d \end{bmatrix}, \tag{4.87}$$

are satisfied, then there exist $l_c \geq\geq 0$, $w_c \geq\geq 0$, $l_d \geq\geq 0$, and $w_d \geq\geq 0$ such that (4.80)–(4.83) hold. Equations (4.85) and (4.86) comprise a set of $3n + m_c + m_d$ linear inequalities with p_i, $i = 1, \ldots, n$, variables, and hence, the feasibility of $p \geq\geq 0$ such that (4.85) and (4.86) hold can be checked by standard linear matrix inequality (LMI) techniques [27].

Finally, we provide sufficient conditions for the case where $\mathcal{G}$ given by (4.76)–(4.79) is lossless with respect to the linear hybrid supply rate $(s_c(u_c, y_c), s_d(u_d, y_d)) = (q_c^T y_c + r_c^T u_c, q_d^T y_d + r_d^T u_d)$.

Theorem 4.8 Let $q_c \in \mathbb{R}^{l_c}$, $r_c \in \mathbb{R}^{m_c}$, $q_d \in \mathbb{R}^{l_d}$, and $r_d \in \mathbb{R}^{m_d}$. Consider the linear impulsive dynamical system $\mathcal{G}$ given by (4.76)–(4.79) and assume that A_c is essentially nonnegative, A_d is nonnegative, $B_c \geq\geq 0$, $B_d \geq\geq 0$, $C_c \geq\geq 0$, $C_d \geq\geq 0$, $D_c \geq\geq 0$, and $D_d \geq\geq 0$. If there exists $p \in \overline{\mathbb{R}}_+^n$ such that

$$0 = x^T(A_c^T p - C_c^T q_c), \quad x \notin \mathcal{Z}_x, \tag{4.88}$$
$$0 = B_c^T p - D_c^T q_c - r_c, \tag{4.89}$$
$$0 = x^T(A_d^T p - p - C_d^T q_d), \quad x \in \mathcal{Z}_x, \tag{4.90}$$
$$0 = B_d^T p - D_d^T q_d - r_d, \tag{4.91}$$

then the linear impulsive dynamical system $\mathcal{G}$ given by (4.76)–(4.79) is lossless with respect to the linear supply rate $(s_c(u_c, y_c), s_d(u_d, y_d)) = (q_c^T y_c + r_c^T u_c, q_d^T y_d + r_d^T u_d)$.

Proof. The proof follows from Theorem 4.6 with $f_c(x) = A_c x$, $G_c(x) = B_c$, $h_c(x) = C_c x$, $J_c(x) = D_c$, $f_d(x) = (A_d - I_n)x$, $G_d(x) = B_d$, $h_d(x) = C_d x$, $J_d(x) = D_d$, and $V_s(x) = p^T x$. □

Chapter Five

Vector Dissipativity Theory for Large-Scale Impulsive
Dynamical Systems

5.1 Introduction

Modern complex dynamical systems[1] are highly interconnected and mutually interdependent, both physically and through a multitude of information and communication network constraints. The sheer size (i.e., dimensionality) and complexity of these large-scale dynamical systems often necessitate a hierarchical decentralized architecture for analyzing and controlling these systems. Specifically, in the analysis and control-system design of complex large-scale dynamical systems it is often desirable to treat the overall system as a collection of interconnected subsystems. The behavior of the composite (i.e., large-scale) system can then be predicted from the behaviors of the individual subsystems and their interconnections. The need for decentralized analysis and control design of large-scale systems is a direct consequence of the physical size and complexity of the dynamical model. In particular, computational complexity may be too large for model analysis while severe constraints on communication links between system sensors, actuators, and processors may render centralized control architectures impractical.

The complexity of modern controlled large-scale dynamical systems is further exacerbated by the use of hierarchial embedded control subsystems within the feedback control system, that is, abstract decision-making units performing logical checks that identify system mode operation and specify the continuous-variable subcontroller to be activated. As discussed in Chapter 1, such systems typically possess a multiechelon hierarchical hybrid decentralized control architecture characterized by continuous-time dynamics at the lower levels of the hierarchy and discrete-time dynamics at the higher levels of the hierarchy. The lower-level units directly interact with the dynamical

---

[1] Here we have in mind large flexible space structures, aerospace systems, electric power systems, network systems, economic systems, and ecological systems, to cite but a few examples.

system to be controlled while the higher-level units receive information from the lower-level units as inputs and provide (possibly discrete) output commands which serve to coordinate and reconcile the (sometimes competing) actions of the lower-level units. The hierarchical controller organization reduces processor cost and controller complexity by breaking up the processing task into relatively small pieces and decomposing the fast and slow control functions. Typically, the higher-level units perform logical checks that determine system mode operation, while the lower-level units execute continuous-variable commands for a given system mode of operation.

An approach to analyzing large-scale dynamical systems was introduced by the pioneering work of Šiljak [151] and involves the notion of *connective stability*. In particular, the large-scale dynamical system is decomposed into a collection of subsystems with local dynamics and uncertain interactions. Then, each subsystem is considered independently so that the stability of each subsystem is combined with the interconnection constraints to obtain a *vector Lyapunov function* for the composite large-scale dynamical system guaranteeing connective stability for the overall system. Vector Lyapunov functions were first introduced by Bellman [19] and Matrosov [119] and further developed in [52, 96, 109, 116–118, 122, 151, 152], with [52, 109, 151, 152] exploiting their utility for analyzing large-scale systems. Extensions of vector Lyapunov function theory that include matrix-valued Lyapunov functions for stability analysis of large-scale dynamical systems appear in the monographs by Martynyuk [117, 118].

The use of vector Lyapunov functions in large-scale system analysis offers a very flexible framework since each component of the vector Lyapunov function can satisfy less rigid requirements as compared to a single scalar Lyapunov function. Weakening the hypothesis on the Lyapunov function enlarges the class of Lyapunov functions that can be used for analyzing the stability of large-scale dynamical systems. In particular, each component of a vector Lyapunov function need not be positive definite with a negative or even negative-semidefinite derivative. The time derivative of the vector Lyapunov function need only satisfy an element-by-element vector inequality involving a vector field of a certain comparison system.

In light of the fact that energy flow modeling arises naturally in large-scale dynamical systems, and vector Lyapunov functions provide a powerful stability analysis framework for these systems, it seems natural that hybrid dissipativity theory [56, 61, 62], on the subsystem level, should play a key role in analyzing large-scale impulsive dynamical systems. Specifically, hybrid dissipativity theory provides a fundamental framework for the analysis and design of impulsive dynamical systems using an input-output description based on system-energy-related considerations. As shown in Chapter 3, the hybrid dissipation hypothesis on impulsive dynamical systems results in a fundamental constraint on their dynamic behavior, wherein a dissipative impulsive dynamical system can deliver only a fraction of its energy to its surroundings and can store only a fraction of the work done to it. Such conservation laws are prevalent in large-scale impulsive dynamical systems such as aerospace, power, network, telecommunications, and transportation systems. Since these systems have numerous input-output properties related to conservation, dissipation, and transport of energy, extending hybrid dissipativity theory to capture conservation and dissipation notions on the subsystem level would provide a natural energy flow model for large-scale impulsive dynamical systems.

Aggregating the dissipativity properties of each of the impulsive subsystems by appropriate storage functions and hybrid supply rates would allow us to study the dissipativity properties of the composite large-scale impulsive system using *vector storage functions* and *vector hybrid supply rates*. Furthermore, since vector Lyapunov functions can be viewed as generalizations of composite energy functions for all of the impulsive subsystems, a generalized notion of hybrid dissipativity, namely, *vector hybrid dissipativity*, with appropriate vector storage functions and vector hybrid supply rates, can be used to construct vector Lyapunov functions for nonlinear feedback large-scale impulsive systems by appropriately combining vector storage functions for the forward and feedback large-scale impulsive systems. Finally, as in classical dynamical system theory, vector dissipativity theory can play a fundamental role in addressing robustness, disturbance rejection, stability of feedback interconnections, and optimality for large-scale impulsive dynamical systems.

In this chapter, we develop vector dissipativity notions for large-scale nonlinear impulsive dynamical systems. In particular, we introduce a generalized definition of dissipativity for large-scale nonlinear impulsive dynamical systems in terms of a *hybrid vector inequality* involving a vector hybrid supply rate, a vector storage function, and an essentially nonnegative, semistable dissipation matrix. Generalized notions of a vector available storage and a vector required supply are also defined and shown to be element-by-element ordered, nonnegative, and finite. On the impulsive subsystem level, the proposed approach provides an energy flow balance over the continuous-time dynamics and the resetting events in terms of the stored subsystem

energy, the supplied subsystem energy, the subsystem energy gained
from all other subsystems independent of the subsystem coupling
strengths, and the subsystem energy dissipated. Furthermore, for
large-scale impulsive dynamical systems decomposed into interconnected impulsive subsystems, dissipativity of the composite impulsive
system is shown to be determined from the dissipativity properties of
the individual impulsive subsystems and the nature of the interconnections. In addition, we develop extended Kalman-Yakubovich-Popov
conditions, in terms of the local impulsive subsystem dynamics and
the interconnection constraints, for characterizing vector dissipativeness via vector storage functions for large-scale impulsive dynamical
systems.

5.2 Vector Dissipativity Theory for Large-Scale Impulsive Dynamical Systems

To develop vector dissipativity notions for large-scale impulsive dynamical systems we consider input/state-dependent impulsive dynamical systems $\mathcal{G}$ of the form

$$\dot{x}(t) = F_c(x(t), u_c(t)), \quad x(t_0) = x_0, \quad (x(t), u_c(t)) \notin \mathcal{Z}, \quad t \geq t_0,$$
$$(5.1)$$
$$\Delta x(t) = F_d(x(t), u_d(t)), \quad (x(t), u_c(t)) \in \mathcal{Z}, \tag{5.2}$$
$$y_c(t) = H_c(x(t), u_c(t)), \quad (x(t), u_c(t)) \notin \mathcal{Z}, \tag{5.3}$$
$$y_d(t) = H_d(x(t), u_d(t)), \quad (x(t), u_c(t)) \in \mathcal{Z}, \tag{5.4}$$

where $x(t) \in \mathcal{D} \subseteq \mathbb{R}^n$, $t \geq t_0$, $u_c(t) \in U_c \subseteq \mathbb{R}^{m_c}$, $u_d(t_k) \in U_d \subseteq \mathbb{R}^{m_d}$,
$y_c(t) \in Y_c \subseteq \mathbb{R}^{l_c}$, $y_d(t_k) \in Y_d \subseteq \mathbb{R}^{l_d}$, $F_c : \mathcal{D} \times U_c \to \mathbb{R}^n$, $F_d : \mathcal{D} \times U_d \to$
$\mathbb{R}^n$, $H_c : \mathcal{D} \times U_c \to Y_c$, $H_d : \mathcal{D} \times U_d \to Y_d$, $\mathcal{D}$ is an open set with $0 \in \mathcal{D}$,
$\mathcal{Z} \subset \mathcal{D} \times U_c$, and $F_c(0,0) = 0$. Here, we assume that $\mathcal{G}$ represents a
large-scale impulsive dynamical system composed of q interconnected
controlled impulsive subsystems $\mathcal{G}_i$ such that, for all $i = 1, \ldots, q$,

$$F_{ci}(x, u_{ci}) = f_{ci}(x_i) + \mathcal{I}_{ci}(x) + G_{ci}(x_i)u_{ci}, \tag{5.5}$$
$$F_{di}(x, u_{di}) = f_{di}(x_i) + \mathcal{I}_{di}(x) + G_{di}(x_i)u_{di}, \tag{5.6}$$
$$H_{ci}(x_i, u_{ci}) = h_{ci}(x_i) + J_{ci}(x_i)u_{ci}, \tag{5.7}$$
$$H_{di}(x_i, u_{di}) = h_{di}(x_i) + J_{di}(x_i)u_{di}, \tag{5.8}$$

where $x_i \in \mathcal{D}_i \subseteq \mathbb{R}^{n_i}$, $u_{ci} \in U_{ci} \subseteq \mathbb{R}^{m_{ci}}$, $u_{di} \in U_{di} \subseteq \mathbb{R}^{m_{di}}$, $y_{ci} \triangleq$
$H_{ci}(x_i, u_{ci}) \in Y_{ci} \subseteq \mathbb{R}^{l_{ci}}$, $y_{di} \triangleq H_{di}(x_i, u_{di}) \in Y_{di} \subseteq \mathbb{R}^{l_{di}}$, $((u_{ci}, u_{di}),$
$(y_{ci}, y_{di}))$ is the hybrid input-output pair for the ith subsystem, f_{ci} :

$\mathbb{R}^{n_i} \to \mathbb{R}^{n_i}$ and $\mathcal{I}_{ci} : \mathcal{D} \to \mathbb{R}^{n_i}$ are Lipschitz continuous and satisfy $f_{ci}(0) = 0$ and $\mathcal{I}_{ci}(0) = 0$, $f_{di} : \mathbb{R}^{n_i} \to \mathbb{R}^{n_i}$ and $\mathcal{I}_{di} : \mathcal{D} \to \mathbb{R}^{n_i}$ are continuous, $G_{ci} : \mathbb{R}^{n_i} \to \mathbb{R}^{n_i \times m_{ci}}$ and $G_{di} : \mathbb{R}^{n_i} \to \mathbb{R}^{n_i \times m_{di}}$ are continuous, $h_{ci} : \mathbb{R}^{n_i} \to \mathbb{R}^{l_{ci}}$ and satisfies $h_{ci}(0) = 0$, $h_{di} : \mathbb{R}^{n_i} \to \mathbb{R}^{l_{di}}$, $J_{ci} : \mathbb{R}^{n_i} \to \mathbb{R}^{l_{ci} \times m_{ci}}$, $J_{di} : \mathbb{R}^{n_i} \to \mathbb{R}^{l_{di} \times m_{di}}$, $\sum_{i=1}^{q} n_i = n$, $\sum_{i=1}^{q} m_{ci} = m_c$, $\sum_{i=1}^{q} m_{di} = m_d$, $\sum_{i=1}^{q} l_{ci} = l_c$, and $\sum_{i=1}^{q} l_{di} = l_d$.

Here, $f_{ci} : \mathcal{D}_i \subseteq \mathbb{R}^{n_i} \to \mathbb{R}^{n_i}$ and $f_{di} : \mathcal{D}_i \subseteq \mathbb{R}^{n_i} \to \mathbb{R}^{n_i}$ define vector fields of each isolated subsystem of (5.1) and (5.2), and $\mathcal{I}_{ci} : \mathcal{D} \to \mathbb{R}^{n_i}$ and $\mathcal{I}_{di} : \mathcal{D} \to \mathbb{R}^{n_i}$ define the structure of the interconnection dynamics of the ith impulsive subsystem with all other impulsive subsystems. Furthermore, for the large-scale dynamical system $\mathcal{G}$ we assume that the required properties for the existence and uniqueness of solutions are satisfied, that is, for each $i \in \{1, \ldots, q\}$, $u_{ci}(\cdot)$ and $u_{di}(\cdot)$ satisfy sufficient regularity conditions such that the system (5.1) and (5.2) has a unique solution forward in time. We define the composite input and composite output for the large-scale impulsive dynamical system $\mathcal{G}$ as $u_c \triangleq [u_{c1}^T, \ldots, u_{cq}^T]^T$, $u_d \triangleq [u_{d1}^T, \ldots, u_{dq}^T]^T$, $y_c \triangleq [y_{c1}^T, \ldots, y_{cq}^T]^T$, and $y_d \triangleq [y_{d1}^T, \ldots, y_{dq}^T]^T$, respectively. In addition, we define $\mathcal{U} \triangleq \mathcal{U}_c \times \mathcal{U}_d$ and $\mathcal{Y} \triangleq \mathcal{Y}_c \times \mathcal{Y}_d$ to be an input and output space, respectively, consisting of left-continuous bounded U-valued and Y-valued functions on the semi-infinite interval $[0, \infty)$.

Definition 5.1 *For the large-scale impulsive dynamical system $\mathcal{G}$ given by (5.1)–(5.4) a vector function $(S_c(u_c, y_c), S_d(u_d, y_d))$, where $S_c(u_c, y_c) \triangleq [s_{c1}(u_{c1}, y_{c1}), \ldots, s_{cq}(u_{cq}, y_{cq})]^T$, $S_d(u_d, y_d) \triangleq [s_{d1}(u_{d1}, y_{d1}), \ldots, s_{dq}(u_{dq}, y_{dq})]^T$, $s_{ci} : U_{ci} \times Y_{ci} \to \mathbb{R}$, and $s_{di} : U_{di} \times Y_{di} \to \mathbb{R}$, $i = 1, \ldots, q$, such that $S_c(0, 0) = 0$ and $S_d(0, 0) = 0$, is called a* vector hybrid supply rate *if it is locally componentwise integrable for all input-output pairs satisfying (5.1)–(5.4), that is, for every $i \in \{1, \ldots, q\}$ and for all input-output pairs $u_{ci} \in U_{ci}$ and $y_{ci} \in Y_{ci}$ satisfying (5.1)–(5.4), $s_{ci}(\cdot, \cdot)$ satisfies $\int_t^{\hat{t}} |s_{ci}(u_{ci}(s), y_{ci}(s))| ds < \infty$, $t, \hat{t} \geq t_0$.*

Note that since all input-output pairs $u_{di}(t_k) \in U_{di}$ and $y_{di}(t_k) \in Y_{di}$ are defined for discrete instants, $s_{di}(\cdot, \cdot)$ in Definition 5.1 satisfies $\sum_{k \in \mathbb{Z}_{[t,\hat{t})}} |s_{di}(u_{di}(t_k), y_{di}(t_k))| < \infty$, where $\mathbb{Z}_{[t,\hat{t})} \triangleq \{k : t \leq t_k < \hat{t}\}$. For the statement of the next definition, recall that a matrix $W \in \mathbb{R}^{q \times q}$ is *semistable* if and only if $\lim_{t \to \infty} e^{Wt}$ exists [25,57], while W is *asymptotically stable* if and only if $\lim_{t \to \infty} e^{Wt} = 0$.

Definition 5.2 *The large-scale impulsive dynamical system $\mathcal{G}$ given by (5.1)–(5.4) is* vector dissipative *(respectively,* exponentially vector dissipative*) with respect to the vector hybrid supply rate (S_c, S_d) if* *there exist a continuous, nonnegative definite vector function* $V_{\mathrm{s}} = [v_{\mathrm{s}1}, \ldots, v_{\mathrm{s}q}]^{\mathrm{T}} : \mathcal{D} \to \overline{\mathbb{R}}_{+}^{q}$, *called a* vector storage function, *and an essentially nonnegative* dissipation matrix $W \in \mathbb{R}^{q \times q}$ *such that* $V_{\mathrm{s}}(0) = 0$, W *is semistable (respectively, asymptotically stable), and the* vector hybrid dissipation inequality

$$V_{\mathrm{s}}(x(T)) \leq\leq e^{W(T-t_0)} V_{\mathrm{s}}(x(t_0)) + \int_{t_0}^{T} e^{W(T-t)} S_{\mathrm{c}}(u_{\mathrm{c}}(t), y_{\mathrm{c}}(t)) \mathrm{d}t$$

$$+ \sum_{k \in \mathbb{Z}_{[t_0,T)}} e^{W(T-t_k)} S_{\mathrm{d}}(u_{\mathrm{d}}(t_k), y_{\mathrm{d}}(t_k)), \quad T \geq t_0, \ (5.9)$$

is satisfied, where $x(t)$, $t \geq t_0$, *is the solution to (5.1)–(5.4) with* $(u_{\mathrm{c}}(t), u_{\mathrm{d}}(t_k)) \in U_{\mathrm{c}} \times U_{\mathrm{d}}$ *and* $x(t_0) = x_0$. *The large-scale impulsive dynamical system* $\mathcal{G}$ *given by (5.1)–(5.4) is* vector lossless *with respect to the vector hybrid supply rate* $(S_{\mathrm{c}}, S_{\mathrm{d}})$ *if the vector hybrid dissipation inequality is satisfied as an equality with* W *semistable.*

Note that if the subsystems $\mathcal{G}_i$ of $\mathcal{G}$ are *disconnected*, that is, $\mathcal{I}_{ci}(x) \equiv 0$ and $\mathcal{I}_{di}(x) \equiv 0$ for all $i = 1, \ldots, q$, and $-W \in \mathbb{R}^{q \times q}$ is diagonal and nonnegative definite, then it follows from Definition 5.2 that each of the disconnected subsystems $\mathcal{G}_i$ is dissipative or exponentially dissipative in the sense of Definition 3.1. A similar remark holds in the case where $q = 1$.

Next, define the *vector available storage* of the large-scale impulsive dynamical system $\mathcal{G}$ by

$$V_{\mathrm{a}}(x_0) \triangleq - \inf_{(u_{\mathrm{c}}(\cdot), u_{\mathrm{d}}(\cdot)), \, T \geq t_0} \left[\int_{t_0}^{T} e^{-W(t-t_0)} S_{\mathrm{c}}(u_{\mathrm{c}}(t), y_{\mathrm{c}}(t)) \mathrm{d}t \right.$$

$$\left. + \sum_{k \in \mathbb{Z}_{[t_0,T)}} e^{-W(t_k-t_0)} S_{\mathrm{d}}(u_{\mathrm{d}}(t_k), y_{\mathrm{d}}(t_k)) \right],$$

$$(5.10)$$

where $x(t)$, $t \geq t_0$, is the solution to (5.1)–(5.4) with $x(t_0) = x_0$ and admissible inputs $(u_{\mathrm{c}}(\cdot), u_{\mathrm{d}}(\cdot)) \in \mathcal{U}_{\mathrm{c}} \times \mathcal{U}_{\mathrm{d}}$. The infimum in (5.10) is taken componentwise which implies that for different elements of $V_{\mathrm{a}}(\cdot)$ the infimum is calculated separately. Note that $V_{\mathrm{a}}(x_0) \geq\geq 0$, $x_0 \in \mathcal{D}$, since $V_{\mathrm{a}}(x_0)$ is the infimum over a set of vectors containing the zero vector $(T = t_0)$. To state the main results of this section recall the definitions of complete reachability and zero-state observability given in Section 3.2.

Theorem 5.1 *Consider the large-scale impulsive dynamical system $\mathcal{G}$ given by (5.1)–(5.4) and assume that $\mathcal{G}$ is completely reachable. Then $\mathcal{G}$ is vector dissipative (respectively, exponentially vector dissipative) with respect to the vector hybrid supply rate (S_c, S_d) if and only if there exist a continuous, nonnegative-definite vector function $V_s : \mathcal{D} \to \overline{\mathbb{R}}_+^q$ and an essentially nonnegative dissipation matrix $W \in \mathbb{R}^{q \times q}$ such that $V_s(0) = 0$, W is semistable (respectively, asymptotically stable), and for all $k \in \overline{\mathbb{Z}}_+$,*

$$V_s(x(\hat{t})) \leq\leq e^{W(\hat{t}-t)} V_s(x(t)) + \int_t^{\hat{t}} e^{W(\hat{t}-s)} S_c(u_c(s), y_c(s)) \mathrm{d}s,$$

$$t_k < t \leq \hat{t} \leq t_{k+1}, \quad (5.11)$$

$$V_s(x(t_k) + F_d(x(t_k), u_d(t_k))) \leq\leq V_s(x(t_k)) + S_d(u_d(t_k), y_d(t_k)).$$

$$(5.12)$$

Alternatively, $\mathcal{G}$ is vector lossless with respect to the vector hybrid supply rate (S_c, S_d) if and only if there exists a continuous, nonnegative-definite vector function $V_s : \mathcal{D} \to \overline{\mathbb{R}}_+^q$ such that (5.11) and (5.12) are satisfied as equalities with W semistable.

Proof. Let $k \in \overline{\mathbb{Z}}_+$ and suppose $\mathcal{G}$ is vector dissipative (respectively, exponentially vector dissipative) with respect to the vector hybrid supply rate (S_c, S_d). Then, there exist a continuous nonnegative-definite vector function $V_s : \mathcal{D} \to \overline{\mathbb{R}}_+^q$ and an essentially nonnegative matrix $W \in \mathbb{R}^{q \times q}$ such that (5.9) holds. Now, since for $t_k < t \leq \hat{t} \leq t_{k+1}$, $\mathbb{Z}_{[t,\hat{t})} = \emptyset$, (5.11) is immediate. Next, it follows from (5.9) that

$$V_s(x(t_k^+)) \leq\leq e^{W(t_k^+ - t_k)} V_s(x(t_k)) + \int_{t_k}^{t_k^+} e^{W(t_k^+ - s)} S_c(u_c(s), y_c(s)) \mathrm{d}s$$

$$+ \sum_{k \in \mathbb{Z}_{[t_k, t_k^+)}} e^{W(t_k^+ - t_k)} S_d(u_d(t_k), y_d(t_k)) \quad (5.13)$$

which, since $\mathbb{Z}_{[t_k, t_k^+)} = k$, implies (5.12).

Conversely, suppose (5.11) and (5.12) hold and let $\hat{t} \geq t \geq t_0$ and $\mathbb{Z}_{[t,\hat{t})} = \{i, i+1, \ldots, j\}$. (Note that if $\mathbb{Z}_{[t,\hat{t})} = \emptyset$ the converse result is a direct consequence of (5.11).) If $\mathbb{Z}_{[t,\hat{t})} \neq \emptyset$, it follows from (5.11) and (5.12) that

$$V_s(x(\hat{t})) - e^{W(\hat{t}-t)} V_s(x(t))$$

$$= V_s(x(\hat{t})) - e^{W(\hat{t}-t_j^+)} V_s(x(t_j^+))$$

$$+ e^{W(\hat{t}-t_j^+)} V_{\mathrm{s}}(x(t_j^+)) - e^{W(\hat{t}-t_{j-1}^+)} V_{\mathrm{s}}(x(t_{j-1}^+))$$

$$+ e^{W(\hat{t}-t_{j-1}^+)} V_{\mathrm{s}}(x(t_{j-1}^+)) - \cdots - e^{W(\hat{t}-t_i^+)} V_{\mathrm{s}}(x(t_i^+))$$

$$+ e^{W(\hat{t}-t_i^+)} V_{\mathrm{s}}(x(t_i^+)) - e^{W(\hat{t}-t)} V_{\mathrm{s}}(x(t))$$

$$= V_{\mathrm{s}}(x(\hat{t})) - e^{W(\hat{t}-t_j)} V_{\mathrm{s}}(x(t_j^+))$$

$$+ e^{W(\hat{t}-t_j)} V_{\mathrm{s}}(x(t_j) + F_{\mathrm{d}}(x(t_j), u_{\mathrm{d}}(t_j))) - e^{W(\hat{t}-t_j)} V_{\mathrm{s}}(x(t_j))$$

$$+ e^{W(\hat{t}-t_j)} V_{\mathrm{s}}(x(t_j)) - e^{W(\hat{t}-t_{j-1}^+)} V_{\mathrm{s}}(x(t_{j-1}^+)) + \cdots$$

$$+ e^{W(\hat{t}-t_i)} V_{\mathrm{s}}(x(t_i) + F_{\mathrm{d}}(x(t_i), u_{\mathrm{d}}(t_i))) - e^{W(\hat{t}-t_i)} V_{\mathrm{s}}(x(t_i))$$

$$+ e^{W(\hat{t}-t_i)} V_{\mathrm{s}}(x(t_i)) - e^{W(\hat{t}-t)} V_{\mathrm{s}}(x(t))$$

$$= V_{\mathrm{s}}(x(\hat{t})) - e^{W(\hat{t}-t_j)} V_{\mathrm{s}}(x(t_j^+))$$

$$+ e^{W(\hat{t}-t_j)} [V_{\mathrm{s}}(x(t_j) + F_{\mathrm{d}}(x(t_j), u_{\mathrm{d}}(t_j))) - V_{\mathrm{s}}(x(t_j))]$$

$$+ e^{W(\hat{t}-t_j)} [V_{\mathrm{s}}(x(t_j)) - e^{W(t_j-t_{j-1})} V_{\mathrm{s}}(x(t_{j-1}^+))] + \cdots$$

$$+ e^{W(\hat{t}-t_i)} [V_{\mathrm{s}}(x(t_i) + F_{\mathrm{d}}(x(t_i), u_{\mathrm{d}}(t_i))) - V_{\mathrm{s}}(x(t_i))]$$

$$+ e^{W(\hat{t}-t_i)} [V_{\mathrm{s}}(x(t_i)) - e^{W(t_i-t)} V_{\mathrm{s}}(x(t))]$$

$$\leq\leq \int_{t_j}^{\hat{t}} e^{W(\hat{t}-s)} S_{\mathrm{c}}(u_{\mathrm{c}}(s), y_{\mathrm{c}}(s)) \mathrm{d}s + e^{W(\hat{t}-t_j)} S_{\mathrm{d}}(u_{\mathrm{d}}(t_j), y_{\mathrm{d}}(t_j))$$

$$+ e^{W(\hat{t}-t_j)} \int_{t_{j-1}}^{t_j} e^{W(t_j-s)} S_{\mathrm{c}}(u_{\mathrm{c}}(s), y_{\mathrm{c}}(s)) \mathrm{d}s + \cdots$$

$$+ e^{W(\hat{t}-t_i)} S_{\mathrm{d}}(u_{\mathrm{d}}(t_i), y_{\mathrm{d}}(t_i))$$

$$+ e^{W(\hat{t}-t_i)} \int_{t}^{t_i} e^{W(t_i-s)} S_{\mathrm{c}}(u_{\mathrm{c}}(s), y_{\mathrm{c}}(s)) \mathrm{d}s$$

$$= \int_{t}^{\hat{t}} e^{W(\hat{t}-s)} S_{\mathrm{c}}(u_{\mathrm{c}}(s), y_{\mathrm{c}}(s)) \mathrm{d}s$$

$$+ \sum_{k \in \mathbb{Z}_{[t,\hat{t})}} e^{W(\hat{t}-t_k)} S_{\mathrm{d}}(u_{\mathrm{d}}(t_k), y_{\mathrm{d}}(t_k)), \tag{5.14}$$

which implies that $\mathcal{G}$ is vector dissipative (respectively, exponentially vector dissipative) with respect to the vector hybrid supply rate (S_{c}, S_{d}).

Finally, similar constructions show that $\mathcal{G}$ is vector lossless with respect to the vector hybrid supply rate ($S_{\mathrm{c}}, S_{\mathrm{d}}$) if and only if (5.11) and (5.12) are satisfied as equalities with W semistable. $\qquad \square$

The following lemma is necessary for the next result.

Lemma 5.1 *Let $W \in \mathbb{R}^{q \times q}$. Then W is essentially nonnegative if and only if $e^{W(t-t_0)}$ is nonnegative for all $t \geq t_0$.*

Proof. The proof of this result appears in [25]. For completeness of exposition, we provide a proof here. If W is essentially nonnegative, then there exists sufficiently large $\alpha > 0$ such that $W_\alpha \triangleq W + \alpha I$ is nonnegative. Hence, $e^{W_\alpha(t-t_0)} = e^{(W+\alpha I)(t-t_0)} \geq\geq 0$, $t \geq t_0$, and hence, $e^{W(t-t_0)} = e^{-\alpha(t-t_0)} e^{W_\alpha(t-t_0)} \geq\geq 0$, $t \geq t_0$.

Conversely, suppose $e^{W(t-t_0)} \geq\geq 0$, $t \geq t_0$, and assume, *ad absurdum*, that there exist i, j such that $i \neq j$ and $W_{(i,j)} < 0$. Now, since $e^{W(t-t_0)} = \sum_{k=0}^{\infty} (k!)^{-1} W^k (t-t_0)^k$, it follows that

$$[e^{W(t-t_0)}]_{(i,j)} = I_{(i,j)} + (t-t_0) W_{(i,j)} + \mathcal{O}((t-t_0)^2). \quad (5.15)$$

Thus, as $t \to t_0$ and $i \neq j$, it follows that $[e^{W(t-t_0)}]_{(i,j)} < 0$ for some t sufficiently close to t_0, which leads to a contradiction. Hence, W is essentially nonnegative. $\square$

Theorem 5.2 *Consider the large-scale impulsive dynamical system $\mathcal{G}$ given by (5.1)–(5.4) and assume that $\mathcal{G}$ is completely reachable. Let $W \in \mathbb{R}^{q \times q}$ be essentially nonnegative and semistable (respectively, asymptotically stable). Then*

$$\int_{t_0}^{T} e^{-W(t-t_0)} S_{\mathrm{c}}(u_{\mathrm{c}}(t), y_{\mathrm{c}}(t)) \mathrm{d}t + \sum_{k \in \mathbb{Z}_{[t_0,T)}} e^{-W(t_k-t_0)} S_{\mathrm{d}}(u_{\mathrm{d}}(t_k), y_{\mathrm{d}}(t_k))$$

$$\geq\geq 0, \quad T \geq t_0, \quad (5.16)$$

for $x(t_0) = 0$ and $(u_{\mathrm{c}}(\cdot), u_{\mathrm{d}}(\cdot)) \in \mathcal{U}_{\mathrm{c}} \times \mathcal{U}_{\mathrm{d}}$ if and only if $V_{\mathrm{a}}(0) = 0$ and $V_{\mathrm{a}}(x)$ is finite for all $x \in \mathcal{D}$. Moreover, if (5.16) holds, then $V_{\mathrm{a}}(x)$, $x \in \mathcal{D}$, is a vector storage function for $\mathcal{G}$, and hence, $\mathcal{G}$ is vector dissipative (respectively, exponentially vector dissipative) with respect to the vector hybrid supply rate $(S_{\mathrm{c}}(u_{\mathrm{c}}, y_{\mathrm{c}}), S_{\mathrm{d}}(u_{\mathrm{d}}, y_{\mathrm{d}}))$.

Proof. Suppose $V_{\mathrm{a}}(0) = 0$ and $V_{\mathrm{a}}(x)$, $x \in \mathcal{D}$, is finite. Then

$$0 = V_{\mathrm{a}}(0) = - \inf_{(u_{\mathrm{c}}(\cdot), u_{\mathrm{d}}(\cdot)), T \geq t_0} \left[\int_{t_0}^{T} e^{-W(t-t_0)} S_{\mathrm{c}}(u_{\mathrm{c}}(t), y_{\mathrm{c}}(t)) \mathrm{d}t \right.$$

$$\left. + \sum_{k \in \mathbb{Z}_{[t_0,T)}} e^{-W(t_k-t_0)} S_{\mathrm{d}}(u_{\mathrm{d}}(t_k), y_{\mathrm{d}}(t_k)) \right], \quad (5.17)$$

which implies (5.16).

Next, suppose (5.16) holds. Then for $x(t_0) = 0$,

$$
- \inf_{(u_c(\cdot),\, u_d(\cdot)),\, T \geq t_0} \left[\int_{t_0}^{T} e^{-W(t-t_0)} S_c(u_c(t), y_c(t)) \mathrm{d}t \right.
$$

$$
\left. + \sum_{k \in \mathbb{Z}_{[t_0,T)}} e^{-W(t_k - t_0)} S_d(u_d(t_k), y_d(t_k)) \right] \leq\leq 0,
$$

(5.18)

which implies that $V_a(0) \leq\leq 0$. However, since $V_a(x_0) \geq\geq 0$, $x_0 \in \mathcal{D}$, it follows that $V_a(0) = 0$. Moreover, since $\mathcal{G}$ is completely reachable it follows that for every $x_0 \in \mathcal{D}$ there exists $\hat{t} > t_0$ and an admissible input $u(\cdot)$ defined on $[t_0, \hat{t}]$ such that $x(\hat{t}) = x_0$.

Now, since (5.16) holds for $x(t_0) = 0$ it follows that for all admissible $(u_c(\cdot), y_c(\cdot)) \in \mathcal{U}_c \times \mathcal{Y}_c$ and $(u_d(\cdot), y_d(\cdot)) \in \mathcal{U}_d \times \mathcal{Y}_d$,

$$
\int_{t_0}^{T} e^{-W(t-t_0)} S_c(u_c(t), y_c(t)) \mathrm{d}t + \sum_{k \in \mathbb{Z}_{[t_0,T)}} e^{-W(t_k - t_0)} S_d(u_d(t_k), y_d(t_k))
$$

$$
\geq\geq 0, \quad T \geq \hat{t}, \quad (5.19)
$$

or, equivalently, multiplying (5.19) by the nonnegative matrix $e^{W(\hat{t}-t_0)}$, $\hat{t} \geq t_0$, (see Lemma 5.1) yields

$$
- \int_{\hat{t}}^{T} e^{-W(t-\hat{t})} S_c(u_c(t), y_c(t)) \mathrm{d}t - \sum_{k \in \mathbb{Z}_{[\hat{t},T)}} e^{-W(t_k - \hat{t})} S_d(u_d(t_k), u_d(t_k))
$$

$$
\leq\leq \int_{t_0}^{\hat{t}} e^{-W(t-\hat{t})} S_c(u_c(t), y_c(t)) \mathrm{d}t
$$

$$
+ \sum_{k \in \mathbb{Z}_{[t_0,\hat{t})}} e^{-W(t_k - \hat{t})} S_d(u_d(t_k), u_d(t_k))
$$

$$
\leq\leq Q(x_0)
$$

$$
<< \infty, \quad T \geq \hat{t}, \quad (u_c(t), u_d(t_k)) \in U_c \times U_d, \quad (5.20)
$$

where $Q : \mathcal{D} \to \mathbb{R}^q$. Hence,

$$
V_a(x_0) = - \inf_{(u_c(\cdot),\, u_d(\cdot)),\, T \geq \hat{t}} \left[\int_{\hat{t}}^{T} e^{-W(t-\hat{t})} S_c(u_c(t), y_c(t)) \mathrm{d}t \right.
$$

$$+ \sum_{k \in \mathbb{Z}_{[\hat{t},T)}} e^{-W(t_k - \hat{t})} S_{\mathrm{d}}(u_{\mathrm{d}}(t_k), u_{\mathrm{d}}(t_k)) \Bigg] \leq\leq Q(x_0) << \infty,$$

$$x_0 \in \mathcal{D}, \qquad (5.21)$$

which implies that $V_{\mathrm{a}}(x_0)$, $x_0 \in \mathcal{D}$, is finite.

Finally, since (5.16) implies that $V_{\mathrm{a}}(0) = 0$ and $V_{\mathrm{a}}(x)$, $x \in \mathcal{D}$, is finite it follows from the definition of the vector available storage that

$$-V_{\mathrm{a}}(x_0) \leq\leq \int_{t_0}^{T} e^{-W(t-t_0)} S_{\mathrm{c}}(u_{\mathrm{c}}(t), y_{\mathrm{c}}(t)) \mathrm{d}t$$

$$+ \sum_{k \in \mathbb{Z}_{[t_0,T)}} e^{-W(t_k - t_0)} S_{\mathrm{d}}(u_{\mathrm{d}}(t_k), u_{\mathrm{d}}(t_k))$$

$$= \int_{t_0}^{t_{\mathrm{f}}} e^{-W(t-t_0)} S_{\mathrm{c}}(u_{\mathrm{c}}(t), y_{\mathrm{c}}(t)) \mathrm{d}t$$

$$+ \sum_{k \in \mathbb{Z}_{[t_0,t_{\mathrm{f}})}} e^{-W(t_k - t_0)} S_{\mathrm{d}}(u_{\mathrm{d}}(t_k), u_{\mathrm{d}}(t_k))$$

$$+ \int_{t_{\mathrm{f}}}^{T} e^{-W(t-t_0)} S_{\mathrm{c}}(u_{\mathrm{c}}(t), y_{\mathrm{c}}(t)) \mathrm{d}t$$

$$+ \sum_{k \in \mathbb{Z}_{[t_{\mathrm{f}},T)}} e^{-W(t_k - t_0)} S_{\mathrm{d}}(u_{\mathrm{d}}(t_k), u_{\mathrm{d}}(t_k)), \quad T \geq t_0. \quad (5.22)$$

Now, multiplying (5.22) by the nonnegative matrix $e^{W(t_{\mathrm{f}} - t_0)}$, $t_{\mathrm{f}} \geq t_0$, (see Lemma 5.1) it follows that

$$e^{W(t_{\mathrm{f}} - t_0)} V_{\mathrm{a}}(x_0) + \int_{t_0}^{t_{\mathrm{f}}} e^{W(t_{\mathrm{f}} - t)} S_{\mathrm{c}}(u_{\mathrm{c}}(t), y_{\mathrm{c}}(t)) \mathrm{d}t$$

$$+ \sum_{k \in \mathbb{Z}_{[t_0,t_{\mathrm{f}})}} e^{W(t_{\mathrm{f}} - t_k)} S_{\mathrm{d}}(u_{\mathrm{d}}(t_k), u_{\mathrm{d}}(t_k))$$

$$\geq\geq - \inf_{(u_{\mathrm{c}}(\cdot), u_{\mathrm{d}}(\cdot)),\, T \geq t_{\mathrm{f}}} \Bigg[\int_{t_{\mathrm{f}}}^{T} e^{-W(t-t_{\mathrm{f}})} S_{\mathrm{c}}(u_{\mathrm{c}}(t), y_{\mathrm{d}}(t)) \mathrm{d}t$$

$$+ \sum_{k \in \mathbb{Z}_{[t_{\mathrm{f}},T)}} e^{-W(t_k - t_{\mathrm{f}})} S_{\mathrm{d}}(u_{\mathrm{d}}(t_k), u_{\mathrm{d}}(t_k)) \Bigg]$$

$$= V_{\mathrm{a}}(x(t_{\mathrm{f}})), \qquad (5.23)$$

which implies that $V_{\mathrm{a}}(x)$, $x \in \mathcal{D}$, is a vector storage function, and hence, $\mathcal{G}$ is vector dissipative (respectively, exponentially vector dissipative) with respect to the vector hybrid supply rate $(S_c(u_c, y_c), S_d$ $(u_d, y_d))$. $\qquad\qquad\qquad\qquad\qquad\qquad\qquad\qquad\qquad\square$

The following lemma is necessary for several of the main results of this chapter.

Lemma 5.2 *Suppose $W \in \mathbb{R}^{q \times q}$ is essentially nonnegative. If W is semistable (respectively, asymptotically stable), then there exist a scalar $\alpha \geq 0$ (respectively, $\alpha > 0$) and a nonnegative vector $p \in \overline{\mathbb{R}}_+^q$, $p \neq 0$, (respectively, positive vector $p \in \mathbb{R}_+^q$) such that*

$$W^{\mathrm{T}}p + \alpha p = 0. \tag{5.24}$$

Proof. Since W is semistable if and only if $\lambda = 0$ or $\operatorname{Re} \lambda < 0$, where $\lambda \in \operatorname{spec}(W)$, and $\operatorname{ind}(W) \leq 1$, it follows from Theorem 4.6 of [21] that $-W^{\mathrm{T}}$ is an M-matrix. Now, recalling that (see [78], p.119) $-W^{\mathrm{T}}$ is an M-matrix if and only if there exist a scalar $\beta > 0$ and an $n \times n$ nonnegative matrix $B \geq\geq 0$ such that $\beta \geq \rho(B)$ and $-W^{\mathrm{T}} = \beta I_q - B$, where $\rho(\cdot)$ denotes the spectral radius, it follows that W^{T} can be written as $W^{\mathrm{T}} = B - \beta I_q$, where $\beta > 0$. Now, since $B \geq\geq 0$, it follows from Theorem 8.3.1 of [77] that $\rho(B) \in \operatorname{spec}(B)$ and there exists $p \geq\geq 0$, $p \neq 0$, such that $Bp = \rho(B)p$. Hence, $W^{\mathrm{T}}p = Bp - \beta p = (\rho(B) - \beta)p = -\alpha p$, where $\alpha \triangleq \beta - \rho(B) \geq 0$, which proves that there exist $p \geq\geq 0$, $p \neq 0$, and $\alpha \geq 0$ such that (5.24) holds. In the case where W is asymptotically stable, the result is a direct consequence of the Perron-Frobenius theorem [21]. $\qquad\square$

It follows from Lemma 5.2 that if $W \in \mathbb{R}^{q \times q}$ is essentially nonnegative and semistable (respectively, asymptotically stable), then there exist a scalar $\alpha \geq 0$ (respectively, $\alpha > 0$) and a nonnegative vector $p \in \overline{\mathbb{R}}_+^q$, $p \neq 0$, (respectively, positive vector $p \in \mathbb{R}_+^q$) such that (5.24) holds. In this case,

$$
\begin{aligned}
p^{\mathrm{T}}e^{Wt} &= p^{\mathrm{T}}[I_q + Wt + \tfrac{1}{2}W^2 t^2 + \cdots] \\
&= p^{\mathrm{T}}[I_q - \alpha t I_q + \tfrac{1}{2}\alpha^2 t^2 I_q + \cdots] \\
&= e^{-\alpha t}p^{\mathrm{T}}, \quad t \in \mathbb{R}.
\end{aligned} \tag{5.25}
$$

Using (5.25), we define the (scalar) *available storage* for the large-scale impulsive dynamical system $\mathcal{G}$ by

$$v_a(x_0) \triangleq - \inf_{(u_c(\cdot), u_d(\cdot)), T \geq t_0} \left[\int_{t_0}^T p^{\mathrm{T}}e^{-W(t-t_0)}S_c(u_c(t), y_c(t))\mathrm{d}t \right.$$

$$+ \sum_{k \in \mathbb{Z}_{[t_0,T)}} p^{\mathrm{T}} e^{-W(t_k-t_0)} S_{\mathrm{d}}(u_{\mathrm{d}}(t_k), y_{\mathrm{d}}(t_k)) \Bigg]$$

$$= - \inf_{(u_{\mathrm{c}}(\cdot), u_{\mathrm{d}}(\cdot)), \, T \geq t_0} \Bigg[\int_{t_0}^{T} e^{\alpha(t-t_0)} s_{\mathrm{c}}(u_{\mathrm{c}}(t), y_{\mathrm{c}}(t)) \mathrm{d}t$$

$$+ \sum_{k \in \mathbb{Z}_{[t_0,T)}} e^{\alpha(t_k-t_0)} s_{\mathrm{d}}(u_{\mathrm{d}}(t_k), y_{\mathrm{d}}(t_k)) \Bigg], \qquad (5.26)$$

where $s_{\mathrm{c}} : U_{\mathrm{c}} \times Y_{\mathrm{c}} \to \mathbb{R}$ and $s_{\mathrm{d}} : U_{\mathrm{d}} \times Y_{\mathrm{d}} \to \mathbb{R}$ defined as $s_{\mathrm{c}}(u_{\mathrm{c}}, y_{\mathrm{c}}) \triangleq p^{\mathrm{T}} S_{\mathrm{c}}(u_{\mathrm{c}}, y_{\mathrm{c}})$ and $s_{\mathrm{d}}(u_{\mathrm{d}}, y_{\mathrm{d}}) \triangleq p^{\mathrm{T}} S_{\mathrm{d}}(u_{\mathrm{d}}, y_{\mathrm{d}})$ form the (scalar) hybrid supply rate $(s_{\mathrm{c}}, s_{\mathrm{d}})$ for the large-scale impulsive dynamical system $\mathcal{G}$. Clearly, $v_{\mathrm{a}}(x) \geq 0$ for all $x \in \mathcal{D}$. As in standard hybrid dissipativity theory developed in Chapter 3, the available storage $v_{\mathrm{a}}(x)$, $x \in \mathcal{D}$, denotes the maximum amount of (scaled) energy that can be extracted from the large-scale impulsive dynamical system $\mathcal{G}$ at any time T.

The following theorem relates vector storage functions and vector hybrid supply rates to scalar storage functions and scalar hybrid supply rates of large-scale impulsive dynamical systems.

Theorem 5.3 *Consider the large-scale impulsive dynamical system* $\mathcal{G}$ *given by (5.1)–(5.4). Suppose* $\mathcal{G}$ *is vector dissipative (respectively, exponentially vector dissipative) with respect to the vector hybrid supply rate* $(S_{\mathrm{c}}(u_{\mathrm{c}}, y_{\mathrm{c}}), S_{\mathrm{d}}(u_{\mathrm{d}}, y_{\mathrm{d}})) : (U_{\mathrm{c}} \times Y_{\mathrm{c}}, U_{\mathrm{d}} \times Y_{\mathrm{d}}) \to \mathbb{R}^{q} \times \mathbb{R}^{q}$ *and with vector storage function* $V_{\mathrm{s}} : \mathcal{D} \to \overline{\mathbb{R}}_{+}^{q}$. *Then there exists* $p \in \mathbb{R}_{+}^{q}$, $p \neq 0$, *(respectively,* $p \in \mathbb{R}_{+}^{q}$*) such that* $\mathcal{G}$ *is dissipative (respectively, exponentially dissipative) with respect to the scalar hybrid supply rate* $(s_{\mathrm{c}}(u_{\mathrm{c}}, y_{\mathrm{c}}), s_{\mathrm{d}}(u_{\mathrm{d}}, y_{\mathrm{d}})) = (p^{\mathrm{T}} S_{\mathrm{c}}(u_{\mathrm{c}}, y_{\mathrm{c}}), p^{\mathrm{T}} S_{\mathrm{d}}(u_{\mathrm{d}}, y_{\mathrm{d}}))$ *and with storage function* $v_{\mathrm{s}}(x) = p^{\mathrm{T}} V_{\mathrm{s}}(x)$, $x \in \mathcal{D}$. *Moreover, in this case* $v_{\mathrm{a}}(x)$, $x \in \mathcal{D}$, *is a storage function for* $\mathcal{G}$ *and*

$$0 \leq v_{\mathrm{a}}(x) \leq v_{\mathrm{s}}(x), \quad x \in \mathcal{D}. \qquad (5.27)$$

Proof. Suppose $\mathcal{G}$ is vector dissipative (respectively, exponentially vector dissipative) with respect to the vector hybrid supply rate $(S_{\mathrm{c}}(u_{\mathrm{c}}, y_{\mathrm{c}}), S_{\mathrm{d}}(u_{\mathrm{d}}, y_{\mathrm{d}}))$. Then there exist an essentially nonnegative, semistable (respectively, asymptotically stable) dissipation matrix W and a vector storage function $V_{\mathrm{s}} : \mathcal{D} \to \overline{\mathbb{R}}_{+}^{q}$ such that the dissipation inequality (5.9) holds. Furthermore, it follows from Lemma 5.2 that there exist $\alpha \geq 0$ (respectively, $\alpha > 0$) and a nonzero vector $p \in \overline{\mathbb{R}}_{+}^{q}$ (respectively, $p \in \mathbb{R}_{+}^{q}$) satisfying (5.24). Hence, premultiplying (5.9) by p^{T} and using (5.25) it follows that

$$e^{\alpha T} v_{\mathrm{s}}(x(T)) \leq e^{\alpha t_0} v_{\mathrm{s}}(x(t_0)) + \int_{t_0}^{T} e^{\alpha t} s_{\mathrm{c}}(u_{\mathrm{c}}(t), y_{\mathrm{c}}(t)) \mathrm{d}t$$

$$+ \sum_{k \in \mathbb{Z}_{[t_0,T)}} e^{\alpha t_k} s_{\mathrm{d}}(u_{\mathrm{d}}(t_k), y_{\mathrm{d}}(t_k)),$$

$$T \geq t_0, \quad (u_{\mathrm{c}}(t), u_{\mathrm{d}}(t_k)) \in U_{\mathrm{c}} \times U_{\mathrm{d}}, \qquad (5.28)$$

where $v_{\mathrm{s}}(x) = p^{\mathrm{T}} V_{\mathrm{s}}(x)$, $x \in \mathcal{D}$, which implies dissipativity (respectively, exponential dissipativity) of $\mathcal{G}$ with respect to the scalar hybrid supply rate $(s_{\mathrm{c}}(u_{\mathrm{c}}, y_{\mathrm{c}}), s_{\mathrm{d}}(u_{\mathrm{d}}, y_{\mathrm{d}}))$ and with storage function $v_{\mathrm{s}}(x)$, $x \in \mathcal{D}$. Moreover, since $v_{\mathrm{s}}(0) = 0$, it follows from (5.28) that for $x(t_0) = 0$,

$$\int_{t_0}^{T} e^{\alpha(t-t_0)} s_{\mathrm{c}}(u_{\mathrm{c}}(t), y_{\mathrm{c}}(t)) \mathrm{d}t + \sum_{k \in \mathbb{Z}_{[t_0,T)}} e^{\alpha(t_k-t_0)} s_{\mathrm{d}}(u_{\mathrm{d}}(t_k), y_{\mathrm{d}}(t_k)) \geq 0,$$

$$T \geq t_0, \quad (u_{\mathrm{c}}(t), u_{\mathrm{d}}(t_k)) \in U_{\mathrm{c}} \times U_{\mathrm{d}}, \qquad (5.29)$$

which, using (5.26), implies that $v_{\mathrm{a}}(0) = 0$. Now, it can be easily shown that $v_{\mathrm{a}}(x)$, $x \in \mathcal{D}$, satisfies (5.28), and hence, the available storage defined by (5.26) is a storage function for $\mathcal{G}$.

Finally, it follows from (5.28) that

$$v_{\mathrm{s}}(x(t_0)) \geq e^{\alpha(T-t_0)} v_{\mathrm{s}}(x(T)) - \int_{t_0}^{T} e^{\alpha(t-t_0)} s_{\mathrm{c}}(u_{\mathrm{c}}(t), y_{\mathrm{c}}(t)) \mathrm{d}t$$

$$- \sum_{k \in \mathbb{Z}_{[t_0,T)}} e^{\alpha(t_k-t_0)} s_{\mathrm{d}}(u_{\mathrm{d}}(t_k), y_{\mathrm{d}}(t_k))$$

$$\geq - \int_{t_0}^{T} e^{\alpha(t-t_0)} s_{\mathrm{c}}(u(t), y(t)) \mathrm{d}t$$

$$- \sum_{k \in \mathbb{Z}_{[t_0,T)}} e^{\alpha(t_k-t_0)} s_{\mathrm{d}}(u_{\mathrm{d}}(t_k), y_{\mathrm{d}}(t_k)),$$

$$T \geq t_0, \quad (u_{\mathrm{c}}(t), u_{\mathrm{d}}(t_k)) \in U_{\mathrm{c}} \times U_{\mathrm{d}}, \qquad (5.30)$$

which implies

$$v_{\mathrm{s}}(x(t_0)) \geq - \inf_{(u_{\mathrm{c}}(\cdot), u_{\mathrm{d}}(\cdot)), T \geq t_0} \left[\int_{t_0}^{T} e^{\alpha(t-t_0)} s_{\mathrm{c}}(u_{\mathrm{c}}(t), y_{\mathrm{c}}(t)) \mathrm{d}t \right.$$

$$\left. + \sum_{k \in \mathbb{Z}_{[t_0,T)}} e^{\alpha(t_k-t_0)} s_{\mathrm{d}}(u_{\mathrm{d}}(t_k), y_{\mathrm{d}}(t_k)) \right]$$

$$= v_{\mathrm{a}}(x(t_0)), \qquad (5.31)$$

and hence, (5.27) holds. □

It follows from Theorem 5.2 that if (5.16) holds for $x(t_0) = 0$, then the vector available storage $V_{\mathrm{a}}(x)$, $x \in \mathcal{D}$, is a vector storage function for $\mathcal{G}$. In this case, it follows from Theorem 5.3 that there exists $p \in \overline{\mathbb{R}}_+^q$, $p \neq 0$, such that $v_{\mathrm{s}}(x) \triangleq p^{\mathrm{T}} V_{\mathrm{a}}(x)$ is a storage function for $\mathcal{G}$ that satisfies (5.28), and hence, by (5.27), $v_{\mathrm{a}}(x) \leq p^{\mathrm{T}} V_{\mathrm{a}}(x)$, $x \in \mathcal{D}$. Furthermore, it is important to note that it follows from Theorem 5.3 that if $\mathcal{G}$ is vector dissipative, then $\mathcal{G}$ can either be (scalar) dissipative or (scalar) exponentially dissipative.

The following theorem provides sufficient conditions guaranteeing that all scalar storage functions defined in terms of vector storage functions, that is, $v_{\mathrm{s}}(x) = p^{\mathrm{T}} V_{\mathrm{s}}(x)$, of a given vector dissipative large-scale impulsive nonlinear dynamical system are positive definite.

Theorem 5.4 *Consider the large-scale impulsive dynamical system $\mathcal{G}$ given by (5.1)–(5.4) and assume that $\mathcal{G}$ is zero-state observable. Furthermore, assume that $\mathcal{G}$ is vector dissipative (respectively, exponentially vector dissipative) with respect to the vector hybrid supply rate $(S_{\mathrm{c}}(u_{\mathrm{c}}, y_{\mathrm{c}}), S_{\mathrm{d}}(u_{\mathrm{d}}, y_{\mathrm{d}}))$ and there exist $\alpha \geq 0$ and $p \in \overline{\mathbb{R}}_+^q$ such that (5.24) holds. In addition, assume that there exist functions $\kappa_{\mathrm{c}i} : Y_{\mathrm{c}i} \to U_{\mathrm{c}i}$ and $\kappa_{\mathrm{d}i} : Y_{\mathrm{d}i} \to U_{\mathrm{d}i}$ such that $\kappa_{\mathrm{c}i}(0) = 0$, $\kappa_{\mathrm{d}i}(0) = 0$, $s_{\mathrm{c}i}(\kappa_{\mathrm{c}i}(y_{\mathrm{c}i}), y_{\mathrm{c}i}) < 0$, $y_{\mathrm{c}i} \neq 0$, and $s_{\mathrm{d}i}(\kappa_{\mathrm{d}i}(y_{\mathrm{d}i}), y_{\mathrm{d}i}) < 0$, $y_{\mathrm{d}i} \neq 0$, for all $i = 1, \ldots, q$. Then for all vector storage functions $V_{\mathrm{s}} : \mathcal{D} \to \overline{\mathbb{R}}_+^q$ the storage function $v_{\mathrm{s}}(x) \triangleq p^{\mathrm{T}} V_{\mathrm{s}}(x)$, $x \in \mathcal{D}$, is positive definite, that is, $v_{\mathrm{s}}(0) = 0$ and $v_{\mathrm{s}}(x) > 0$, $x \in \mathcal{D}$, $x \neq 0$.*

Proof. It follows from Theorem 5.3 that $v_{\mathrm{a}}(x)$, $x \in \mathcal{D}$, is a storage function for $\mathcal{G}$ that satisfies (5.28). Next, suppose, *ad absurdum*, that there exists $x \in \mathcal{D}$ such that $v_{\mathrm{a}}(x) = 0$, $x \neq 0$. Then it follows from the definition of $v_{\mathrm{a}}(x)$, $x \in \mathcal{D}$, that for $x(t_0) = x$,

$$\int_{t_0}^{T} e^{\alpha(t-t_0)} s_{\mathrm{c}}(u_{\mathrm{c}}(t), y_{\mathrm{c}}(t)) \mathrm{d}t + \sum_{k \in \mathbb{Z}_{[t_0, T)}} e^{\alpha(t_k - t_0)} s_{\mathrm{d}}(u_{\mathrm{d}}(t_k), y_{\mathrm{d}}(t_k)) \geq 0,$$

$$T \geq t_0, \quad (u_{\mathrm{c}}(t), u_{\mathrm{d}}(t_k)) \in U_{\mathrm{c}} \times U_{\mathrm{d}}. \quad (5.32)$$

However, for $u_{\mathrm{c}i} = \kappa_{\mathrm{c}i}(y_{\mathrm{c}i})$ and $u_{\mathrm{d}i} = \kappa_{\mathrm{d}i}(y_{\mathrm{d}i})$ we have $s_{\mathrm{c}i}(\kappa_{\mathrm{c}i}(y_{\mathrm{c}i}), y_{\mathrm{c}i}) < 0$, $s_{\mathrm{d}i}(\kappa_{\mathrm{d}i}(y_{\mathrm{d}i}), y_{\mathrm{d}i}) < 0$, $y_{\mathrm{c}i} \neq 0$, $y_{\mathrm{d}i} \neq 0$ for all $i = 1, \ldots, q$, and since $p >> 0$, it follows that $y_{\mathrm{c}i}(t) = 0$, $t_k < t \leq t_{k+1}$, $y_{\mathrm{d}i}(t_k) = 0$, $k \in \overline{\mathbb{Z}}_+$, $i = 1, \ldots, q$, which further implies that $u_{\mathrm{c}i}(t) = 0$, $t_k < t \leq t_{k+1}$, and $u_{\mathrm{d}i}(t_k) = 0$, $k \in \overline{\mathbb{Z}}_+$, $i = 1, \ldots, q$. Since $\mathcal{G}$ is zero-state observable it follows that $x = 0$, and hence, $v_{\mathrm{a}}(x) = 0$ if and only if $x = 0$. The result now follows from (5.27). Finally, for the exponentially vector dissipative case it follows from Lemma 5.2 that $p >> 0$, with the rest of the proof being identical to that above. $\square$

Next, we introduce the concept of *vector required supply* of a large-scale impulsive dynamical system. Specifically, define the vector required supply of the large-scale impulsive dynamical system $\mathcal{G}$ by

$$V_{\mathrm{r}}(x_0) \triangleq \inf_{(u_{\mathrm{c}}(\cdot),u_{\mathrm{d}}(\cdot)),\, T \leq t_0} \left[\int_T^{t_0} e^{-W(t-t_0)} S_{\mathrm{c}}(u_{\mathrm{c}}(t), y_{\mathrm{c}}(t)) \mathrm{d}t \right.$$

$$\left. + \sum_{k \in \mathbb{Z}_{[T,t_0)}} e^{-W(t_k-t_0)} S_{\mathrm{d}}(u_{\mathrm{d}}(t_k), y_{\mathrm{d}}(t_k)) \right],$$

(5.33)

where $x(t)$, $t \geq T$, is the solution to (5.1)–(5.4) with $x(T) = 0$ and $x(t_0) = x_0$. Note that since, with $x(t_0) = 0$, the infimum in (5.33) is the zero vector it follows that $V_{\mathrm{r}}(0) = 0$. Moreover, since $\mathcal{G}$ is completely reachable it follows that $V_{\mathrm{r}}(x) << \infty$, $x \in \mathcal{D}$. Using the notion of the vector required supply we present necessary and sufficient conditions for vector dissipativity of a large-scale impulsive dynamical system with respect to a vector hybrid supply rate.

Theorem 5.5 *Consider the large-scale impulsive dynamical system $\mathcal{G}$ given by (5.1)–(5.4) and assume that $\mathcal{G}$ is completely reachable. Then $\mathcal{G}$ is vector dissipative (respectively, exponentially vector dissipative) with respect to the vector hybrid supply rate $(S_{\mathrm{c}}(u_{\mathrm{c}}, y_{\mathrm{c}}), S_{\mathrm{d}}(u_{\mathrm{d}}, y_{\mathrm{d}}))$ if and only if*

$$0 \leq\leq V_{\mathrm{r}}(x) << \infty, \qquad x \in \mathcal{D}. \tag{5.34}$$

Moreover, if (5.34) holds, then $V_{\mathrm{r}}(x)$, $x \in \mathcal{D}$, is a vector storage function for $\mathcal{G}$. Finally, if the vector available storage $V_{\mathrm{a}}(x)$, $x \in \mathcal{D}$, is a vector storage function for $\mathcal{G}$, then

$$0 \leq\leq V_{\mathrm{a}}(x) \leq\leq V_{\mathrm{r}}(x) << \infty, \qquad x \in \mathcal{D}. \tag{5.35}$$

Proof. Suppose (5.34) holds and let $x(t)$, $t \in \mathbb{R}$, satisfy (5.1)–(5.4) with admissible inputs $(u_{\mathrm{c}}(t), u_{\mathrm{d}}(t_k)) \in U_{\mathrm{c}} \times U_{\mathrm{d}}$, $t \in \mathbb{R}$, $k \in \overline{\mathbb{Z}}_+$, and $x(t_0) = x_0$. Then, it follows from the definition of $V_{\mathrm{r}}(\cdot)$ that for $T \leq t_{\mathrm{f}} \leq t_0$, $u_{\mathrm{c}}(\cdot) \in \mathcal{U}_{\mathrm{c}}$, and $u_{\mathrm{d}}(\cdot) \in \mathcal{U}_{\mathrm{d}}$,

$$V_{\mathrm{r}}(x_0) \leq\leq \int_T^{t_0} e^{-W(t-t_0)} S_{\mathrm{c}}(u_{\mathrm{c}}(t), y_{\mathrm{c}}(t)) \mathrm{d}t$$

$$+ \sum_{k \in \mathbb{Z}_{[T,t_0)}} e^{-W(t_k - t_0)} S_{\mathrm{d}}(u_{\mathrm{d}}(t_k), y_{\mathrm{d}}(t_k))$$

$$= \int_T^{t_f} e^{-W(t - t_0)} S_{\mathrm{c}}(u_{\mathrm{c}}(t), y_{\mathrm{c}}(t)) \mathrm{d}t$$

$$+ \sum_{k \in \mathbb{Z}_{[T,t_f)}} e^{-W(t_k - t_0)} S_{\mathrm{d}}(u_{\mathrm{d}}(t_k), y_{\mathrm{d}}(t_k))$$

$$+ \int_{t_f}^{t_0} e^{-W(t - t_0)} S_{\mathrm{c}}(u_{\mathrm{c}}(t), y_{\mathrm{c}}(t)) \mathrm{d}t$$

$$+ \sum_{k \in \mathbb{Z}_{[t_f,t_0)}} e^{-W(t_k - t_0)} S_{\mathrm{d}}(u_{\mathrm{d}}(t_k), y_{\mathrm{d}}(t_k)), \qquad (5.36)$$

and hence,

$$V_{\mathrm{r}}(x_0) \leq\leq e^{W(t_0 - t_f)} \inf_{(u_{\mathrm{c}}(\cdot), u_{\mathrm{d}}(\cdot)), \, T \leq t_f} \left[\int_T^{t_f} e^{-W(t - t_f)} S_{\mathrm{c}}(u_{\mathrm{c}}(t), y_{\mathrm{c}}(t)) \mathrm{d}t \right.$$

$$\left. + \sum_{k \in \mathbb{Z}_{[T,t_f)}} e^{-W(t_k - t_f)} S_{\mathrm{d}}(u_{\mathrm{d}}(t_k), y_{\mathrm{d}}(t_k)) \right]$$

$$+ \int_{t_f}^{t_0} e^{-W(t - t_0)} S_{\mathrm{c}}(u_{\mathrm{c}}(t), y_{\mathrm{c}}(t)) \mathrm{d}t$$

$$+ \sum_{k \in \mathbb{Z}_{[t_f,t_0)}} e^{-W(t_k - t_0)} S_{\mathrm{d}}(u_{\mathrm{d}}(t_k), y_{\mathrm{d}}(t_k))$$

$$= e^{W(t_0 - t_f)} V_{\mathrm{r}}(x(t_f)) + \int_{t_f}^{t_0} e^{-W(t - t_0)} S_{\mathrm{c}}(u_{\mathrm{c}}(t), y_{\mathrm{c}}(t)) \mathrm{d}t$$

$$+ \sum_{k \in \mathbb{Z}_{[t_f,t_0)}} e^{-W(t_k - t_0)} S_{\mathrm{d}}(u_{\mathrm{d}}(t_k), y_{\mathrm{d}}(t_k)), \qquad (5.37)$$

which shows that $V_{\mathrm{r}}(x)$, $x \in \mathcal{D}$, is a vector storage function for $\mathcal{G}$, and hence, $\mathcal{G}$ is vector dissipative with respect to the vector hybrid supply rate $(S_{\mathrm{c}}(u_{\mathrm{c}}, y_{\mathrm{c}}), S_{\mathrm{d}}(u_{\mathrm{d}}, y_{\mathrm{d}}))$.

Conversely, suppose that $\mathcal{G}$ is vector dissipative with respect to the vector hybrid supply rate $(S_{\mathrm{c}}(u_{\mathrm{c}}, y_{\mathrm{c}}), S_{\mathrm{d}}(u_{\mathrm{d}}, y_{\mathrm{d}}))$. Then there exists a nonnegative vector storage function $V_{\mathrm{s}}(x)$, $x \in \mathcal{D}$, such that $V_{\mathrm{s}}(0) = 0$. Since $\mathcal{G}$ is completely reachable it follows that for $x(t_0) = x_0$ there exist $T < t_0$ and $u_{\mathrm{c}}(t)$, $t \in [T, t_0]$, and $u_{\mathrm{d}}(t_k)$, $k \in \mathbb{Z}_{[T,t_0]}$, such that $x(T) = 0$. Hence, it follows from the vector hybrid dissipation inequality (5.9) that

$$0 \leq\leq V_{\mathrm{s}}(x(t_0)) \leq\leq e^{W(t_0-T)} V_{\mathrm{s}}(x(T)) + \int_T^{t_0} e^{W(t_0-t)} S_{\mathrm{c}}(u_{\mathrm{c}}(t), y_{\mathrm{c}}(t)) \mathrm{d}t$$

$$+ \sum_{k \in \mathbb{Z}_{[T,t_0)}} e^{W(t_0-t_k)} S_{\mathrm{d}}(u_{\mathrm{d}}(t_k), y_{\mathrm{d}}(t_k)), \qquad (5.38)$$

which implies that for all $T \leq t_0$, $u_{\mathrm{c}}(t) \in U_{\mathrm{c}}$, and $u_{\mathrm{d}}(t_k) \in U_{\mathrm{d}}$,

$$0 \leq\leq \int_T^{t_0} e^{W(t_0-t)} S_{\mathrm{c}}(u_{\mathrm{c}}(t), y_{\mathrm{c}}(t)) \mathrm{d}t$$

$$+ \sum_{k \in \mathbb{Z}_{[T,t_0)}} e^{W(t_0-t_k)} S_{\mathrm{d}}(u_{\mathrm{d}}(t_k), y_{\mathrm{d}}(t_k)) \qquad (5.39)$$

or, equivalently,

$$0 \leq\leq \inf_{(u_{\mathrm{c}}(\cdot), u_{\mathrm{d}}(\cdot)), T \leq t_0} \left[\int_T^{t_0} e^{W(t_0-t)} S_{\mathrm{c}}(u_{\mathrm{c}}(t), y_{\mathrm{c}}(t)) \mathrm{d}t \right.$$

$$\left. + \sum_{k \in \mathbb{Z}_{[T,t_0)}} e^{W(t_0-t_k)} S_{\mathrm{d}}(u_{\mathrm{d}}(t_k), y_{\mathrm{d}}(t_k)) \right]$$

$$= V_{\mathrm{r}}(x_0). \qquad (5.40)$$

Since, by complete reachability, $V_{\mathrm{r}}(x) << \infty$, $x \in \mathcal{D}$, it follows that (5.34) holds.

Finally, suppose that $V_{\mathrm{a}}(x)$, $x \in \mathcal{D}$, is a vector storage function. Then for $x(T) = 0$, $x(t_0) = x_0$, $u_{\mathrm{c}}(t) \in U_{\mathrm{c}}$, and $u_{\mathrm{d}}(t_k) \in U_{\mathrm{d}}$, it follows that

$$V_{\mathrm{a}}(x(t_0)) \leq\leq e^{W(t_0-T)} V_{\mathrm{a}}(x(T)) + \int_T^{t_0} e^{W(t_0-t)} S_{\mathrm{c}}(u_{\mathrm{c}}(t), y_{\mathrm{c}}(t)) \mathrm{d}t$$

$$+ \sum_{k \in \mathbb{Z}_{[T,t_0)}} e^{W(t_0-t_k)} S_{\mathrm{d}}(u_{\mathrm{d}}(t_k), y_{\mathrm{d}}(t_k)), \qquad (5.41)$$

which implies that

$$0 \leq\leq V_{\mathrm{a}}(x(t_0)) \leq\leq \inf_{(u_{\mathrm{c}}(\cdot), u_{\mathrm{d}}(\cdot)), T \leq t_0} \left[\int_T^{t_0} e^{W(t_0-t)} S_{\mathrm{c}}(u_{\mathrm{c}}(t), y_{\mathrm{c}}(t)) \mathrm{d}t \right.$$

$$\left. + \sum_{k \in \mathbb{Z}_{[T,t_0)}} e^{W(t_0-t_k)} S_{\mathrm{d}}(u_{\mathrm{d}}(t_k), y_{\mathrm{d}}(t_k)) \right]$$

$$= V_{\mathrm{r}}(x(t_0)), \quad x \in \mathcal{D}. \qquad (5.42)$$

Since $x(t_0) = x_0 \in \mathcal{D}$ is arbitrary and, by complete reachability, $V_r(x) << \infty$, $x \in \mathcal{D}$, (5.42) implies (5.35). $\qquad\qquad\square$

The next result is a direct consequence of Theorems 5.2 and 5.5.

Proposition 5.1 *Consider the large-scale impulsive dynamical system $\mathcal{G}$ given by (5.1)–(5.4) and assume $\mathcal{G}$ is completely reachable. Let $M = \text{diag}\,[\mu_1, \ldots, \mu_q]$ be such that $0 \le \mu_i \le 1$, $i = 1, \ldots, q$. If $V_a(x)$, $x \in \mathcal{D}$, and $V_r(x)$, $x \in \mathcal{D}$, are vector storage functions for $\mathcal{G}$, then*

$$V_s(x) = MV_a(x) + (I_q - M)V_r(x), \quad x \in \mathcal{D}, \qquad (5.43)$$

is a vector storage function for $\mathcal{G}$.

Proof. First note that $M \ge\ge 0$ and $I_q - M \ge\ge 0$ if and only if $M = \text{diag}\,[\mu_1, \ldots, \mu_q]$ and $\mu_i \in [0, 1]$, $i = 1, \ldots, q$. Now, the result is a direct consequence of the complete reachability of $\mathcal{G}$ along with vector hybrid dissipation inequality (5.9) by noting that if $V_a(x)$ and $V_r(x)$ satisfy (5.9), then $V_s(x)$ satisfies (5.9). $\qquad\qquad\square$

Next, recall that if $\mathcal{G}$ is vector dissipative (respectively, exponentially vector dissipative), then there exist $p \in \overline{\mathbb{R}}_+^q$, $p \ne 0$, and $\alpha \ge 0$ (respectively, $p \in \mathbb{R}_+^q$ and $\alpha > 0$) such that (5.24) and (5.25) hold. Now, define the (scalar) *required supply* for the large-scale impulsive dynamical system $\mathcal{G}$ by

$$v_r(x_0) \overset{\triangle}{=} \inf_{(u_c(\cdot), u_d(\cdot)),\, T \le t_0} \left[\int_T^{t_0} p^T e^{-W(t-t_0)} S_c(u_c(t), y_c(t)) dt \right.$$

$$\left. + \sum_{k \in \mathbb{Z}_{[T, t_0)}} e^{-W(t_k - t_0)} S_d(u_d(t_k), y_d(t_k)) \right]$$

$$= \inf_{(u_c(\cdot), u_d(\cdot)),\, T \le t_0} \left[\int_T^{t_0} e^{\alpha(t-t_0)} s_c(u_c(t), y_c(t)) dt \right.$$

$$\left. + \sum_{k \in \mathbb{Z}_{[T, t_0)}} e^{\alpha(t_k - t_0)} s_d(u_d(t_k), y_d(t_k)) \right], \quad x_0 \in \mathcal{D}, \quad (5.44)$$

where $s_c(u_c, y_c) = p^T S_c(u_c, y_c)$, $s_d(u_d, y_d) = p^T S_d(u_d, y_d)$, and $x(t)$, $t \ge T$, is the solution to (5.1)–(5.4) with $x(T) = 0$ and $x(t_0) = x_0$. It follows from (5.44) that the required supply of a large-scale impulsive dynamical system is the minimum amount of generalized energy which can be delivered to the large-scale system in order to transfer it from an initial state $x(T) = 0$ to a given state $x(t_0) = x_0$. Using the same arguments as in case of the vector required supply, it follows that $v_r(0) = 0$ and $v_r(x) < \infty$, $x \in \mathcal{D}$.

Next, using the notion of the required supply, we show that all storage functions of the form $v_s(x) = p^T V_s(x)$, where $p \in \mathbb{R}^q_+$, $p \neq 0$, are bounded from above by the required supply and bounded from below by the available storage. Hence, a dissipative large-scale impulsive dynamical system can deliver to its surroundings only a fraction of all of its stored subsystem energies and can store only a fraction of the work done to all of its subsystems.

Corollary 5.1 *Consider the large-scale impulsive dynamical system $\mathcal{G}$ given by (5.1)–(5.4). Assume that $\mathcal{G}$ is vector dissipative with respect to the vector hybrid supply rate $(S_c(u_c, y_c), S_d(u_d, y_d))$ and with vector storage function $V_s : \mathcal{D} \to \overline{\mathbb{R}}^q_+$. Then $v_r(x)$, $x \in \mathcal{D}$, is a storage function for $\mathcal{G}$. Moreover, if $v_s(x) \triangleq p^T V_s(x)$, $x \in \mathcal{D}$, where $p \in \mathbb{R}^q_+$, $p \neq 0$, then*

$$0 \leq v_a(x) \leq v_s(x) \leq v_r(x) < \infty, \quad x \in \mathcal{D}. \qquad (5.45)$$

Proof. It follows from Theorem 5.3 that if $\mathcal{G}$ is vector dissipative with respect to the vector hybrid supply rate $(S_c(u_c, y_c), S_d(u_d, y_d))$ and with a vector storage function $V_s : \mathcal{D} \to \overline{\mathbb{R}}^q_+$, then there exists $p \in \mathbb{R}^q_+$, $p \neq 0$, such that $\mathcal{G}$ is dissipative with respect to the hybrid supply rate $(s_c(u_c, y_c), s_d(u_d, y_d)) = (p^T S_c(u_c, y_c), p^T S_d(u_d, y_d))$ and with storage function $v_s(x) = p^T V_s(x)$, $x \in \mathcal{D}$. Hence, it follows from (5.28), with $x(T) = 0$ and $x(t_0) = x_0$, that

$$\int_T^{t_0} e^{\alpha(t-t_0)} s_c(u_c(t), y_c(t)) dt + \sum_{k \in \mathbb{Z}_{[T,t_0)}} e^{\alpha(t_k-t_0)} s_d(u_d(t_k), y_d(t_k)) \geq 0,$$

$$T \leq t_0, \quad (u_c, u_d) \in \mathcal{U}_c \times \mathcal{U}_d, \qquad (5.46)$$

which implies that $v_r(x_0) \geq 0$, $x_0 \in \mathcal{D}$.

Furthermore, it is easy to see from the definition of the required supply that $v_r(x)$, $x \in \mathcal{D}$, satisfies the dissipation inequality (5.28). Hence, $v_r(x)$, $x \in \mathcal{D}$, is a storage function for $\mathcal{G}$. Moreover, it follows from the dissipation inequality (5.28), with $x(T) = 0$, $x(t_0) = x_0$, $u_c(t) \in \mathcal{U}_c$, and $u_d(t_k) \in \mathcal{U}_d$, that

$$e^{\alpha t_0} v_s(x(t_0)) \leq e^{\alpha T} v_s(x(T)) + \int_T^{t_0} e^{\alpha t} s_c(u_c(t), y_c(t)) dt$$

$$+ \sum_{k \in \mathbb{Z}_{[T,t_0)}} e^{\alpha t_k} s_{\mathrm{d}}(u_{\mathrm{d}}(t_k), y_{\mathrm{d}}(t_k))$$

$$= \int_T^{t_0} e^{\alpha t} s_{\mathrm{c}}(u_{\mathrm{c}}(t), y_{\mathrm{c}}(t)) \mathrm{d}t$$

$$+ \sum_{k \in \mathbb{Z}_{[T,t_0)}} e^{\alpha t_k} s_{\mathrm{d}}(u_{\mathrm{d}}(t_k), y_{\mathrm{d}}(t_k)), \tag{5.47}$$

which implies that

$$v_{\mathrm{s}}(x(t_0)) \leq \inf_{(u_{\mathrm{c}}(\cdot), u_{\mathrm{d}}(\cdot)),\, T \leq t_0} \left[\int_T^{t_0} e^{\alpha(t-t_0)} s_{\mathrm{c}}(u_{\mathrm{c}}(t), y_{\mathrm{c}}(t)) \mathrm{d}t \right.$$

$$\left. + \sum_{k \in \mathbb{Z}_{[T,t_0)}} e^{\alpha(t_k - t_0)} s_{\mathrm{d}}(u_{\mathrm{d}}(t_k), y_{\mathrm{d}}(t_k)) \right]$$

$$= v_{\mathrm{r}}(x(t_0)). \tag{5.48}$$

Finally, it follows from Theorem 5.3 that $v_{\mathrm{a}}(x)$, $x \in \mathcal{D}$, is a storage function for $\mathcal{G}$, and hence, using (5.27) and (5.48), (5.45) holds. $\quad\square$

It follows from Theorem 5.5 that if $\mathcal{G}$ is vector dissipative with respect to the vector hybrid supply rate $(S_{\mathrm{c}}(u_{\mathrm{c}}, y_{\mathrm{c}}), S_{\mathrm{d}}(u_{\mathrm{d}}, y_{\mathrm{d}}))$, then $V_{\mathrm{r}}(x)$, $x \in \mathcal{D}$, is a vector storage function for $\mathcal{G}$ and, by Theorem 5.3, there exists $p \in \mathbb{R}_+^q$, $p \neq 0$, such that $v_{\mathrm{s}}(x) \triangleq p^{\mathrm{T}} V_{\mathrm{r}}(x)$, $x \in \mathcal{D}$, is a storage function for $\mathcal{G}$ satisfying (5.28). Hence, it follows from Corollary 5.1 that $p^{\mathrm{T}} V_{\mathrm{r}}(x) \leq v_{\mathrm{r}}(x)$, $x \in \mathcal{D}$. The next result relates the vector (respectively, scalar) available storage and the vector (respectively, scalar) required supply for vector lossless large-scale impulsive dynamical systems.

Theorem 5.6 *Consider the large-scale impulsive dynamical system $\mathcal{G}$ given by (5.1)–(5.4). Assume that $\mathcal{G}$ is completely reachable to and from the origin. If $\mathcal{G}$ is vector lossless with respect to the vector hybrid supply rate $(S_{\mathrm{c}}(u_{\mathrm{c}}, y_{\mathrm{c}}), S_{\mathrm{d}}(u_{\mathrm{d}}, y_{\mathrm{d}}))$ and $V_{\mathrm{a}}(x)$, $x \in \mathcal{D}$, is a vector storage function, then $V_{\mathrm{a}}(x) = V_{\mathrm{r}}(x)$, $x \in \mathcal{D}$. Moreover, if $V_{\mathrm{s}}(x)$, $x \in \mathcal{D}$, is a vector storage function, then all (scalar) storage functions of the form $v_{\mathrm{s}}(x) = p^{\mathrm{T}} V_{\mathrm{s}}(x)$, $x \in \mathcal{D}$, where $p \in \mathbb{R}_+^q$, $p \neq 0$, are given by*

$$v_{\mathrm{s}}(x_0) = v_{\mathrm{a}}(x_0) = v_{\mathrm{r}}(x_0) = - \int_{t_0}^{T_+} e^{\alpha(t-t_0)} s_{\mathrm{c}}(u_{\mathrm{c}}(t), y_{\mathrm{c}}(t)) \mathrm{d}t$$

$$- \sum_{k \in \mathbb{Z}_{[t_0, T_+)}} e^{\alpha(t_k - t_0)} s_{\mathrm{d}}(u_{\mathrm{d}}(t_k), y_{\mathrm{d}}(t_k))$$

$$= \int_{T_-}^{t_0} e^{\alpha(t-t_0)} s_{\mathrm{c}}(u_{\mathrm{c}}(t), y_{\mathrm{c}}(t)) \mathrm{d}t$$

$$+ \sum_{k \in \mathbb{Z}_{[T_-, t_0)}} e^{\alpha(t_k - t_0)} s_{\mathrm{d}}(u_{\mathrm{d}}(t_k), y_{\mathrm{d}}(t_k)),$$

$$(5.49)$$

where $x(t)$, $t \geq t_0$, is the solution to (5.1)–(5.4) with $u_{\mathrm{c}}(\cdot) \in \mathcal{U}_{\mathrm{c}}$, $u_{\mathrm{d}}(\cdot) \in \mathcal{U}_{\mathrm{d}}$, $x(t_0) = x_0 \in \mathcal{D}$, $s_{\mathrm{c}}(u_{\mathrm{c}}, y_{\mathrm{c}}) = p^{\mathrm{T}} S_{\mathrm{c}}(u_{\mathrm{c}}, y_{\mathrm{c}})$, and $s_{\mathrm{d}}(u_{\mathrm{d}}, y_{\mathrm{d}}) = p^{\mathrm{T}} S_{\mathrm{d}}(u_{\mathrm{d}}, y_{\mathrm{d}})$, for any $T_+ > t_0$ and $T_- < t_0$ such that $x(T_+) = 0$ and $x(T_-) = 0$.

Proof. Suppose $\mathcal{G}$ is vector lossless with respect to the vector hybrid supply rate $(S_{\mathrm{c}}(u_{\mathrm{c}}, y_{\mathrm{c}}), S_{\mathrm{d}}(u_{\mathrm{d}}, y_{\mathrm{d}}))$. Since $\mathcal{G}$ is completely reachable to and from the origin it follows that for every $x_0 = x(t_0) \in \mathcal{D}$ there exist $T_+ > t_0$, $T_- < t_0$, $u_{\mathrm{c}}(t) \in U_{\mathrm{c}}$, and $u_{\mathrm{d}}(t_k) \in U_{\mathrm{d}}$, $t \in [T_-, T_+]$, $k \in \mathbb{Z}_{[T_-, T_+]}$, such that $x(T_-) = 0$, $x(T_+) = 0$, and $x(t_0) = x_0$. Now, it follows from the dissipation inequality (5.9) which is satisfied as an equality that

$$0 = \int_{T_-}^{T_+} e^{W(T_+ - t)} S_{\mathrm{c}}(u_{\mathrm{c}}(t), y_{\mathrm{c}}(t)) \mathrm{d}t$$

$$+ \sum_{k \in \mathbb{Z}_{[T_-, T_+)}} e^{W(T_+ - t_k)} S_{\mathrm{d}}(u_{\mathrm{d}}(t_k), y_{\mathrm{d}}(t_k)), \qquad (5.50)$$

or, equivalently,

$$0 = \int_{T_-}^{T_+} e^{-W(t - t_0)} S_{\mathrm{c}}(u_{\mathrm{c}}(t), y_{\mathrm{c}}(t)) \mathrm{d}t$$

$$+ \sum_{k \in \mathbb{Z}_{[T_-, T_+)}} e^{-W(t_k - t_0)} S_{\mathrm{d}}(u_{\mathrm{d}}(t_k), y_{\mathrm{d}}(t_k))$$

$$= \int_{T_-}^{t_0} e^{-W(t - t_0)} S_{\mathrm{c}}(u_{\mathrm{c}}(t), y_{\mathrm{c}}(t)) \mathrm{d}t$$

$$+ \sum_{k \in \mathbb{Z}_{[T_-, t_0)}} e^{-W(t_k - t_0)} S_{\mathrm{d}}(u_{\mathrm{d}}(t_k), y_{\mathrm{d}}(t_k))$$

$$+ \int_{t_0}^{T_+} e^{-W(t - t_0)} S_{\mathrm{c}}(u_{\mathrm{c}}(t), y_{\mathrm{c}}(t)) \mathrm{d}t$$

$$+ \sum_{k \in \mathbb{Z}_{[t_0, T_+)}} e^{-W(t_k - t_0)} S_{\mathrm{d}}(u_{\mathrm{d}}(t_k), y_{\mathrm{d}}(t_k))$$

$$\geq \geq \inf_{(u_c(\cdot), u_d(\cdot)), T_- \leq t_0} \left[\int_{T_-}^{t_0} e^{-W(t-t_0)} S_c(u_c(t), y_c(t)) dt \right.$$

$$\left. + \sum_{k \in \mathbb{Z}_{[T_-, t_0)}} e^{-W(t_k - t_0)} S_d(u_d(t_k), y_d(t_k)) \right]$$

$$+ \inf_{(u_c(\cdot), u_d(\cdot)), T_+ \geq t_0} \left[\int_{t_0}^{T_+} e^{-W(t-t_0)} S_c(u_c(t), y_c(t)) dt \right.$$

$$\left. + \sum_{k \in \mathbb{Z}_{[t_0, T_+)}} e^{-W(t_k - t_0)} S_d(u_d(t_k), y_d(t_k)) \right]$$

$$= V_r(x_0) - V_a(x_0), \tag{5.51}$$

which implies that $V_r(x_0) \leq \leq V_a(x_0)$, $x_0 \in \mathcal{D}$. However, it follows from Theorem 5.5 that if $\mathcal{G}$ is vector dissipative and $V_a(x)$, $x \in \mathcal{D}$, is a vector storage function, then $V_a(x) \leq \leq V_r(x)$, $x \in \mathcal{D}$, which along with (5.51) implies that $V_a(x) = V_r(x)$, $x \in \mathcal{D}$.

Next, since $\mathcal{G}$ is vector lossless there exist a nonzero vector $p \in \overline{\mathbb{R}}_+^q$ and a scalar $\alpha \geq 0$ satisfying (5.24). Now, it follows from (5.50) that

$$0 = \int_{T_-}^{T_+} p^T e^{-W(t-t_0)} S_c(u_c(t), y_c(t)) dt$$

$$+ \sum_{k \in \mathbb{Z}_{[T_-, T_+)}} p^T e^{-W(t_k - t_0)} S_d(u_d(t_k), y_d(t_k))$$

$$= \int_{T_-}^{T_+} e^{\alpha(t-t_0)} s_c(u_c(t), y_c(t)) dt$$

$$+ \sum_{k \in \mathbb{Z}_{[T_-, T_+)}} e^{\alpha(t_k - t_0)} s_d(u_d(t_k), y_d(t_k))$$

$$= \int_{T_-}^{t_0} e^{\alpha(t-t_0)} s_c(u_c(t), y_c(t)) dt$$

$$+ \sum_{k \in \mathbb{Z}_{[T_-, t_0)}} e^{\alpha(t_k - t_0)} s_d(u_d(t_k), y_d(t_k))$$

$$+ \int_{t_0}^{T_+} e^{\alpha(t-t_0)} s_c(u_c(t), y_c(t)) dt$$

$$+ \sum_{k \in \mathbb{Z}_{[t_0, T_+)}} e^{\alpha(t_k - t_0)} s_d(u_d(t_k), y_d(t_k))$$

$$\geq \inf_{(u_{\mathrm{c}}(\cdot),u_{\mathrm{d}}(\cdot)),\, T_{-}\leq t_0} \left[\int_{T_{-}}^{t_0} e^{\alpha(t-t_0)} s_{\mathrm{c}}(u_{\mathrm{c}}(t),y_{\mathrm{c}}(t))\mathrm{d}t \right.$$

$$\left. + \sum_{k\in\mathbb{Z}_{[T_{-},t_0)}} e^{\alpha(t_k-t_0)} s_{\mathrm{d}}(u_{\mathrm{d}}(t_k),y_{\mathrm{d}}(t_k)) \right]$$

$$+ \inf_{(u_{\mathrm{c}}(\cdot),u_{\mathrm{d}}(\cdot)),\, T_{+}\geq t_0} \left[\int_{t_0}^{T_{+}} e^{\alpha(t-t_0)} s_{\mathrm{c}}(u_{\mathrm{c}}(t),y_{\mathrm{c}}(t))\mathrm{d}t \right.$$

$$\left. + \sum_{k\in\mathbb{Z}_{[t_0,T_{+})}} e^{\alpha(t_k-t_0)} s_{\mathrm{d}}(u_{\mathrm{d}}(t_k),y_{\mathrm{d}}(t_k)) \right]$$

$$= v_{\mathrm{r}}(x_0) - v_{\mathrm{a}}(x_0), \quad x_0 \in \mathcal{D}, \tag{5.52}$$

which along with (5.45) implies that for any (scalar) storage function of the form $v_{\mathrm{s}}(x) = p^{\mathrm{T}} V_{\mathrm{s}}(x)$, $x \in \mathcal{D}$, the equality $v_{\mathrm{a}}(x) = v_{\mathrm{s}}(x) = v_{\mathrm{r}}(x)$, $x \in \mathcal{D}$, holds. Moreover, since $\mathcal{G}$ is vector lossless the inequalities (5.28) and (5.47) are satisfied as equalities and

$$v_{\mathrm{s}}(x_0) = - \int_{t_0}^{T_{+}} e^{\alpha(t-t_0)} s_{\mathrm{c}}(u_{\mathrm{c}}(t),y_{\mathrm{c}}(t))\mathrm{d}t$$

$$- \sum_{k\in\mathbb{Z}_{[t_0,T_{+})}} e^{\alpha(t_k-t_0)} s_{\mathrm{d}}(u_{\mathrm{d}}(t_k),y_{\mathrm{d}}(t_k))$$

$$= \int_{T_{-}}^{t_0} e^{\alpha(t-t_0)} s_{\mathrm{c}}(u_{\mathrm{c}}(t),y_{\mathrm{c}}(t))\mathrm{d}t$$

$$+ \sum_{k\in\mathbb{Z}_{[T_{-},t_0)}} e^{\alpha(t_k-t_0)} s_{\mathrm{d}}(u_{\mathrm{d}}(t_k),y_{\mathrm{d}}(t_k)), \tag{5.53}$$

where $x(t)$, $t \geq t_0$, is the solution to (5.1)–(5.4) with $u_{\mathrm{c}}(t) \in U_{\mathrm{c}}$, $u_{\mathrm{d}}(t_k) \in U_{\mathrm{d}}$, $x(T_{-}) = 0$, $x(T_{+}) = 0$, and $x(t_0) = x_0 \in \mathcal{D}$. □

The next proposition presents a characterization for vector dissipativity of large-scale impulsive dynamical systems in the case where $V_{\mathrm{s}}(\cdot)$ is continuously differentiable.

Proposition 5.2 *Consider the large-scale impulsive dynamical system $\mathcal{G}$ given by (5.1)–(5.4), assume $V_{\mathrm{s}} = [v_{\mathrm{s}1},\ldots,v_{\mathrm{s}q}]^{\mathrm{T}} : \mathcal{D} \to \overline{\mathbb{R}}_{+}^{q}$ is a continuously differentiable vector storage function for $\mathcal{G}$, and assume $\mathcal{G}$ is completely reachable. Then $\mathcal{G}$ is vector dissipative with respect to* *the vector hybrid supply rate* $(S_c(u_c, y_c), S_d(u_d, y_d))$ *if and only if*

$$\dot{V}_s(x(t)) \leq\leq WV_s(x(t)) + S_c(u_c(t), y_c(t)), \quad t_k < t \leq t_{k+1}, \quad (5.54)$$
$$V_s(x(t_k) + F_d(x(t_k), u_d(t_k))) \leq\leq V_s(x(t_k)) + S_d(u_d(t_k), y_d(t_k)),$$
$$k \in \overline{\mathbb{Z}}_+, \quad (5.55)$$

where $\dot{V}_s(x(t))$ *denotes the total time derivative of each component of* $V_s(\cdot)$ *along the state trajectories* $x(t), t_k < t \leq t_{k+1},$ *of* $\mathcal{G}.$

Proof. Suppose $\mathcal{G}$ is vector dissipative with respect to the vector hybrid supply rate $(S_c(u_c, y_c), S_d(u_d, y_d))$ and with a continuously differentiable vector storage function $V_s : \mathcal{D} \to \overline{\mathbb{R}}_+^q$. Then, with $T = \hat{t}$ and $t_0 = t$, it follows from (5.11) that there exists a nonnegative vector function $l(t, \hat{t}, x_0, u_c(\cdot)) \geq\geq 0, t_{k+1} \geq \hat{t} \geq t > t_k, x_0 \in \mathcal{D}, u_c(\cdot) \in \mathcal{U}_c,$ such that

$$V_s(x(\hat{t})) = e^{W(\hat{t}-t)}V_s(x(t)) + \int_t^{\hat{t}} e^{W(\hat{t}-\sigma)}S_c(u_c(\sigma), y_c(\sigma))d\sigma$$
$$- l(t, \hat{t}, x_0, u_c(\cdot)), \quad (5.56)$$

or, equivalently,

$$e^{-W\hat{t}}V_s(x(\hat{t})) - e^{-Wt}V_s(x(t)) = \int_t^{\hat{t}} e^{-W\sigma}S_c(u_c(\sigma), y_c(\sigma))d\sigma$$
$$- e^{-W\hat{t}}l(t, \hat{t}, x_0, u_c(\cdot)). \quad (5.57)$$

Now, dividing (5.57) by $\hat{t}-t$ and letting $\hat{t} \to t^+$, (5.57) is equivalent to

$$\frac{d}{d\sigma}\left[e^{-W\sigma}V_s(x(\sigma))\right]\Big|_{\sigma=t} = e^{-Wt}S_c(u_c(t), y_c(t))$$
$$- e^{-Wt} \lim_{\hat{t}\to t^+} \frac{l(t, \hat{t}, x_0, u_c(\cdot))}{\hat{t} - t}, \quad (5.58)$$

where the limit in (5.58) exists since $V_s(\cdot)$ is assumed to be continuously differentiable. Next, premultiplying (5.58) by e^{Wt}, $t \geq 0$, yields

$$\dot{V}_s(x(t)) - WV_s(x(t)) = S_c(u_c(t), y_c(t)) - \lim_{\hat{t}\to t^+} \frac{l(t, \hat{t}, x_0, u_c(\cdot))}{\hat{t} - t}, \quad (5.59)$$

which, since $\lim_{\hat{t}\to t^+} \frac{l(t,\hat{t},x_0,u_c(\cdot))}{\hat{t}-t} \geq\geq 0$ and t is arbitrary, gives (5.54). Inequality (5.55) is a restatement of (5.12).

The converse is immediate from Theorem 5.1. $\qquad\square$

Recall that if a disconnected subsystem $\mathcal{G}_i$ (i.e., $\mathcal{I}_{ci}(x) \equiv 0$ and $\mathcal{I}_{di}(x) \equiv 0, i \in \{1, \ldots, q\}$) of a large-scale impulsive dynamical system $\mathcal{G}$ is exponentially dissipative (respectively, dissipative) with respect to a hybrid supply rate $(s_{ci}(u_{ci}, y_{ci}), s_{di}(u_{di}, y_{di}))$, then there exist a storage function $v_{si} : \mathbb{R}^{n_i} \to \overline{\mathbb{R}}_+$ and a constant $\varepsilon_i > 0$ (respectively, $\varepsilon_i = 0$), $i = 1, \ldots, q$, such that the dissipation inequality

$$e^{\varepsilon_i T} v_{si}(x(T)) \leq e^{\varepsilon_i t_0} v_{si}(x(t_0)) + \int_{t_0}^{T} e^{\varepsilon_i t} s_{ci}(u_{ci}(t), y_{ci}(t)) \mathrm{d}t$$

$$+ \sum_{k \in \mathbb{Z}_{[t_0, T)}} e^{\varepsilon_i t_k} s_{di}(u_{di}(t_k), y_{di}(t_k)), \quad T \geq t_0, \quad (5.60)$$

holds. In the case where $v_{si} : \mathbb{R}^{n_i} \to \overline{\mathbb{R}}_+$ is continuously differentiable and $\mathcal{G}$ is completely reachable, (5.60) yields

$$v'_{si}(x_i)(f_{ci}(x_i) + G_{ci}(x_i)u_{ci}) \leq -\varepsilon_i v_{si}(x_i) + s_{ci}(u_{ci}, y_{ci}),$$
$$x \notin \mathcal{Z}_i, \quad u_{ci} \in U_{ci}, \quad (5.61)$$
$$v_{si}(x_i + f_{di}(x_i) + G_{di}(x_i)u_{di}) \leq v_{si}(x_i) + s_{di}(u_{di}, y_{di}),$$
$$x \in \mathcal{Z}_i, \quad u_{di} \in U_{di}, \quad (5.62)$$

where $\mathcal{Z}_i \triangleq \mathbb{R}^{n_1} \times \cdots \times \mathbb{R}^{n_{i-1}} \times \mathcal{Z}_{x_i} \times \mathbb{R}^{n_{i+1}} \times \cdots \times \mathbb{R}^q \subset \mathbb{R}^n$ and $\mathcal{Z}_{x_i} \subset \mathbb{R}^{n_i}, i = 1, \ldots, q$. The next result relates exponential dissipativity with respect to a scalar hybrid supply rate of each disconnected subsystem $\mathcal{G}_i$ of $\mathcal{G}$ with vector dissipativity (or, possibly, exponential vector dissipativity) of $\mathcal{G}$ with respect to a vector hybrid supply rate.

Proposition 5.3 *Consider the large-scale impulsive dynamical system $\mathcal{G}$ given by (5.1)–(5.4) with $\mathcal{Z}_x = \cup_{i=1}^{q} \mathcal{Z}_i$. Assume that $\mathcal{G}$ is completely reachable and each disconnected subsystem $\mathcal{G}_i$ of $\mathcal{G}$ is exponentially dissipative with respect to the hybrid supply rate $(s_{ci}(u_{ci}, y_{ci}), s_{di}(u_{di}, y_{di}))$ and with a continuously differentiable storage function $v_{si} : \mathbb{R}^{n_i} \to \overline{\mathbb{R}}_+, i = 1, \ldots, q$. Furthermore, assume that the interconnection functions $\mathcal{I}_{ci} : \mathcal{D} \to \mathbb{R}^{n_i}$ and $\mathcal{I}_{di} : \mathcal{D} \to \mathbb{R}^{n_i}, i = 1, \ldots, q$, of $\mathcal{G}$ are such that*

$$v'_{si}(x_i)\mathcal{I}_{ci}(x) \leq \sum_{j=1}^{q} \xi_{ij}(x)v_{sj}(x_j), \quad x \notin \mathcal{Z}_x, \quad (5.63)$$
$$v_{si}(x_i + f_{di}(x_i) + \mathcal{I}_{di}(x) + G_{di}(x_i)u_{di}) \leq v_{si}(x_i + f_{di}(x_i)$$
$$+ G_{di}(x_i)u_{di}), \quad x \in \mathcal{Z}_x, \quad u_{di} \in U_{di}, \quad i = 1, \ldots, q, \quad (5.64)$$

where $\xi_{ij} : \mathcal{D} \to \mathbb{R}, i, j = 1, \ldots, q$, are given bounded functions. If $W \in \mathbb{R}^{q \times q}$ is semistable (respectively, asymptotically stable), with

$$W_{(i,j)} = \begin{cases} -\varepsilon_i + \alpha_{ii}, & i = j, \\ \alpha_{ij}, & i \neq j, \end{cases} \quad (5.65)$$

where $\varepsilon_i > 0$ and $\alpha_{ij} \triangleq \max\{0, \sup_{x \in \mathcal{D}} \xi_{ij}(x)\}$, for all $i, j = 1, \ldots, q$, then $\mathcal{G}$ is vector dissipative (respectively, exponentially vector dissipative) with respect to the vector hybrid supply rate $(S_c(u_c, y_c), S_d(u_d, y_d)$ $) \triangleq ([s_{c1}(u_{c1}, y_{c1}), \ldots, s_{cq}(u_{cq}, y_{cq})]^T, [s_{d1}(u_{d1}, y_{d1}), \ldots, s_{dq}(u_{dq}, y_{dq})]^T$ $)$ and with vector storage function $V_s(x) \triangleq [v_{s1}(x_1), \ldots, v_{sq}(x_q)]^T$, $x \in \mathcal{D}$.

Proof. Since each disconnected subsystem $\mathcal{G}_i$ of $\mathcal{G}$ is exponentially dissipative with respect to the hybrid supply rate $s_{ci}(u_{ci}, y_{ci})$, $i = 1, \ldots, q$, it follows from (5.61)–(5.64) that, for all $u_{ci} \in U_{ci}$ and $i = 1, \ldots, q$,

$$\dot{v}_{si}(x_i(t)) = v'_{si}(x_i(t))[f_{ci}(x_i(t)) + \mathcal{I}_{ci}(x(t)) + G_{ci}(x_i(t))u_{ci}(t)]$$

$$\leq -\varepsilon_i v_{si}(x_i(t)) + s_{ci}(u_{ci}(t), y_{ci}(t)) + \sum_{j=1}^{q} \xi_{ij}(x(t))v_{sj}(x_j(t))$$

$$\leq -\varepsilon_i v_{si}(x_i(t)) + s_{ci}(u_{ci}(t), y_{ci}(t)) + \sum_{j=1}^{q} \alpha_{ij} v_{sj}(x_j(t)),$$

$$t_k < t \leq t_{k+1}, \quad (5.66)$$

and

$$v_{si}(x_i(t_k) + f_{di}(x_i(t_k)) + \mathcal{I}_{di}(x(t_k)) + G_{di}(x_i(t_k))u_{di}(t_k))$$
$$\leq v_{si}(x_i(t_k) + f_{di}(x_i(t_k)) + G_{di}(x_i(t_k))u_{di}(t_k))$$
$$\leq v_{si}(x_i(t_k)) + s_{di}(u_{di}(t_k), y_{di}(t_k)), \quad k \in \overline{\mathbb{Z}}_+. \quad (5.67)$$

Now, the result follows from Proposition 5.2 by noting that for all subsystems $\mathcal{G}_i$ of $\mathcal{G}$,

$$\dot{V}_s(x(t)) \leq\leq W V_s(x(t)) + S_c(u_c(t), y_c(t)), \quad t_k < t \leq t_{k+1},$$
$$u_c(t) \in U_c, \quad (5.68)$$
$$V_s(x(t_k) + F_d(x(t_k), u_d(t_k))) \leq\leq V_s(x(t_k)) + S_d(u_d(t_k), y_d(t_k)),$$
$$k \in \overline{\mathbb{Z}}_+, \quad u_d(t_k) \in U_d, \quad (5.69)$$

where W is essentially nonnegative and, by assumption, semistable (respectively, asymptotically stable), and the vector function $V_s(x) \triangleq [v_{s1}(x_1), \ldots, v_{sq}(x_q)]^T$, for all $x \in \mathcal{D}$, is a vector storage function for $\mathcal{G}$. $\square$

As a special case of vector dissipativity theory we can analyze the stability of large-scale impulsive dynamical systems. Specifically, assume that the large-scale impulsive dynamical system $\mathcal{G}$ is vector dissipative (respectively, exponentially vector dissipative) with respect to the vector hybrid supply rate $(S_c(u_c, y_c), S_d(u_d, y_d))$ and with a continuously differentiable vector storage function $V_s : \mathcal{D} \to \overline{\mathbb{R}}_+^q$. Moreover, assume that the conditions of Theorem 5.4 are satisfied. Then it follows from Proposition 5.2, with $u_c(t) \equiv 0$, $u_d(t_k) \equiv 0$, $y_c(t) \equiv 0$, and $y_d(t_k) \equiv 0$, that

$$\dot{V}_s(x(t)) \leq\leq W V_s(x(t)), \quad t_k < t \leq t_{k+1} \tag{5.70}$$
$$V_s(x(t_k) + f_d(x(t_k)) + \mathcal{I}_d(x(t_k))) \leq\leq V_s(x(t_k)), \quad k \in \overline{\mathbb{Z}}_+, \tag{5.71}$$

where $x(t)$, $t \geq t_0$, is a solution to (5.1)–(5.4) with $x(t_0) = x_0$, $u_c(t) \equiv 0$, and $u_d(t_k) \equiv 0$. Now, it follows from Theorem 2.11, with $w_c(z) = W z$ and $w_d(z) = 0$, that the zero solution $x(t) \equiv 0$ to (5.1)–(5.4), with $u_c(t) \equiv 0$ and $u_d(t_k) \equiv 0$, is Lyapunov (respectively, asymptotically) stable.

More generally, the problem of control system design for large-scale impulsive dynamical systems can be addressed within the framework of vector dissipativity theory. In particular, suppose that there exists a continuously differentiable vector function $V_s : \mathcal{D} \to \overline{\mathbb{R}}_+^q$ such that $V_s(0) = 0$ and

$$\dot{V}_s(x(t)) \leq\leq \mathcal{F}_c(V_s(x(t)), u_c(t)), \quad t_k < t \leq t_{k+1}, \quad u_c(t) \in U_c, \tag{5.72}$$
$$V_s(x(t_k) + F_d(x(t_k), u_d(t_k))) \leq\leq V_s(x(t_k)), \quad k \in \overline{\mathbb{Z}}_+, \quad u_d(t_k) \in U_d, \tag{5.73}$$

where $\mathcal{F}_c : \overline{\mathbb{R}}_+^q \times \mathbb{R}^{m_c} \to \mathbb{R}^q$ and $\mathcal{F}_c(0,0) = 0$. Then the control system design problem for a large-scale impulsive dynamical system reduces to constructing a hybrid *energy* feedback control law (ϕ_c, ϕ_d) : $\overline{\mathbb{R}}_+^q \times \overline{\mathbb{R}}_+^q \to U_c \times U_d$ of the form

$$u_c = \phi_c(V_s(x)) \triangleq [\phi_{c1}^T(V_s(x)), \dots, \phi_{cq}^T(V_s(x))]^T, \quad x \notin \mathcal{Z}_x, \tag{5.74}$$
$$u_d = \phi_d(V_s(x)) \triangleq [\phi_{d1}^T(V_s(x)), \dots, \phi_{dq}^T(V_s(x))]^T, \quad x \in \mathcal{Z}_x, \tag{5.75}$$

where $\phi_{ci} : \overline{\mathbb{R}}_+^q \to U_{ci}$, $\phi_{ci}(0) = 0$, $\phi_{di} : \overline{\mathbb{R}}_+^q \to U_{di}$, $i = 1, \dots, q$, such that the zero solution $z(t) \equiv 0$ to the comparison system

$$\dot{z}(t) = w_c(z(t)), \quad z(t_0) = V_s(x(t_0)), \quad t \geq t_0, \tag{5.76}$$

is rendered asymptotically stable, where $w_c(z) \triangleq \mathcal{F}_c(z, \phi_c(z))$ is of class $\mathcal{W}$, and Assumptions A1 and A2 hold. In this case, if there exists $p \in \mathbb{R}_+^q$ such that $v_s(x) \triangleq p^T V_s(x)$, $x \in \mathcal{D}$, is positive definite, then it follows from Theorem 2.11 that the zero solution $x(t) \equiv 0$ to

(5.1)–(5.4), with u_c and u_d given by (5.74) and (5.75), respectively, is asymptotically stable.

As can be seen from the above discussion, using an energy feedback control architecture and exploiting the comparison system within the control design for large-scale impulsive dynamical systems can significantly reduce the dimensionality of a control synthesis problem in terms of a number of states that need to be stabilized. It should be noted, however, that for stability analysis of large-scale impulsive dynamical systems the comparison system need not be linear as implied by (5.70). A nonlinear comparison system would still guarantee stability of a large-scale impulsive dynamical system provided that the conditions of Theorem 2.11 are satisfied.

5.3 Extended Kalman-Yakubovich-Popov Conditions for Large-Scale Impulsive Dynamical Systems

In this section, we show that vector dissipativeness (respectively, exponential vector dissipativeness) of a large-scale impulsive dynamical system $\mathcal{G}$ of the form (5.1)–(5.4) can be characterized in terms of the local subsystem functions $f_{ci}(\cdot)$, $G_{ci}(\cdot)$, $h_{ci}(\cdot)$, $J_{ci}(\cdot)$, $f_{di}(\cdot)$, $G_{di}(\cdot)$, $h_{di}(\cdot)$, and $J_{di}(\cdot)$, along with the interconnection structures $\mathcal{I}_{ci}(\cdot)$ and $\mathcal{I}_{di}(\cdot)$ for $i = 1, \ldots, q$. For the results in this section we consider the special case of dissipative systems with quadratic vector hybrid supply rates and set $\mathcal{D} = \mathbb{R}^n$, $U_{ci} = \mathbb{R}^{m_{ci}}$, $U_{di} = \mathbb{R}^{m_{di}}$, $Y_{ci} = \mathbb{R}^{l_{ci}}$, and $Y_{di} = \mathbb{R}^{l_{di}}$. Furthermore, we assume that $\mathcal{Z} = \mathcal{Z}_x \times \mathbb{R}^{m_c}$, where $\mathcal{Z}_x \subset \mathcal{D}$, so that resetting occurs only when $x(t)$ intersects $\mathcal{Z}_x$. Specifically, let $R_{ci} \in \mathbb{S}^{m_{ci}}$, $S_{ci} \in \mathbb{R}^{l_{ci} \times m_{ci}}$, $Q_{ci} \in \mathbb{S}^{l_{ci}}$, $R_{di} \in \mathbb{S}^{m_{di}}$, $S_{di} \in \mathbb{R}^{l_{di} \times m_{di}}$, and $Q_{di} \in \mathbb{S}^{l_{di}}$ be given, and assume $S_c(u_c, y_c)$ is such that $s_{ci}(u_{ci}, y_{ci}) = y_{ci}^T Q_{ci} y_{ci} + 2 y_{ci}^T S_{ci} u_{ci} + u_{ci}^T R_{ci} u_{ci}$ and $S_d(u_d, y_d)$ is such that $s_{di}(u_{di}, y_{di}) = y_{di}^T Q_{di} y_{di} + 2 y_{di}^T S_{di} u_{di} + u_{di}^T R_{di} u_{di}$, $i = 1, \ldots, q$. Furthermore, for the remainder of this chapter we assume that there exists a continuously differentiable vector storage function $V_s(x)$, $x \in \mathbb{R}^n$, for the large-scale impulsive dynamical system $\mathcal{G}$.

For the statement of the next result recall that $x = [x_1^T, \ldots, x_q^T]^T$, $u_c = [u_{c1}^T, \ldots, u_{cq}^T]^T$, $y_c = [y_{c1}^T, \ldots, y_{cq}^T]^T$, $u_d = [u_{d1}^T, \ldots, u_{dq}^T]^T$, $y_d = [y_{d1}^T, \ldots, y_{dq}^T]^T$, $x_i \in \mathbb{R}^{n_i}$, $u_{ci} \in \mathbb{R}^{m_{ci}}$, $y_{ci} \in \mathbb{R}^{l_{ci}}$, $u_{di} \in \mathbb{R}^{m_{di}}$, $y_{di} \in \mathbb{R}^{l_{di}}$, $i = 1, \ldots, q$, $\sum_{i=1}^q n_i = n$, $\sum_{i=1}^q m_{ci} = m_c$, $\sum_{i=1}^q m_{di} = m_d$, $\sum_{i=1}^q l_{ci} = l_c$, and $\sum_{i=1}^q l_{di} = l_d$. Furthermore, for (5.1)–(5.4) define $\mathcal{F}_c : \mathbb{R}^n \to \mathbb{R}^n$, $G_c : \mathbb{R}^n \to \mathbb{R}^{n \times m_c}$, $h_c : \mathbb{R}^n \to \mathbb{R}^{l_c}$, $J_c : \mathbb{R}^n \to \mathbb{R}^{l_c \times m_c}$, $\mathcal{F}_d : \mathbb{R}^n \to \mathbb{R}^n$, $G_d : \mathbb{R}^n \to \mathbb{R}^{n \times m_d}$, $h_d : \mathbb{R}^n \to \mathbb{R}^{l_d}$, and $J_{\mathrm{d}} : \mathbb{R}^n \to \mathbb{R}^{l_{\mathrm{d}} \times m_{\mathrm{d}}}$ by $\mathcal{F}_{\mathrm{c}}(x) \triangleq [\mathcal{F}_{\mathrm{c1}}^{\mathrm{T}}(x), \ldots, \mathcal{F}_{\mathrm{cq}}^{\mathrm{T}}(x)]^{\mathrm{T}}$, $\mathcal{F}_{\mathrm{d}}(x) \triangleq$ $[\mathcal{F}_{\mathrm{d1}}^{\mathrm{T}}(x), \ldots, \mathcal{F}_{\mathrm{dq}}^{\mathrm{T}}(x)]^{\mathrm{T}}$, where $\mathcal{F}_{\mathrm{c}i}(x) \triangleq f_{\mathrm{c}i}(x_i) + \mathcal{I}_{\mathrm{c}i}(x)$, $\mathcal{F}_{\mathrm{d}i}(x) \triangleq f_{\mathrm{d}i}(x_i)$ $+ \mathcal{I}_{\mathrm{d}i}(x)$, $i = 1, \ldots, q$, $G_{\mathrm{c}}(x) \triangleq \mathrm{diag}[G_{\mathrm{c1}}(x_1), \ldots, G_{\mathrm{cq}}(x_q)]$, $G_{\mathrm{d}}(x) \triangleq$ $\mathrm{diag}[G_{\mathrm{d1}}(x_1), \ldots, G_{\mathrm{dq}}(x_q)]$, $h_{\mathrm{c}}(x) \triangleq [h_{\mathrm{c1}}^{\mathrm{T}}(x_1), \ldots, h_{\mathrm{cq}}^{\mathrm{T}}(x_q)]^{\mathrm{T}}$, $h_{\mathrm{d}}(x) \triangleq$ $[h_{\mathrm{d1}}^{\mathrm{T}}(x_1), \ldots, h_{\mathrm{dq}}^{\mathrm{T}}(x_q)]^{\mathrm{T}}$, $J_{\mathrm{c}}(x) \triangleq \mathrm{diag}\,[J_{\mathrm{c1}}(x_1), \ldots, J_{\mathrm{cq}}(x_q)]$, and $J_{\mathrm{d}}(x)$ $\triangleq \mathrm{diag}[J_{\mathrm{d1}}(x_1), \ldots, J_{\mathrm{dq}}(x_q)]$. Moreover, for all $i = 1, \ldots, q$, define $\hat{R}_{\mathrm{c}i} \in \mathbb{S}^{m_{\mathrm{c}}}$, $\hat{S}_{\mathrm{c}i} \in \mathbb{R}^{l_{\mathrm{c}} \times m_{\mathrm{c}}}$, $\hat{Q}_{\mathrm{c}i} \in \mathbb{S}^{l_{\mathrm{c}}}$, $\hat{R}_{\mathrm{d}i} \in \mathbb{S}^{m_{\mathrm{d}}}$, $\hat{S}_{\mathrm{d}i} \in \mathbb{R}^{l_{\mathrm{d}} \times m_{\mathrm{d}}}$, and $\hat{Q}_{\mathrm{d}i} \in \mathbb{S}^{l_{\mathrm{d}}}$ such that each of these block matrices consists of zero blocks except, respectively, for the matrix blocks $R_{\mathrm{c}i} \in \mathbb{S}^{m_{\mathrm{c}i}}$, $S_{\mathrm{c}i} \in \mathbb{R}^{l_{\mathrm{c}i} \times m_{\mathrm{c}i}}$, $Q_{\mathrm{c}i} \in \mathbb{S}^{l_{\mathrm{c}i}}$, $R_{\mathrm{d}i} \in \mathbb{S}^{m_{\mathrm{d}i}}$, $S_{\mathrm{d}i} \in \mathbb{R}^{l_{\mathrm{d}i} \times m_{\mathrm{d}i}}$, and $Q_{\mathrm{d}i} \in \mathbb{S}^{l_{\mathrm{d}i}}$ on (i, i) position.

The next result introduces a more general definition of vector dissipativity involving an underlying nonlinear comparison system.

Definition 5.3 *The large-scale impulsive dynamical system $\mathcal{G}$ given by (5.1)–(5.4) is vector dissipative (respectively, exponentially vector dissipative) with respect to the vector hybrid supply rate $(S_{\mathrm{c}}(u_{\mathrm{c}}, y_{\mathrm{c}}), S_{\mathrm{d}}(u_{\mathrm{d}}, y_{\mathrm{d}}))$ if there exist a continuous, nonnegative definite vector function $V_{\mathrm{s}} = [v_{\mathrm{s1}}, \ldots, v_{\mathrm{sq}}]^{\mathrm{T}} : \mathcal{D} \to \overline{\mathbb{R}}_+^q$, called a vector storage function, and a class $\mathcal{W}$ function $w_{\mathrm{c}} : \overline{\mathbb{R}}_+^q \to \mathbb{R}^q$ such that $V_{\mathrm{s}}(0) = 0$, $w_{\mathrm{c}}(0) = 0$, the zero solution $z(t) \equiv 0$ to the comparison system*

$$\dot{z}(t) = w_{\mathrm{c}}(z(t)), \quad z(t_0) = z_0, \quad t \geq t_0, \tag{5.77}$$

is Lyapunov (respectively, asymptotically) stable, and the vector hybrid dissipation inequality

$$V_{\mathrm{s}}(x(T)) \leq\leq V_{\mathrm{s}}(x(t_0)) + \int_{t_0}^{T} w_{\mathrm{c}}(V_{\mathrm{s}}(x(t)))\mathrm{d}t + \int_{t_0}^{T} S_{\mathrm{c}}(u_{\mathrm{c}}(t), y_{\mathrm{c}}(t))\mathrm{d}t$$
$$+ \sum_{k \in \mathbb{Z}_{[t_0, T)}} S_{\mathrm{d}}(u_{\mathrm{d}}(t_k), y_{\mathrm{d}}(t_k)), \quad T \geq t_0, \tag{5.78}$$

is satisfied, where $x(t)$, $t \geq t_0$, is the solution to (5.1)–(5.4) with $u_{\mathrm{c}}(t) \in U_{\mathrm{c}}$ and $u_{\mathrm{d}}(t_k) \in U_{\mathrm{d}}$. The large-scale impulsive dynamical system $\mathcal{G}$ given by (5.1)–(5.4) is vector lossless with respect to the vector hybrid supply rate $(S_{\mathrm{c}}(u_{\mathrm{c}}, y_{\mathrm{c}}), S_{\mathrm{d}}(u_{\mathrm{d}}, y_{\mathrm{d}}))$ if the vector hybrid dissipation inequality is satisfied as an equality with the zero solution $z(t) \equiv 0$ to (5.77) being Lyapunov stable.

If $\mathcal{G}$ is completely reachable and $V_{\mathrm{s}}(\cdot)$ is continuously differentiable, then (5.78) can be equivalently written as

$$\dot{V}_{\mathrm{s}}(x(t)) \leq\leq w_{\mathrm{c}}(V_{\mathrm{s}}(x(t))) + S_{\mathrm{c}}(u_{\mathrm{c}}(t), y_{\mathrm{c}}(t)), \quad t_k < t \leq t_{k+1}, \tag{5.79}$$

$$V_{\mathrm{s}}(x(t_k) + F_{\mathrm{d}}(x(t_k), u_{\mathrm{d}}(t_k))) \leq\leq V_{\mathrm{s}}(x(t_k)) + S_{\mathrm{d}}(u_{\mathrm{d}}(t_k), y_{\mathrm{d}}(t_k)),$$
$$k \in \overline{\mathbb{Z}}_+, \quad (5.80)$$

with $u_{\mathrm{c}}(t) \in U_{\mathrm{c}}$ and $u_{\mathrm{d}}(t_k) \in U_{\mathrm{d}}$. If in Definition 5.3 the function $w_{\mathrm{c}} : \overline{\mathbb{R}}_+^q \to \mathbb{R}^q$ is such that $w_{\mathrm{c}}(z) = Wz$, where $W \in \mathbb{R}^{q \times q}$, then W is essentially nonnegative and Definition 5.3 collapses to Definition 5.2.

Theorem 5.7 *Consider the large-scale impulsive dynamical system $\mathcal{G}$ given by (5.1)–(5.4). Let $R_{\mathrm{c}i} \in \mathbb{S}^{m_{\mathrm{c}i}}$, $S_{\mathrm{c}i} \in \mathbb{R}^{l_{\mathrm{c}i} \times m_{\mathrm{c}i}}$, $Q_{\mathrm{c}i} \in \mathbb{S}^{l_{\mathrm{c}i}}$, $R_{\mathrm{d}i} \in \mathbb{S}^{m_{\mathrm{d}i}}$, $S_{\mathrm{d}i} \in \mathbb{R}^{l_{\mathrm{d}i} \times m_{\mathrm{d}i}}$, and $Q_{\mathrm{d}i} \in \mathbb{S}^{l_{\mathrm{d}i}}$, $i = 1, \ldots, q$. Then $\mathcal{G}$ is vector dissipative (respectively, exponentially vector dissipative) with respect to the quadratic hybrid supply rate $(S_{\mathrm{c}}(u_{\mathrm{c}}, y_{\mathrm{c}}), S_{\mathrm{d}}(u_{\mathrm{d}}, y_{\mathrm{d}}))$, where $s_{\mathrm{c}i}(u_{\mathrm{c}i}, y_{\mathrm{c}i}) = y_{\mathrm{c}i}^{\mathrm{T}} Q_{\mathrm{c}i} y_{\mathrm{c}i} + 2 y_{\mathrm{c}i}^{\mathrm{T}} S_{\mathrm{c}i} u_{\mathrm{c}i} + u_{\mathrm{c}i}^{\mathrm{T}} R_{\mathrm{c}i} u_{\mathrm{c}i}$ and $s_{\mathrm{d}i}(u_{\mathrm{d}i}, y_{\mathrm{d}i}) = y_{\mathrm{d}i}^{\mathrm{T}} Q_{\mathrm{d}i} y_{\mathrm{d}i} + 2 y_{\mathrm{d}i}^{\mathrm{T}} S_{\mathrm{d}i} u_{\mathrm{d}i} + u_{\mathrm{d}i}^{\mathrm{T}} R_{\mathrm{d}i} u_{\mathrm{d}i}$, $i = 1, \ldots, q$, if there exist functions $V_{\mathrm{s}} = [v_{\mathrm{s}1}, \ldots, v_{\mathrm{s}q}]^{\mathrm{T}} : \mathbb{R}^n \to \overline{\mathbb{R}}_+^q$, $w_{\mathrm{c}} = [w_{\mathrm{c}1}, \ldots, w_{\mathrm{c}q}]^{\mathrm{T}} : \overline{\mathbb{R}}_+^q \to \mathbb{R}^q$, $\ell_{\mathrm{c}i} : \mathbb{R}^n \to \mathbb{R}^{s_{\mathrm{c}i}}$, $\mathcal{Z}_{\mathrm{c}i} : \mathbb{R}^n \to \mathbb{R}^{s_{\mathrm{c}i} \times m_{\mathrm{c}}}$, $\ell_{\mathrm{d}i} : \mathbb{R}^n \to \mathbb{R}^{s_{\mathrm{d}i}}$, $\mathcal{Z}_{\mathrm{d}i} : \mathbb{R}^n \to \mathbb{R}^{s_{\mathrm{d}i} \times m_{\mathrm{d}}}$, $P_{1i} : \mathbb{R}^n \to \mathbb{R}^{1 \times m_{\mathrm{d}}}$, and $P_{2i} : \mathbb{R}^n \to \mathbb{N}^{m_{\mathrm{d}}}$ such that $v_{\mathrm{s}i}(\cdot)$ is continuously differentiable, $v_{\mathrm{s}i}(0) = 0$, $i = 1, \ldots, q$, $w_{\mathrm{c}}(\cdot) \in \mathcal{W}$, $w_{\mathrm{c}}(0) = 0$, the zero solution $z(t) \equiv 0$ to (5.77) is Lyapunov (respectively, asymptotically) stable,*

$$v_{\mathrm{s}i}(x + \mathcal{F}_{\mathrm{d}}(x) + G_{\mathrm{d}}(x)u_{\mathrm{d}}) = v_{\mathrm{s}i}(x + \mathcal{F}_{\mathrm{d}}(x)) + P_{1i}(x)u_{\mathrm{d}} + u_{\mathrm{d}}^{\mathrm{T}} P_{2i}(x)u_{\mathrm{d}},$$
$$x \in \mathcal{Z}_x, \quad u_{\mathrm{d}} \in \mathbb{R}^{m_{\mathrm{d}}}, \quad (5.81)$$

and, for all $i = 1, \ldots, q$,

$$0 = v_{\mathrm{s}i}'(x)\mathcal{F}_{\mathrm{c}}(x) - h_{\mathrm{c}}^{\mathrm{T}}(x)\hat{Q}_{\mathrm{c}i} h_{\mathrm{c}}(x) - w_{\mathrm{c}i}(V_{\mathrm{s}}(x)) + \ell_{\mathrm{c}i}^{\mathrm{T}}(x)\ell_{\mathrm{c}i}(x),$$
$$x \notin \mathcal{Z}_x, \quad (5.82)$$

$$0 = \tfrac{1}{2} v_{\mathrm{s}i}'(x)G_{\mathrm{c}}(x) - h_{\mathrm{c}}^{\mathrm{T}}(x)(\hat{S}_{\mathrm{c}i} + \hat{Q}_{\mathrm{c}i} J_{\mathrm{c}}(x)) + \ell_{\mathrm{c}i}^{\mathrm{T}}(x)\mathcal{Z}_{\mathrm{c}i}(x), \quad x \notin \mathcal{Z}_x,$$
$$(5.83)$$

$$0 = \hat{R}_{\mathrm{c}i} + J_{\mathrm{c}}^{\mathrm{T}}(x)\hat{S}_{\mathrm{c}i} + \hat{S}_{\mathrm{c}i}^{\mathrm{T}} J_{\mathrm{c}}(x) + J_{\mathrm{c}}^{\mathrm{T}}(x)\hat{Q}_{\mathrm{c}i} J_{\mathrm{c}}(x) - \mathcal{Z}_{\mathrm{c}i}^{\mathrm{T}}(x)\mathcal{Z}_{\mathrm{c}i}(x),$$
$$x \notin \mathcal{Z}_x, \quad (5.84)$$

$$0 = v_{\mathrm{s}i}(x + \mathcal{F}_{\mathrm{d}}(x)) - h_{\mathrm{d}}^{\mathrm{T}}(x)\hat{Q}_{\mathrm{d}i} h_{\mathrm{d}}(x) - v_{\mathrm{s}i}(x) + \ell_{\mathrm{d}i}^{\mathrm{T}}(x)\ell_{\mathrm{d}i}(x), \quad x \in \mathcal{Z}_x,$$
$$(5.85)$$

$$0 = \tfrac{1}{2} P_{1i}(x) - h_{\mathrm{d}}^{\mathrm{T}}(x)(\hat{S}_{\mathrm{d}i} + \hat{Q}_{\mathrm{d}i} J_{\mathrm{d}}(x)) + \ell_{\mathrm{d}i}^{\mathrm{T}}(x)\mathcal{Z}_{\mathrm{d}i}(x), \quad x \in \mathcal{Z}_x, \quad (5.86)$$

$$0 = \hat{R}_{\mathrm{d}i} + J_{\mathrm{d}}^{\mathrm{T}}(x)\hat{S}_{\mathrm{d}i} + \hat{S}_{\mathrm{d}i}^{\mathrm{T}} J_{\mathrm{d}}(x) + J_{\mathrm{d}}^{\mathrm{T}}(x)\hat{Q}_{\mathrm{d}i} J_{\mathrm{d}}(x) - P_{2i}(x)$$
$$- \mathcal{Z}_{\mathrm{d}i}^{\mathrm{T}}(x)\mathcal{Z}_{\mathrm{d}i}(x), \quad x \in \mathcal{Z}_x. \quad (5.87)$$

Proof. Suppose that there exist functions $v_{\mathrm{s}i} : \mathbb{R}^n \to \overline{\mathbb{R}}_+$, $\ell_{\mathrm{c}i} : \mathbb{R}^n \to \mathbb{R}^{s_{\mathrm{c}i}}$, $\mathcal{Z}_{\mathrm{c}i} : \mathbb{R}^n \to \mathbb{R}^{s_{\mathrm{c}i} \times m_{\mathrm{c}}}$, $\ell_{\mathrm{d}i} : \mathbb{R}^n \to \mathbb{R}^{s_{\mathrm{d}i}}$, $\mathcal{Z}_{\mathrm{d}i} : \mathbb{R}^n \to \mathbb{R}^{s_{\mathrm{d}i} \times m_{\mathrm{d}}}$,

$w_c : \overline{\mathbb{R}}_+^q \to \mathbb{R}^q$, $P_{1i} : \mathbb{R}^n \to \mathbb{R}^{1 \times m_d}$, and $P_{2i} : \mathbb{R}^n \to \mathbb{N}^{m_d}$ such that $v_{si}(\cdot)$ is continuously differentiable and nonnegative definite, $v_{si}(0) = 0$, $i = 1, \ldots, q$, $w_c(0) = 0$, $w_c(\cdot) \in \mathcal{W}$, the zero solution $z(t) \equiv 0$ to (5.77) is Lyapunov (respectively, asymptotically) stable, and (5.81)–(5.87) are satisfied. Then for any $u_c(t) \in \mathbb{R}^{m_c}$, $t, \hat{t} \in \mathbb{R}$, $t_k < t \le \hat{t} \le t_{k+1}$, $k \in \overline{\mathbb{Z}}_+$, and $i = 1, \ldots, q$, it follows from (5.82)–(5.84) that

$$
\begin{aligned}
\int_t^{\hat{t}} s_{ci}(u_{ci}(\sigma), y_{ci}(\sigma)) \mathrm{d}\sigma &= \int_t^{\hat{t}} [u_c^T(\sigma) \hat{R}_{ci} u_c(\sigma) + 2 y_c^T(\sigma) \hat{S}_{ci} u_c(\sigma) \\
&\quad + y_c^T(\sigma) \hat{Q}_{ci} y_c(\sigma)] \mathrm{d}\sigma \\
&= \int_t^{\hat{t}} [h_c^T(x(\sigma)) \hat{Q}_{ci} h_c(x(\sigma)) \\
&\quad + 2 h_c^T(x(\sigma))(\hat{S}_{ci} + \hat{Q}_{ci} J_c(x(\sigma))) u_c(\sigma) \\
&\quad + u_c^T(\sigma)(J_c^T(x(\sigma)) \hat{Q}_{ci} J_c(x(\sigma)) + J_c^T(x(\sigma)) \hat{S}_{ci} \\
&\quad + \hat{S}_{ci}^T J_c(x(\sigma)) + \hat{R}_{ci}) u_c(\sigma)] \mathrm{d}\sigma \\
&= \int_t^{\hat{t}} [v_{si}'(x(\sigma))(\mathcal{F}_c(x(\sigma)) + G_c(x(\sigma)) u_c(\sigma)) \\
&\quad + \ell_{ci}^T(x(\sigma)) \ell_{ci}(x(\sigma)) \\
&\quad + 2 \ell_{ci}^T(x(\sigma)) \mathcal{Z}_{ci}(x(\sigma)) u_c(\sigma) \\
&\quad + u_c^T(\sigma) \mathcal{Z}_{ci}^T(x(\sigma)) \mathcal{Z}_{ci}(x(\sigma)) u_c(\sigma) \\
&\quad - w_{ci}(V_s(x(\sigma)))] \mathrm{d}\sigma \\
&= \int_t^{\hat{t}} [\dot{v}_{si}(x(\sigma)) \\
&\quad + [\ell_{ci}(x(\sigma)) + \mathcal{Z}_{ci}(x(\sigma)) u_c(\sigma)]^T [\ell_{ci}(x(\sigma)) \\
&\quad + \mathcal{Z}_{ci}(x(\sigma)) u_c(\sigma)] - w_{ci}(V_s(x(\sigma)))] \mathrm{d}\sigma \\
&\ge v_{si}(x(\hat{t})) - v_{si}(x(t)) - \int_t^{\hat{t}} w_{ci}(V_s(x(\sigma))) \mathrm{d}\sigma,
\end{aligned}
$$

(5.88)

where $x(\sigma)$, $\sigma \in (t_k, t_{k+1}]$, satisfies (5.1).

Next, for any $u_d(t_k) \in \mathbb{R}^{m_d}$, $t_k \in \mathbb{R}$, and $k \in \overline{\mathbb{Z}}_+$, it follows from (5.81) and (5.85)–(5.87) that

$$
\begin{aligned}
v_{si}(x &+ \mathcal{F}_d(x) + G_d(x) u_d) - v_{si}(x) \\
&= v_{si}(x + \mathcal{F}_d(x)) - v_{si}(x) + P_{1i}(x) u_d + u_d^T P_{2i}(x) u_d \\
&= h_d^T(x) \hat{Q}_{di} h_d(x) - \ell_{di}^T(x) \ell_{di}(x) + 2[h_d^T(x)(\hat{Q}_{di} J_d(x) \\
&\quad + \hat{S}_{di}) - \ell_{di}^T(x) \mathcal{Z}_{di}(x)] u_d + u_d^T [\hat{R}_{di} + \hat{S}_{di}^T J_d(x)
\end{aligned}
$$

$$+ J_d^T(x)\hat{S}_{di} + J_d^T(x)\hat{Q}_{di}J_d(x) - \mathcal{Z}_{di}^T(x)\mathcal{Z}_{di}(x)]u_d$$
$$= s_{di}(u_{di}, y_{di}) - [\ell_{di}(x) + \mathcal{Z}_{di}(x)u_d]^T[\ell_{di}(x) + \mathcal{Z}_{di}(x)u_d]$$
$$\leq s_{di}(u_{di}, y_{di}). \tag{5.89}$$

Now, using (5.88) and (5.89) the result is immediate with vector storage function $V_s(x) = [v_{s1}(x), \ldots, v_{sq}(x)]^T$, $x \in \mathbb{R}^n$. □

Using (5.82)–(5.87) it follows that for $T \geq t_0 \geq 0$, $k \in \mathbb{Z}_{[t_0,T)}$, and $i = 1, \ldots, q$,

$$\int_{t_0}^T s_{ci}(u_{ci}(t), y_{ci}(t))dt + \int_{t_0}^T w_{ci}(V_s(x(t)))dt$$
$$+ \sum_{k \in \mathbb{Z}_{[t_0,T)}} s_{di}(u_d(t_k), y_d(t_k))$$
$$= v_{si}(x(T)) - v_{si}(x(t_0))$$
$$+ \int_{t_0}^T [\ell_{ci}(x(t)) + \mathcal{Z}_{ci}(x(t))u_c(t)]^T[\ell_{ci}(x(t)) + \mathcal{Z}_{ci}(x(t))u_c(t)]dt$$
$$+ \sum_{k \in \mathbb{Z}_{[t_0,T)}} [\ell_{di}(x(t_k)) + \mathcal{Z}_{di}(x(t_k))u_d(t_k)]^T[\ell_{di}(x(t_k))$$
$$+ \mathcal{Z}_{di}(x(t_k))u_d(t_k)], \tag{5.90}$$

where $V_s(x) = [v_{s1}(x), \ldots, v_{sq}(x)]^T$, $x \in \mathbb{R}^n$, which can be interpreted as a *generalized energy* balance equation for the ith impulsive subsystem of $\mathcal{G}$, where $v_{si}(x(T)) - v_{si}(x(t_0))$ is the stored or accumulated generalized energy of the ith impulsive subsystem; the two path-dependent terms on the left are, respectively, the external supplied energy to the ith subsystem over the continuous-time dynamics and the energy gained over the continuous-time dynamics by the ith subsystem from the net energy flow between all subsystems due to subsystem coupling; the last discrete term on the left corresponds to the external supplied energy to the ith subsystem at the resetting instants; the second path-dependent term on the right corresponds to the dissipated energy from the ith impulsive subsystem over the continuous-time dynamics; and the last discrete term on the right corresponds to the dissipated energy from the ith impulsive subsystem at the resetting instants.

Equivalently, (5.90) can be rewritten as

$$\dot{v}_{si}(x(t)) = s_{ci}(u_{ci}(t), y_{ci}(t)) + w_{ci}(V_s(x(t)))$$
$$- [\ell_{ci}(x(t)) + \mathcal{Z}_{ci}(x(t))u_c(t)]^T[\ell_{ci}(x(t)) + \mathcal{Z}_{ci}(x(t))u_c(t)],$$

$$t_k < t \le t_{k+1}, \quad i = 1, \ldots, q, \quad (5.91)$$

$$v_{si}(x(t_k) + \mathcal{F}_d(x(t_k)) + G_d(x(t_k))u_d(t_k)) - v_{si}(x(t_k))$$
$$= s_{di}(u_d(t_k), y_d(t_k)) - [\ell_{di}(x(t_k))$$
$$+ \mathcal{Z}_{di}(x(t_k))u_d(t_k)]^T[\ell_{di}(x(t_k)) + \mathcal{Z}_{di}(x(t_k))u_d(t_k)],$$
$$k \in \overline{\mathbb{Z}}_+, \quad (5.92)$$

which yields a set of q generalized energy conservation equations for the large-scale impulsive dynamical system $\mathcal{G}$. Specifically, (5.91) shows that the rate of change in generalized energy, or generalized power, over the time interval $t \in (t_k, t_{k+1}]$ for the ith subsystem of $\mathcal{G}$ is equal to the generalized system power input to the ith subsystem plus the instantaneous rate of energy supplied to the ith subsystem from the net energy flow between all subsystems minus the internal generalized system power dissipated from the ith subsystem; (5.92) shows that the change of energy at the resetting times t_k, $k \in \overline{\mathbb{Z}}_+$, is equal to the external generalized system supplied energy at the resetting times minus the generalized dissipated energy at the resetting times.

Note that if $\mathcal{G}$, with $(u_c(t), u_d(t_k)) \equiv (0, 0)$, is vector dissipative (respectively, exponentially vector dissipative) with respect to the quadratic hybrid supply rate, and $Q_{ci} \le 0$ and $Q_{di} \le 0$, $i = 1, \ldots, q$, then it follows from the vector hybrid dissipation inequality that for all $k \in \overline{\mathbb{Z}}_+$,

$$\dot{V}_s(x(t)) \le\le w_c(V_s(x(t))) + S_c(0, y_c(t)) \le\le w_c(V_s(x(t))),$$
$$t_k < t \le t_{k+1}, \quad (5.93)$$
$$V_s(x(t_k) + \mathcal{F}_d(x(t_k))) - V_s(x(t_k)) \le\le S_d(0, y_d(t_k)) \le\le 0, \quad (5.94)$$

where $S_c(0, y_c) = [s_{c1}(0, y_{c1}), \ldots, s_{cq}(0, y_{cq})]^T$, $S_d(0, y_d) = [s_{d1}(0, y_{d1}), \ldots, s_{dq}(0, y_{dq})]^T$, $s_{ci}(0, y_{ci}(t)) = y_{ci}^T(t)Q_{ci}y_{ci}(t) \le 0$, $s_{di}(0, y_{di}(t_k)) = y_{di}^T(t_k)Q_{di}y_{di}(t_k) \le 0$, $t_k < t \le t_{k+1}$, $k \in \overline{\mathbb{Z}}_+$, $i = 1, \ldots, q$, and $x(t)$, $t \ge t_0$, is the solution to (5.1)–(5.4) with $(u_c(t), u_d(t_k)) \equiv (0, 0)$. If, in addition, there exists $p \in \mathbb{R}_+^q$ such that $p^T V_s(x)$, $x \in \mathbb{R}^n$, is positive definite, then it follows from Theorem 2.11 that the undisturbed $((u_c(t), u_d(t_k)) \equiv (0, 0))$ large-scale impulsive dynamical system (5.1)–(5.4) is Lyapunov (respectively, asymptotically) stable.

Next, we extend the notions of passivity and nonexpansivity to vector passivity and vector nonexpansivity.

Definition 5.4 *The large-scale impulsive dynamical system $\mathcal{G}$ given by (5.1)–(5.4) with $m_{ci} = l_{ci}$, $m_{di} = l_{di}$, $i = 1, \ldots, q$, is vector passive (respectively, vector exponentially passive) if it is vector dissipa-*

tive (respectively, exponentially vector dissipative) with respect to the vector hybrid supply rate $(S_c(u_c, y_c), S_d(u_d, y_d))$, *where* $s_{ci}(u_{ci}, y_{ci}) = 2y_{ci}^T u_{ci}$ *and* $s_{di}(u_{di}, y_{di}) = 2y_{di}^T u_{di}$, $i = 1, \ldots, q$.

Definition 5.5 *The large-scale impulsive dynamical system* $\mathcal{G}$ *given by (5.1)–(5.4) is vector nonexpansive (respectively, vector exponentially nonexpansive) if it is vector dissipative (respectively, exponentially vector dissipative) with respect to the vector hybrid supply rate* $(S_c(u_c, y_c), S_d(u_d, y_d))$, *where* $s_{ci}(u_{ci}, y_{ci}) = \gamma_{ci}^2 u_{ci}^T u_{ci} - y_{ci}^T y_{ci}$ *and* s_{di} $(u_{di}, y_{di}) = \gamma_{di}^2 u_{di}^T u_{di} - y_{di}^T y_{di}$, $i = 1, \ldots, q$, *and* $\gamma_{ci} > 0$, $\gamma_{di} > 0$, $i = 1, \ldots, q$, *are given.*

Note that a mixed vector passive-nonexpansive formulation of $\mathcal{G}$ can also be considered. Specifically, one can consider large-scale impulsive dynamical systems $\mathcal{G}$ which are vector dissipative with respect to vector hybrid supply rates $(S_c(u_c, y_c), S_d(u_d, y_d))$, where $s_{ci}(u_{ci}, y_{ci}) = 2y_{ci}^T u_{ci}$, $s_{di}(u_{di}, y_{di}) = 2y_{di}^T u_{di}$, $i \in \mathbb{Z}_p$, $s_{cj}(u_{cj}, y_{cj}) = \gamma_{cj}^2 u_{cj}^T u_{cj} - y_{cj}^T y_{cj}$, $\gamma_{cj} > 0$, $s_{dj}(u_{dj}, y_{dj}) = \gamma_{dj}^2 u_{dj}^T u_{dj} - y_{dj}^T y_{dj}$, $\gamma_{dj} > 0$, $j \in \mathbb{Z}_{ne}$, $\mathbb{Z}_p \cap \mathbb{Z}_{ne} = \emptyset$, and $\mathbb{Z}_p \cup \mathbb{Z}_{ne} = \{1, \ldots, q\}$. Furthermore, hybrid supply rates for vector input strict passivity, vector output strict passivity, and vector input-output strict passivity generalizing the dissipativity notions given in [74] can also be considered. However, for simplicity of exposition we do not do so here.

The next result presents constructive sufficient conditions guaranteeing vector dissipativity of $\mathcal{G}$ with respect to a quadratic hybrid supply rate for the case where the vector storage function $V_s(x)$, $x \in \mathbb{R}^n$, is component decoupled, that is, $V_s(x) = [v_{s1}(x_1), \ldots, v_{sq}(x_q)]^T$, $x \in \mathbb{R}^n$.

Theorem 5.8 *Consider the large-scale impulsive dynamical system* $\mathcal{G}$ *given by (5.1)–(5.4). Assume that there exist functions* $V_s = [v_{s1}, \ldots, v_{sq}]^T : \mathbb{R}^n \to \overline{\mathbb{R}}_+^q$, $w_c = [w_{c1}, \ldots, w_{cq}]^T : \overline{\mathbb{R}}_+^q \to \mathbb{R}^q$, $\ell_{ci} : \mathbb{R}^n \to \mathbb{R}^{s_{ci}}$, $\mathcal{Z}_{ci} : \mathbb{R}^n \to \mathbb{R}^{s_{ci} \times m_{ci}}$, $\ell_{di} : \mathbb{R}^n \to \mathbb{R}^{s_{di}}$, $\mathcal{Z}_{di} : \mathbb{R}^n \to \mathbb{R}^{s_{di} \times m_{di}}$, $P_{1i} : \mathbb{R}^n \to \mathbb{R}^{1 \times m_{di}}$, *and* $P_{2i} : \mathbb{R}^n \to \mathbb{N}^{m_{di}}$ *such that* $V_s(x) = [v_{s1}(x_1), \ldots, v_{sq}(x_q)]^T$, $v_{si}(\cdot)$ *is continuously differentiable,* $v_{si}(0) = 0$, $i = 1, \ldots, q$, $w_c(\cdot) \in \mathcal{W}$, $w_c(0) = 0$, *the zero solution* $z(t) \equiv 0$ *to (5.77) is Lyapunov (respectively, asymptotically) stable, and, for all* $x \in \mathbb{R}^n$ *and* $i = 1, \ldots, q$,

$$0 \le v_{si}(x_i + \mathcal{F}_{di}(x)) - v_{si}(x_i + \mathcal{F}_{di}(x) + G_{di}(x_i)u_{di}) + P_{1i}(x)u_{di}$$
$$+ u_{di}^T P_{2i}(x)u_{di}, \quad x \in \mathcal{Z}_x, \quad u_{di} \in \mathbb{R}^{m_{di}}, \qquad (5.95)$$
$$0 \ge v'_{si}(x_i)\mathcal{F}_{ci}(x) - h_{ci}^T(x_i)Q_{ci}h_{ci}(x_i) - w_{ci}(V_s(x)) + \ell_{ci}^T(x_i)\ell_{ci}(x_i),$$
$$x \notin \mathcal{Z}_x, \quad (5.96)$$

$$0 = \tfrac{1}{2}v'_{si}(x_i)G_{ci}(x_i) - h_{ci}^{\mathrm{T}}(x_i)(S_{ci} + Q_{ci}J_{ci}(x_i)) + \ell_{ci}^{\mathrm{T}}(x_i)\mathcal{Z}_{ci}(x_i),$$
$$x \notin \mathcal{Z}_x, \quad (5.97)$$

$$0 \leq R_{ci} + J_{ci}^{\mathrm{T}}(x_i)S_{ci} + S_{ci}^{\mathrm{T}}J_{ci}(x_i) + J_{ci}^{\mathrm{T}}(x_i)Q_{ci}J_{ci}(x_i) - \mathcal{Z}_{ci}^{\mathrm{T}}(x_i)\mathcal{Z}_{ci}(x_i),$$
$$x \notin \mathcal{Z}_x, \quad (5.98)$$

$$0 \geq v_{si}(x_i + \mathcal{F}_{di}(x)) - h_{di}^{\mathrm{T}}(x_i)Q_{di}h_{di}(x_i) - v_{si}(x_i) + \ell_{di}^{\mathrm{T}}(x_i)\ell_{di}(x_i),$$
$$x \in \mathcal{Z}_x, \quad (5.99)$$

$$0 = \tfrac{1}{2}P_{1i}(x) - h_{di}^{\mathrm{T}}(x_i)(S_{di} + Q_{di}J_{di}(x_i)) + \ell_{di}^{\mathrm{T}}(x_i)\mathcal{Z}_{di}(x_i), \quad x \in \mathcal{Z}_x,$$
$$(5.100)$$

$$0 \leq R_{di} + J_{di}^{\mathrm{T}}(x_i)S_{di} + S_{di}^{\mathrm{T}}J_{di}(x_i) + J_{di}^{\mathrm{T}}(x_i)Q_{di}J_{di}(x_i) - P_{2i}(x)$$
$$- \mathcal{Z}_{di}^{\mathrm{T}}(x_i)\mathcal{Z}_{di}(x_i), \quad x \in \mathcal{Z}_x. \quad (5.101)$$

Then $\mathcal{G}$ is vector dissipative (respectively, exponentially vector dissipative) with respect to the vector hybrid supply rate $(S_c(u_c, y_c), S_d(u_d, y_d))$, where $s_{ci}(u_{ci}, y_{ci}) = u_{ci}^{\mathrm{T}}R_{ci}u_{ci} + 2y_{ci}^{\mathrm{T}}S_{ci}u_{ci} + y_{ci}^{\mathrm{T}}Q_{ci}y_{ci}$ and $s_{di}(u_{di}, y_{di}) = u_{di}^{\mathrm{T}}R_{di}u_{di} + 2y_{di}^{\mathrm{T}}S_{di}u_{di} + y_{di}^{\mathrm{T}}Q_{di}y_{di}$, $i = 1, \ldots, q$.

Proof. For any admissible input $u_c(t) = [u_{c1}^{\mathrm{T}}(t), \ldots, u_{cq}^{\mathrm{T}}(t)]^{\mathrm{T}}$ such that $u_{ci}(t) \in \mathbb{R}^{m_{ci}}$, $t, \hat{t} \in \mathbb{R}$, $t_k < t \leq \hat{t} \leq t_{k+1}$, $k \in \overline{\mathbb{Z}}_+$, and $i = 1, \ldots, q$, it follows from (5.96)–(5.98) that

$$\int_t^{\hat{t}} s_{ci}(u_{ci}(\sigma), y_{ci}(\sigma))\mathrm{d}\sigma = \int_t^{\hat{t}} [u_{ci}^{\mathrm{T}}(\sigma)R_{ci}u_{ci}(\sigma) + 2y_{ci}^{\mathrm{T}}(\sigma)S_{ci}u_{ci}(\sigma)$$
$$+ y_{ci}^{\mathrm{T}}(\sigma)Q_{ci}y_{ci}(\sigma)]\mathrm{d}\sigma$$
$$= \int_t^{\hat{t}} [h_{ci}^{\mathrm{T}}(x_i(\sigma))Q_{ci}h_{ci}(x_i(\sigma))$$
$$+ 2h_{ci}^{\mathrm{T}}(x_i(\sigma))(S_{ci} + Q_{ci}J_{ci}(x_i(\sigma)))$$
$$u_{ci}(\sigma) + u_{ci}^{\mathrm{T}}(\sigma)(J_{ci}^{\mathrm{T}}(x_i(\sigma))Q_{ci}J_{ci}(x_i(\sigma))$$
$$+ J_{ci}^{\mathrm{T}}(x_i(\sigma))S_{ci} + S_{ci}^{\mathrm{T}}J_{ci}(x_i(\sigma))$$
$$+ R_{ci})u_{ci}(\sigma)]\mathrm{d}\sigma$$
$$\geq \int_t^{\hat{t}} [v'_{si}(x_i(\sigma))[\mathcal{F}_{ci}(x(\sigma)) + G_{ci}(x_i(\sigma))u_{ci}(\sigma)]$$
$$+ \ell_{ci}^{\mathrm{T}}(x_i(\sigma))\ell_{ci}(x_i(\sigma))$$
$$+ 2\ell_{ci}^{\mathrm{T}}(x_i(\sigma))\mathcal{Z}_{ci}(x_i(\sigma))u_{ci}(\sigma)$$
$$+ u_{ci}^{\mathrm{T}}(\sigma)\mathcal{Z}_{ci}^{\mathrm{T}}(x_i(\sigma))\mathcal{Z}_{ci}(x_i(\sigma))u_{ci}(\sigma)$$
$$- w_{ci}(V_s(x(\sigma)))]\mathrm{d}\sigma$$
$$= \int_t^{\hat{t}} [\dot{v}_{si}(x_i(\sigma)) + [\ell_{ci}(x_i(\sigma))$$

$$+\mathcal{Z}_{ci}(x_i(\sigma))u_{ci}(\sigma)]^{\mathrm{T}}[\ell_{ci}(x_i(\sigma))$$
$$+\mathcal{Z}_{ci}(x_i(\sigma))u_{ci}(\sigma)] - w_{ci}(V_{\mathrm{s}}(x(\sigma)))]\mathrm{d}\sigma$$
$$\geq v_{si}(x_i(\hat{t})) - v_{si}(x_i(t)) - \int_t^{\hat{t}} w_{ci}(V_{\mathrm{s}}(x(\sigma)))\mathrm{d}\sigma,$$
$$(5.102)$$

where $x(\sigma)$, $t_k < \sigma \leq t_{k+1}$, satisfies (5.1).

Next, for any admissible input $u_{\mathrm{d}}(t_k) = [u_{\mathrm{d}1}^{\mathrm{T}}(t_k), \ldots, u_{\mathrm{d}q}^{\mathrm{T}}(t_k)]^{\mathrm{T}}$ such that $u_{\mathrm{d}i}(t_k) \in \mathbb{R}^{m_{\mathrm{d}i}}$, $t_k \in \mathbb{R}$, $k \in \overline{\mathbb{Z}}_+$, and $i = 1, \ldots, q$, it follows from (5.95) and (5.99)–(5.101) that

$$
\begin{aligned}
s_{\mathrm{d}i}(u_{\mathrm{d}i}(t_k), y_{\mathrm{d}i}(t_k)) =& \; u_{\mathrm{d}i}^{\mathrm{T}}(t_k)R_{\mathrm{d}i}u_{\mathrm{d}i}(t_k) + 2y_{\mathrm{d}i}^{\mathrm{T}}(t_k)S_{\mathrm{d}i}u_{\mathrm{d}i}(t_k) \\
&+ y_{\mathrm{d}i}^{\mathrm{T}}(t_k)Q_{\mathrm{d}i}y_{\mathrm{d}i}(t_k) \\
=& \; h_{\mathrm{d}i}^{\mathrm{T}}(x_i(t_k))Q_{\mathrm{d}i}h_{\mathrm{d}i}(x_i(t_k)) \\
&+ 2h_{\mathrm{d}i}^{\mathrm{T}}(x_i(t_k))(S_{\mathrm{d}i} + Q_{\mathrm{d}i}J_{\mathrm{d}i}(x_i(t_k)))u_{\mathrm{d}i}(t_k) \\
&+ u_{\mathrm{d}i}^{\mathrm{T}}(t_k)(J_{\mathrm{d}i}^{\mathrm{T}}(x_i(t_k))Q_{\mathrm{d}i}J_{\mathrm{d}i}(x_i(t_k)) \\
&+ J_{\mathrm{d}i}^{\mathrm{T}}(x_i(t_k))S_{\mathrm{d}i} + S_{\mathrm{d}i}^{\mathrm{T}}J_{\mathrm{d}i}(x_i(t_k)) + R_{\mathrm{d}i})u_{\mathrm{d}i}(t_k) \\
\geq& \; v_{si}(x_i(t_k) + \mathcal{F}_{\mathrm{d}i}(x(t_k))) + P_{1i}(x(t_k))u_{\mathrm{d}i}(t_k) \\
&+ \ell_{\mathrm{d}i}^{\mathrm{T}}(x_i(t_k))\ell_{\mathrm{d}i}(x_i(t_k)) \\
&+ 2\ell_{\mathrm{d}i}^{\mathrm{T}}(x_i(t_k))\mathcal{Z}_{\mathrm{d}i}(x_i(t_k))u_{\mathrm{d}i}(t_k) \\
&+ u_{\mathrm{d}i}^{\mathrm{T}}(t_k)P_{2i}(x(t_k))u_{\mathrm{d}i}(t_k) \\
&+ u_{\mathrm{d}i}^{\mathrm{T}}(t_k)\mathcal{Z}_{\mathrm{d}i}^{\mathrm{T}}(x_i(t_k))\mathcal{Z}_{\mathrm{d}i}(x_i(t_k))u_{\mathrm{d}i}(t_k) \\
&- v_{si}(x_i(t_k)) \\
\geq& \; v_{si}(x_i(t_k) + \mathcal{F}_{\mathrm{d}i}(x(t_k)) + G_{\mathrm{d}i}(x_i(t_k))u_{\mathrm{d}i}(t_k)) \\
&+ [\ell_{\mathrm{d}i}(x_i(t_k)) + \mathcal{Z}_{\mathrm{d}i}(x_i(t_k))u_{\mathrm{d}i}(t_k)]^{\mathrm{T}}[\ell_{\mathrm{d}i}(x_i(t_k)) \\
&+ \mathcal{Z}_{\mathrm{d}i}(x_i(t_k))u_{\mathrm{d}i}(t_k)] - v_{si}(x_i(t_k)) \\
\geq& \; v_{si}(x_i(t_k) + \mathcal{F}_{\mathrm{d}i}(x(t_k)) + G_{\mathrm{d}i}(x_i(t_k))u_{\mathrm{d}i}(t_k)) \\
&- v_{si}(x_i(t_k)), \qquad\qquad\qquad\qquad\qquad\qquad\qquad (5.103)
\end{aligned}
$$

where $x(t_k)$, $k \in \overline{\mathbb{Z}}_+$, satisfies (5.2). Now, the result follows from (5.102) and (5.103) with vector storage function $V_{\mathrm{s}}(x) = [v_{s1}(x_1), \ldots, v_{sq}(x_q)]^{\mathrm{T}}$, $x \in \mathbb{R}^n$. $\qquad\square$

Finally, we provide necessary and sufficient conditions for the case where the large-scale impulsive dynamical system $\mathcal{G}$ is vector lossless with respect to a quadratic hybrid supply rate.

Theorem 5.9 *Consider the large-scale impulsive dynamical system $\mathcal{G}$ given by (5.1)–(5.4). Let $R_{ci} \in \mathbb{S}^{m_{ci}}$, $S_{ci} \in \mathbb{R}^{l_{ci} \times m_{ci}}$, $Q_{ci} \in \mathbb{S}^{l_{ci}}$,*

$R_{di} \in \mathbb{S}^{m_{di}}$, $S_{di} \in \mathbb{R}^{l_{di} \times m_{di}}$, and $Q_{di} \in \mathbb{S}^{l_{di}}$, $i = 1, \ldots, q$. Then $\mathcal{G}$ is vector lossless with respect to the quadratic hybrid supply rate $(S_c(u_c, y_c), S_d(u_d, y_d))$, where $s_{ci}(u_{ci}, y_{ci}) = u_{ci}^T R_{ci} u_{ci} + 2 y_{ci}^T S_{ci} u_{ci} + y_{ci}^T Q_{ci} y_{ci}$ and $s_{di}(u_{di}, y_{di}) = u_{di}^T R_{di} u_{di} + 2 y_{di}^T S_{di} u_{di} + y_{di}^T Q_{di} y_{di}$, $i = 1, \ldots, q$, if and only if there exist functions $V_s = [v_{s1}, \ldots, v_{sq}]^T : \mathbb{R}^n \to \overline{\mathbb{R}}_+^q$, $P_{1i} : \mathbb{R}^n \to \mathbb{R}^{1 \times m_d}$, $P_{2i} : \mathbb{R}^n \to \mathbb{N}^{m_d}$, and $w_c = [w_{c1}, \ldots, w_{cq}]^T : \overline{\mathbb{R}}_+^q \to \mathbb{R}^q$ such that $v_{si}(\cdot)$ is continuously differentiable, $v_{si}(0) = 0$, $i = 1, \ldots, q$, $w_c \in \mathcal{W}$, $w_c(0) = 0$, the zero solution $z(t) \equiv 0$ to (5.77) is Lyapunov stable, and, for all $x \in \mathbb{R}^n$, $i = 1, \ldots, q$, (5.81) holds and

$$0 = v_{si}'(x)\mathcal{F}_c(x) - h_c^T(x)\hat{Q}_{ci}h_c(x) - w_{ci}(V_s(x)), \quad x \notin \mathcal{Z}_x, \qquad (5.104)$$

$$0 = \tfrac{1}{2}v_{si}'(x)G_c(x) - h_c^T(x)(\hat{S}_{ci} + \hat{Q}_{ci}J_c(x)), \quad x \notin \mathcal{Z}_x, \qquad (5.105)$$

$$0 = \hat{R}_{ci} + J_c^T(x)\hat{S}_{ci} + \hat{S}_{ci}^T J_c(x) + J_c^T(x)\hat{Q}_{ci}J_c(x), \quad x \notin \mathcal{Z}_x, \qquad (5.106)$$

$$0 = v_{si}(x + \mathcal{F}_d(x)) - h_d^T(x)\hat{Q}_{di}h_d(x) - v_{si}(x), \quad x \in \mathcal{Z}_x, \qquad (5.107)$$

$$0 = \tfrac{1}{2}P_{1i}(x) - h_d^T(x)(\hat{S}_{di} + \hat{Q}_{di}J_d(x)), \quad x \in \mathcal{Z}_x, \qquad (5.108)$$

$$0 = \hat{R}_{di} + J_d^T(x)\hat{S}_{di} + \hat{S}_{di}^T J_d(x) + J_d^T(x)\hat{Q}_{di}J_d(x) - P_{2i}(x), \quad x \in \mathcal{Z}_x. \qquad (5.109)$$

Proof. Sufficiency follows as in the proof of Theorem 5.7. To show necessity, suppose that $\mathcal{G}$ is lossless with respect to the quadratic hybrid supply rate $(S_c(u_c, y_c), S_d(u_d, y_d))$. Then, there exist continuous functions $V_s = [v_{s1}, \ldots, v_{sq}]^T : \mathbb{R}^n \to \overline{\mathbb{R}}_+^q$ and $w_c = [w_{c1}, \ldots, w_{cq}]^T : \overline{\mathbb{R}}_+^q \to \mathbb{R}^q$ such that $V_s(0) = 0$, the zero solution $z(t) \equiv 0$ to (5.77) is Lyapunov stable, and for all $k \in \overline{\mathbb{Z}}_+$, $i = 1, \ldots, q$,

$$v_{si}(x(\hat{t})) - v_{si}(x(t)) = \int_t^{\hat{t}} s_{ci}(u_{ci}(\sigma), y_{ci}(\sigma))\mathrm{d}\sigma + \int_t^{\hat{t}} w_{ci}(V_s(x(\sigma)))\mathrm{d}\sigma,$$
$$t_k < t \le \hat{t} \le t_{k+1}, \qquad (5.110)$$

and

$$v_{si}(x(t_k) + \mathcal{F}_d(x(t_k)) + G_d(x(t_k))u_d(t_k)) = v_{si}(x(t_k))$$
$$+ s_{di}(u_{di}(t_k), y_{di}(t_k)). \qquad (5.111)$$

Now, dividing (5.110) by $\hat{t} - t^+$ and letting $\hat{t} \to t^+$, (5.110) is equivalent to

$$\dot{v}_{si}(x(t)) = v_{si}'(x(t))[\mathcal{F}_c(x(t)) + G_c(x(t))u_c(t)]$$
$$= s_{ci}(u_{ci}(t), y_{ci}(t)) + w_{ci}(V_s(x(t))), \quad t_k < t \le t_{k+1}. \quad (5.112)$$

Next, with $t = t_0$, it follows from (5.112) that

$$v'_{si}(x_0)[\mathcal{F}_c(x_0) + G_c(x_0)u_c(t_0)] = s_{ci}(u_{ci}(t_0), y_{ci}(t_0)) + w_{ci}(V_s(x_0)),$$
$$x_0 \notin \mathcal{Z}_x, \quad u_c(t_0) \in \mathbb{R}^{m_c}. \quad (5.113)$$

Since $x_0 \notin \mathcal{Z}_x$ is arbitrary, it follows that

$$
\begin{aligned}
v'_{si}(x)[\mathcal{F}_c(x) + G_c(x)u_c] &= w_{ci}(V_s(x)) + u_c^{\mathrm{T}}\hat{R}_{ci}u_c + 2y_c^{\mathrm{T}}\hat{S}_{ci}u_c \\
&\quad + y_c^{\mathrm{T}}\hat{Q}_{ci}y_c \\
&= w_{ci}(V_s(x)) + h_c^{\mathrm{T}}(x)\hat{Q}_{ci}h_c(x) \\
&\quad + 2h_c^{\mathrm{T}}(x)(\hat{Q}_{ci}J_c(x) + \hat{S}_{ci})u_c \\
&\quad + u_c^{\mathrm{T}}(\hat{R}_{ci} + \hat{S}_{ci}^{\mathrm{T}}J_c(x) + J_c^{\mathrm{T}}(x)\hat{S}_{ci} \\
&\quad + J_c^{\mathrm{T}}(x)\hat{Q}_{ci}J_c(x))u_c, \quad x \in \mathbb{R}^n, \quad u_c \in \mathbb{R}^{m_c}.
\end{aligned}
$$
$$(5.114)$$

Now, equating coefficients of equal powers yields (5.104)–(5.106).

Next, it follows from (5.111) with $k = 1$ that

$$
\begin{aligned}
v_{si}(x(t_1) + \mathcal{F}_d(x(t_1)) + G_d(x(t_1))u_d(t_1)) &= v_{si}(x(t_1)) \\
&\quad + s_{di}(u_{di}(t_1), y_{di}(t_1)).
\end{aligned}
$$
$$(5.115)$$

Now, since the continuous-time dynamics (5.1) are Lipschitz, it follows that for arbitrary $x \in \mathcal{Z}_x$ there exists $x_0 \notin \mathcal{Z}_x$ such that $x(t_1) = x$. Hence, it follows from (5.115) that

$$
\begin{aligned}
v_{si}(x + \mathcal{F}_d(x) + G_d(x)u_d) &= v_{si}(x) + u_d^{\mathrm{T}}\hat{R}_{di}u_d \\
&\quad + 2y_d^{\mathrm{T}}\hat{S}_{di}u_d + y_d^{\mathrm{T}}\hat{Q}_{di}y_d \\
&= v_{si}(x) + h_d^{\mathrm{T}}(x)\hat{Q}_{di}h_d(x) \\
&\quad + 2h_d^{\mathrm{T}}(x)(\hat{Q}_{di}J_d(x) + \hat{S}_{di})u_d \\
&\quad + u_d^{\mathrm{T}}(\hat{R}_{di} + \hat{S}_{di}^{\mathrm{T}}J_d(x) + J_d^{\mathrm{T}}(x)\hat{S}_{di} \\
&\quad + J_d^{\mathrm{T}}(x)\hat{Q}_{di}J_d(x))u_d, \\
&\qquad x \in \mathbb{R}^n, \quad u_d \in \mathbb{R}^{m_d}. \quad (5.116)
\end{aligned}
$$

Since the right-hand side of (5.116) is quadratic in u_d it follows that $v_{si}(x + \mathcal{F}_d(x) + G_d(x)u_d)$ is quadratic in u_d, and hence, there exist $P_{1i} : \mathbb{R}^n \to \mathbb{R}^{1 \times m_d}$ and $P_{2i} : \mathbb{R}^n \to \mathbb{N}^{m_d}$, $i = 1, \dots, q$, such that

$$v_{si}(x + \mathcal{F}_d(x) + G_d(x)u_d) = v_{si}(x + \mathcal{F}_d(x)) + P_{1i}(x)u_d + u_d^{\mathrm{T}}P_{2i}(x)u_d,$$
$$x \in \mathbb{R}^n, \quad u_d \in \mathbb{R}^{m_d}. \quad (5.117)$$

Now, using (5.117) and equating coefficients of equal powers in (5.116) yields (5.107)–(5.109). $\qquad\qquad\square$

5.4 Specialization to Large-Scale Linear Impulsive Dynamical Systems

In this section, we specialize the results of Section 5.3 to the case of large-scale linear impulsive dynamical systems. Specifically, we assume that $w_c(\cdot) \in \mathcal{W}$ is linear so that $w_c(z) = Wz$, where $W \in \mathbb{R}^{q \times q}$ is essentially nonnegative, and consider the large-scale linear impulsive dynamical system $\mathcal{G}$ given by

$$\dot{x}(t) = A_c x(t) + B_c u_c(t), \quad x(t) \notin \mathcal{Z}_x, \tag{5.118}$$

$$\Delta x(t) = (A_d - I_n) x(t) + B_d u_d(t), \quad x(t) \in \mathcal{Z}_x, \tag{5.119}$$

$$y_c(t) = C_c x(t) + D_c u_c(t), \quad x(t) \notin \mathcal{Z}_x, \tag{5.120}$$

$$y_d(t) = C_d x(t) + D_d u_d(t), \quad x(t) \in \mathcal{Z}_x, \tag{5.121}$$

where $A_c \in \mathbb{R}^{n \times n}$ is partitioned as $A_c \triangleq [A_{cij}], i, j = 1, \ldots, q, A_{cij} \in \mathbb{R}^{n_i \times n_j}, \sum_{i=1}^{q} n_i = n, B_c = \text{block–diag}[B_{c1}, \ldots, B_{cq}], C_c = \text{block–diag}[C_{c1}, \ldots, C_{cq}], D_c = \text{block–diag}[D_{c1}, \ldots, D_{cq}], B_{ci} \in \mathbb{R}^{n_i \times m_{ci}}, C_{ci} \in \mathbb{R}^{l_{ci} \times n_i}, D_{ci} \in \mathbb{R}^{l_{ci} \times m_{ci}}, A_d \in \mathbb{R}^{n \times n}$ is partitioned as $A_d \triangleq [A_{dij}], i, j = 1, \ldots, q, A_{dij} \in \mathbb{R}^{n_i \times n_j}, B_d = \text{block–diag}[B_{d1}, \ldots, B_{dq}], C_d = \text{block–diag}[C_{d1}, \ldots, C_{dq}], D_d = \text{block–diag}[D_{d1}, \ldots, D_{dq}], B_{di} \in \mathbb{R}^{n_i \times m_{di}}, C_{di} \in \mathbb{R}^{l_{di} \times n_i}, D_{di} \in \mathbb{R}^{l_{di} \times m_{di}}$, and $i = 1, \ldots, q$.

Theorem 5.10 *Consider the large-scale linear impulsive dynamical system $\mathcal{G}$ given by (5.118)–(5.121). Let $R_{ci} \in \mathbb{S}^{m_{ci}}, S_{ci} \in \mathbb{R}^{l_{ci} \times m_{ci}}, Q_{ci} \in \mathbb{S}^{l_{ci}}, R_{di} \in \mathbb{S}^{m_{di}}, S_{di} \in \mathbb{R}^{l_{di} \times m_{di}}, Q_{di} \in \mathbb{S}^{l_{di}}, i = 1, \ldots, q$. Then $\mathcal{G}$ is vector dissipative (respectively, exponentially vector dissipative) with respect to the vector hybrid supply rate $(S_c(u_c, y_c), S_d(u_d, y_d))$, where $s_{ci}(u_{ci}, y_{ci}) = u_{ci}^T R_{ci} u_{ci} + 2 y_{ci}^T S_{ci} u_{ci} + y_{ci}^T Q_{ci} y_{ci}$ and $s_{di}(u_{di}, y_{di}) = u_{di}^T R_{di} u_{di} + 2 y_{di}^T S_{di} u_{di} + y_{di}^T Q_{di} y_{di}, i = 1, \ldots, q$, if there exist $W \in \mathbb{R}^{q \times q}, P_i \in \mathbb{N}^n, L_{ci} \in \mathbb{R}^{s_{ci} \times n}, Z_{ci} \in \mathbb{R}^{s_{ci} \times m_c}, L_{di} \in \mathbb{R}^{s_{di} \times n}$, and $Z_{di} \in \mathbb{R}^{s_{di} \times m_d}, i = 1, \ldots, q$, such that W is essentially nonnegative and semistable (respectively, asymptotically stable), and, for all $i = 1, \ldots, q$,*

$$0 = x^T (A_c^T P_i + P_i A_c - C_c^T \hat{Q}_{ci} C_c - \sum_{j=1}^{q} W_{(i,j)} P_j + L_{ci}^T L_{ci}) x, \quad x \notin \mathcal{Z}_x,$$

$$\tag{5.122}$$

$$0 = x^{\mathrm{T}}(P_i B_{\mathrm{c}} - C_{\mathrm{c}}^{\mathrm{T}}(\hat{S}_{ci} + \hat{Q}_{ci} D_{\mathrm{c}}) + L_{ci}^{\mathrm{T}} Z_{ci}), \quad x \notin \mathcal{Z}_x, \tag{5.123}$$

$$0 = \hat{R}_{ci} + D_{\mathrm{c}}^{\mathrm{T}} \hat{S}_{ci} + \hat{S}_{ci}^{\mathrm{T}} D_{\mathrm{c}} + D_{\mathrm{c}}^{\mathrm{T}} \hat{Q}_{ci} D_{\mathrm{c}} - Z_{ci}^{\mathrm{T}} Z_{ci}, \tag{5.124}$$

$$0 = x^{\mathrm{T}}(A_{\mathrm{d}}^{\mathrm{T}} P_i A_{\mathrm{d}} - C_{\mathrm{d}}^{\mathrm{T}} \hat{Q}_{di} C_{\mathrm{d}} - P_i + L_{di}^{\mathrm{T}} L_{di})x, \quad x \in \mathcal{Z}_x, \tag{5.125}$$

$$0 = x^{\mathrm{T}}(A_{\mathrm{d}}^{\mathrm{T}} P_i B_{\mathrm{d}} - C_{\mathrm{d}}^{\mathrm{T}}(\hat{S}_{di} + \hat{Q}_{di} D_{\mathrm{d}}) + L_{di}^{\mathrm{T}} Z_{di}), \quad x \in \mathcal{Z}_x, \tag{5.126}$$

$$0 = \hat{R}_{di} + D_{\mathrm{d}}^{\mathrm{T}} \hat{S}_{di} + \hat{S}_{di}^{\mathrm{T}} D_{\mathrm{d}} + D_{\mathrm{d}}^{\mathrm{T}} \hat{Q}_{di} D_{\mathrm{d}} - B_{\mathrm{d}}^{\mathrm{T}} P_i B_{\mathrm{d}} - Z_{di}^{\mathrm{T}} Z_{di}. \tag{5.127}$$

Proof. The proof follows from Theorem 5.7 with $\mathcal{F}_{\mathrm{c}}(x) = A_{\mathrm{c}} x$, $G_{\mathrm{c}}(x) = B_{\mathrm{c}}$, $h_{\mathrm{c}}(x) = C_{\mathrm{c}} x$, $J_{\mathrm{c}}(x) = D_{\mathrm{c}}$, $w_{\mathrm{c}}(r) = Wr$, $\ell_{ci}(x) = L_{ci} x$, $\mathcal{Z}_{ci}(x) = Z_{ci}$, $\mathcal{F}_{\mathrm{d}}(x) = A_{\mathrm{d}} x$, $G_{\mathrm{d}}(x) = B_{\mathrm{d}}$, $h_{\mathrm{d}}(x) = C_{\mathrm{d}} x$, $J_{\mathrm{d}}(x) = D_{\mathrm{d}}$, $\ell_{di}(x) = L_{di} x$, $\mathcal{Z}_{di}(x) = Z_{di}$, $P_{1i}(x) = 2x^{\mathrm{T}} A_{\mathrm{d}}^{\mathrm{T}} P_i B_{\mathrm{d}}$, $P_{2i}(x) = B_{\mathrm{d}}^{\mathrm{T}} P_i B_{\mathrm{d}}$, and $v_{si}(x) = x^{\mathrm{T}} P_i x$, $i = 1, \ldots, q$. □

Note that (5.122)–(5.127) are implied by

$$\begin{bmatrix} \mathcal{A}_{ci} & \mathcal{B}_{ci} \\ \mathcal{B}_{ci}^{\mathrm{T}} & \mathcal{C}_{ci} \end{bmatrix} = - \begin{bmatrix} L_{ci}^{\mathrm{T}} \\ Z_{ci}^{\mathrm{T}} \end{bmatrix} \begin{bmatrix} L_{ci} & Z_{ci} \end{bmatrix} \leq 0, \tag{5.128}$$

$$\begin{bmatrix} \mathcal{A}_{di} & \mathcal{B}_{di} \\ \mathcal{B}_{di}^{\mathrm{T}} & \mathcal{C}_{di} \end{bmatrix} = - \begin{bmatrix} L_{di}^{\mathrm{T}} \\ Z_{di}^{\mathrm{T}} \end{bmatrix} \begin{bmatrix} L_{di} & Z_{di} \end{bmatrix} \leq 0, \quad i = 1, \ldots, q, \tag{5.129}$$

where, for all $i = 1, \ldots, q$,

$$\mathcal{A}_{ci} = A_{\mathrm{c}}^{\mathrm{T}} P_i + P_i A_{\mathrm{c}} - C_{\mathrm{c}}^{\mathrm{T}} \hat{Q}_{ci} C_{\mathrm{c}} - \sum_{j=1}^{q} W_{(i,j)} P_j, \tag{5.130}$$

$$\mathcal{B}_{ci} = P_i B_{\mathrm{c}} - C_{\mathrm{c}}^{\mathrm{T}}(\hat{S}_{ci} + \hat{Q}_{ci} D_{\mathrm{c}}), \tag{5.131}$$

$$\mathcal{C}_{ci} = -(\hat{R}_{ci} + D_{\mathrm{c}}^{\mathrm{T}} \hat{S}_{ci} + \hat{S}_{ci}^{\mathrm{T}} D_{\mathrm{c}} + D_{\mathrm{c}}^{\mathrm{T}} \hat{Q}_{ci} D_{\mathrm{c}}), \tag{5.132}$$

$$\mathcal{A}_{di} = A_{\mathrm{d}}^{\mathrm{T}} P_i A_{\mathrm{d}} - C_{\mathrm{d}}^{\mathrm{T}} \hat{Q}_{di} C_{\mathrm{d}} - P_i, \tag{5.133}$$

$$\mathcal{B}_{di} = A_{\mathrm{d}}^{\mathrm{T}} P_i B_{\mathrm{d}} - C_{\mathrm{d}}^{\mathrm{T}}(\hat{S}_{di} + \hat{Q}_{di} D_{\mathrm{d}}), \tag{5.134}$$

$$\mathcal{C}_{di} = -(\hat{R}_{di} + D_{\mathrm{d}}^{\mathrm{T}} \hat{S}_{di} + \hat{S}_{di}^{\mathrm{T}} D_{\mathrm{d}} + D_{\mathrm{d}}^{\mathrm{T}} \hat{Q}_{di} D_{\mathrm{d}} - B_{\mathrm{d}}^{\mathrm{T}} P_i B_{\mathrm{d}}). \tag{5.135}$$

Hence, vector dissipativity of large-scale linear impulsive dynamical systems with respect to quadratic hybrid supply rates can be characterized via (cascade) linear matrix inequalities (LMIs) [27]. A similar remark holds for Theorem 5.11 below.

The next result presents sufficient conditions guaranteeing vector dissipativity of $\mathcal{G}$ with respect to a quadratic hybrid supply rate in case where the vector storage function is component decoupled.

Theorem 5.11 *Consider the large-scale linear impulsive dynamical system $\mathcal{G}$ given by (5.118)–(5.121). Let $R_{ci} \in \mathbb{S}^{m_{ci}}$, $S_{ci} \in \mathbb{R}^{l_{ci} \times m_{ci}}$, $Q_{ci} \in \mathbb{S}^{l_{ci}}$, $R_{di} \in \mathbb{S}^{m_{di}}$, $S_{di} \in \mathbb{R}^{l_{di} \times m_{di}}$, and $Q_{di} \in \mathbb{S}^{l_{di}}$, $i = 1, \ldots, q$, be* *given. Assume there exist matrices* $W \in \mathbb{R}^{q \times q}$, $P_i \in \mathbb{N}^{n_i}$, $L_{cii} \in \mathbb{R}^{s_{cii} \times n_i}$, $Z_{cii} \in \mathbb{R}^{s_{cii} \times m_{ci}}$, $L_{dii} \in \mathbb{R}^{s_{dii} \times n_i}$, $Z_{dii} \in \mathbb{R}^{s_{dii} \times m_{ci}}$, $i = 1, \ldots, q$, $L_{cij} \in \mathbb{R}^{s_{cij} \times n_i}$, $Z_{cij} \in \mathbb{R}^{s_{cij} \times n_j}$, $L_{dij} \in \mathbb{R}^{s_{dij} \times n_i}$, *and* $Z_{dij} \in \mathbb{R}^{s_{dij} \times n_j}$, $i, j = 1, \ldots, q$, $i \neq j$, *such that* W *is essentially nonnegative and semistable (respectively, asymptotically stable), and, for all* $i = 1, \ldots, q$,

$$0 \geq x_i^{\mathrm{T}} \left(A_{cii}^{\mathrm{T}} P_i + P_i A_{cii} - C_{ci}^{\mathrm{T}} Q_{ci} C_{ci} - W_{(i,i)} P_i + L_{cii}^{\mathrm{T}} L_{cii} \right.$$

$$\left. - \sum_{j=1, j \neq i}^{q} L_{cij}^{\mathrm{T}} L_{cij} \right) x_i, \quad x \notin \mathcal{Z}_x, \tag{5.136}$$

$$0 = x_i^{\mathrm{T}} (P_i B_{ci} - C_{ci}^{\mathrm{T}} S_{ci} - C_{ci}^{\mathrm{T}} Q_{ci} D_{ci} + L_{cii}^{\mathrm{T}} Z_{cii}), \quad x \notin \mathcal{Z}_x \tag{5.137}$$

$$0 \leq R_{ci} + D_{ci}^{\mathrm{T}} S_{ci} + S_{ci}^{\mathrm{T}} D_{ci} + D_{ci}^{\mathrm{T}} Q_{ci} D_{ci} - Z_{cii}^{\mathrm{T}} Z_{cii}, \tag{5.138}$$

$$0 \geq x_i^{\mathrm{T}} \left(A_{dii}^{\mathrm{T}} P_i A_{dii} - C_{di}^{\mathrm{T}} Q_{di} C_{di} - P_i + L_{dii}^{\mathrm{T}} L_{dii} \right.$$

$$\left. + \sum_{j=1, j \neq i}^{q} L_{dij}^{\mathrm{T}} L_{dij} \right) x_i, \quad x \in \mathcal{Z}_x, \tag{5.139}$$

$$0 = x_i^{\mathrm{T}} (A_{dii}^{\mathrm{T}} P_i B_{di} - C_{di}^{\mathrm{T}} S_{di} - C_{di}^{\mathrm{T}} Q_{di} D_{di} + L_{dii}^{\mathrm{T}} Z_{dii}), \quad x \in \mathcal{Z}_x, \tag{5.140}$$

$$0 \leq R_{di} + D_{di}^{\mathrm{T}} S_{di} + S_{di}^{\mathrm{T}} D_{di} + D_{di}^{\mathrm{T}} Q_{di} D_{di} - B_{di}^{\mathrm{T}} P_i B_{di} - Z_{dii}^{\mathrm{T}} Z_{dii}, \tag{5.141}$$

and for $j = 1, \ldots, q, l = 1, \ldots, q, j \neq i, l \neq i$,

$$0 = x_i^{\mathrm{T}} (P_i A_{cij} + L_{cij}^{\mathrm{T}} Z_{cij}), \quad x \notin \mathcal{Z}_x, \tag{5.142}$$

$$0 \leq x_j^{\mathrm{T}} (W_{(i,j)} P_j - Z_{cij}^{\mathrm{T}} Z_{cij}) x_j, \quad x \notin \mathcal{Z}_x, \tag{5.143}$$

$$0 = x_j^{\mathrm{T}} (A_{dij}^{\mathrm{T}} P_i B_{di}), \quad x \in \mathcal{Z}_x, \tag{5.144}$$

$$0 \geq x_j^{\mathrm{T}} (A_{dij}^{\mathrm{T}} P_i A_{dil}) x_l, \quad x \in \mathcal{Z}_x, \tag{5.145}$$

$$0 \geq x_i^{\mathrm{T}} (A_{dii}^{\mathrm{T}} P_i A_{dij} + L_{dij}^{\mathrm{T}} Z_{dij}) x_j, \quad x \in \mathcal{Z}_x, \tag{5.146}$$

$$0 = x_j^{\mathrm{T}} (Z_{dij}^{\mathrm{T}} Z_{dij}) x_j, \quad x \in \mathcal{Z}_x. \tag{5.147}$$

Then $\mathcal{G}$ *is vector dissipative (respectively, exponentially vector dissipative) with respect to the vector hybrid supply rate* $(S_{\mathrm{c}}(u_{\mathrm{c}}, y_{\mathrm{c}}), S_{\mathrm{d}}(u_{\mathrm{d}}, y_{\mathrm{d}}))$, *where* $s_{ci}(u_{ci}, y_{ci}) = u_{ci}^{\mathrm{T}} R_{ci} u_{ci} + 2 y_{ci}^{\mathrm{T}} S_{ci} u_{ci} + y_{ci}^{\mathrm{T}} Q_{ci} y_{ci}$ *and* $s_{di}(u_{di}, y_{di}) = u_{di}^{\mathrm{T}} R_{di} u_{di} + 2 y_{di}^{\mathrm{T}} S_{di} u_{di} + y_{di}^{\mathrm{T}} Q_{di} y_{di}$, $i = 1, \ldots, q$.

Proof. Since P_i is nonnegative definite, the function $v_{si}(x_i) \stackrel{\triangle}{=} x_i^{\mathrm{T}} P_i x_i$, $x_i \in \mathbb{R}^{n_i}$, is nonnegative definite and $v_{si}(0) = 0$. Moreover, since $v_{si}(\cdot)$ is continuously differentiable it follows from (5.136)–(5.147) that for all $u_{ci}(t) \in \mathbb{R}^{m_{ci}}$, $u_{di}(t_k) \in \mathbb{R}^{m_{di}}$, $i = 1, \dots, q$, and $t_k < t \le t_{k+1}$, $k \in \overline{\mathbb{Z}}_+$,

$$
\dot{v}_{si}(x_i(t))
$$

$$
= 2x_i^{\mathrm{T}}(t) P_i \left[\sum_{j=1}^{q} A_{cij} x_j(t) + B_{ci} u_{ci}(t) \right]
$$

$$
\le x_i^{\mathrm{T}}(t) \left[W_{(i,i)} P_i + C_{ci}^{\mathrm{T}} Q_{ci} C_{ci} - L_{cii}^{\mathrm{T}} L_{cii} - \sum_{j=1,\, j\ne i}^{q} L_{cij}^{\mathrm{T}} L_{cij} \right] x_i(t)
$$

$$
- \sum_{j=1,\, j\ne i}^{q} 2x_i^{\mathrm{T}}(t) L_{cij}^{\mathrm{T}} Z_{cij} x_j(t)
$$

$$
+ 2x_i^{\mathrm{T}}(t) C_{ci}^{\mathrm{T}} S_{ci} u_{ci}(t) + 2x_i^{\mathrm{T}}(t) C_{ci}^{\mathrm{T}} Q_{ci} D_{ci} u_{ci}(t)
$$

$$
- 2x_i^{\mathrm{T}}(t) L_{cii}^{\mathrm{T}} Z_{cii} u_{ci}(t) + \sum_{j=1,\, j\ne i}^{q} x_j^{\mathrm{T}}(t) [W_{(i,j)} P_j - Z_{cij}^{\mathrm{T}} Z_{cij}] x_j(t)
$$

$$
+ u_{ci}^{\mathrm{T}}(t) R_{ci} u_{ci}(t) + 2u_{ci}^{\mathrm{T}}(t) D_{ci}^{\mathrm{T}} S_{ci} u_{ci}(t) + u_{ci}^{\mathrm{T}}(t) D_{ci}^{\mathrm{T}} Q_{ci} D_{ci} u_{ci}(t)
$$

$$
- u_{ci}^{\mathrm{T}}(t) Z_{cii}^{\mathrm{T}} Z_{cii} u_{ci}(t)
$$

$$
= \sum_{j=1}^{q} W_{(i,j)} v_{sj}(x_j(t)) + u_{ci}^{\mathrm{T}}(t) R_{ci} u_{ci}(t) + 2y_{ci}^{\mathrm{T}}(t) S_{ci} u_{ci}(t)
$$

$$
+ y_{ci}^{\mathrm{T}}(t) Q_{ci} y_{ci}(t)
$$

$$
- [L_{cii} x_i(t) + Z_{cii} u_{ci}(t)]^{\mathrm{T}} [L_{cii} x_i(t) + Z_{cii} u_{ci}(t)]
$$

$$
- \sum_{j=1,\, j\ne i}^{q} (L_{cij} x_i(t) + Z_{cij} x_j(t))^{\mathrm{T}} (L_{cij} x_i(t) + Z_{cij} x_j(t))
$$

$$
\le s_{ci}(u_{ci}(t), y_{ci}(t)) + \sum_{j=1}^{q} W_{(i,j)} v_{sj}(x_j(t)). \tag{5.148}
$$

Furthermore,

$$
v_{si} \left(\sum_{j=1}^{q} A_{dij} x_j(t_k) + B_{di} u_{di}(t_k) \right) = \left[\sum_{j=1}^{q} A_{dij} x_j(t_k) + B_{di} u_{di}(t_k) \right]^{\mathrm{T}}
$$

$$
\cdot P_i \left[\sum_{j=1}^{q} A_{\mathrm{d}ij} x_j(t_k) + B_{\mathrm{d}i} u_{\mathrm{d}i}(t_k) \right]
$$

$$
\leq x_i^{\mathrm{T}}(t_k) \left[P_i + C_{\mathrm{d}i}^{\mathrm{T}} Q_{\mathrm{d}i} C_{\mathrm{d}i} - L_{\mathrm{d}ii}^{\mathrm{T}} L_{\mathrm{d}ii} - \sum_{j=1, j\neq i}^{q} L_{\mathrm{d}ij}^{\mathrm{T}} L_{\mathrm{d}ij} \right] x_i(t_k)
$$

$$
- \sum_{j=1, j\neq i}^{q} 2 x_i^{\mathrm{T}}(t_k) L_{\mathrm{d}ij}^{\mathrm{T}} Z_{\mathrm{d}ij} x_j(t_k) + 2 x_i^{\mathrm{T}}(t_k) C_{\mathrm{d}i}^{\mathrm{T}} S_{\mathrm{d}i} u_{\mathrm{d}i}(t_k)
$$

$$
+ 2 x_i^{\mathrm{T}}(t_k) C_{\mathrm{d}i}^{\mathrm{T}} Q_{\mathrm{d}i} D_{\mathrm{d}i} u_{\mathrm{d}i}(t_k)
$$

$$
- 2 x_i^{\mathrm{T}}(t_k) L_{\mathrm{d}ii}^{\mathrm{T}} Z_{\mathrm{d}ii} u_{\mathrm{d}i}(t_k) - \sum_{j=1, j\neq i}^{q} x_j^{\mathrm{T}}(t_k) Z_{\mathrm{d}ij}^{\mathrm{T}} Z_{\mathrm{d}ij} x_j(t_k)
$$

$$
+ u_{\mathrm{d}i}^{\mathrm{T}}(t_k) R_{\mathrm{d}i} u_{\mathrm{d}i}(t_k) + 2 u_{\mathrm{d}i}^{\mathrm{T}}(t_k) D_{\mathrm{d}i}^{\mathrm{T}} S_{\mathrm{d}i} u_{\mathrm{d}i}(t_k)
$$

$$
+ u_{\mathrm{d}i}^{\mathrm{T}}(t_k) D_{\mathrm{d}i}^{\mathrm{T}} Q_{\mathrm{d}i} D_{\mathrm{d}i} u_{\mathrm{d}i}(t_k) - u_{\mathrm{d}i}^{\mathrm{T}}(t_k) Z_{\mathrm{d}ii}^{\mathrm{T}} Z_{\mathrm{d}ii} u_{\mathrm{d}i}(t_k)
$$

$$
= v_{\mathrm{s}i}(x_i(t_k)) + u_{\mathrm{d}i}^{\mathrm{T}}(t_k) R_{\mathrm{d}i} u_{\mathrm{d}i}(t_k) + 2 y_{\mathrm{d}i}^{\mathrm{T}}(t_k) S_{\mathrm{d}i} u_{\mathrm{d}i}(t_k)
$$

$$
+ y_{\mathrm{d}i}^{\mathrm{T}}(t_k) Q_{\mathrm{d}i} y_{\mathrm{d}i}(t_k) - [L_{\mathrm{d}ii} x_i(t_k) + Z_{\mathrm{d}ii} u_{\mathrm{d}i}(t_k)]^{\mathrm{T}} [L_{\mathrm{d}ii} x_i(t_k)
$$

$$
+ Z_{\mathrm{d}ii} u_{\mathrm{d}i}(t_k)] - \sum_{j=1, j\neq i}^{q} [L_{\mathrm{d}ij} x_i(t_k)
$$

$$
+ Z_{\mathrm{d}ij} x_j(t_k)]^{\mathrm{T}} [L_{\mathrm{d}ij} x_i(t_k) + Z_{\mathrm{d}ij} x_j(t_k)]
$$

$$
\leq s_{\mathrm{d}i}(u_{\mathrm{d}i}(t_k), y_{\mathrm{d}i}(t_k)) + v_{\mathrm{s}i}(x_i(t_k)). \tag{5.149}
$$

Writing (5.148) and (5.149) in vector form yields

$$
\dot{V}_{\mathrm{s}}(x) \leq\leq W V_{\mathrm{s}}(x) + S_{\mathrm{c}}(u_{\mathrm{c}}, y_{\mathrm{c}}), \quad u_{\mathrm{c}} \in \mathbb{R}^{m_{\mathrm{c}}}, \quad x \notin \mathcal{Z}_x, \tag{5.150}
$$

$$
V_{\mathrm{s}}(A_{\mathrm{d}} x + B_{\mathrm{d}} u_{\mathrm{d}}) \leq\leq V_{\mathrm{s}}(x) + S_{\mathrm{d}}(u_{\mathrm{d}}, y_{\mathrm{d}}), \quad u_{\mathrm{d}} \in \mathbb{R}^{m_{\mathrm{d}}}, \quad x \in \mathcal{Z}_x, \tag{5.151}
$$

where $V_{\mathrm{s}}(x) \triangleq [v_{\mathrm{s}1}(x_1), \ldots, v_{\mathrm{s}q}(x_q)]^{\mathrm{T}}$, $x \in \mathbb{R}^n$. Now, it follows from Definition 5.3 that $\mathcal{G}$ is vector dissipative (respectively, exponentially vector dissipative) with respect to the vector hybrid supply rate $(S_{\mathrm{c}}(u_{\mathrm{c}}, y_{\mathrm{c}}), S_{\mathrm{d}}(u_{\mathrm{d}}, y_{\mathrm{d}}))$ and with vector storage function $V_{\mathrm{s}}(x)$, $x \in \mathbb{R}^n$. $\qquad \square$

Chapter Six

Stability and Feedback Interconnections of
Dissipative Impulsive Dynamical Systems

6.1 Introduction

In Chapters 2 and 3 stability and dissipativity theory for nonlinear impulsive dynamical systems was developed. Using the concepts of dissipativity and exponential dissipativity for impulsive systems, in this chapter we develop feedback interconnection stability results for nonlinear impulsive dynamical systems. The feedback system can be impulsive, nonlinear, and either dynamic or static. General stability criteria are given for Lyapunov, asymptotic, and exponential stability of feedback impulsive systems. In the case of quadratic supply rates involving net system power and input-output energy, these results generalize the positivity and small-gain theorems [53, 74, 142, 165, 171] to the case of nonlinear impulsive dynamical systems. In particular, we show that if the nonlinear impulsive dynamical systems $\mathcal{G}$ and $\mathcal{G}_c$ are dissipative (respectively, exponentially dissipative) with respect to quadratic supply rates corresponding to net system power, or weighted input and output energy, then the negative feedback interconnection of $\mathcal{G}$ and $\mathcal{G}_c$ is Lyapunov (respectively, asymptotically) stable.

6.2 Stability of Feedback Interconnections of Dissipative Impulsive Dynamical Systems

In this section, we consider feedback interconnections of dissipative impulsive dynamical systems. Specifically, using the notion of dissipative and exponentially dissipative impulsive dynamical systems introduced in Chapter 3, with appropriate storage functions and supply rates, we construct Lyapunov functions for interconnected impulsive dynamical systems by appropriately combining storage functions for each subsystem. Here, we restrict our attention to input/state-dependent impulsive dynamical systems. Analogous results, with the exception of results requiring the impulsive invariance principle, hold for time-dependent impulsive dynamical systems.

We begin by considering the nonlinear dynamical system $\mathcal{G}$ given by

$$\dot{x}(t) = f_{\mathrm{c}}(x(t)) + G_{\mathrm{c}}(x(t))u_{\mathrm{c}}(t), \quad x(0) = x_0, \quad (x(t), u_{\mathrm{c}}(t)) \notin \mathcal{Z},$$
$$(6.1)$$
$$\Delta x(t) = f_{\mathrm{d}}(x(t)) + G_{\mathrm{d}}(x(t))u_{\mathrm{d}}(t), \quad (x(t), u_{\mathrm{c}}(t)) \in \mathcal{Z}, \qquad (6.2)$$
$$y_{\mathrm{c}}(t) = h_{\mathrm{c}}(x(t)) + J_{\mathrm{c}}(x(t))u_{\mathrm{c}}(t), \quad (x(t), u_{\mathrm{c}}(t)) \notin \mathcal{Z}, \qquad (6.3)$$
$$y_{\mathrm{d}}(t) = h_{\mathrm{d}}(x(t)) + J_{\mathrm{d}}(x(t))u_{\mathrm{d}}(t), \quad (x(t), u_{\mathrm{c}}(t)) \in \mathcal{Z}, \qquad (6.4)$$

where $t \geq 0$, $x(t) \in \mathcal{D} \subseteq \mathbb{R}^n$, $\mathcal{D}$ is an open set with $0 \in \mathcal{D}$, $\Delta x(t) \triangleq x(t^+) - x(t)$, $u_{\mathrm{c}}(t) \in U_{\mathrm{c}} \subseteq \mathbb{R}^{m_{\mathrm{c}}}$, $u_{\mathrm{d}}(t_k) \in U_{\mathrm{d}} \subseteq \mathbb{R}^{m_{\mathrm{d}}}$, t_k denotes the kth instant of time at which $(x(t), u_{\mathrm{c}}(t))$ intersects $\mathcal{Z}$ for a particular trajectory $x(t)$ and input $u_{\mathrm{c}}(t)$, $y_{\mathrm{c}}(t) \in \mathbb{R}^{l_{\mathrm{c}}}$, $y_{\mathrm{d}}(t_k) \in \mathbb{R}^{l_{\mathrm{d}}}$, $f_{\mathrm{c}} : \mathcal{D} \to \mathbb{R}^n$ is Lipschitz continuous on $\mathcal{D}$ and satisfies $f_{\mathrm{c}}(0) = 0$, $G_{\mathrm{c}} : \mathcal{D} \to \mathbb{R}^{n \times m_{\mathrm{c}}}$, $f_{\mathrm{d}} : \mathcal{D} \to \mathbb{R}^n$ is continuous on $\mathcal{D}$, $G_{\mathrm{d}} : \mathcal{D} \to \mathbb{R}^{n \times m_{\mathrm{d}}}$, $h_{\mathrm{c}} : \mathcal{D} \to \mathbb{R}^{l_{\mathrm{c}}}$ and satisfies $h_{\mathrm{c}}(0) = 0$, $J_{\mathrm{c}} : \mathcal{D} \to \mathbb{R}^{l_{\mathrm{c}} \times m_{\mathrm{c}}}$, $h_{\mathrm{d}} : \mathcal{D} \to \mathbb{R}^{l_{\mathrm{d}}}$, $J_{\mathrm{d}} : \mathcal{D} \to \mathbb{R}^{l_{\mathrm{d}} \times m_{\mathrm{d}}}$, and $\mathcal{Z} \triangleq \mathcal{Z}_x \times \mathcal{Z}_{u_{\mathrm{c}}}$, where $\mathcal{Z}_x \subset \mathcal{D}$ and $\mathcal{Z}_{u_{\mathrm{c}}} \subset U_{\mathrm{c}}$, is the resetting set.

Furthermore, consider the impulsive nonlinear feedback system $\mathcal{G}_{\mathrm{c}}$ given by

$$\dot{x}_{\mathrm{c}}(t) = f_{\mathrm{cc}}(x_{\mathrm{c}}(t)) + G_{\mathrm{cc}}(u_{\mathrm{cc}}(t), x_{\mathrm{c}}(t))u_{\mathrm{cc}}(t), \quad x_{\mathrm{c}}(0) = x_{\mathrm{c}0},$$
$$(x_{\mathrm{c}}(t), u_{\mathrm{cc}}(t)) \notin \mathcal{Z}_{\mathrm{c}}, \quad (6.5)$$
$$\Delta x_{\mathrm{c}}(t) = f_{\mathrm{dc}}(x_{\mathrm{c}}(t)) + G_{\mathrm{dc}}(u_{\mathrm{dc}}(t), x_{\mathrm{c}}(t))u_{\mathrm{dc}}(t), \quad (x_{\mathrm{c}}(t), u_{\mathrm{cc}}(t)) \in \mathcal{Z}_{\mathrm{c}},$$
$$(6.6)$$
$$y_{\mathrm{cc}}(t) = h_{\mathrm{cc}}(x_{\mathrm{c}}(t)) + J_{\mathrm{cc}}(u_{\mathrm{cc}}(t), x_{\mathrm{c}}(t))u_{\mathrm{cc}}(t), \quad (x_{\mathrm{c}}(t), u_{\mathrm{cc}}(t)) \notin \mathcal{Z}_{\mathrm{c}},$$
$$(6.7)$$
$$y_{\mathrm{dc}}(t) = h_{\mathrm{dc}}(x_{\mathrm{c}}(t)) + J_{\mathrm{dc}}(u_{\mathrm{dc}}(t), x_{\mathrm{c}}(t))u_{\mathrm{dc}}(t), \quad (x_{\mathrm{c}}(t), u_{\mathrm{cc}}(t)) \in \mathcal{Z}_{\mathrm{c}},$$
$$(6.8)$$

where $t \geq 0$, $\Delta x_{\mathrm{c}}(t) = x_{\mathrm{c}}(t^+) - x_{\mathrm{c}}(t)$, $x_{\mathrm{c}}(t) \in \mathbb{R}^{n_{\mathrm{c}}}$, $u_{\mathrm{cc}}(t) \in U_{\mathrm{cc}} \subseteq \mathbb{R}^{m_{\mathrm{cc}}}$, $u_{\mathrm{dc}}(t_k) \in U_{\mathrm{dc}} \subseteq \mathbb{R}^{m_{\mathrm{dc}}}$, $y_{\mathrm{cc}}(t) \in \mathbb{R}^{l_{\mathrm{cc}}}$, $y_{\mathrm{dc}}(t_k) \in \mathbb{R}^{l_{\mathrm{dc}}}$, $f_{\mathrm{cc}} : \mathbb{R}^{n_{\mathrm{c}}} \to \mathbb{R}^{n_{\mathrm{c}}}$ is Lipschitz continuous on $\mathbb{R}^{n_{\mathrm{c}}}$ and satisfies $f_{\mathrm{cc}}(0) = 0$, $G_{\mathrm{cc}} : \mathbb{R}^{m_{\mathrm{cc}}} \times \mathbb{R}^{n_{\mathrm{c}}} \to \mathbb{R}^{n_{\mathrm{c}} \times m_{\mathrm{cc}}}$, $f_{\mathrm{dc}} : \mathbb{R}^{n_{\mathrm{c}}} \to \mathbb{R}^{n_{\mathrm{c}}}$ is continuous on $\mathbb{R}^{n_{\mathrm{c}}}$, $G_{\mathrm{dc}} : \mathbb{R}^{m_{\mathrm{dc}}} \times \mathbb{R}^{n_{\mathrm{c}}} \to \mathbb{R}^{n_{\mathrm{c}} \times m_{\mathrm{dc}}}$, $J_{\mathrm{cc}} : \mathbb{R}^{m_{\mathrm{cc}}} \times \mathbb{R}^{n_{\mathrm{c}}} \to \mathbb{R}^{l_{\mathrm{cc}} \times m_{\mathrm{cc}}}$, $h_{\mathrm{cc}} : \mathbb{R}^{n_{\mathrm{c}}} \to \mathbb{R}^{l_{\mathrm{cc}}}$ and satisfies $h_{\mathrm{cc}}(0) = 0$, $J_{\mathrm{dc}} : \mathbb{R}^{m_{\mathrm{dc}}} \times \mathbb{R}^{n_{\mathrm{c}}} \to \mathbb{R}^{l_{\mathrm{dc}} \times m_{\mathrm{dc}}}$, $h_{\mathrm{dc}} : \mathbb{R}^{n_{\mathrm{c}}} \to \mathbb{R}^{l_{\mathrm{dc}}}$, $m_{\mathrm{cc}} = l_{\mathrm{c}}$, $m_{\mathrm{dc}} = l_{\mathrm{d}}$, $l_{\mathrm{cc}} = m_{\mathrm{c}}$, $l_{\mathrm{dc}} = m_{\mathrm{d}}$, and $\mathcal{Z}_{\mathrm{c}} \triangleq \mathcal{Z}_{\mathrm{c}x_{\mathrm{c}}} \times \mathcal{Z}_{\mathrm{c}u_{\mathrm{cc}}}$, where $\mathcal{Z}_{\mathrm{c}x_{\mathrm{c}}} \subset \mathbb{R}^{n_{\mathrm{c}}}$ and $\mathcal{Z}_{\mathrm{c}u_{\mathrm{cc}}} \subset U_{\mathrm{cc}}$, is such that Assumptions A1 and A2 of Chapter 2 hold. Note that with the feedback interconnection given by Figure 6.1, $(u_{\mathrm{cc}}, u_{\mathrm{dc}}) = (y_{\mathrm{c}}, y_{\mathrm{d}})$ and $(y_{\mathrm{cc}}, y_{\mathrm{dc}}) = (-u_{\mathrm{c}}, -u_{\mathrm{d}})$.

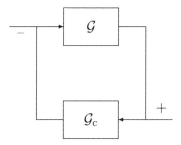

Figure 6.1 Feedback interconnection of $\mathcal{G}$ and $\mathcal{G}_c$.

Furthermore, note that, for generality, we allow the feedback system $\mathcal{G}_c$ to be of dimension n_c which may be less than the plant order n.

Even though the input-output pairs of the feedback interconnection shown on Figure 6.1 consist of two-vector inputs/two-vector outputs, at any given instant of time a single-vector input/single-vector output is active. Here, we assume that the negative feedback interconnection of $\mathcal{G}$ and $\mathcal{G}_c$ is well posed, that is, $\det[I_{m_c} + J_{cc}(y_c, x_c)J_c(x)] \neq 0$ and $\det[I_{m_d} + J_{dc}(y_d, x_c)J_d(x)] \neq 0$ for all y_c, y_d, x, and x_c. The following results give sufficient conditions for Lyapunov, asymptotic, and exponential stability of the negative feedback interconnection given by Figure 6.1. In contrast to Chapter 2, in this chapter we represent the resetting time $\tau_k(x_0)$ for a state-dependent impulsive dynamical system by t_k. This minor abuse of notation considerably simplifies the presentation.

For the results of this section we define the closed-loop resetting set $\tilde{\mathcal{Z}}_{\tilde{x}} \triangleq \mathcal{Z}_x \times \mathcal{Z}_{cx_c} \cup \{(x, x_c) : (\mathcal{F}_{cc}(x), \mathcal{F}_c(x_c)) \in \mathcal{Z}_{cu_{cc}} \times \mathcal{Z}_{u_c}\}$, where $\mathcal{F}_{cc}(\cdot)$ and $\mathcal{F}_c(\cdot)$ are functions of x and x_c arising from the algebraic loops due to u_{cc} and u_c, respectively. Note that since the feedback interconnection of $\mathcal{G}$ and $\mathcal{G}_c$ is well posed, it follows that $\tilde{\mathcal{Z}}_{\tilde{x}}$ is well defined and depends on the closed-loop states $\tilde{x} \triangleq [x^T\ x_c^T]^T$. In the special case where $J_c(x) \equiv 0$ and $J_{cc}(u_{cc}, x_c) \equiv 0$ it follows that $\tilde{\mathcal{Z}}_{\tilde{x}} = \mathcal{Z}_x \times \mathcal{Z}_{cx_c} \cup \{(x, x_c) : (h_c(x), h_{cc}(x_c)) \in \mathcal{Z}_{cu_{cc}} \times \mathcal{Z}_{u_c}\}$. Furthermore, note that in the case where $\mathcal{Z} = \emptyset$, that is, the plant is a continuous-time dynamical system without any resetting, it follows that $\tilde{\mathcal{Z}}_{\tilde{x}} = \mathcal{Z}_{cx_c} \cup \{(x, x_c) : h_c(x) \in \mathcal{Z}_{cu_{cc}}\}$, and hence, knowledge of x_c and y_c is sufficient to determine whether or not the closed-loop state vector is in the set $\tilde{\mathcal{Z}}_{\tilde{x}}$. Here we assume that the solution $s(t, \tilde{x}_0)$ of the dynamical system resulting from the feedback interconnection of $\mathcal{G}$ and $\mathcal{G}_c$ is such that Assumption 2.1 is satisfied. For the statement of the results of this section let $\mathcal{T}^{c}_{x_0,u_c}$ denote the set of resetting times of $\mathcal{G}$, let $\mathcal{T}_{x_0,u_c}$ denote the complement of $\mathcal{T}^{c}_{x_0,u_c}$, that is, $[0,\infty)\backslash\mathcal{T}^{c}_{x_0,u_c}$, let $\mathcal{T}^{c}_{x_{c0},u_{cc}}$ denote the set of resetting times of $\mathcal{G}_c$, and let $\mathcal{T}_{x_{c0},u_{cc}}$ denote the complement of $\mathcal{T}^{c}_{x_{c0},u_{cc}}$, that is, $[0,\infty)\backslash\mathcal{T}^{c}_{x_{c0},u_{cc}}$.

Theorem 6.1 *Consider the closed-loop system consisting of the non-linear impulsive dynamical systems $\mathcal{G}$ given by (6.1)–(6.4) and $\mathcal{G}_c$ given by (6.5)–(6.8) with input-output pairs $(u_c, u_d; y_c, y_d)$ and $(u_{cc}, u_{dc}; y_{cc}, y_{dc})$, respectively, and with $(u_{cc}, u_{dc}) = (y_c, y_d)$ and $(y_{cc}, y_{dc}) = (-u_c, -u_d)$. Assume $\mathcal{G}$ and $\mathcal{G}_c$ are zero-state observable, dissipative with respect to the hybrid supply rates $(s_c(u_c, y_c), s_d(u_d, y_d))$ and $(s_{cc}(u_{cc}, y_{cc}), s_{dc}(u_{dc}, y_{dc}))$, respectively, and with continuously differentiable positive definite, radially unbounded storage functions $V_s(\cdot)$ and $V_{sc}(\cdot)$, respectively, such that $V_s(0) = 0$ and $V_{sc}(0) = 0$. Furthermore, assume there exists a scalar $\sigma > 0$ such that $s_c(u_c, y_c) + \sigma s_{cc}(u_{cc}, y_{cc}) \le 0$ and $s_d(u_d, y_d) + \sigma s_{dc}(u_{dc}, y_{dc}) \le 0$. Then the following statements hold:*

i) *The negative feedback interconnection of $\mathcal{G}$ and $\mathcal{G}_c$ is Lyapunov stable.*

ii) *If $\mathcal{G}$ is strongly zero-state observable, $\mathcal{G}_c$ is exponentially dissipative with respect to the hybrid supply rate $(s_{cc}(u_{cc}, y_{cc}), s_{dc}(u_{dc}, y_{dc}))$, and $\mathrm{rank}[G_{cc}(u_{cc}, 0)] = m_{cc}$, $u_{cc} \in U_{cc}$, then the negative feedback interconnection of $\mathcal{G}$ and $\mathcal{G}_c$ is globally asymptotically stable.*

iii) *If $\mathcal{G}$ and $\mathcal{G}_c$ are exponentially dissipative with respect to the supply rates $(s_c(u_c, y_c), s_d(u_d, y_d))$ and $(s_{cc}(u_{cc}, y_{cc}), s_{dc}(u_{dc}, y_{dc}))$, respectively, and $V_s(\cdot)$ and $V_{sc}(\cdot)$ are such that there exist constants α, α_c, β, $\beta_c > 0$ such that*

$$\alpha\|x\|^2 \le V_s(x) \le \beta\|x\|^2, \quad x \in \mathbb{R}^n, \tag{6.9}$$
$$\alpha_c\|x_c\|^2 \le V_{sc}(x_c) \le \beta_c\|x_c\|^2, \quad x_c \in \mathbb{R}^{n_c}, \tag{6.10}$$

then the negative feedback interconnection of $\mathcal{G}$ and $\mathcal{G}_c$ is globally exponentially stable.

Proof. Let $\tilde{\mathcal{T}}^c \triangleq \mathcal{T}^{c}_{x_0,u_c} \cup \mathcal{T}^{c}_{x_{c0},u_{cc}}$ and $t_k \in \tilde{\mathcal{T}}^c$, $k \in \overline{\mathbb{Z}}_+$. First, note that it follows from Assumptions A1 and A2 of Chapter 2 that the resetting times $t_k (= \tau_k(\tilde{x}_0))$ for the feedback system are well defined and distinct for every closed-loop system trajectory.

i) Consider the Lyapunov function candidate $V(x, x_c) = V_s(x) + \sigma V_{sc}(x_c)$. Now, the corresponding Lyapunov derivative of $V(x, x_c)$ along the state trajectories $(x(t), x_c(t))$, $t \in (t_k, t_{k+1}]$, is given by

$$
\begin{aligned}
\dot{V}(x(t), x_c(t)) &= \dot{V}_s(x(t)) + \sigma \dot{V}_{sc}(x_c(t)) \\
&\le s_c(u_c(t), y_c(t)) + \sigma s_{cc}(u_{cc}(t), y_{cc}(t)) \\
&\le 0, \quad (x(t), x_c(t)) \notin \tilde{\mathcal{Z}}_{\tilde{x}},
\end{aligned} \tag{6.11}
$$

and the Lyapunov difference of $V(x, x_c)$ at the resetting times t_k, $k \in \overline{\mathbb{Z}}_+$, is given by

$$
\begin{aligned}
\Delta V(x(t_k), x_c(t_k)) &= \Delta V_s(x(t_k)) + \sigma \Delta V_{sc}(x_c(t_k) \\
&\le s_d(u_d(t_k), y_d(t_k)) + \sigma s_{dc}(u_{dc}(t_k), y_{dc}(t_k)) \\
&\le 0, \quad (x(t_k), x_c(t_k)) \in \tilde{\mathcal{Z}}_{\tilde{x}}.
\end{aligned} \tag{6.12}
$$

Now, Lyapunov stability of the negative feedback interconnection of $\mathcal{G}$ and $\mathcal{G}_c$ follows as a direct consequence of Theorem 2.1.

ii) Next, if $\mathcal{G}_c$ is exponentially dissipative it follows that for some scalar $\varepsilon_{cc} > 0$,

$$
\begin{aligned}
\dot{V}(x(t), x_c(t)) &= \dot{V}_s(x(t)) + \sigma \dot{V}_{sc}(x_c(t)) \\
&\le -\sigma \varepsilon_{cc} V_{sc}(x_c(t)) + s_c(u_c(t), y_c(t)) + \sigma s_{cc}(u_{cc}(t), y_{cc}(t)) \\
&\le -\sigma \varepsilon_{cc} V_{sc}(x_c(t)), \quad (x(t), x_c(t)) \notin \tilde{\mathcal{Z}}_{\tilde{x}}, \ t_k < t \le t_{k+1},
\end{aligned} \tag{6.13}
$$

and

$$
\begin{aligned}
\Delta V(x(t_k), x_c(t_k)) &= \Delta V_s(x(t_k)) + \sigma \Delta V_{sc}(x_c(t_k)) \\
&\le s_d(u_d(t_k), y_d(t_k)) + \sigma s_{dc}(u_{dc}(t_k), y_{dc}(t_k)) \\
&\le 0, \quad (x(t_k), x_c(t_k)) \in \tilde{\mathcal{Z}}_{\tilde{x}}, \ k \in \overline{\mathbb{Z}}_+.
\end{aligned} \tag{6.14}
$$

Let $\mathcal{R} \triangleq \{(x, x_c) \in \mathbb{R}^n \times \mathbb{R}^{n_c} : (x, x_c) \notin \tilde{\mathcal{Z}}_{\tilde{x}}, \dot{V}(x, x_c) = 0\} \cup \{(x, x_c) \in \mathbb{R}^n \times \mathbb{R}^{n_c} : (x, x_c) \in \tilde{\mathcal{Z}}_{\tilde{x}}, \Delta V(x, x_c) = 0\}$, where $\dot{V}(x, x_c)$ and $\Delta V(x, x_c)$ denote the total derivative and difference of $V(x, x_c)$ of the closed-loop system for all $(x, x_c) \notin \tilde{\mathcal{Z}}_{\tilde{x}}$ and $(x, x_c) \in \tilde{\mathcal{Z}}_{\tilde{x}}$, respectively. Since $V_{sc}(x_c)$ is positive definite, note that $\dot{V}(x, x_c) = 0$ for all $(x, x_c) \in \mathbb{R}^n \times \mathbb{R}^{n_c} \backslash \tilde{\mathcal{Z}}_{\tilde{x}}$ only if $x_c = 0$. Now, since rank$[G_{cc}(u_{cc}, 0)] = m_{cc}$, $u_{cc} \in U_{cc}$, it follows that on every invariant set $\mathcal{M}$ contained in $\mathcal{R}$, $u_{cc}(t) = y_c(t) \equiv 0$, and hence, $y_{cc}(t) \equiv -u_c(t) \equiv 0$ so that $\dot{x}(t) = f_c(x(t))$. Now, since $\mathcal{G}$ is strongly zero-state observable it follows that $\mathcal{R} = \{(0, 0)\} \cup \{(x, x_c) \in \mathbb{R}^n \times \mathbb{R}^{n_c} : (x, x_c) \in \tilde{\mathcal{Z}}_{\tilde{x}}, \Delta V(x, x_c) = 0\}$ contains no solution other than the trivial solution $(x(t), x_c(t)) \equiv (0, 0)$.

Hence, it follows from Theorem 2.3 that $(x(t), x_c(t)) \to \mathcal{M} = \{(0,0)\}$ as $t \to \infty$. Now, global asymptotic stability of the negative feedback interconnection of $\mathcal{G}$ and $\mathcal{G}_c$ follows from the fact that $V_s(\cdot)$ and $V_{sc}(\cdot)$ are, by assumption, radially unbounded.

iii) Finally, if $\mathcal{G}$ and $\mathcal{G}_c$ are exponentially dissipative and (6.9) and (6.10) hold, it follows that that for all $t \in (t_k, t_{k+1}]$,

$$
\begin{aligned}
\dot{V}(x(t), x_c(t)) &= \dot{V}_s(x(t)) + \sigma \dot{V}_{sc}(x_c(t)) \\
&\leq -\varepsilon_c V_s(x(t)) - \sigma \varepsilon_{cc} V_{sc}(x_c(t)) + s_c(u_c(t), y_c(t)) \\
&\quad + \sigma s_{cc}(u_{cc}(t), y_{cc}(t)) \\
&\leq -\min\{\varepsilon_c, \varepsilon_{cc}\} V(x(t), x_c(t)), \quad (x(t), x_c(t)) \notin \tilde{\mathcal{Z}}_{\tilde{x}},
\end{aligned}
\tag{6.15}
$$

and $\Delta V(x(t_k), x_c(t_k))$, $(x(t_k), x_c(t_k)) \in \tilde{\mathcal{Z}}_{\tilde{x}}$, $k \in \overline{\mathbb{Z}}_+$, satisfies (6.14). Now, Theorem 2.1 implies that the negative feedback interconnection of $\mathcal{G}$ and $\mathcal{G}_c$ is globally exponentially stable. $\qquad\square$

The next result presents Lyapunov, asymptotic, and exponential stability of dissipative feedback systems with quadratic supply rates.

Theorem 6.2 *Let* $Q_c \in \mathbb{S}^{l_c}$, $S_c \in \mathbb{R}^{l_c \times m_c}$, $R_c \in \mathbb{S}^{m_c}$, $Q_d \in \mathbb{S}^{l_d}$, $S_d \in \mathbb{R}^{l_d \times m_d}$, $R_d \in \mathbb{S}^{m_d}$, $Q_{cc} \in \mathbb{S}^{l_{cc}}$, $S_{cc} \in \mathbb{R}^{l_{cc} \times m_{cc}}$, $R_{cc} \in \mathbb{S}^{m_{cc}}$, $Q_{dc} \in \mathbb{S}^{l_{dc}}$, $S_{dc} \in \mathbb{R}^{l_{dc} \times m_{dc}}$, *and* $R_{dc} \in \mathbb{S}^{m_{dc}}$. *Consider the closed-loop system consisting of the nonlinear impulsive dynamical systems* $\mathcal{G}$ *given by (6.1)–(6.4) and* $\mathcal{G}_c$ *given by (6.5)–(6.8), and assume* $\mathcal{G}$ *and* $\mathcal{G}_c$ *are zero-state observable. Furthermore, assume* $\mathcal{G}$ *is dissipative with respect to the quadratic hybrid supply rate* $(s_c(u_c, y_c), s_d(u_d, y_d)) = (y_c^{\mathrm{T}} Q_c y_c + 2 y_c^{\mathrm{T}} S_c u_c + u_c^{\mathrm{T}} R_c u_c, y_d^{\mathrm{T}} Q_d y_d + 2 y_d^{\mathrm{T}} S_d u_d + u_d^{\mathrm{T}} R_d u_d)$ *and has a radially unbounded storage function* $V_s(\cdot)$, *and* $\mathcal{G}_c$ *is dissipative with respect to the quadratic hybrid supply rate* $(s_{cc}(u_{cc}, y_{cc}), s_{dc}(u_{dc}, y_{dc})) = (y_{cc}^{\mathrm{T}} Q_{cc} y_{cc} + 2 y_{cc}^{\mathrm{T}} S_{cc} u_{cc} + u_{cc}^{\mathrm{T}} R_{cc} u_{cc}, y_{dc}^{\mathrm{T}} Q_{dc} y_{dc} + 2 y_{dc}^{\mathrm{T}} S_{dc} u_{dc} + u_{dc}^{\mathrm{T}} R_{dc} u_{dc})$ *and has a radially unbounded storage function* $V_{sc}(\cdot)$. *Finally, assume there exists a scalar* $\sigma > 0$ *such that*

$$
\hat{Q}_c \triangleq \begin{bmatrix} Q_c + \sigma R_{cc} & -S_c + \sigma S_{cc}^{\mathrm{T}} \\ -S_c^{\mathrm{T}} + \sigma S_{cc} & R_c + \sigma Q_{cc} \end{bmatrix} \leq 0,
\tag{6.16}
$$

$$
\hat{Q}_d \triangleq \begin{bmatrix} Q_d + \sigma R_{dc} & -S_d + \sigma S_{dc}^{\mathrm{T}} \\ -S_d^{\mathrm{T}} + \sigma S_{dc} & R_d + \sigma Q_{dc} \end{bmatrix} \leq 0.
\tag{6.17}
$$

Then the following statements hold:

i) *The negative feedback interconnection of* $\mathcal{G}$ *and* $\mathcal{G}_c$ *is Lyapunov stable.*

ii) *If $\mathcal{G}$ is strongly zero-state observable, $\mathcal{G}_c$ is exponentially dissipative with respect to the hybrid supply rate $(s_{cc}(u_{cc}, y_{cc}), s_{dc}(u_{dc}, y_{dc}))$, and $\mathrm{rank}[G_{cc}(u_{cc}, 0)] = m_{cc}, u_{cc} \in U_{cc}$, then the negative feedback interconnection of $\mathcal{G}$ and $\mathcal{G}_c$ is globally asymptotically stable.*

iii) *If $\mathcal{G}$ and $\mathcal{G}_c$ are exponentially dissipative with respect to the supply rates $(s_c(u_c, y_c), s_d(u_d, y_d))$ and $(s_{cc}(u_{cc}, y_{cc}), s_{dc}(u_{dc}, y_{dc}))$, respectively, and there exist constants $\alpha, \alpha_c, \beta, \beta_c > 0$ such that (6.9) and (6.10) hold, then the negative feedback interconnection of $\mathcal{G}$ and $\mathcal{G}_c$ is globally exponentially stable.*

iv) *If $\hat{Q}_c < 0$ and $\hat{Q}_d < 0$, then the negative feedback interconnection of $\mathcal{G}$ and $\mathcal{G}_c$ is globally asymptotically stable.*

Proof. Statements i)–iii) are a direct consequence of Theorem 6.1 by noting

$$s_c(u_c, y_c) + \sigma s_{cc}(u_{cc}, y_{cc}) = \begin{bmatrix} y_c \\ y_{cc} \end{bmatrix}^T \hat{Q}_c \begin{bmatrix} y_c \\ y_{cc} \end{bmatrix}, \qquad (6.18)$$

$$s_d(u_d, y_d) + \sigma s_{dc}(u_{dc}, y_{dc}) = \begin{bmatrix} y_d \\ y_{dc} \end{bmatrix}^T \hat{Q}_d \begin{bmatrix} y_d \\ y_{dc} \end{bmatrix}, \qquad (6.19)$$

and hence, $s_c(u_c, y_c) + \sigma s_{cc}(u_{cc}, y_{cc}) \le 0$ and $s_d(u_d, y_d) + \sigma s_{dc}(u_{dc}, y_{dc}) \le 0$.

To show iv) consider the Lyapunov function candidate $V(x, x_c) = V_s(x) + \sigma V_{sc}(x)$. Noting that $u_{cc} = y_c$ and $y_{cc} = -u_c$, it follows that the corresponding Lyapunov derivative satisfies

$$\begin{aligned}
\dot{V}(x(t), x_c(t)) &= \dot{V}_s(x(t)) + \sigma \dot{V}_{sc}(x_c(t)) \\
&\le s_c(u_c(t), y_c(t)) + \sigma s_{cc}(u_{cc}(t), y_{cc}(t)) \\
&= y_c^T(t) Q_c y_c(t) + 2 y_c^T(t) S_c u_c(t) + u_c^T(t) R_c u_c(t) \\
&\quad + \sigma [y_{cc}^T(t) Q_{cc} y_{cc}(t) + 2 y_{cc}^T(t) S_{cc} u_{cc}(t) \\
&\quad + u_{cc}^T(t) R_{cc} u_{cc}(t)] \\
&= \begin{bmatrix} y_c(t) \\ y_{cc}(t) \end{bmatrix}^T \hat{Q}_c \begin{bmatrix} y_c(t) \\ y_{cc}(t) \end{bmatrix} \\
&\le 0, \qquad (x(t), x_c(t)) \notin \tilde{Z}_{\tilde{x}}, \quad t_k < t \le t_{k+1}, \quad (6.20)
\end{aligned}$$

and, similarly, the Lyapunov difference satisfies

$$\Delta V(x(t_k), x_c(t_k)) = \begin{bmatrix} y_d(t_k) \\ y_{dc}(t_k) \end{bmatrix}^T \hat{Q}_d \begin{bmatrix} y_d(t_k) \\ y_{dc}(t_k) \end{bmatrix}$$

$$\le 0, \quad (x(t_k), x_c(t_k)) \in \tilde{Z}_{\tilde{x}}, \quad k \in \overline{\mathbb{Z}}_+, \qquad (6.21)$$

which implies that the negative feedback interconnection of $\mathcal{G}$ and $\mathcal{G}_c$ is Lyapunov stable.

Next, let $\mathcal{R} \triangleq \{(x, x_c) \in \mathbb{R}^n \times \mathbb{R}^{n_c} : (x, x_c) \notin \tilde{\mathcal{Z}}_{\tilde{x}}, \dot{V}(x, x_c) = 0\} \cup \{(x, x_c) \in \mathbb{R}^n \times \mathbb{R}^{n_c} : (x, x_c) \in \tilde{\mathcal{Z}}_{\tilde{x}}, \Delta V(x, x_c) = 0\}$, where $\dot{V}(x, x_c)$ and $\Delta V(x, x_c)$ denote the total derivative and difference of $V(x, x_c)$ of the closed-loop system for all $(x, x_c) \notin \tilde{\mathcal{Z}}_{\tilde{x}}$ and $(x, x_c) \in \tilde{\mathcal{Z}}_{\tilde{x}}$, respectively. Now, note that $\dot{V}(x, x_c) = 0$ for all $(x, x_c) \in \mathbb{R}^n \times \mathbb{R}^{n_c} \backslash \tilde{\mathcal{Z}}_{\tilde{x}}$ if and only if $(y_c, y_{cc}) = (0, 0)$ and $\Delta V(x, x_c) = 0$ for all $(x, x_c) \in \tilde{\mathcal{Z}}_{\tilde{x}}$ if and only if $(y_d, y_{dc}) = (0, 0)$. Since $\mathcal{G}$ and $\mathcal{G}_c$ are zero-state observable it follows that $\mathcal{R} \triangleq \{(x, x_c) \in \mathbb{R}^n \times \mathbb{R}^{n_c} : (x, x_c) \notin \tilde{\mathcal{Z}}_{\tilde{x}}, (h_c(x), h_{cc}(x_c)) = (0, 0)\} \cup \{(x, x_c) \in \tilde{\mathcal{Z}}_{\tilde{x}}, (h_d(x), h_{dc}(x_c)) = (0, 0)\}$ which contains no solution other than the trivial solution $(x(t), x_c(t)) \equiv (0, 0)$. Hence, it follows from Theorem 2.3 that $(x(t), x_c(t)) \rightarrow \mathcal{M} = \{(0, 0)\}$ as $t \rightarrow \infty$. Finally, global asymptotic stability follows from the fact that $V_s(\cdot)$ and $V_{sc}(\cdot)$ are, by assumption, radially unbounded. $\square$

The following result generalizes the classical positivity and small-gain theorems to the case of impulsive systems. For this result note that if a nonlinear dynamical system $\mathcal{G}$ is dissipative (respectively, exponentially dissipative) with respect to the hybrid supply rate $(s_c(u_c, y_c), s_d(u_d, y_d)) = (2u_c^T y_c, 2u_d^T y_d)$, then, with $(\kappa_c(y_c), \kappa_d(y_d)) = (-k_c y_c, -k_d y_d)$, where $k_c, k_d > 0$, it follows that $(s_c(u_c, y_c), s_d(u_d, y_d)) = (-2k_c y_c^T y_c, -2k_d y_d^T y_d) < (0, 0)$, $(y_c, y_d) \neq (0, 0)$. Alternatively, if a nonlinear dynamical system $\mathcal{G}$ is dissipative (respectively, exponentially dissipative) with respect to the hybrid supply rate $(s_c(u_c, y_c), s_d(u_d, y_d)) = (\gamma_c^2 u_c^T u_c - y_c^T y_c, \gamma_d^2 u_d^T u_d - y_d^T y_d)$, where $\gamma_c, \gamma_d > 0$, then, with $(\kappa_c(y_c), \kappa_d(y_d)) = (0, 0)$, it follows that $(s_c(u_c, y_c), s_d(u_d, y_d)) = (-y_c^T y_c, -y_d^T y_d) < (0, 0)$, $(y_c, y_d) \neq (0, 0)$. Hence, if $\mathcal{G}$ is zero-state observable it follows from Theorem 3.3 that all storage functions of $\mathcal{G}$ are positive definite.

Corollary 6.1 *Consider the closed-loop system consisting of the non-linear impulsive dynamical systems $\mathcal{G}$ given by (6.1)–(6.4) and $\mathcal{G}_c$ given by (6.5)–(6.8). Assume $\mathcal{G}$ and $\mathcal{G}_c$ are zero-state observable. Then the following statements hold:*

   i) If $\mathcal{G}$ is passive and strongly zero-state observable, $\mathcal{G}_c$ is exponentially passive, and $\text{rank}[G_{cc}(u_{cc}, 0)] = m_{cc}$, $u_{cc} \in U_{cc}$, then the negative feedback interconnection of $\mathcal{G}$ and $\mathcal{G}_c$ is asymptotically stable.

   ii) If $\mathcal{G}$ and $\mathcal{G}_c$ are exponentially passive with storage functions $V_s(\cdot)$ and $V_{sc}(\cdot)$, respectively, such that (6.9) and (6.10) hold, then the *negative feedback interconnection of $\mathcal{G}$ and $\mathcal{G}_{\mathrm{c}}$ is exponentially stable.*

iii) If $\mathcal{G}$ is nonexpansive with gains $\gamma_{\mathrm{c}}, \gamma_{\mathrm{d}} > 0$ and strongly zero-state observable, $\mathcal{G}_{\mathrm{c}}$ is exponentially nonexpansive with gains $\gamma_{\mathrm{cc}} > 0$, $\gamma_{\mathrm{dc}} > 0$, $\mathrm{rank}[G_{\mathrm{cc}}(u_{\mathrm{cc}}, 0)] = m_{\mathrm{cc}}$, $u_{\mathrm{cc}} \in U_{\mathrm{cc}}$, $\gamma_{\mathrm{c}}\gamma_{\mathrm{cc}} \leq 1$, and $\gamma_{\mathrm{d}}\gamma_{\mathrm{dc}} \leq 1$, then the negative feedback interconnection of $\mathcal{G}$ and $\mathcal{G}_{\mathrm{c}}$ is asymptotically stable.

iv) If $\mathcal{G}$ and $\mathcal{G}_{\mathrm{c}}$ are exponentially nonexpansive with storage functions $V_{\mathrm{s}}(\cdot)$ and $V_{\mathrm{sc}}(\cdot)$, respectively, such that (6.9) and (6.10) hold, and with gains $\gamma_{\mathrm{c}}, \gamma_{\mathrm{d}} > 0$ and $\gamma_{\mathrm{cc}}, \gamma_{\mathrm{dc}} > 0$, respectively, such that $\gamma_{\mathrm{c}}\gamma_{\mathrm{cc}} \leq 1$ and $\gamma_{\mathrm{d}}\gamma_{\mathrm{dc}} \leq 1$, then the negative feedback interconnection of $\mathcal{G}$ and $\mathcal{G}_{\mathrm{c}}$ is exponentially stable.

Proof. The proof is a direct consequence of Theorem 6.2. Specifically, statements i) and ii) follow from Theorem 6.2 with $Q_{\mathrm{c}} = 0$, $Q_{\mathrm{d}} = 0$, $Q_{\mathrm{cc}} = 0$, $Q_{\mathrm{dc}} = 0$, $S_{\mathrm{c}} = I_{m_{\mathrm{c}}}$, $S_{\mathrm{d}} = I_{m_{\mathrm{d}}}$, $S_{\mathrm{cc}} = I_{m_{\mathrm{cc}}}$, $S_{\mathrm{dc}} = I_{m_{\mathrm{dc}}}$, $R_{\mathrm{c}} = 0$, $R_{\mathrm{d}} = 0$, $R_{\mathrm{cc}} = 0$, and $R_{\mathrm{dc}} = 0$. Statements iii) and iv) follow from Theorem 6.2 with $Q_{\mathrm{c}} = -I_{l_{\mathrm{c}}}$, $Q_{\mathrm{d}} = -I_{l_{\mathrm{d}}}$, $Q_{\mathrm{cc}} = -I_{l_{\mathrm{cc}}}$, $Q_{\mathrm{dc}} = -I_{l_{\mathrm{dc}}}$, $S_{\mathrm{c}} = 0$, $S_{\mathrm{d}} = 0$, $S_{\mathrm{cc}} = 0$, $S_{\mathrm{dc}} = 0$, and $R_{\mathrm{c}} = \gamma_{\mathrm{c}}^2 I_{m_{\mathrm{c}}}$, $R_{\mathrm{d}} = \gamma_{\mathrm{d}}^2 I_{m_{\mathrm{d}}}$, $R_{\mathrm{cc}} = \gamma_{\mathrm{cc}}^2 I_{m_{\mathrm{cc}}}$, and $R_{\mathrm{dc}} = \gamma_{\mathrm{dc}}^2 I_{m_{\mathrm{dc}}}$. $\square$

Global asymptotic stability of the negative feedback interconnection of $\mathcal{G}$ and $\mathcal{G}_{\mathrm{c}}$ is also guaranteed if the nonlinear impulsive system $\mathcal{G}$ is input strict passive (respectively, output strict passive) and the nonlinear impulsive compensator $\mathcal{G}_{\mathrm{c}}$ is input strict passive (respectively, output strict passive) [74].

6.3 Hybrid Controllers for Combustion Systems

In this section, we apply the concepts developed in Section 6.2 to the control of thermoacoustic instabilities in combustion processes. Engineering applications involving steam and gas turbines and jet and ramjet engines for power generation and propulsion technology involve combustion processes. Due to the inherent coupling between several intricate physical phenomena in these processes involving acoustics, thermodynamics, fluid mechanics, and chemical kinetics, the dynamic behavior of combustion systems is characterized by highly complex nonlinear models [9, 10, 43, 86]. The unstable dynamic coupling between heat release in combustion processes generated by reacting mixtures releasing chemical energy and unsteady motions in the combustor develop acoustic pressure and velocity oscillations which can severely impact operating conditions and system performance. These pressure oscillations, known as *thermoacoustic instabilities*, often lead to high vibration levels causing mechanical failures, high levels of acoustic noise, high burn rates, and even component melting. Hence, the need for active control to mitigate combustion-induced pressure instabilities is critical.

To design hybrid controllers for combustion systems we concentrate on a two-mode, nonlinear time-averaged combustion model with nonlinearities present due to the second-order gas dynamics. This model is developed in [43] and is given by

$$\dot{x}_1(t) = \alpha_1 x_1(t) + \theta_1 x_2(t) - \beta(x_1(t)x_3(t) + x_2(t)x_4(t)) + u_{s1}(t),$$
$$x_1(0) = x_{10}, \quad (6.22)$$
$$\dot{x}_2(t) = -\theta_1 x_1(t) + \alpha_1 x_2(t) + \beta(x_2(t)x_3(t) - x_1(t)x_4(t)) + u_{s2}(t),$$
$$x_2(0) = x_{20}, \quad (6.23)$$
$$\dot{x}_3(t) = \alpha_2 x_3(t) + \theta_2 x_4(t) + \beta(x_1^2(t) - x_2^2(t)) + u_{s3}(t), \quad x_3(0) = x_{30},$$
$$(6.24)$$
$$\dot{x}_4(t) = -\theta_2 x_3(t) + \alpha_2 x_4(t) + 2\beta x_1(t)x_2(t) + u_{s4}(t), \quad x_4(0) = x_{40},$$
$$(6.25)$$

where $\alpha_1, \alpha_2 \in \mathbb{R}$ represent growth/decay constants, $\theta_1, \theta_2 \in \mathbb{R}$ represent frequency shift constants, $\beta = ((\gamma + 1)/8\gamma)\omega_1$, where γ denotes the ratio of specific heats, ω_1 is the frequency of the fundamental mode, and u_{si}, $i = 1, \ldots, 4$, are control input signals. For the data parameters $\alpha_1 = 5$, $\alpha_2 = -55$, $\theta_1 = 4$, $\theta_2 = 32$, $\gamma = 1.4$, $\omega_1 = 1$, and $x_0 = [1\ 1\ 1\ 1]^{\mathrm{T}}$, the open-loop (i.e., $u_{si}(t) \equiv 0, i = 1, \ldots, 4$) dynamics (6.22)–(6.25) result in a limit cycle instability. Figures 6.2, 6.3, and 6.4 show, respectively, the phase portrait, state response, and *plant energy*

$$V_s(x) \triangleq x_1^2 + x_2^2 + x_3^2 + x_4^2 \qquad (6.26)$$

versus time.

To design a stabilizing time-dependent hybrid controller for (6.22)–(6.25) we first design a continuous-time control law

$$u_s = -K_s x + u_c, \qquad (6.27)$$

where $K_s \triangleq \mathrm{diag}[k_{s1}, k_{s2}, k_{s3}, k_{s4}]$, $x \triangleq [x_1, x_2, x_3, x_4]^{\mathrm{T}}$, $u_s \triangleq [u_{s1}, u_{s2}, u_{s3}, u_{s4}]^{\mathrm{T}}$, and $u_c \triangleq [u_{c1}, u_{c2}, u_{c3}, u_{c4}]^{\mathrm{T}}$. In this case, (6.22)–

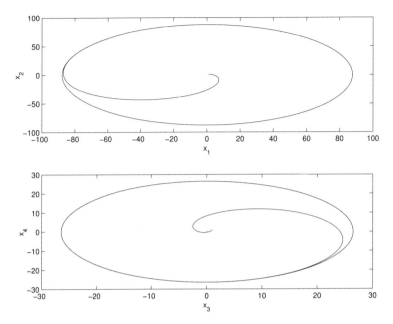

Figure 6.2 Phase portrait.

(6.25) are given by (6.1) and (6.2) with $\mathcal{Z} = \emptyset$ and

$$f_c(x) = \begin{bmatrix} \alpha_1 x_1 + \theta_1 x_2 - \beta(x_1 x_3 + x_2 x_4) - k_{s1} x_1 \\ -\theta_1 x_1 + \alpha_1 x_2 + \beta(x_2 x_3 - x_1 x_4) - k_{s2} x_2 \\ \alpha_2 x_3 + \theta_2 x_4 + \beta(x_1^2 - x_2^2) - k_{s3} x_3 \\ -\theta_2 x_3 + \alpha_2 x_4 + 2\beta x_1 x_2 - k_{s4} x_4 \end{bmatrix}, \quad G_c(x) = I_4,$$

$$(6.28)$$

$$f_d(x) = 0, \qquad G_d(x) = 0. \tag{6.29}$$

Now, with $y_c = x$, $k_{s1} = k_{s2} = \alpha_1$, and $k_{s3} = k_{s4} = 0$, it follows that (6.1) and (6.3), with $f_c(x)$ and $G_c(x)$ given by (6.28) and $h_c(x) = x$ and $J_c(x) = 0$, is passive with input u_c, output y_c, and plant energy function, or storage function, $V_s(x)$. Hence, $V_s'(x) f_c(x) \leq 0$, $x \in \mathbb{R}^4$. Furthermore, (6.1) and (6.3), with $f_c(x)$ and $G_c(x)$ given by (6.28) and $h_c(x) = x$ and $J_c(x) = 0$, is zero-state observable. Figures 6.5, 6.6, and 6.7 show, respectively, the phase portrait, state response, and plant energy of the controlled system (6.1) and (6.3) with $u_s = -K_s x + u_c$ and $u_c \equiv 0$.

To improve the performance of the above controller, we use the flexibility in u_c to design a hybrid controller. Specifically, consider the hybrid controller emulating the plant structure given by (6.5)–

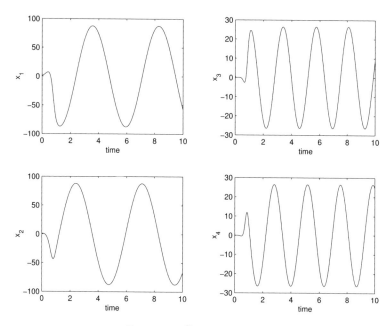

Figure 6.3 State response.

(6.8) with $\mathcal{S}_c = \mathcal{T} \times \mathbb{R}^{n_c} \times \mathbb{R}^{m_{cc}}$,

$$f_{cc}(x_c) = \begin{bmatrix} \alpha_1 x_{c1} + \theta_1 x_{c2} - \beta(x_{c1}x_{c3} + x_{c2}x_{c4}) - k_{c1}x_{c1} \\ -\theta_1 x_{c1} + \alpha_1 x_{c2} + \beta(x_{c2}x_{c3} - x_{c1}x_{c4}) - k_{c2}x_{c2} \\ \alpha_2 x_{c3} + \theta_2 x_{c4} + \beta(x_{c1}^2 - x_{c2}^2) - k_{c3}x_{c3} \\ -\theta_2 x_{c3} + \alpha_2 x_{c4} + 2\beta x_{c1}x_{c2} - k_{c4}x_{c4} \end{bmatrix}, (6.30)$$

$$G_{cc}(x_c) = I_4, \tag{6.31}$$

$$f_{dc}(x_c) = \begin{bmatrix} -x_{c1} \\ -x_{c2} \\ -x_{c3} \\ -x_{c4} \end{bmatrix}, \qquad G_{dc}(x_c) = 0, \tag{6.32}$$

$$h_{cc}(x_c) = -[x_{c1},\, x_{c2},\, x_{c3},\, x_{c4}]^{\mathrm{T}}, \tag{6.33}$$

$$J_{cc}(x_c) = 0,\ h_{dc}(x_c) = 0,\ J_{dc}(x_c) = 0, \tag{6.34}$$

where $k_{c1} > \alpha_1$, $k_{c2} > \alpha_1$, $k_{c3} > \alpha_2$, and $k_{c4} > \alpha_2$. It can be easily shown using the results of Chapter 3 that the hybrid controller (6.5)–(6.8) with dynamics given by (6.30)–(6.34), resetting set $\mathcal{S}_c =$

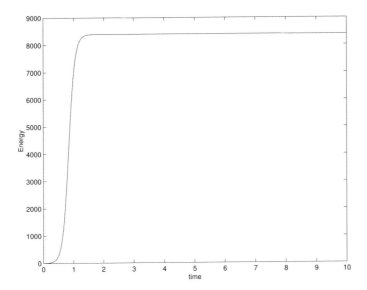

Figure 6.4 Plant energy versus time.

$\mathcal{T} \times \mathbb{R}^{n_c} \times \mathbb{R}^{m_{cc}}$, input y_c, and output $-u_c$ is exponentially passive with *controller energy*, or storage function, $V_{sc}(x_c) \triangleq x_{c1}^2 + x_{c2}^2 + x_{c3}^2 + x_{c4}^2$. Hence, $V_{sc}'(x_c) f_{cc}(x_c) \leq -\varepsilon V_{sc}(x_c)$, $x_c \in \mathbb{R}^4$, where $\varepsilon = \min\{\alpha_1 - k_{c1}, \alpha_1 - k_{c2}, \alpha_2 - k_{c3}, \alpha_2 - k_{c4}\}$. Furthermore, note that rank $[G_{cc}(0)] = 4$. Hence, stability of the closed-loop system (6.1), (6.3), and (6.5)–(6.8) is guaranteed by Theorem 6.1. Finally, we note that the *total energy* of the closed-loop system (6.1), (6.3), and (6.5)–(6.8) is given by

$$V(\tilde{x}) \triangleq V_s(x) + V_{sc}(x_c) = x_1^2 + x_2^2 + x_3^2 + x_4^2 + x_{c1}^2 + x_{c2}^2 + x_{c3}^2 + x_{c4}^2, \tag{6.35}$$

where $\tilde{x} \triangleq [x^{\mathrm{T}} \ x_c^{\mathrm{T}}]^{\mathrm{T}}$.

The effect of the resetting law (6.6) with $f_{dc}(x_c)$ and $G_{dc}(x_c)$ given by (6.32), is to cause all the controller states to be instantaneously reset to zero, that is, the resetting law (6.6) implies $V_{sc}(x_c + \Delta x_c) = 0$. The closed-loop resetting law is thus given by

$$\Delta \tilde{x} = \begin{bmatrix} 0 & 0 & 0 & 0 & -x_{c1} & -x_{c2} & -x_{c3} & -x_{c4} \end{bmatrix}^{\mathrm{T}}. \tag{6.36}$$

Note that since

$$\tilde{x} + \Delta \tilde{x} = \begin{bmatrix} x_1 & x_2 & x_3 & x_4 & 0 & 0 & 0 & 0 \end{bmatrix}^{\mathrm{T}}, \tag{6.37}$$

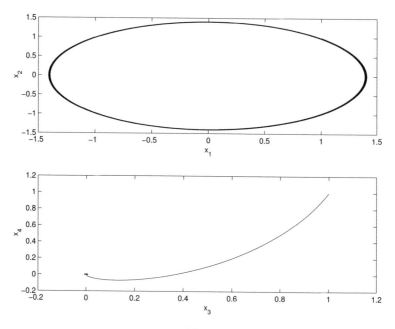

Figure 6.5 Phase portrait.

it follows that

$$V(\tilde{x} + \Delta\tilde{x}) = V_s(x) \tag{6.38}$$

and

$$V(\tilde{x} + \Delta\tilde{x}) - V(\tilde{x}) = -V_{sc}(x_c) \le 0. \tag{6.39}$$

Now, from (6.39) it follows that the resetting law (6.6) causes the total energy to instantaneously decrease by an amount equal to the accumulated controller energy.

To illustrate the dynamic behavior of the closed-loop system, let $\alpha_1 = 5$, $\alpha_2 = -55$, $k_{s1} = \alpha_1$, $k_{s2} = \alpha_1$, $k_{s3} = 0$, $k_{s4} = 0$, $k_{c1} = \alpha_1 + 0.1$, $k_{c2} = \alpha_1 + 0.1$, $k_{c3} = 0$, $k_{c4} = 0$, and $\mathcal{T} = \{2, 4, 6, \ldots\}$, so that the controller resets periodically with a period of 2 seconds. The response of the controlled system (6.1) and (6.3) with the resetting controller (6.5)–(6.8) and initial condition $x_0 = [1\ 1\ 1\ 1\ 0\ 0\ 0\ 0]^{\mathrm{T}}$ is shown in Figure 6.8. Note that the control force versus time is discontinuous at the resetting times. A comparison of the plant energy, control energy, and total energy is given in Figure 6.9.

In this example the resetting times were chosen arbitrarily. However, with the same choice of controller parameters we can choose a resetting time to achieve finite-time stabilization. Specifically, this

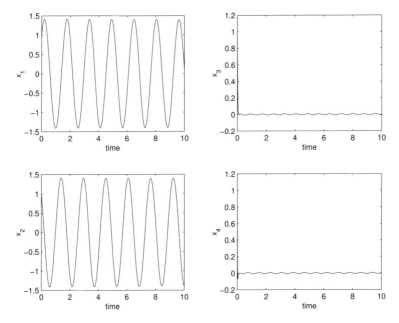

Figure 6.6 State response.

resetting time will correspond to the time at which all of the energy of the plant is drawn to the controller. This resetting time can be obtained from the energy history of the closed-loop system without resetting. In particular, the time instant when the plant and controller energy interchange is such that the plant energy is at zero corresponds to the resetting time that achieves finite-time stabilization. For this example, finite-time stability is achieved by choosing the resetting instant at $t = 1.6223$ sec.

Next, we describe the mathematical setting and design of an input/state-dependent resetting controller. We consider the plant and resetting controller as given above with $\mathcal{S}_c = [0, \infty) \times \mathcal{Z}_{cx_c} \times \mathcal{Z}_{cu_{cc}}$, where

$$\mathcal{Z}_{cx_c} \times \mathcal{Z}_{cu_{cc}} = \{(x_c, u_{cc}) : f_{dc}(x_c) \neq 0 \text{ and } V'_{sc}(x_c)[f_{cc}(x_c) + G_{cc}(x_c)u_{cc}] \leq 0\}. \tag{6.40}$$

The resetting set (6.40) is thus defined to be the set of all controller states and input points that represent nonincreasing control energy, except for those points that satisfy $f_{dc}(x_c) = 0$. The states x_c that satisfy $f_{dc}(x_c) = 0$ are states that do not change under the action of the resetting law, and hence, we need to exclude these states from

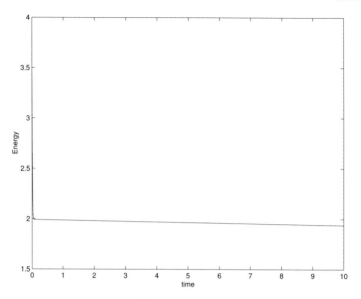

Figure 6.7 Plant energy versus time.

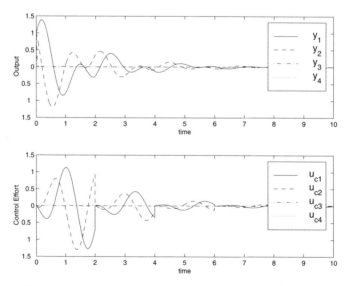

Figure 6.8 Time-dependent resetting controller: Output and control effort versus
time.

the resetting set to ensure that Assumption A1 of Chapter 2 is not
violated.

For the four-state, time-averaged combustion system given by (6.1)
and (6.3) with $\mathcal{Z} = \emptyset$, dynamics (6.28), and output $h_c(x) = x$, the

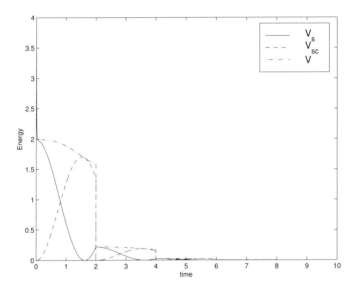

Figure 6.9 Time-dependent resetting controller: Plant, controller, and total energy.

input/state-dependent resetting set (6.40) becomes

$$
\begin{aligned}
\mathcal{Z}_{cx_c} \times \mathcal{Z}_{u_{cc}} = \{(x_c, u_{cc}) : \ & f_{dc}(x_c) \neq 0 \text{ and } 2u_{cc1}(\alpha_1 x_{c1} + \theta_1 x_{c2} \\
& - \beta(x_{c1} x_{c3} + x_{c2} x_{c4}) - k_{c1} x_{c1} + u_{cc1}) \\
& + 2u_{cc2}(-\theta_1 x_{c1} + \alpha_1 x_{c2} + \beta(x_{c2} x_{c3} - x_{c1} x_{c4}) - k_{c2} x_{c2} \\
& + u_{cc2}) + 2u_{cc3}(\alpha_2 x_{c3} + \theta_2 x_{c4} + \beta(x_{c1}^2 - x_{c2}^2) \\
& - k_{c3} x_{c3} + u_{cc3}) + 2u_{cc4}(-\theta_2 x_{c3} + \alpha_2 x_{c4} + 2\beta x_{c1} x_{c2} \\
& - k_{c4} x_{c4} + u_{cc4})) \leq 0\},
\end{aligned}
\tag{6.41}
$$

where u_{cci}, $i = 1, \ldots, 4$, represents the ith component of u_{cc}. Now, it can be shown that Assumptions A1 and A2 of Chapter 2 are satisfied using straightforward calculations. Furthermore, since the resetting controller given by (6.31) and (6.32) is exponentially passive for $\mathcal{S}_c = [0, \infty) \times \mathbb{R}^{n_c} \times \mathbb{R}^{m_{cc}}$, it follows that the resetting controller is exponentially passive for $\mathcal{S}_c = [0, \infty) \times \mathcal{Z}_{cx_c} \times \mathcal{Z}_{cu_{cc}}$. Hence, asymptotic stability of the closed-loop system (6.1), (6.3), and (6.5)–(6.8) is guaranteed by Theorem 6.1. Finally, note that knowledge of x_c and y_c is sufficient to determine whether or not the closed-loop state vector $\tilde{x}$ is in the resetting set $\tilde{\mathcal{Z}}_{\tilde{x}}$, where

$$
\begin{aligned}
\tilde{\mathcal{Z}}_{\tilde{x}} = \mathcal{Z}_{cx_c} \cup \{(x, x_c) : h_c(x) \in U_{cc}\} = \{\tilde{x} : \ & f_{dc}(x_c) \neq 0 \text{ and} \\
& V_{sc}'(x_c)[f_{cc}(x_c) + G_{cc}(x_c)h_c(x)] \leq 0\}.
\end{aligned}
\tag{6.42}
$$

To illustrate the dynamics behavior of the closed-loop system we again choose $\alpha_1 = 5$, $\alpha_2 = -55$, $k_{s1} = \alpha_1$, $k_{s2} = \alpha_1$, $k_{s3} = 0$, $k_{s4} = 0$, $k_{c1} = \alpha_1 + 0.1$, $k_{c2} = \alpha_1 + 0.1$, $k_{c3} = 0$, and $k_{c4} = 0$, with initial condition $x_0 = [1\ 1\ 1\ 1\ 0\ 0\ 0\ 0]^{\mathrm{T}}$. The response of controlled system (6.1) and (6.3), with dynamics (6.28) and $h_c(x) = x$, and the state-dependent resetting controller given by (6.5)–(6.8), with dynamics (6.31)–(6.34) and resetting set (6.41), is given in Figure 6.10. The total energy, plant energy, and controller energy versus time are shown in Figure 6.11. Note that the proposed input/state-dependent resetting controller achieves finite-time stabilization.

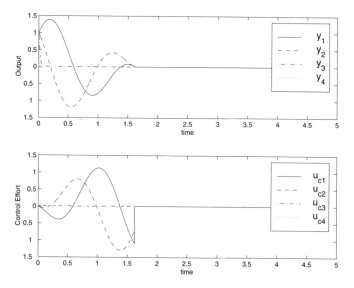

Figure 6.10 State-dependent resetting controller: Output and control effort versus time.

6.4 Feedback Interconnections of Nonlinear Impulsive Nonnegative Dynamical Systems

In this section, we consider stability of feedback interconnections of impulsive nonnegative dynamical systems. Specifically, using concepts of dissipativity and exponential dissipativity for impulsive nonnegative dynamical systems, we develop feedback interconnection stability results for nonlinear nonnegative impulsive dynamical systems. In particular, general stability criteria are given for Lyapunov and asymptotic stability of feedback interconnections of impulsive nonnegative systems. These results can be viewed as a generalization of the

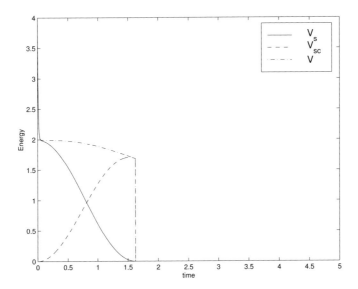

Figure 6.11 State-dependent resetting controller: Plant, controller, and total energy.

positivity and the small gain theorems [74] to impulsive nonnegative systems with linear supply rates involving net input-output system flux.

We begin by considering the nonlinear impulsive hybrid dynamical system $\mathcal{G}$ given by (6.1)–(6.4) where $f_c(\cdot)$ is essentially nonnegative, $f_d(\cdot)$ is such that $x + f_d(x)$ is nonnegative for all $x \in \overline{\mathbb{R}}^n_+$, and $G_c(\cdot)$, $G_d(\cdot)$, $h_c(\cdot)$, $h_d(\cdot)$, $J_c(\cdot)$, and $J_d(\cdot)$ are nonnegative functions. Furthermore, consider the nonlinear impulsive nonnegative feedback system $\mathcal{G}_c$ given by

$$\dot{x}_c(t) = f_{cc}(x_c(t)) + G_{cc}(x_c(t))u_{cc}(t), \quad x_c(0) = x_{c0},$$
$$(x_c(t), u_{cc}(t)) \notin \mathcal{Z}_c, \quad (6.43)$$
$$\Delta x_c(t) = f_{dc}(x_c(t)) + G_{dc}(x_c(t))u_{dc}(t), \quad (x_c(t), u_{cc}(t)) \in \mathcal{Z}_c, \quad (6.44)$$
$$y_{cc}(t) = h_{cc}(x_c(t)), \quad (x_c(t), u_{cc}(t)) \notin \mathcal{Z}_c, \quad (6.45)$$
$$y_{dc}(t) = h_{dc}(x_c(t)), \quad (x_c(t), u_{cc}(t)) \in \mathcal{Z}_c, \quad (6.46)$$

where $t \geq 0$, $x_c(t) \in \overline{\mathbb{R}}^{n_c}_+$, $\Delta x_c(t) \triangleq x_c(t^+) - x_c(t)$, $u_{cc}(t) \in U_{cc} \subseteq \overline{\mathbb{R}}^{m_{cc}}_+$, $u_{dc}(t_k) \in U_{dc} \subseteq \overline{\mathbb{R}}^{m_{dc}}_+$, t_k denotes the kth instant of time at which $(x_c(t), u_{cc}(t))$ intersects $\mathcal{Z}_c \subset \overline{\mathbb{R}}^{n_c}_+ \times U_{cc}$ for a particular trajectory $x_c(t)$ and input $u_{cc}(t)$, $y_{cc}(t) \in Y_{cc} \subseteq \overline{\mathbb{R}}^{l_{cc}}_+$, $y_{dc}(t_k) \in Y_{dc} \subseteq \overline{\mathbb{R}}^{l_{dc}}_+$, $f_{cc} : \mathbb{R}^{n_c} \to \mathbb{R}^{n_c}$ is Lipschitz continuous and is essentially nonnegative, $G_{cc} : \mathbb{R}^{n_c} \to \mathbb{R}^{n_c \times m_{cc}}$ and satisfies $G_{cc}(x_c) \geq\geq 0$, $x_c \in \overline{\mathbb{R}}^{n_c}_+$,

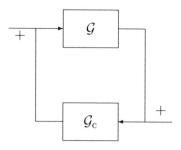

Figure 6.12 Feedback interconnection of $\mathcal{G}$ and $\mathcal{G}_c$.

$f_{\mathrm{dc}} : \mathbb{R}^{n_c} \to \mathbb{R}^{n_c}$ is continuous and is such that $x_c + f_{\mathrm{dc}}(x_c)$ is nonnegative for all $x_c \in \overline{\mathbb{R}}_+^{n_c}$, $G_{\mathrm{dc}} : \mathbb{R}^{n_c} \to \mathbb{R}^{n_c \times m_{\mathrm{dc}}}$ and satisfies $G_{\mathrm{dc}}(x_c) \geq\geq 0$, $x_c \in \overline{\mathbb{R}}_+^{n_c}$, $h_{\mathrm{cc}} : \mathbb{R}^{n_c} \to \mathbb{R}^{l_{cc}}$ and satisfies $h_{\mathrm{cc}}(x_c) \geq\geq 0$, $x_c \in \overline{\mathbb{R}}_+^{n_c}$, $h_{\mathrm{dc}} : \mathbb{R}^{n_c} \to \mathbb{R}^{l_{\mathrm{dc}}}$ and satisfies $h_{\mathrm{dc}}(x_c) \geq\geq 0$, $x_c \in \overline{\mathbb{R}}_+^{n_c}$, $m_{\mathrm{cc}} = l_c$, $m_{\mathrm{dc}} = l_d$, $l_{cc} = m_c$, $l_{\mathrm{dc}} = m_d$, and $\mathcal{Z}_c \triangleq \mathcal{Z}_{cx_c} \times \mathcal{Z}_{cu_{cc}} \subset \overline{\mathbb{R}}_+^{n_c} \times U_{cc}$.

Here, we assume that $u_{\mathrm{cc}}(\cdot)$ and $u_{\mathrm{dc}}(\cdot)$ are restricted to the class of *admissible* inputs consisting of measurable functions such that $(u_{\mathrm{cc}}(t),$ $u_{\mathrm{dc}}(t_k)) \in U_{cc} \times U_{\mathrm{dc}}$ for all $t \geq 0$ and $k \in \mathbb{Z}_{[0,t)} \triangleq \{k : 0 \leq t_k < t\}$, where the constraint set $U_{cc} \times U_{\mathrm{dc}}$ is given with $(0,0) \in U_{cc} \times U_{\mathrm{dc}}$. Furthermore, we assume that the set $\mathcal{Z}_c = \{(x_c, u_{\mathrm{cc}}) : \mathcal{X}_c(x_c, u_{\mathrm{cc}}) = 0\}$, where $\mathcal{X}_c : \overline{\mathbb{R}}_+^{n_c} \times U_{cc} \to \mathbb{R}$. In addition, we assume that the system functions $f_{\mathrm{cc}}(\cdot)$, $f_{\mathrm{dc}}(\cdot)$, $G_{\mathrm{cc}}(\cdot)$, $G_{\mathrm{dc}}(\cdot)$, $h_{\mathrm{cc}}(\cdot)$, and $h_{\mathrm{dc}}(\cdot)$ are continuous mappings. Finally, for the nonlinear dynamical system (6.43) we assume that the required properties for the existence and uniqueness of solutions are satisfied such that (6.43) has a unique solution for all $t \in \mathbb{R}$ [14,93]. Note that with the positive feedback interconnection given by Figure 6.12, $(u_{\mathrm{cc}}, u_{\mathrm{dc}}) = (y_c, y_d)$ and $(y_{\mathrm{cc}}, y_{\mathrm{dc}}) = (u_c, u_d)$. Furthermore, even though the input-output pairs of the feedback interconnection shown on Figure 6.12 consist of two-vector inputs/two-vector outputs, at any given instant of time a single-vector input/single-vector output is active.

Next, we define the closed-loop resetting set

$$\tilde{\mathcal{Z}}_{\tilde{x}} \triangleq \mathcal{Z}_x \times \mathcal{Z}_{cx_c} \cup \{(x, x_c) : (h_c(x) + J_c(x)h_{\mathrm{cc}}(x_c), h_{\mathrm{cc}}(x_c)) \in \mathcal{Z}_{cu_{cc}} \times \mathcal{Z}_{u_c}\}.$$
$$(6.47)$$

Note that since the positive feedback interconnection of $\mathcal{G}$ and $\mathcal{G}_c$ is well posed, it follows that $\tilde{\mathcal{Z}}_{\tilde{x}}$ is well defined and depends on the closed-loop states $\tilde{x} \triangleq [x^{\mathrm{T}} x_c^{\mathrm{T}}]^{\mathrm{T}}$. As in Section 6.2, here we assume that the solution $s(t, \tilde{x}_0)$ to the dynamical system resulting from the feedback interconnection of $\mathcal{G}$ and $\mathcal{G}_c$ is such that Assumption 2.1 is satisfied.

The following theorem gives sufficient conditions for Lyapunov and asymptotic stability of the positive feedback interconnection given by Figure 6.12. For the statement of this result recall the definitions of $\mathcal{T}_{x0,u_c}$, $\mathcal{T}^c_{x0,u_c}$, $\mathcal{T}_{xc0,u_{cc}}$, and $\mathcal{T}^c_{xc0,u_{cc}}$.

Theorem 6.3 *Let* $q_c \in \mathbb{R}^{l_c}$, $r_c \in \mathbb{R}^{m_c}$, $q_d \in \mathbb{R}^{l_d}$, $r_d \in \mathbb{R}^{m_d}$, $q_{cc} \in \mathbb{R}^{l_{cc}}$, $r_{cc} \in \mathbb{R}^{m_{cc}}$, $q_{dc} \in \mathbb{R}^{l_{dc}}$, *and* $r_{dc} \in \mathbb{R}^{m_{dc}}$. *Consider the nonlinear impulsive nonnegative dynamical systems* $\mathcal{G}$ *and* $\mathcal{G}_c$ *given by (4.35)–(4.38) and (6.43)–(6.46), respectively. Assume* $\mathcal{G}$ *and* $\mathcal{G}_c$ *are dissipative with respect to the linear hybrid supply rates* $(q_c^T y_c + r_c^T u_c, q_d^T y_d + r_d^T u_d)$ *and* $(q_{cc}^T y_{cc} + r_{cc}^T u_{cc}, q_{dc}^T y_{dc} + r_{dc}^T u_{dc})$, *and with continuously differentiable, positive-definite storage functions* $V_s(\cdot)$ *and* $V_{sc}(\cdot)$, *respectively, such that* $V_s(0) = 0$ *and* $V_{sc}(0) = 0$. *Furthermore, assume there exists a scalar* $\sigma > 0$ *such that* $q_c + \sigma r_{cc} \leq\leq 0$, $r_c + \sigma q_{cc} \leq\leq 0$, $q_d + \sigma r_{dc} \leq\leq 0$, *and* $r_d + \sigma q_{dc} \leq\leq 0$. *Then the following statements hold:*

i) *The positive feedback interconnection of* $\mathcal{G}$ *and* $\mathcal{G}_c$ *is Lyapunov stable.*

ii) *If* $\mathcal{G}$ *and* $\mathcal{G}_c$ *are strongly zero-state observable and* $q_c + \sigma r_{cc} << 0$ *and* $r_c + \sigma q_{cc} << 0$, *then the positive feedback interconnection of* $\mathcal{G}$ *and* $\mathcal{G}_c$ *is asymptotically stable.*

iii) *If* $\mathcal{G}$ *is strongly zero-state observable,* $\mathcal{G}_c$ *is exponentially dissipative with respect to the linear hybrid supply rate* $(q_{cc}^T y_{cc} + r_{cc}^T u_{cc}, q_{dc}^T y_{dc} + r_{dc}^T u_{dc})$, *and* rank $G_{cc}(0) = m_{cc}$, *then the positive feedback interconnection of* $\mathcal{G}$ *and* $\mathcal{G}_c$ *is asymptotically stable.*

iv) *If* $\mathcal{G}$ *and* $\mathcal{G}_c$ *are exponentially dissipative with respect to the linear hybrid supply rates* $(q_c^T y_c + r_c^T u_c, q_d^T y_d + r_d^T u_d)$ *and* $(q_{cc}^T y_{cc} + r_{cc}^T u_{cc}, q_{dc}^T y_{dc} + r_{dc}^T u_{dc})$, *then the positive feedback interconnection of* $\mathcal{G}$ *and* $\mathcal{G}_c$ *is asymptotically stable.*

Proof. Let $\tilde{\mathcal{T}}^c \triangleq \mathcal{T}^c_{x0,u_c} \cup \mathcal{T}^c_{xc0,u_{cc}}$ and $t_k \in \tilde{\mathcal{T}}^c, k \in \overline{\mathbb{Z}}_+$. Note that it follows from Assumptions A1 and A2 that the resetting times $t_k = \tau_k(\tilde{x}_0)$ for the feedback system are well defined and distinct for every closed-loop trajectory. Furthermore, note that the positive feedback interconnection of $\mathcal{G}$ and $\mathcal{G}_c$ is defined by the closed-loop impulsive dynamics given by

$$\begin{bmatrix} \dot{x}(t) \\ \dot{x}_c(t) \end{bmatrix} = \begin{bmatrix} f_c(x(t)) + G_c(x(t))h_{cc}(x_c(t)) \\ f_{cc}(x_c(t)) + G_{cc}(x_c(t))h_c(x(t)) \end{bmatrix}$$

$$+ \begin{bmatrix} 0 \\ G_{cc}(x_c(t))J_c(x(t))h_{cc}(x_c(t)) \end{bmatrix}, \quad (x(t), x_c(t)) \notin \tilde{\mathcal{Z}}_{\tilde{x}},$$

(6.48)

$$\begin{bmatrix} \Delta x(t) \\ \Delta x_c(t) \end{bmatrix} = \begin{bmatrix} f_d(x(t)) + G_d(x(t))h_{dc}(x_c(t)) \\ f_{dc}(x_c(t)) + G_{dc}(x_c(t))h_d(x(t)) \end{bmatrix}$$
$$+ \begin{bmatrix} 0 \\ G_{dc}(x_c(t))J_d(x(t))h_{dc}(x_c(t)) \end{bmatrix}, \quad (x(t), x_c(t)) \in \tilde{\mathcal{Z}}_{\tilde{x}},$$

(6.49)

which implies that

$$\tilde{f}_c(\tilde{x}) \triangleq \begin{bmatrix} f_c(x) + G_c(x)h_{cc}(x_c) \\ f_{cc}(x_c) + G_{cc}(x_c)h_c(x) + G_{cc}(x_c)J_c(x)h_{cc}(x_c) \end{bmatrix}$$

(6.50)

is essentially nonnegative and

$$\tilde{x} + \tilde{f}_d(\tilde{x}) \triangleq \begin{bmatrix} x + f_d(x) + G_d(x)h_{dc}(x_c) \\ x_c + f_{dc}(x_c) + G_{dc}(x_c)h_d(x) + G_{dc}(x_c)J_d(x)h_{dc}(x_c) \end{bmatrix}$$

(6.51)

is nonnegative. Hence, it follows from Proposition 4.1 that $\overline{\mathbb{R}}_+^n \times \overline{\mathbb{R}}_+^{n_c}$ is an invariant set with respect to the impulsive closed-loop system (6.48) and (6.49), and hence, $x(t) \geq\geq 0$, $x_c(t) \geq\geq 0$, $u_c(t) = y_{cc}(t) \geq\geq 0$, $u_d(t_k) = y_{dc}(t_k) \geq\geq 0$, $y_c(t) = u_{cc}(t) \geq\geq 0$, and $y_d(t_k) = u_{dc}(t_k) \geq\geq 0$, $t \geq 0$, $k \in \mathbb{Z}_+$.

$i)$ Consider the Lyapunov function candidate $V(x, x_c) = V_s(x) + \sigma V_{sc}(x_c)$. Now, the corresponding Lyapunov derivative of $V(x, x_c)$ along the state trajectories $(x(t), x_c(t))$, $t \in (t_k, t_{k+1}]$, is given by

$$\dot{V}(x(t), x_c(t)) = \dot{V}_s(x(t)) + \sigma \dot{V}_{sc}(x_c(t)) \leq q_c^{\mathrm{T}} y_c + r_c^{\mathrm{T}} u_c$$
$$+ \sigma(q_{cc}^{\mathrm{T}} y_{cc} + r_{cc}^{\mathrm{T}} u_{cc})$$
$$\leq 0, \quad (x(t), x_c(t)) \notin \tilde{\mathcal{Z}}_{\tilde{x}},$$

(6.52)

and the Lyapunov difference of $V(x, x_c)$ at the resetting times t_k, $k \in \mathbb{Z}_+$, is given by

$$\Delta V(x(t_k), x_c(t_k)) = \Delta V_s(x(t_k)) + \sigma \Delta V_{sc}(x_c(t_k)) \leq q_d^{\mathrm{T}} y_d + r_d^{\mathrm{T}} u_d$$
$$+ \sigma(q_{dc}^{\mathrm{T}} y_{dc} + r_{dc}^{\mathrm{T}} u_{dc})$$
$$\leq 0, \quad (x(t), x_c(t)) \in \tilde{\mathcal{Z}}_{\tilde{x}}.$$

(6.53)

Now, Lyapunov stability of the positive feedback interconnection of $\mathcal{G}$ and $\mathcal{G}_c$ follows as a direct consequence of Theorem 4.1.

$ii)$ With $V(x, x_c) = V_s(x) + \sigma V_{sc}(x_c)$, Lyapunov stability follows from $i)$. Furthermore, if $q_c + \sigma r_{cc} << 0$ and $r_c + \sigma q_{cc} << 0$, then, using the strong zero-state observability assumption, it follows from (6.52) and (6.53) that the largest invariant set contained in

$$\mathcal{R} \triangleq \{(x, x_c) \in \mathcal{D} \times \mathbb{R}^{n_c} : (x, x_c) \notin \tilde{\mathcal{Z}}_{\tilde{x}}, \ \dot{V}(x, x_c) = 0)\}$$
$$\cup \{(x, x_c) \in \mathcal{D} \times \mathbb{R}^{n_c} : (x, x_c) \in \tilde{\mathcal{Z}}_{\tilde{x}}, \ \Delta V(x, x_c) = 0\} \quad (6.54)$$

is given by $\mathcal{M} = \{(0,0)\}$. Hence, asymptotic stability of the closed-loop system follows from Theorem 4.2.

iii) If $\mathcal{G}_c$ is exponentially dissipative it follows that for some scalar $\varepsilon_{cc} > 0$,

$$\dot{V}(x(t), x_c(t)) = \dot{V}_s(x(t)) + \sigma \dot{V}_{sc}(x_c(t))$$
$$\leq -\sigma \varepsilon_{cc} V_{sc}(x_c(t)) + q_c^T y_c + r_c^T u_c + \sigma(q_{cc}^T y_{cc} + r_{cc}^T u_{cc})$$
$$\leq -\sigma \varepsilon_{cc} V_{sc}(x_c(t)) \leq 0, \quad (x(t), x_c(t)) \notin \tilde{\mathcal{Z}}_{\tilde{x}}, \quad (6.55)$$

and the Lyapunov difference $\Delta V(x(t_k), x_c(t_k))$, $k \in \overline{\mathbb{Z}}_+$, at the resetting times for the closed-loop system satisfies (6.53). Since $V_{sc}(x_c)$ is positive definite, note that $\dot{V}(x, x_c) = 0$ for all $(x, x_c) \notin \tilde{\mathcal{Z}}_{\tilde{x}}$ only if $x_c = 0$. Furthermore, since rank $G_{cc}(0) = m_{cc}$, it follows that on every invariant set $\mathcal{M}$ contained in $\mathcal{R}$ given by (6.54), $u_{cc}(t) = y_c(t) \equiv 0$, and hence, $y_{cc}(t) = u_c(t) \equiv 0$ so that $\dot{x}(t) = f_c(x(t))$. Now, since $\mathcal{G}$ is strongly zero-state observable it follows that $\mathcal{R} = \{(0,0)\} \cup \{(x, x_c) \in \overline{\mathbb{R}}_+^n \times \overline{\mathbb{R}}_+^{n_c} : (x, x_c) \in \tilde{\mathcal{Z}}_{\tilde{x}}, \ \Delta V(x, x_c) = 0\}$ contains no solution other than the trivial solution $(x(t), x_c(t)) \equiv (0,0)$. Hence, it follows from Theorem 4.2 that the closed-loop system is asymptotically stable.

iv) Finally, if $\mathcal{G}$ and $\mathcal{G}_c$ are exponentially dissipative it follows that

$$\dot{V}(x(t), x_c(t)) = \dot{V}_s(x(t)) + \sigma \dot{V}_{sc}(x_c(t))$$
$$\leq -\varepsilon_c V_s(x(t)) - \sigma \varepsilon_{cc} V_{sc}(x_c(t)) + q_c^T y_c + r_c^T u_c$$
$$+ \sigma(q_{cc}^T y_{cc} + r_{cc}^T u_{cc})$$
$$\leq -\min\{\varepsilon_c, \varepsilon_{cc}\} V(x(t), x_c(t)), \quad (x(t), x_c(t)) \notin \tilde{\mathcal{Z}}_{\tilde{x}}, \quad (6.56)$$

and $\Delta V(x(t_k), x_c(t_k))$, $(x(t), x_c(t)) \in \tilde{\mathcal{Z}}_{\tilde{x}}$, $k \in \overline{\mathbb{Z}}_+$, satisfies (6.53). Now, Theorem 4.1 implies that the positive feedback interconnection of $\mathcal{G}$ and $\mathcal{G}_c$ is asymptotically stable. $\square$

Theorem 6.3 also holds for the more general architecture of the feedback system $\mathcal{G}_c$, wherein $y_{cc} = h_{cc}(x_c) + J_{cc}(x_c)u_{cc}$ and $y_{dc} = h_{dc}(x_c) + J_{dc}(x_c)u_{dc}$, where $J_{cc} : \mathbb{R}^{n_c} \rightarrow \mathbb{R}^{l_{cc} \times m_{cc}}$, $J_{dc} : \mathbb{R}^{n_c} \rightarrow \mathbb{R}^{l_{dc} \times m_{dc}}$, $J_{cc}(x_c) \geq\geq 0$, $x_c \notin \mathcal{Z}_c$, and $J_{dc}(x_c) \geq\geq 0$, $x_c \in \mathcal{Z}_c$. In this case, however, we assume that the positive feedback interconnection of $\mathcal{G}$

and $\mathcal{G}_c$ is well posed, that is, $\det[I_{m_c} - J_{cc}(x_c)J_c(x)] \neq 0$, $(x, x_c) \notin \tilde{\mathcal{Z}}_{\tilde{x}}$, and $\det[I_{m_d} - J_{dc}(x_c)J_d(x)] \neq 0$, $(x, x_c) \in \tilde{\mathcal{Z}}_{\tilde{x}}$.

The following corollary to Theorem 6.3 addresses linear hybrid supply rates of the form $(s_c(u_c, y_c), s_d(u_d, y_d)) = (\mathbf{e}^T u_c - \mathbf{e}^T y_c, \mathbf{e}^T u_d - \mathbf{e}^T y_d)$ and $(s_{cc}(u_{cc}, y_{cc}), s_{dc}(u_{dc}, y_{dc})) = (\mathbf{e}^T u_{cc} - \mathbf{e}^T y_{cc}, \mathbf{e}^T u_{dc} - \mathbf{e}^T y_{dc})$.

Corollary 6.2 *Consider the nonlinear impulsive nonnegative dynamical systems $\mathcal{G}$ and $\mathcal{G}_c$ given by (4.35)–(4.38) and (6.43)–(6.46), respectively. Assume $\mathcal{G}$ is nonaccumulative with a continuously differentiable, positive-definite storage function $V_s(\cdot)$, and $\mathcal{G}_c$ is exponentially nonaccumulative with a continuously differentiable, positive-definite storage function $V_{sc}(\cdot)$. Then the following statements hold:*

i) If $\mathcal{G}$ is strongly zero-state observable and $\operatorname{rank} G_{cc}(0) = m_{cc}$, then the positive feedback interconnection of $\mathcal{G}$ and $\mathcal{G}_c$ is asymptotically stable.

ii) If $\mathcal{G}$ is exponentially nonaccumulative, then the positive feedback interconnection of $\mathcal{G}$ and $\mathcal{G}_c$ is asymptotically stable.

Proof. The proof is a direct consequence of *iii)* and *iv)* of Theorem 6.3 with $\sigma = 1$, $q_c = -r_{cc} = -\mathbf{e}$, $q_d = -r_{dc} = -\mathbf{e}$, $r_c = -q_{cc} = \mathbf{e}$, and $r_d = -q_{dc} = \mathbf{e}$. $\qquad\qquad\qquad\qquad\qquad\qquad\qquad\qquad\qquad\square$

6.5 Stability of Feedback Interconnections of Large-Scale Impulsive Dynamical Systems

In this section, we use the concepts of vector dissipativity and vector storage functions as candidate vector Lyapunov functions to develop feedback interconnection stability results of large-scale impulsive dynamical systems. General stability criteria are given for Lyapunov and asymptotic stability of feedback large-scale impulsive dynamical systems. Specifically, we consider input/state-dependent impulsive large-scale dynamical systems $\mathcal{G}$ of the form

$$\dot{x}(t) = F_c(x(t), u_c(t)), \quad x(t_0) = x_0, \quad (x(t), u_c(t)) \notin \mathcal{Z}, \quad t \geq t_0, \tag{6.57}$$

$$\Delta x(t) = F_d(x(t), u_d(t)), \quad (x(t), u_c(t)) \in \mathcal{Z}, \tag{6.58}$$

$$y_c(t) = H_c(x(t), u_c(t)), \quad (x(t), u_c(t)) \notin \mathcal{Z}, \tag{6.59}$$

$$y_d(t) = H_d(x(t), u_d(t)), \quad (x(t), u_c(t)) \in \mathcal{Z}, \tag{6.60}$$

where $x(t) \in \mathcal{D} \subseteq \mathbb{R}^n$, $t \geq t_0$, $u_{\mathrm{c}}(t) \in U_{\mathrm{c}} \subseteq \mathbb{R}^{m_{\mathrm{c}}}$, $u_{\mathrm{d}}(t_k) \in U_{\mathrm{d}} \subseteq \mathbb{R}^{m_{\mathrm{d}}}$, $y_{\mathrm{c}}(t) \in Y_{\mathrm{c}} \subseteq \mathbb{R}^{l_{\mathrm{c}}}$, $y_{\mathrm{d}}(t_k) \in Y_{\mathrm{d}} \subseteq \mathbb{R}^{l_{\mathrm{d}}}$, $F_{\mathrm{c}} : \mathcal{D} \times U_{\mathrm{c}} \to \mathbb{R}^n$, $F_{\mathrm{d}} : \mathcal{D} \times U_{\mathrm{d}} \to \mathbb{R}^n$, $H_{\mathrm{c}} : \mathcal{D} \times U_{\mathrm{c}} \to Y_{\mathrm{c}}$, $H_{\mathrm{d}} : \mathcal{D} \times U_{\mathrm{d}} \to Y_{\mathrm{d}}$, $\mathcal{D}$ is an open set with $0 \in \mathcal{D}$, $\mathcal{Z} \subset \mathcal{D} \times U_{\mathrm{c}}$, and $F_{\mathrm{c}}(0,0) = 0$.

Here, we assume that $\mathcal{G}$ represents a large-scale impulsive dynamical system composed of q interconnected controlled impulsive subsystems $\mathcal{G}_i$ such that, for all $i = 1, \ldots, q$,

$$F_{\mathrm{c}i}(x, u_{\mathrm{c}i}) = f_{\mathrm{c}i}(x_i) + \mathcal{I}_{\mathrm{c}i}(x) + G_{\mathrm{c}i}(x_i)u_{\mathrm{c}i}, \qquad (6.61)$$

$$F_{\mathrm{d}i}(x, u_{\mathrm{d}i}) = f_{\mathrm{d}i}(x_i) + \mathcal{I}_{\mathrm{d}i}(x) + G_{\mathrm{d}i}(x_i)u_{\mathrm{d}i}, \qquad (6.62)$$

$$H_{\mathrm{c}i}(x_i, u_{\mathrm{c}i}) = h_{\mathrm{c}i}(x_i) + J_{\mathrm{c}i}(x_i)u_{\mathrm{c}i}, \qquad (6.63)$$

$$H_{\mathrm{d}i}(x_i, u_{\mathrm{d}i}) = h_{\mathrm{d}i}(x_i) + J_{\mathrm{d}i}(x_i)u_{\mathrm{d}i}, \qquad (6.64)$$

where $x_i \in \mathcal{D}_i \subseteq \mathbb{R}^{n_i}$, $u_{\mathrm{c}i} \in U_{\mathrm{c}i} \subseteq \mathbb{R}^{m_{\mathrm{c}i}}$, $u_{\mathrm{d}i} \in U_{\mathrm{d}i} \subseteq \mathbb{R}^{m_{\mathrm{d}i}}$, $y_{\mathrm{c}i} \triangleq H_{\mathrm{c}i}(x_i, u_{\mathrm{c}i}) \in Y_{\mathrm{c}i} \subseteq \mathbb{R}^{l_{\mathrm{c}i}}$, $y_{\mathrm{d}i} \triangleq H_{\mathrm{d}i}(x_i, u_{\mathrm{d}i}) \in Y_{\mathrm{d}i} \subseteq \mathbb{R}^{l_{\mathrm{d}i}}$, $((u_{\mathrm{c}i}, u_{\mathrm{d}i})$, $(y_{\mathrm{c}i}, y_{\mathrm{d}i}))$ is the hybrid input-output pair for the ith subsystem, $f_{\mathrm{c}i} : \mathbb{R}^{n_i} \to \mathbb{R}^{n_i}$ and $\mathcal{I}_{\mathrm{c}i} : \mathcal{D} \to \mathbb{R}^{n_i}$ are Lipschitz continuous and satisfy $f_{\mathrm{c}i}(0) = 0$ and $\mathcal{I}_{\mathrm{c}i}(0) = 0$, $f_{\mathrm{d}i} : \mathbb{R}^{n_i} \to \mathbb{R}^{n_i}$ and $\mathcal{I}_{\mathrm{d}i} : \mathcal{D} \to \mathbb{R}^{n_i}$ are continuous, $G_{\mathrm{c}i} : \mathbb{R}^{n_i} \to \mathbb{R}^{n_i \times m_{\mathrm{c}i}}$ and $G_{\mathrm{d}i} : \mathbb{R}^{n_i} \to \mathbb{R}^{n_i \times m_{\mathrm{d}i}}$ are continuous, $h_{\mathrm{c}i} : \mathbb{R}^{n_i} \to \mathbb{R}^{l_{\mathrm{c}i}}$ and satisfies $h_{\mathrm{c}i}(0) = 0$, $h_{\mathrm{d}i} : \mathbb{R}^{n_i} \to \mathbb{R}^{l_{\mathrm{d}i}}$, $J_{\mathrm{c}i} : \mathbb{R}^{n_i} \to \mathbb{R}^{l_{\mathrm{c}i} \times m_{\mathrm{c}i}}$, $J_{\mathrm{d}i} : \mathbb{R}^{n_i} \to \mathbb{R}^{l_{\mathrm{d}i} \times m_{\mathrm{d}i}}$, $\sum_{i=1}^q n_i = n$, $\sum_{i=1}^q m_{\mathrm{c}i} = m_{\mathrm{c}}$, $\sum_{i=1}^q m_{\mathrm{d}i} = m_{\mathrm{d}}$, $\sum_{i=1}^q l_{\mathrm{c}i} = l_{\mathrm{c}}$, and $\sum_{i=1}^q l_{\mathrm{d}i} = l_{\mathrm{d}}$. Furthermore, for the large-scale dynamical system $\mathcal{G}$ we assume that the required properties for the existence and uniqueness of solutions are satisfied, that is, for each $i \in \{1, \ldots, q\}$, $u_{\mathrm{c}i}(\cdot)$ and $u_{\mathrm{d}i}(\cdot)$ satisfy sufficient regularity conditions such that the system (6.57) and (6.58) has a unique solution forward in time. We define the composite input and composite output for the large-scale impulsive dynamical system $\mathcal{G}$ as $u_{\mathrm{c}} \triangleq [u_{\mathrm{c}1}^{\mathrm{T}}, \ldots, u_{\mathrm{c}q}^{\mathrm{T}}]^{\mathrm{T}}$, $u_{\mathrm{d}} \triangleq [u_{\mathrm{d}1}^{\mathrm{T}}, \ldots, u_{\mathrm{d}q}^{\mathrm{T}}]^{\mathrm{T}}$, $y_{\mathrm{c}} \triangleq [y_{\mathrm{c}1}^{\mathrm{T}}, \ldots, y_{\mathrm{c}q}^{\mathrm{T}}]^{\mathrm{T}}$, and $y_{\mathrm{d}} \triangleq [y_{\mathrm{d}1}^{\mathrm{T}}, \ldots, y_{\mathrm{d}q}^{\mathrm{T}}]^{\mathrm{T}}$, respectively.

Next, we consider a dynamical large-scale impulsive feedback system $\mathcal{G}_{\mathrm{c}}$ given by

$$\dot{x}_{\mathrm{c}}(t) = F_{\mathrm{cc}}(x_{\mathrm{c}}(t), u_{\mathrm{cc}}(t)), \quad x_{\mathrm{c}}(t_0) = x_{\mathrm{c}0}, \quad (x_{\mathrm{c}}(t), u_{\mathrm{cc}}(t)) \notin \mathcal{Z}_{\mathrm{c}}, \qquad (6.65)$$

$$\Delta x_{\mathrm{c}}(t) = F_{\mathrm{dc}}(x_{\mathrm{c}}(t), u_{\mathrm{dc}}(t)), \quad (x_{\mathrm{c}}(t), u_{\mathrm{cc}}(t)) \in \mathcal{Z}_{\mathrm{c}}, \qquad (6.66)$$

$$y_{\mathrm{cc}}(t) = H_{\mathrm{cc}}(x_{\mathrm{c}}(t), u_{\mathrm{cc}}(t)), \quad (x_{\mathrm{c}}(t), u_{\mathrm{cc}}(t)) \notin \mathcal{Z}_{\mathrm{c}}, \qquad (6.67)$$

$$y_{\mathrm{dc}}(t) = H_{\mathrm{dc}}(x_{\mathrm{c}}(t), u_{\mathrm{dc}}(t)), \quad (x_{\mathrm{c}}(t), u_{\mathrm{cc}}(t)) \in \mathcal{Z}_{\mathrm{c}}, \qquad (6.68)$$

where $F_{\mathrm{cc}} : \mathbb{R}^{n_{\mathrm{c}}} \times U_{\mathrm{cc}} \to \mathbb{R}^{n_{\mathrm{c}}}$, $F_{\mathrm{dc}} : \mathbb{R}^{n_{\mathrm{c}}} \times U_{\mathrm{dc}} \to \mathbb{R}^{n_{\mathrm{c}}}$, $H_{\mathrm{cc}} : \mathbb{R}^{n_{\mathrm{c}}} \times U_{\mathrm{cc}} \to Y_{\mathrm{cc}}$, $H_{\mathrm{dc}} : \mathbb{R}^{n_{\mathrm{c}}} \times U_{\mathrm{dc}} \to Y_{\mathrm{dc}}$, $F_{\mathrm{cc}} \triangleq [F_{\mathrm{cc}1}^{\mathrm{T}}, \ldots, F_{\mathrm{cc}q}^{\mathrm{T}}]^{\mathrm{T}}$, $F_{\mathrm{dc}} \triangleq$

$[F_{\mathrm{dc}1}^{\mathrm{T}}, \ldots, F_{\mathrm{dc}q}^{\mathrm{T}}]^{\mathrm{T}}$, $H_{\mathrm{cc}} \triangleq [H_{\mathrm{cc}1}^{\mathrm{T}}, \ldots, H_{\mathrm{cc}q}^{\mathrm{T}}]^{\mathrm{T}}$, $H_{\mathrm{dc}} \triangleq [H_{\mathrm{dc}1}^{\mathrm{T}}, \ldots, H_{\mathrm{dc}q}^{\mathrm{T}}]^{\mathrm{T}}$, $U_{\mathrm{cc}} \subseteq \mathbb{R}^{l_{\mathrm{c}}}$, $U_{\mathrm{dc}} \subseteq \mathbb{R}^{l_{\mathrm{d}}}$, $Y_{\mathrm{cc}} \subseteq \mathbb{R}^{m_{\mathrm{c}}}$, $Y_{\mathrm{dc}} \subseteq \mathbb{R}^{m_{\mathrm{d}}}$. Moreover, for all $i = 1, \ldots, q$, we assume that

$$F_{\mathrm{cc}i}(x_{\mathrm{c}}, u_{\mathrm{cc}i}) = f_{\mathrm{cc}i}(x_{\mathrm{c}i}) + \mathcal{I}_{\mathrm{cc}i}(x_{\mathrm{c}}) + G_{\mathrm{cc}i}(x_{\mathrm{c}i})u_{\mathrm{cc}i}, \qquad (6.69)$$

$$F_{\mathrm{dc}i}(x_{\mathrm{c}}, u_{\mathrm{dc}i}) = f_{\mathrm{dc}i}(x_{\mathrm{c}i}) + \mathcal{I}_{\mathrm{dc}i}(x_{\mathrm{c}}) + G_{\mathrm{dc}i}(x_{\mathrm{c}i})u_{\mathrm{dc}i}, \qquad (6.70)$$

$$H_{\mathrm{cc}i}(x_{\mathrm{c}i}, u_{\mathrm{cc}i}) = h_{\mathrm{cc}i}(x_{\mathrm{c}i}) + J_{\mathrm{cc}i}(x_{\mathrm{c}i})u_{\mathrm{cc}i}, \qquad (6.71)$$

$$H_{\mathrm{dc}i}(x_{\mathrm{c}i}, u_{\mathrm{dc}i}) = h_{\mathrm{dc}i}(x_{\mathrm{c}i}) + J_{\mathrm{dc}i}(x_{\mathrm{c}i})u_{\mathrm{dc}i}, \qquad (6.72)$$

where $u_{\mathrm{cc}i} \in U_{\mathrm{cc}i} \subseteq \mathbb{R}^{l_{\mathrm{c}i}}$, $u_{\mathrm{dc}i} \in U_{\mathrm{dc}i} \subseteq \mathbb{R}^{l_{\mathrm{d}i}}$, $y_{\mathrm{cc}i} \triangleq H_{\mathrm{cc}i}(x_{\mathrm{c}i}, u_{\mathrm{cc}i}) \in Y_{\mathrm{cc}i} \subseteq \mathbb{R}^{m_{\mathrm{c}i}}$, $y_{\mathrm{dc}i} \triangleq H_{\mathrm{dc}i}(x_{\mathrm{c}i}, u_{\mathrm{dc}i}) \in Y_{\mathrm{dc}i} \subseteq \mathbb{R}^{m_{\mathrm{d}i}}$, $f_{\mathrm{cc}i} : \mathbb{R}^{n_{\mathrm{c}i}} \to \mathbb{R}^{n_{\mathrm{c}i}}$ and $\mathcal{I}_{\mathrm{cc}i} : \mathbb{R}^{n_{\mathrm{c}}} \to \mathbb{R}^{n_{\mathrm{c}i}}$ satisfy $f_{\mathrm{cc}i}(0) = 0$ and $\mathcal{I}_{\mathrm{cc}i}(0) = 0$, $f_{\mathrm{dc}i} : \mathbb{R}^{n_{\mathrm{c}i}} \to \mathbb{R}^{n_{\mathrm{c}i}}$, $\mathcal{I}_{\mathrm{dc}i} : \mathbb{R}^{n_{\mathrm{c}}} \to \mathbb{R}^{n_{\mathrm{c}i}}$, $G_{\mathrm{cc}i} : \mathbb{R}^{n_{\mathrm{c}i}} \to \mathbb{R}^{n_{\mathrm{c}i} \times l_{\mathrm{c}i}}$, $G_{\mathrm{dc}i} : \mathbb{R}^{n_{\mathrm{c}i}} \to \mathbb{R}^{n_{\mathrm{c}i} \times l_{\mathrm{d}i}}$, $h_{\mathrm{cc}i} : \mathbb{R}^{n_{\mathrm{c}i}} \to \mathbb{R}^{m_{\mathrm{c}i}}$ and satisfies $h_{\mathrm{cc}i}(0) = 0$, $h_{\mathrm{dc}i} : \mathbb{R}^{n_{\mathrm{c}i}} \to \mathbb{R}^{m_{\mathrm{d}i}}$, $J_{\mathrm{cc}i} : \mathbb{R}^{n_{\mathrm{c}i}} \to \mathbb{R}^{m_{\mathrm{c}i} \times l_{\mathrm{c}i}}$, $J_{\mathrm{dc}i} : \mathbb{R}^{n_{\mathrm{c}i}} \to \mathbb{R}^{m_{\mathrm{d}i} \times l_{\mathrm{d}i}}$, and $\sum_{i=1}^{q} n_{\mathrm{c}i} = n_{\mathrm{c}}$. Furthermore, we define the composite input and composite output for the system $\mathcal{G}_{\mathrm{c}}$ as $u_{\mathrm{cc}} \triangleq [u_{\mathrm{cc}1}^{\mathrm{T}}, \ldots, u_{\mathrm{cc}q}^{\mathrm{T}}]^{\mathrm{T}}$, $u_{\mathrm{dc}} \triangleq [u_{\mathrm{dc}1}^{\mathrm{T}}, \ldots, u_{\mathrm{dc}q}^{\mathrm{T}}]^{\mathrm{T}}$, $y_{\mathrm{cc}} \triangleq [y_{\mathrm{cc}1}^{\mathrm{T}}, \ldots, y_{\mathrm{cc}q}^{\mathrm{T}}]^{\mathrm{T}}$, and $y_{\mathrm{dc}} \triangleq [y_{\mathrm{dc}1}^{\mathrm{T}}, \ldots, y_{\mathrm{dc}q}^{\mathrm{T}}]^{\mathrm{T}}$, respectively. In this case, $U_{\mathrm{cc}} = U_{\mathrm{cc}1} \times \cdots \times U_{\mathrm{cc}q}$, $U_{\mathrm{dc}} = U_{\mathrm{dc}1} \times \cdots \times U_{\mathrm{dc}q}$, $Y_{\mathrm{cc}} = Y_{\mathrm{cc}1} \times \cdots \times Y_{\mathrm{cc}q}$, and $Y_{\mathrm{dc}} = Y_{\mathrm{dc}1} \times \cdots \times Y_{\mathrm{dc}q}$. Note that with the negative feedback interconnection given by Figure 6.13, $(u_{\mathrm{cc}}, u_{\mathrm{dc}}) = (y_{\mathrm{c}}, y_{\mathrm{d}})$ and $(y_{\mathrm{cc}}, y_{\mathrm{dc}}) = (-u_{\mathrm{c}}, -u_{\mathrm{d}})$.

We assume that the negative feedback interconnection of $\mathcal{G}$ and $\mathcal{G}_{\mathrm{c}}$ is well posed, that is, $\det[I_{m_{\mathrm{c}i}} + J_{\mathrm{cc}i}(x_{\mathrm{c}i})J_{\mathrm{c}i}(x_i)] \neq 0$, $\det[I_{m_{\mathrm{d}i}} + J_{\mathrm{dc}i}(x_{\mathrm{c}i})J_{\mathrm{d}i}(x_i)] \neq 0$ for all $x_i \in \mathbb{R}^{n_i}$, $x_{\mathrm{c}i} \in \mathbb{R}^{n_{\mathrm{c}i}}$, and $i = 1, \ldots, q$. Next, we assume that $\mathcal{Z}_{\mathrm{c}} \triangleq \mathcal{Z}_{\mathrm{c}x_{\mathrm{c}}} \times \mathcal{Z}_{\mathrm{cu}_{\mathrm{cc}}} = \{(x_{\mathrm{c}}, u_{\mathrm{cc}}) : \mathcal{X}_{\mathrm{c}}(x_{\mathrm{c}}, u_{\mathrm{cc}}) = 0\}$, where $\mathcal{X}_{\mathrm{c}} : \mathbb{R}^{n_{\mathrm{c}}} \times U_{\mathrm{cc}} \to \mathbb{R}$, and define the closed-loop resetting set

$$\tilde{\mathcal{Z}}_{\tilde{x}} \triangleq \mathcal{Z}_x \times \mathcal{Z}_{\mathrm{c}x_{\mathrm{c}}} \cup \{(x, x_{\mathrm{c}}) : (\mathcal{L}_{\mathrm{cc}}(x, x_{\mathrm{c}}), \mathcal{L}_{\mathrm{c}}(x, x_{\mathrm{c}})) \in \mathcal{Z}_{\mathrm{cu}_{\mathrm{cc}}} \times \mathcal{Z}_{u_{\mathrm{c}}}\},$$
$$(6.73)$$

where $\mathcal{L}_{\mathrm{cc}}(\cdot, \cdot)$ and $\mathcal{L}_{\mathrm{c}}(\cdot, \cdot)$ are functions of x and x_{c} arising from the algebraic loops due to u_{cc} and u_{c}, respectively. Note that since the feedback interconnection of $\mathcal{G}$ and $\mathcal{G}_{\mathrm{c}}$ is well posed, it follows that $\tilde{\mathcal{Z}}_{\tilde{x}}$ is well defined and depends on the closed-loop states $\tilde{x} \triangleq [x^{\mathrm{T}} x_{\mathrm{c}}^{\mathrm{T}}]^{\mathrm{T}}$. Furthermore, we assume that for the large-scale systems $\mathcal{G}$ and $\mathcal{G}_{\mathrm{c}}$, the conditions of Theorem 5.4 are satisfied, that is, if $V_{\mathrm{s}}(x)$, $x \in \mathbb{R}^n$, and $V_{\mathrm{cs}}(x_{\mathrm{c}})$, $x_{\mathrm{c}} \in \mathbb{R}^{n_{\mathrm{c}}}$, are vector storage functions for $\mathcal{G}$ and $\mathcal{G}_{\mathrm{c}}$, respectively, then there exist $p \in \mathbb{R}_+^q$ and $p_{\mathrm{c}} \in \mathbb{R}_+^q$ such that the functions $v_{\mathrm{s}}(x) = p^{\mathrm{T}} V_{\mathrm{s}}(x)$, $x \in \mathbb{R}^n$, and $v_{\mathrm{cs}}(x_{\mathrm{c}}) = p_{\mathrm{c}}^{\mathrm{T}} V_{\mathrm{cs}}(x_{\mathrm{c}})$, $x_{\mathrm{c}} \in \mathbb{R}^{n_{\mathrm{c}}}$, are positive definite. The following result gives sufficient conditions for

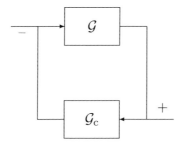

Figure 6.13 Feedback interconnection of large-scale systems $\mathcal{G}$ and $\mathcal{G}_c$.

Lyapunov and asymptotic stability of the feedback interconnection given by Figure 6.13.

Theorem 6.4 *Consider the large-scale impulsive dynamical systems $\mathcal{G}$ and $\mathcal{G}_c$ given by (6.57)–(6.60) and (6.65)–(6.68), respectively. Assume that $\mathcal{G}$ and $\mathcal{G}_c$ are vector dissipative with respect to the vector hybrid supply rates $(S_c(u_c, y_c), S_d(u_d, y_d))$ and $(S_{cc}(u_{cc}, y_{cc}), S_{dc}(u_{dc}, y_{dc}))$, and with continuously differentiable vector storage functions $V_s(\cdot)$ and $V_{cs}(\cdot)$, and dissipation matrices $W \in \mathbb{R}^{q \times q}$ and $W_c \in \mathbb{R}^{q \times q}$, respectively.*

i) *If there exists $\Sigma \triangleq \operatorname{diag}[\sigma_1, \ldots, \sigma_q] > 0$ such that $S_c(u_c, y_c) + \Sigma S_{cc}(u_{cc}, y_{cc}) \leq\leq 0$, $S_d(u_d, y_d) + \Sigma S_{dc}(u_{dc}, y_{dc}) \leq\leq 0$, and $\tilde{W} \in \mathbb{R}^{q \times q}$ is semistable (respectively, asymptotically stable), where $\tilde{W}_{(i,j)} \triangleq \max\{W_{(i,j)}, (\Sigma W_c \Sigma^{-1})_{(i,j)}\} = \max\{W_{(i,j)}, \frac{\sigma_i}{\sigma_j} W_{c(i,j)}\}$, $i, j = 1, \ldots, q$, then the negative feedback interconnection of $\mathcal{G}$ and $\mathcal{G}_c$ is Lyapunov (respectively, asymptotically) stable.*

ii) *Let $Q_{ci} \in \mathbb{S}^{l_{ci}}$, $S_{ci} \in \mathbb{R}^{l_{ci} \times m_{ci}}$, $R_{ci} \in \mathbb{S}^{m_{ci}}$, $Q_{di} \in \mathbb{S}^{l_{di}}$, $S_{di} \in \mathbb{R}^{l_{di} \times m_{di}}$, $R_{di} \in \mathbb{S}^{m_{di}}$, $Q_{cci} \in \mathbb{S}^{m_{ci}}$, $S_{cci} \in \mathbb{R}^{m_{ci} \times l_{ci}}$, $R_{cci} \in \mathbb{S}^{l_{ci}}$, $Q_{dci} \in \mathbb{S}^{m_{di}}$, $S_{dci} \in \mathbb{R}^{m_{di} \times l_{di}}$, and $R_{dci} \in \mathbb{S}^{l_{di}}$, and suppose $S_c(u_c, y_c) = [s_{c1}(u_{c1}, y_{c1}), \ldots, s_{cq}(u_{cq}, y_{cq})]^T$, $S_d(u_d, y_d) = [s_{d1}(u_{d1}, y_{d1}), \ldots, s_{dq}(u_{dq}, y_{dq})]^T$, $S_{cc}(u_{cc}, y_{cc}) = [s_{cc1}(u_{cc1}, y_{cc1}), \ldots, s_{ccq}(u_{ccq}, y_{ccq})]^T$, and $S_{dc}(u_{dc}, y_{dc}) = [s_{dc1}(u_{dc1}, y_{dc1}), \ldots, s_{dcq}(u_{dcq}, y_{dcq})]^T$, where $s_{ci}(u_{ci}, y_{ci}) = u_{ci}^T R_{ci} u_{ci} + 2 y_{ci}^T S_{ci} u_{ci} + y_{ci}^T Q_{ci} y_{ci}$, $s_{di}(u_{di}, y_{di}) = u_{di}^T R_{di} u_{di} + 2 y_{di}^T S_{di} u_{di} + y_{di}^T Q_{di} y_{di}$, $s_{cci}(u_{cci}, y_{cci}) = u_{cci}^T R_{cci} u_{cci} + 2 y_{cci}^T S_{cci} u_{cci} + y_{cci}^T Q_{cci} y_{cci}$, and $s_{dci}(u_{dci}, y_{dci}) = u_{dci}^T R_{dci} u_{dci} + 2 y_{dci}^T S_{dci} u_{dci} + y_{dci}^T Q_{dci} y_{dci}$, $i = 1, \ldots, q$. If there exists $\Sigma \triangleq \operatorname{diag}[\sigma_1, \ldots, \sigma_q] > 0$ such that for all $i =$*

$$1, \ldots, q,$$

$$\tilde{Q}_{\mathrm{c}i} \triangleq \begin{bmatrix} Q_{\mathrm{c}i} + \sigma_i R_{\mathrm{cc}i} & -S_{\mathrm{c}i} + \sigma_i S_{\mathrm{cc}i}^{\mathrm{T}} \\ -S_{\mathrm{c}i}^{\mathrm{T}} + \sigma_i S_{\mathrm{cc}i} & R_{\mathrm{c}i} + \sigma_i Q_{\mathrm{cc}i} \end{bmatrix} \leq 0, \qquad (6.74)$$

$$\tilde{Q}_{\mathrm{d}i} \triangleq \begin{bmatrix} Q_{\mathrm{d}i} + \sigma_i R_{\mathrm{dc}i} & -S_{\mathrm{d}i} + \sigma_i S_{\mathrm{dc}i}^{\mathrm{T}} \\ -S_{\mathrm{d}i}^{\mathrm{T}} + \sigma_i S_{\mathrm{dc}i} & R_{\mathrm{d}i} + \sigma_i Q_{\mathrm{dc}i} \end{bmatrix} \leq 0, \qquad (6.75)$$

and $\tilde{W} \in \mathbb{R}^{q \times q}$ is semistable (respectively, asymptotically stable), where $\tilde{W}_{(i,j)} \triangleq \max\{W_{(i,j)}, (\Sigma W_{\mathrm{c}} \Sigma^{-1})_{(i,j)}\} = \max\{W_{(i,j)}, \frac{\sigma_i}{\sigma_j} \cdot W_{\mathrm{c}(i,j)}\}$, $i, j = 1, \ldots, q$, then the negative feedback interconnection of $\mathcal{G}$ and $\mathcal{G}_{\mathrm{c}}$ is Lyapunov (respectively, asymptotically) stable.

Proof. Let $\tilde{\mathcal{T}}^{\mathrm{c}} \triangleq \mathcal{T}_{x_0, u_{\mathrm{c}}}^{\mathrm{c}} \cup \mathcal{T}_{x_{\mathrm{c}0}, u_{\mathrm{cc}}}^{\mathrm{c}}$ and $t_k \in \tilde{\mathcal{T}}^{\mathrm{c}}$, $k \in \overline{\mathbb{Z}}_+$. First, note that it follows from Assumptions A1 and A2 that the resetting times $t_k(= \tau_k(\tilde{x}_0))$ for the feedback system are well defined and distinct for every closed-loop trajectory.

i) Consider the vector Lyapunov function candidate $V(x, x_{\mathrm{c}}) = V_{\mathrm{s}}(x) + \Sigma V_{\mathrm{cs}}(x_{\mathrm{c}})$, $(x, x_{\mathrm{c}}) \in \mathbb{R}^n \times \mathbb{R}^{n_{\mathrm{c}}}$, and note that the corresponding vector Lyapunov derivative of $V(x, x_{\mathrm{c}})$ along the state trajectories $(x(t), x_{\mathrm{c}}(t))$, $t \in (t_k, t_{k+1})$, is given by

$$\begin{aligned}
\dot{V}(x(t), x_{\mathrm{c}}(t)) &= \dot{V}_{\mathrm{s}}(x(t)) + \Sigma \dot{V}_{\mathrm{cs}}(x_{\mathrm{c}}(t)) \\
&\leq\leq S_{\mathrm{c}}(u_{\mathrm{c}}(t), y_{\mathrm{c}}(t)) + \Sigma S_{\mathrm{cc}}(u_{\mathrm{cc}}(t), y_{\mathrm{cc}}(t)) + W V_{\mathrm{s}}(x(t)) \\
&\quad + \Sigma W_{\mathrm{c}} V_{\mathrm{cs}}(x_{\mathrm{c}}(t)) \\
&\leq\leq W V_{\mathrm{s}}(x(t)) + \Sigma W_{\mathrm{c}} \Sigma^{-1} \Sigma V_{\mathrm{cs}}(x_{\mathrm{c}}(t)) \\
&\leq\leq \tilde{W}(V_{\mathrm{s}}(x(t)) + \Sigma V_{\mathrm{cs}}(x_{\mathrm{c}}(t))) \\
&= \tilde{W} V(x(t), x_{\mathrm{c}}(t)), \qquad (x(t), x_{\mathrm{c}}(t)) \notin \tilde{\mathcal{Z}}_{\tilde{x}}, \qquad (6.76)
\end{aligned}$$

and the Lyapunov difference of $V(x, x_{\mathrm{c}})$ at the resetting times t_k, $k \in \overline{\mathbb{Z}}_+$, is given by

$$\begin{aligned}
\Delta V(x(t_k), x_{\mathrm{c}}(t_k)) &= \Delta V_{\mathrm{s}}(x(t_k)) + \Sigma \Delta V_{\mathrm{cs}}(x_{\mathrm{c}}(t_k)) \\
&\leq\leq S_{\mathrm{d}}(u_{\mathrm{d}}(t_k), y_{\mathrm{d}}(t_k)) + \Sigma S_{\mathrm{dc}}(u_{\mathrm{dc}}(t_k), y_{\mathrm{dc}}(t_k)) \\
&\leq\leq 0, \qquad (x(t), x_{\mathrm{c}}(t)) \in \tilde{\mathcal{Z}}_{\tilde{x}}. \qquad (6.77)
\end{aligned}$$

Next, since for $V_{\mathrm{s}}(x)$, $x \in \mathbb{R}^n$, and $V_{\mathrm{cs}}(x_{\mathrm{c}})$, $x_{\mathrm{c}} \in \mathbb{R}^{n_{\mathrm{c}}}$, there exist, by assumption, $p \in \mathbb{R}_+^q$ and $p_{\mathrm{c}} \in \mathbb{R}_+^q$ such that the functions $v_{\mathrm{s}}(x) = p^{\mathrm{T}} V_{\mathrm{s}}(x)$, $x \in \mathbb{R}^n$, and $v_{\mathrm{cs}}(x_{\mathrm{c}}) = p_{\mathrm{c}}^{\mathrm{T}} V_{\mathrm{cs}}(x_{\mathrm{c}})$, $x_{\mathrm{c}} \in \mathbb{R}^{n_{\mathrm{c}}}$, are positive definite, and noting that $v_{\mathrm{cs}}(x_{\mathrm{c}}) \leq \max_{i=1,\ldots,q}\{p_{\mathrm{c}i}\} \mathrm{e}^{\mathrm{T}} V_{\mathrm{cs}}(x_{\mathrm{c}})$, where $p_{\mathrm{c}i}$ is the ith element of p_{c} and $\mathrm{e} \triangleq [1, \ldots, 1]^{\mathrm{T}}$, it follows that

$\mathbf{e}^{\mathrm{T}}V_{\mathrm{cs}}(x_{\mathrm{c}})$, $x_{\mathrm{c}} \in \mathbb{R}^{n_{\mathrm{c}}}$, is positive definite. Now, since

$$\min_{i=1,\dots,q} \{p_i\sigma_i\}\mathbf{e}^{\mathrm{T}}V_{\mathrm{cs}}(x_{\mathrm{c}}) \leq p^{\mathrm{T}}\Sigma V_{\mathrm{cs}}(x_{\mathrm{c}}), \qquad (6.78)$$

it follows that $p^{\mathrm{T}}\Sigma V_{\mathrm{cs}}(x_{\mathrm{c}})$, $x_{\mathrm{c}} \in \mathbb{R}^{n_{\mathrm{c}}}$, is positive definite. Hence, the function $v(x, x_{\mathrm{c}}) \triangleq p^{\mathrm{T}}V(x, x_{\mathrm{c}})$, $(x, x_{\mathrm{c}}) \in \mathbb{R}^{n} \times \mathbb{R}^{n_{\mathrm{c}}}$, is positive definite. Now, the result is a direct consequence of Theorem 2.11.

$ii)$ The proof follows from $i)$ by noting that, for all $i = 1, \dots, q$,

$$s_{\mathrm{c}i}(u_{\mathrm{c}i}, y_{\mathrm{c}i}) + \sigma_i s_{\mathrm{cc}i}(u_{\mathrm{cc}i}, y_{\mathrm{cc}i}) = \begin{bmatrix} y_{\mathrm{c}} \\ y_{\mathrm{cc}} \end{bmatrix}^{\mathrm{T}} \tilde{Q}_{\mathrm{c}i} \begin{bmatrix} y_{\mathrm{c}} \\ y_{\mathrm{cc}} \end{bmatrix}, \qquad (6.79)$$

$$s_{\mathrm{d}i}(u_{\mathrm{d}i}, y_{\mathrm{d}i}) + \sigma_i s_{\mathrm{dc}i}(u_{\mathrm{dc}i}, y_{\mathrm{dc}i}) = \begin{bmatrix} y_{\mathrm{d}} \\ y_{\mathrm{dc}} \end{bmatrix}^{\mathrm{T}} \tilde{Q}_{\mathrm{d}i} \begin{bmatrix} y_{\mathrm{d}} \\ y_{\mathrm{dc}} \end{bmatrix}, \qquad (6.80)$$

and hence, $S_{\mathrm{c}}(u_{\mathrm{c}}, y_{\mathrm{c}}) + \Sigma S_{\mathrm{cc}}(u_{\mathrm{cc}}, y_{\mathrm{cc}}) \leq\leq 0$ and $S_{\mathrm{d}}(u_{\mathrm{d}}, y_{\mathrm{d}}) + \Sigma S_{\mathrm{dc}}(u_{\mathrm{dc}}, y_{\mathrm{dc}}) \leq\leq 0$. $\qquad \square$

For the next result note that if the large-scale impulsive dynamical system $\mathcal{G}$ is vector dissipative with respect to the vector hybrid supply rate $(S_{\mathrm{c}}(u_{\mathrm{c}}, y_{\mathrm{c}}), S_{\mathrm{d}}(u_{\mathrm{d}}, y_{\mathrm{d}}))$, where $s_{\mathrm{c}i}(u_{\mathrm{c}i}, y_{\mathrm{c}i}) = 2y_{\mathrm{c}i}^{\mathrm{T}}u_{\mathrm{c}i}$ and $s_{\mathrm{d}i}(u_{\mathrm{d}i}, y_{\mathrm{d}i}) = 2y_{\mathrm{d}i}^{\mathrm{T}}u_{\mathrm{d}i}$, $i = 1, \dots, q$, then, with $\kappa_{\mathrm{c}i}(y_{\mathrm{c}i}) = -\kappa_{\mathrm{c}i}y_{\mathrm{c}i}$ and $\kappa_{\mathrm{d}i}(y_{\mathrm{d}i}) = -\kappa_{\mathrm{d}i}y_{\mathrm{d}i}$, where $\kappa_{\mathrm{c}i} > 0$, $\kappa_{\mathrm{d}i} > 0$, $i = 1, \dots, q$, it follows that $s_{\mathrm{c}i}(\kappa_{\mathrm{c}i}(y_{\mathrm{c}i}), y_{\mathrm{c}i}) = -2\kappa_{\mathrm{c}i}y_{\mathrm{c}i}^{\mathrm{T}}y_{\mathrm{c}i} < 0$ and $s_{\mathrm{d}i}(\kappa_{\mathrm{d}i}(y_{\mathrm{d}i}), y_{\mathrm{d}i}) = -2\kappa_{\mathrm{d}i}y_{\mathrm{d}i}^{\mathrm{T}}y_{\mathrm{d}i} < 0$, $y_{\mathrm{c}i} \neq 0$, $y_{\mathrm{d}i} \neq 0$, $i = 1, \dots, q$. Alternatively, if $\mathcal{G}$ is vector dissipative with respect to the vector hybrid supply rate $(S_{\mathrm{c}}(u_{\mathrm{c}}, y_{\mathrm{c}}), S_{\mathrm{d}}(u_{\mathrm{d}}, y_{\mathrm{d}}))$, where $s_{\mathrm{c}i}(u_{\mathrm{c}i}, y_{\mathrm{c}i}) = \gamma_{\mathrm{c}i}^2 u_{\mathrm{c}i}^{\mathrm{T}}u_{\mathrm{c}i} - y_{\mathrm{c}i}^{\mathrm{T}}y_{\mathrm{c}i}$ and $s_{\mathrm{d}i}(u_{\mathrm{d}i}, y_{\mathrm{d}i}) = \gamma_{\mathrm{d}i}^2 u_{\mathrm{d}i}^{\mathrm{T}}u_{\mathrm{d}i} - y_{\mathrm{d}i}^{\mathrm{T}}y_{\mathrm{d}i}$, where $\gamma_{\mathrm{c}i} > 0$, $\gamma_{\mathrm{d}i} > 0$, $i = 1, \dots, q$, then, with $\kappa_{\mathrm{c}i}(y_{\mathrm{c}i}) = 0$ and $\kappa_{\mathrm{d}i}(y_{\mathrm{d}i}) = 0$, it follows that $s_{\mathrm{c}i}(\kappa_{\mathrm{c}i}(y_{\mathrm{c}i}), y_{\mathrm{c}i}) = -y_{\mathrm{c}i}^{\mathrm{T}}y_{\mathrm{c}i} < 0$ and $s_{\mathrm{d}i}(\kappa_{\mathrm{d}i}(y_{\mathrm{d}i}), y_{\mathrm{d}i}) = -y_{\mathrm{d}i}^{\mathrm{T}}y_{\mathrm{d}i} < 0$, $y_{\mathrm{c}i} \neq 0$, $y_{\mathrm{d}i} \neq 0$, $i = 1, \dots, q$. Hence, if $\mathcal{G}$ is zero-state observable and the dissipation matrix W is such that there exist $\alpha \geq 0$ and $p \in \mathbb{R}_+^q$ such that (5.24) holds, then it follows from Theorem 5.4 that (scalar) storage functions of the form $v_{\mathrm{s}}(x) = p^{\mathrm{T}}V_{\mathrm{s}}(x)$, $x \in \mathbb{R}^n$, where $V_{\mathrm{s}}(\cdot)$ is a vector storage function for $\mathcal{G}$, are positive definite. If $\mathcal{G}$ is exponentially vector dissipative, then p is positive.

Corollary 6.3 *Consider the large-scale impulsive dynamical systems $\mathcal{G}$ and $\mathcal{G}_{\mathrm{c}}$ given by (6.57)–(6.60) and (6.65)–(6.68), respectively. Assume that $\mathcal{G}$ and $\mathcal{G}_{\mathrm{c}}$ are zero-state observable and the dissipation matrices $W \in \mathbb{R}^{q \times q}$ and $W_{\mathrm{c}} \in \mathbb{R}^{q \times q}$ are such that there exist, respectively, $\alpha \geq 0$, $p \in \mathbb{R}_+^q$, $\alpha_{\mathrm{c}} \geq 0$, and $p_{\mathrm{c}} \in \mathbb{R}_+^q$ such that (5.24) is satisfied. Then the following statements hold:*

i) *If $\mathcal{G}$ and $\mathcal{G}_{\mathrm{c}}$ are vector passive and $\tilde{W} \in \mathbb{R}^{q \times q}$ is asymptotically stable, where $\tilde{W}_{(i,j)} \triangleq \max\{W_{(i,j)}, W_{\mathrm{c}(i,j)}\}$, $i, j = 1, \ldots, q$, then the negative feedback interconnection of $\mathcal{G}$ and $\mathcal{G}_{\mathrm{c}}$ is asymptotically stable.*

ii) *If $\mathcal{G}$ and $\mathcal{G}_{\mathrm{c}}$ are vector nonexpansive and $\tilde{W} \in \mathbb{R}^{q \times q}$ is asymptotically stable, where $\tilde{W}_{(i,j)} \triangleq \max\{W_{(i,j)}, W_{\mathrm{c}(i,j)}\}$, $i, j = 1, \ldots, q$, then the negative feedback interconnection of $\mathcal{G}$ and $\mathcal{G}_{\mathrm{c}}$ is asymptotically stable.*

Proof. The proof is a direct consequence of Theorem 6.4. Specifically, statement *i)* follows from Theorem 6.4 with $R_{\mathrm{c}i} = 0$, $S_{\mathrm{c}i} = I_{m_{\mathrm{c}i}}$, $Q_{\mathrm{c}i} = 0$, $R_{\mathrm{d}i} = 0$, $S_{\mathrm{d}i} = I_{m_{\mathrm{d}i}}$, $Q_{\mathrm{d}i} = 0$, $R_{\mathrm{cc}i} = 0$, $S_{\mathrm{cc}i} = I_{m_{\mathrm{c}i}}$, $Q_{\mathrm{cc}i} = 0$, $R_{\mathrm{dc}i} = 0$, $S_{\mathrm{dc}i} = I_{m_{\mathrm{d}i}}$, $Q_{\mathrm{dc}i} = 0$, $i = 1, \ldots, q$, and $\Sigma = I_q$. Statement *ii)* follows from Theorem 6.4 with $R_{\mathrm{c}i} = \gamma_{\mathrm{c}i}^2 I_{m_{\mathrm{c}i}}$, $S_{\mathrm{c}i} = 0$, $Q_{\mathrm{c}i} = -I_{l_{\mathrm{c}i}}$, $R_{\mathrm{d}i} = \gamma_{\mathrm{d}i}^2 I_{m_{\mathrm{d}i}}$, $S_{\mathrm{d}i} = 0$, $Q_{\mathrm{d}i} = -I_{l_{\mathrm{d}i}}$, $R_{\mathrm{cc}i} = \gamma_{\mathrm{cc}i}^2 I_{l_{\mathrm{c}i}}$, $S_{\mathrm{cc}i} = 0$, $Q_{\mathrm{cc}i} = -I_{m_{\mathrm{c}i}}$, $R_{\mathrm{dc}i} = \gamma_{\mathrm{dc}i}^2 I_{l_{\mathrm{d}i}}$, $S_{\mathrm{dc}i} = 0$, $Q_{\mathrm{dc}i} = -I_{m_{\mathrm{d}i}}$, $i = 1, \ldots, q$, and $\Sigma = I_q$. $\square$

Chapter Seven

Energy-Based Control for Impulsive Port-Controlled Hamiltonian Systems

7.1 Introduction

In a recent series of papers [136–138] a passivity-based control framework for port-controlled Hamiltonian systems is established. Specifically, the authors in [136–138] develop a controller design methodology that achieves stabilization via system passivation. In particular, the interconnection and damping matrix functions of the port-controlled Hamiltonian system are shaped so that the physical (Hamiltonian) system structure is preserved at the closed-loop level, and the closed-loop energy function is equal to the difference between the physical energy of the system and the energy supplied by the controller. Since the Hamiltonian structure is preserved at the closed-loop level, the passivity-based controller is *robust* with respect to unmodeled passive dynamics. Furthermore, passivity-based control architectures are extremely appealing since the control action has a clear *physical* energy interpretation which can considerably simplify controller implementation.

Modern complex engineering systems involve multiple modes of operation, placing stringent demands on controller design and implementation of increasing complexity. As discussed in Chapter 1, such systems typically possess a multiechelon hierarchical hybrid control architecture characterized by continuous-time dynamics at the lower levels of the hierarchy and discrete-time dynamics at the higher levels of hierarchy. The mathematical description of many of these systems can be characterized by impulsive differential equations. Furthermore, since certain dynamical systems such as telecommunications, transportation, biological, physiological, power, and network systems involve high-level, abstract hierarchies with input-output properties related to conservation, dissipation, and transport of mass and/or energy, these systems can be modeled as *impulsive port-controlled Hamiltonian systems.*

In this chapter, we use the stability and dissipativity framework for impulsive dynamical systems developed in Chapters 2 and 3 to extend the results in [136–138] to nonlinear impulsive port-controlled Hamiltonian systems. Specifically, we develop an energy-based hybrid feedback control framework for nonlinear impulsive port-controlled Hamiltonian systems that preserves the physical hybrid Hamiltonian structure at the closed-loop level. In particular, we present sufficient conditions for hybrid feedback stabilization that preserve the physical hybrid Hamiltonian structure at the closed-loop level while providing a shaped Hamiltonian energy function as a Lyapunov function for the closed-loop impulsive system. These sufficient conditions consist of a hybrid system of two partial differential equations involving the continuous-time dynamics and the resetting (discrete-time) dynamics. We emphasize that our approach is constructive in nature providing a hybrid system of partial differential equations whose solutions, when they exist, characterize the set of all desired shaped Hamiltonian energy functions that can be assigned while preserving the hybrid Hamiltonian structure at the closed-loop system level.

Unlike the passivity-based control framework developed in [136–138] for port-controlled Hamiltonian systems with continuous flows, our approach does not achieve stabilization via hybrid system passivation in the sense of [56, 61]. However, under certain conditions on the open and closed-loop dissipation matrix functions, the closed-loop energy function over the continuous-time trajectories is equal to the difference between the physical energy of the hybrid system and the energy supplied by hybrid controller. Furthermore, the closed-loop energy function at the resetting instants is nonincreasing.

7.2 Impulsive Port-Controlled Hamiltonian Systems

In this section, we introduce nonlinear impulsive port-controlled Hamiltonian systems. We begin by considering an *input/state-dependent* impulsive port-controlled Hamiltonian system $\mathcal{G}$ given by

$$\dot{x}(t) = [\mathcal{J}_c(x(t)) - \mathcal{R}_c(x(t))]\left(\frac{\partial\mathcal{H}}{\partial x}(x(t))\right)^{\mathrm{T}} + G_c(x(t))u_c(t),$$

$$x(0) = x_0, \quad (x(t), u_c(t)) \notin \mathcal{Z}, \quad (7.1)$$

$$\Delta x(t) = [\mathcal{J}_d(x(t)) - \mathcal{R}_d(x(t))]\left(\frac{\partial\mathcal{H}}{\partial x}(x(t))\right)^{\mathrm{T}} + G_d(x(t))u_d(t),$$

$$(x(t), u_c(t)) \in \mathcal{Z}, \quad (7.2)$$

$$y_c(t) = h_c(x(t)) + J_c(x(t))u_c(t), \quad (x(t), u_c(t)) \notin \mathcal{Z}, \quad (7.3)$$

$$y_{\mathrm{d}}(t) = h_{\mathrm{d}}(x(t)) + J_{\mathrm{d}}(x(t))u_{\mathrm{d}}(t), \quad (x(t), u_{\mathrm{c}}(t)) \in \mathcal{Z}, \qquad (7.4)$$

where $t \geq 0$, $x(t) \in \mathcal{D} \subseteq \mathbb{R}^n$, $\mathcal{D}$ is an open set, $\Delta x(t) \triangleq x(t^+) - x(t)$, $u_{\mathrm{c}}(t) \in \mathcal{U}_{\mathrm{c}} \subseteq \mathbb{R}^{m_{\mathrm{c}}}$, $u_{\mathrm{d}}(t_k) \in \mathcal{U}_{\mathrm{d}} \subseteq \mathbb{R}^{m_{\mathrm{d}}}$, t_k denotes the kth instant of time at which $(x(t), u_{\mathrm{c}}(t))$ intersects $\mathcal{Z}$ for a particular trajectory $x(t)$ and input $u_{\mathrm{c}}(t)$, $y_{\mathrm{c}}(t) \in Y_{\mathrm{c}} \subseteq \mathbb{R}^{l_{\mathrm{c}}}$, $y_{\mathrm{d}}(t_k) \in Y_{\mathrm{d}} \subseteq \mathbb{R}^{l_{\mathrm{d}}}$, $\mathcal{H} : \mathcal{D} \to \mathbb{R}$ is a continuously differentiable *Hamiltonian function* for the impulsive system (7.1)–(7.4), $\mathcal{J}_{\mathrm{c}} : \mathcal{D} \to \mathbb{R}^{n \times n}$ is such that $\mathcal{J}_{\mathrm{c}}(x) = -\mathcal{J}_{\mathrm{c}}^{\mathrm{T}}(x)$, $\mathcal{R}_{\mathrm{c}} : \mathcal{D} \to \mathbb{S}^n$ is such that $\mathcal{R}_{\mathrm{c}}(x) \geq 0$, $x \in \mathcal{D}$, $[\mathcal{J}_{\mathrm{c}}(x) - \mathcal{R}_{\mathrm{c}}(x)] \left(\frac{\partial \mathcal{H}}{\partial x}(x)\right)^{\mathrm{T}}$, $x \in \mathcal{D}$, is Lipschitz continuous, $G_{\mathrm{c}} : \mathcal{D} \to \mathbb{R}^{n \times m_{\mathrm{c}}}$, $\mathcal{J}_{\mathrm{d}} : \mathcal{D} \to \mathbb{R}^{n \times n}$ is such that $\mathcal{J}_{\mathrm{d}}(x) = -\mathcal{J}_{\mathrm{d}}^{\mathrm{T}}(x)$, $\mathcal{R}_{\mathrm{d}} : \mathcal{D} \to \mathbb{S}^n$ is such that $\mathcal{R}_{\mathrm{d}}(x) \geq 0$, $x \in \mathcal{D}$, $[\mathcal{J}_{\mathrm{d}}(x) - \mathcal{R}_{\mathrm{d}}(x)] \left(\frac{\partial \mathcal{H}}{\partial x}(x)\right)^{\mathrm{T}}$, $x \in \mathcal{D}$, is continuous, $G_{\mathrm{d}} : \mathcal{D} \to \mathbb{R}^{n \times m_{\mathrm{c}}}$, $h_{\mathrm{c}} : \mathcal{D} \to \mathbb{R}^{l_{\mathrm{c}}}$, $J_{\mathrm{c}} : \mathcal{D} \to \mathbb{R}^{l_{\mathrm{c}} \times m_{\mathrm{c}}}$, $h_{\mathrm{d}} : \mathcal{D} \to \mathbb{R}^{l_{\mathrm{d}}}$, $J_{\mathrm{d}} : \mathcal{D} \to \mathbb{R}^{l_{\mathrm{d}} \times m_{\mathrm{d}}}$, and $\mathcal{Z} \triangleq (\mathcal{Z}_x \times \mathcal{U}_{\mathrm{c}}) \cup (\mathbb{R}^n \times \mathcal{Z}_{u_{\mathrm{c}}}) \subset \mathcal{D} \times \mathcal{U}_{\mathrm{c}}$ is the resetting set.

The skew-symmetric matrix functions $\mathcal{J}_{\mathrm{c}}(x)$ and $\mathcal{J}_{\mathrm{d}}(x)$, $x \in \mathcal{D}$, capture the internal hybrid system interconnection structure, the input matrix functions $G_{\mathrm{c}}(x)$ and $G_{\mathrm{d}}(x)$, $x \in \mathcal{D}$, capture hybrid interconnections with the environment, and the symmetric nonnegative-definite matrix functions $\mathcal{R}_{\mathrm{c}}(x)$ and $\mathcal{R}_{\mathrm{d}}(x)$, $x \in \mathcal{D}$, capture hybrid system dissipation. Here, we assume that $u_{\mathrm{c}}(\cdot)$ and $u_{\mathrm{d}}(\cdot)$ are restricted to the class of *admissible* inputs consisting of measurable functions such that $(u_{\mathrm{c}}(t), u_{\mathrm{d}}(t_k)) \in \mathcal{U}_{\mathrm{c}} \times \mathcal{U}_{\mathrm{d}}$ for all $t \geq 0$ and $k \in \mathbb{Z}_{[0,t)} \triangleq \{k : 0 \leq t_k < t\}$. We denote the solution to (7.1) and (7.2) with initial condition $x_0 \in \mathcal{D}$ by $s(t, x_0)$, $t \geq 0$, and the set of the resetting times $t_k \equiv \tau_k(x_0)$ for a particular trajectory $s(\cdot, x_0)$ by $[0, \infty) \backslash \mathcal{T}_{x_0, u_{\mathrm{c}}} \triangleq \{t_1, t_2, \ldots\}$, where $\mathcal{T}_{x_0, u_{\mathrm{c}}}$ is a dense subset of the semi-infinite interval $[0, \infty)$ such that $\mathcal{T}_{x_0, u_{\mathrm{c}}}^{\mathrm{c}} \triangleq [0, \infty) \backslash \mathcal{T}_{x_0, u_{\mathrm{c}}}$ is (finitely or infinitely) countable. For notational convenience we write $\mathcal{T}$ and $\mathcal{T}^{\mathrm{c}}$ for $\mathcal{T}_{x_0, u_{\mathrm{c}}}$ and $\mathcal{T}_{x_0, u_{\mathrm{c}}}^{\mathrm{c}}$, respectively.

Note that the solution $x(t)$, $t \geq 0$, of (7.1) and (7.2) is left-continuous. Furthermore, as shown in Chapter 2, if the resetting set is such that it removes $x(t_k)$ from the resetting set and if no trajectory can intersect the interior of $\mathcal{Z}$, then the resetting times t_k, $k \in \overline{\mathbb{Z}}_+$, are well defined and distinct. Since the resetting times are well defined and distinct, and since the solution to (7.1) exists and is unique, it follows that the solution of the impulsive port-controlled Hamiltonian system (7.1) and (7.2) also exists and is unique over a forward time interval. However, as discussed in Chapter 2, the analysis of impulsive dynamical systems can be quite involved. In particular, such systems can exhibit Zenoness and beating, as well as confluence. Furthermore, due to Zeno solutions, *not* every bounded solution of an impulsive dynamical system over a forward time interval can be extended to infinity. Here, we assume that Assumptions A1 and A2 established in Chapter 2 hold, and hence we allow for the possibility of confluence and Zeno solutions, however, we preclude the possibility of beating.

It is important to note that in our impulsive system formulation (7.1) and (7.2) we assume that the impact model dynamics (7.2) is Hamiltonian. For mechanical systems with collisions this is without loss of generality. To see this, let $x = [q^{\mathrm{T}}, \dot{q}^{\mathrm{T}}]^{\mathrm{T}}$, where $q \in \mathbb{R}^{\hat{n}}$ represents generalized positions and $\dot{q} \in \mathbb{R}^{\hat{n}}$ represents generalized velocities, and $\hat{n} = \frac{n}{2}$, and note that the impact dynamics are given by

$$q(t_k^+) = q(t_k), \tag{7.5}$$

$$\left[\frac{\partial T}{\partial \dot{q}}(t_k^+) \right]^{\mathrm{T}} = \mathcal{I}(q(t_k), \dot{q}(t_k)) \left[\frac{\partial T}{\partial \dot{q}}(t_k) \right]^{\mathrm{T}}, \tag{7.6}$$

where $T(q, \dot{q}) = \frac{1}{2}\dot{q}^{\mathrm{T}} M(q)\dot{q}$ is the system kinetic energy, $M(q) > 0$, $q \in \mathbb{R}^{\hat{n}}$, is the system inertia matrix function, $\mathcal{I} : \mathbb{R}^{\hat{n}} \times \mathbb{R}^{\hat{n}} \to \mathbb{R}^{\hat{n} \times \hat{n}}$ is an impact matrix function, and t_k, t_k^+ are the instants before and after collisions, respectively.

The impact function $\mathcal{I}(\cdot, \cdot)$ can be quite difficult to characterize since solid impacts can involve stress waves, expansions in colliding solids, and reflections from solid boundaries. To capture the dynamics of these waves it is often necessary to use partial differential equations. For an additional discussion on impact dynamics see [33, 34]. However, assuming that across a collision event the generalized system velocities change according to the law of conservation of momentum, and the generalized velocities account for the loss of kinetic energy in a collision, (7.5) and (7.6) can be rewritten as

$$
\begin{bmatrix} \Delta q(t_k) \\ \Delta \dot{q}(t_k) \end{bmatrix} = \begin{bmatrix} 0 & 0 \\ 0 & M^{-1}(q)\mathcal{I}(q, \dot{q})M(q) - I_{\hat{n}} \end{bmatrix} \begin{bmatrix} q(t_k) \\ \dot{q}(t_k) \end{bmatrix}
$$

$$
= \begin{bmatrix} 0 & 0 \\ 0 & M^{-1}(q)\mathcal{I}(q, \dot{q}) - M^{-1}(q) \end{bmatrix} \begin{bmatrix} \left(\frac{\partial \mathcal{H}}{\partial q}(q(t_k), \dot{q}(t_k)) \right)^{\mathrm{T}} \\ \left(\frac{\partial \mathcal{H}}{\partial \dot{q}}(q(t_k), \dot{q}(t_k)) \right)^{\mathrm{T}} \end{bmatrix}, \tag{7.7}
$$

where $\mathcal{H}(q, \dot{q}) = T(q, \dot{q}) + V(q)$ denotes the total system energy and $V(q)$ is the system potential energy.

Next, note that the matrix function $M^{-1}(q)\mathcal{I}(q, \dot{q}) - M^{-1}(q)$, $(q, \dot{q}) \in \mathbb{R}^{\hat{n}} \times \mathbb{R}^{\hat{n}}$, can be represented as a sum of a skew-symmetric matrix function and a negative-semidefinite matrix function if and only if

$$(\mathcal{I}^{\mathrm{T}}(q, \dot{q}) - I_{\hat{n}})M^{-1}(q) + M^{-1}(q)(\mathcal{I}(q, \dot{q}) - I_{\hat{n}}) \leq 0,$$
$$(q, \dot{q}) \in \mathbb{R}^{\hat{n}} \times \mathbb{R}^{\hat{n}}. \quad (7.8)$$

Now, assuming that the kinetic energy after the impact is less than or equal to the kinetic energy before the impact, that is,

$$T(q(t_k^+), \dot{q}(t_k^+)) \leq T(q(t_k), \dot{q}(t_k)), \quad (7.9)$$

it follows from (7.7) and (7.9), since (7.9) holds for arbitrary $q, \dot{q} \in \mathbb{R}^{\hat{n}}$, that

$$M(q)\mathcal{I}^{\mathrm{T}}(q, \dot{q})M^{-1}(q)\mathcal{I}(q, \dot{q})M(q) \leq M(q), \quad (q, \dot{q}) \in \mathbb{R}^{\hat{n}} \times \mathbb{R}^{\hat{n}}, \quad (7.10)$$

which is equivalent to

$$\sigma_{\max}[M^{-\frac{1}{2}}(q)\mathcal{I}(q, \dot{q})M^{\frac{1}{2}}(q)] \leq 1, \quad (q, \dot{q}) \in \mathbb{R}^{\hat{n}} \times \mathbb{R}^{\hat{n}}, \quad (7.11)$$

where $\sigma_{\max}(\cdot)$ denotes the maximum singular value. Now, it follows from (7.11) that

$$M^{-\frac{1}{2}}(q)\mathcal{I}(q, \dot{q})M^{\frac{1}{2}}(q) + M^{\frac{1}{2}}(q)\mathcal{I}^{\mathrm{T}}(q, \dot{q})M^{-\frac{1}{2}}(q)$$
$$\leq \sigma_{\max}[M^{-\frac{1}{2}}(q)\mathcal{I}(q, \dot{q})M^{\frac{1}{2}}(q)$$
$$+ M^{\frac{1}{2}}(q)\mathcal{I}^{\mathrm{T}}(q, \dot{q})M^{-\frac{1}{2}}(q)]I_{\hat{n}}$$
$$\leq 2\sigma_{\max}[M^{-\frac{1}{2}}(q)\mathcal{I}(q, \dot{q})M^{\frac{1}{2}}(q)]I_{\hat{n}}$$
$$\leq 2I_{\hat{n}}, \quad (q, \dot{q}) \in \mathbb{R}^{\hat{n}} \times \mathbb{R}^{\hat{n}}, \quad (7.12)$$

where $(\cdot)^{1/2}$ denotes the (unique) positive-definite square root. Hence,

$$M^{-\frac{1}{2}}(q)\mathcal{I}(q, \dot{q})M^{\frac{1}{2}}(q) + M^{\frac{1}{2}}(q)\mathcal{I}^{\mathrm{T}}(q, \dot{q})M^{-\frac{1}{2}}(q) - 2I_{\hat{n}} \leq 0,$$
$$(q, \dot{q}) \in \mathbb{R}^{\hat{n}} \times \mathbb{R}^{\hat{n}}, \quad (7.13)$$

which is equivalent to (7.8), and hence, the impact dynamics (7.7) can be written in a Hamiltonian form

$$\Delta x(t_k) = [\mathcal{J}(x(t_k)) - \mathcal{R}(x(t_k))] \left(\frac{\partial \mathcal{H}}{\partial x}(x(t_k)) \right)^{\mathrm{T}}. \quad (7.14)$$

Finally, we note that $G_{\mathrm{d}}(x)u_{\mathrm{d}}$ in (7.2) provides the additional flexibility of including an impulsive control to the impact dynamics. See [168] for additional details.

Assuming that the Hamiltonian energy function $\mathcal{H}(\cdot)$ is lower bounded, it can be shown (with an additional structural constraint on $\mathcal{H}(\cdot)$)

that impulsive port-controlled Hamiltonian systems provide a hybrid energy balance in terms of the stored or accumulated energy, hybrid supplied system energy, dissipated energy over the continuous-time dynamics, and dissipated energy at the resetting instants. To see this, let the hybrid inputs and hybrid outputs be dual (conjugated) variables so that $y_c(t) = G_c^T(x(t)) \left(\frac{\partial \mathcal{H}}{\partial x}(x(t))\right)^T$, $(x(t), u_c(t)) \notin \mathcal{Z}$, $y_d(t) = G_d^T(x(t)) \left(\frac{\partial \mathcal{H}}{\partial x}(x(t))\right)^T$, $(x(t), u_c(t)) \in \mathcal{Z}$, and assume $\mathcal{H}(\cdot)$ is such that[1]

$$\mathcal{H}\left(x + [\mathcal{J}_d(x) - \mathcal{R}_d(x)]\left(\frac{\partial \mathcal{H}}{\partial x}(x)\right)^T + G_d(x)u_d\right) = \mathcal{H}(x)$$

$$+ \frac{\partial \mathcal{H}}{\partial x}(x) [\mathcal{J}_d(x) - \mathcal{R}_d(x)]\left(\frac{\partial \mathcal{H}}{\partial x}(x)\right)^T + \left(\frac{\partial \mathcal{H}}{\partial x}(x)\right) G_d(x)u_d,$$

$$x \in \mathcal{D}, \quad u_d \in U_d. \qquad (7.15)$$

Now, computing the rate of change of the Hamiltonian along the system state trajectories $x(t)$, $t \in (t_k, t_{k+1}]$, and the Hamiltonian difference at the resetting times t_k, $k \in \overline{\mathbb{Z}}_+$, yields the set of energy conservation equations given by[2]

$$\dot{\mathcal{H}}(x(t)) = u_c^T(t)y_c(t) - \frac{\partial \mathcal{H}}{\partial x}(x(t))\mathcal{R}_c(x(t))\left(\frac{\partial \mathcal{H}}{\partial x}(x(t))\right)^T,$$

$$t_k < t \leq t_{k+1}, \qquad (7.16)$$

$$\Delta\mathcal{H}(x(t_k)) \triangleq \mathcal{H}(x(t_k^+)) - \mathcal{H}(x(t_k)) = u_d^T(t_k)y_d(t_k)$$

$$- \frac{\partial \mathcal{H}}{\partial x}(x(t_k))\mathcal{R}_d(x(t_k))\left(\frac{\partial \mathcal{H}}{\partial x}(x(t_k))\right)^T, \quad k \in \overline{\mathbb{Z}}_+. \qquad (7.17)$$

Equation (7.16) shows that the rate of change in energy, or power, over the time interval $t \in (t_k, t_{k+1}]$ is equal to the system power input minus the internal system power dissipated, while (7.17) shows that the change of energy at the resetting times t_k, $k \in \overline{\mathbb{Z}}_+$, is equal to the supplied system energy at the resetting times minus the dissipated energy at the resetting times. Using Theorem 3.2, (7.16) and (7.17)

---

[1]The structural constraint on the Hamiltonian given by (7.15) is natural for nonnegative and compartmental dynamical systems where the state vector is restricted to the nonnegative orthant of the state space [57, 63]. For these systems the Hamiltonian represents the total mass/energy in the system and is a *linear* function of the state. For details, see Section 4.3.

[2]Note that (7.16) holds even if $\mathcal{H}(\cdot)$ does *not* satisfy the structural constraint (7.15).

can be equivalently written as

$$\mathcal{H}(x(t)) - \mathcal{H}(x(0)) = \int_0^t u_c^T(s) y_c(s) ds + \sum_{k \in \mathbb{Z}_{[0,t)}} u_d^T(t_k) y_d(t_k)$$

$$- \int_0^t \frac{\partial \mathcal{H}}{\partial x}(x(s)) \mathcal{R}_c(x(s)) \left(\frac{\partial \mathcal{H}}{\partial x}(x(s)) \right)^T ds$$

$$- \sum_{k \in \mathbb{Z}_{[0,t)}} \frac{\partial \mathcal{H}}{\partial x}(x(t_k)) \mathcal{R}_d(x(t_k)) \left(\frac{\partial \mathcal{H}}{\partial x}(x(t_k)) \right)^T,$$

$$t \geq 0. \quad (7.18)$$

Equation (7.18) shows that the stored or accumulated system energy is equal to the energy supplied to the system via the hybrid external inputs u_c and u_d minus the energy dissipated over the continuous-time dynamics and the resetting instants. Since $\mathcal{R}_c(x)$ and $\mathcal{R}_d(x)$ are nonnegative definite for all $x \in \mathcal{D}$, it follows from (7.18) that

$$- \left[\int_0^t u_c^T(s) y_c(s) ds + \sum_{k \in \mathbb{Z}_{[0,t)}} u_d^T(t_k) y_d(t_k) \right] \leq \mathcal{H}(x(0)), \quad (7.19)$$

which shows that the energy that can be extracted from the impulsive port-controlled Hamiltonian system through the hybrid input-output ports is less than or equal to the initial energy stored in the system. Hence, impulsive port-controlled Hamiltonian systems with the structural constraint (7.15) are passive systems in the sense of Definition 3.5.

7.3 Energy-Based Hybrid Feedback Control

In this section, we present an energy-based hybrid feedback control framework for nonlinear impulsive port-controlled Hamiltonian systems that preserves the Hamiltonian structure at the closed-loop level. In particular, we obtain constructive sufficient conditions for feedback stabilization of an arbitrary equilibrium point in $\mathcal{D}$ that provide a shaped energy function for the closed-loop system while preserving a hybrid Hamiltonian structure at the closed-loop level. To address the energy-based hybrid feedback control problem let $\phi_c : \mathcal{D} \to U_c$ and $\phi_d : \mathcal{D} \to U_d$. If $(u_c(t), u_d(t_k)) = (\phi_c(x(t)), \phi_d(x(t_k)))$, then $(u_c(\cdot), u_d(\cdot))$ is a *hybrid feedback control*. Note that with the hybrid feedback control law $(u_c(t), u_d(t_k)) = (\phi_c(x(t)), \phi_d(x(t_k)))$, the resetting set $\mathcal{Z}$ can be equivalently rewritten as a state-dependent manifold $\mathcal{Z} = \mathcal{Z}_x \cup \{x \in \mathcal{D} : \phi_c(x) \in \mathcal{Z}_{u_c}\}$. To state the main result of this section, we assume that the impulsive closed-loop system (7.1) and (7.2) with hybrid feedback controller $(u_c(t), u_d(t_k)) = (\phi_c(x(t)), \phi_d(x(t_k)))$ is such that Assumption 2.1 holds. For the closed-loop system (7.1) and (7.2) this assumption takes the following form.

Assumption 7.1 *Consider the impulsive port-controlled Hamiltonian dynamical system $\mathcal{G}$ given by (7.1) and (7.2) with hybrid feedback controller $(u_c(t), u_d(t_k)) = (\phi_c(x(t)), \phi_d(x(t_k)))$, and let $s(t, x_0)$, $t \geq 0$, denote the solution to (7.1) and (7.2) with initial condition x_0. Then for every $x_0 \in \mathcal{D}$, there exists a dense subset $\mathcal{T}_{x_0} \subseteq [0, \infty)$ such that $[0, \infty) \backslash \mathcal{T}_{x_0}$ is (finitely or infinitely) countable and for every $\varepsilon > 0$ and $t \in \mathcal{T}_{x_0}$, there exists $\delta(\varepsilon, x_0, t) > 0$ such that if $\|x_0 - y\| < \delta(\varepsilon, x_0, t)$, $y \in \mathcal{D}$, then $\|s(t, x_0) - s(t, y)\| < \varepsilon$.*

As shown in Chapter 2, sufficient conditions that guarantee that the trajectories of the closed-loop nonlinear impulsive dynamical system satisfy Assumption 7.1 are Lipshitz continuity of the continuous-time, closed-loop dynamics and the existence of a continuously differentiable function $\mathcal{X} : \mathcal{D} \to \mathbb{R}$ such that the resetting set is given by $\mathcal{Z} = \{x \in \mathcal{D} : \mathcal{X}(x) = 0\}$, where $\mathcal{X}'(x) \neq 0, x \in \mathcal{Z}$, and $\frac{\partial \mathcal{X}(x)}{\partial x} \left([\mathcal{J}_c(x) - \mathcal{R}_c(x)] \left(\frac{\partial \mathcal{H}}{\partial x}(x) \right)^{\mathrm{T}} + G_c(x)\phi_c(x) \right) \neq 0, x \in \mathcal{Z}$. The last condition ensures that the solution $s(t, x_0)$, $t \geq 0$, of the closed-loop system is not tangent to the resetting set $\mathcal{Z}$ for all initial conditions $x_0 \in \mathcal{D}$.

Next, we provide constructive sufficient conditions for energy-based hybrid feedback control of impulsive port-controlled Hamiltonian systems. Here, we restrict our attention to the *state-dependent* impulsive port-controlled Hamiltonian system (7.1)–(7.4), that is, $\mathcal{Z}_{u_c} = U_c$. Specifically, we seek hybrid feedback controllers $(u_c(t), u_d(t_k)) = (\phi_c (x(t)), \phi_d(x(t_k)))$, where $\phi_c : \mathcal{D} \to U_c$ and $\phi_d : \mathcal{Z}_x \to U_d$, such that the closed-loop system has the form

$$\dot{x}(t) = [\mathcal{J}_c(x(t)) - \mathcal{R}_c(x(t))] \left(\frac{\partial \mathcal{H}}{\partial x}(x(t)) \right)^{\mathrm{T}} + G_c(x(t))\phi_c(x(t))$$

$$= [\mathcal{J}_{cs}(x(t)) - \mathcal{R}_{cs}(x(t))] \left(\frac{\partial \mathcal{H}_s}{\partial x}(x(t)) \right)^{\mathrm{T}},$$

$$x(0) = x_0, \quad x(t) \notin \mathcal{Z}_x, \quad (7.20)$$

$$\Delta x(t) = [\mathcal{J}_d(x(t)) - \mathcal{R}_d(x(t))] \left(\frac{\partial \mathcal{H}}{\partial x}(x(t)) \right)^{\mathrm{T}} + G_d(x(t))\phi_d(x(t))$$

$$= [\mathcal{J}_{\mathrm{ds}}(x(t)) - \mathcal{R}_{\mathrm{ds}}(x(t))] \left(\frac{\partial \mathcal{H}_{\mathrm{s}}}{\partial x}(x(t)) \right)^{\mathrm{T}}, \quad x(t) \in \mathcal{Z}_x, \quad (7.21)$$

where $\mathcal{H}_{\mathrm{s}} : \mathcal{D} \to \mathbb{R}$ is a *shaped Hamiltonian function* for the closed-loop system (7.20) and (7.21), $\mathcal{J}_{\mathrm{cs}} : \mathcal{D} \to \mathbb{R}^{n \times n}$ is a shaped interconnection matrix function for the continuous-time closed-loop system and satisfies $\mathcal{J}_{\mathrm{cs}}(x) = -\mathcal{J}_{\mathrm{cs}}^{\mathrm{T}}(x)$, $\mathcal{R}_{\mathrm{cs}} : \mathcal{D} \to \mathbb{S}^n$ is a shaped dissipation matrix function for the continuous-time closed-loop system and satisfies $\mathcal{R}_{\mathrm{cs}}(x) \geq 0$, $x \in \mathcal{D}$, $\mathcal{J}_{\mathrm{ds}} : \mathcal{Z}_x \to \mathbb{R}^{n \times n}$ is a shaped interconnection matrix function for the closed-loop resetting dynamics and satisfies $\mathcal{J}_{\mathrm{ds}}(x) = -\mathcal{J}_{\mathrm{ds}}^{\mathrm{T}}(x)$, and $\mathcal{R}_{\mathrm{ds}} : \mathcal{Z}_x \to \mathbb{S}^n$ is a shaped dissipation matrix function for the closed-loop resetting dynamics and satisfies $\mathcal{R}_{\mathrm{ds}}(x) \geq 0$, $x \in \mathcal{Z}_x$.

Theorem 7.1 *Consider the nonlinear impulsive port-controlled Hamiltonian system given by (7.1) and (7.2). Assume there exist functions* $\phi_{\mathrm{c}} : \mathcal{D} \to U_{\mathrm{c}}$, $\phi_{\mathrm{d}} : \mathcal{Z}_x \to U_{\mathrm{d}}$, $\mathcal{H}_{\mathrm{s}}$, $\mathcal{H}_{\mathrm{c}} : \mathcal{D} \to \mathbb{R}$, $\mathcal{J}_{\mathrm{cs}}$, $\mathcal{J}_{\mathrm{ca}} : \mathcal{D} \to \mathbb{R}^{n \times n}$, $\mathcal{R}_{\mathrm{cs}}$, $\mathcal{R}_{\mathrm{ca}} : \mathcal{D} \to \mathbb{R}^{n \times n}$, $\mathcal{J}_{\mathrm{ds}}$, $\mathcal{J}_{\mathrm{da}} : \mathcal{Z}_x \to \mathbb{R}^{n \times n}$, and $\mathcal{R}_{\mathrm{ds}}$, $\mathcal{R}_{\mathrm{da}} : \mathcal{Z}_x \to \mathbb{R}^{n \times n}$ such that $\mathcal{H}_{\mathrm{s}}(x) = \mathcal{H}(x) + \mathcal{H}_{\mathrm{c}}(x)$ is continuously differentiable, $\mathcal{J}_{\mathrm{cs}}(x) = \mathcal{J}_{\mathrm{c}}(x) + \mathcal{J}_{\mathrm{ca}}(x)$, $\mathcal{J}_{\mathrm{cs}}(x) = -\mathcal{J}_{\mathrm{cs}}^{\mathrm{T}}(x)$, $\mathcal{R}_{\mathrm{cs}}(x) = \mathcal{R}_{\mathrm{c}}(x) + \mathcal{R}_{\mathrm{ca}}(x)$, $\mathcal{R}_{\mathrm{cs}}(x) = \mathcal{R}_{\mathrm{cs}}^{\mathrm{T}}(x) \geq 0$, $x \in \mathcal{D}$, $\mathcal{J}_{\mathrm{ds}}(x) = \mathcal{J}_{\mathrm{d}}(x) + \mathcal{J}_{\mathrm{da}}(x)$, $\mathcal{J}_{\mathrm{ds}}(x) = -\mathcal{J}_{\mathrm{ds}}^{\mathrm{T}}(x)$, $\mathcal{R}_{\mathrm{ds}}(x) = \mathcal{R}_{\mathrm{d}}(x) + \mathcal{R}_{\mathrm{da}}(x)$, $\mathcal{R}_{\mathrm{ds}}(x) = \mathcal{R}_{\mathrm{ds}}^{\mathrm{T}}(x) \geq 0$, $x \in \mathcal{Z}_x$, and*

$$\mathcal{H}_{\mathrm{s}}(x + y) = \mathcal{H}_{\mathrm{s}}(x) + \frac{\partial \mathcal{H}_{\mathrm{s}}}{\partial x}(x)y + \frac{1}{2}y^{\mathrm{T}}\frac{\partial^2 \mathcal{H}_{\mathrm{s}}}{\partial x^2}(x)y, \quad x, y \in \mathcal{D}, \quad (7.22)$$

$$\frac{\partial \mathcal{H}_{\mathrm{c}}}{\partial x}(x_{\mathrm{e}}) = -\frac{\partial \mathcal{H}}{\partial x}(x_{\mathrm{e}}), \quad x_{\mathrm{e}} \in \mathcal{D}, \quad (7.23)$$

$$\frac{\partial^2 \mathcal{H}_{\mathrm{c}}}{\partial x^2}(x_{\mathrm{e}}) > -\frac{\partial^2 \mathcal{H}}{\partial x^2}(x_{\mathrm{e}}), \quad x_{\mathrm{e}} \in \mathcal{D}, \quad (7.24)$$

$$\frac{\partial \mathcal{H}_{\mathrm{s}}}{\partial x}(x) [\mathcal{J}_{\mathrm{ds}}(x) - \mathcal{R}_{\mathrm{ds}}(x)]$$
$$+ \frac{1}{2}(\mathcal{J}_{\mathrm{ds}}(x) - \mathcal{R}_{\mathrm{ds}}(x))^{\mathrm{T}}\frac{\partial^2 \mathcal{H}_{\mathrm{s}}}{\partial x^2}(x)(\mathcal{J}_{\mathrm{ds}}(x) - \mathcal{R}_{\mathrm{ds}}(x))\left(\frac{\partial \mathcal{H}_{\mathrm{s}}}{\partial x}(x)\right)^{\mathrm{T}} \leq 0,$$
$$x \in \mathcal{Z}_x, \quad (7.25)$$

$$[\mathcal{J}_{\mathrm{cs}}(x) - \mathcal{R}_{\mathrm{cs}}(x)]\left(\frac{\partial \mathcal{H}_{\mathrm{c}}}{\partial x}(x)\right)^{\mathrm{T}} = -[\mathcal{J}_{\mathrm{ca}}(x) - \mathcal{R}_{\mathrm{ca}}(x)]\left(\frac{\partial \mathcal{H}}{\partial x}(x)\right)^{\mathrm{T}}$$
$$+ G_{\mathrm{c}}(x)\phi_{\mathrm{c}}(x), \quad x \notin \mathcal{Z}_x, \quad (7.26)$$

$$[\mathcal{J}_{\mathrm{ds}}(x) - \mathcal{R}_{\mathrm{ds}}(x)]\left(\frac{\partial \mathcal{H}_{\mathrm{c}}}{\partial x}(x)\right)^{\mathrm{T}} = -[\mathcal{J}_{\mathrm{da}}(x) - \mathcal{R}_{\mathrm{da}}(x)]\left(\frac{\partial \mathcal{H}}{\partial x}(x)\right)^{\mathrm{T}}$$
$$+ G_{\mathrm{d}}(x)\phi_{\mathrm{d}}(x), \quad x \in \mathcal{Z}_x. \quad (7.27)$$

Then the equilibrium solution $x(t) \equiv x_{\mathrm{e}}$ of the closed-loop system (7.20) and (7.21) is Lyapunov stable. If, in addition, $\mathcal{D}_{\mathrm{c}} \subseteq \mathcal{D}$ is a compact positively invariant set with respect to (7.20) and (7.21), *and the largest invariant set contained in*

$$\mathcal{R} \triangleq \left\{ x \in \mathcal{D}_c : x \notin \mathcal{Z}_x, \frac{\partial \mathcal{H}_s}{\partial x}(x)\mathcal{R}_{cs}(x) \left(\frac{\partial \mathcal{H}_s}{\partial x}(x) \right)^{\mathrm{T}} = 0 \right\}$$

$$\cup \left\{ x \in \mathcal{D}_c : x \in \mathcal{Z}_x, \frac{\partial \mathcal{H}_s}{\partial x}(x)[\mathcal{J}_{ds}(x) - \mathcal{R}_{ds}(x) \right.$$

$$+ \frac{1}{2}(\mathcal{J}_{ds}(x) - \mathcal{R}_{ds}(x))^{\mathrm{T}} \frac{\partial^2 \mathcal{H}_s}{\partial x^2}(x)(\mathcal{J}_{ds}(x) - \mathcal{R}_{ds}(x))] \left(\frac{\partial \mathcal{H}_s}{\partial x}(x) \right)^{\mathrm{T}}$$

$$\left. = 0 \right\} \tag{7.28}$$

is $\mathcal{M} = \{x_e\}$, *then the equilibrium solution* $x(t) \equiv x_e$ *of the closed-loop system (7.20) and (7.21) is locally asymptotically stable and* $\mathcal{D}_c$ *is a subset of the domain of attraction of (7.20) and (7.21).*

Proof. First, note that for $\mathcal{Z} = \mathcal{Z}_x$, it follows from Assumptions A1 and A2 that the resetting times t_k $(= \tau_k(x_0))$ are well defined and distinct for every trajectory of (7.20) and (7.21). Conditions (7.26) and (7.27) imply that with hybrid feedback controller $(u_c(t), u_d(t_k)) = (\phi_c(x(t)), \phi_d(x(t_k)))$, the closed-loop system (7.1) and (7.2) has a Hamiltonian structure given by (7.20) and (7.21). Furthermore, it follows from (7.22)–(7.24) that the energy function $\mathcal{H}_s(\cdot)$ has a global minimum at $x = x_e$. Hence, $x = x_e$ is an equilibrium point of the closed-loop system.

Next, consider the Lyapunov function candidate for the closed-loop system (7.20) and (7.21) given by $V(x) = \mathcal{H}_s(x) - \mathcal{H}_s(x_e)$. Now, the corresponding Lyapunov derivative of $V(x)$ along the closed-loop state trajectories $x(t)$, $t \in (t_k, t_{k+1}]$, is given by

$$\dot{V}(x(t)) = \dot{\mathcal{H}}_s(x(t)) = -\frac{\partial \mathcal{H}_s}{\partial x}(x(t))\mathcal{R}_{cs}(x(t)) \left(\frac{\partial \mathcal{H}_s}{\partial x}(x(t)) \right)^{\mathrm{T}} \le 0,$$

$$x(t) \notin \mathcal{Z}_x, \tag{7.29}$$

and the Lyapunov difference of $V(x)$ at the resetting times t_k, $k \in \overline{\mathbb{Z}}_+$, is given by

$$\Delta V(x(t)) \triangleq V(x(t^+)) - V(x(t))$$

$$= \mathcal{H}_s \left(x(t) + (\mathcal{J}_{ds}(x(t)) - \mathcal{R}_{ds}(x(t))) \left(\frac{\partial \mathcal{H}_s}{\partial x}(x(t)) \right)^{\mathrm{T}} \right)$$

$$- \mathcal{H}_s(x(t)), \quad x(t) \in \mathcal{Z}_x, \tag{7.30}$$

or, equivalently, using (7.22) and (7.25),

$$
\begin{aligned}
\Delta V(x(t)) = \frac{\partial \mathcal{H}_{\mathrm{s}}}{\partial x}(x(t)) \Bigg[&\mathcal{J}_{\mathrm{ds}}(x(t)) - \mathcal{R}_{\mathrm{ds}}(x(t)) \\
&+ \frac{1}{2}(\mathcal{J}_{\mathrm{ds}}(x(t)) - \mathcal{R}_{\mathrm{ds}}(x(t)))^{\mathrm{T}} \frac{\partial^2 \mathcal{H}_{\mathrm{s}}}{\partial x^2}(x(t)) \\
&\cdot (\mathcal{J}_{\mathrm{ds}}(x(t)) - \mathcal{R}_{\mathrm{ds}}(x(t))) \Bigg] \left(\frac{\partial \mathcal{H}_{\mathrm{s}}}{\partial x}(x(t)) \right)^{\mathrm{T}} \\
\leq 0, \quad x(t) \in \mathcal{Z}_x.
\end{aligned} \tag{7.31}
$$

Thus, it follows from Theorem 2.1 that the equilibrium solution $x(t) \equiv x_{\mathrm{e}}$ of (7.20) and (7.21) is Lyapunov stable. Asymptotic stability of the closed-loop system follows immediately from Theorem 2.3. $\quad\square$

Theorem 7.1 presents constructive sufficient conditions for hybrid feedback stabilization that preserve the physical hybrid Hamiltonian structure at the closed-loop level while providing a shaped Hamiltonian energy function as a Lyapunov function for the closed-loop system. These sufficient conditions consist of a system of two partial differential equations parameterized by the auxiliary energy function ($\mathcal{H}_{\mathrm{c}}$), the auxiliary interconnection matrix functions ($\mathcal{J}_{\mathrm{ca}}, \mathcal{J}_{\mathrm{da}}$), and auxiliary dissipation matrix functions ($\mathcal{R}_{\mathrm{ca}}, \mathcal{R}_{\mathrm{da}}$). The solutions of (7.26) and (7.27) characterize the set of all desired shaped energy functions that can be assigned while preserving the system hybrid Hamiltonian structure at the closed-loop level.

To apply Theorem 7.1, we fix the structure of the interconnection ($\mathcal{J}_{\mathrm{cs}}, \mathcal{J}_{\mathrm{ds}}$) and dissipation ($\mathcal{R}_{\mathrm{cs}}, \mathcal{R}_{\mathrm{ds}}$) matrix functions and solve for the closed-loop energy function $\mathcal{H}_{\mathrm{s}}$. Although in this case solving (7.25)–(7.27) appears formidable, the equations are, in fact, quite tractable since the partial differential equations are parameterized via the interconnection and dissipation matrix functions which can be chosen by the control designer to satisfy system physical constraints. Alternatively, we can fix the shaped Hamiltonian $\mathcal{H}_{\mathrm{s}}$ and solve for the interconnection and dissipation matrix functions. In this case, we do not need to solve a set of partial differential equations, but rather a set of algebraic equations.

If rank $G_{\mathrm{c}}(x) = m_{\mathrm{c}}$, $x \notin \mathcal{Z}_x$, rank $G_{\mathrm{d}}(x) = m_{\mathrm{d}}$, $x \in \mathcal{Z}_x$, rank$[G_{\mathrm{c}}(x)$ $b_{\mathrm{c}}(x)] = $ rank $G_{\mathrm{c}}(x) = m_{\mathrm{c}}$, $x \notin \mathcal{Z}_x$, and rank$[G_{\mathrm{d}}(x) \; b_{\mathrm{d}}(x)] = $ rank

$G_{\mathrm{d}}(x) = m_{\mathrm{d}}$, $x \in \mathcal{Z}_x$, where

$$b_{\mathrm{c}}(x) = [\mathcal{J}_{\mathrm{cs}}(x) - \mathcal{R}_{\mathrm{cs}}(x)] \left(\frac{\partial \mathcal{H}_{\mathrm{c}}}{\partial x}(x) \right)^{\mathrm{T}}$$

$$+ [\mathcal{J}_{\mathrm{ca}}(x) - \mathcal{R}_{\mathrm{ca}}(x)] \left(\frac{\partial \mathcal{H}}{\partial x}(x) \right)^{\mathrm{T}}, \qquad (7.32)$$

$$b_{\mathrm{d}}(x) = [\mathcal{J}_{\mathrm{ds}}(x) - \mathcal{R}_{\mathrm{ds}}(x)] \left(\frac{\partial \mathcal{H}_{\mathrm{c}}}{\partial x}(x) \right)^{\mathrm{T}}$$

$$+ [\mathcal{J}_{\mathrm{da}}(x) - \mathcal{R}_{\mathrm{da}}(x)] \left(\frac{\partial \mathcal{H}}{\partial x}(x) \right)^{\mathrm{T}}, \qquad (7.33)$$

then an explicit expression for the stabilizing hybrid feedback controller satisfying (7.26) and (7.27) is given by $\phi_{\mathrm{c}}(x) = (G_{\mathrm{c}}^{\mathrm{T}}(x)G_{\mathrm{c}}(x))^{-1}$ $\cdot G_{\mathrm{c}}^{\mathrm{T}}(x)b_{\mathrm{c}}(x)$, $x \notin \mathcal{Z}_x$, and $\phi_{\mathrm{d}}(x) = (G_{\mathrm{d}}^{\mathrm{T}}(x)G_{\mathrm{d}}(x))^{-1}G_{\mathrm{d}}^{\mathrm{T}}(x)\, b_{\mathrm{d}}(x)$, $x \in \mathcal{Z}_x$. Alternatively, if $\mathrm{rank}[G_{\mathrm{c}}(x)\, b_{\mathrm{c}}(x)] = \mathrm{rank}\, G_{\mathrm{c}}(x) < m_{\mathrm{c}}$, $x \notin \mathcal{Z}_x$, and $\mathrm{rank}[G_{\mathrm{d}}(x)\, b_{\mathrm{d}}(x)] = \mathrm{rank}\, G_{\mathrm{d}}(x) < m_{\mathrm{d}}$, $x \in \mathcal{Z}_x$, then the hybrid feedback controller $\phi_{\mathrm{c}}(x) = G_{\mathrm{c}}^{+}(x)b_{\mathrm{c}}(x) + [I_{m_{\mathrm{c}}} - G_{\mathrm{c}}^{+}(x)G_{\mathrm{c}}(x)]z_{\mathrm{c}}$, where $(\cdot)^{+}$ denotes the Moore-Penrose generalized inverse, $z_{\mathrm{c}} \in \mathbb{R}^{m_{\mathrm{c}}}$, and $x \notin \mathcal{Z}_x$, and $\phi_{\mathrm{d}}(x) = G_{\mathrm{d}}^{+}(x)b_{\mathrm{d}}(x) + [I_{m_{\mathrm{d}}} - G_{\mathrm{d}}^{+}(x)G_{\mathrm{d}}(x)]z_{\mathrm{d}}$, where $z_{\mathrm{d}} \in \mathbb{R}^{m_{\mathrm{d}}}$ and $x \in \mathcal{Z}_x$, satisfies (7.26) and (7.27).

If for a certain class of systems the discrete-time dynamics cannot be written in a port-controlled Hamiltonian form, then we can simply take

$$\Delta x(t_k) = f_{\mathrm{d}}(x(t_k)) + G_{\mathrm{d}}(x(t_k))u_{\mathrm{d}}(t_k), \qquad (7.34)$$

where $f_{\mathrm{d}} : \mathcal{Z}_x \rightarrow \mathbb{R}^n$ is continuous. In this case, Theorem 7.1 holds with (7.22), (7.25), and (7.27) replaced by the more general condition

$$\Delta \mathcal{H}_{\mathrm{s}}(x) = \mathcal{H}_{\mathrm{s}}(x + f_{\mathrm{d}}(x) + G_{\mathrm{d}}(x)\phi_{\mathrm{d}}(x)) - \mathcal{H}_{\mathrm{s}}(x) \le 0, \quad x \in \mathcal{Z}_x, \qquad (7.35)$$

for the shaped Hamiltonian function for the closed-loop system. Furthermore, the set given in (7.28) becomes

$$\mathcal{R} \triangleq \{ x \in \mathcal{D}_{\mathrm{c}} : x \notin \mathcal{Z}_x,\ \dot{\mathcal{H}}_{\mathrm{s}}(x) = 0 \} \cup \{ x \in \mathcal{D}_{\mathrm{c}} : x \in \mathcal{Z}_x,\ \Delta \mathcal{H}_{\mathrm{s}}(x) = 0 \}. \qquad (7.36)$$

Under certain conditions on the system dissipation, the hybrid energy-based controller given by Theorem 7.1 provides an energy balance over the continuous-time trajectories of the controlled system.

To see this, let $\mathcal{R}_{\mathrm{ca}}(x) \equiv 0$ and $\mathcal{R}_{\mathrm{c}}(x) \left(\frac{\partial \mathcal{H}_{\mathrm{c}}}{\partial x}(x)\right)^{\mathrm{T}} = 0$, $x \in \mathcal{D}$. In this case, the continuous-time closed-loop dynamics are given by

$$\dot{x}(t) = [\mathcal{J}_{\mathrm{cs}}(x(t)) - \mathcal{R}_{\mathrm{c}}(x(t))] \left(\frac{\partial \mathcal{H}_{\mathrm{s}}}{\partial x}(x(t))\right)^{\mathrm{T}}, \qquad x(0) = x_0,$$

$$x(t) \notin \mathcal{Z}_x. \quad (7.37)$$

Along the continuous-time trajectories $x(t)$, $t \in (t_k, t_{k+1}]$, it follows that

$$\begin{aligned}
\dot{\mathcal{H}}_{\mathrm{s}}(x(t)) &= -\frac{\partial \mathcal{H}_{\mathrm{s}}}{\partial x}(x(t)) \mathcal{R}_{\mathrm{c}}(x(t)) \left(\frac{\partial \mathcal{H}_{\mathrm{s}}}{\partial x}(x(t))\right)^{\mathrm{T}} \\
&= -\left(\frac{\partial \mathcal{H}}{\partial x}(x(t)) + \frac{\partial \mathcal{H}_{\mathrm{c}}}{\partial x}(x(t))\right) \mathcal{R}_{\mathrm{c}}(x(t)) \\
&\quad \cdot \left(\frac{\partial \mathcal{H}}{\partial x}(x(t)) + \frac{\partial \mathcal{H}_{\mathrm{c}}}{\partial x}(x(t))\right)^{\mathrm{T}} \\
&= -\frac{\partial \mathcal{H}}{\partial x}(x(t)) \mathcal{R}_{\mathrm{c}}(x(t)) \left(\frac{\partial \mathcal{H}}{\partial x}(x(t))\right)^{\mathrm{T}}, \qquad t_k < t \le t_{k+1},
\end{aligned}$$

$$(7.38)$$

or, equivalently, using (7.16),

$$\dot{\mathcal{H}}_{\mathrm{s}}(x(t)) = \dot{\mathcal{H}}(x(t)) - u_{\mathrm{c}}^{\mathrm{T}}(t) y_{\mathrm{c}}(t), \qquad t_k < t \le t_{k+1}. \quad (7.39)$$

Now, integrating (7.39) yields

$$\mathcal{H}_{\mathrm{s}}(x(t)) = \mathcal{H}(x(t)) - \int_{\hat{t}}^{t} u_{\mathrm{c}}^{\mathrm{T}}(s) y_{\mathrm{c}}(s) \mathrm{d}s + \kappa(\hat{t}), \qquad t_k < \hat{t} \le t \le t_{k+1},$$

$$(7.40)$$

where $\kappa(\hat{t}) \triangleq \mathcal{H}_{\mathrm{s}}(x(\hat{t})) - \mathcal{H}(x(\hat{t}))$, which shows that the closed-loop energy function $\mathcal{H}_{\mathrm{s}}(\cdot)$ over the continuous-time trajectories is equal to the difference between the physical energy $\mathcal{H}(\cdot)$ of the hybrid system and the energy supplied by the hybrid controller modulo $\kappa(\cdot)$. Furthermore, it follows from (7.25) that at the resetting times t_k, $\Delta \mathcal{H}_{\mathrm{s}}(x(t_k)) \le 0$, $k \in \overline{\mathbb{Z}}_+$.

7.4 Energy-Based Hybrid Dynamic Compensation via the Energy-Casimir Method

In this section, we consider energy-based hybrid dynamic control for impulsive port-controlled Hamiltonian systems, wherein energy shaping is achieved by combining the physical energy of the plant and the emulated energy of the controller. For systems with continuous flows, this approach has been extensively studied in [133,134] to design Euler-Lagrange controllers for potential energy shaping of mechanical systems.

We begin by considering the input/state-dependent hybrid port-controlled Hamiltonian system $\mathcal{G}$ given by (7.1)–(7.4) with $m_c = l_c$, $m_d = l_d$, and the hybrid outputs y_c and y_d replaced by

$$y_c(t) = G_c^T(x(t)) \left(\frac{\partial \mathcal{H}}{\partial x}(x(t)) \right)^T, \quad (x(t), u_c(t)) \notin \mathcal{Z}, \quad (7.41)$$

$$y_d(t) = G_d^T(x(t)) \left(\frac{\partial \mathcal{H}}{\partial x}(x(t)) \right)^T, \quad (x(t), u_c(t)) \in \mathcal{Z}. \quad (7.42)$$

Furthermore, we consider the input/state-dependent impulsive port-controlled Hamiltonian feedback control system $\mathcal{G}_c$ given by

$$\dot{x}_c(t) = [\mathcal{J}_{cc}(x_c(t)) - \mathcal{R}_{cc}(x_c(t))] \left(\frac{\partial \mathcal{H}_c}{\partial x_c}(x_c(t)) \right)^T + G_{cc}(x_c(t)) u_{cc}(t),$$

$$x_c(0) = x_{c0}, \quad (x_c(t), u_{cc}(t)) \notin \mathcal{Z}_c, \quad (7.43)$$

$$\Delta x_c(t) = [\mathcal{J}_{dc}(x_c(t)) - \mathcal{R}_{dc}(x_c(t))] \left(\frac{\partial \mathcal{H}_c}{\partial x_c}(x_c(t)) \right)^T + G_{dc}(x_c(t)) u_{dc}(t),$$

$$(x_c(t), u_{cc}(t)) \in \mathcal{Z}_c, \quad (7.44)$$

$$y_{cc}(t) = G_{cc}^T(x_c(t)) \left(\frac{\partial \mathcal{H}_c}{\partial x_c}(x_c(t)) \right)^T, \quad (x_c(t), u_{cc}(t)) \notin \mathcal{Z}_c, \quad (7.45)$$

$$y_{dc}(t) = G_{dc}^T(x_c(t)) \left(\frac{\partial \mathcal{H}_c}{\partial x_c}(x_c(t)) \right)^T, \quad (x_c(t), u_{cc}(t)) \in \mathcal{Z}_c, \quad (7.46)$$

where $t \geq 0$, $x_c(t) \in \mathbb{R}^{n_c}$, $u_{cc}(t) \in U_{cc} \subseteq \mathbb{R}^{m_{cc}}$, $u_{dc}(t_k) \in U_{dc} \subseteq \mathbb{R}^{m_{dc}}$, $y_{cc}(t) \in Y_{cc} \subseteq \mathbb{R}^{l_{cc}}$, $y_{dc}(t_k) \in Y_{dc} \subseteq \mathbb{R}^{l_{dc}}$, $m_{cc} = l_{cc}$, $m_{dc} = l_{dc}$, $\mathcal{H}_c : \mathbb{R}^{n_c} \to \mathbb{R}$ is a continuously differentiable Hamiltonian function of the feedback control system $\mathcal{G}_c$, $\mathcal{J}_{cc} : \mathbb{R}^{n_c} \to \mathbb{R}^{n_c \times n_c}$ is such that $\mathcal{J}_{cc}(x_c) = -\mathcal{J}_{cc}^T(x_c)$, $\mathcal{R}_{cc} : \mathbb{R}^{n_c} \to \mathbb{S}^{n_c}$ is such that $\mathcal{R}_{cc}(x_c) \geq 0$, $x_c \in \mathbb{R}^{n_c}$, $[\mathcal{J}_{cc}(x_c) - \mathcal{R}_{cc}(x_c)] \left(\frac{\partial \mathcal{H}_c}{\partial x_c}(x_c) \right)^T$, $x_c \in \mathbb{R}^{n_c}$, is Lipschitz continuous, $G_{cc} : \mathbb{R}^{n_c} \to \mathbb{R}^{n_c \times m_{cc}}$, $\mathcal{J}_{dc} : \mathbb{R}^{n_c} \to \mathbb{R}^{n_c \times n_c}$ is such that $\mathcal{J}_{dc}(x_c) = -\mathcal{J}_{dc}^T(x_c)$, $\mathcal{R}_{dc} : \mathbb{R}^{n_c} \to \mathbb{S}^{n_c}$ is such that $\mathcal{R}_{dc}(x_c) \geq 0$, $x_c \in \mathbb{R}^{n_c}$, $[\mathcal{J}_{dc}(x_c) - \mathcal{R}_{dc}(x_c)] \left(\frac{\partial \mathcal{H}_c}{\partial x_c}(x_c) \right)^T x_c \in \mathbb{R}^{n_c}$, is continuous, $G_{dc} : \mathbb{R}^{n_c} \to \mathbb{R}^{n_c \times m_{dc}}$, $m_{cc} = l_c$, $m_{dc} = l_d$, $l_{cc} = m_c$, $l_{dc} = m_d$, and $\mathcal{Z}_c \triangleq (\mathcal{Z}_{cx_c} \times U_{cc}) \cup (\mathbb{R}^{n_c} \times \mathcal{Z}_{cu_{cc}}) \subset \mathbb{R}^{n_c} \times U_{cc}$ is the resetting set for the system $\mathcal{G}_c$.

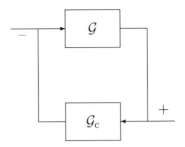

Figure 7.1 Negative feedback interconnection of port-controlled Hamiltonian systems $\mathcal{G}$ and $\mathcal{G}_c$.

Here, we assume that $u_{cc}(\cdot)$ and $u_{dc}(\cdot)$ are restricted to the class of admissible inputs consisting of measurable functions such that $(u_{cc}(t),$ $u_{dc}(t_k)) \in U_{cc} \times U_{dc}$ for all $t \geq 0$ and $k \in \mathbb{Z}_{[0,t)} \triangleq \{k : 0 \leq t_k < t\}$. Finally, for the nonlinear dynamical system (7.43) we assume that the required properties for the existence and uniqueness of solutions are satisfied such that (7.43) has a unique solution for all $t \in \mathbb{R}$ [14, 93]. Note that with the negative feedback interconnection given by Figure 7.1, $(u_{cc}, u_{dc}) = (y_c, y_d)$ and $(y_{cc}, y_{dc}) = (-u_c, -u_d)$. Furthermore, even though the input-output pairs of the feedback interconnection shown on Figure 7.1 consist of two-vector inputs/two-vector outputs, at any given instant of time a single-vector input/single-vector output is active.

Next, we define the closed-loop resetting set

$$
\tilde{\mathcal{Z}}_{\tilde{x}} \triangleq \left\{ \mathcal{Z}_x \cup \left(\left\{ x \in \mathcal{D} : G_c^{\mathrm{T}}(x) \left(\frac{\partial \mathcal{H}}{\partial x}(x) \right)^{\mathrm{T}} \in \mathcal{Z}_{cu_{cc}} \right\} \times \mathbb{R}^{n_c} \right) \right.
$$

$$
\cup \left\{ (\mathcal{D} \times \mathcal{Z}_{cx_c}) \cup \left\{ x_c \in \mathbb{R}^{n_c} : -G_{cc}^{\mathrm{T}}(x_c) \left(\frac{\partial \mathcal{H}_c}{\partial x_c}(x_c) \right)^{\mathrm{T}} \right. \right.
$$

$$
\left. \left. \left. \in \mathcal{Z}_{u_c} \right\} \right\}. \quad (7.47)
$$

Note that since the negative feedback interconnection of $\mathcal{G}$ and $\mathcal{G}_c$ is well posed, it follows that $\tilde{\mathcal{Z}}_{\tilde{x}}$ is well defined and depends on the closed-loop states $\tilde{x} \triangleq [x^{\mathrm{T}} x_c^{\mathrm{T}}]^{\mathrm{T}}$. Let $\mathcal{T}_{x0,u_c}^c$ denote the set of resetting times of $\mathcal{G}$, let $\mathcal{T}_{x0,u_c}$ denote the complement of $\mathcal{T}_{x0,u_c}^c$, that is, $\mathcal{T}_{x0,u_c} = [0, \infty) \backslash \mathcal{T}_{x0,u_c}^c$, let $\mathcal{T}_{xc0,u_{cc}}^c$ denote the set of resetting times of $\mathcal{G}_c$ and let $\mathcal{T}_{xc0,u_{cc}}$ denote the complement of $\mathcal{T}_{xc0,u_{cc}}^c$, that is, $\mathcal{T}_{xc0,u_{cc}} = [0, \infty) \backslash$

$\mathcal{T}^{c}_{x_{c0},u_{cc}}$. Furthermore, let $\tilde{\mathcal{T}}^{c} \triangleq \mathcal{T}^{c}_{x_{0},u_{c}} \cup \mathcal{T}^{c}_{x_{c0},u_{cc}}$ and $t_{k} \in \tilde{\mathcal{T}}^{c}, k \in \overline{\mathbb{Z}}_{+}$, so that $t_{k} = \tau_{k}(\tilde{x}_{0})$ denoting the resetting times for the feedback system are well defined and distinct for every closed-loop trajectory. As in Section 7.3, here we assume that the solution $s(t,\tilde{x}_{0})$ to the dynamical system resulting from the feedback interconnection of $\mathcal{G}$ and $\mathcal{G}_{c}$ is such that Assumption 7.1 is satisfied.

With the feedback interconnection given by $(u_{cc},u_{dc}) = (y_{c},y_{d})$ and $(y_{cc},y_{dc}) = (-u_{c},-u_{d})$, the closed-loop dynamics can be written in hybrid Hamiltonian form given by

$$
\dot{\tilde{x}}(t) = \left(\begin{bmatrix} \mathcal{J}_{c}(x(t)) & -G_{c}(x(t))G_{cc}^{T}(x_{c}(t)) \\ G_{cc}(x_{c}(t))G_{c}^{T}(x(t)) & \mathcal{J}_{cc}(x_{c}(t)) \end{bmatrix} \right.
$$
$$
\left. - \begin{bmatrix} \mathcal{R}_{c}(x(t)) & 0 \\ 0 & \mathcal{R}_{cc}(x_{c}(t)) \end{bmatrix} \right) \begin{bmatrix} \left(\frac{\partial \mathcal{H}}{\partial x}(x(t)) \right)^{T} \\ \left(\frac{\partial \mathcal{H}_{c}}{\partial x_{c}}(x_{c}(t)) \right)^{T} \end{bmatrix},
$$
$$
\tilde{x}(0) = \tilde{x}_{0}, \quad \tilde{x}(t) \notin \tilde{\mathcal{Z}}_{\tilde{x}}, \qquad (7.48)
$$

$$
\Delta\tilde{x}(t) = \left(\begin{bmatrix} \mathcal{J}_{d}(x(t)) & -G_{d}(x(t))G_{dc}^{T}(x_{c}(t)) \\ G_{dc}(x_{c}(t))G_{d}^{T}(x(t)) & \mathcal{J}_{dc}(x_{c}(t)) \end{bmatrix} \right.
$$
$$
\left. - \begin{bmatrix} \mathcal{R}_{d}(x(t)) & 0 \\ 0 & \mathcal{R}_{dc}(x_{c}(t)) \end{bmatrix} \right) \begin{bmatrix} \left(\frac{\partial \mathcal{H}}{\partial x}(x(t)) \right)^{T} \\ \left(\frac{\partial \mathcal{H}_{c}}{\partial x_{c}}(x_{c}(t)) \right)^{T} \end{bmatrix},
$$
$$
\tilde{x}(t) \in \tilde{\mathcal{Z}}_{\tilde{x}}. \qquad (7.49)
$$

It can be seen from (7.48) and (7.49) that by relating the controller state variables x_{c} to the plant state variables x, one can shape the Hamiltonian function $\mathcal{H}(\cdot) + \mathcal{H}_{c}(\cdot)$ so as to preserve the hybrid Hamiltonian structure under dynamic feedback for part of the closed-loop system associated with the plant dynamics. Since the closed-loop impulsive dynamical system (7.48) and (7.49) is Hamiltonian involving skew-symmetric interconnection matrix function terms and nonnegative definite dissipation matrix function terms, we can establish the existence of energy-Casimir functions [26, 160] (i.e., dynamical invariants) that are independent of the closed-loop Hamiltonian and relate the controller states to the plant states. Since energy-Casimir functions are composed of integrals of motion, it follows that these functions are constant along the trajectories of the closed-loop system (7.48) and (7.49). Furthermore, since the controller Hamiltonian $\mathcal{H}_{c}(\cdot)$ can be assigned, the energy-Casimir method can be used to construct suitable Lyapunov functions for the closed-loop system.

To proceed, consider the candidate vector energy-Casimir function

$E : \mathcal{D} \times \mathbb{R}^{n_c} \to \mathbb{R}^{n_c}$, where $E(\cdot, \cdot)$ is continuously differentiable and has the form

$$E(x, x_c) = x_c - F(x), \quad x \in \mathcal{D}, \quad x_c \in \mathbb{R}^{n_c}, \qquad (7.50)$$

where $F : \mathcal{D} \to \mathbb{R}^{n_c}$ is continuously differentiable and satisfies

$$F(x + y) = F(x) + \frac{\partial F}{\partial x}(x)y, \quad x, y \in \mathcal{D}. \qquad (7.51)$$

To ensure that the candidate vector energy-Casimir function $E(\cdot, \cdot)$ is constant along the trajectories of (7.48) and (7.49) we require that

$$\dot{E}(x(t), x_c(t)) = \dot{x}_c(t) - \frac{\partial F}{\partial x}(x(t))\dot{x}(t) = 0, \quad \tilde{x}(t) \notin \tilde{\mathcal{Z}}_{\tilde{x}}, \ (7.52)$$

$$\begin{aligned} \Delta E(x(t), x_c(t)) &\triangleq E(x(t^+), x_c(t^+)) - E(x(t), x_c(t)) \\ &= \Delta x_c(t) - F(x(t^+)) + F(x(t)) \\ &= 0, \quad \tilde{x}(t) \in \tilde{\mathcal{Z}}_{\tilde{x}}. \end{aligned} \qquad (7.53)$$

Using (7.51), we can arrive at a set of *sufficient* conditions which guarantee that (7.52) and (7.53) hold. Specifically, it follows from (7.48) and (7.49) that (7.52) and (7.53) can be rewritten as

$$\dot{E}(x(t), x_c(t)) = \left[\begin{array}{c} \left[G_{cc}(x_c)G_c^T(x) - \frac{\partial F}{\partial x}(x)(\mathcal{J}_c(x) - \mathcal{R}_c(x)) \right]^T \\ \left[\mathcal{J}_{cc}(x_c) - \mathcal{R}_{cc}(x_c) + \frac{\partial F}{\partial x}(x)G_c(x)G_{cc}^T(x_c) \right]^T \end{array} \right]^T$$
$$\cdot \left[\begin{array}{c} \left(\frac{\partial \mathcal{H}}{\partial x}(x(t)) \right)^T \\ \left(\frac{\partial \mathcal{H}_c}{\partial x_c}(x_c(t)) \right)^T \end{array} \right], \quad \tilde{x}(t) \notin \tilde{\mathcal{Z}}_{\tilde{x}}, \qquad (7.54)$$

$$\Delta E(x(t), x_c(t)) = \left[\begin{array}{c} \left[G_{dc}(x_c)G_d^T(x) - \frac{\partial F}{\partial x}(x)(\mathcal{J}_d(x) - \mathcal{R}_d(x)) \right]^T \\ \left[\mathcal{J}_{dc}(x_c) - \mathcal{R}_{dc}(x_c) + \frac{\partial F}{\partial x}(x)G_d(x)G_{dc}^T(x_c) \right]^T \end{array} \right]^T$$
$$\cdot \left[\begin{array}{c} \left(\frac{\partial \mathcal{H}}{\partial x}(x(t)) \right)^T \\ \left(\frac{\partial \mathcal{H}_c}{\partial x_c}(x_c(t)) \right)^T \end{array} \right], \quad \tilde{x}(t) \in \tilde{\mathcal{Z}}_{\tilde{x}}. \qquad (7.55)$$

Hence, a set of sufficient conditions such that (7.52) and (7.53) hold are given by

$$G_{cc}(x_c)G_c^T(x) - \frac{\partial F}{\partial x}(x)(\mathcal{J}_c(x) - \mathcal{R}_c(x)) = 0, \quad x \in \mathcal{D}, \quad x_c \in \mathbb{R}^{n_c}, \qquad (7.56)$$

$$\mathcal{J}_{cc}(x_c) - \mathcal{R}_{cc}(x_c) + \frac{\partial F}{\partial x}(x)G_c(x)G_{cc}^T(x_c) = 0, \quad x \in \mathcal{D}, \quad x_c \in \mathbb{R}^{n_c},$$

$$(7.57)$$

$$G_{\mathrm{dc}}(x_{\mathrm{c}})G_{\mathrm{d}}^{\mathrm{T}}(x) - \frac{\partial F}{\partial x}(x)(\mathcal{J}_{\mathrm{d}}(x) - \mathcal{R}_{\mathrm{d}}(x)) = 0, \quad x \in \mathcal{D}, \quad x_{\mathrm{c}} \in \mathbb{R}^{n_{\mathrm{c}}},$$

$$(7.58)$$

$$\mathcal{J}_{\mathrm{dc}}(x_{\mathrm{c}}) - \mathcal{R}_{\mathrm{dc}}(x_{\mathrm{c}}) + \frac{\partial F}{\partial x}(x)G_{\mathrm{d}}(x)G_{\mathrm{dc}}^{\mathrm{T}}(x_{\mathrm{c}}) = 0, \quad x \in \mathcal{D}, \quad x_{\mathrm{c}} \in \mathbb{R}^{n_{\mathrm{c}}}.$$

$$(7.59)$$

The following proposition summarizes the above results.

Proposition 7.1 *Consider the feedback interconnection of the port-controlled Hamiltonian systems* $\mathcal{G}$ *and* $\mathcal{G}_{\mathrm{c}}$ *given by (7.1), (7.2), (7.41), (7.42), and (7.43)–(7.46), respectively. If there exists a continuously differentiable function* $F : \mathcal{D} \to \mathbb{R}^{n_{\mathrm{c}}}$ *satisfying (7.51) and, for all* $(x, x_{\mathrm{c}}) \in \mathcal{D} \times \mathbb{R}^{n_{\mathrm{c}}}$,

$$\frac{\partial F}{\partial x}(x)\mathcal{J}_{\mathrm{c}}(x)\left(\frac{\partial F}{\partial x}(x)\right)^{\mathrm{T}} - \mathcal{J}_{\mathrm{cc}}(x_{\mathrm{c}}) = 0, \qquad (7.60)$$

$$\mathcal{R}_{\mathrm{cc}}(x_{\mathrm{c}}) = 0, \qquad (7.61)$$

$$\mathcal{R}_{\mathrm{c}}(x)\left(\frac{\partial F}{\partial x}(x)\right)^{\mathrm{T}} = 0, \qquad (7.62)$$

$$\frac{\partial F}{\partial x}(x)\mathcal{J}_{\mathrm{c}}(x) - G_{\mathrm{cc}}(x_{\mathrm{c}})G_{\mathrm{c}}^{\mathrm{T}}(x) = 0, \qquad (7.63)$$

$$\frac{\partial F}{\partial x}(x)\mathcal{J}_{\mathrm{d}}(x)\left(\frac{\partial F}{\partial x}(x)\right)^{\mathrm{T}} - \mathcal{J}_{\mathrm{dc}}(x_{\mathrm{c}}) = 0, \qquad (7.64)$$

$$\mathcal{R}_{\mathrm{dc}}(x_{\mathrm{c}}) = 0, \qquad (7.65)$$

$$\mathcal{R}_{\mathrm{d}}(x)\left(\frac{\partial F}{\partial x}(x)\right)^{\mathrm{T}} = 0, \qquad (7.66)$$

$$\frac{\partial F}{\partial x}(x)\mathcal{J}_{\mathrm{d}}(x) - G_{\mathrm{dc}}(x_{\mathrm{c}})G_{\mathrm{d}}^{\mathrm{T}}(x) = 0, \qquad (7.67)$$

then

$$E(\tilde{x}(t)) = x_{\mathrm{c}}(t) - F(x(t)) = c, \quad t \geq 0, \qquad (7.68)$$

where $c \in \mathbb{R}^{n_{\mathrm{c}}}$ *and* $\tilde{x}(t) = [x^{\mathrm{T}}(t), x_{\mathrm{c}}^{\mathrm{T}}(t)]^{\mathrm{T}}$, *satisfies (7.48) and (7.49).*

Proof. Postmultiplying (7.56) by $\left(\frac{\partial F}{\partial x}(x)\right)^{\mathrm{T}}$, it follows from (7.56) and (7.57) that

$$\frac{\partial F}{\partial x}(x)[\mathcal{J}_{\mathrm{c}}(x) - \mathcal{R}_{\mathrm{c}}(x)]\left(\frac{\partial F}{\partial x}(x)\right)^{\mathrm{T}} = \mathcal{J}_{\mathrm{cc}}(x_{\mathrm{c}}) + \mathcal{R}_{\mathrm{cc}}(x_{\mathrm{c}}), \quad x \in \mathcal{D},$$

$$x_{\mathrm{c}} \in \mathbb{R}^{n_{\mathrm{c}}}. \qquad (7.69)$$

Next, using the fact that the sum of a skew-symmetric and symmetric matrix is zero if and only if the individual matrices are zero, it follows that (7.69) is equivalent to

$$\frac{\partial F}{\partial x}(x)\mathcal{J}_c(x)\left(\frac{\partial F}{\partial x}(x)\right)^{\mathrm{T}} - \mathcal{J}_{cc}(x_c) = 0, \quad x \in \mathcal{D}, \quad x_c \in \mathbb{R}^{n_c}, \quad (7.70)$$

$$\mathcal{R}_{cc}(x_c) + \frac{\partial F}{\partial x}(x)\mathcal{R}_c(x)\left(\frac{\partial F}{\partial x}(x)\right)^{\mathrm{T}} = 0, \quad x \in \mathcal{D}, \quad x_c \in \mathbb{R}^{n_c}. \quad (7.71)$$

Now, since $\mathcal{R}_c(x) \geq 0$, $x \in \mathcal{D}$, and $\mathcal{R}_{cc}(x_c) \geq 0$, $x \in \mathbb{R}^{n_c}$, it follows that (7.70) and (7.71) are equivalent to (7.60)–(7.62). Hence, it follows that (7.56) can be rewritten as (7.63). Analogously, it can be shown that (7.58) and (7.59) are equivalent to (7.64)–(7.67). The equivalence between (7.56)–(7.59) and (7.60)–(7.67) proves the result.   □

Note that conditions (7.60)–(7.67) are necessary and sufficient for (7.56)–(7.59) to hold, which, in turn, provide sufficient conditions for guaranteeing that the vector energy-Casimir function $E(\cdot, \cdot)$ is constant along the trajectories of the closed-loop system (7.48) and (7.49). The constant vector $c \in \mathbb{R}^{n_c}$ in (7.68) depends on the initial conditions for the plant and controller states.

For the statement of the next result we consider the feedback interconnection of the two state-dependent impulsive port-controlled Hamiltonian systems $\mathcal{G}$ and $\mathcal{G}_c$ given by (7.1), (7.2), (7.41), (7.42), and (7.43)–(7.46), where $\mathcal{Z}_{u_c} = U_c$ and $\mathcal{Z}_{cu_{cc}} = U_{cc}$, respectively. In this case the resetting set for the closed-loop system is given by $\mathcal{Z}_{\tilde{x}} \triangleq (\mathcal{Z}_x \times \mathbb{R}^{n_c}) \cup (\mathcal{D} \times \mathcal{Z}_{cx_c})$. Furthermore, if conditions (7.60)–(7.67) are satisfied, then the controller state variables along the trajectories of the closed-loop system given by (7.48) and (7.49) can be represented in terms of the plant state variables as $x_c(t) = F(x(t)) + c$, $t \geq 0$, $x(t) \in \mathcal{D}$, $c \in \mathbb{R}^{n_c}$. Hence, the resetting set $\mathcal{Z}_{\tilde{x}}$ for the closed-loop system can be redefined as $\hat{\mathcal{Z}}_x = \mathcal{Z}_x \cup \{x \in \mathcal{D} : F(x) + c \in \mathcal{Z}_{cx_c}\}$. In this case, it follows that the continuous-time closed-loop system associated with the plant dynamics is given by

$$\dot{x}(t) = [\mathcal{J}_c(x(t)) - \mathcal{R}_c(x(t))]\left(\frac{\partial \mathcal{H}}{\partial x}(x(t))\right)^{\mathrm{T}}$$

$$-G_c(x(t))G_{cc}^{\mathrm{T}}(x_c(t))\left(\frac{\partial \mathcal{H}_c}{\partial x_c}(x_c(t))\right)^{\mathrm{T}}$$

$$= [\mathcal{J}_c(x(t)) - \mathcal{R}_c(x(t))]\left(\frac{\partial \mathcal{H}}{\partial x}(x(t)) + \frac{\partial \mathcal{H}_c}{\partial x_c}(x_c(t))\frac{\partial F}{\partial x}(x(t))\right)^{\mathrm{T}}$$

$$= [\mathcal{J}_{\mathrm{c}}(x(t)) - \mathcal{R}_{\mathrm{c}}(x(t))] \left(\frac{\partial \mathcal{H}_{\mathrm{s}}}{\partial x}(x(t)) \right)^{\mathrm{T}}, \quad x(0) = x_0,$$

$$x(t) \notin \hat{\mathcal{Z}}_x, \qquad (7.72)$$

and, similarly, the resetting closed-loop system associated with the plant dynamics is given by

$$\Delta x(t) = [\mathcal{J}_{\mathrm{d}}(x(t)) - \mathcal{R}_{\mathrm{d}}(x(t))] \left(\frac{\partial \mathcal{H}_{\mathrm{s}}}{\partial x}(x(t)) \right)^{\mathrm{T}}, \quad x(t) \in \hat{\mathcal{Z}}_x, \quad (7.73)$$

where $\mathcal{H}_{\mathrm{s}}(x) = \mathcal{H}(x) + \mathcal{H}_{\mathrm{c}}(F(x) + c)$, $x \in \mathcal{D}$, is the *shaped* Hamiltonian function for the closed-loop system (7.72) and (7.73).

Next, we use the existence of the vector energy-Casimir function to construct stabilizing hybrid dynamic controllers that guarantee that the impulsive closed-loop system associated with the impulsive plant dynamics preserves the hybrid Hamiltonian structure without the need for solving a set of partial differential equations.

Theorem 7.2 *Consider the feedback interconnection of the state-dependent impulsive port-controlled Hamiltonian systems $\mathcal{G}$ and $\mathcal{G}_{\mathrm{c}}$ given by (7.1), (7.2), (7.41), (7.42), and (7.43)–(7.46), respectively. Assume there exists a continuously differentiable function $F : \mathcal{D} \to \mathbb{R}^{n_{\mathrm{c}}}$ satisfying (7.51) such that conditions (7.60)–(7.67) hold for all $(x, x_{\mathrm{c}}) \in \mathcal{D} \times \mathbb{R}^{n_{\mathrm{c}}}$, and assume that the Hamiltonian function $\mathcal{H}_{\mathrm{c}} : \mathbb{R}^{n_{\mathrm{c}}} \to \mathbb{R}$ of the hybrid feedback controller $\mathcal{G}_{\mathrm{c}}$ is such that $\mathcal{H}_{\mathrm{s}} : \mathcal{D} \to \mathbb{R}$ is given by $\mathcal{H}_{\mathrm{s}}(x) = \mathcal{H}(x) + \mathcal{H}_{\mathrm{c}}(F(x) + c)$, $x \in \mathcal{D}$, and condition (7.22) holds. If*

$$\frac{\partial \mathcal{H}_{\mathrm{c}}}{\partial x}(F(x_{\mathrm{e}}) + c) = -\frac{\partial \mathcal{H}}{\partial x}(x_{\mathrm{e}}), \quad x_{\mathrm{e}} \in \mathcal{D}, \qquad (7.74)$$

$$\frac{\partial^2 \mathcal{H}_{\mathrm{c}}}{\partial x^2}(F(x_{\mathrm{e}}) + c) > -\frac{\partial^2 \mathcal{H}}{\partial x^2}(x_{\mathrm{e}}), \quad x_{\mathrm{e}} \in \mathcal{D}, \qquad (7.75)$$

$$\frac{\partial \mathcal{H}_{\mathrm{s}}}{\partial x}(x) \left[\mathcal{J}_{\mathrm{d}}(x) - \mathcal{R}_{\mathrm{d}}(x) \right.$$

$$\left. + \frac{1}{2}(\mathcal{J}_{\mathrm{d}}(x) - \mathcal{R}_{\mathrm{d}}(x))^{\mathrm{T}} \frac{\partial^2 \mathcal{H}_{\mathrm{s}}}{\partial x^2}(x)(\mathcal{J}_{\mathrm{d}}(x) - \mathcal{R}_{\mathrm{d}}(x)) \right] \left(\frac{\partial \mathcal{H}_{\mathrm{s}}}{\partial x}(x) \right)^{\mathrm{T}} \leq 0,$$

$$x \in \hat{\mathcal{Z}}_x, \qquad (7.76)$$

then the equilibrium solution $x(t) \equiv x_{\mathrm{e}}$ of the system (7.72) and (7.73) is Lyapunov stable. If, in addition, $\mathcal{D}_{\mathrm{c}} \subseteq \mathcal{D}$ is a compact positively *invariant set with respect to (7.72) and (7.73) and the largest invariant set contained in*

$$\mathcal{R} \triangleq \left\{ x \in \mathcal{D}_{\mathrm{c}} : x \notin \hat{\mathcal{Z}}_x, \frac{\partial \mathcal{H}_{\mathrm{s}}}{\partial x}(x) \mathcal{R}_{\mathrm{c}}(x) \left(\frac{\partial \mathcal{H}_{\mathrm{s}}}{\partial x}(x) \right)^{\mathrm{T}} = 0 \right\}$$

$$\cup \left\{ x \in \mathcal{D}_{\mathrm{c}} : x \in \hat{\mathcal{Z}}_x, \frac{\partial \mathcal{H}_{\mathrm{s}}}{\partial x}(x) \left[\mathcal{J}_{\mathrm{d}}(x) - \mathcal{R}_{\mathrm{d}}(x) \right. \right.$$

$$\left. + \frac{1}{2} (\mathcal{J}_{\mathrm{d}}(x) - \mathcal{R}_{\mathrm{d}}(x))^{\mathrm{T}} \frac{\partial^2 \mathcal{H}_{\mathrm{s}}}{\partial x^2}(x)(\mathcal{J}_{\mathrm{d}}(x) - \mathcal{R}_{\mathrm{d}}(x)) \right] \left(\frac{\partial \mathcal{H}_{\mathrm{s}}}{\partial x}(x) \right)^{\mathrm{T}}$$

$$\left. = 0 \right\} \tag{7.77}$$

is $\mathcal{M} = \{x_{\mathrm{e}}\}$, then the equilibrium solution $x(t) \equiv x_{\mathrm{e}}$ of the closed-loop system (7.72) and (7.73) is locally asymptotically stable.

Proof. Conditions (7.51) and (7.60)–(7.67) imply that the closed-loop dynamics of the impulsive port-controlled Hamiltonian system $\mathcal{G}$, and the hybrid controller $\mathcal{G}_{\mathrm{c}}$ associated with the plant states can be written in the form given by (7.72) and (7.73). Now, using identical arguments as in the proof of Theorem 7.1, conditions (7.74) and (7.75) guarantee the existence of the Lyapunov function candidate $V(x) = \mathcal{H}_{\mathrm{s}}(x) - \mathcal{H}_{\mathrm{s}}(x_{\mathrm{e}})$, $x \in \mathcal{D}$, which, along with (7.76), guarantees Lyapunov stability of the equilibrium solution $x(t) \equiv x_{\mathrm{e}}$ of the closed-loop system (7.72) and (7.73). Asymptotic stability of $x(t) \equiv x_{\mathrm{e}}$ follows from Theorem 2.3. $\qquad\square$

As in the static hybrid controller case, the hybrid dynamic controller given by Theorem 7.2 also provides an energy balance interpretation over the continuous-time trajectories of the controlled system. To see this, note that since by (7.61) $\mathcal{R}_{\mathrm{cc}}(x_{\mathrm{c}}) = 0$, $x_{\mathrm{c}} \in \mathbb{R}^{n_{\mathrm{c}}}$, it follows that the controller Hamiltonian $\mathcal{H}_{\mathrm{c}}(\cdot)$ satisfies

$$\dot{\mathcal{H}}_{\mathrm{c}}(F(x(t)) + c) = y_{\mathrm{cc}}^{\mathrm{T}}(t) y_{\mathrm{c}}(t) = -u_{\mathrm{c}}^{\mathrm{T}}(t) y_{\mathrm{c}}(t), \quad t_k < t \leq t_{k+1}. \tag{7.78}$$

Now, it follows that

$$\dot{\mathcal{H}}_{\mathrm{s}}(x(t)) = \dot{\mathcal{H}}(x(t)) + \dot{\mathcal{H}}_{\mathrm{c}}(F(x(t)) + c) = \dot{\mathcal{H}}(x(t)) - u_{\mathrm{c}}^{\mathrm{T}}(t) y_{\mathrm{c}}(t),$$
$$t_k < t \leq t_{k+1}, \tag{7.79}$$

which yields (7.40). Moreover, it follows from (7.76) that at the resetting times t_k, $\Delta \mathcal{H}_{\mathrm{s}}(x(t_k)) \leq 0$, $k \in \overline{\mathbb{Z}}_+$.

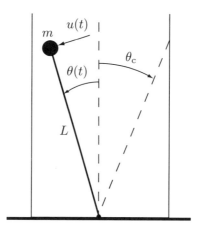

Figure 7.2 Constrained inverted pendulum.

7.5 Energy-Based Hybrid Control Design

In this section, we apply the proposed energy-based hybrid control design framework to two examples. For the first example, consider the constrained inverted pendulum shown in Figure 7.2, where $m = 1\,\text{kg}$ and $L = 1\,\text{m}$. In the case where $|\theta(t)| < \theta_{\text{c}}$, the system is governed by the dynamic equation of motion

$$\ddot{\theta}(t) - g \sin \theta(t) = u(t), \quad \theta(0) = \theta_0, \quad \dot{\theta}(0) = \dot{\theta}_0, \quad t \geq 0, \quad (7.80)$$

where g denotes the gravitational acceleration and $u(\cdot)$ is a (thruster) control force. At the instant of collision with the vertical constraint $|\theta(t)| = \theta_{\text{c}}$, the system resets according to the resetting law

$$\theta(t_k^+) = \theta(t_k), \qquad \dot{\theta}(t_k^+) = -e\dot{\theta}(t_k), \qquad (7.81)$$

where $e \in [0, 1)$ is the coefficient of restitution.

Defining $x_1 = \theta$ and $x_2 = \dot{\theta}$, we can rewrite the continuous-time dynamics and the resetting dynamics in state space form (7.1) and (7.2) with $x \triangleq [x_1, x_2]^{\text{T}}$,

$$\mathcal{J}_{\text{c}}(x) = \begin{bmatrix} 0 & 1 \\ -1 & 0 \end{bmatrix}, \quad \mathcal{R}_{\text{c}}(x) = 0, \quad G_{\text{c}}(x) = \begin{bmatrix} 0 \\ -1 \end{bmatrix},$$

$$\mathcal{J}_{\text{d}}(x) = 0, \quad \mathcal{R}_{\text{d}}(x) = \begin{bmatrix} 0 & 0 \\ 0 & (1+e) \end{bmatrix}, \quad G_{\text{d}}(x) = 0,$$

$$\mathcal{D} = \{x \in \mathbb{R}^2 : |x_1| \leq \theta_0\}, \ \mathcal{Z}_x = \{x \in \mathbb{R}^2 : x_1 = \theta_{\text{c}}, \ x_2 > 0\} \cup \{x \in \mathbb{R}^2 :$$

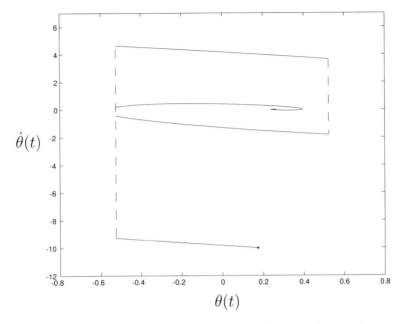

$\dot\theta(t)$

$\theta(t)$

Figure 7.3 Phase portrait of the constrained inverted pendulum.

$x_1 = -\theta_c$, $x_2 < 0\}$, and Hamiltonian function $\mathcal{H}(\cdot)$ corresponding to the total energy in the system given by $\mathcal{H}(x) = \frac{x_2^2}{2} + g\cos x_1$.

Next, to stabilize the equilibrium point $x_e = [\theta_e, 0]^T$, where $|\theta_e| < \theta_c$, we assign the shaped Hamiltonian $\mathcal{H}_s(x) = \frac{x_2^2}{2} + \frac{1}{2}(x_1 - \theta_e)^2$ function for the closed-loop system. Furthermore, we set $\mathcal{J}_{ca}(x) = 0$ and

$$ \mathcal{R}_{ca}(x) = \begin{bmatrix} 0 & 0 \\ 0 & 1 \end{bmatrix}, \qquad x \in \mathcal{D}, \quad x \notin \mathcal{Z}_x. $$

In this case, it follows from (7.26) that the continuous-time feedback controller is given by $u = \phi_c(x) = x_2 + (x_1 - \theta_e) + g\sin x_1$, $x \in \mathcal{D}$, $x \notin \mathcal{Z}_x$. Note that since $G_d(x) = 0$, it is impossible to shape the resetting dynamics. Next, note that $\dot{\mathcal{H}}_s(x) = -x_2^2 \leq 0$, $x \in \mathcal{D}$, $x \notin \mathcal{Z}_x$, and $\Delta\mathcal{H}_s(x) = \frac{e^2 x_2^2}{2} - \frac{x_2^2}{2} \leq 0$, $x \in \mathcal{Z}_x$. Hence, $\mathcal{R} \triangleq \{x \in \mathcal{D} : x \notin \mathcal{Z}_x, \dot{\mathcal{H}}_s = 0\} \cup \{x \in \mathcal{D} : x \in \mathcal{Z}_x, \Delta\mathcal{H}_s = 0\} = \{x \in \mathcal{D} : x_2 = 0\}$. Finally, since for every $x \in \mathcal{R}$, $\dot{x}_2 \neq 0$ if and only if $x_1 \neq \theta_e$, it follows that the largest invariant set contained in $\mathcal{R}$ is given by $\mathcal{M} = \{x_e\}$, and hence, the equilibrium solution $x(t) \equiv [\theta_e, 0]^T$ is asymptotically stable.

With $e = 0.5$, $\theta_c = 30°$, and $\theta_e = 15°$, Figure 7.3 shows the phase portrait of the impulsive port-controlled Hamiltonian system. Figure

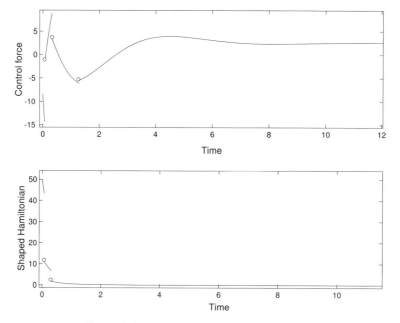

Figure 7.4 Control force and shaped Hamiltonian versus time.

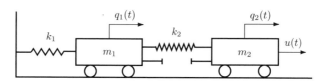

Figure 7.5 Two-mass system with constraint buffers.

7.4 shows the control force versus time and the shaped Hamiltonian versus time. Note that the control force and shaped Hamiltonian versus time are discontinuous at the resetting times.

For the next example, consider the two-mass, two-spring system with buffer constraints of length $\frac{L}{2}$ shown in Figure 7.5. A control force $u(\cdot)$ acts on mass 2 with the goal to stabilize the position of the second mass. Between collisions the system dynamics, with state variables defined in Figure 7.5, are given by

$$m_1\ddot{q}_1(t) + (k_1 + k_2)q_1(t) - k_2 q_2(t) = 0, \quad q_1(0) = q_{01}, \quad \dot{q}_1(0) = \dot{q}_{01},$$
$$t \geq 0, \quad (7.82)$$
$$m_2\ddot{q}_2(t) - k_2 q_1(t) + k_2 q_2(t) = u(t), \quad q_2(0) = q_{02}, \quad \dot{q}_2(0) = \dot{q}_{02}.$$
$$(7.83)$$

At the instant of a collision, the velocities of the masses change according to the law of conservation of linear momentum and the loss of kinetic energy due to a collision so that

$$m_1 \dot{q}_1(t_k^+) + m_2 \dot{q}_2(t_k^+) = m_1 \dot{q}_1(t_k) + m_2 \dot{q}_2(t_k), \qquad (7.84)$$
$$\dot{q}_1(t_k^+) - \dot{q}_2(t_k^+) = -e(\dot{q}_1(t_k) - \dot{q}_2(t_k)), \qquad (7.85)$$

where $e \in [0, 1)$ is the coefficient of restitution. Solving (7.84) and (7.85) for $\dot{q}_1(t_k^+)$ and $\dot{q}_2(t_k^+)$, the resetting dynamics are given by

$$\Delta \dot{q}_1(t_k) = \dot{q}_1(t_k^+) - \dot{q}_1(t_k) = -\frac{(1+e)m_2}{m_1 + m_2}(\dot{q}_1(t_k) - \dot{q}_2(t_k)), \quad (7.86)$$

$$\Delta \dot{q}_2(t_k) = \dot{q}_2(t_k^+) - \dot{q}_2(t_k) = \frac{(1+e)m_1}{m_1 + m_2}(\dot{q}_1(t_k) - \dot{q}_2(t_k)). \quad (7.87)$$

Defining $x_1 = q_1$, $x_2 = \dot{q}_1$, $x_3 = q_2$, and $x_4 = \dot{q}_2$, we can rewrite (7.82), (7.83), (7.86), and (7.87) in state space form (7.1) and (7.2) with $x = [x_1, x_2, x_3, x_4]^T$, $\mathcal{R}_c(x) = 0$, $G_c(x) = [0, 0, 0, 1]^T$, $\mathcal{J}_d(x) = 0$, $G_d(x) = 0$, $u_c = \frac{u}{m_2}$,

$$\mathcal{J}_c(x) = \begin{bmatrix} 0 & \frac{1}{m_1} & 0 & 0 \\ -\frac{1}{m_1} & 0 & 0 & 0 \\ 0 & 0 & 0 & \frac{1}{m_2} \\ 0 & 0 & -\frac{1}{m_2} & 0 \end{bmatrix},$$

$$\mathcal{R}_d(x) = \begin{bmatrix} 0 & 0 & 0 & 0 \\ 0 & \frac{(1+e)m_2}{m_1(m_1+m_2)} & 0 & -\frac{(1+e)}{(m_1+m_2)} \\ 0 & 0 & 0 & 0 \\ 0 & -\frac{(1+e)}{(m_1+m_2)} & 0 & \frac{(1+e)m_1}{m_2(m_1+m_2)} \end{bmatrix},$$

$\mathcal{D} = \mathbb{R}^4$, $\mathcal{Z}_x = \{x \in \mathbb{R}^4 : x_1 - x_3 = L, x_2 > x_4\}$, and Hamiltonian function $\mathcal{H}(\cdot)$ corresponding to the total energy in the system given by $\mathcal{H}(x) = \frac{m_1 x_2^2}{2} + \frac{m_2 x_4^2}{2} + \frac{k_1 x_1^2}{2} + \frac{k_2(x_3 - x_1)^2}{2}$.

Next, to stabilize the equilibrium point $x_e = [x_{1e}, 0, x_{3e}, 0]^T$, where $x_{1e} = \frac{k_2}{(k_1 + k_2)} x_{3e}$, with a steady-state control value of

$$u_{c\,ss} = \frac{k_1 k_2}{m_2(k_1 + k_2)} x_{3e},$$

we assign the shaped Hamiltonian function $\mathcal{H}_s(x) = \frac{m_1 x_2^2}{2} + \frac{m_2 x_4^2}{2} + \frac{k_1 x_1^2}{2} + \frac{k_2(x_3 - x_1)^2}{2} - \frac{k_1 k_2}{(k_1 + k_2)} x_{3e} x_3$ for the closed-loop system. Furthermore, we set $\mathcal{J}_{\mathrm{ca}}(x) \equiv 0$ and

$$
\mathcal{R}_{\mathrm{ca}}(x) = \begin{bmatrix} 0 & 0 & 0 & 0 \\ 0 & 0 & 0 & 0 \\ 0 & 0 & 0 & 0 \\ 0 & 0 & 0 & \frac{1}{m_2} \end{bmatrix}.
$$

In this case, it follows from (7.26) that the continuous-time feedback controller is given by $u_{\mathrm{c}} = \phi_{\mathrm{c}}(x) = \frac{k_1 k_2}{m_2 (k_1 + k_2)} x_{3\mathrm{e}} - x_4$, $x \in \mathcal{D}$, $x \notin \mathcal{Z}_x$. Note that since $G_{\mathrm{d}}(x) = 0$, we can only shape the continuous-time dynamics. Next, note that $\dot{\mathcal{H}}_{\mathrm{s}}(x) = -m_2 x_4^2 \leq 0$, $x \in \mathcal{D}$, $x \notin \mathcal{Z}_x$, and $\Delta \mathcal{H}_{\mathrm{s}}(x) = \frac{(e^2-1)m_1 m_2 (x_2 - x_4)^2}{2(m_1 + m_2)} \leq 0$, $x \in \mathcal{Z}_x$. Hence, $\mathcal{R} \triangleq \{x \in \mathcal{D} : x \notin \mathcal{Z}_x, \dot{\mathcal{H}}_{\mathrm{s}} = 0\} \cup \{x \in \mathcal{D} : x \in \mathcal{Z}_x, \Delta \mathcal{H}_{\mathrm{s}} = 0\} = \{x \in \mathcal{D} : x \notin \mathcal{Z}_x, x_4 = 0\} \cup \varnothing$. Now, if $\mathcal{M} \subseteq \mathcal{R}$ is the largest invariant set contained in $\mathcal{R}$, then for any $x_0 \in \mathcal{M}$, $x_4(t) \equiv 0$, which implies that $x_1(t) - x_3(t) + \frac{k_1}{k_1+k_2} x_{3\mathrm{e}} = 0$ and $\dot{x}_3(t) = 0$, $t \geq 0$. In this case, it follows that $\dot{x}_1(t) = 0$, and hence, $\dot{x}_2(t) = 0$, $t \geq 0$. Hence, the only point that belongs to $\mathcal{M}$ is $x_{\mathrm{e}} = [\frac{k_2}{(k_1+k_2)} x_{3\mathrm{e}}, 0, x_{3\mathrm{e}}, 0]^{\mathrm{T}}$, which implies that x_{e} is an asymptotically stable equilibrium point of the closed-loop system.

With $m_1 = 1.5\,\mathrm{kg}$, $m_2 = 0.8\,\mathrm{kg}$, $k_1 = 0.1\,\mathrm{N/m}$, $k_2 = 0.3\,\mathrm{N/m}$, $L = 0.4\,\mathrm{m}$, $x_{3\mathrm{e}} = 3\,\mathrm{m}$, and $e = 0.5$, Figure 7.6 shows the phase portrait of x_2 versus x_4 of the impulsive port-controlled Hamiltonian system. Figures 7.7 and 7.8 show, respectively, the positions and velocities of the masses versus time. Finally, Figure 7.9 shows the control force versus time and the shaped Hamiltonian versus time.

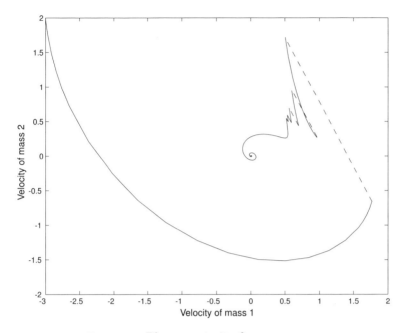

Figure 7.6 Phase portrait of x_2 versus x_4.

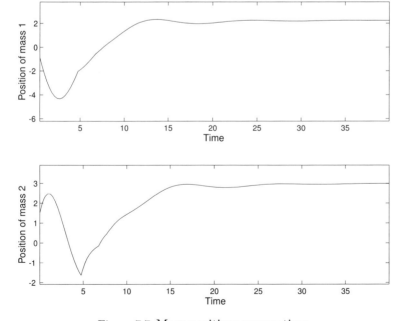

Figure 7.7 Mass positions versus time.

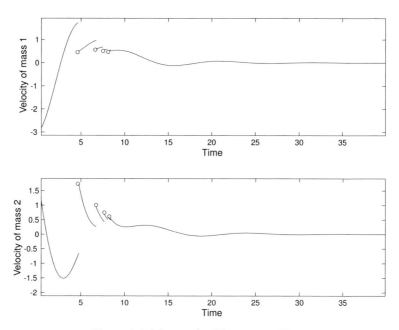

Figure 7.8 Mass velocities versus time.

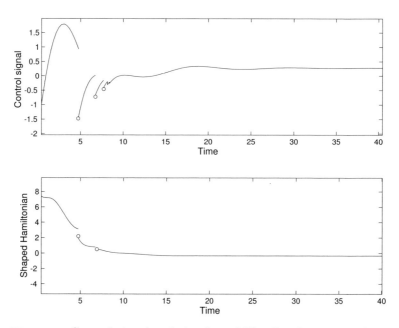

Figure 7.9 Control signal and the shaped Hamiltonian versus time.

Chapter Eight

Energy and Entropy-Based Hybrid Stabilization for Nonlinear Dynamical Systems

8.1 Introduction

Energy is a concept that underlies our understanding of all physical phenomena and is a measure of the ability of a dynamical system to produce changes (motion) in its own system state as well as changes in the system states of its surroundings. In control engineering, dissipativity theory [165], which encompasses passivity theory, provides a fundamental framework for the analysis and control design of dynamical systems using an input-output system description based on system-energy-related considerations [107, 134, 160]. The notion of energy here refers to abstract energy notions for which a physical system energy interpretation is not necessary. As noted in Chapter 3, the dissipation hypothesis on dynamical systems results in a fundamental constraint on their dynamic behavior, wherein a dissipative dynamical system can only deliver a fraction of its energy to its surroundings and can only store a fraction of the work done to it. Thus, dissipativity theory provides a powerful framework for the analysis and control design of dynamical systems based on generalized energy considerations by exploiting the notion that numerous physical systems have certain input-output properties related to conservation, dissipation, and transport of energy. Such conservation laws are prevalent in dynamical systems such as mechanical systems, fluid systems, electromechanical systems, electrical systems, combustion systems, structural vibration systems, biological systems, physiological systems, power systems, telecommunications systems, and economic systems, to cite but a few examples.

Energy-based control for Euler-Lagrange dynamical systems and Hamiltonian dynamical systems based on passivity notions has received considerable attention in the literature [131, 133–135, 150, 159]. This controller design technique achieves system stabilization by shaping the energy of the closed-loop system which involves the physical system energy and the controller emulated energy. Specifically, *en-*

ergy shaping is achieved by modifying the system potential energy in such a way so that the shaped potential energy function for the closed-loop system possesses a unique global minimum at a desired equilibrium point. Next, damping is *injected* via feedback control modifying the system dissipation to guarantee asymptotic stability of the closed-loop system. A central feature of this energy-based stabilization approach is that the Lagrangian system form is preserved at the closed-loop system level. Furthermore, the control action has a clear physical energy interpretation, wherein the total energy of the closed-loop Euler-Lagrange system corresponds to the difference between the physical system energy and the emulated energy supplied by the controller.

More recently, a passivity-based control framework for port-controlled Hamiltonian systems is established in [136–138, 143, 160]. Specifically, the authors in [136–138] develop a controller design methodology that achieves stabilization via system passivation. In particular, the interconnection and damping matrix functions of the port-controlled Hamiltonian system are shaped so that the physical (Hamiltonian) system structure is preserved at the closed-loop level, and the closed-loop energy function is equal to the difference between the physical energy of the system and the energy supplied by the controller. Since the Hamiltonian structure is preserved at the closed-loop level, the passivity-based controller is *robust* with respect to unmodeled passive dynamics. Furthermore, passivity-based control architectures are extremely appealing since the control action has a clear *physical* energy interpretation which can considerably simplify controller implementation.

In this chapter, we develop a novel energy-dissipating hybrid control framework for Lagrangian, port-controlled Hamiltonian, and dissipative dynamical systems. These dynamical systems cover a very broad spectrum of applications including mechanical, electrical, electromechanical, structural, biological, and power systems. The fixed-order, energy-based hybrid controller is a hybrid controller that emulates an approximately lossless hybrid dynamical system and exploits the feature that the states of the dynamic controller may be reset to enhance the overall energy dissipation in the closed-loop system. An important feature of the hybrid controller is that its structure can be associated with an energy function. In a mechanical Euler-Lagrange system, positions typically correspond to elastic deformations, which contribute to the potential energy of the system, whereas velocities typically correspond to momenta, which contribute to the kinetic energy of the system. On the other hand, while our energy-based hybrid controller has dynamical states that emulate the motion of a physical lossless system, these states only "exist" as numerical representations inside the processor. Consequently, while one can associate an *emulated energy* with these states, this energy is merely a mathematical construct and does not correspond to any physical form of energy.

The concept of an energy-based hybrid controller can be viewed as a feedback control technique that exploits the coupling between a physical dynamical system and an energy-based controller to efficiently remove energy from the physical system. Specifically, if a dissipative or lossless plant is at high energy level, and a lossless feedback controller at a low energy level is attached to it, then energy will generally tend to flow from the plant into the controller, decreasing the plant energy and increasing the controller energy [90]. Of course, emulated energy, and not physical energy, is accumulated by the controller. Conversely, if the attached controller is at a high energy level and a plant is at a low energy level, then energy can flow from the controller to the plant, since a controller can generate real, physical energy to effect the required energy flow. Hence, if and when the controller states coincide with a high emulated energy level, then we can *reset* these states to remove the emulated energy so that the emulated energy is not returned to the plant. In this case, the overall closed-loop system consisting of the plant and the controller possesses discontinuous flows since it combines logical switchings with continuous dynamics, leading to impulsive differential equations. Within the context of vibration control using resetting virtual absorbers, these ideas were first explored in [35].

8.2 Hybrid Control and Impulsive Dynamical Systems

In this section, we consider continuous-time nonlinear dynamical systems of the form

$$\dot{x}_{\mathrm{p}}(t) = f_{\mathrm{p}}(x_{\mathrm{p}}(t), u(t)), \quad x_{\mathrm{p}}(0) = x_{\mathrm{p}0}, \quad t \geq 0, \tag{8.1}$$
$$y(t) = h_{\mathrm{p}}(x_{\mathrm{p}}(t)), \tag{8.2}$$

where $t \geq 0$, $x_{\mathrm{p}}(t) \in \mathcal{D}_{\mathrm{p}} \subseteq \mathbb{R}^{n_{\mathrm{p}}}$, $\mathcal{D}_{\mathrm{p}}$ is an open set with $0 \in \mathcal{D}_{\mathrm{p}}$, $u(t) \in \mathbb{R}^{m}$, $f_{\mathrm{p}} : \mathcal{D}_{\mathrm{p}} \times \mathbb{R}^{m} \to \mathbb{R}^{n_{\mathrm{p}}}$ is smooth (i.e., infinitely differentiable) on $\mathcal{D}_{\mathrm{p}} \times \mathbb{R}^{m}$ and satisfies $f_{\mathrm{p}}(0, 0) = 0$, and $h_{\mathrm{p}} : \mathcal{D}_{\mathrm{p}} \to \mathbb{R}^{l}$ is continuous and satisfies $h_{\mathrm{p}}(0) = 0$. Furthermore, we consider hybrid (resetting) dynamic controllers of the form

$$\dot{x}_{\mathrm{c}}(t) = f_{\mathrm{cc}}(x_{\mathrm{c}}(t), y(t)), \quad x_{\mathrm{c}}(0) = x_{\mathrm{c}0}, \quad (x_{\mathrm{c}}(t), y(t)) \notin \mathcal{Z}_{\mathrm{c}}, \tag{8.3}$$

$$\Delta x_{\mathrm{c}}(t) = f_{\mathrm{dc}}(x_{\mathrm{c}}(t), y(t)), \quad (x_{\mathrm{c}}(t), y(t)) \in \mathcal{Z}_{\mathrm{c}}, \tag{8.4}$$

$$u(t) = h_{\mathrm{cc}}(x_{\mathrm{c}}(t), y(t)), \tag{8.5}$$

where $t \geq 0$, $x_{\mathrm{c}}(t) \in \mathcal{D}_{\mathrm{c}} \subseteq \mathbb{R}^{n_{\mathrm{c}}}$, $\mathcal{D}_{\mathrm{c}}$ is an open set with $0 \in \mathcal{D}_{\mathrm{c}}$, $\Delta x_{\mathrm{c}}(t) \triangleq x_{\mathrm{c}}(t^{+}) - x_{\mathrm{c}}(t)$, $f_{\mathrm{cc}} : \mathcal{D}_{\mathrm{c}} \times \mathbb{R}^{l} \to \mathbb{R}^{n_{\mathrm{c}}}$ is smooth on $\mathcal{D}_{\mathrm{c}} \times \mathbb{R}^{l}$ and satisfies $f_{\mathrm{cc}}(0,0) = 0$, $h_{\mathrm{cc}} : \mathcal{D}_{\mathrm{c}} \times \mathbb{R}^{l} \to \mathbb{R}^{m}$ is continuous and satisfies $h_{\mathrm{cc}}(0,0) = 0$, $f_{\mathrm{dc}} : \mathcal{D}_{\mathrm{c}} \times \mathbb{R}^{l} \to \mathbb{R}^{n_{\mathrm{c}}}$ is continuous, and $\mathcal{Z}_{\mathrm{c}} \subset \mathcal{D}_{\mathrm{c}} \times \mathbb{R}^{l}$ is the resetting set. Note that, for generality, we allow the hybrid dynamic controller to be of fixed dimension n_{c} which may be less than the plant order n_{p}.

The equations of motion for the closed-loop dynamical system (8.1)–(8.5) have the form

$$\dot{x}(t) = f_{\mathrm{c}}(x(t)), \quad x(0) = x_0, \quad x(t) \notin \mathcal{Z}, \tag{8.6}$$

$$\Delta x(t) = f_{\mathrm{d}}(x(t)), \quad x(t) \in \mathcal{Z}, \tag{8.7}$$

where

$$x \triangleq \begin{bmatrix} x_{\mathrm{p}} \\ x_{\mathrm{c}} \end{bmatrix} \in \mathbb{R}^{n}, \quad f_{\mathrm{c}}(x) \triangleq \begin{bmatrix} f_{\mathrm{p}}(x_{\mathrm{p}}, h_{\mathrm{cc}}(x_{\mathrm{c}}, h_{\mathrm{p}}(x_{\mathrm{p}}))) \\ f_{\mathrm{cc}}(x_{\mathrm{c}}, h_{\mathrm{p}}(x_{\mathrm{p}})) \end{bmatrix}, \tag{8.8}$$

$$f_{\mathrm{d}}(x) \triangleq \begin{bmatrix} 0 \\ f_{\mathrm{dc}}(x_{\mathrm{c}}, h_{\mathrm{p}}(x_{\mathrm{p}})) \end{bmatrix}, \tag{8.9}$$

and $\mathcal{Z} \triangleq \{x \in \mathcal{D} : (x_{\mathrm{c}}, h_{\mathrm{p}}(x_{\mathrm{p}})) \in \mathcal{Z}_{\mathrm{c}}\}$, with $n \triangleq n_{\mathrm{p}} + n_{\mathrm{c}}$ and $\mathcal{D} \triangleq \mathcal{D}_{\mathrm{p}} \times \mathcal{D}_{\mathrm{c}}$. Note that although the closed-loop state vector consists of plant states and controller states, it is clear from (8.9) that only those states associated with the controller are reset. For convenience, we use the notation $s(t, x_0)$ to denote the solution $x(t)$ of (8.6) and (8.7) at time $t \geq 0$ with initial condition $x(0) = x_0$.

For a particular closed-loop trajectory $x(t)$, we let $t_k \triangleq \tau_k(x_0)$ denote the kth instant of time at which $x(t)$ intersects $\mathcal{Z}$, and we denote the resetting times by t_k. Thus, the trajectory of the closed-loop system (8.6) and (8.7) from the initial condition $x(0) = x_0$ is given by $s(t, x_0)$ for $t \geq 0$, and note that $s(\cdot, x_0)$ is continuous everywhere except at the resetting times t_k for $k = 1, 2, \ldots$. To ensure the well-posedness of the resetting times, we make the following additional assumptions. These assumptions represent the specialization of A1 and A2 of Chapter 2 to a state-dependent resetting set $\mathcal{Z}$.

A1. If $x \in \overline{\mathcal{Z}} \backslash \mathcal{Z}$, then there exists $\varepsilon > 0$ such that, for all $0 < \delta < \varepsilon$, $s(\delta, x) \notin \mathcal{Z}$.

A2. If $x \in \mathcal{Z}$, then $x + f_{\mathrm{d}}(x) \notin \mathcal{Z}$.

Since the resetting times are well defined and distinct, and since the solution to (8.6) exists and is unique, it follows that the solution of the impulsive dynamical system (8.6) and (8.7) also exists and is unique over a forward time interval. Here we assume that if the solution to (8.6) and (8.7) is Zeno, then it is convergent and the continuous and discrete parts of the state converge to a unique value at the Zeno time.

As shown in Chapter 2, Assumption 2.1 is key in guaranteeing invariance of positive limit sets for state-dependent impulsive dynamical systems. However, as can be seen from the proof of Theorem 2.2, in order to guarantee invariance of positive limit sets for state-dependent impulsive dynamical systems, the quasi-continuous dependence property need only be satisfied for trajectories starting on the positive limit set. However, since it is generally difficult to verify that the quasi-continuous dependence property holds for trajectories on the positive limit set, in Assumption 2.1 we assume that the quasi-continuous dependence property holds for every trajectory in $\mathcal{D}$. In practice, however, this assumption can be restrictive. In this chapter, we weaken Assumption 2.1 by assuming point-wise continuous dependence and show that this weakened version of Assumption 2.1 is sufficient for guaranteeing invariance of positive limit sets for a special class of state-dependent impulsive dynamical systems. Specifically, we consider impulsive dynamical systems of the form (8.6) and (8.7) for which $x_0 \in \mathcal{Z}$ implies that $x_0 + f_{\mathrm{d}}(x_0) \in \overline{\mathcal{Z}} \backslash \mathcal{Z}$. In this case, it will be shown that the positive limit sets of all trajectories of (8.6) and (8.7) lie on $\overline{\mathcal{Z}} \backslash \mathcal{Z}$. Hence, we need only assume quasi-continuous dependence for trajectories starting outside $\mathcal{Z}$.

Assumption 8.1 *Consider the impulsive dynamical system (8.6) and (8.7), and let $s(t, x_0)$, $t \geq 0$, denote the solution to (8.6) and (8.7) with initial condition x_0. Then for every $x_0 \notin \mathcal{Z}$ and every $\varepsilon > 0$ and $t \neq t_k$, there exists $\delta(\varepsilon, x_0, t) > 0$ such that if $\|x_0 - z\| < \delta(\varepsilon, x_0, t)$, $z \in \mathcal{D}$, then $\|s(t, x_0) - s(t, z)\| < \varepsilon$.*

The following result provides a generalization of the results given in Section 2.3 for establishing sufficient conditions for guaranteeing that the impulsive dynamical system (8.6) and (8.7) satisfies Assumption 8.1.

Proposition 8.1 *Consider the impulsive dynamical system $\mathcal{G}$ given by (8.6) and (8.7). Assume that Assumptions A1 and A2 hold, $\tau_1(\cdot)$ is continuous at every $x \notin \overline{\mathcal{Z}}$ such that $0 < \tau_1(x) < \infty$, and if $x \in \mathcal{Z}$, then $x + f_{\mathrm{d}}(x) \in \overline{\mathcal{Z}} \backslash \mathcal{Z}$. Furthermore, for every $x \in \overline{\mathcal{Z}} \backslash \mathcal{Z}$ such that $0 < \tau_1(x) < \infty$, assume that the following statements hold:*

i) If a sequence $\{x_i\}_{i=1}^{\infty} \in \mathcal{D}$ is such that $\lim_{i\to\infty} x_i = x$ and $\lim_{i\to\infty} \tau_1(x_i)$ exists, then either $f_{\mathrm{d}}(x) = 0$ and $\lim_{i\to\infty} \tau_1(x_i) = 0$, or $\lim_{i\to\infty} \tau_1(x_i) = \tau_1(x)$.

ii) If a sequence $\{x_i\}_{i=1}^{\infty} \in \overline{\mathcal{Z}}\backslash\mathcal{Z}$ is such that $\lim_{i\to\infty} x_i = x$ and $\lim_{i\to\infty} \tau_1(x_i)$ exists, then $\lim_{i\to\infty} \tau_1(x_i) = \tau_1(x)$.

Then $\mathcal{G}$ satisfies Assumption 8.1.

Proof. Let $x_0 \in \overline{\mathcal{Z}}\backslash\mathcal{Z}$ and let $\{x_i\}_{i=1}^{\infty} \in \mathcal{D}$ be such that $f_{\mathrm{d}}(x_0) = 0$ and $\lim_{i\to\infty} \tau_1(x_i) = 0$ hold. Define $z_i \triangleq s(\tau_1(x_i), x_i) + f_{\mathrm{d}}(s(\tau_1(x_i), x_i))$ $= \psi(\tau_1(x_i), x_i) + f_{\mathrm{d}}(\psi(\tau_1(x_i), x_i))$, $i = 1, 2, \ldots$, where $\psi(t, x_0)$ denotes the solution to the continuous-time dynamics (8.6), and note that, since $f_{\mathrm{d}}(x_0) = 0$ and $\lim_{i\to\infty} \tau_1(x_i) = 0$, it follows that $\lim_{i\to\infty} z_i = x_0$. Hence, since by assumption $z_i \in \overline{\mathcal{Z}}\backslash\mathcal{Z}$, $i = 1, 2, \ldots$, it follows from *ii)* that $\lim_{i\to\infty} \tau_1(z_i) = \tau_1(x_0)$, or, equivalently, $\lim_{i\to\infty} \tau_2(x_i) = \tau_1(x_0)$. Similarly, it can be shown that $\lim_{i\to\infty} \tau_{k+1}(x_i) = \tau_k(x_0)$, $k = 2, 3, \ldots$. Next, note that

$$\lim_{i\to\infty} s(\tau_2(x_i), x_i) = \lim_{i\to\infty} \psi(\tau_2(x_i) - \tau_1(x_i), s(\tau_1(x_i), x_i)$$
$$+ f_{\mathrm{d}}(s(\tau_1(x_i), x_i)))$$
$$= \psi(\tau_1(x_0), x_0)$$
$$= s(\tau_1(x_0), x_0).$$

Now, using mathematical induction it can be shown that $\lim_{i\to\infty} s(\tau_{k+1}(x_i), x_i) = s(\tau_k(x_0), x_0)$, $k = 2, 3, \ldots$.

Next, let $k \in \{1, 2, \ldots\}$ and let $t \in (\tau_k(x_0), \tau_{k+1}(x_0))$. Since $\lim_{i\to\infty} \tau_{k+1}(x_i) = \tau_k(x_0)$, it follows that there exists $I \in \{1, 2, \ldots\}$ such that $\tau_{k+1}(x_i) < t$ and $\tau_{k+2}(x_i) > t$ for all $i > I$. Hence, it follows that for every $t \in (\tau_k(x_0), \tau_{k+1}(x_0))$,

$$\lim_{i\to\infty} s(t, x_i) = \lim_{i\to\infty} \psi(t - \tau_{k+1}(x_i), s(\tau_{k+1}(x_i), x_i)$$
$$+ f_{\mathrm{d}}(s(\tau_{k+1}(x_i), x_i)))$$
$$= \psi(t - \tau_k(x_0), s(\tau_k(x_0), x_0) + f_{\mathrm{d}}(s(\tau_k(x_0), x_0)))$$
$$= s(t, x_0).$$

Alternatively, if $x_0 \in \overline{\mathcal{Z}}\backslash\mathcal{Z}$ is such that $\lim_{i\to\infty} \tau_1(x_i) = \tau_1(x_0)$ for $\{x_i\}_{i=1}^{\infty} \in \overline{\mathcal{Z}}\backslash\mathcal{Z}$, then using identical arguments as above, it can be shown that $\lim_{i\to\infty} s(t, x_i) = s(t, x_0)$ for every $t \in (\tau_k(x_0), \tau_{k+1}(x_0))$, $k = 1, 2, \ldots$.

Finally, let $x_0 \notin \overline{\mathcal{Z}}$, $0 < \tau_1(x_0) < \infty$, and assume $\tau_1(\cdot)$ is continuous. In this case, it follows from the definition of $\tau_1(x_0)$ that for every

$x_0 \notin \overline{\mathcal{Z}}$ and $t \in (\tau_1(x_0), \tau_2(x_0)]$,

$$s(t, x_0) = \psi(t - \tau_1(x_0), s(\tau_1(x_0), x_0) + f_d(s(\tau_1(x_0), x_0))). \qquad (8.10)$$

Since $\psi(\cdot, \cdot)$ is continuous in both its arguments, $\tau_1(\cdot)$ is continuous at x_0, and $f_d(\cdot)$ is continuous, it follows that $s(t, \cdot)$ is continuous at x_0 for every $t \in (\tau_1(x_0), \tau_2(x_0))$. Next, for every sequence $\{x_i\}_{i=1}^{\infty} \in \mathcal{D}$ such that $\lim_{i \to \infty} x_i = x_0$, it follows that $\lim_{i \to \infty} s(\tau_1(x_i), x_i) = \lim_{i \to \infty} \psi(\tau_1(x_i), x_i) = \psi(\tau_1(x_0), x_0) = s(\tau_1(x_0), x_0)$. Furthermore, note that by assumption $z_i \triangleq s(\tau_1(x_i), x_i) + f_d(s(\tau_1(x_i), x_i)) \in \overline{\mathcal{Z}} \backslash \mathcal{Z}$, $i = 0, 1, \dots$. Hence, it follows that for all $t \in (\tau_k(z_0), \tau_{k+1}(z_0))$, $k = 1, 2, \dots$, $\lim_{i \to \infty} s(t, z_i) = s(t, z_0)$, or, equivalently, for all $t \in (\tau_k(x_0), \tau_{k+1}(x_0))$, $k = 2, 3, \dots$, $\lim_{i \to \infty} s(t, x_i) = s(t, x_0)$, which proves the result. $\qquad \square$

Proposition 8.1 presents a generalization of Proposition 2.1 to the case where the resetting set $\mathcal{Z}$ is not necessarily closed. This generalization is key in developing energy-based and entropy-based hybrid controllers. The following result provides sufficient conditions for establishing continuity of $\tau_1(\cdot)$ at $x_0 \notin \overline{\mathcal{Z}}$ and *sequential continuity* of $\tau_1(\cdot)$ at $x_0 \in \overline{\mathcal{Z}} \backslash \mathcal{Z}$, that is, $\lim_{i \to \infty} \tau_1(x_i) = \tau_1(x_0)$ for $\{x_i\}_{i=1}^{\infty} \notin \mathcal{Z}$ and $\lim_{i \to \infty} x_i = x_0$. For this result, the following definition is needed. First, however, recall that the *Lie derivative* of a smooth function $\mathcal{X} : \mathcal{D} \to \mathbb{R}$ along the vector field of the continuous-time dynamics $f_c(x)$ is given by $L_{f_c} \mathcal{X}(x) \triangleq \frac{d}{dt} \mathcal{X}(\psi(t, x))|_{t=0} = \frac{\partial \mathcal{X}(x)}{\partial x} f_c(x)$, and the *zeroth* and *higher-order Lie derivatives* are, respectively, defined by $L_{f_c}^0 \mathcal{X}(x) \triangleq \mathcal{X}(x)$ and $L_{f_c}^k \mathcal{X}(x) \triangleq L_{f_c}(L_{f_c}^{k-1} \mathcal{X}(x))$, where $k \geq 1$.

Definition 8.1 Let $\mathcal{M} \triangleq \{x \in \mathcal{D} : \mathcal{X}(x) = 0\}$, *where* $\mathcal{X} : \mathcal{D} \to \mathbb{R}$ *is an infinitely differentiable function. A point* $x \in \mathcal{M}$ *such that* $f_c(x) \neq 0$ *is* transversal *to (8.6) if there exists* $k \in \{1, 2, \dots\}$ *such that*

$$L_{f_c}^r \mathcal{X}(x) = 0, \quad r = 0, \dots, 2k - 2, \quad L_{f_c}^{2k-1} \mathcal{X}(x) \neq 0. \qquad (8.11)$$

Proposition 8.2 *Consider the impulsive dynamical system (8.6) and (8.7). Let* $\mathcal{X} : \mathcal{D} \to \mathbb{R}$ *be an infinitely differentiable function such that* $\mathcal{Z} = \{x \in \mathcal{D} : \mathcal{X}(x) = 0\}$, *and assume that every* $x \in \mathcal{Z}$ *is transversal to (8.6). Then at every* $x_0 \notin \overline{\mathcal{Z}}$ *such that* $0 < \tau_1(x_0) < \infty$, $\tau_1(\cdot)$ *is continuous. Furthermore, if* $x_0 \in \overline{\mathcal{Z}} \backslash \mathcal{Z}$ *is such that* $\tau_1(x_0) \in (0, \infty)$ *and* $\{x_i\}_{i=1}^{\infty} \in \overline{\mathcal{Z}} \backslash \mathcal{Z}$ *or* $\lim_{i \to \infty} \tau_1(x_i) > 0$, *where* $\{x_i\}_{i=1}^{\infty} \notin \overline{\mathcal{Z}}$ *is such that* $\lim_{i \to \infty} x_i = x_0$ *and* $\lim_{i \to \infty} \tau_1(x_i)$ *exists, then* $\lim_{i \to \infty} \tau_1(x_i) = \tau_1(x_0)$.

Proof. Let $x_0 \notin \overline{\mathcal{Z}}$ be such that $0 < \tau_1(x_0) < \infty$. It follows from the definition of $\tau_1(\cdot)$ that $s(t, x_0) = \psi(t, x_0)$, $t \in [0, \tau_1(x_0)]$, $\mathcal{X}(s(t, x_0)) \neq 0$, $t \in (0, \tau_1(x_0))$, and $\mathcal{X}(s(\tau_1(x_0), x_0)) = 0$. Without loss of generality, let $\mathcal{X}(s(t, x_0)) > 0$, $t \in (0, \tau_1(x_0))$. Since $\hat{x} \triangleq \psi(\tau_1(x_0), x_0) \in \mathcal{Z}$ is transversal to (8.6), it follows that there exists $\theta > 0$ such that $\mathcal{X}(\psi(t, \hat{x})) > 0$, $t \in [-\theta, 0)$, and $\mathcal{X}(\psi(t, \hat{x})) < 0$, $t \in (0, \theta]$. (This fact can be easily shown by expanding $\mathcal{X}(\psi(t, x))$ via a Taylor series expansion about $\hat{x}$ and using the fact that $\hat{x}$ is transversal to (8.6).) Hence, $\mathcal{X}(\psi(t, x_0)) > 0$, $t \in [\hat{t}_1, \tau_1(x_0))$, and $\mathcal{X}(\psi(t, x_0)) < 0$, $t \in (\tau_1(x_0), \hat{t}_2]$, where $\hat{t}_1 \triangleq \tau_1(x_0) - \theta$ and $\hat{t}_2 \triangleq \tau_1(x_0) + \theta$.

Next, let $\varepsilon \triangleq \min\{\mathcal{X}(\psi(\hat{t}_1, x_0)), \mathcal{X}(\psi(\hat{t}_2, x_0))\}$. Now, it follows from the continuity of $\mathcal{X}(\cdot)$ and the continuous dependence of $\psi(\cdot, \cdot)$ on the system initial conditions that there exists $\delta > 0$ such that

$$\sup_{0 \le t \le \hat{t}_2} |\mathcal{X}(\psi(t, x)) - \mathcal{X}(\psi(t, x_0))| < \varepsilon, \quad x \in \mathcal{B}_\delta(x_0), \quad (8.12)$$

which implies that $\mathcal{X}(\psi(\hat{t}_1, x)) > 0$ and $\mathcal{X}(\psi(\hat{t}_2, x)) < 0$, $x \in \mathcal{B}_\delta(x_0)$. Hence, it follows that $\hat{t}_1 < \tau_1(x) < \hat{t}_2$, $x \in \mathcal{B}_\delta(x_0)$. The continuity of $\tau_1(\cdot)$ at x_0 now follows immediately by noting that θ can be chosen arbitrarily small.

Finally, let $x_0 \in \overline{\mathcal{Z}} \backslash \mathcal{Z}$ be such that $\lim_{i \to \infty} x_i = x_0$ for some sequence $\{x_i\}_{i=1}^\infty \in \overline{\mathcal{Z}} \backslash \mathcal{Z}$. Then using similar arguments as above it can be shown that $\lim_{i \to \infty} \tau_1(x_i) = \tau_1(x_0)$. Alternatively, if $x_0 \in \overline{\mathcal{Z}} \backslash \mathcal{Z}$ is such that $\lim_{i \to \infty} x_i = x_0$ and $\lim_{i \to \infty} \tau_1(x_i) > 0$ for some sequence $\{x_i\}_{i=1}^\infty \notin \mathcal{Z}$, then it follows that there exists sufficiently small $\hat{t} > 0$ and $I \in \mathbb{Z}_+$ such that $s(\hat{t}, x_i) = \psi(\hat{t}, x_i)$, $i = I, I+1, \ldots$, which implies that $\lim_{i \to \infty} s(\hat{t}, x_i) = s(\hat{t}, x_0)$. Next, define $z_i \triangleq \psi(\hat{t}, x_i)$, $i = 0, 1, \ldots$, so that $\lim_{i \to \infty} z_i = z_0$ and note that it follows from the transversality assumption that $z_0 \notin \overline{\mathcal{Z}}$, which implies that $\tau_1(\cdot)$ is continuous at z_0. Hence, $\lim_{i \to \infty} \tau_1(z_i) = \tau_1(z_0)$. The result now follows by noting that $\tau_1(x_i) = \hat{t} + \tau_1(z_i)$, $i = 1, 2, \ldots$. $\square$

Note that if $x_0 \notin \mathcal{Z}$ is such that $\lim_{i \to \infty} \tau_1(x_i) \neq \tau_1(x_0)$ for some sequence $\{x_i\}_{i=1}^\infty \notin \mathcal{Z}$, then it follows from Proposition 8.2 that $\lim_{i \to \infty} \tau_1(x_i) = 0$. The next result characterizes impulsive dynamical system limit sets for impulsive dynamical systems satisfying the *weak* quasi-continuous dependence Assumption 8.1 in terms of continuously differentiable functions. In particular, we show that the system trajectories of a state-dependent impulsive dynamical system converge to an invariant set contained in a union of level surfaces characterized by the continuous-time system dynamics and the resetting system dynamics. For the next set of results we assume that $f_c(\cdot)$, $f_d(\cdot)$, and $\mathcal{Z}$ are such that the dynamical system $\mathcal{G}$ given by (8.6) and (8.7) satisfies Assumptions A1, A2, and 8.1, and $\mathcal{Z} \cap \{x : f_d(x) = x\}$ is empty.

Theorem 8.1 *Consider the impulsive dynamical system (8.6) and (8.7), assume $\mathcal{D}_{ci} \subset \mathcal{D}$ is a compact positively invariant set with respect to (8.6) and (8.7), assume that if $x_0 \in \mathcal{Z}$ then $x_0 + f_d(x_0) \in \overline{\mathcal{Z}}\backslash\mathcal{Z}$, and assume that there exists a continuously differentiable function $V : \mathcal{D}_{ci} \to \mathbb{R}$ such that*

$$V'(x)f_c(x) \leq 0, \quad x \in \mathcal{D}_{ci}, \quad x \notin \mathcal{Z}, \tag{8.13}$$
$$V(x + f_d(x)) \leq V(x), \quad x \in \mathcal{D}_{ci}, \quad x \in \mathcal{Z}. \tag{8.14}$$

Let $\mathcal{R} \triangleq \{x \in \mathcal{D}_{ci} : x \notin \mathcal{Z}, V'(x)f_c(x) = 0\} \cup \{x \in \mathcal{D}_{ci} : x \in \mathcal{Z}, V(x + f_d(x)) = V(x)\}$ and let $\mathcal{M}$ denote the largest invariant set contained in $\mathcal{R}$. If $x_0 \in \mathcal{D}_{ci}$, then $x(t) \to \mathcal{M}$ as $t \to \infty$. Furthermore, if $0 \in \overset{\circ}{\mathcal{D}}_{ci}$, $V(0) = 0$, $V(x) > 0$, $x \neq 0$, and the set $\mathcal{R}$ contains no invariant set other than the set $\{0\}$, then the zero solution $x(t) \equiv 0$ to (8.6) and (8.7) is asymptotically stable and $\mathcal{D}_{ci}$ is a subset of the domain of attraction of (8.6) and (8.7).

Proof. The proof is similar to the proof of Theorem 2.3 and, hence, is omitted. $\square$

Setting $\mathcal{D} = \mathbb{R}^n$ and requiring $V(x) \to \infty$ as $\|x\| \to \infty$ in Theorem 8.1, it follows that the zero solution $x(t) \equiv 0$ to (8.6) and (8.7) is globally asymptotically stable. A similar remark holds for Theorem 8.2 below.

Theorem 8.2 *Consider the impulsive dynamical system (8.6) and (8.7), assume $\mathcal{D}_{ci} \subset \mathcal{D}$ is a compact positively invariant set with respect to (8.6) and (8.7) such that $0 \in \overset{\circ}{\mathcal{D}}_{ci}$, assume that if $x_0 \in \mathcal{Z}$ then $x_0 + f_d(x_0) \in \overline{\mathcal{Z}}\backslash\mathcal{Z}$, and assume that for all $x_0 \in \mathcal{D}_{ci}$, $x_0 \neq 0$, there exists $\tau \geq 0$ such that $x(\tau) \in \mathcal{Z}$, where $x(t)$, $t \geq 0$, denotes the solution to (8.6) and (8.7) with the initial condition x_0. Furthermore, assume there exists a continuously differentiable function $V : \mathcal{D}_{ci} \to \mathbb{R}$ such that $V(0) = 0$, $V(x) > 0$, $x \neq 0$,*

$$V'(x)f_c(x) \leq 0, \quad x \in \mathcal{D}_{ci}, \quad x \notin \mathcal{Z}, \tag{8.15}$$
$$V(x + f_d(x)) < V(x), \quad x \in \mathcal{D}_{ci}, \quad x \in \mathcal{Z}. \tag{8.16}$$

Then the zero solution $x(t) \equiv 0$ to (8.6) and (8.7) is asymptotically stable and $\mathcal{D}_{ci}$ is a subset of the domain of attraction of (8.6) and (8.7).

Proof. The proof is identical to the proof of Corollary 2.3 with Theorem 8.1 invoked in place of Corollary 2.1. □

8.3 Hybrid Control Design for Dissipative Dynamical Systems

In this section, we present a hybrid controller design framework for dissipative dynamical systems [165]. Specifically, we consider nonlinear dynamical systems $\mathcal{G}_p$ of the form

$$\dot{x}_p(t) = f_p(x_p(t), u(t)), \quad x_p(0) = x_{p0}, \quad t \geq 0, \qquad (8.17)$$
$$y(t) = h_p(x_p(t)), \qquad (8.18)$$

where $t \geq 0$, $x_p(t) \in \mathcal{D}_p \subseteq \mathbb{R}^{n_p}$, $\mathcal{D}_p$ is an open set with $0 \in \mathcal{D}_p$, $u(t) \in \mathbb{R}^m$, $y(t) \in \mathbb{R}^l$, $f_p : \mathcal{D}_p \times \mathbb{R}^m \to \mathbb{R}^{n_p}$ is smooth on $\mathcal{D}_p \times \mathbb{R}^m$ and satisfies $f_p(0,0) = 0$, and $h_p : \mathcal{D}_p \to \mathbb{R}^l$ is continuous and satisfies $h_p(0) = 0$. Furthermore, for the nonlinear dynamical system $\mathcal{G}_p$ we assume that the required properties for the existence and uniqueness of solutions are satisfied, that is, $u(\cdot)$ satisfies sufficient regularity conditions such that (8.17) has a unique solution forward in time.

Next, we consider hybrid resetting dynamic controllers $\mathcal{G}_c$ of the form

$$\dot{x}_c(t) = f_{cc}(x_c(t), y(t)), \quad x_c(0) = x_{c0}, \quad (x_c(t), y(t)) \notin \mathcal{Z}_c, \quad (8.19)$$
$$\Delta x_c(t) = \eta(y(t)) - x_c(t), \quad (x_c(t), y(t)) \in \mathcal{Z}_c, \qquad (8.20)$$
$$y_c(t) = h_{cc}(x_c(t), y(t)), \qquad (8.21)$$

where $x_c(t) \in \mathcal{D}_c \subseteq \mathbb{R}^{n_c}$, $\mathcal{D}_c$ is an open set with $0 \in \mathcal{D}_c$, $y(t) \in \mathbb{R}^l$, $y_c(t) \in \mathbb{R}^m$, $f_{cc} : \mathcal{D}_c \times \mathbb{R}^l \to \mathbb{R}^{n_c}$ is smooth on $\mathcal{D}_c \times \mathbb{R}^l$ and satisfies $f_{cc}(0,0) = 0$, $\eta : \mathbb{R}^l \to \mathcal{D}_c$ is continuous and satisfies $\eta(0) = 0$, and $h_{cc} : \mathcal{D}_c \times \mathbb{R}^l \to \mathbb{R}^m$ is continuous and satisfies $h_{cc}(0,0) = 0$.

Recall that for the dynamical system $\mathcal{G}_p$ given by (8.17) and (8.18), a function $s(u, y)$, where $s : \mathbb{R}^m \times \mathbb{R}^l \to \mathbb{R}$ is such that $s(0,0) = 0$, is called a *supply rate* [165] if it is locally integrable for all input-output pairs satisfying (8.17) and (8.18), that is, for all input-output pairs $u(\cdot) \in \mathcal{U}$ and $y(\cdot) \in \mathcal{Y}$ satisfying (8.17) and (8.18), $s(\cdot, \cdot)$ satisfies $\int_t^{\hat{t}} |s(u(\sigma), y(\sigma))| d\sigma < \infty$, $t, \hat{t} \geq 0$. Here, $\mathcal{U}$ and $\mathcal{Y}$ are input and output spaces, respectively, that are assumed to be closed under the shift operator. Furthermore, we assume that $\mathcal{G}_p$ is *dissipative with respect to the supply rate* $s(u, y)$, and hence, there exists a continuous, nonnegative-definite *storage function* $V_s : \mathcal{D}_p \to \overline{\mathbb{R}}_+$ such that $V_s(0) =$

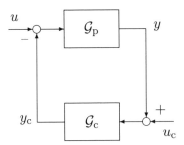

Figure 8.1 Feedback interconnection of $\mathcal{G}_p$ and $\mathcal{G}_c$.

0 and

$$V_s(x_p(t)) = V_s(x_p(t_0)) + \int_{t_0}^{t} [s(u(\sigma), y(\sigma)) - d(x_p(\sigma))]d\sigma, \quad t \geq t_0,$$

for all $t_0, t \geq 0$, where $x_p(t)$, $t \geq t_0$, is the solution to (8.17) with $u(\cdot) \in \mathcal{U}$ and $d : \mathcal{D}_p \rightarrow \overline{\mathbb{R}}_+$ is a continuous, nonnegative-definite dissipation rate function. In addition, we assume that the nonlinear dynamical system $\mathcal{G}_p$ is *completely reachable* [165] and *zero-state observable* [165], and there exists a function $\kappa : \mathbb{R}^l \rightarrow \mathbb{R}^m$ such that $\kappa(0) = 0$ and $s(\kappa(y), y) < 0$, $y \neq 0$, so that all storage functions $V_s(x_p)$, $x_p \in \mathcal{D}_p$, of $\mathcal{G}_p$ are positive definite [75]. Finally, we assume that $V_s(\cdot)$ is continuously differentiable.

Consider the negative feedback interconnection of $\mathcal{G}_p$ and $\mathcal{G}_c$ given in Figure 8.1 such that $y = u_c$ and $u = -y_c$. In this case, the closed-loop system $\mathcal{G}$ is given by

$$\dot{x}(t) = f_c(x(t)), \quad x(0) = x_0, \quad x(t) \notin \mathcal{Z}, \quad t \geq 0, \quad (8.22)$$
$$\Delta x(t) = f_d(x(t)), \quad x(t) \in \mathcal{Z}, \quad (8.23)$$

where $t \geq 0$, $x(t) \triangleq [x_p^T(t), x_c^T(t)]^T$, $\mathcal{Z} = \{x \in \mathcal{D} : (x_c, h_p(x_p)) \in \mathcal{Z}_c\}$,

$$f_c(x) = \begin{bmatrix} f_p(x_p, h_{cc}(x_c, -h_p(x_p))) \\ f_{cc}(x_c, h_p(x_p)) \end{bmatrix}, \quad f_d(x) = \begin{bmatrix} 0 \\ \eta(h_p(x_p)) - x_c \end{bmatrix}.$$
$$(8.24)$$

Assume that there exists an infinitely differentiable function $V_c : \mathcal{D}_c \times \mathbb{R}^l \rightarrow \overline{\mathbb{R}}_+$ such that $V_c(x_c, y) \geq 0$, $x_c \in \mathcal{D}_c$, $y \in \mathbb{R}^l$, and $V_c(x_c, y) = 0$ if and only if $x_c = \eta(y)$ and

$$\dot{V}_c(x_c(t), y(t)) = s_c(u_c(t), y_c(t)), \quad (x_c(t), y(t)) \notin \mathcal{Z}, \quad t \geq 0, \quad (8.25)$$

where $s_c : \mathbb{R}^l \times \mathbb{R}^m \to \mathbb{R}$ is such that $s_c(0,0) = 0$ and is locally integrable for all input-output pairs satisfying (8.19)–(8.21).

We associate with the plant a positive-definite, continuously differentiable function $V_p(x_p) \triangleq V_s(x_p)$, which we will refer to as the *plant energy*. Furthermore, we associate with the controller a nonnegative-definite, infinitely differentiable function $V_c(x_c, y)$ called the controller *emulated energy*. Finally, we associate with the closed-loop system the function

$$V(x) \triangleq V_p(x_p) + V_c(x_c, h_p(x_p)), \tag{8.26}$$

called the *total energy*.

Next, we construct the resetting set for the closed-loop system $\mathcal{G}$ in the following form

$$\mathcal{Z} = \{(x_p, x_c) \in \mathcal{D}_p \times \mathcal{D}_c : L_{f_c} V_c(x_c, h_p(x_p)) = 0$$

$$\text{and } V_c(x_c, h_p(x_p)) > 0\}. \tag{8.27}$$

The resetting set $\mathcal{Z}$ is thus defined to be the set of all points in the closed-loop state space that correspond to decreasing controller emulated energy. By resetting the controller states, the plant energy can never increase after the first resetting event. Furthermore, if the continuous-time dynamics of the closed-loop system are lossless and the closed-loop system total energy is conserved between resetting events, then a decrease in plant energy is accompanied by a corresponding increase in emulated energy. Hence, this approach allows the plant energy to flow to the controller, where it increases the emulated energy but does not allow the emulated energy to flow back to the plant after the first resetting event. This energy-dissipating hybrid controller effectively enforces a one-way energy transfer between the plant and the controller after the first resetting event. For practical implementation, knowledge of x_c and y is sufficient to determine whether or not the closed-loop state vector is in the set $\mathcal{Z}$.

The next theorem gives sufficient conditions for asymptotic stability of the closed-loop system $\mathcal{G}$ using state-dependent hybrid controllers.

Theorem 8.3 *Consider the closed-loop hybrid dynamical system $\mathcal{G}$ given by (8.22) and (8.23) with the resetting set $\mathcal{Z}$ given by (8.27). Assume that $\mathcal{D}_{ci} \subset \mathcal{D}$ is a compact positively invariant set with respect to $\mathcal{G}$ such that $0 \in \overset{\circ}{\mathcal{D}}_{ci}$, assume that $\mathcal{G}_p$ is lossless with respect to the supply rate $s(u,y)$ (i.e., $d(x_p) \equiv 0$) and with a positive definite, continuously differentiable storage function $V_p(x_p)$, $x_p \in \mathcal{D}_p$,*

and assume there exists a smooth (i.e., infinitely differentiable) func-
tion $V_c : \mathcal{D}_c \times \mathbb{R}^l \to \overline{\mathbb{R}}_+$ *such that* $V_c(x_c, y) \geq 0$, $x_c \in \mathcal{D}_c$, $y \in \mathbb{R}^l$, *and*
$V_c(x_c, y) = 0$ *if and only if* $x_c = \eta(y)$ *and (8.25) holds. Furthermore,*
assume that every $x_0 \in \overline{\mathcal{Z}}$ *is transversal to (8.22) and*

$$s(u, y) + s_c(u_c, y_c) = 0, \quad x \notin \mathcal{Z}, \tag{8.28}$$

where $y = u_c = h_p(x_p)$, $u = -y_c = -h_{cc}(x_c, h_p(x_p))$, *and* $\mathcal{Z}$ *is*
given by (8.27). Then the zero solution $x(t) \equiv 0$ *to the closed-*
loop system $\mathcal{G}$ *is asymptotically stable. In addition, the total energy*
function $V(x)$ *of* $\mathcal{G}$ *given by (8.26) is strictly decreasing across re-*
setting events. Alternatively, assume $\mathcal{G}_p$ *is dissipative with respect*
to the supply rate $s(u, y)$ *and the largest invariant set contained in*
$\mathcal{R} \triangleq \{(x_p, x_c) \in \mathcal{D}_{ci} : d(x_p) = 0\}$ *is* $\mathcal{M} = \{(0, 0)\}$. *Then the zero*
solution $x(t) \equiv 0$ *to* $\mathcal{G}$ *is asymptotically stable. Finally, if* $\mathcal{D}_p = \mathbb{R}^{n_p}$,
$\mathcal{D}_c = \mathbb{R}^{n_c}$, *and* $V(\cdot)$ *is radially unbounded, then the above asymptotic*
stability results are global.

Proof. First we consider the case where $\mathcal{G}_p$ is lossless with respect
to the supply rate $s(u, y)$. Note that since $V_c(x_c, y) \geq 0$, $x_c \in \mathcal{D}_c$,
$y \in \mathbb{R}^l$, it follows that

$$\begin{aligned} \overline{\mathcal{Z}} &= \{(x_p, x_c) \in \mathcal{D}_p \times \mathcal{D}_c : L_{f_c} V_c(x_c, h_p(x_p)) = 0 \\ &\quad \text{and } V_c(x_c, h_p(x_p)) \geq 0\} \\ &= \{(x_p, x_c) \in \mathcal{D}_p \times \mathcal{D}_c : \mathcal{X}(x) = 0\}, \end{aligned} \tag{8.29}$$

where $\mathcal{X}(x) = L_{f_c} V_c(x_c, h_p(x_p))$. Next, we show that if the transversality condition (8.11) holds, then Assumptions A1, A2, and 8.1 hold,
and, for every $x_0 \in \mathcal{D}_{ci}$, there exists $\tau \geq 0$ such that $x(\tau) \in \mathcal{Z}$. Note
that if $x_0 \in \overline{\mathcal{Z}} \backslash \mathcal{Z}$, that is, $V_c(x_c(0), h_p(x_p(0))) = 0$ and $L_{f_c} V_c(x_c(0), h_p$
$(x_p(0))) = 0$, it follows from the transversality condition that there
exists $\delta > 0$ such that for all $t \in (0, \delta]$, $L_{f_c} V_c(x_c(t), h_p(x_p(t))) \neq 0$.
Hence, since $V_c(x_c(t), h_p(x_p(t))) = V_c(x_c(0), h_p(x_p(0))) + t L_{f_c} V_c(x_c(\tau)$
$, h_p(x_p(\tau)))$ for some $\tau \in (0, t]$ and $V_c(x_c, y) \geq 0$, $x_c \in \mathcal{D}_c$, $y \in \mathbb{R}^l$,
it follows that $V_c(x_c(t), h_p(x_p(t))) > 0$, $t \in (0, \delta]$, which implies
that Assumption A1 is satisfied. Furthermore, if $x \in \mathcal{Z}$ then, since
$V_c(x_c, y) = 0$ if and only if $x_c = \eta(y)$, it follows from (8.23) that
$x + f_d(x) \in \overline{\mathcal{Z}} \backslash \mathcal{Z}$. Hence, Assumption A2 holds.

Next, consider the set

$$\mathcal{M}_\gamma \triangleq \{x \in \mathcal{D}_{ci} : V_c(x_c, h_p(x_p)) = \gamma\}, \tag{8.30}$$

where $\gamma \geq 0$. It follows from the transversality condition that for
every $\gamma \geq 0$, $\mathcal{M}_\gamma$ does not contain any nontrivial trajectory of $\mathcal{G}$. To

see this, suppose, *ad absurdum*, there exists a nontrivial trajectory $x(t) \in \mathcal{M}_\gamma$, $t \geq 0$, for some $\gamma \geq 0$. In this case, it follows that $\frac{d^k}{dt^k} V_c(x_c(t), h_p(x_p(t))) = L_{f_c}^k V_c(x_c(t), h_p(x_p(t))) \equiv 0$, $k = 1, 2, \ldots$, which contradicts the transversality condition.

Next, we show that for every $x_0 \notin \mathcal{Z}$, $x_0 \neq 0$, there exists $\tau > 0$ such that $x(\tau) \in \mathcal{Z}$. To see this, suppose, *ad absurdum*, $x(t) \notin \mathcal{Z}$, $t \geq 0$, which implies that

$$\frac{d}{dt} V_c(x_c(t), h_p(x_p(t))) \neq 0, \quad t \geq 0, \tag{8.31}$$

or

$$V_c(x_c(t), h_p(x_p(t))) = 0, \quad t \geq 0. \tag{8.32}$$

If (8.31) holds, then it follows that $V_c(x_c(t), h_p(x_p(t)))$ is a (decreasing or increasing) monotonic function of time. Hence, $V_c(x_c(t), h_p(x_p(t))) \to \gamma$ as $t \to \infty$, where $\gamma \geq 0$ is a constant, which implies that the positive limit set of the closed-loop system is contained in $\mathcal{M}_\gamma$ for some $\gamma \geq 0$, and hence, is a contradiction. Similarly, if (8.32) holds then $\mathcal{M}_0$ contains a nontrivial trajectory of $\mathcal{G}$ also leading to a contradiction. Hence, for every $x_0 \notin \mathcal{Z}$, there exists $\tau > 0$ such that $x(\tau) \in \mathcal{Z}$. Thus, it follows that for every $x_0 \notin \mathcal{Z}$, $0 < \tau_1(x_0) < \infty$. Now, it follows from Proposition 8.2 that $\tau_1(\cdot)$ is continuous at $x_0 \notin \overline{\mathcal{Z}}$. Furthermore, for all $x_0 \in \overline{\mathcal{Z}} \backslash \mathcal{Z}$ and for every sequence $\{x_i\}_{i=1}^\infty \in \overline{\mathcal{Z}} \backslash \mathcal{Z}$ converging to $x_0 \in \overline{\mathcal{Z}} \backslash \mathcal{Z}$, it follows from the transversality condition and Proposition 8.2 that $\lim_{i \to \infty} \tau_1(x_i) = \tau_1(x_0)$. Next, let $x_0 \in \overline{\mathcal{Z}} \backslash \mathcal{Z}$ and let $\{x_i\}_{i=1}^\infty \in \mathcal{D}_{ci}$ be such that $\lim_{i \to \infty} x_i = x_0$ and $\lim_{i \to \infty} \tau_1(x_i)$ exists. In this case, it follows from Proposition 8.2 that either $\lim_{i \to \infty} \tau_1(x_i) = 0$ or $\lim_{i \to \infty} \tau_1(x_i) = \tau_1(x_0)$. Furthermore, since $x_0 \in \overline{\mathcal{Z}} \backslash \mathcal{Z}$ corresponds to the case where $V_c(x_{c0}, h_p(x_{p0})) = 0$, it follows that $x_{c0} = \eta(h_p(x_{p0}))$, and hence, $f_d(x_0) = 0$. Now, it follows from Proposition 8.1 that Assumption 8.1 holds.

To show that the zero solution $x(t) \equiv 0$ to $\mathcal{G}$ is asymptotically stable, consider the Lyapunov function candidate corresponding to the total energy function $V(x)$ given by (8.26). Since $\mathcal{G}_p$ is lossless with respect to the supply rate $s(u, y)$, and (8.25) and (8.28) hold, it follows that

$$\dot{V}(x(t)) = s(u(t), y(t)) + s_c(u_c(t), y_c(t)) = 0, \quad x(t) \notin \mathcal{Z}. \tag{8.33}$$

Furthermore, it follows from (8.24) and (8.27) that

$$\Delta V(x(t_k)) = V_c(x_c(t_k^+), h_p(x_p(t_k^+))) - V_c(x_c(t_k), h_p(x_p(t_k)))$$

$$
\begin{aligned}
&= V_c(\eta(h_p(x_p(t_k))), h_p(x_p(t_k))) - V_c(x_c(t_k), h_p(x_p(t_k))) \\
&= -V_c(x_c(t_k), h_p(x_p(t_k))) \\
&< 0, \quad x(t_k) \in \mathcal{Z}, \quad k \in \overline{\mathbb{Z}}_+.
\end{aligned}
\tag{8.34}
$$

Thus, it follows from Theorem 8.2 that the zero solution $x(t) \equiv 0$ to $\mathcal{G}$ is asymptotically stable. If $\mathcal{D}_p = \mathbb{R}^{n_p}$, $\mathcal{D}_c = \mathbb{R}^{n_c}$, and $V(\cdot)$ is radially unbounded, then global asymptotic stability is immediate.

If $\mathcal{G}_p$ is dissipative with respect to the supply rate $s(u, y)$ and for every $x_0 \notin \mathcal{Z}$, $x_0 = 0$, there exists $\tau > 0$ such that $x(\tau) \in \mathcal{Z}$, then the proof is identical to the proof for the lossless case. Alternatively, if there exists $k_{max} \geq 0$ such that $k \leq k_{max}$, that is, the closed-loop system trajectory intersects the resetting set $\mathcal{Z}$ a finite number of times, then the closed-loop impulsive dynamical system possesses a continuous flow for all $t > t_{k_{max}}$. In this case, since the largest invariant set contained in $\mathcal{R}$ is $\{(0, 0)\}$, closed-loop asymptotic stability of $\mathcal{G}$ follows from standard Lyapunov and invariant set arguments. Finally, if $\mathcal{D}_p = \mathbb{R}^{n_p}$, $\mathcal{D}_c = \mathbb{R}^{n_c}$, and $V(\cdot)$ is radially unbounded, then global asymptotic stability is immediate. $\square$

If $V_c = V_c(x_c, y)$ is only a function of x_c and $V_c(x_c)$ is a positive-definite function, then we can choose $\eta(y) \equiv 0$. In this case, $V_c(x_c) = 0$ if and only if $x_c = 0$, and hence, Theorem 8.3 specializes to the case of a negative feedback interconnection of a dissipative dynamical system $\mathcal{G}_p$ and a hybrid lossless controller $\mathcal{G}_c$.

In the proof of Theorem 8.3, we assume that $x_0 \notin \mathcal{Z}$ for $x_0 \neq 0$. This proviso is necessary since it may be possible to reset the states of the closed-loop system to the origin, in which case $x(s) = 0$ for a finite value of s. In this case, for $t > s$, we have $V(x(t)) = V(x(s)) = V(0) = 0$. This situation does not present a problem, however, since reaching the origin in finite time is a stronger condition than reaching the origin as $t \to \infty$.

Finally, we specialize the hybrid controller design framework just presented to *port-controlled Hamiltonian systems*. Specifically, consider the port-controlled Hamiltonian system given by

$$
\dot{x}_p(t) = [\mathcal{J}_p(x_p(t)) - \mathcal{R}_p(x_p(t))] \left(\frac{\partial \mathcal{H}_p}{\partial x_p}(x_p(t)) \right)^{\mathrm{T}} + G_p(x_p(t)) u(t),
$$
$$
x_p(0) = x_{p0}, \quad t \geq 0, \quad (8.35)
$$
$$
y(t) = G_p^{\mathrm{T}}(x_p(t)) \left(\frac{\partial \mathcal{H}_p}{\partial x_p}(x_p(t)) \right)^{\mathrm{T}},
\tag{8.36}
$$

where $x_p(t) \in \mathcal{D}_p \subseteq \mathbb{R}^{n_p}$, $\mathcal{D}_p$ is an open set with $0 \in \mathcal{D}_p$, $u(t) \in \mathbb{R}^m$,

$y(t) \in \mathbb{R}^m$, $\mathcal{H}_{\mathrm{p}} : \mathcal{D}_{\mathrm{p}} \to \mathbb{R}$ is an infinitely differentiable Hamiltonian function for the system (8.35) and (8.36), $\mathcal{J}_{\mathrm{p}} : \mathcal{D}_{\mathrm{p}} \to \mathbb{R}^{n_{\mathrm{p}} \times n_{\mathrm{p}}}$ is such that $\mathcal{J}_{\mathrm{p}}(x_{\mathrm{p}}) = -\mathcal{J}_{\mathrm{p}}^{\mathrm{T}}(x_{\mathrm{p}})$, $\mathcal{R}_{\mathrm{p}} : \mathcal{D}_{\mathrm{p}} \to \mathbb{R}^{n_{\mathrm{p}} \times n_{\mathrm{p}}}$ is such that $\mathcal{R}_{\mathrm{p}}(x_{\mathrm{p}}) = \mathcal{R}_{\mathrm{p}}^{\mathrm{T}}(x_{\mathrm{p}}) \geq 0$, $x_{\mathrm{p}} \in \mathcal{D}_{\mathrm{p}}$, $[\mathcal{J}_{\mathrm{p}}(x_{\mathrm{p}}) - \mathcal{R}_{\mathrm{p}}(x_{\mathrm{p}})](\frac{\partial \mathcal{H}_{\mathrm{p}}}{\partial x_{\mathrm{p}}}(x_{\mathrm{p}}))^{\mathrm{T}}$, $x_{\mathrm{p}} \in \mathcal{D}_{\mathrm{p}}$, is smooth on $\mathcal{D}_{\mathrm{p}}$, and $G_{\mathrm{p}} : \mathcal{D}_{\mathrm{p}} \to \mathbb{R}^{n_{\mathrm{p}} \times m}$. The skew-symmetric matrix function $\mathcal{J}_{\mathrm{p}}(x_{\mathrm{p}})$, $x_{\mathrm{p}} \in \mathcal{D}_{\mathrm{p}}$, captures the internal system interconnection structure and the symmetric nonnegative definite matrix function $\mathcal{R}_{\mathrm{p}}(x_{\mathrm{p}})$, $x_{\mathrm{p}} \in \mathcal{D}_{\mathrm{p}}$, captures system dissipation. Furthermore, we assume that $\mathcal{H}_{\mathrm{p}}(0) = 0$ and $\mathcal{H}_{\mathrm{p}}(x_{\mathrm{p}}) > 0$ for all $x_{\mathrm{p}} \neq 0$ and $x_{\mathrm{p}} \in \mathcal{D}_{\mathrm{p}}$.

Next, consider the fixed-order, energy-based hybrid controller

$$\dot{x}_{\mathrm{c}}(t) = \mathcal{J}_{\mathrm{cc}}(x_{\mathrm{c}}(t)) \left(\frac{\partial \mathcal{H}_{\mathrm{c}}}{\partial x_{\mathrm{c}}}(x_{\mathrm{c}}(t)) \right)^{\mathrm{T}} + G_{\mathrm{cc}}(x_{\mathrm{c}}(t)) y(t), \quad x_{\mathrm{c}}(0) = x_{\mathrm{c}0},$$
$$(x_{\mathrm{p}}(t), x_{\mathrm{c}}(t)) \notin \mathcal{Z}, \quad (8.37)$$

$$\Delta x_{\mathrm{c}}(t) = -x_{\mathrm{c}}(t), \quad (x_{\mathrm{p}}(t), x_{\mathrm{c}}(t)) \in \mathcal{Z}, \quad (8.38)$$

$$u(t) = -G_{\mathrm{cc}}^{\mathrm{T}}(x_{\mathrm{c}}(t)) \left(\frac{\partial \mathcal{H}_{\mathrm{c}}}{\partial x_{\mathrm{c}}}(x_{\mathrm{c}}(t)) \right)^{\mathrm{T}}, \quad (8.39)$$

where $t \geq 0$, $x_{\mathrm{c}}(t) \in \mathcal{D}_{\mathrm{c}} \subseteq \mathbb{R}^{n_{\mathrm{c}}}$, $\mathcal{D}_{\mathrm{c}}$ is an open set with $0 \in \mathcal{D}_{\mathrm{c}}$, $\Delta x_{\mathrm{c}}(t) \triangleq x_{\mathrm{c}}(t^+) - x_{\mathrm{c}}(t)$, $\mathcal{H}_{\mathrm{c}} : \mathcal{D}_{\mathrm{c}} \to \mathbb{R}$ is an infinitely differentiable Hamiltonian function for (8.37), $\mathcal{J}_{\mathrm{cc}} : \mathcal{D}_{\mathrm{c}} \to \mathbb{R}^{n_{\mathrm{c}} \times n_{\mathrm{c}}}$ is such that $\mathcal{J}_{\mathrm{cc}}(x_{\mathrm{c}}) = -\mathcal{J}_{\mathrm{cc}}^{\mathrm{T}}(x_{\mathrm{c}})$, $x_{\mathrm{c}} \in \mathcal{D}_{\mathrm{c}}$, $\mathcal{J}_{\mathrm{cc}}(x_{\mathrm{c}})(\frac{\partial \mathcal{H}_{\mathrm{c}}}{\partial x_{\mathrm{c}}}(x_{\mathrm{c}}))^{\mathrm{T}}$, $x_{\mathrm{c}} \in \mathcal{D}_{\mathrm{c}}$, is smooth on $\mathcal{D}_{\mathrm{c}}$, $G_{\mathrm{cc}} : \mathcal{D}_{\mathrm{c}} \to \mathbb{R}^{n_{\mathrm{c}} \times m}$, and the resetting set $\mathcal{Z} \subset \mathcal{D}_{\mathrm{p}} \times \mathcal{D}_{\mathrm{c}}$ is given by

$$\mathcal{Z} \triangleq \left\{ (x_{\mathrm{p}}, x_{\mathrm{c}}) \in \mathcal{D}_{\mathrm{p}} \times \mathcal{D}_{\mathrm{c}} : \frac{\mathrm{d}}{\mathrm{d}t} \mathcal{H}_{\mathrm{c}}(x_{\mathrm{c}}) = 0 \text{ and } \mathcal{H}_{\mathrm{c}}(x_{\mathrm{c}}) > 0 \right\}, \quad (8.40)$$

where $\frac{\mathrm{d}}{\mathrm{d}t} \mathcal{H}_{\mathrm{c}}(x_{\mathrm{c}}(t)) \triangleq \lim_{\tau \to t^-} \frac{1}{t-\tau} [\mathcal{H}_{\mathrm{c}}(x_{\mathrm{c}}(t)) - \mathcal{H}_{\mathrm{c}}(x_{\mathrm{c}}(\tau))]$ whenever the limit on the right-hand side exists. Here, we assume that $\mathcal{H}_{\mathrm{c}}(0) = 0$ and $\mathcal{H}_{\mathrm{c}}(x_{\mathrm{c}}) > 0$ for all $x_{\mathrm{c}} \neq 0$ and $x_{\mathrm{c}} \in \mathcal{D}_{\mathrm{c}}$.

Note that $\mathcal{H}_{\mathrm{p}}(x_{\mathrm{p}})$, $x_{\mathrm{p}} \in \mathcal{D}_{\mathrm{p}}$, is the plant energy and $\mathcal{H}_{\mathrm{c}}(x_{\mathrm{c}})$, $x_{\mathrm{c}} \in \mathcal{D}_{\mathrm{c}}$, is the controller emulated energy. Furthermore, the closed-loop system energy is given by $\mathcal{H}(x_{\mathrm{p}}, x_{\mathrm{c}}) \triangleq \mathcal{H}_{\mathrm{p}}(x_{\mathrm{p}}) + \mathcal{H}_{\mathrm{c}}(x_{\mathrm{c}})$. Next, note that the total energy function $\mathcal{H}(x_{\mathrm{p}}, x_{\mathrm{c}})$ along the trajectories of the closed-loop dynamics (8.35)–(8.39) satisfies

$$\frac{\mathrm{d}}{\mathrm{d}t} \mathcal{H}(x_{\mathrm{p}}(t), x_{\mathrm{c}}(t)) = -\frac{\partial \mathcal{H}_{\mathrm{p}}}{\partial x_{\mathrm{p}}}(x_{\mathrm{p}}(t)) \mathcal{R}_{\mathrm{p}}(x_{\mathrm{p}}(t)) \left(\frac{\partial \mathcal{H}_{\mathrm{p}}}{\partial x_{\mathrm{p}}}(x_{\mathrm{p}}(t)) \right)^{\mathrm{T}} \leq 0,$$
$$(x_{\mathrm{p}}(t), x_{\mathrm{c}}(t)) \notin \mathcal{Z}, \quad t_k < t \leq t_{k+1}, \quad (8.41)$$

$$\Delta \mathcal{H}(x_{\mathrm{p}}(t_k), x_{\mathrm{c}}(t_k)) = -\mathcal{H}_{\mathrm{c}}(x_{\mathrm{c}}(t_k)), \quad (x_{\mathrm{p}}(t_k), x_{\mathrm{c}}(t_k)) \in \mathcal{Z}, \quad k \in \overline{\mathbb{Z}}_+. \tag{8.42}$$

Here, we assume that every $(x_{\mathrm{p}0}, x_{\mathrm{c}0}) \in \overline{\mathcal{Z}}$ is transversal to the closed-loop dynamical system given by (8.35)–(8.39). Furthermore, we assume $\mathcal{D}_{\mathrm{ci}} \subset \mathcal{D}_{\mathrm{p}} \times \mathcal{D}_{\mathrm{c}}$ is a compact positively invariant set with respect to the closed-loop dynamical system (8.35)–(8.39) such that $0 \in \overset{\circ}{\mathcal{D}}_{\mathrm{ci}}$. In this case, it follows from Theorem 8.3, with $V_{\mathrm{s}}(x_{\mathrm{p}}) = \mathcal{H}_{\mathrm{p}}(x_{\mathrm{p}})$, $V_{\mathrm{c}}(x_{\mathrm{c}}, y) = \mathcal{H}_{\mathrm{c}}(x_{\mathrm{c}})$, $s(u, y) = u^{\mathrm{T}} y$, and $s_{\mathrm{c}}(u_{\mathrm{c}}, y_{\mathrm{c}}) = u_{\mathrm{c}}^{\mathrm{T}} y_{\mathrm{c}}$, that if $R_{\mathrm{p}}(x_{\mathrm{p}}) \equiv 0$, then the zero solution $(x_{\mathrm{p}}(t), x_{\mathrm{c}}(t)) \equiv (0, 0)$ to the closed-loop system (8.35)–(8.39), with $\mathcal{Z}$ given by (8.40), is asymptotically stable. Alternatively, if $R_{\mathrm{p}}(x_{\mathrm{p}}) \neq 0$, $x_{\mathrm{p}} \in \mathcal{D}_{\mathrm{p}}$, and the largest invariant set contained in

$$\mathcal{R} \triangleq \left\{ (x_{\mathrm{p}}, x_{\mathrm{c}}) \in \mathcal{D}_{\mathrm{ci}} : \frac{\partial \mathcal{H}_{\mathrm{p}}}{\partial x_{\mathrm{p}}}(x_{\mathrm{p}}) R_{\mathrm{p}}(x_{\mathrm{p}}) \left(\frac{\partial \mathcal{H}_{\mathrm{p}}}{\partial x_{\mathrm{p}}}(x_{\mathrm{p}}) \right)^{\mathrm{T}} = 0 \right\} \tag{8.43}$$

is $\mathcal{M} = \{(0,0)\}$, then the zero solution $(x_{\mathrm{p}}(t), x_{\mathrm{c}}(t)) \equiv (0, 0)$ of the closed-loop system (8.35)–(8.39), with $\mathcal{Z}$ given by (8.40), is asymptotically stable.

8.4 Lagrangian and Hamiltonian Dynamical Systems

Consider the governing equations of motion of an $\hat{n}_{\mathrm{p}}$-degree-of-freedom dynamical system given by the *Euler-Lagrange* equation

$$\frac{\mathrm{d}}{\mathrm{d}t} \left[\frac{\partial \mathcal{L}}{\partial \dot{q}}(q(t), \dot{q}(t)) \right]^{\mathrm{T}} - \left[\frac{\partial \mathcal{L}}{\partial q}(q(t), \dot{q}(t)) \right]^{\mathrm{T}} + \left[\frac{\partial \mathcal{R}}{\partial \dot{q}}(\dot{q}(t)) \right]^{\mathrm{T}} = u(t),$$
$$q(0) = q_0, \quad \dot{q}(0) = \dot{q}_0, \tag{8.44}$$

where $t \geq 0$, $q \in \mathbb{R}^{\hat{n}_{\mathrm{p}}}$ represents the generalized system positions, $\dot{q} \in \mathbb{R}^{\hat{n}_{\mathrm{p}}}$ represents the generalized system velocities, $\mathcal{L} : \mathbb{R}^{\hat{n}_{\mathrm{p}}} \times \mathbb{R}^{\hat{n}_{\mathrm{p}}} \rightarrow \mathbb{R}$ denotes the system Lagrangian given by $\mathcal{L}(q, \dot{q}) = T(q, \dot{q}) - U(q)$, where $T : \mathbb{R}^{\hat{n}_{\mathrm{p}}} \times \mathbb{R}^{\hat{n}_{\mathrm{p}}} \rightarrow \mathbb{R}$ is the system kinetic energy and $U : \mathbb{R}^{\hat{n}_{\mathrm{p}}} \rightarrow \mathbb{R}$ is the system potential energy, $\mathcal{R} : \mathbb{R}^{\hat{n}_{\mathrm{p}}} \rightarrow \mathbb{R}$ represents the Rayleigh dissipation function satisfying $\frac{\partial \mathcal{R}}{\partial \dot{q}}(\dot{q})\dot{q} \geq 0$, $\dot{q} \in \mathbb{R}^{\hat{n}_{\mathrm{p}}}$, and $u \in \mathbb{R}^{\hat{n}_{\mathrm{p}}}$ is the vector of generalized control forces acting on the system. Furthermore, let $\mathcal{H} : \mathbb{R}^{\hat{n}_{\mathrm{p}}} \times \mathbb{R}^{\hat{n}_{\mathrm{p}}} \rightarrow \mathbb{R}$ denote the *Legendre transformation* of the Lagrangian function $\mathcal{L}(q, \dot{q})$ with respect to the generalized velocity $\dot{q}$ defined by

$$\mathcal{H}(q, p) \triangleq \dot{q}^{\mathrm{T}} p - \mathcal{L}(q, \dot{q}), \tag{8.45}$$

where p denotes the vector of generalized momenta given by

$$p(q, \dot{q}) = \left[\frac{\partial \mathcal{L}}{\partial \dot{q}}(q, \dot{q}) \right]^{\mathrm{T}}, \qquad (8.46)$$

where the map from the generalized velocities $\dot{q}$ to the generalized momenta p is assumed to be *bijective* (i.e., one-to-one and onto). Now, if $\mathcal{H}(q, p)$ is lower bounded, then we can always shift $\mathcal{H}(q, p)$ so that, with a minor abuse of notation, $\mathcal{H}(q, p) \geq 0$, $(q, p) \in \mathbb{R}^{\hat{n}_{\mathrm{P}}} \times \mathbb{R}^{\hat{n}_{\mathrm{P}}}$. In this case, using (8.44) and the fact that

$$\frac{\mathrm{d}}{\mathrm{d}t}[\mathcal{L}(q, \dot{q})] = \frac{\partial \mathcal{L}}{\partial q}(q, \dot{q})\dot{q} + \frac{\partial \mathcal{L}}{\partial \dot{q}}(q, \dot{q})\ddot{q}, \qquad (8.47)$$

it follows that

$$\frac{\mathrm{d}}{\mathrm{d}t}\mathcal{H}(q, p) = u^{\mathrm{T}}\dot{q} - \frac{\partial \mathcal{R}}{\partial \dot{q}}(\dot{q})\dot{q}. \qquad (8.48)$$

Next, we transform the Euler-Lagrange equations to the *Hamiltonian* equations of motion. To reduce the Euler-Lagrange equations (8.44) to a Hamiltonian system of equations consider the Legendre transformation $\mathcal{H}(q, p)$ given by (8.45) and note that it follows from (8.44)–(8.46) that

$$\dot{q}(t) = \left[\frac{\partial \mathcal{H}}{\partial p}(q(t), p(t)) \right]^{\mathrm{T}}, \quad q(0) = q_0, \quad t \geq 0, \qquad (8.49)$$

$$\dot{p}(t) = -\left[\frac{\partial \mathcal{H}}{\partial q}(q(t), p(t)) \right]^{\mathrm{T}} - \left[\frac{\partial \mathcal{R}}{\partial \dot{q}}(\dot{q}(t)) \right]^{\mathrm{T}} + u(t), \quad p(0) = p_0, \qquad (8.50)$$

where $p \in \mathbb{R}^{\hat{n}_{\mathrm{P}}}$, $q \in \mathbb{R}^{\hat{n}_{\mathrm{P}}}$, and $\mathcal{H}(\cdot, \cdot)$ is a lower bounded *Hamiltonian function*. These equations provide a fundamental structure of the mathematical description of numerous physical dynamical systems by capturing energy conservation and energy dissipation, as well as internal interconnection structural properties of physical dynamical systems. It is well known that the Hamiltonian system dynamics (8.49) and (8.50) is equivalent to the Lagrangian system dynamics (8.44). Thus, a stabilizing controller for (8.49) and (8.50) with output $y = [h_1^{\mathrm{T}}(q), h_2^{\mathrm{T}}(\dot{q})]^{\mathrm{T}}$ serves as a stabilizing controller for (8.44) with the same output. Hence, in the next section we only consider controller designs for Lagrangian systems of the form (8.44) since these controllers can be equivalently applied, via a suitable transformation, to Hamiltonian systems of the form (8.49) and (8.50).

8.5 Hybrid Control Design for Euler-Lagrange Systems

In this section, we present a hybrid feedback control framework for Euler-Lagrange dynamical systems. Specifically, consider the Lagrangian system (8.44) with outputs

$$y = \begin{bmatrix} h_1(q) \\ h_2(\dot{q}) \end{bmatrix} = \begin{bmatrix} h_1(q) \\ h_2\left(\frac{\partial \mathcal{H}}{\partial p}(q,p)\right) \end{bmatrix}, \tag{8.51}$$

where $h_1 : \mathbb{R}^{\hat{n}_{\mathrm{p}}} \to \mathbb{R}^{l_1}$ and $h_2 : \mathbb{R}^{\hat{n}_{\mathrm{p}}} \to \mathbb{R}^{l-l_1}$ are continuously differentiable, $h_1(0) = 0$, $h_2(0) = 0$, and $h_1(q) \not\equiv 0$. We assume that the system kinetic energy is such that $T(q,\dot{q}) = \frac{1}{2}\dot{q}^{\mathrm{T}}[\frac{\partial T}{\partial \dot{q}}(q,\dot{q})]^{\mathrm{T}}$, $T(q,0) = 0$, and $T(q,\dot{q}) > 0$, $\dot{q} \neq 0$, $\dot{q} \in \mathbb{R}^{\hat{n}_{\mathrm{p}}}$. We also assume that the system potential energy $U(\cdot)$ is such that $U(0) = 0$ and $U(q) > 0$, $q \neq 0$, $q \in \mathcal{D}_{\mathrm{q}} \subseteq \mathbb{R}^{\hat{n}_{\mathrm{p}}}$, which implies that $\mathcal{H}(q,p) = T(q,\dot{q}) + U(q) > 0$, $(q,\dot{q}) \neq 0$, $(q,\dot{q}) \in \mathcal{D}_{\mathrm{q}} \times \mathbb{R}^{\hat{n}_{\mathrm{p}}}$.

Next, consider the energy-based hybrid controller

$$\frac{\mathrm{d}}{\mathrm{dt}}\left[\frac{\partial \mathcal{L}_{\mathrm{c}}}{\partial \dot{q}_{\mathrm{c}}}(q_{\mathrm{c}}(t),\dot{q}_{\mathrm{c}}(t),y_q(t))\right]^{\mathrm{T}} - \left[\frac{\partial \mathcal{L}_{\mathrm{c}}}{\partial q_{\mathrm{c}}}(q_{\mathrm{c}}(t),\dot{q}_{\mathrm{c}}(t),y_q(t))\right]^{\mathrm{T}} = 0,$$

$$q_{\mathrm{c}}(0) = q_{\mathrm{c}0}, \quad \dot{q}_{\mathrm{c}}(0) = \dot{q}_{\mathrm{c}0}, \quad (q_{\mathrm{c}}(t),\dot{q}_{\mathrm{c}}(t),y(t)) \notin \mathcal{Z}_{\mathrm{c}}, \tag{8.52}$$

$$\begin{bmatrix} \Delta q_{\mathrm{c}}(t) \\ \Delta \dot{q}_{\mathrm{c}}(t) \end{bmatrix} = \begin{bmatrix} \eta(y_q(t)) - q_{\mathrm{c}}(t) \\ -\dot{q}_{\mathrm{c}}(t) \end{bmatrix}, \quad (q_{\mathrm{c}}(t),\dot{q}_{\mathrm{c}}(t),y(t)) \in \mathcal{Z}_{\mathrm{c}}, \tag{8.53}$$

$$u(t) = \left[\frac{\partial \mathcal{L}_{\mathrm{c}}}{\partial q}(q_{\mathrm{c}}(t),\dot{q}_{\mathrm{c}}(t),y_q(t))\right]^{\mathrm{T}}, \tag{8.54}$$

where $t \geq 0$, $q_{\mathrm{c}} \in \mathbb{R}^{\hat{n}_{\mathrm{c}}}$ represents virtual controller positions, $\dot{q}_{\mathrm{c}} \in \mathbb{R}^{\hat{n}_{\mathrm{c}}}$ represents virtual controller velocities, $y_q \triangleq h_1(q)$, $\mathcal{L}_{\mathrm{c}} : \mathbb{R}^{\hat{n}_{\mathrm{c}}} \times \mathbb{R}^{\hat{n}_{\mathrm{c}}} \times \mathbb{R}^{l_1} \to \mathbb{R}$ denotes the controller Lagrangian given by

$$\mathcal{L}_{\mathrm{c}}(q_{\mathrm{c}},\dot{q}_{\mathrm{c}},y_q) \triangleq T_{\mathrm{c}}(q_{\mathrm{c}},\dot{q}_{\mathrm{c}}) - U_{\mathrm{c}}(q_{\mathrm{c}},y_q), \tag{8.55}$$

where $T_{\mathrm{c}} : \mathbb{R}^{\hat{n}_{\mathrm{c}}} \times \mathbb{R}^{\hat{n}_{\mathrm{c}}} \to \mathbb{R}$ is the controller kinetic energy and $U_{\mathrm{c}} : \mathbb{R}^{\hat{n}_{\mathrm{c}}} \times \mathbb{R}^{l_1} \to \mathbb{R}$ is the controller potential energy, $\eta(\cdot)$ is a continuously differentiable function such that $\eta(0) = 0$, $\mathcal{Z}_{\mathrm{c}} \subset \mathbb{R}^{\hat{n}_{\mathrm{c}}} \times \mathbb{R}^{\hat{n}_{\mathrm{c}}} \times \mathbb{R}^l$ is the resetting set, $\Delta q_{\mathrm{c}}(t) \triangleq q_{\mathrm{c}}(t^+) - q_{\mathrm{c}}(t)$, and $\Delta \dot{q}_{\mathrm{c}}(t) \triangleq \dot{q}_{\mathrm{c}}(t^+) - \dot{q}_{\mathrm{c}}(t)$. We assume that the controller kinetic energy $T_{\mathrm{c}}(q_{\mathrm{c}},\dot{q}_{\mathrm{c}})$ is such that $T_{\mathrm{c}}(q_{\mathrm{c}},\dot{q}_{\mathrm{c}}) = \frac{1}{2}\dot{q}_{\mathrm{c}}^{\mathrm{T}}[\frac{\partial T_{\mathrm{c}}}{\partial \dot{q}_{\mathrm{c}}}(q_{\mathrm{c}},\dot{q}_{\mathrm{c}})]^{\mathrm{T}}$, with $T_{\mathrm{c}}(q_{\mathrm{c}},0) = 0$ and $T_{\mathrm{c}}(q_{\mathrm{c}},\dot{q}_{\mathrm{c}}) > 0$, $\dot{q}_{\mathrm{c}} \neq 0$, $\dot{q}_{\mathrm{c}} \in \mathbb{R}^{\hat{n}_{\mathrm{c}}}$. Furthermore, we assume that $U_{\mathrm{c}}(\eta(y_q),y_q) = 0$ and $U_{\mathrm{c}}(q_{\mathrm{c}},y_q) > 0$ for $q_{\mathrm{c}} \neq \eta(y_q)$, $q_{\mathrm{c}} \in \mathcal{D}_{q_{\mathrm{c}}} \subseteq \mathbb{R}^{\hat{n}_{\mathrm{c}}}$.

As in Section 8.3, note that $V_p(q, \dot{q}) \triangleq T(q, \dot{q}) + U(q)$ is the plant energy and $V_c(q_c, \dot{q}_c, y_q) \triangleq T_c(q_c, \dot{q}_c) + U_c(q_c, y_q)$ is the controller emulated energy. Finally,

$$V(q, \dot{q}, q_c, \dot{q}_c) \triangleq V_p(q, \dot{q}) + V_c(q_c, \dot{q}_c, y_q), \tag{8.56}$$

is the total energy of the closed-loop system. It is important to note that the Lagrangian dynamical system (8.44) is *not* dissipative with outputs y_q or y. Next, we study the behavior of the total energy function $V(q, \dot{q}, q_c, \dot{q}_c)$ along the trajectories of the closed-loop system dynamics. For the closed-loop system, we define our resetting set as

$$\mathcal{Z} \triangleq \{(q, \dot{q}, q_c, \dot{q}_c) : (q_c, \dot{q}_c, y) \in \mathcal{Z}_c\}. \tag{8.57}$$

Note that

$$\frac{\mathrm{d}}{\mathrm{d}t} V_p(q, \dot{q}) = \frac{\mathrm{d}}{\mathrm{d}t} \mathcal{H}(q, p) = u^T \dot{q} - \frac{\partial \mathcal{R}}{\partial \dot{q}}(\dot{q})\dot{q}, \quad (q, \dot{q}, q_c, \dot{q}_c) \notin \mathcal{Z}. \tag{8.58}$$

To obtain an expression for $\frac{\mathrm{d}}{\mathrm{d}t} V_c(q_c, \dot{q}_c, y_q)$ when $(q, \dot{q}, q_c, \dot{q}_c) \notin \mathcal{Z}$, define the controller Hamiltonian by

$$\mathcal{H}_c(q_c, \dot{q}_c, p_c, y_q) \triangleq \dot{q}_c^T p_c - \mathcal{L}_c(q_c, \dot{q}_c, y_q), \tag{8.59}$$

where the virtual controller momentum p_c is given by $p_c(q_c, \dot{q}_c, y_q) = \left[\frac{\partial \mathcal{L}_c}{\partial \dot{q}_c}(q_c, \dot{q}_c, y_q)\right]^T$. Next, note that the controller (8.52) and (8.54) can be written in Hamiltonian form. Specifically, it follows from (8.52) and (8.59) that

$$\dot{p}_c(t) = -\left[\frac{\partial \mathcal{H}_c}{\partial q_c}(q_c(t), \dot{q}_c(t), p_c(t), y_q(t))\right]^T,$$
$$(q(t), \dot{q}(t), q_c(t), \dot{q}_c(t)) \notin \mathcal{Z}, \tag{8.60}$$

$$\dot{q}_c(t) = \left[\frac{\partial \mathcal{H}_c}{\partial p_c}(q_c(t), \dot{q}_c(t), p_c(t), y_q(t))\right]^T,$$
$$(q(t), \dot{q}(t), q_c(t), \dot{q}_c(t)) \notin \mathcal{Z}, \tag{8.61}$$

$$u(t) = -\left[\frac{\partial \mathcal{H}_c}{\partial q}(q_c(t), \dot{q}_c(t), p_c(t), y_q(t))\right]^T, \tag{8.62}$$

where $\mathcal{H}_c(q_c, \dot{q}_c, p_c, y_q) = T_c(q_c, \dot{q}_c) + U_c(q_c, y_q)$. Now, it follows from (8.52) and the structure of $T_c(q_c, \dot{q}_c)$ that, for $t \in (t_k, t_{k+1}]$,

$$0 = \frac{\mathrm{d}}{\mathrm{d}t} \left[p_c(q_c(t), \dot{q}_c(t), y_q(t))\right]^T \dot{q}_c(t) - \frac{\partial \mathcal{L}_c}{\partial q_c}(q_c(t), \dot{q}_c(t), y_q(t))\dot{q}_c(t)$$

$$= \frac{\mathrm{d}}{\mathrm{d}t} \left[p_{\mathrm{c}}^{\mathrm{T}}(q_{\mathrm{c}}(t), \dot{q}_{\mathrm{c}}(t), y_q(t))\dot{q}_{\mathrm{c}}(t) \right] - p_{\mathrm{c}}^{\mathrm{T}}(q_{\mathrm{c}}(t), \dot{q}_{\mathrm{c}}(t), y_q(t))\ddot{q}_{\mathrm{c}}(t)$$

$$+ \frac{\partial \mathcal{L}_{\mathrm{c}}}{\partial \dot{q}_{\mathrm{c}}}(q_{\mathrm{c}}(t), \dot{q}_{\mathrm{c}}(t), y_q(t))\ddot{q}_{\mathrm{c}}(t) + \frac{\partial \mathcal{L}_{\mathrm{c}}}{\partial q}(q_{\mathrm{c}}(t), \dot{q}_{\mathrm{c}}(t), y_q(t))\dot{q}(t)$$

$$- \frac{\mathrm{d}}{\mathrm{d}t} \mathcal{L}_{\mathrm{c}}(q_{\mathrm{c}}(t), \dot{q}_{\mathrm{c}}(t), y_q(t))$$

$$= \frac{\mathrm{d}}{\mathrm{d}t}[p_{\mathrm{c}}^{\mathrm{T}}(q_{\mathrm{c}}(t), \dot{q}_{\mathrm{c}}(t), y_q(t))\dot{q}_{\mathrm{c}}(t) - \mathcal{L}_{\mathrm{c}}(q_{\mathrm{c}}(t), \dot{q}_{\mathrm{c}}(t), y_q(t))]$$

$$+ \frac{\partial \mathcal{L}_{\mathrm{c}}}{\partial q}(q_{\mathrm{c}}(t), \dot{q}_{\mathrm{c}}(t), y_q(t))\dot{q}(t)$$

$$= \frac{\mathrm{d}}{\mathrm{d}t} \mathcal{H}_{\mathrm{c}}(q_{\mathrm{c}}(t), \dot{q}_{\mathrm{c}}(t), p_{\mathrm{c}}(t), y_q(t)) + \frac{\partial \mathcal{L}_{\mathrm{c}}}{\partial q}(q_{\mathrm{c}}(t), \dot{q}_{\mathrm{c}}(t), y_q(t))\dot{q}(t)$$

$$= \frac{\mathrm{d}}{\mathrm{d}t} V_{\mathrm{c}}(q_{\mathrm{c}}(t), \dot{q}_{\mathrm{c}}(t), y_q(t)) + \frac{\partial \mathcal{L}_{\mathrm{c}}}{\partial q}(q_{\mathrm{c}}(t), \dot{q}_{\mathrm{c}}(t), y_q(t))\dot{q}(t),$$

$$(q(t), \dot{q}(t), q_{\mathrm{c}}(t), \dot{q}_{\mathrm{c}}(t)) \notin \mathcal{Z}. \quad (8.63)$$

Hence,

$$\frac{\mathrm{d}}{\mathrm{d}t} V(q(t), \dot{q}(t), q_{\mathrm{c}}(t), \dot{q}_{\mathrm{c}}(t)) = u(t)^{\mathrm{T}}\dot{q}(t) - \frac{\partial \mathcal{L}_{\mathrm{c}}}{\partial q}(q_{\mathrm{c}}(t), \dot{q}_{\mathrm{c}}(t), y_q(t))\dot{q}(t)$$

$$- \frac{\partial \mathcal{R}}{\partial \dot{q}}(\dot{q}(t))\dot{q}(t)$$

$$= - \frac{\partial \mathcal{R}}{\partial \dot{q}}(\dot{q}(t))\dot{q}(t)$$

$$\leq 0, \quad (q(t), \dot{q}(t), q_{\mathrm{c}}(t), \dot{q}_{\mathrm{c}}(t)) \notin \mathcal{Z},$$

$$t_k < t \leq t_{k+1}, \quad (8.64)$$

which implies that the total energy of the closed-loop system between resetting events is nonincreasing. Alternatively, if $\mathcal{R}(\dot{q}) \equiv 0$, then $\frac{\mathrm{d}}{\mathrm{d}t} V(q, \dot{q}, q_{\mathrm{c}}, \dot{q}_{\mathrm{c}}) = 0$, $(q, \dot{q}, q_{\mathrm{c}}, \dot{q}_{\mathrm{c}}) \notin \mathcal{Z}$, which implies that the total energy of the closed-loop system is conserved between resetting events. The total energy difference across resetting events is given by

$$\Delta V(q(t_k), \dot{q}(t_k), q_{\mathrm{c}}(t_k), \dot{q}_{\mathrm{c}}(t_k)) = T_{\mathrm{c}}(q_{\mathrm{c}}(t_k^+), \dot{q}_{\mathrm{c}}(t_k^+)) + U_{\mathrm{c}}(q_{\mathrm{c}}(t_k^+), y_q(t_k))$$

$$- V_{\mathrm{c}}(q_{\mathrm{c}}(t_k), \dot{q}_{\mathrm{c}}(t_k), y_q(t_k))$$

$$= T_{\mathrm{c}}(\eta(y_q(t_k)), 0)$$

$$+ U_{\mathrm{c}}(\eta(y_q(t_k)), y_q(t_k))$$

$$- V_{\mathrm{c}}(q_{\mathrm{c}}(t_k), \dot{q}_{\mathrm{c}}(t_k), y_q(t_k))$$

$$= - V_{\mathrm{c}}(q_{\mathrm{c}}(t_k), \dot{q}_{\mathrm{c}}(t_k), y_q(t_k)),$$

$$< 0, \quad (q(t_k), \dot{q}(t_k), q_{\mathrm{c}}(t_k), \dot{q}_{\mathrm{c}}(t_k)) \in \mathcal{Z},$$

$$k \in \overline{\mathbb{Z}}_+, \quad (8.65)$$

which implies that the resetting law (8.53) ensures the total energy decrease across resetting events by an amount equal to the accumulated emulated energy.

Here, we concentrate on an energy-dissipating state-dependent resetting controller that affects a one-way energy transfer between the plant and the controller. Specifically, consider the closed-loop system (8.44), (8.51)–(8.54), where $\mathcal{Z}$ is defined by

$$\mathcal{Z} \triangleq \left\{ (q, \dot{q}, q_{\mathrm{c}}, \dot{q}_{\mathrm{c}}) : \frac{\mathrm{d}}{\mathrm{d}t} V_{\mathrm{c}}(q_{\mathrm{c}}, \dot{q}_{\mathrm{c}}, y_q) = 0 \text{ and } V_{\mathrm{c}}(q_{\mathrm{c}}, \dot{q}_{\mathrm{c}}, y_q) > 0 \right\}. \tag{8.66}$$

Since $y_q = h_1(q)$ and

$$\frac{\mathrm{d}}{\mathrm{d}t} V_{\mathrm{c}}(q_{\mathrm{c}}, \dot{q}_{\mathrm{c}}, y_q) = - \left[\frac{\partial \mathcal{L}_{\mathrm{c}}}{\partial q}(q_{\mathrm{c}}, \dot{q}_{\mathrm{c}}, y_q) \right] \dot{q} = \left[\frac{\partial U_{\mathrm{c}}}{\partial q}(q_{\mathrm{c}}, y_q) \right] \dot{q}, \tag{8.67}$$

it follows that (8.66) can be equivalently rewritten as

$$\mathcal{Z} = \left\{ (q, \dot{q}, q_{\mathrm{c}}, \dot{q}_{\mathrm{c}}) : \left[\frac{\partial U_{\mathrm{c}}}{\partial q}(q_{\mathrm{c}}, h_1(q)) \right] \dot{q} = 0 \text{ and } V_{\mathrm{c}}(q_{\mathrm{c}}, \dot{q}_{\mathrm{c}}, h_1(q)) > 0 \right\}. \tag{8.68}$$

Once again, for practical implementation, knowledge of q_{c}, $\dot{q}_{\mathrm{c}}$, and y_q is sufficient to determine whether or not the closed-loop state vector is in the set $\mathcal{Z}$.

The next theorem gives sufficient conditions for stabilization of Euler-Lagrange dynamical systems using state-dependent hybrid controllers. For this result define the closed-loop system states $x \triangleq [q^{\mathrm{T}}, \dot{q}^{\mathrm{T}}, q_{\mathrm{c}}^{\mathrm{T}}, \dot{q}_{\mathrm{c}}^{\mathrm{T}}]^{\mathrm{T}}$.

Theorem 8.4 *Consider the closed-loop dynamical system $\mathcal{G}$ given by (8.44), (8.51)–(8.54), with $\frac{\partial \mathcal{R}(\dot{q})}{\partial \dot{q}} \dot{q} \equiv 0$ and the resetting set $\mathcal{Z}$ given by (8.66). Assume that $\mathcal{D}_{\mathrm{ci}} \subset \mathcal{D}_{\mathrm{q}} \times \mathbb{R}^{\hat{n}_{\mathrm{p}}} \times \mathcal{D}_{\mathrm{q_c}} \times \mathbb{R}^{\hat{n}_{\mathrm{c}}}$ is a compact positively invariant set with respect to $\mathcal{G}$ such that $0 \in \overset{\circ}{\mathcal{D}}_{\mathrm{ci}}$. Furthermore, assume that the transversality condition (8.11) holds with $\mathcal{X}(x) = \frac{\mathrm{d}}{\mathrm{d}t} V_{\mathrm{c}}(q_{\mathrm{c}}, \dot{q}_{\mathrm{c}}, y_q)$. Then the zero solution $x(t) \equiv 0$ to $\mathcal{G}$ is asymptotically stable. In addition, the total energy function $V(x)$ of $\mathcal{G}$ given by (8.56) is strictly decreasing across resetting events. Alternatively, if $\frac{\partial \mathcal{R}(\dot{q})}{\partial \dot{q}} \dot{q} \neq 0$, $\dot{q} \in \mathbb{R}^{\hat{n}_{\mathrm{p}}}$, and the largest invariant set contained in*

$$\mathcal{R} \triangleq \left\{ x \in \mathcal{D}_{\mathrm{ci}} : \frac{\partial \mathcal{R}(\dot{q})}{\partial \dot{q}} \dot{q} = 0 \right\} \tag{8.69}$$

is $\mathcal{M} = \{0\}$, then the zero solution $x(t) \equiv 0$ to $\mathcal{G}$ is asymptotically stable. Finally, if $\mathcal{D}_q = \mathbb{R}^{\hat{n}_p}$, $\mathcal{D}_{qc} = \mathbb{R}^{\hat{n}_c}$, and the total energy function $V(x)$ is radially unbounded, then the above asymptotic stability results are global.

Proof. The proof is a direct consequence of Theorem 8.3 with $V_p(x_p) = V_p(q, \dot{q})$, $V_c(x_c, y) = V_c(q_c, \dot{q}_c, y_q)$, $y = u_c = x_p$, $u = -y_c = \frac{\partial \mathcal{L}_c}{\partial q}$, $s(u, y) = u^T \rho(y)$, $s_c(u_c, y_c) = y_c^T \rho(u_c)$, where $\rho(y) = \rho\left(\begin{bmatrix} q \\ \dot{q} \end{bmatrix}\right) = \dot{q}$, and $\eta(y)$ replaced by $\begin{bmatrix} \eta(y_q) \\ 0 \end{bmatrix}$. $\square$

8.6 Thermodynamic Stabilization

In this section, we use the recently developed notion of system thermodynamics [65] to develop thermodynamically consistent hybrid controllers for lossless dynamical systems. Specifically, since our energy-based hybrid controller architecture involves the exchange of energy with conservation laws describing transfer, accumulation, and dissipation of energy between the controller and the plant, we construct a modified hybrid controller that guarantees that the closed-loop system is consistent with basic thermodynamic principles after the first resetting event. To develop thermodynamically consistent hybrid controllers consider the closed-loop system $\mathcal{G}$ given by (8.22) and (8.23), with $\mathcal{Z}$ given by

$$\mathcal{Z} \triangleq \left\{ x \in \mathcal{D} : \phi(x)(V_p(x) - V_c(x)) = 0 \text{ and } V_c(x) > 0 \right\}, \quad (8.70)$$

where $\phi(x) \triangleq -\dot{V}_c(x)$, $x \notin \mathcal{Z}$. It follows from (8.33) that $\phi(\cdot)$ is the net energy flow from the plant to the controller, and hence, we refer to $\phi(\cdot)$ as the *net energy flow* function.

We assume that the energy flow function $\phi(x)$ is infinitely differentiable and the transversality condition (8.11) holds with $\mathcal{X}(x) = \phi(x)(V_p(x) - V_c(x))$. To ensure a thermodynamically consistent energy flow between the plant and controller after the first resetting event, the controller resetting logic must be designed in such a way so as to satisfy three key thermodynamic axioms on the closed-loop system level. Namely, between resettings the energy flow function $\phi(\cdot)$ must satisfy the following two axioms [64, 65]:

Axiom i) For the *connectivity matrix* $\mathcal{C} \in \mathbb{R}^{2 \times 2}$ [65, p. 56] associated with the closed-loop system $\mathcal{G}$ defined by

$$\mathcal{C}_{(i,j)} \triangleq \begin{cases} 0, & \text{if } \phi(x(t)) \equiv 0, \\ 1, & \text{otherwise,} \end{cases} \quad i \neq j, \quad i, j = 1, 2, \quad t \geq t_1^+, \quad (8.71)$$

and

$$\mathcal{C}_{(i,i)} = -\mathcal{C}_{(k,i)}, \quad i \neq k, \quad i, k = 1, 2, \quad (8.72)$$

rank $\mathcal{C} = 1$, and for $\mathcal{C}_{(i,j)} = 1$, $i \neq j$, $\phi(x(t)) = 0$ if and only if $V_p(x(t)) = V_c(x(t))$, $x(t) \notin \mathcal{Z}$, $t \geq t_1^+$.

Axiom ii) $\phi(x(t))(V_p(x(t)) - V_c(x(t))) \leq 0$, $x(t) \notin \mathcal{Z}$, $t \geq t_1^+$.

Furthermore, across resettings the energy difference between the plant and the controller must satisfy the following axiom [68, 69]:

Axiom iii) $[V_p(x + f_d(x)) - V_c(x + f_d(x))][V_p(x) - V_c(x)] \geq 0$, $x \in \mathcal{Z}$.

The fact that $\phi(x(t)) = 0$ if and only if $V_p(x(t)) = V_c(x(t))$, $x(t) \notin \mathcal{Z}$, $t \geq t_1^+$, implies that the plant and the controller are *connected*; alternatively, $\phi(x(t)) \equiv 0$, $t \geq t_1^+$, implies that the plant and the controller are *disconnected*. Axiom i) implies that if the energies in the plant and the controller are equal, then energy exchange between the plant and controller is not possible unless a resetting event occurs. This statement is consistent with the *zeroth law of thermodynamics*, which postulates that temperature equality is a necessary and sufficient condition for thermal equilibrium of an isolated system. Axiom ii) implies that energy flows from a more energetic system to a less energetic system and is consistent with the *second law of thermodynamics*, which states that heat (energy) must flow in the direction of lower temperatures. Finally, Axiom iii) implies that the energy difference between the plant and the controller across resetting instants is monotonic, that is, $[V_p(x(t_k^+)) - V_c(x(t_k^+))][V_p(x(t_k)) - V_c(x(t_k))] \geq 0$ for all $V_p(x) \neq V_c(x)$, $x \in \mathcal{Z}$, $k \in \overline{\mathbb{Z}}_+$.

With the resetting law given by (8.70), it follows that the closed-loop dynamical system $\mathcal{G}$ satisfies Axioms i)–iii) for all $t \geq t_1$. To see this, note that since $\phi(x) \not\equiv 0$, the connectivity matrix $\mathcal{C}$ is given by

$$\mathcal{C} = \begin{bmatrix} -1 & 1 \\ 1 & -1 \end{bmatrix}, \quad (8.73)$$

and hence, rank $\mathcal{C} = 1$. The second condition in Axiom i) need not be satisfied since the case where $\phi(x) = 0$ or $V_p(x) = V_c(x)$ corresponds to a resetting instant. Furthermore, it follows from the definition of the resetting set (8.70) that Axiom ii) is satisfied for the closed-loop system for all $t \geq t_1^+$. Finally, since $V_c(x + f_d(x)) = 0$ and $V_p(x + f_d(x)) = V_p(x)$, $x \in \mathcal{Z}$, it follows from the definition of the resetting set that

$$[V_p(x + f_d(x)) - V_c(x + f_d(x))][V_p(x) - V_c(x)]$$
$$= V_p(x)[V_p(x) - V_c(x)] \geq 0, \quad x \in \mathcal{Z}, \quad (8.74)$$

and hence, Axiom iii) is satisfied across resettings. Hence, the closed-loop system $\mathcal{G}$ is thermodynamically consistent after the first resetting event in the sense of [64, 65, 68, 69].

Next, we give a hybrid definition of entropy for the closed-loop system $\mathcal{G}$ that generalizes the continuous-time and discrete-time entropy definitions established in [64, 65, 68, 69].

Definition 8.2 *For the impulsive closed-loop system $\mathcal{G}$ given by (8.22) and (8.23), a function $S : \overline{\mathbb{R}}_+^2 \to \mathbb{R}$ satisfying*

$$S(E(x(T))) \geq S(E(x(t_1))) - \int_{t_1}^{T} \frac{d(x_p(t))}{c + V_p(x(t))} dt$$

$$-\frac{1}{c} \sum_{k \in \mathbb{Z}_{[t_1, T)}} V_c(x(t_k)), \quad T \geq t_1, \quad (8.75)$$

where $k \in \mathbb{Z}_{[t_1, T)} \triangleq \{k : t_1 \leq t_k < T\}$, $E \triangleq [V_p, V_c]^T$, $d : \mathcal{D}_p \to \mathbb{R}$ is a continuous, nonnegative-definite dissipation rate function, $c > 0$, is called an entropy *function of $\mathcal{G}$.*

The next result gives necessary and sufficient conditions for establishing the existence of an entropy function of $\mathcal{G}$ over an interval $t \in (t_k, t_{k+1}]$ involving the consecutive resetting times t_k and t_{k+1}, $k \in \mathbb{Z}_+$.

Theorem 8.5 *For the impulsive closed-loop system $\mathcal{G}$ given by (8.22) and (8.23), a function $S : \overline{\mathbb{R}}_+^2 \to \mathbb{R}$ is an entropy function of $\mathcal{G}$ if and only if*

$$S(E(x(\hat{t}))) \geq S(E(x(t))) - \int_t^{\hat{t}} \frac{d(x_p(s))}{c + V_p(x(s))} ds,$$

$$t_k < t \leq \hat{t} \leq t_{k+1}, \quad (8.76)$$

$$S(E(x(t_k)) + f_d(x(t_k)))) \geq S(E(x(t_k))) - \frac{V_c(x(t_k))}{c}, \quad k \in \mathbb{Z}_+.$$

$$(8.77)$$

Proof. Let $k \in \mathbb{Z}_+$ and suppose $S(E)$ is an entropy function of $\mathcal{G}$. Then, (8.75) holds. Now, since for $t_k < t \le \hat{t} \le t_{k+1}$, $\mathbb{Z}_{[t,\hat{t})} = \emptyset$, (8.76) is immediate. Next, note that

$$S(E(x(t_k^+))) \ge S(E(x(t_k))) - \int_{t_k}^{t_k^+} \frac{d(x_\mathrm{p}(s))}{c + V_\mathrm{p}(x(s))} \mathrm{d}s - \frac{V_\mathrm{c}(x(t_k))}{c},$$

$$(8.78)$$

which, since $\mathbb{Z}_{[t_k, t_k^+)} = k$, implies (8.77).

Conversely, suppose (8.76) and (8.77) hold, and let $\hat{t} \ge t \ge t_1$ and $\mathbb{Z}_{[t,\hat{t})} = \{i, i+1, \ldots, j\}$. (Note that if $\mathbb{Z}_{[t,\hat{t})} = \emptyset$ the converse result is a direct consequence of (8.76).) If $\mathbb{Z}_{[t,\hat{t})} \ne \emptyset$, it follows from (8.76) and (8.77) that

$$
\begin{aligned}
S(E(x(\hat{t}))) &- S(E(x(t))) \\
&= S(E(x(\hat{t}))) - S(E(x(t_j^+))) \\
&\quad + \sum_{m=0}^{j-i-1} S(E(x(t_{j-m}^+))) - S(E(x(t_{j-m-1}^+))) \\
&\quad + S(E(x(t_i^+))) - S(E(x(t))) \\
&= S(E(x(\hat{t}))) - S(E(x(t_j^+))) \\
&\quad + \sum_{m=0}^{j-i} S(E(x(t_{j-m}) + f_\mathrm{d}(x(t_{j-m})))) - S(E(x(t_{j-m}))) \\
&\quad + \sum_{m=0}^{j-i-1} S(E(x(t_{j-m}))) - S(E(x(t_{j-m-1}^+))) \\
&\quad + S(E(x(t_i))) - S(E(x(t))) \\
&\ge - \int_{t_j^+}^{\hat{t}} \frac{d(x_\mathrm{p}(s))}{c + V_\mathrm{p}(x(s))} \mathrm{d}s - \frac{1}{c} \sum_{m=0}^{j-i} V_\mathrm{c}(x(t_{j-m})) \\
&\quad - \sum_{m=0}^{j-i-1} \int_{t_{j-m-1}^+}^{t_{j-m}} \frac{d(x_\mathrm{p}(s))}{c + V_\mathrm{p}(x(s))} \mathrm{d}s - \int_t^{t_i} \frac{d(x_\mathrm{p}(s))}{c + V_\mathrm{p}(x(s))} \mathrm{d}s \\
&= - \int_t^{\hat{t}} \frac{d(x_\mathrm{p}(s))}{c + V_\mathrm{p}(x(s))} \mathrm{d}s - \frac{1}{c} \sum_{k \in \mathbb{Z}_{[t,\hat{t})}} V_\mathrm{c}(x(t_k)), \quad (8.79)
\end{aligned}
$$

which implies that $S(E)$ is an entropy function of $\mathcal{G}$. □

The next theorem establishes the existence of an entropy function for the impulsive closed-loop system $\mathcal{G}$.

Theorem 8.6 *Consider the impulsive closed-loop system $\mathcal{G}$ given by (8.22) and (8.23), with $\mathcal{Z}$ given by (8.70). Then the function S : $\overline{\mathbb{R}}_+^2 \to \mathbb{R}$ given by*

$$S(E) = \log_e(c + V_\mathrm{p}) + \log_e(c + V_\mathrm{c}) - 2\log_e c, \quad E \in \overline{\mathbb{R}}_+^2, \quad (8.80)$$

where $c > 0$, is a continuously differentiable entropy function of $\mathcal{G}$. In addition,

$$\dot{S}(E(x(t))) > -\frac{d(x_\mathrm{p}(t))}{c + V_\mathrm{p}(x(t))}, \quad x(t) \notin \mathcal{Z}, \quad t_k < t \le t_{k+1}, \quad (8.81)$$

$$-\frac{V_\mathrm{c}(x(t_k))}{c} < \Delta S(E(x(t_k))) < -\frac{V_\mathrm{c}(x(t_k))}{c + V_\mathrm{c}(x(t_k))}, \quad x(t_k) \in \mathcal{Z},$$
$$k \in \mathbb{Z}_+. \quad (8.82)$$

Proof. Since $\dot{V}_\mathrm{p}(x(t)) = \phi(x(t)) - d(x_\mathrm{p}(t))$ and $\dot{V}_\mathrm{c}(x(t)) = -\phi(x(t))$, $x(t) \notin \mathcal{Z}$, $t \in (t_k, t_{k+1}]$, $k \in \mathbb{Z}_+$, it follows that

$$\dot{S}(E(x(t))) = \frac{\phi(x(t))(V_\mathrm{c}(x(t)) - V_\mathrm{p}(x(t)))}{(c + V_\mathrm{p}(x(t)))(c + V_\mathrm{c}(x(t)))} - \frac{d(x_\mathrm{p}(t))}{c + V_\mathrm{p}(x(t))}$$

$$> -\frac{d(x_\mathrm{p}(t))}{c + V_\mathrm{p}(x(t))}, \quad x(t) \notin \mathcal{Z}, \quad t_k < t \le t_{k+1}. \quad (8.83)$$

Furthermore, since $V_\mathrm{c}(x(t_k)) + f_\mathrm{d}(x(t_k))) = 0$ and $V_\mathrm{p}(x(t_k) + f_\mathrm{d}(x(t_k))) = V_\mathrm{p}(x(t_k))$, $x(t_k) \in \mathcal{Z}$, $k \in \mathbb{Z}_+$, it follows that

$$\Delta S(E(x(t_k))) = \log_e\left[1 - \frac{V_\mathrm{c}(x(t_k))}{c + V_\mathrm{c}(x(t_k))}\right] > -\frac{V_\mathrm{c}(x(t_k))}{c}, \quad x(t_k) \in \mathcal{Z},$$
$$k \in \mathbb{Z}_+, \quad (8.84)$$

and

$$\Delta S(E(x(t_k))) = \log_e\left[1 - \frac{V_\mathrm{c}(x(t_k))}{c + V_\mathrm{c}(x(t_k))}\right] < -\frac{V_\mathrm{c}(x(t_k))}{c + V_\mathrm{c}(x(t_k))},$$
$$x(t_k) \in \mathcal{Z}, \quad k \in \mathbb{Z}_+, \quad (8.85)$$

where in (8.84) and (8.85) we used the fact that $\frac{x}{1+x} < \log_e(1+x) < x$, $x > -1$, $x \neq 0$. The result is now an immediate consequence of Theorem 8.5. $\qquad\square$

Note that in the absence of energy dissipation into the environment (i.e., $d(x_\mathrm{p}(x)) \equiv 0$) it follows from (8.81) that the entropy of the closed-loop system strictly increases between resetting events, which is consistent with thermodynamic principles. This is not surprising since in this case the closed-loop system is *adiabatically isolated* (i.e., the system does not exchange energy (heat) with the environment) and the total energy of the closed-loop system is conserved between resetting events. Alternatively, it follows from (8.82) that the entropy of the closed-loop system strictly decreases across resetting events since the total energy strictly decreases at each resetting instant, and hence, energy is not conserved across resetting events.

Using Theorem 8.6, the resetting set $\mathcal{Z}$ given by (8.70) can be rewritten as

$$\mathcal{Z} \triangleq \left\{ x \in \mathcal{D} : \frac{d}{dt} S(E(x)) + \frac{d(x_\mathrm{p})}{c + V_\mathrm{p}(x)} = 0 \right.$$

$$\left. \text{and } V_\mathrm{c}(x) > 0 \right\}, \quad (8.86)$$

where $\mathcal{X}(x) \triangleq \frac{d}{dt} S(E(x)) + \frac{d(x_\mathrm{p})}{c + V_\mathrm{p}(x)}$ is a continuously differentiable function that defines the resetting set as its zero level set. The resetting set (8.70) or, equivalently, (8.86) is motivated by thermodynamic principles and guarantees that the energy of the closed-loop system is always flowing from regions of higher to lower energies after the first resetting event, which is consistent with the second law of thermodynamics. As shown in Theorem 8.6, this guarantees the existence of an entropy function $S(E)$ for the closed-loop system that satisfies the Clausius-type inequality (8.81) between resetting events. If $\phi(x) = 0$ or $V_\mathrm{p}(x) = V_\mathrm{c}(x)$, then inequality (8.81) would be subverted, and hence, we reset the compensator states in order to ensure that the second law of thermodynamics is not violated.

Finally, if $\mathcal{D}_\mathrm{ci} \subset \mathcal{D}$ is a compact positively invariant set with respect to the closed-loop dynamical system $\mathcal{G}$ given by (8.22) and (8.23) such that $0 \in \overset{\circ}{\mathcal{D}}_\mathrm{ci}$, $d(x_\mathrm{p}) \equiv 0$, and the transversality condition (8.11) holds with $\mathcal{X}(x) = \frac{d}{dt} S(E(x)) + \frac{d(x_\mathrm{p})}{c + V_\mathrm{p}(x)}$, then it follows from Theorem 8.3 that the zero solution $x(t) \equiv 0$ of the closed-loop system $\mathcal{G}$, with resetting set $\mathcal{Z}$ given by (8.70), is asymptotically stable. Alternatively, if $d(x_\mathrm{p}) \neq 0$, $x_\mathrm{p} \in \mathcal{D}_\mathrm{p}$, and the largest invariant set contained in $\mathcal{R} \triangleq \{ x \in \mathcal{D}_\mathrm{ci} : d(x_\mathrm{p}) = 0 \}$ is $\{0\}$, then the zero solution $x(t) \equiv 0$ of the closed-loop system $\mathcal{G}$ is asymptotically stable. Furthermore, in this case, the hybrid controller (8.52) and (8.53), with resetting set (8.70), is a *thermodynamically stabilizing* compensator. Analogous thermodynamically stabilizing compensators can be constructed for port-controlled Hamiltonian and Euler-Lagrange dynamical systems.

8.7 Energy-Dissipating Hybrid Control Design

In this section, we apply the energy dissipating hybrid controller synthesis framework developed in Sections 8.5 and 8.6 to two examples. For the first example, consider the vector second-order nonlinear *Lienard* system given by

$$\ddot{q}(t) + f(q(t)) = u(t), \quad q(0) = q_0, \quad \dot{q}(0) = \dot{q}_0, \quad t \geq 0, \quad (8.87)$$

$$y(t) = \begin{bmatrix} C_1 q(t) \\ C_2 \dot{q}(t) \end{bmatrix}, \qquad (8.88)$$

where $q \in \mathbb{R}^{\hat{n}_{\mathrm{p}}}$, $f : \mathbb{R}^{\hat{n}_{\mathrm{p}}} \to \mathbb{R}^{\hat{n}_{\mathrm{p}}}$ is infinitely differentiable, $f(q) = 0$ if and only if $q = 0$, $C_1 \in \mathbb{R}^{l_1 \times \hat{n}_{\mathrm{p}}}$, $C_2 \in \mathbb{R}^{(l-l_1) \times \hat{n}_{\mathrm{p}}}$, and

$$\frac{\partial f_i}{\partial q_j} = \frac{\partial f_j}{\partial q_i}, \quad i, j = 1, \ldots, \hat{n}_{\mathrm{p}}. \qquad (8.89)$$

The plant energy of the system is given by

$$
\begin{aligned}
V_{\mathrm{p}}(q, \dot{q}) &= T(q, \dot{q}) + U(q) \\
&= \frac{1}{2}\dot{q}^{\mathrm{T}}\dot{q} + \int_{0,\,\mathrm{path}}^{q} f^{\mathrm{T}}(\sigma) \mathrm{d}\sigma \\
&= \frac{1}{2}\dot{q}^{\mathrm{T}}\dot{q} + \int_{0,\,\mathrm{path}}^{q} \sum_{i=1}^{\hat{n}_{\mathrm{p}}} f_i(\sigma) \mathrm{d}\sigma_i \\
&= \frac{1}{2}\dot{q}^{\mathrm{T}}\dot{q} + \int_{0}^{q_1} f_1(\sigma_1, 0, \ldots, 0) \mathrm{d}\sigma_1 + \int_{0}^{q_2} f_2(q_1, \sigma_2, 0, \ldots, 0) \mathrm{d}\sigma_2 \\
&\quad + \cdots + \int_{0}^{q_{\hat{n}_{\mathrm{p}}}} f_{\hat{n}_{\mathrm{p}}}(q_1, q_2, \ldots, q_{\hat{n}_{\mathrm{p}}-1}, \sigma_{\hat{n}_{\mathrm{p}}}) \mathrm{d}\sigma_{\hat{n}_{\mathrm{p}}}, \qquad (8.90)
\end{aligned}
$$

where $T(q, \dot{q}) = \frac{1}{2}\dot{q}^{\mathrm{T}}\dot{q}$ and $U(q) = \int_{0,\,\mathrm{path}}^{q} f^{\mathrm{T}}(\sigma)\mathrm{d}\sigma$. Note that the path integral in (8.90) is taken over any path joining the origin to $q \in \mathbb{R}^{\hat{n}_{\mathrm{p}}}$. Furthermore, the path integral in (8.90) is well defined since $f(\cdot)$ is such that $\frac{\partial f}{\partial q}$ is symmetric, and hence, $f(\cdot)$ is a gradient of a real-valued function [8, Theorem 10-37]. Here, we assume that $U(0) = 0$ and $U(q) > 0$ for $q \neq 0$, $q \in \mathbb{R}^{\hat{n}_{\mathrm{p}}}$. Note that defining $p \triangleq \dot{q}$ and

$$\mathcal{H}(q, p) \triangleq \frac{1}{2}p^{\mathrm{T}}p + \int_{0,\,\mathrm{path}}^{q} f^{\mathrm{T}}(\sigma)\mathrm{d}\sigma, \qquad (8.91)$$

it follows that (8.87) can be written in Hamiltonian form

$$\dot{q}(t) = \left[\frac{\partial \mathcal{H}}{\partial p}(q(t), p(t))\right]^{\mathrm{T}}, \quad q(0) = q_0, \quad t \geq 0, \qquad (8.92)$$

$$\dot{p}(t) = -\left[\frac{\partial \mathcal{H}}{\partial q}(q(t), p(t))\right]^{\mathrm{T}} + u, \quad p(0) = p_0. \qquad (8.93)$$

To design a state-dependent hybrid controller for the Lienard system (8.87), let $C_1 = C_2 = I_{\hat{n}_{\mathrm{p}}}$, let

$$T_{\mathrm{c}}(q_{\mathrm{c}}, \dot{q}_{\mathrm{c}}) = \frac{1}{2}\dot{q}_{\mathrm{c}}^{\mathrm{T}}\dot{q}_{\mathrm{c}}, \qquad (8.94)$$

$$U_{\mathrm{c}}(q_{\mathrm{c}}, q) = \int_{0, \,\mathrm{path}}^{q_{\mathrm{c}}-q} g^{\mathrm{T}}(\sigma)\mathrm{d}\sigma, \qquad (8.95)$$

where $q_{\mathrm{c}} \in \mathbb{R}^{\hat{n}_{\mathrm{p}}}$, $g : \mathbb{R}^{\hat{n}_{\mathrm{p}}} \to \mathbb{R}^{\hat{n}_{\mathrm{p}}}$ is infinitely differentiable, $g(x) = 0$ if and only if $x = 0$, and $g'(0)$ is positive definite, and let

$$\frac{\partial g_i}{\partial x_j} = \frac{\partial g_j}{\partial x_i}, \quad i, j = 1, \ldots, \hat{n}_{\mathrm{p}}, \qquad (8.96)$$

so that

$$\mathcal{L}_{\mathrm{c}}(q_{\mathrm{c}}, \dot{q}_{\mathrm{c}}, q) = \frac{1}{2}\dot{q}_{\mathrm{c}}^{\mathrm{T}}\dot{q}_{\mathrm{c}} - \int_{0, \,\mathrm{path}}^{q_{\mathrm{c}}-q} g^{\mathrm{T}}(\sigma)\mathrm{d}\sigma. \qquad (8.97)$$

Here, we assume that $\int_{0, \,\mathrm{path}}^{x} g^{\mathrm{T}}(\sigma)\mathrm{d}\sigma > 0$ for all $x \neq 0$, $x \in \mathbb{R}^{\hat{n}_{\mathrm{p}}}$. In this case, the state-dependent hybrid controller has the form

$$\ddot{q}_{\mathrm{c}}(t) + g(q_{\mathrm{c}}(t) - q(t)) = 0, \quad (q(t), \dot{q}(t), q_{\mathrm{c}}(t), \dot{q}_{\mathrm{c}}(t)) \notin \mathcal{Z}, \quad t \geq 0, \qquad (8.98)$$

$$\begin{bmatrix} \Delta q_{\mathrm{c}}(t) \\ \Delta \dot{q}_{\mathrm{c}}(t) \end{bmatrix} = \begin{bmatrix} q(t) - q_{\mathrm{c}}(t) \\ -\dot{q}_{\mathrm{c}}(t) \end{bmatrix}, \quad (q(t), \dot{q}(t), q_{\mathrm{c}}(t), \dot{q}_{\mathrm{c}}(t)) \in \mathcal{Z}, \quad t \geq 0, \qquad (8.99)$$

$$u(t) = g(q_{\mathrm{c}}(t) - q(t)), \qquad (8.100)$$

with the resetting set (8.66) taking the form

$$\mathcal{Z} = \left\{ (q, \dot{q}, q_{\mathrm{c}}, \dot{q}_{\mathrm{c}}) : [g(q_{\mathrm{c}} - q)]^{\mathrm{T}}\dot{q} = 0 \text{ and } \begin{bmatrix} q - q_{\mathrm{c}} \\ -\dot{q}_{\mathrm{c}} \end{bmatrix} \neq 0 \right\}. \qquad (8.101)$$

Here, we consider the case where $\hat{n}_{\mathrm{p}} = \frac{n_{\mathrm{p}}}{2} = 1$. To show that A1 holds in this case, we show that upon reaching a nonequilibrium point $x(t) \triangleq [q(t), \dot{q}(t), q_{\mathrm{c}}(t), \dot{q}_{\mathrm{c}}(t)]^{\mathrm{T}} \notin \mathcal{Z}$ that is in the closure of $\mathcal{Z}$, the continuous-time dynamics $\dot{x} = f_c(x)$ remove $x(t)$ from $\overline{\mathcal{Z}}$, and hence, necessarily move the trajectory a finite distance away from $\mathcal{Z}$. If $x(t) \notin \mathcal{Z}$ is an equilibrium point, then $x(s) \notin \mathcal{Z}$, $s \geq t$, which is also consistent with A1.

The closure of $\mathcal{Z}$ is given by

$$\overline{\mathcal{Z}} = \{(q, \dot{q}, q_c, \dot{q}_c) : [g(q_c - q)]\dot{q} \geq 0\}. \tag{8.102}$$

Furthermore, the points x^* satisfying $[q^* - q_c^*, -\dot{q}_c^*]^{\mathrm{T}} = 0$ have the form

$$x^* \triangleq [\; q \quad \dot{q} \quad q \quad 0 \;]^{\mathrm{T}}, \tag{8.103}$$

that is, $q_c = q$ and $\dot{q}_c = 0$. It follows that $x^* \notin \mathcal{Z}$, although $x^* \in \overline{\mathcal{Z}}$.

To show that the continuous-time dynamics $\dot{x} = f_c(x)$ remove x^* from $\overline{\mathcal{Z}}$, note that

$$\frac{\mathrm{d}}{\mathrm{d}t} V_{\mathrm{p}}(q, \dot{q}) = [g(q_c - q)]\dot{q} \tag{8.104}$$

and

$$\frac{\mathrm{d}^2}{\mathrm{d}t^2} V_{\mathrm{p}}(q, \dot{q}) = \ddot{q}[g(q_c - q)] + \dot{q}[g'(q_c - q)](\dot{q}_c - \dot{q}), \tag{8.105}$$

$$\frac{\mathrm{d}^3}{\mathrm{d}t^3} V_{\mathrm{p}}(q, \dot{q}) = q^{(3)}[g(q_c - q)] + [g'(q_c - q)](\dot{q}\ddot{q}_c + 2\dot{q}_c\ddot{q} - 3\dot{q}\ddot{q})$$
$$+ [g''(q_c - q)](\dot{q}_c - \dot{q})^2 \dot{q}, \tag{8.106}$$

$$\frac{\mathrm{d}^4}{\mathrm{d}t^4} V_{\mathrm{p}}(q, \dot{q}) = q^{(4)}[g(q_c - q)] + [g'(q_c - q)](3\dot{q}_c q^{(3)} - 4\dot{q}q^{(3)} + 3\ddot{q}\ddot{q}_c$$
$$+ \dot{q}q_c^{(3)} - 3\ddot{q}^2)$$
$$+ [g''(q_c - q)](3\dot{q}\dot{q}_c\ddot{q}_c + 3\dot{q}_c^2\ddot{q} - 9\dot{q}\dot{q}_c\ddot{q} - 3\dot{q}^2\ddot{q}_c + 6\dot{q}^2\ddot{q})$$
$$+ g^{(3)}(q_c - q)(\dot{q}_c - \dot{q})^3 \dot{q}, \tag{8.107}$$

where $g^{(n)}(t) \triangleq \frac{\mathrm{d}^n g(t)}{\mathrm{d}t^n}$. Since

$$\left.\frac{\mathrm{d}^2}{\mathrm{d}t^2} V_{\mathrm{p}}(q, \dot{q})\right|_{x=x^*} = -g'(0)\dot{q}^2, \tag{8.108}$$

it follows that if $\dot{q} \neq 0$, then the continuous-time dynamics $\dot{x} = f_c(x)$ remove x^* from $\overline{\mathcal{Z}}$. If $\dot{q} = 0$, then it follows from (8.105)–(8.107) that

$$\left.\frac{\mathrm{d}^2}{\mathrm{d}t^2} V_{\mathrm{p}}(q, \dot{q})\right|_{x=x^*, \dot{q}=0} = 0, \tag{8.109}$$

$$\left.\frac{\mathrm{d}^3}{\mathrm{d}t^3} V_{\mathrm{p}}(q, \dot{q})\right|_{x=x^*, \dot{q}=0} = 0, \tag{8.110}$$

$$\frac{\mathrm{d}^4}{\mathrm{d}t^4} V_{\mathrm{p}}(q, \dot{q}) \bigg|_{x=x^*, \dot{q}=0} = -3g'(0)\ddot{q}^2, \qquad (8.111)$$

where in the evaluation of (8.110) and (8.111) we use the fact that if $q_{\mathrm{c}} = q$ and $\dot{q}_{\mathrm{c}} = 0$, then $\ddot{q}_{\mathrm{c}} = 0$, which follows immediately from the continuous-time dynamics. Since if $\dot{q} = 0$ and $\ddot{q} \neq 0$, then the lowest-order nonzero time derivative of $V_{\mathrm{p}}(x_{\mathrm{p}})$ is negative, it follows that the continuous-time dynamics remove x^* from $\overline{\mathcal{Z}}$. However, if $\dot{q} = 0$ and $\ddot{q} = 0$, then it follows from the continuous-time dynamics that x^* is necessarily an equilibrium point, in which case the trajectory never again enters $\mathcal{Z}$. Therefore, we can conclude that A1 is indeed valid for this system. Also, since $f_{\mathrm{d}}(x + f_{\mathrm{d}}(x)) = 0$, it follows from (8.101) that if $x \in \mathcal{Z}$, then $x + f_{\mathrm{d}}(x) \notin \mathcal{Z}$, and thus A2 holds.

For thermodynamic stabilization, the resetting set (8.70) is given by

$$\mathcal{Z} = \left\{ (q, \dot{q}, q_{\mathrm{c}}, \dot{q}_{\mathrm{c}}) : \dot{q}^{\mathrm{T}}[g(q_{\mathrm{c}} - q)][V_{\mathrm{p}}(q, \dot{q}) - V_{\mathrm{c}}(q_{\mathrm{c}}, \dot{q}_{\mathrm{c}}, q)] = 0 \right.$$

$$\left. \text{and} \ \begin{bmatrix} q - q_{\mathrm{c}} \\ -\dot{q}_{\mathrm{c}} \end{bmatrix} \neq 0 \right\}. \qquad (8.112)$$

Furthermore, the entropy function $S(E)$ is given by

$$S(E) = \log_e[1 + V_{\mathrm{p}}(q, \dot{q})] + \log_e[1 + V_{\mathrm{c}}(q_{\mathrm{c}}, \dot{q}_{\mathrm{c}}, q)]. \qquad (8.113)$$

To illustrate the behavior of the closed-loop impulsive dynamical system, let $\hat{n}_{\mathrm{p}} = \frac{n_{\mathrm{p}}}{2} = 1$, $f(x) = x + x^3$, and $g(x) = 3x$ with initial conditions $q(0) = 0$, $\dot{q}(0) = 1$, $q_{\mathrm{c}}(0) = 0$, and $\dot{q}_{\mathrm{c}}(0) = 0$. For this system, the transversality condition is sufficiently complex that we have been unable to show it analytically. This condition was verified numerically, and hence, Assumption 8.1 holds. Figure 8.2 shows the controlled plant position and velocity states versus time, while 8.3 shows the virtual position and velocity compensator states versus time. Figure 8.4 shows the control force versus time. Note that the compensator states are the only states that reset. Furthermore, the control force versus time is discontinuous at the resetting times. A comparison of the plant energy, controller energy, and total energy is shown in Figure 8.5. Figures 8.6–8.9 show analogous representations for the thermodynamically stabilizing compensator. Finally, Figure 8.10 shows the closed-loop system entropy versus time. Note that the entropy of the closed-loop system strictly increases between resetting events.

As our next example, we consider the rotational/translational proof-mass actuator (RTAC) nonlinear system studied in [36]. The system

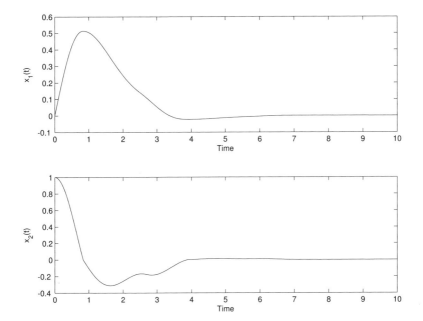

Figure 8.2 Plant position and velocity versus time.

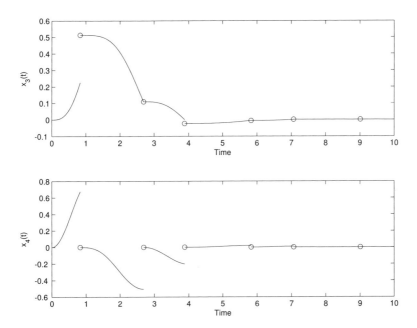

Figure 8.3 Controller position and velocity versus time.

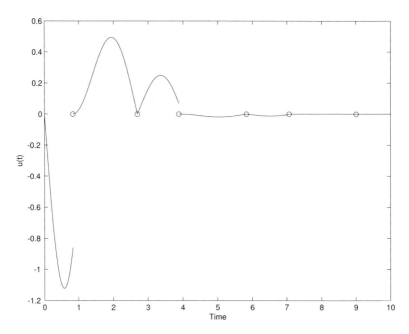

Figure 8.4 Control signal versus time.

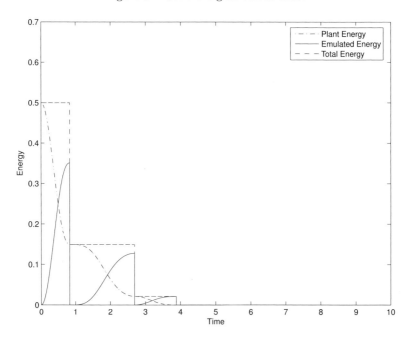

Figure 8.5 Plant, emulated, and total energy versus time.

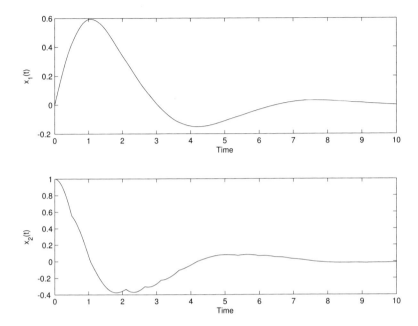

Figure 8.6 Plant position and velocity versus time for thermodynamic controller.

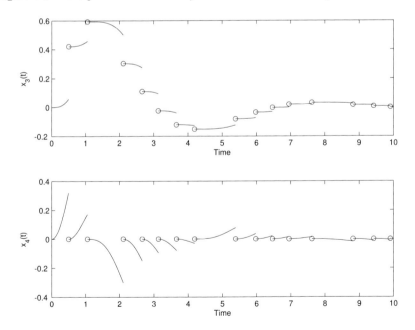

Figure 8.7 Controller position and velocity versus time for thermodynamic controller.

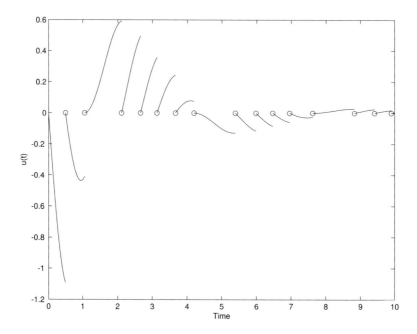

Figure 8.8 Control signal versus time for thermodynamic controller.

(see Figure 8.11) involves an eccentric rotational inertia, which acts as a proof-mass actuator mounted on a translational oscillator. The oscillator cart of mass M is connected to a fixed support via a linear spring of stiffness k. The cart is constrained to one-dimensional motion and the rotational proof-mass actuator consists of a mass m and mass moment of inertia I located a distance e from the center of mass of the cart. In Figure 8.11, N denotes the control torque applied to the proof mass. Since the motion is constrained to the horizontal plane the gravitational forces are not considered in the dynamic analysis.

Letting q, $\dot{q}$, θ, and $\dot{\theta}$ denote the translational position and velocity of the cart and the angular position and velocity of the rotational proof mass, respectively, and using the energy function

$$V_s(q, \dot{q}, \theta, \dot{\theta}) = \frac{1}{2}[kq^2 + (M + m)\dot{q}^2 + (I + me^2)\dot{\theta}^2 + 2me\dot{q}\dot{\theta}\cos\theta],$$

$$(8.114)$$

the nonlinear dynamic equations of motion are given by

$$(M + m)\ddot{q} + kq = -me(\ddot{\theta}\cos\theta - \dot{\theta}^2\sin\theta), \qquad (8.115)$$

$$(I + me^2)\ddot{\theta} = -me\ddot{q}\cos\theta + N, \qquad (8.116)$$

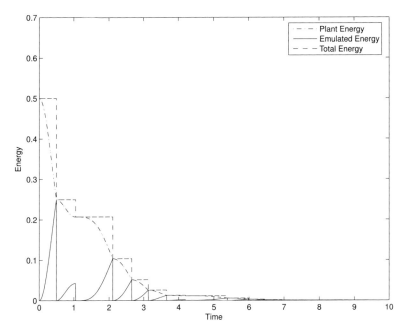

Figure 8.9 Plant, emulated, and total energy versus time for thermodynamic controller.

with problem data given in Table 8.1 and output $y = [\theta, \dot{\theta}]^{\mathrm{T}}$. The physical configuration of the system necessitates the constraint $|q| \leq 0.025\,\mathrm{m}$. In addition, the control torque is limited by $|N| \leq 0.100\,\mathrm{N}$ m [36]. With the normalization

$$\xi \triangleq \sqrt{\frac{M+m}{I+me^2}}q, \quad \tau \triangleq \sqrt{\frac{k}{M+m}}t, \quad u \triangleq \frac{M+m}{k(I+me^2)}N, \quad (8.117)$$

the equations of motion become

$$\ddot{\xi} + \xi = \varepsilon(\dot{\theta}^2 \sin\theta - \ddot{\theta}\cos\theta), \tag{8.118}$$
$$\ddot{\theta} = -\varepsilon\ddot{\xi}\cos\theta + u, \tag{8.119}$$

where ξ is the normalized cart position and u represents the non-dimensionalized control torque. In the normalized equations (8.118) and (8.119), the symbol $(\dot{\cdot})$ represents differentiation with respect to the normalized time τ and the parameter ε represents the coupling between the translational and rotational motions and is defined by

$$\varepsilon \triangleq \frac{me}{\sqrt{(I+me^2)(M+m)}}. \tag{8.120}$$

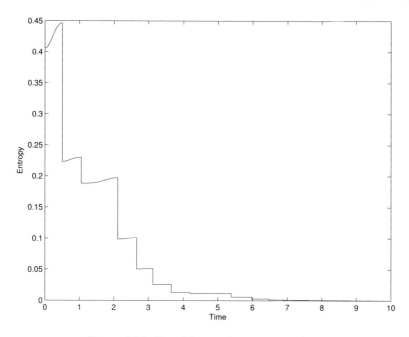

Figure 8.10 Closed-loop entropy versus time.

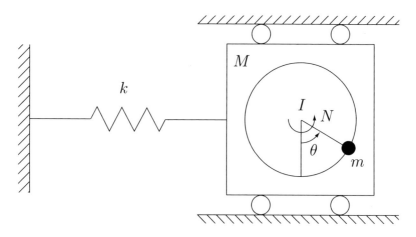

Figure 8.11 Rotational/translational proof-mass actuator.

Since the plant energy function (8.114) is not positive definite in $\mathbb{R}^4$, we first design a control law $u = -k_\theta \theta + \hat{u}$, where $k_\theta > 0$, with associated positive definite normalized plant energy function given by

$$V_{\mathrm{s}}(\xi, \dot{\xi}, \theta, \dot{\theta}) = \frac{1}{2}\xi^2 + \frac{1}{2}\dot{\xi}^2 + \frac{1}{2}k_\theta \theta^2 + \frac{1}{2}\dot{\theta}^2 + \varepsilon \dot{\xi}\dot{\theta}\cos\theta. \quad (8.121)$$

Description	Parameter	Value	Units
Cart mass	M	1.3608	kg
Arm mass	m	0.096	kg
Arm eccentricity	e	0.0592	m
Arm inertia	I	0.0002175	kg m^2
Spring stiffness	k	186.3	N/m
Coupling parameter	ε	0.200	—

Table 8.1 Problem data for the RTAC [36].

To design a state-dependent hybrid controller for (8.118) and (8.119), let $n_c = 1$, $V_c(\xi_c, \dot{\xi}_c, \theta) = \frac{1}{2} m_c \dot{\xi}_c^2 + \frac{1}{2} k_c (\xi_c - \theta)^2$, $\mathcal{L}_c(\xi_c, \dot{\xi}_c, \theta) = \frac{1}{2} m_c \dot{\xi}_c^2 - \frac{1}{2} k_c (\xi_c - \theta)^2$, $y_q = \theta$, and $\eta(y_q) = y_q$, where $m_c > 0$ and $k_c > 0$. Then the state-dependent hybrid controller has the form

$$m_c \ddot{\xi}_c + k_c (\xi_c - \theta) = 0, \quad (\xi_c, \dot{\xi}_c, \theta, \dot{\theta}) \notin \mathcal{Z}, \qquad (8.122)$$

$$\begin{bmatrix} \Delta \xi_c \\ \Delta \dot{\xi}_c \end{bmatrix} = \begin{bmatrix} \theta - \xi_c \\ -\dot{\xi}_c \end{bmatrix}, \quad (\xi_c, \dot{\xi}_c, \theta, \dot{\theta}) \in \mathcal{Z}, \qquad (8.123)$$

$$\hat{u} = k_c (\xi_c - \theta), \qquad (8.124)$$

with the resetting set (8.66) taking the form

$$\mathcal{Z} = \left\{ (\xi_c, \dot{\xi}_c, \theta, \dot{\theta}) \in \mathbb{R}^4 : k_c \dot{\theta}(\xi_c - \theta) = 0 \text{ and } \begin{bmatrix} \theta - \xi_c \\ -\dot{\xi}_c \end{bmatrix} \neq 0 \right\}.$$
$$(8.125)$$

To show that A1 holds, we show that upon reaching a nonequilibrium point $x(\tau) \triangleq [\xi(\tau), \dot{\xi}(\tau), \theta(\tau), \dot{\theta}(\tau), \xi_c(\tau), \dot{\xi}_c(\tau)]^{\mathrm{T}} \notin \mathcal{Z}$ that is in the closure of $\mathcal{Z}$, the continuous-time dynamics $\dot{x} = f_c(x)$ remove $x(\tau)$ from $\overline{\mathcal{Z}}$, and thus necessarily move the trajectory a finite distance away from $\mathcal{Z}$. If $x(\tau) \notin \mathcal{Z}$ is an equilibrium point, then $x(s) \notin \mathcal{Z}$, $s \geq \tau$, which is also consistent with A1.

The closure of $\mathcal{Z}$ is given by

$$\overline{\mathcal{Z}} = \left\{ (\xi_c, \dot{\xi}_c, \theta, \dot{\theta}) : k_c \dot{\theta}(\xi_c - \theta) \geq 0 \right\}. \qquad (8.126)$$

Furthermore, the points x^* satisfying $[\theta^* - \xi_c^*, -\dot{\xi}_c^*]^{\mathrm{T}} = 0$ have the form

$$x^* \triangleq \begin{bmatrix} \xi & \dot{\xi} & \theta & \dot{\theta} & \theta & 0 \end{bmatrix}^{\mathrm{T}}, \qquad (8.127)$$

that is, $\xi_c = \theta$ and $\dot{\xi}_c = 0$. It follows that $x^* \notin \mathcal{Z}$, although $x^* \in \overline{\mathcal{Z}}$.

To show that the continuous-time dynamics $\dot{x} = f_{\mathrm{c}}(x)$ remove x^* from $\overline{\mathcal{Z}}$, note that

$$\frac{\mathrm{d}}{\mathrm{d}\tau}V_{\mathrm{s}}(\xi, \dot{\xi}, \theta, \dot{\theta}) = k_{\mathrm{c}}\dot{\theta}(\xi_{\mathrm{c}} - \theta) \tag{8.128}$$

and

$$\frac{\mathrm{d}^2}{\mathrm{d}\tau^2}V_{\mathrm{s}}(\xi, \dot{\xi}, \theta, \dot{\theta}) = k_{\mathrm{c}}\ddot{\theta}(\xi_{\mathrm{c}} - \theta) + k_{\mathrm{c}}\dot{\theta}(\dot{\xi}_{\mathrm{c}} - \dot{\theta}), \tag{8.129}$$

$$\frac{\mathrm{d}^3}{\mathrm{d}\tau^3}V_{\mathrm{s}}(\xi, \dot{\xi}, \theta, \dot{\theta}) = k_{\mathrm{c}}\theta^{(3)}(\xi_{\mathrm{c}} - \theta) + 2k_{\mathrm{c}}\ddot{\theta}(\dot{\xi}_{\mathrm{c}} - \dot{\theta}) + k_{\mathrm{c}}\dot{\theta}(\ddot{\xi}_{\mathrm{c}} - \ddot{\theta}), \tag{8.130}$$

$$\frac{\mathrm{d}^4}{\mathrm{d}\tau^4}V_{\mathrm{s}}(\xi, \dot{\xi}, \theta, \dot{\theta}) = k_{\mathrm{c}}\theta^{(4)}(\xi_{\mathrm{c}} - \theta) + 3k_{\mathrm{c}}\theta^{(3)}(\dot{\xi}_{\mathrm{c}} - \dot{\theta}) + 3k_{\mathrm{c}}\ddot{\theta}(\ddot{\xi}_{\mathrm{c}} - \ddot{\theta})$$
$$+ k_{\mathrm{c}}\dot{\theta}(\xi_{\mathrm{c}}^{(3)} - \theta^{(3)}), \tag{8.131}$$

where $g^{(n)}(\tau) \triangleq \frac{\mathrm{d}^n g(\tau)}{\mathrm{d}\tau^n}$. Since

$$\left.\frac{\mathrm{d}^2}{\mathrm{d}\tau^2}V_{\mathrm{s}}(\xi, \dot{\xi}, \theta, \dot{\theta})\right|_{x=x^*} = -k_{\mathrm{c}}\dot{\theta}^2, \tag{8.132}$$

it follows that if $\dot{\theta} \neq 0$, then the continuous-time dynamics $\dot{x} = f_{\mathrm{c}}(x)$ remove x^* from $\overline{\mathcal{Z}}$. If $\dot{\theta} = 0$, then it follows from (8.129)–(8.131) that

$$\left.\frac{\mathrm{d}^2}{\mathrm{d}\tau^2}V_{\mathrm{s}}(\xi, \dot{\xi}, \theta, \dot{\theta})\right|_{x=x^*, \dot{\theta}=0} = 0, \tag{8.133}$$

$$\left.\frac{\mathrm{d}^3}{\mathrm{d}\tau^3}V_{\mathrm{s}}(\xi, \dot{\xi}, \theta, \dot{\theta})\right|_{x=x^*, \dot{\theta}=0} = 0, \tag{8.134}$$

$$\left.\frac{\mathrm{d}^4}{\mathrm{d}\tau^4}V_{\mathrm{s}}(\xi, \dot{\xi}, \theta, \dot{\theta})\right|_{x=x^*, \dot{\theta}=0} = -3k_{\mathrm{c}}\ddot{q}^2, \tag{8.135}$$

where in the evaluation of (8.134) and (8.135) we use the fact that if $\xi_{\mathrm{c}} = \theta$ and $\dot{\xi}_{\mathrm{c}} = 0$, then $\ddot{\xi}_{\mathrm{c}} = 0$, which follows immediately from the continuous-time dynamics. Since if $\dot{\theta} = 0$ and $\ddot{\theta} \neq 0$, then the lowest-order nonzero time derivative of $\dot{V}_{\mathrm{s}}(\xi, \dot{\xi}, \theta, \dot{\theta})$ is negative, it follows that the continuous-time dynamics remove x^* from $\overline{\mathcal{Z}}$. However, if $\dot{\theta} = 0$ and $\ddot{\theta} = 0$, then it follows from the continuous-time dynamics that x^* is necessarily an equilibrium point, in which case the trajectory never again enters $\mathcal{Z}$. Therefore, we can conclude that A1 is indeed valid for this system. Also, since $f_{\mathrm{d}}(x + f_{\mathrm{d}}(x)) = 0$, it follows from (8.125) that if $x \in \mathcal{Z}$, then $x + f_{\mathrm{d}}(x) \notin \mathcal{Z}$, and thus A2 holds.

For thermodynamic stabilization, the output y is modified as $y = [\xi, \dot{\xi}, \theta, \dot{\theta}]^{\mathrm{T}}$ and the resetting set (8.70) is given by

$$\mathcal{Z} = \Big\{ (\xi, \dot{\xi}, \theta, \dot{\theta}, \xi_{\mathrm{c}}, \dot{\xi}_{\mathrm{c}}) \in \mathbb{R}^6 : k_{\mathrm{c}}\dot{\theta}(\xi_{\mathrm{c}} - \theta)[V_{\mathrm{s}}(\xi, \dot{\xi}, \theta, \dot{\theta}) - V_{\mathrm{c}}(\xi_{\mathrm{c}}, \dot{\xi}_{\mathrm{c}}, \theta)]$$

$$= 0 \text{ and } \begin{bmatrix} \theta - \xi_c \\ -\dot{\xi}_c \end{bmatrix} \neq 0 \Bigg\}. \qquad (8.136)$$

Furthermore, the entropy function $S(E)$ is given by

$$S(E) = \log_e[1 + V_s(\xi, \dot{\xi}, \theta, \dot{\theta})] + \log_e[1 + V_c(\xi_c, \dot{\xi}_c, \theta)]. \quad (8.137)$$

To illustrate the behavior of the closed-loop impulsive dynamical system, let $m_c = 0.2$, $k_c = 1$, and $k_\theta = 1$ with initial conditions $\xi(0) = 1$, $\dot{\xi}(0) = 0$, $\theta(0) = 0$, $\dot{\theta}(0) = 0$, $\xi_c(0) = 0$, and $\dot{\xi}_c(0) = 0$. For thermodynamic stabilization, the initial conditions are given by $\xi(0) = 0.6$, $\dot{\xi}(0) = 0$, $\theta(0) = 0$, $\dot{\theta}(0) = 0$, $\xi_c(0) = 0.8$, and $\dot{\xi}_c(0) = 0$. For this system, the transversality condition is sufficiently complex that we have been unable to show it analytically. This condition was verified numerically, and hence, Assumption 8.1 holds. Figures 8.12 and 8.13 show the translational position of the cart and the angular position of the rotational proof mass versus time. Figure 8.14 shows the control torque versus time. Note that the compensator states are the only states that reset. Furthermore, the control torque versus time is discontinuous at the resetting times. A comparison of the plant energy, control energy, and total energy is shown in Figure 8.15. Figures 8.16–8.19 show analogous representations for the thermodynamically stabilizing compensator. Finally, Figure 8.20 shows the closed-loop system entropy versus time. Note that the entropy of the closed-loop system strictly increases between resetting events.

Our final example considers the design of a hybrid controller for the combustion system we considered in Section 6.3. Recall that this model is given by

$$\dot{x}_1(t) = \alpha_1 x_1(t) + \theta_1 x_2(t) - \beta(x_1(t)x_3(t) + x_2(t)x_4(t)) + u_1(t),$$
$$x_1(0) = x_{10}, \quad t \geq 0, \quad (8.138)$$
$$\dot{x}_2(t) = -\theta_1 x_1(t) + \alpha_1 x_2(t) + \beta(x_2(t)x_3(t) - x_1(t)x_4(t)) + u_2(t),$$
$$x_2(0) = x_{20}, \quad (8.139)$$
$$\dot{x}_3(t) = \alpha_2 x_3(t) + \theta_2 x_4(t) + \beta(x_1^2(t) - x_2^2(t)) + u_3(t), \quad x_3(0) = x_{30},$$
$$(8.140)$$
$$\dot{x}_4(t) = -\theta_2 x_3(t) + \alpha_2 x_4(t) + 2\beta x_1(t)x_2(t) + u_4(t), \quad x_4(0) = x_{40},$$
$$(8.141)$$

where $x \triangleq [x_1, x_2, x_3, x_4]^T \in \mathbb{R}^4$ is the plant state, $u \triangleq [u_1, u_2, u_3, u_4]^T \in \mathbb{R}^4$ is the control input, $i = 1, \ldots, 4$, $\alpha_1, \alpha_2 \in \mathbb{R}$ represent growth/decay constants, $\theta_1, \theta_2 \in \mathbb{R}$ represent frequency shift constants, $\beta = ((\gamma + 1)/8\gamma)\omega_1$, where γ denotes the ratio of specific heats, ω_1 is frequency of the fundamental mode, and u_i, $i = 1, \ldots, 4$, are control input signals. For the data parameters $\alpha_1 = 5$, $\alpha_2 = -55$, $\theta_1 = 4$, $\theta_2 = 32$, $\gamma = 1.4$, $\omega_1 = 1$, and $x(0) = [1, 1, 1, 1]^T$, the open-loop ($u_i(t) \equiv 0$, $i = 1, 2, 3, 4$) dynamics (8.138)–(8.141) result in a limit cycle instability. In addition, with the plant energy defined by $V_p(x) \triangleq \frac{1}{2}(x_1^2 + x_2^2 + x_3^2 + x_4^2)$, (8.138)–(8.141) is dissipative with respect to the supply rate $\hat{u}^T y$, where $\hat{u} \triangleq [u_1 + \alpha_1 x_1, u_2 + \alpha_1 x_2, u_3, u_4]^T$ and $y \triangleq x$.

Next, consider the reduced-order dynamic compensator given by (8.19)–(8.21) with $f_{cc}(x_c, y) = A_c x_c + B_c y$, $\eta(y) = 0$, $h_{cc}(x_c, y) = B_c^T x_c$, where $x_c \triangleq [x_{c1}, x_{c2}]^T \in \mathbb{R}^2$,

$$A_c = \begin{bmatrix} 0 & 1 \\ -1 & 0 \end{bmatrix}, \quad B_c = \begin{bmatrix} 0 & 0 & 0 & 0 \\ 4 & 0 & 0 & 0 \end{bmatrix}, \quad (8.142)$$

and controller energy given by $V_c(x_c) = \frac{1}{2}x_c^T x_c$. Furthermore, the resetting set (8.27) is given by $\mathcal{Z} = \{(x, x_c) : x_c^T B_c x = 0, \ x_c \neq 0\}$.

To illustrate the behavior of the closed-loop impulsive dynamical system, we choose the initial condition $x_c(0) = [0, 0]^T$. For this system a straightforward, but lengthy, calculation shows that A1 and A2 hold. However, the transversality condition is sufficiently complex that we have been unable to show it analytically. This condition was verified numerically and Assumption 8.1 appears to hold. Figure 8.21 shows the state trajectories of the plant versus time, while Figure 8.22 shows the state trajectories of the compensator versus time. Figure 8.23 shows the control inputs u_1 and u_2 versus time. Note that the compensator states are the only states that reset. Furthermore, the control force versus time is discontinuous at the resetting times. A comparison of the plant energy, controller energy, and total energy is shown in Figure 8.24. Note that the proposed energy-based hybrid controller achieves *finite-time stabilization*.

Next, we consider the case where $\alpha_1 = 0$ and $\alpha_2 = 0$, that is, there is no decay or growth in the system. The other system parameters remain as before. In this case, the system is lossless with respect to the supply rate $u^T y$. For this problem we consider an entropy-based hybrid dynamic compensator given by (8.19)–(8.21) with $f_{cc}(x_c, y) = A_c x_c + B_c y$, $\eta(y) = 0$, $h_{cc}(x_c, y) = B_c^T x_c$, where $x_c \triangleq [x_{c1}, x_{c2}, x_{c3}, x_{c4}]^T \in \mathbb{R}^4$,

$$A_c = \begin{bmatrix} 0 & 1 & 0 & 0 \\ -1 & 0 & 0 & 0 \\ 0 & 0 & 0 & 1 \\ 0 & 0 & -1 & 0 \end{bmatrix}, \quad B_c = \begin{bmatrix} 0 & -30 & 0 & 0 \\ 30 & 0 & 0 & 0 \\ 0 & 0 & 60 & 0 \\ 0 & 0 & 0 & 0 \end{bmatrix}, \quad (8.143)$$

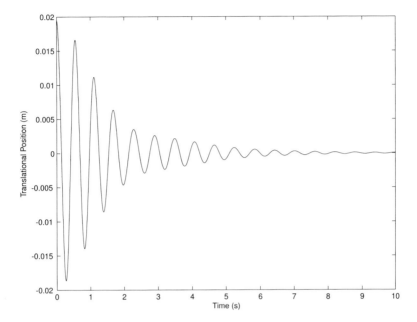

Figure 8.12 Translational position of the cart versus time.

and controller energy given by $V_c(x_c) = \frac{1}{2}x_c^T x_c$. Furthermore, the entropy function $S(E)$ is given by $S(E) = \log_e[1 + V_p(x)] + \log_e[1 + V_c(x_c)]$, and the resetting set (8.70) is given by

$$\mathcal{Z} = \left\{ (x, x_c) : x_c^T B_c x[V_c(x_c) - V_p(x)] = 0, \ x_c \neq 0 \right\}.$$

To illustrate the behavior of the closed-loop impulsive dynamical system, we choose initial condition $x_c(0) = [0, 0, 0, 0]^T$. Straightforward calculations show that Assumptions A1 and A2 hold. However, the transversality condition is sufficiently complex that we have been unable to show it analytically. This was verified numerically, and hence, Assumption 8.1 appears to hold. Figure 8.25 shows the state trajectories of the plant versus time, while Figure 8.26 shows the state trajectories of the compensator versus time. Figure 8.27 shows the control input versus time. Note that the compensator states are the only states that reset. Furthermore, the control force versus time is discontinuous at the resetting times. A comparison of the plant energy, controller energy, and total energy is shown in Figure 8.28. Finally, Figure 8.29 shows the closed-loop system entropy versus time. Note that the entropy of the closed-loop system strictly increases between resetting events.

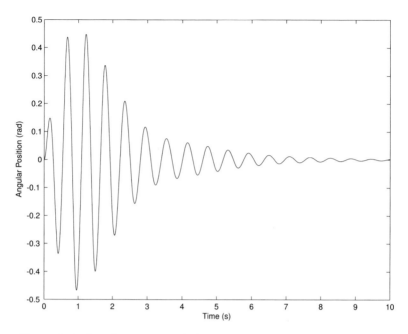

Figure 8.13 Angular position of the rotational proof mass versus time.

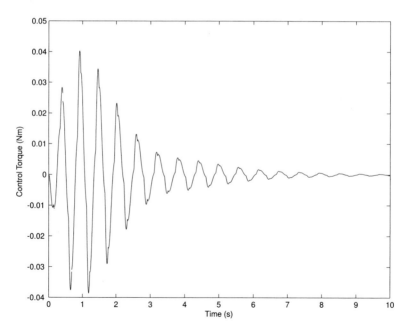

Figure 8.14 Control torque versus time.

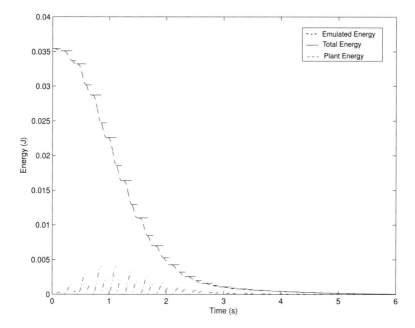

Figure 8.15 Plant, emulated, and total energy versus time.

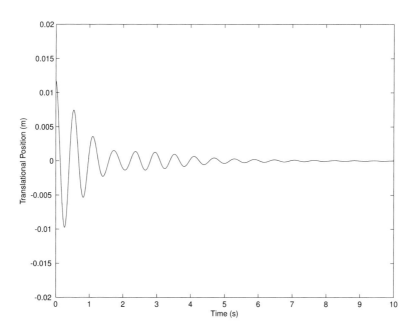

Figure 8.16 Translational position of the cart versus time for thermodynamic controller.

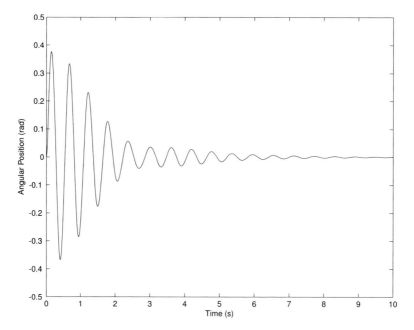

Figure 8.17 Angular position of the rotational proof mass versus time for
            thermodynamic controller.

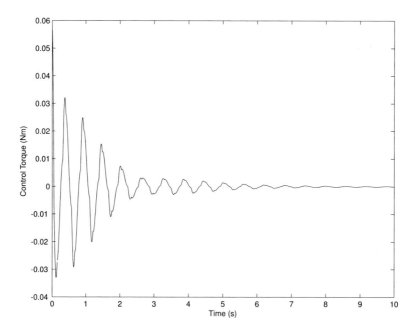

Figure 8.18 Control torque versus time for thermodynamic controller.

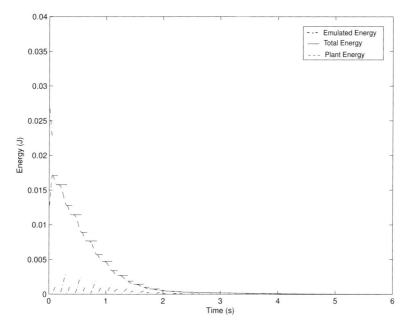

Figure 8.19 Plant, emulated, and total energy versus time for thermodynamic controller.

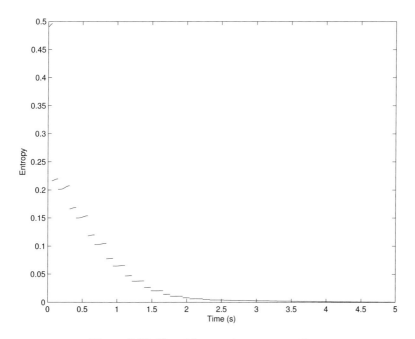

Figure 8.20 Closed-loop entropy versus time.

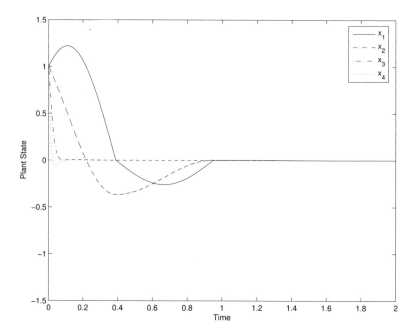

Figure 8.21 Plant state trajectories versus time.

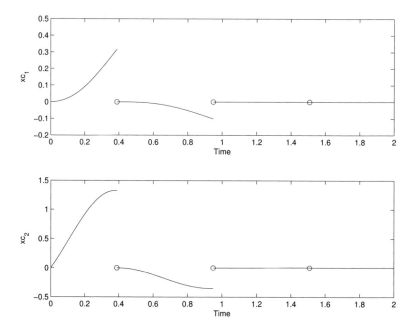

Figure 8.22 Compensator state trajectories versus time.

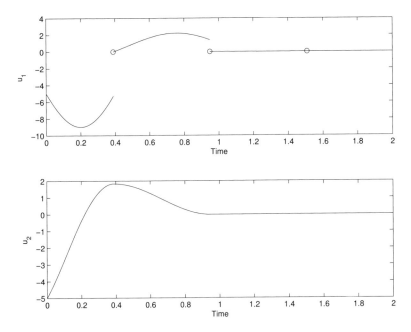

Figure 8.23 Control signal versus time.

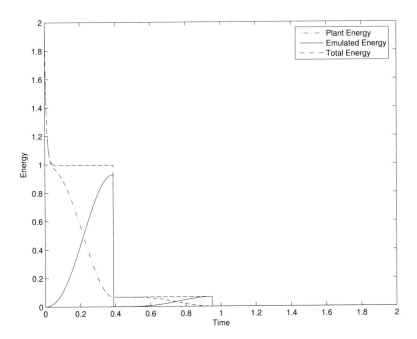

Figure 8.24 Plant, emulated, and total energy versus time.

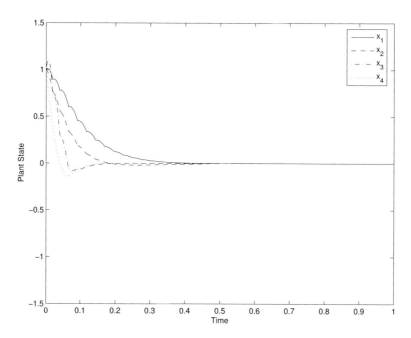

Figure 8.25 Plant state trajectories versus time for thermodynamic controller.

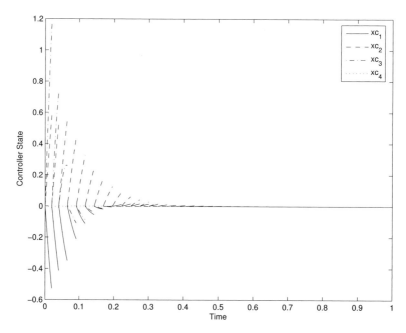

Figure 8.26 Compensator state trajectories versus time for thermodynamic controller.

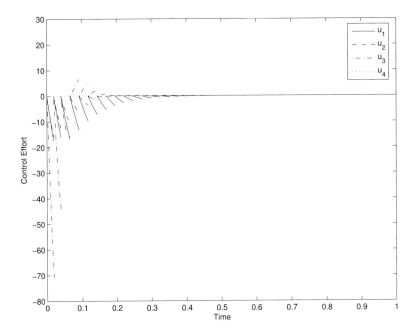

Figure 8.27 Control input versus time for thermodynamic controller.

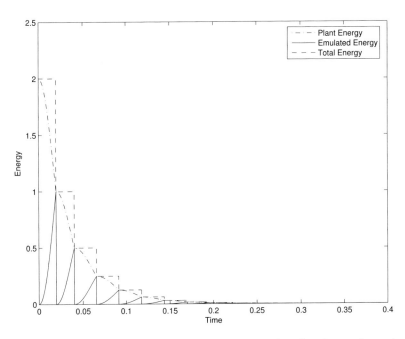

Figure 8.28 Plant, emulated, and total energy versus time for thermodynamic controller.

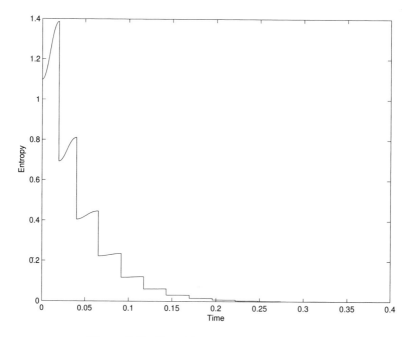

Figure 8.29 Closed-loop entropy versus time.

8.8 Energy-Dissipating Hybrid Control for Impulsive Dynamical Systems

In this section, we extend the results of Section 8.3 to lossless impulsive dynamical systems. Extensions to dissipative impulsive dynamical systems can be trivially addressed as in Section 8.3. We begin by considering the controlled impulsive dynamical systems of the form

$$\dot{x}_{\mathrm{p}}(t) = f_{\mathrm{cp}}(x_{\mathrm{p}}(t), u_{\mathrm{c}}(t)), \quad x_{\mathrm{p}}(0) = x_{\mathrm{p}0}, \quad (x_{\mathrm{p}}(t), u_{\mathrm{c}}(t)) \notin \mathcal{Z}_{\mathrm{p}},$$
$$(8.144)$$
$$\Delta x_{\mathrm{p}}(t) = f_{\mathrm{dp}}(x_{\mathrm{p}}(t), u_{\mathrm{d}}(t)), \quad (x_{\mathrm{p}}(t), u_{\mathrm{c}}(t)) \in \mathcal{Z}_{\mathrm{p}}, \qquad (8.145)$$
$$y(t) = h_{\mathrm{p}}(x_{\mathrm{p}}(t)), \qquad (8.146)$$

where $t \geq 0$, $x_{\mathrm{p}}(t) \in \mathcal{D}_{\mathrm{p}} \subseteq \mathbb{R}^{n_{\mathrm{p}}}$, $\mathcal{D}_{\mathrm{p}}$ is an open set with $0 \in \mathcal{D}_{\mathrm{p}}$, $\Delta x_{\mathrm{p}}(t) \triangleq x_{\mathrm{p}}(t^{+}) - x_{\mathrm{p}}(t)$, $u_{\mathrm{c}}(t) \in \mathbb{R}^{m_{\mathrm{c}}}$, $u_{\mathrm{d}}(t) \in \mathbb{R}^{m_{\mathrm{d}}}$, $f_{\mathrm{cp}} : \mathcal{D}_{\mathrm{p}} \times \mathbb{R}^{m_{\mathrm{c}}} \rightarrow \mathbb{R}^{n_{\mathrm{p}}}$ is smooth on $\mathcal{D}_{\mathrm{p}}$ and satisfies $f_{\mathrm{cp}}(0,0) = 0$, $f_{\mathrm{dp}} : \mathcal{D}_{\mathrm{p}} \times \mathbb{R}^{m_{\mathrm{d}}} \rightarrow \mathbb{R}^{n_{\mathrm{p}}}$ is continuous, $h_{\mathrm{p}} : \mathcal{D}_{\mathrm{p}} \rightarrow \mathbb{R}^{l}$ is continuous and satisfies $h_{\mathrm{p}}(0) = 0$, and $\mathcal{Z}_{\mathrm{p}} \subset \mathcal{D}_{\mathrm{p}} \times \mathbb{R}^{m_{\mathrm{c}}}$ is the resetting set. Furthermore, we consider hybrid (resetting) dynamic controllers of the form

$$\dot{x}_{\mathrm{c}}(t) = f_{\mathrm{cc}}(x_{\mathrm{c}}(t), y(t)), \quad x_{\mathrm{c}}(0) = x_{\mathrm{c}0}, \quad (x_{\mathrm{c}}(t), y(t)) \notin \mathcal{Z}_{\mathrm{c}}, \quad (8.147)$$

$$\Delta x_{\mathrm{c}}(t) = f_{\mathrm{dc}}(x_{\mathrm{c}}(t), y(t)), \quad (x_{\mathrm{c}}(t), y(t)) \in \mathcal{Z}_{\mathrm{c}}, \tag{8.148}$$
$$u_{\mathrm{c}}(t) = h_{\mathrm{cc}}(x_{\mathrm{c}}(t), y(t)), \tag{8.149}$$
$$u_{\mathrm{d}}(t) = h_{\mathrm{dc}}(x_{\mathrm{c}}(t), y(t)), \tag{8.150}$$

where $t \geq 0$, $x_{\mathrm{c}}(t) \in \mathcal{D}_{\mathrm{c}} \subseteq \mathbb{R}^{n_{\mathrm{c}}}$, $\mathcal{D}_{\mathrm{c}}$ is an open set with $0 \in \mathcal{D}_{\mathrm{c}}$, $\Delta x_{\mathrm{c}}(t) \triangleq x_{\mathrm{c}}(t^+) - x_{\mathrm{c}}(t)$, $f_{\mathrm{cc}} : \mathcal{D}_{\mathrm{c}} \times \mathbb{R}^l \rightarrow \mathbb{R}^{n_{\mathrm{c}}}$ is smooth on $\mathcal{D}_{\mathrm{c}}$ and satisfies $f_{\mathrm{cc}}(0,0) = 0$, $f_{\mathrm{dc}} : \mathcal{D}_{\mathrm{c}} \times \mathbb{R}^l \rightarrow \mathbb{R}^{n_{\mathrm{c}}}$ is continuous, $h_{\mathrm{cc}} : \mathcal{D}_{\mathrm{c}} \times \mathbb{R}^l \rightarrow \mathbb{R}^{m_{\mathrm{c}}}$ is continuous and satisfies $h_{\mathrm{cc}}(0,0) = 0$, $h_{\mathrm{dc}} : \mathcal{D}_{\mathrm{c}} \times \mathbb{R}^l \rightarrow \mathbb{R}^{m_{\mathrm{d}}}$ is continuous, and $\mathcal{Z}_{\mathrm{c}} \subset \mathcal{D}_{\mathrm{c}} \times \mathbb{R}^l$ is the resetting set.

The equations of motion for the closed-loop impulsive dynamical system (8.144)–(8.150) have the form

$$\dot{x}(t) = f_{\mathrm{c}}(x(t)), \quad x(0) = x_0, \quad x(t) \notin \mathcal{Z}, \tag{8.151}$$
$$\Delta x(t) = f_{\mathrm{d}}(x(t)), \quad x(t) \in \mathcal{Z}, \tag{8.152}$$

where

$$x \triangleq \begin{bmatrix} x_{\mathrm{p}} \\ x_{\mathrm{c}} \end{bmatrix} \in \mathbb{R}^n, \quad f_{\mathrm{c}}(x) \triangleq \begin{bmatrix} f_{\mathrm{cp}}(x_{\mathrm{p}}, h_{\mathrm{cc}}(x_{\mathrm{c}}, h_{\mathrm{p}}(x_{\mathrm{p}}))) \\ f_{\mathrm{cc}}(x_{\mathrm{c}}, h_{\mathrm{p}}(x_{\mathrm{p}})) \end{bmatrix}, \tag{8.153}$$

$$f_{\mathrm{d}}(x) \triangleq \begin{bmatrix} f_{\mathrm{dp}}(x_{\mathrm{p}}, h_{\mathrm{dc}}(x_{\mathrm{c}}, h_{\mathrm{p}}(x_{\mathrm{p}}))) \chi_{\mathcal{Z}_1}(x) \\ f_{\mathrm{dc}}(x_{\mathrm{c}}, h_{\mathrm{p}}(x_{\mathrm{p}})) \chi_{\mathcal{Z}_2}(x) \end{bmatrix},$$

$$\chi_{\mathcal{Z}_i}(x) \triangleq \begin{cases} 1, & x \in \mathcal{Z}_i, \\ 0, & x \notin \mathcal{Z}_i, \end{cases} \quad i = 1, 2, \tag{8.154}$$

and $\mathcal{Z} \triangleq \mathcal{Z}_1 \cup \mathcal{Z}_2$, $\mathcal{Z}_1 \triangleq \{x \in \mathcal{D} : (x_{\mathrm{p}}, h_{\mathrm{cc}}(x_{\mathrm{c}}, h_{\mathrm{p}}(x_{\mathrm{p}}))) \in \mathcal{Z}_{\mathrm{p}}\}$, $\mathcal{Z}_2 \triangleq \{x \in \mathcal{D} : (x_{\mathrm{c}}, h_{\mathrm{p}}(x_{\mathrm{p}})) \in \mathcal{Z}_{\mathrm{c}}\}$, with $n \triangleq n_{\mathrm{p}} + n_{\mathrm{c}}$ and $\mathcal{D} \triangleq \mathcal{D}_{\mathrm{p}} \times \mathcal{D}_{\mathrm{c}}$. To ensure well-posedness of the resetting times, we assume that Assumptions A1 and A2 hold.

The following definition and proposition is a generalization of Definition 8.1 and Proposition 8.2.

Definition 8.3 *Let* $\mathcal{M} \triangleq \{x \in \mathcal{D} : \mathcal{X}_{\mathrm{p}}(x) = 0\} \cup \{x \in \mathcal{D} : \mathcal{X}_{\mathrm{c}}(x) = 0\}$, *where* $\mathcal{X}_{\mathrm{p}} : \mathcal{D} \rightarrow \mathbb{R}$ *and* $\mathcal{X}_{\mathrm{c}} : \mathcal{D} \rightarrow \mathbb{R}$ *are infinitely differentiable functions. A point* $x \in \mathcal{M}$ *such that* $f_{\mathrm{c}}(x) \neq 0$ *is transversal to (8.151) if there exist* $k_{\mathrm{p}} \in \{1, 2, \ldots\}$ *and* $k_{\mathrm{c}} \in \{1, 2, \ldots\}$ *such that*

$$L_{f_{\mathrm{c}}}^r \mathcal{X}_{\mathrm{p}}(x) = 0, \quad r = 0, \ldots, 2k_{\mathrm{p}} - 2, \quad L_{f_{\mathrm{c}}}^{2k_{\mathrm{p}}-1} \mathcal{X}_{\mathrm{p}}(x) \neq 0, \tag{8.155}$$
$$L_{f_{\mathrm{c}}}^r \mathcal{X}_{\mathrm{c}}(x) = 0, \quad r = 0, \ldots, 2k_{\mathrm{c}} - 2, \quad L_{f_{\mathrm{c}}}^{2k_{\mathrm{c}}-1} \mathcal{X}_{\mathrm{c}}(x) \neq 0. \tag{8.156}$$

Proposition 8.3 *Consider the impulsive dynamical system (8.151) and (8.152). Let* $\mathcal{X}_{\mathrm{p}} : \mathcal{D} \rightarrow \mathbb{R}$ *and* $\mathcal{X}_{\mathrm{c}} : \mathcal{D} \rightarrow \mathbb{R}$ *be infinitely differentiable functions such that* $\overline{\mathcal{Z}} = \{x \in \mathcal{D} : \mathcal{X}_{\mathrm{p}}(x) = 0\} \cup \{x \in \mathcal{D} : \mathcal{X}_{\mathrm{c}}(x) = $

$0\}$, *and assume every* $x \in \overline{\mathcal{Z}}$ *is transversal to (8.151). Then at every* $x_0 \notin \overline{\mathcal{Z}}$ *such that* $0 < \tau_1(x_0) < \infty$, $\tau_1(\cdot)$ *is continuous. Furthermore, if* $x_0 \in \overline{\mathcal{Z}} \backslash \mathcal{Z}$ *is such that* $\tau_1(x_0) \in (0, \infty)$ *and* $\{x_i\}_{i=1}^{\infty} \in \overline{\mathcal{Z}} \backslash \mathcal{Z}$ *or* $\lim_{i \to \infty} \tau_1(x_i) > 0$, *where* $\{x_i\}_{i=1}^{\infty} \notin \overline{\mathcal{Z}}$ *is such that* $\lim_{i \to \infty} x_i = x_0$ *and* $\lim_{i \to \infty} \tau_1(x_i)$ *exists, then* $\lim_{i \to \infty} \tau_1(x_i) = \tau_1(x_0)$.

Proof. The proof is similar to the proof of Proposition 8.2 and, hence, is omitted. $\qquad\qquad\qquad\qquad\qquad\qquad\qquad\qquad\qquad\square$

Next, we present a hybrid controller design framework for lossless impulsive dynamical systems. Specifically, we consider impulsive dynamical systems $\mathcal{G}_p$ of the form given by (8.144)–(8.146) where $u(\cdot)$ satisfies sufficient regularity conditions such that (8.144) has a unique solution between the resetting times. Furthermore, we consider hybrid resetting dynamic controllers $\mathcal{G}_c$ of the form

$$\dot{x}_c(t) = f_{cc}(x_c(t), y(t)), \quad x_c(0) = x_{c0}, \quad (x_c(t), y(t)) \notin \mathcal{Z}_c, \quad (8.157)$$

$$\Delta x_c(t) = \eta(y(t)) - x_c(t), \quad (x_c(t), y(t)) \in \mathcal{Z}_c, \quad\quad\quad\quad (8.158)$$

$$y_{cc}(t) = h_{cc}(x_c(t), y(t)), \quad\quad\quad\quad\quad\quad\quad\quad\quad\quad\quad (8.159)$$

$$y_{dc}(t) = h_{dc}(x_c(t), y(t)), \quad\quad\quad\quad\quad\quad\quad\quad\quad\quad\quad (8.160)$$

where $x_c(t) \in \mathcal{D}_c \subseteq \mathbb{R}^{n_c}$, $\mathcal{D}_c$ is an open set with $0 \in \mathcal{D}_c$, $y(t) \in \mathbb{R}^l$, $y_{cc}(t) \in \mathbb{R}^{m_c}$, $y_{dc}(t) \in \mathbb{R}^{m_d}$, $f_{cc} : \mathcal{D}_c \times \mathbb{R}^l \to \mathbb{R}^{n_c}$ is smooth on $\mathcal{D}_c$ and satisfies $f_{cc}(0,0) = 0$, $\eta : \mathbb{R}^l \to \mathcal{D}_c$ is continuous and satisfies $\eta(0) = 0$, $h_{cc} : \mathcal{D}_c \times \mathbb{R}^l \to \mathbb{R}^{m_c}$ is continuous and satisfies $h_{cc}(0,0) = 0$, and $h_{dc} : \mathcal{D}_c \times \mathbb{R}^l \to \mathbb{R}^{m_d}$ is continuous.

We assume that $\mathcal{G}_p$ is lossless with respect to the hybrid supply rate $(s_c(u_c, y), s_d(u_d, y))$, and hence, there exists a continuous, nonnegative-definite storage function $V_s : \mathcal{D}_p \to \mathbb{R}$ such that $V_s(0) = 0$ and

$$V_s(x_p(t)) = V_s(x_p(t_0)) + \int_{t_0}^{t} s_c(u_c(\sigma), y(\sigma)) d\sigma$$

$$+ \sum_{k \in \mathbb{Z}_{[t, t_0)}} s_d(u_d(t_k), y(t_k)), \quad t \geq t_0, \quad (8.161)$$

for all $t_0, t \geq 0$, where $x_p(t)$, $t \geq t_0$, is the solution to (8.144) and (8.145) with $(u_c, u_d) \in \mathcal{U}_c \times \mathcal{U}_d$. Equivalently, it follows from Theorem 3.2 that over the interval $t \in (t_k, t_{k+1}]$, (8.161) can be written as

$$V_s(x_p(\hat{t})) - V_s(x_p(t)) = \int_{t}^{\hat{t}} s_c(u_c(\sigma), y(\sigma)) d\sigma, \quad t_k < t \leq \hat{t} \leq t_{k+1},$$

$$k \in \overline{\mathbb{Z}}_+, \quad (8.162)$$

$$V_s(x_p(t_k) + f_{dp}(x_p(t_k), u_d(t_k))) - V_s(x_p(t_k)) = s_d(u_d(t_k), y(t_k)),$$
$$k \in \overline{\mathbb{Z}}_+. \quad (8.163)$$

In addition, we assume that the nonlinear impulsive dynamical system $\mathcal{G}_p$ is completely reachable and zero-state observable, and there exist functions $\kappa_c : \mathbb{R}^l \rightarrow \mathbb{R}^{m_c}$ and $\kappa_d : \mathbb{R}^l \rightarrow \mathbb{R}^{m_d}$ such that $\kappa_c(0) = 0$, $\kappa_d(0) = 0$, $s_c(\kappa_c(y), y) < 0$, $y \neq 0$, and $s_d(\kappa_d(y), y) < 0$, $y \neq 0$, so that all storage functions $V_s(x_p)$, $x_p \in \mathcal{D}_p$, of $\mathcal{G}_p$ are positive definite. Finally, we assume that $V_s(\cdot)$ is continuously differentiable.

Next, consider the negative feedback interconnection of $\mathcal{G}_p$ and $\mathcal{G}_c$ given by $y = u_{cc}$ and $(u_c, u_d) = (-y_{cc}, -y_{dc})$. In this case, the closed-loop system $\mathcal{G}$ is given by

$$\dot{x}(t) = f_c(x(t)), \quad x(0) = x_0, \quad x(t) \notin \mathcal{Z}, \quad t \geq 0, \quad (8.164)$$
$$\Delta x(t) = f_d(x(t)), \quad x(t) \in \mathcal{Z}, \quad (8.165)$$

where $t \geq 0$, $x(t) \triangleq [x_p^T(t), x_c^T(t)]^T$, $\mathcal{Z} \triangleq \mathcal{Z}_1 \cup \mathcal{Z}_2$, $\mathcal{Z}_1 \triangleq \{x \in \mathcal{D} : (x_p, -h_{cc}(x_c, h_p(x_p))) \in \mathcal{Z}_p\}$, $\mathcal{Z}_2 \triangleq \{x \in \mathcal{D} : (x_c, h_p(x_p)) \in \mathcal{Z}_c\}$,

$$f_c(x) \triangleq \begin{bmatrix} f_{cp}(x_p, -h_{cc}(x_c, h_p(x_p))) \\ f_{cc}(x_c, h_p(x_p)) \end{bmatrix},$$
$$f_d(x) \triangleq \begin{bmatrix} f_{dp}(x_p, -h_{dc}(x_c, h_p(x_p)))\chi_{\mathcal{Z}_1}(x) \\ (\eta(h_p(x_p)) - x_c)\chi_{\mathcal{Z}_2}(x) \end{bmatrix}. \quad (8.166)$$

Assume that there exists an infinitely differentiable function $V_c : \mathcal{D}_c \times \mathbb{R}^l \rightarrow \overline{\mathbb{R}}_+$ such that $V_c(x_c, y) \geq 0$, $x_c \in \mathcal{D}_c$, $y \in \mathbb{R}^l$, $V_c(x_c, y) = 0$ if and only if $x_c = \eta(y)$, and

$$\dot{V}_c(x_c(t), y(t)) = s_{cc}(u_{cc}(t), y_{cc}(t)), \quad (x_c(t), y(t)) \notin \mathcal{Z}_c, \quad t \geq 0, \quad (8.167)$$

where $s_{cc} : \mathbb{R}^l \times \mathbb{R}^{m_c} \rightarrow \mathbb{R}$ is such that $s_{cc}(0, 0) = 0$ and is locally integrable for all input-output pairs satisfying (8.157)–(8.160).

As in Section 8.3, we associate with the plant a positive-definite, continuously differentiable function $V_p(x_p) \triangleq V_s(x_p)$, which we will refer to as the plant energy. Furthermore, we associate with the controller a nonnegative-definite, infinitely-differentiable function $V_c(x_c, y)$ called the controller emulated energy. Finally, we associate with the closed-loop system the function

$$V(x) \triangleq V_p(x_p) + V_c(x_c, h_p(x_p)), \quad (8.168)$$

called the total energy.

Next, we construct the resetting set for $\mathcal{G}_{\mathrm{c}}$ in the following form

$$
\begin{aligned}
\mathcal{Z}_2 &= \{(x_{\mathrm{p}}, x_{\mathrm{c}}) \in \mathcal{D}_{\mathrm{p}} \times \mathcal{D}_{\mathrm{c}} : L_{f_{\mathrm{c}}} V_{\mathrm{c}}(x_{\mathrm{c}}, h_{\mathrm{p}}(x_{\mathrm{p}})) = 0 \\
&\quad \text{and } V_{\mathrm{c}}(x_{\mathrm{c}}, h_{\mathrm{p}}(x_{\mathrm{p}})) > 0\} \\
&= \{(x_{\mathrm{p}}, x_{\mathrm{c}}) \in \mathcal{D}_{\mathrm{p}} \times \mathcal{D}_{\mathrm{c}} : s_{\mathrm{cc}}(h_{\mathrm{p}}(x_{\mathrm{p}}), h_{\mathrm{cc}}(x_{\mathrm{c}}, h_{\mathrm{p}}(x_{\mathrm{p}}))) = 0 \\
&\quad \text{and } V_{\mathrm{c}}(x_{\mathrm{c}}, h_{\mathrm{p}}(x_{\mathrm{p}})) > 0\} \, .
\end{aligned}
\tag{8.169}
$$

The resetting set $\mathcal{Z}_2$ is thus defined to be the set of all points in the closed-loop state space that correspond to decreasing controller emulated energy. By resetting the controller states, the plant energy can never increase after the first resetting event. Furthermore, if the closed-loop system total energy is conserved between resetting events, then a decrease in plant energy is accompanied by a corresponding increase in emulated energy. Hence, this approach allows the plant energy to flow to the controller, where it increases the emulated energy but does not allow the emulated energy to flow back to the plant after the first resetting event. For practical implementation, knowledge of x_{c} and y is sufficient to determine whether or not the closed-loop state vector is in the set $\mathcal{Z}_2$. The next theorem gives sufficient conditions for asymptotic stability of the closed-loop system $\mathcal{G}$ using state-dependent hybrid controllers.

Theorem 8.7 *Consider the closed-loop impulsive dynamical system $\mathcal{G}$ given by (8.164) and (8.165) with the resetting set $\mathcal{Z}_2$ given by (8.169). Assume that $\mathcal{D}_{\mathrm{ci}} \subset \mathcal{D}$ is a compact positively invariant set with respect to $\mathcal{G}$ such that $0 \in \overset{\circ}{\mathcal{D}}_{\mathrm{ci}}$, assume that if $x_0 \in \mathcal{Z}_1$ then $x_0 + f_{\mathrm{d}}(x_0) \in \overline{\mathcal{Z}}_1 \backslash \mathcal{Z}_1$, and if $x_0 \in \overline{\mathcal{Z}}_1 \backslash \mathcal{Z}_1$, then $f_{\mathrm{dp}}(x_{\mathrm{p}0}, -h_{\mathrm{dc}}(x_{\mathrm{c}0}, h_{\mathrm{p}}(x_{\mathrm{p}0}))) = 0$, where $\overline{\mathcal{Z}}_1 = \{x \in \mathcal{D} : \mathcal{X}_{\mathrm{p}}(x) = 0\}$ with an infinitely differentiable function $\mathcal{X}_{\mathrm{p}}(\cdot)$, and assume that $\mathcal{G}_{\mathrm{p}}$ is lossless with respect to the hybrid supply rate $(s_{\mathrm{c}}(u_{\mathrm{c}}, y), s_{\mathrm{d}}(u_{\mathrm{d}}, y))$ and with a positive-definite, continuously differentiable storage function $V_{\mathrm{p}}(x_{\mathrm{p}})$, $x_{\mathrm{p}} \in \mathcal{D}_{\mathrm{p}}$. In addition, assume there exists a smooth function $V_{\mathrm{c}} : \mathcal{D}_{\mathrm{c}} \times \mathbb{R}^l \to \overline{\mathbb{R}}_+$ such that $V_{\mathrm{c}}(x_{\mathrm{c}}, y) \geq 0$, $x_{\mathrm{c}} \in \mathcal{D}_{\mathrm{c}}$, $y \in \mathbb{R}^l$, $V_{\mathrm{c}}(x_{\mathrm{c}}, y) = 0$ if and only if $x_{\mathrm{c}} = \eta(y)$, and (8.167) holds. Furthermore, assume that every $x_0 \in \overline{\mathcal{Z}}$ is transversal to (8.164) with $\mathcal{X}_{\mathrm{c}}(x) = L_{f_{\mathrm{c}}} V_{\mathrm{c}}(x_{\mathrm{c}}, h_{\mathrm{p}}(x_{\mathrm{p}}))$, and*

$$
\begin{aligned}
s_{\mathrm{c}}(u_{\mathrm{c}}, y) + s_{\mathrm{cc}}(u_{\mathrm{cc}}, y_{\mathrm{cc}}) &= 0, \quad x \notin \mathcal{Z}, \tag{8.170} \\
s_{\mathrm{d}}(u_{\mathrm{d}}, y) &< 0, \quad x \in \mathcal{Z}_1, \tag{8.171}
\end{aligned}
$$

where $y = u_{\mathrm{cc}} = h_{\mathrm{p}}(x_{\mathrm{p}})$, $u_{\mathrm{c}} = -y_{\mathrm{cc}} = -h_{\mathrm{cc}}(x_{\mathrm{c}}, h_{\mathrm{p}}(x_{\mathrm{p}}))$, and $y_{\mathrm{d}} = -y_{\mathrm{dc}} = -h_{\mathrm{dc}}(x_{\mathrm{c}}, h_{\mathrm{p}}(x_{\mathrm{p}}))$. Then the zero solution $x(t) \equiv 0$ to the closed-loop system $\mathcal{G}$ is asymptotically stable. In addition, the total energy function $V(x)$ of $\mathcal{G}$ given by (8.168) is strictly decreasing *across resetting events. Finally, if $\mathcal{D}_p = \mathbb{R}^{n_p}$, $\mathcal{D}_c = \mathbb{R}^{n_c}$, and $V(\cdot)$ is radially unbounded, then the zero solution $x(t) \equiv 0$ to $\mathcal{G}$ is globally asymptotically stable.*

Proof. First, note that since $V_c(x_c, y) \geq 0$, $x_c \in \mathcal{D}_c$, $y \in \mathbb{R}^l$, it follows that

$$\overline{\mathcal{Z}} = \overline{\mathcal{Z}}_1 \cup \{(x_p, x_c) \in \mathcal{D}_p \times \mathcal{D}_c : L_{f_c} V_c(x_c, h_p(x_p)) = 0$$
$$\text{and } V_c(x_c, h_p(x_p)) \geq 0\}$$
$$= \overline{\mathcal{Z}}_1 \cup \{(x_p, x_c) \in \mathcal{D}_p \times \mathcal{D}_c : \mathcal{X}_c(x) = 0\}, \qquad (8.172)$$

where $\mathcal{X}_c(x) = L_{f_c} V_c(x_c, h_p(x_p))$. Next, we show that if the transversality condition (8.155) and (8.156) holds, then Assumptions A1, A2, and 8.1 hold, and, for every $x_0 \in \mathcal{D}_{ci}$, there exists $\tau \geq 0$ such that $x(\tau) \in \mathcal{Z}$. Note that if $x_0 \in \overline{\mathcal{Z}} \backslash \mathcal{Z}$, that is, $\mathcal{X}_p(x(0)) = 0$ or $V_c(x_c(0), h_p(x(0))) = 0$ and $L_{f_c} V_c(x_c(0), h_p(x_p(0))) = 0$, it follows from the transversality condition that there exists $\delta > 0$ such that for all $t \in (0, \delta]$, $\mathcal{X}_p(x(t)) \neq 0$ and $L_{f_c} V_c(x_c(t), h_p(x_p(t))) \neq 0$. Hence, since $V_c(x_c(t), h_p(x_p(t))) = V_c(x_c(0), h_p(x_p(0))) + t L_{f_c} V_c(x_c(\tau), h_p(x_p(\tau)))$ for some $\tau \in (0, t]$ and $V_c(x_c, y) \geq 0$, $x_c \in \mathcal{D}_c$, $y \in \mathbb{R}^l$, it follows that $V_c(x_c(t), h_p(x_p(t))) > 0$, $t \in (0, \delta]$, which implies that A1 is satisfied. Furthermore, if $x \in \mathcal{Z}$ then, since $V_c(x_c, y) = 0$ if and only if $x_c = \eta(y)$, it follows from (8.165) that $x + f_d(x) \in \overline{\mathcal{Z}}_2 \backslash \mathcal{Z}_2$, and hence, $x + f_d(x) \in \overline{\mathcal{Z}} \backslash \mathcal{Z}$. Hence, A2 holds. Assumption 8.1 and the fact that for every $x_0 \notin \mathcal{Z}$, $x_0 \neq 0$, there exists $\tau > 0$ such that $x(\tau) \in \mathcal{Z}$ follow as in the proof of Theorem 8.3 with Proposition 8.3 invoked in the place of Proposition 8.2.

To show that the zero solution $x(t) \equiv 0$ to $\mathcal{G}$ is asymptotically stable, consider the Lyapunov function candidate corresponding to the total energy function $V(x)$ given by (8.168). Since $\mathcal{G}_p$ is lossless with respect to the hybrid supply rate $(s_c(u_c, y), s_d(u_d, y))$ and (8.167) and (8.170) hold, it follows that

$$\dot{V}(x(t)) = s_c(u_c(t), y(t)) + s_{cc}(u_{cc}(t), y_{cc}(t)) = 0, \quad x(t) \notin \mathcal{Z}. \qquad (8.173)$$

Furthermore, it follows from (8.163), (8.166), and (8.169) that

$$\Delta V(x(t_k)) = V_p(x_p(t_k^+)) - V_p(x_p(t_k))$$
$$+ V_c(x_c(t_k^+), h_p(x_p(t_k^+))) - V_c(x_c(t_k), h_p(x_p(t_k)))$$
$$= s_d(u_d(t_k), y(t_k)) \chi_{\mathcal{Z}_1}(x(t_k))$$
$$+ [V_c(\eta(h_p(x_p(t_k))), h_p(x_p(t_k)))$$

$$-V_{\mathrm{c}}(x_{\mathrm{c}}(t_k), h_{\mathrm{p}}(x_{\mathrm{p}}(t_k)))]\chi_{\mathcal{Z}_2}(x(t_k))$$
$$= s_{\mathrm{d}}(u_{\mathrm{d}}(t_k), y(t_k))\chi_{\mathcal{Z}_1}(x(t_k))$$
$$\qquad -V_{\mathrm{c}}(x_{\mathrm{c}}(t_k), h_{\mathrm{p}}(x_{\mathrm{p}}(t_k)))\chi_{\mathcal{Z}_2}(x(t_k))$$
$$< 0, \quad x(t_k) \in \mathcal{Z}, \quad k \in \overline{\mathbb{Z}}_+. \qquad (8.174)$$

Thus, it follows from Theorem 8.2 that the zero solution $x(t) \equiv 0$ to $\mathcal{G}$ is asymptotically stable. Finally, if $\mathcal{D}_{\mathrm{p}} = \mathbb{R}^{n_{\mathrm{p}}}$, $\mathcal{D}_{\mathrm{c}} = \mathbb{R}^{n_{\mathrm{c}}}$, and $V(\cdot)$ is radially unbounded, then global asymptotic stability is immediate. $\qquad\square$

It is important to note that Theorem 8.7 also holds for the case where (8.171) is replaced by $s_{\mathrm{d}}(u_{\mathrm{d}}, y) \leq 0$, $x \in \mathcal{Z}_1$. In this case, it can be shown using similar arguments as in the proof of Theorem 8.3 that for every $x_0 \notin \mathcal{Z}$, $x_0 \neq 0$, there exists $\tau > 0$ such that $x(\tau) \in \mathcal{Z}_2$.

Finally, we specialize the hybrid controller design framework just presented to impulsive port-controlled Hamiltonian systems. Specifically, consider the state-dependent impulsive port-controlled Hamiltonian system given by

$$\dot{x}_{\mathrm{p}}(t) = \mathcal{J}_{\mathrm{cp}}(x_{\mathrm{p}}(t))\left(\frac{\partial \mathcal{H}_{\mathrm{p}}}{\partial x_{\mathrm{p}}}(x_{\mathrm{p}}(t))\right)^{\mathrm{T}} + G_{\mathrm{p}}(x_{\mathrm{p}}(t))u_{\mathrm{c}}(t),$$
$$x_{\mathrm{p}}(0) = x_{\mathrm{p}0}, \quad (x_{\mathrm{p}}(t), u_{\mathrm{c}}(t)) \notin \mathcal{Z}_{\mathrm{p}}, \quad (8.175)$$
$$\Delta x_{\mathrm{p}}(t) = \mathcal{J}_{\mathrm{dp}}(x_{\mathrm{p}}(t))\left(\frac{\partial \mathcal{H}_{\mathrm{p}}}{\partial x_{\mathrm{p}}}(x_{\mathrm{p}}(t))\right)^{\mathrm{T}} + G_{\mathrm{p}}(x_{\mathrm{p}}(t))u_{\mathrm{d}}(t),$$
$$(x_{\mathrm{p}}(t), u_{\mathrm{c}}(t)) \in \mathcal{Z}_{\mathrm{p}}, \quad (8.176)$$
$$y(t) = G_{\mathrm{p}}^{\mathrm{T}}(x_{\mathrm{p}}(t))\left(\frac{\partial \mathcal{H}_{\mathrm{p}}}{\partial x_{\mathrm{p}}}(x_{\mathrm{p}}(t))\right)^{\mathrm{T}}, \quad (8.177)$$

where $t \geq 0$, $x_{\mathrm{p}}(t) \in \mathcal{D}_{\mathrm{p}} \subseteq \mathbb{R}^{n_{\mathrm{p}}}$, $\mathcal{D}_{\mathrm{p}}$ is an open set with $0 \in \mathcal{D}_{\mathrm{p}}$, $u_{\mathrm{c}}(t) \in \mathbb{R}^m$, $u_{\mathrm{d}}(t) \in \mathbb{R}^m$, $y(t) \in \mathbb{R}^m$, $\mathcal{H}_{\mathrm{p}} : \mathcal{D}_{\mathrm{p}} \to \mathbb{R}$ is an infinitely differentiable Hamiltonian function for the system (8.175)–(8.177), $\mathcal{J}_{\mathrm{cp}} : \mathcal{D}_{\mathrm{p}} \to \mathbb{R}^{n_{\mathrm{p}} \times n_{\mathrm{p}}}$ is such that $\mathcal{J}_{\mathrm{cp}}(x_{\mathrm{p}}) = -\mathcal{J}_{\mathrm{cp}}^{\mathrm{T}}(x_{\mathrm{p}})$, $x_{\mathrm{p}} \in \mathcal{D}_{\mathrm{p}}$, $\mathcal{J}_{\mathrm{cp}}(x_{\mathrm{p}})(\frac{\partial \mathcal{H}_{\mathrm{p}}}{\partial x_{\mathrm{p}}}(x_{\mathrm{p}}))^{\mathrm{T}}$, $x_{\mathrm{p}} \in \mathcal{D}_{\mathrm{p}}$, is Lipschitz continuous on $\mathcal{D}_{\mathrm{p}}$, $G_{\mathrm{p}} : \mathcal{D}_{\mathrm{p}} \to \mathbb{R}^{n_{\mathrm{p}} \times m}$, $\mathcal{J}_{\mathrm{dp}} : \mathcal{D}_{\mathrm{p}} \to \mathbb{R}^{n_{\mathrm{p}} \times n_{\mathrm{p}}}$ is such that $\mathcal{J}_{\mathrm{dp}}(x_{\mathrm{p}}) = -\mathcal{J}_{\mathrm{dp}}^{\mathrm{T}}(x_{\mathrm{p}})$, $x_{\mathrm{p}} \in \mathcal{D}_{\mathrm{p}}$, $\mathcal{J}_{\mathrm{dp}}(x_{\mathrm{p}})(\frac{\partial \mathcal{H}_{\mathrm{p}}}{\partial x_{\mathrm{p}}}(x_{\mathrm{p}}))^{\mathrm{T}}$, $x_{\mathrm{p}} \in \mathcal{D}_{\mathrm{p}}$, is continuous on $\mathcal{D}_{\mathrm{p}}$, and $\mathcal{Z}_{\mathrm{p}} \triangleq \mathcal{Z}_{x_{\mathrm{p}}} \times \mathcal{Z}_{u_{\mathrm{c}}} \subset \mathcal{D}_{\mathrm{p}} \times \mathbb{R}^m$ is the resetting set. Furthermore, assume $\mathcal{H}_{\mathrm{p}}(\cdot)$ is such that

$$\mathcal{H}_{\mathrm{p}}\left(x_{\mathrm{p}} + \mathcal{J}_{\mathrm{dp}}(x_{\mathrm{p}})\left(\frac{\partial \mathcal{H}_{\mathrm{p}}}{\partial x_{\mathrm{p}}}(x_{\mathrm{p}})\right)^{\mathrm{T}} + G_{\mathrm{p}}(x_{\mathrm{p}})u_{\mathrm{d}}\right)$$

$$= \mathcal{H}_{\mathrm{p}}(x_{\mathrm{p}}) + \frac{\partial \mathcal{H}_{\mathrm{p}}}{\partial x_{\mathrm{p}}}(x_{\mathrm{p}})G_{\mathrm{p}}(x_{\mathrm{p}})u_{\mathrm{d}}, \quad x_{\mathrm{p}} \in \mathcal{D}_{\mathrm{p}}, \quad u_{\mathrm{d}} \in \mathbb{R}^m. \quad (8.178)$$

Finally, we assume that $\mathcal{H}_{\mathrm{p}}(0) = 0$ and $\mathcal{H}_{\mathrm{p}}(x_{\mathrm{p}}) > 0$ for all $x_{\mathrm{p}} \neq 0$ and $x_{\mathrm{p}} \in \mathcal{D}_{\mathrm{p}}$.

Next, consider the fixed-order, energy-based hybrid controller

$$\dot{x}_{\mathrm{c}}(t) = \mathcal{J}_{\mathrm{cc}}(x_{\mathrm{c}}(t)) \left(\frac{\partial \mathcal{H}_{\mathrm{c}}}{\partial x_{\mathrm{c}}}(x_{\mathrm{c}}(t)) \right)^{\mathrm{T}} + G_{\mathrm{cc}}(x_{\mathrm{c}}(t))y(t),$$

$$x_{\mathrm{c}}(0) = x_{\mathrm{c}0}, \quad (x_{\mathrm{c}}(t), y(t)) \notin \mathcal{Z}_{\mathrm{c}}, \quad (8.179)$$

$$\Delta x_{\mathrm{c}}(t) = -x_{\mathrm{c}}(t), \quad (x_{\mathrm{c}}(t), y(t)) \in \mathcal{Z}_{\mathrm{c}}, \quad (8.180)$$

$$u_{\mathrm{c}}(t) = -G_{\mathrm{cc}}^{\mathrm{T}}(x_{\mathrm{c}}(t)) \left(\frac{\partial \mathcal{H}_{\mathrm{c}}}{\partial x_{\mathrm{c}}}(x_{\mathrm{c}}(t)) \right)^{\mathrm{T}}, \quad (8.181)$$

$$u_{\mathrm{d}}(t) = -G_{\mathrm{p}}^{\mathrm{T}}(x_{\mathrm{p}}(t)) \left(\frac{\partial \mathcal{H}_{\mathrm{p}}}{\partial x_{\mathrm{p}}}(x_{\mathrm{p}}(t)) \right)^{\mathrm{T}}, \quad (8.182)$$

where $t \geq 0$, $x_{\mathrm{c}}(t) \in \mathcal{D}_{\mathrm{c}} \subseteq \mathbb{R}^{n_{\mathrm{c}}}$, $\mathcal{D}_{\mathrm{c}}$ is an open set with $0 \in \mathcal{D}_{\mathrm{c}}$, $\Delta x_{\mathrm{c}}(t) \triangleq x_{\mathrm{c}}(t^+) - x_{\mathrm{c}}(t)$, $\mathcal{H}_{\mathrm{c}} : \mathcal{D}_{\mathrm{c}} \to \mathbb{R}$ is an infinitely differentiable Hamiltonian function for (8.179), $\mathcal{J}_{\mathrm{cc}} : \mathcal{D}_{\mathrm{c}} \to \mathbb{R}^{n_{\mathrm{c}} \times n_{\mathrm{c}}}$ is such that $\mathcal{J}_{\mathrm{cc}}(x_{\mathrm{c}}) = -\mathcal{J}_{\mathrm{cc}}^{\mathrm{T}}(x_{\mathrm{c}})$, $x_{\mathrm{c}} \in \mathcal{D}_{\mathrm{c}}$, $\mathcal{J}_{\mathrm{cc}}(x_{\mathrm{c}})(\frac{\partial \mathcal{H}_{\mathrm{c}}}{\partial x_{\mathrm{c}}}(x_{\mathrm{c}}))^{\mathrm{T}}$, $x_{\mathrm{c}} \in \mathcal{D}_{\mathrm{c}}$, is Lipschitz continuous on $\mathcal{D}_{\mathrm{c}}$, $G_{\mathrm{cc}} : \mathcal{D}_{\mathrm{c}} \to \mathbb{R}^{n_{\mathrm{c}} \times m}$, and the resetting set $\mathcal{Z}_{\mathrm{c}} \subset \mathcal{D}_{\mathrm{p}} \times \mathcal{D}_{\mathrm{c}}$ given by

$$\mathcal{Z}_{\mathrm{c}} \triangleq \left\{ (x_{\mathrm{p}}, x_{\mathrm{c}}) \in \mathcal{D}_{\mathrm{p}} \times \mathcal{D}_{\mathrm{c}} : \frac{\mathrm{d}}{\mathrm{d}t}\mathcal{H}_{\mathrm{c}}(x_{\mathrm{c}}) = 0 \text{ and } \mathcal{H}_{\mathrm{c}}(x_{\mathrm{c}}) > 0 \right\}. \quad (8.183)$$

Finally, we assume that $\mathcal{H}_{\mathrm{c}}(0) = 0$ and $\mathcal{H}_{\mathrm{c}}(x_{\mathrm{c}}) > 0$ for all $x_{\mathrm{c}} \neq 0$ and $x_{\mathrm{c}} \in \mathcal{D}_{\mathrm{c}}$.

Note that $\mathcal{H}_{\mathrm{p}}(x_{\mathrm{p}})$, $x_{\mathrm{p}} \in \mathcal{D}_{\mathrm{p}}$, is the plant energy and $\mathcal{H}_{\mathrm{c}}(x_{\mathrm{c}})$, $x_{\mathrm{c}} \in \mathcal{D}_{\mathrm{c}}$, is the controller emulated energy. Furthermore, the closed-loop system energy is given by $\mathcal{H}(x_{\mathrm{p}}, x_{\mathrm{c}}) \triangleq \mathcal{H}_{\mathrm{p}}(x_{\mathrm{p}}) + \mathcal{H}_{\mathrm{c}}(x_{\mathrm{c}})$. The resetting set $\mathcal{Z}$ is given by $\mathcal{Z} \triangleq \mathcal{Z}_1 \cup \mathcal{Z}_2$, where

$$\mathcal{Z}_1 \triangleq \left\{ (x_{\mathrm{p}}, x_{\mathrm{c}}) \in \mathcal{D}_{\mathrm{p}} \times \mathcal{D}_{\mathrm{c}} : \left(x_{\mathrm{p}}, -G_{\mathrm{cc}}^{\mathrm{T}}(x_{\mathrm{c}}) \left(\frac{\partial \mathcal{H}_{\mathrm{c}}}{\partial x_{\mathrm{c}}}(x_{\mathrm{c}}) \right)^{\mathrm{T}} \right) \in \mathcal{Z}_{\mathrm{p}} \right\},$$

$$(8.184)$$

$$\mathcal{Z}_2 \triangleq \left\{ (x_{\mathrm{p}}, x_{\mathrm{c}}) \in \mathcal{D}_{\mathrm{p}} \times \mathcal{D}_{\mathrm{c}} : \left(x_{\mathrm{c}}, G_{\mathrm{p}}^{\mathrm{T}}(x_{\mathrm{p}}) \left(\frac{\partial \mathcal{H}_{\mathrm{p}}}{\partial x_{\mathrm{p}}}(x_{\mathrm{p}}) \right)^{\mathrm{T}} \right) \in \mathcal{Z}_{\mathrm{c}} \right\}.$$

$$(8.185)$$

Here, we assume that $\overline{\mathcal{Z}}_1 = \{(x_\mathrm{p}, x_\mathrm{c}) \in \mathcal{D}_\mathrm{p} \times \mathcal{D}_\mathrm{c} : \mathcal{X}_1(x_\mathrm{p}, x_\mathrm{c}) = 0\}$. Furthermore, if $(x_\mathrm{p}, x_\mathrm{c}) \in \mathcal{Z}_1$ then $x_\mathrm{p} + \mathcal{J}_{\mathrm{dp}}(x_\mathrm{p})(\frac{\partial \mathcal{H}_\mathrm{p}}{\partial x_\mathrm{p}}(x_\mathrm{p}))^\mathrm{T} - G_\mathrm{p}(x_\mathrm{p})G_\mathrm{p}^\mathrm{T}(x_\mathrm{p})(\frac{\partial \mathcal{H}_\mathrm{p}}{\partial x_\mathrm{p}}(x_\mathrm{p}))^\mathrm{T} \in \overline{\mathcal{Z}}_1 \backslash \mathcal{Z}_1$, and if $(x_\mathrm{p}, x_\mathrm{c}) \in \overline{\mathcal{Z}}_1 \backslash \mathcal{Z}_1$ then $\mathcal{J}_{\mathrm{dp}}(x_\mathrm{p})(\frac{\partial \mathcal{H}_\mathrm{p}}{\partial x_\mathrm{p}}(x_\mathrm{p}))^\mathrm{T} - G_\mathrm{p}(x_\mathrm{p})G_\mathrm{p}^\mathrm{T}(x_\mathrm{p})(\frac{\partial \mathcal{H}_\mathrm{p}}{\partial x_\mathrm{p}}(x_\mathrm{p}))^\mathrm{T} = 0$. Finally, we assume that

$$\mathcal{Z}_1 \cap \left\{ (x_\mathrm{p}, x_\mathrm{c}) \in \mathcal{D}_\mathrm{p} \times \mathcal{D}_\mathrm{c} : G_\mathrm{p}^\mathrm{T}(x_\mathrm{p}) \left(\frac{\partial \mathcal{H}_\mathrm{p}}{\partial x_\mathrm{p}}(x_\mathrm{p}) \right)^\mathrm{T} = 0 \right\} = \varnothing. \quad (8.186)$$

Next, note that the total energy function $\mathcal{H}(x_\mathrm{p}, x_\mathrm{c})$ along the trajectories of the closed-loop dynamics (8.175)–(8.185) satisfies

$$\frac{\mathrm{d}}{\mathrm{d}t}\mathcal{H}(x_\mathrm{p}(t), x_\mathrm{c}(t)) = 0, \quad (x_\mathrm{p}(t), x_\mathrm{c}(t)) \notin \mathcal{Z}, \quad (8.187)$$

$$\Delta\mathcal{H}(x_\mathrm{p}(t_k), x_\mathrm{c}(t_k)) = -\frac{\partial \mathcal{H}_\mathrm{p}}{\partial x_\mathrm{p}}(x_\mathrm{p}(t_k))G_\mathrm{p}(x_\mathrm{p}(t_k))G_\mathrm{p}^\mathrm{T}(x_\mathrm{p}(t_k))$$
$$\left(\frac{\partial \mathcal{H}_\mathrm{p}}{\partial x_\mathrm{p}}(x_\mathrm{p}(t_k)) \right)^\mathrm{T} \chi_{\mathcal{Z}_1}(x_\mathrm{p}(t_k), x_\mathrm{c}(t_k))$$
$$-\mathcal{H}_\mathrm{c}(x_\mathrm{c}(t_k))\chi_{\mathcal{Z}_2}(x_\mathrm{p}(t_k), x_\mathrm{c}(t_k)),$$
$$(x_\mathrm{p}(t_k), x_\mathrm{c}(t_k)) \in \mathcal{Z}, \quad k \in \overline{\mathbb{Z}}_+. \quad (8.188)$$

Here, we assume that every $(x_{\mathrm{p}0}, x_{\mathrm{c}0}) \in \overline{\mathcal{Z}}$ is transversal to the closed-loop dynamical system given by (8.175)–(8.185) with $\mathcal{X}_\mathrm{p}(x_\mathrm{p}, x_\mathrm{c}) = \mathcal{X}_1(x_\mathrm{p}, x_\mathrm{c})$ and $\mathcal{X}_\mathrm{c}(x_\mathrm{p}, x_\mathrm{c}) = \frac{\mathrm{d}}{\mathrm{d}t}\mathcal{H}_\mathrm{c}(x_\mathrm{c})$. Furthermore, we assume $\mathcal{D}_{\mathrm{ci}} \subset \mathcal{D}_\mathrm{p} \times \mathcal{D}_\mathrm{c}$ is a compact positively invariant set with respect to the closed-loop dynamical system (8.175)–(8.185), such that $0 \in \overset{\circ}{\mathcal{D}}_{\mathrm{ci}}$. In this case, it follows from Theorem 8.7, with $V_\mathrm{s}(x_\mathrm{p}) = \mathcal{H}_\mathrm{p}(x_\mathrm{p})$, $V_\mathrm{c}(x_\mathrm{c}, y) = \mathcal{H}_\mathrm{c}(x_\mathrm{c})$, $s_\mathrm{c}(u_\mathrm{c}, y) = u_\mathrm{c}^\mathrm{T} y$, $s_\mathrm{d}(u_\mathrm{d}, y) = u_\mathrm{d}^\mathrm{T} y$, and $s_\mathrm{cc}(u_\mathrm{cc}, y_\mathrm{cc}) = u_\mathrm{cc}^\mathrm{T} y_\mathrm{cc}$, that that the zero solution $(x_\mathrm{p}(t), x_\mathrm{c}(t)) \equiv (0, 0)$ to the closed-loop system (8.175)–(8.185) is asymptotically stable.

8.9 Hybrid Control Design for Nonsmooth Euler-Lagrange Systems

In this section, we present a hybrid feedback control framework for nonsmooth Euler-Lagrange dynamical systems. Consider the governing equations of motion of an $\hat{n}_\mathrm{p}$-degree-of-freedom dynamical system given by the *hybrid Euler-Lagrange* equation

$$\frac{\mathrm{d}}{\mathrm{d}t}\left[\frac{\partial \mathcal{L}}{\partial \dot{q}}(q(t),\dot{q}(t))\right]^{\mathrm{T}} - \left[\frac{\partial \mathcal{L}}{\partial q}(q(t),\dot{q}(t))\right]^{\mathrm{T}} = u_{\mathrm{c}}(t),$$

$$q(0) = q_0, \quad \dot{q}(0) = \dot{q}_0, \quad (q(t),\dot{q}(t)) \notin \mathcal{Z}_{\mathrm{p}}, \quad (8.189)$$

$$\begin{bmatrix} \Delta q(t) \\ \Delta \dot{q}(t) \end{bmatrix} = \begin{bmatrix} P(q(t)) - q(t) \\ Q(\dot{q}(t)) - \dot{q}(t) \end{bmatrix}, \quad (q(t),\dot{q}(t)) \in \mathcal{Z}_{\mathrm{p}}, \quad (8.190)$$

with outputs

$$y = \begin{bmatrix} h_1(q) \\ h_2(\dot{q}) \end{bmatrix}, \quad (8.191)$$

where $t \geq 0$, $q \in \mathbb{R}^{\hat{n}_{\mathrm{P}}}$ represents the generalized system positions, $\dot{q} \in \mathbb{R}^{\hat{n}_{\mathrm{P}}}$ represents the generalized system velocities, $\mathcal{L} : \mathbb{R}^{\hat{n}_{\mathrm{P}}} \times \mathbb{R}^{\hat{n}_{\mathrm{P}}} \to \mathbb{R}$ denotes the system Lagrangian given by $\mathcal{L}(q,\dot{q}) = T(q,\dot{q}) - U(q)$, where $T : \mathbb{R}^{\hat{n}_{\mathrm{P}}} \times \mathbb{R}^{\hat{n}_{\mathrm{P}}} \to \mathbb{R}$ is the system kinetic energy and $U : \mathbb{R}^{\hat{n}_{\mathrm{P}}} \to \mathbb{R}$ is the system potential energy, $u_{\mathrm{c}} \in \mathbb{R}^{\hat{n}_{\mathrm{P}}}$ is the vector of generalized control forces acting on the system, $\mathcal{Z}_{\mathrm{p}} \subset \mathbb{R}^{\hat{n}_{\mathrm{P}}} \times \mathbb{R}^{\hat{n}_{\mathrm{P}}}$ is the resetting set such that the closure of $\mathcal{Z}_{\mathrm{p}}$ is given by

$$\overline{\mathcal{Z}}_{\mathrm{p}} \triangleq \{(q,\dot{q}) : H(q,\dot{q}) = 0\}, \quad (8.192)$$

where $H : \mathbb{R}^{\hat{n}_{\mathrm{P}}} \times \mathbb{R}^{\hat{n}_{\mathrm{P}}} \to \mathbb{R}$ is an infinitely differentiable function. Here, $P : \mathbb{R}^{\hat{n}_{\mathrm{P}}} \to \mathbb{R}^{\hat{n}_{\mathrm{P}}}$ and $Q : \mathbb{R}^{\hat{n}_{\mathrm{P}}} \to \mathbb{R}^{\hat{n}_{\mathrm{P}}}$ are continuous functions such that if $(q,\dot{q}) \in \mathcal{Z}_{\mathrm{p}}$, then $(P(q),Q(\dot{q})) \in \overline{\mathcal{Z}}_{\mathrm{p}} \backslash \mathcal{Z}_{\mathrm{p}}$, and if $(q,\dot{q}) \in \overline{\mathcal{Z}}_{\mathrm{p}} \backslash \mathcal{Z}_{\mathrm{p}}$, then $(P(q),Q(\dot{q})) = (q,\dot{q})$, $T(P(q),Q(\dot{q})) + U(P(q)) < T(q,\dot{q}) + U(q)$, $(q,\dot{q}) \in \mathcal{Z}_{\mathrm{p}}$, and $h_1 : \mathbb{R}^{\hat{n}_{\mathrm{P}}} \to \mathbb{R}^{l_1}$ and $h_2 : \mathbb{R}^{\hat{n}_{\mathrm{P}}} \to \mathbb{R}^{l-l_1}$ are continuously differentiable functions such that $h_1(0) = 0$, $h_2(0) = 0$, and $h_1(q) \not\equiv 0$. We assume that the system kinetic energy is such that $T(q,\dot{q}) = \frac{1}{2}\dot{q}^{\mathrm{T}}[\frac{\partial T}{\partial \dot{q}}(q,\dot{q})]^{\mathrm{T}}$, $T(q,0) = 0$, and $T(q,\dot{q}) > 0$, $\dot{q} \neq 0$, $\dot{q} \in \mathbb{R}^{\hat{n}_{\mathrm{P}}}$.

Furthermore, let $\mathcal{H} : \mathbb{R}^{\hat{n}_{\mathrm{P}}} \times \mathbb{R}^{\hat{n}_{\mathrm{P}}} \to \mathbb{R}$ denote the Legendre transformation of the Lagrangian function $\mathcal{L}(q,\dot{q})$ with respect to the generalized velocity $\dot{q}$ defined by $\mathcal{H}(q,p) \triangleq \dot{q}^{\mathrm{T}}p - \mathcal{L}(q,\dot{q})$, where p denotes the vector of generalized momenta given by

$$p(q,\dot{q}) = \left[\frac{\partial \mathcal{L}}{\partial \dot{q}}(q,\dot{q})\right]^{\mathrm{T}}, \quad (8.193)$$

where the map from the generalized velocities $\dot{q}$ to the generalized momenta p is assumed to be bijective. Now, if $\mathcal{H}(q,p)$ is lower bounded, then we can always shift $\mathcal{H}(q,p)$ so that, with a minor abuse of notation, $\mathcal{H}(q,p) \geq 0$, $(q,p) \in \mathbb{R}^{\hat{n}_{\mathrm{P}}} \times \mathbb{R}^{\hat{n}_{\mathrm{P}}}$. In this case, using (8.189) and the fact that

$$\frac{\mathrm{d}}{\mathrm{d}t}[\mathcal{L}(q,\dot{q})] = \frac{\partial \mathcal{L}}{\partial q}(q,\dot{q})\dot{q} + \frac{\partial \mathcal{L}}{\partial \dot{q}}(q,\dot{q})\ddot{q}, \quad (q,\dot{q}) \notin \mathcal{Z}_{\mathrm{p}}, \quad (8.194)$$

it follows that $\frac{\mathrm{d}}{\mathrm{d}t}\mathcal{H}(q,p) = u_{\mathrm{c}}^{\mathrm{T}}\dot{q}$, $(q,\dot{q}) \notin \mathcal{Z}_{\mathrm{p}}$. We also assume that the system potential energy $U(\cdot)$ is such that $U(0) = 0$ and $U(q) > 0$, $q \neq 0$, $q \in \mathcal{D}_{\mathrm{q}} \subseteq \mathbb{R}^{\hat{n}_{\mathrm{P}}}$, which implies that $\mathcal{H}(q,p) = T(q,\dot{q}) + U(q) > 0$, $(q,\dot{q}) \neq 0$, $(q,\dot{q}) \in \mathcal{D}_{\mathrm{q}} \times \mathbb{R}^{\hat{n}_{\mathrm{P}}}$.

Next, consider the energy-based hybrid controller

$$\frac{\mathrm{d}}{\mathrm{d}t}\left[\frac{\partial \mathcal{L}_{\mathrm{c}}}{\partial \dot{q}_{\mathrm{c}}}(q_{\mathrm{c}}(t), \dot{q}_{\mathrm{c}}(t), y_{q}(t))\right]^{\mathrm{T}} - \left[\frac{\partial \mathcal{L}_{\mathrm{c}}}{\partial q_{\mathrm{c}}}(q_{\mathrm{c}}(t), \dot{q}_{\mathrm{c}}(t), y_{q}(t))\right]^{\mathrm{T}} = 0,$$

$$q_{\mathrm{c}}(0) = q_{\mathrm{c}0}, \quad \dot{q}_{\mathrm{c}}(0) = \dot{q}_{\mathrm{c}0}, \quad (q_{\mathrm{c}}(t), \dot{q}_{\mathrm{c}}(t), y(t)) \notin \mathcal{Z}_{\mathrm{c}}, \quad (8.195)$$

$$\begin{bmatrix} \Delta q_{\mathrm{c}}(t) \\ \Delta \dot{q}_{\mathrm{c}}(t) \end{bmatrix} = \begin{bmatrix} \eta(y_{q}(t)) - q_{\mathrm{c}}(t) \\ -\dot{q}_{\mathrm{c}}(t) \end{bmatrix}, \quad (q_{\mathrm{c}}(t), \dot{q}_{\mathrm{c}}(t), y(t)) \in \mathcal{Z}_{\mathrm{c}}, \quad (8.196)$$

$$u_{\mathrm{c}}(t) = \left[\frac{\partial \mathcal{L}_{\mathrm{c}}}{\partial q}(q_{\mathrm{c}}(t), \dot{q}_{\mathrm{c}}(t), y_{q}(t))\right]^{\mathrm{T}}, \quad (8.197)$$

where $t \geq 0$, $q_{\mathrm{c}} \in \mathbb{R}^{\hat{n}_{\mathrm{c}}}$ represents virtual controller positions, $\dot{q}_{\mathrm{c}} \in \mathbb{R}^{\hat{n}_{\mathrm{c}}}$ represents virtual controller velocities, $y_{q} \triangleq h_{1}(q)$, $\mathcal{L}_{\mathrm{c}} : \mathbb{R}^{\hat{n}_{\mathrm{c}}} \times \mathbb{R}^{\hat{n}_{\mathrm{c}}} \times \mathbb{R}^{l_{1}} \to \mathbb{R}$ denotes the controller Lagrangian given by $\mathcal{L}_{\mathrm{c}}(q_{\mathrm{c}}, \dot{q}_{\mathrm{c}}, y_{q}) \triangleq T_{\mathrm{c}}(q_{\mathrm{c}}, \dot{q}_{\mathrm{c}}) - U_{\mathrm{c}}(q_{\mathrm{c}}, y_{q})$, where $T_{\mathrm{c}} : \mathbb{R}^{\hat{n}_{\mathrm{c}}} \times \mathbb{R}^{\hat{n}_{\mathrm{c}}} \to \mathbb{R}$ is the controller kinetic energy, $U_{\mathrm{c}} : \mathbb{R}^{\hat{n}_{\mathrm{c}}} \times \mathbb{R}^{l_{1}} \to \mathbb{R}$ is the controller potential energy, $\eta(\cdot)$ is a continuously differentiable function such that $\eta(0) = 0$, $\mathcal{Z}_{\mathrm{c}} \subset \mathbb{R}^{\hat{n}_{\mathrm{c}}} \times \mathbb{R}^{\hat{n}_{\mathrm{c}}} \times \mathbb{R}^{l}$ is the resetting set, $\Delta q_{\mathrm{c}}(t) \triangleq q_{\mathrm{c}}(t^{+}) - q_{\mathrm{c}}(t)$, and $\Delta \dot{q}_{\mathrm{c}}(t) \triangleq \dot{q}_{\mathrm{c}}(t^{+}) - \dot{q}_{\mathrm{c}}(t)$. We assume that the controller kinetic energy $T_{\mathrm{c}}(q_{\mathrm{c}}, \dot{q}_{\mathrm{c}})$ is such that $T_{\mathrm{c}}(q_{\mathrm{c}}, \dot{q}_{\mathrm{c}}) = \frac{1}{2}\dot{q}_{\mathrm{c}}^{\mathrm{T}}[\frac{\partial T_{\mathrm{c}}}{\partial \dot{q}_{\mathrm{c}}}(q_{\mathrm{c}}, \dot{q}_{\mathrm{c}})]^{\mathrm{T}}$, with $T_{\mathrm{c}}(q_{\mathrm{c}}, 0) = 0$ and $T_{\mathrm{c}}(q_{\mathrm{c}}, \dot{q}_{\mathrm{c}}) > 0$, $\dot{q}_{\mathrm{c}} \neq 0$, $\dot{q}_{\mathrm{c}} \in \mathbb{R}^{\hat{n}_{\mathrm{c}}}$. Furthermore, we assume that $U_{\mathrm{c}}(\eta(y_{q}), y_{q}) = 0$ and $U_{\mathrm{c}}(q_{\mathrm{c}}, y_{q}) > 0$ for $q_{\mathrm{c}} \neq \eta(y_{q})$, $q_{\mathrm{c}} \in \mathcal{D}_{q_{\mathrm{c}}} \subseteq \mathbb{R}^{\hat{n}_{\mathrm{c}}}$.

As in Section 8.5, note that $V_{\mathrm{p}}(q,\dot{q}) \triangleq T(q,\dot{q}) + U(q)$ is the plant energy and $V_{\mathrm{c}}(q_{\mathrm{c}}, \dot{q}_{\mathrm{c}}, y_{q}) \triangleq T_{\mathrm{c}}(q_{\mathrm{c}}, \dot{q}_{\mathrm{c}}) + U_{\mathrm{c}}(q_{\mathrm{c}}, y_{q})$ is the controller emulated energy. Furthermore, $V(q,\dot{q}, q_{\mathrm{c}}, \dot{q}_{\mathrm{c}}) \triangleq V_{\mathrm{p}}(q,\dot{q}) + V_{\mathrm{c}}(q_{\mathrm{c}}, \dot{q}_{\mathrm{c}}, y_{q})$ is the total energy of the closed-loop system. It is important to note that the Lagrangian dynamical system (8.189) is not lossless with outputs y_{q} or y. Next, we study the behavior of the total energy function $V(q,\dot{q}, q_{\mathrm{c}}, \dot{q}_{\mathrm{c}})$ along the trajectories of the closed-loop system dynamics. For the closed-loop system, we define our resetting set as $\mathcal{Z} \triangleq \mathcal{Z}_{1} \cup \mathcal{Z}_{2}$, where $\mathcal{Z}_{1} \triangleq \{(q,\dot{q}, q_{\mathrm{c}}, \dot{q}_{\mathrm{c}}) : (q,\dot{q}) \in \mathcal{Z}_{\mathrm{p}}\}$ and $\mathcal{Z}_{2} \triangleq \{(q,\dot{q}, q_{\mathrm{c}}, \dot{q}_{\mathrm{c}}) : (q_{\mathrm{c}}, \dot{q}_{\mathrm{c}}, y) \in \mathcal{Z}_{\mathrm{c}}\}$. Note that

$$\frac{\mathrm{d}}{\mathrm{d}t}V_{\mathrm{p}}(q,\dot{q}) = \frac{\mathrm{d}}{\mathrm{d}t}\mathcal{H}(q,p) = u_{\mathrm{c}}^{\mathrm{T}}\dot{q}, \quad (q,\dot{q}, q_{\mathrm{c}}, \dot{q}_{\mathrm{c}}) \notin \mathcal{Z}. \quad (8.198)$$

To obtain an expression for $\frac{d}{dt}V_c(q_c, \dot{q}_c, y_q)$ when $(q, \dot{q}, q_c, \dot{q}_c) \notin \mathcal{Z}$, define the controller Hamiltonian by

$$\mathcal{H}_c(q_c, \dot{q}_c, p_c, y_q) \triangleq \dot{q}_c^T p_c - \mathcal{L}_c(q_c, \dot{q}_c, y_q), \qquad (8.199)$$

where the virtual controller momentum p_c is given by $p_c(q_c, \dot{q}_c, y_q) = \left[\frac{\partial \mathcal{L}_c}{\partial \dot{q}_c}(q_c, \dot{q}_c, y_q)\right]^T$. Then $\mathcal{H}_c(q_c, \dot{q}_c, p_c, y_q) = T_c(q_c, \dot{q}_c) + U_c(q_c, y_q)$. Now, it follows from (8.195) and the structure of $T_c(q_c, \dot{q}_c)$ that, for $t \in (t_k, t_{k+1}]$,

$$0 = \frac{d}{dt}[p_c(q_c(t), \dot{q}_c(t), y_q(t))]^T \dot{q}_c(t) - \frac{\partial \mathcal{L}_c}{\partial q_c}(q_c(t), \dot{q}_c(t), y_q(t))\dot{q}_c(t)$$

$$= \frac{d}{dt}\left[p_c^T(q_c(t), \dot{q}_c(t), y_q(t))\dot{q}_c(t)\right] - p_c^T(q_c(t), \dot{q}_c(t), y_q(t))\ddot{q}_c(t)$$

$$+ \frac{\partial \mathcal{L}_c}{\partial \dot{q}_c}(q_c(t), \dot{q}_c(t), y_q(t))\ddot{q}_c(t) + \frac{\partial \mathcal{L}_c}{\partial q}(q_c(t), \dot{q}_c(t), y_q(t))\dot{q}(t)$$

$$- \frac{d}{dt}\mathcal{L}_c(q_c(t), \dot{q}_c(t), y_q(t))$$

$$= \frac{d}{dt}[p_c^T(q_c(t), \dot{q}_c(t), y_q(t))\dot{q}_c(t) - \mathcal{L}_c(q_c(t), \dot{q}_c(t), y_q(t))]$$

$$+ \frac{\partial \mathcal{L}_c}{\partial q}(q_c(t), \dot{q}_c(t), y_q(t))\dot{q}(t)$$

$$= \frac{d}{dt}\mathcal{H}_c(q_c(t), \dot{q}_c(t), p_c(t), y_q(t)) + \frac{\partial \mathcal{L}_c}{\partial q}(q_c(t), \dot{q}_c(t), y_q(t))\dot{q}(t)$$

$$= \frac{d}{dt}V_c(q_c(t), \dot{q}_c(t), y_q(t)) + \frac{\partial \mathcal{L}_c}{\partial q}(q_c(t), \dot{q}_c(t), y_q(t))\dot{q}(t),$$

$$(q(t), \dot{q}(t), q_c(t), \dot{q}_c(t)) \notin \mathcal{Z}. \qquad (8.200)$$

Hence,

$$\frac{d}{dt}V(q(t), \dot{q}(t), q_c(t), \dot{q}_c(t)) = u_c^T(t)\dot{q}(t) - \frac{\partial \mathcal{L}_c}{\partial q}(q_c(t), \dot{q}_c(t), y_q(t))\dot{q}(t)$$

$$= 0, \quad (q(t), \dot{q}(t), q_c(t), \dot{q}_c(t)) \notin \mathcal{Z},$$

$$t_k < t \le t_{k+1}, \qquad (8.201)$$

which implies that the total energy of the closed-loop system between resetting events is conserved.

The total energy difference across resetting events is given by

$$\Delta V(q(t_k), \dot{q}(t_k), q_c(t_k), \dot{q}_c(t_k)) = V_p(q(t_k^+), \dot{q}(t_k^+)) - V_p(q(t_k), \dot{q}(t_k))$$

$$+ T_c(q_c(t_k^+), \dot{q}_c(t_k^+))$$

$$+ U_c(q_c(t_k^+), y_q(t_k))$$

$$-V_\mathrm{c}(q_\mathrm{c}(t_k), \dot{q}_\mathrm{c}(t_k), y_q(t_k))$$
$$= [V_\mathrm{p}(P(q(t_k)), Q(\dot{q}(t_k)))$$
$$-V_\mathrm{p}(q(t_k), \dot{q}(t_k))]$$
$$\cdot \chi_{\mathcal{Z}_1}(q(t_k), \dot{q}(t_k), q_\mathrm{c}(t_k), \dot{q}_\mathrm{c}(t_k))$$
$$-V_\mathrm{c}(q_\mathrm{c}(t_k), \dot{q}_\mathrm{c}(t_k), y_q(t_k))$$
$$\cdot \chi_{\mathcal{Z}_2}(q(t_k), \dot{q}(t_k), q_\mathrm{c}(t_k), \dot{q}_\mathrm{c}(t_k))$$
$$< 0, \quad (q(t_k), \dot{q}(t_k), q_\mathrm{c}(t_k), \dot{q}_\mathrm{c}(t_k)) \in \mathcal{Z},$$
$$k \in \overline{\mathbb{Z}}_+, \quad (8.202)$$

which implies that the resetting law (8.196) ensures the total energy decrease across resetting events.

Here, we concentrate on an energy dissipating state-dependent resetting controller that affects a one-way energy transfer between the plant and the controller. Specifically, consider the closed-loop system (8.189)–(8.197), where $\mathcal{Z}_\mathrm{c}$ is defined by

$$\mathcal{Z}_\mathrm{c} \triangleq \left\{ (q, \dot{q}, q_\mathrm{c}, \dot{q}_\mathrm{c}) : \frac{\mathrm{d}}{\mathrm{d}t} V_\mathrm{c}(q_\mathrm{c}, \dot{q}_\mathrm{c}, y_q) = 0 \text{ and } V_\mathrm{c}(q_\mathrm{c}, \dot{q}_\mathrm{c}, y_q) > 0 \right\}.$$
$$(8.203)$$

Since $y_q = h_1(q)$ and

$$\frac{\mathrm{d}}{\mathrm{d}t} V_\mathrm{c}(q_\mathrm{c}, \dot{q}_\mathrm{c}, y_q) = - \left[\frac{\partial \mathcal{L}_\mathrm{c}}{\partial q}(q_\mathrm{c}, \dot{q}_\mathrm{c}, y_q) \right] \dot{q} = \left[\frac{\partial U_\mathrm{c}}{\partial q}(q_\mathrm{c}, y_q) \right] \dot{q},$$
$$(q_\mathrm{c}, \dot{q}_\mathrm{c}, y) \notin \mathcal{Z}_\mathrm{c}, \quad (8.204)$$

it follows that (8.203) can be equivalently rewritten as

$$\mathcal{Z}_\mathrm{c} = \left\{ (q, \dot{q}, q_\mathrm{c}, \dot{q}_\mathrm{c}) : \left[\frac{\partial U_\mathrm{c}}{\partial q}(q_\mathrm{c}, h_1(q)) \right] \dot{q} = 0 \right.$$
$$\left. \text{and } V_\mathrm{c}(q_\mathrm{c}, \dot{q}_\mathrm{c}, h_1(q)) > 0 \right\}. \quad (8.205)$$

Once again, for practical implementation, knowledge of q_c, $\dot{q}_\mathrm{c}$, and y is often sufficient to determine whether or not the closed-loop state vector is in the set $\mathcal{Z}_\mathrm{c}$.

The next theorem gives sufficient conditions for stabilization of non-smooth Euler-Lagrange dynamical systems using state-dependent hybrid controllers. For this result define the closed-loop system states $x \triangleq [q^\mathrm{T}, \dot{q}^\mathrm{T}, q_\mathrm{c}^\mathrm{T}, \dot{q}_\mathrm{c}^\mathrm{T}]^\mathrm{T}$.

Theorem 8.8 *Consider the closed-loop dynamical system $\mathcal{G}$ given by (8.189)–(8.197), with the resetting set $\mathcal{Z}_c$ given by (8.203). Assume that $\mathcal{D}_{ci} \subset \mathcal{D}_q \times \mathbb{R}^{\hat{n}_p} \times \mathcal{D}_{qc} \times \mathbb{R}^{\hat{n}_c}$ is a compact positively invariant set with respect to $\mathcal{G}$ such that $0 \in \overset{\circ}{\mathcal{D}}_{ci}$. Furthermore, assume that the transversality condition (8.155) and (8.156) holds with $\mathcal{X}_p(x) = H(q, \dot{q})$ and $\mathcal{X}_c(x) = \frac{d}{dt} V_c(q_c, \dot{q}_c, y_q)$. Then the zero solution $x(t) \equiv 0$ to $\mathcal{G}$ is asymptotically stable. In addition, the total energy function $V(x)$ of $\mathcal{G}$ is strictly decreasing across resetting events. Finally, if $\mathcal{D}_q = \mathbb{R}^{\hat{n}_p}$, $\mathcal{D}_{qc} = \mathbb{R}^{\hat{n}_c}$, and the total energy function $V(x)$ is radially unbounded, then the zero solution $x(t) \equiv 0$ to $\mathcal{G}$ is globally asymptotically stable.*

Proof. The proof is similar to the proof of Theorem 8.7 with $V_p(x_p) = V_p(q, \dot{q})$, $V_c(x_c, y) = V_c(q_c, \dot{q}_c, y_q)$, $y = u_{cc} = x_p$, $u_c = -y_{cc} = \frac{\partial \mathcal{L}_c}{\partial q}$, $s_c(u_c, y) = u_c^T \rho(y)$, $s_d(u_d, y) = 0$, $V_p(P(q), Q(\dot{q})) - V_p(q, \dot{q}) < 0$, $(q, \dot{q}) \in \mathcal{Z}_p$, $s_{cc}(u_{cc}, y_{cc}) = y_{cc}^T \rho(u_c)$, where $\rho(y) = \rho\left(\begin{bmatrix} q \\ \dot{q} \end{bmatrix}\right) = \dot{q}$, $\eta(y)$ replaced by $\begin{bmatrix} \eta(y_q) \\ 0 \end{bmatrix}$, and noting that (8.201) and (8.202) hold. $\qquad\qquad\square$

8.10 Hybrid Control Design for Impact Mechanics

In this section, we apply the energy-dissipating hybrid controller synthesis framework presented in Section 9.9 to the constrained inverted pendulum shown in Figure 8.30, where $m = 1$ kg and $L = 1$ m. In the case where $|\theta(t)| < \theta_c \leq \frac{\pi}{2}$, the system is governed by the dynamic equation of motion

$$\ddot{\theta}(t) - g \sin \theta(t) = u_c(t), \quad \theta(0) = \theta_0, \quad \dot{\theta}(0) = \dot{\theta}_0, \quad t \geq 0, \quad (8.206)$$

where g denotes the gravitational acceleration and $u_c(\cdot)$ is a (thruster) control force. At the instant of collision with the vertical constraint $|\theta(t)| = \theta_c$, the system resets according to the resetting law

$$\theta(t_k^+) = \theta(t_k), \quad \dot{\theta}(t_k^+) = -e\dot{\theta}(t_k), \quad (8.207)$$

where $e \in [0, 1)$ is the coefficient of restitution. Defining $q = \theta$ and $\dot{q} = \dot{\theta}$, we can rewrite the continuous-time dynamics (8.206) and resetting dynamics (8.207) in Lagrangian form (8.189) and (8.190) with $\mathcal{L}(q, \dot{q}) = \frac{1}{2}\dot{q}^2 - g \cos q$, $P(q) = q$, $Q(\dot{q}) = -e\dot{q}$, and $\mathcal{Z}_p = \{(q, \dot{q}) \in \mathbb{R}^2 : q = \theta_c, \dot{q} > 0\} \cup \{(q, \dot{q}) \in \mathbb{R}^2 : q = -\theta_c, \dot{q} < 0\}$.

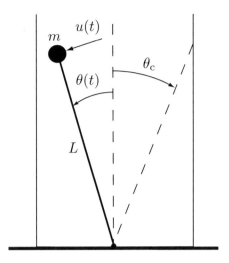

Figure 8.30 Constrained inverted pendulum.

Next, to stabilize the equilibrium point $(q_e, \dot{q}_e) = (0,0)$, consider the hybrid dynamic compensator

$$\ddot{q}_c(t) + k_c q_c(t) = k_c q(t), \quad q_c(0) = q_{c0}, \quad (q(t), \dot{q}(t), q_c(t), \dot{q}_c(t)) \notin \mathcal{Z}_c,$$
$$t \geq 0, \quad (8.208)$$

$$\begin{bmatrix} \Delta q_c(t) \\ \Delta \dot{q}_c(t) \end{bmatrix} = \begin{bmatrix} q(t) - q_c(t) \\ -\dot{q}_c(t) \end{bmatrix}, \quad (q(t), \dot{q}(t), q_c(t), \dot{q}_c(t)) \in \mathcal{Z}_c, (8.209)$$

$$u_c(t) = -k_p q + k_c(q_c(t) - q(t)), \quad (8.210)$$

where $k_p > g$ and $k_c > 0$, with the resetting set (8.203) taking the form

$$\mathcal{Z}_c = \left\{ (q, \dot{q}, q_c, \dot{q}_c) : k_c(q_c - q)\dot{q} = 0 \text{ and } \begin{bmatrix} q - q_c \\ -\dot{q}_c \end{bmatrix} \neq 0 \right\}. \quad (8.211)$$

To illustrate the behavior of the closed-loop impulsive dynamical system, let $\theta_c = \frac{\pi}{6}$, $g = 9.8$, $e = 0.5$, $k_p = 9.9$, and $k_c = 2$ with initial conditions $q(0) = 0$, $\dot{q}(0) = 1$, $q_c(0) = 0$, and $\dot{q}_c(0) = 0$. For this system a straightforward, but lengthy, calculation shows that Assumptions A1, A2, and 8.1 hold. Figure 8.31 shows the phase portrait of the closed-loop impulsive dynamical system with $x_1 = q$ and $x_2 = \dot{q}$. Figure 8.32 shows the controlled plant position and velocity states versus time, while Figure 8.33 shows the controller position and velocity versus time. Figure 8.34 shows the control force versus time. Note that for this example the plant velocity and the controller velocity are the only states that reset. Furthermore, in this

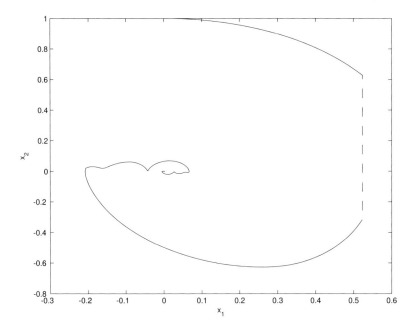

Figure 8.31 Phase portrait of the constraint inverted pendulum.

case, the control force is continuous since the plant position and the controller position are continuous functions of time.

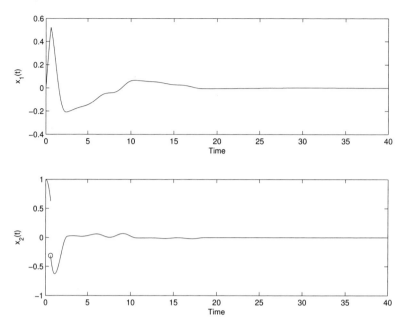

Figure 8.32 Plant position and velocity versus time.

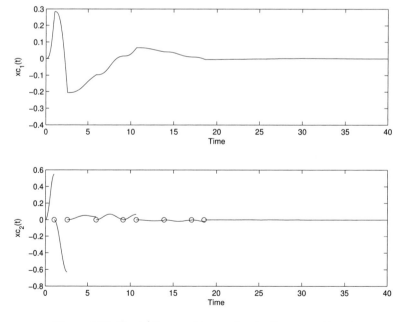

Figure 8.33 Controller position and velocity versus time.

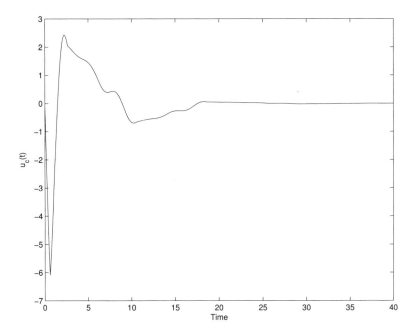

Figure 8.34 Control signal versus time.

Chapter Nine

Optimal Control for Impulsive Dynamical Systems

9.1 Introduction

In this chapter, we consider a hybrid feedback optimal control problem over an infinite horizon involving a hybrid nonlinear-nonquadratic performance functional. The performance functional involves a continuous-time cost for addressing performance of the continuous-time system dynamics and a discrete-time cost for addressing performance at the resetting instants. Furthermore, the hybrid cost functional can be evaluated in closed form as long as the nonlinear-nonquadratic cost functional considered is related in a specific way to an underlying Lyapunov function that guarantees asymptotic stability of the nonlinear closed-loop impulsive system. This Lyapunov function is shown to be a solution of a steady-state, hybrid Hamilton-Jacobi-Bellman equation, and hence, guarantees both optimality and stability of the feedback controlled impulsive dynamical system. The overall framework provides the foundation for extending linear-quadratic feedback control methods to nonlinear impulsive dynamical systems. We note that the optimal control framework for impulsive dynamical systems developed herein is quite different from the quasi-variational inequality methods for impulsive and hybrid control developed in the literature (e.g., [16–18, 30]). Specifically, quasi-variational methods do not guarantee asymptotic stability via Lyapunov functions and do not necessarily yield feedback controllers. In contrast, the proposed approach provides hybrid *feedback* controllers guaranteeing closed-loop stability via an underlying Lyapunov function.

9.2 Impulsive Optimal Control

In this section, we consider an optimal control problem for nonlinear impulsive dynamical systems involving a notion of optimality with respect to a hybrid nonlinear-nonquadratic performance functional. Specifically, we consider the following impulsive optimal control problem.

Impulsive Optimal Control Problem. Consider the nonlinear impulsive controlled system given by

$$\dot{x}(t) = F_c(x(t), u_c(t), t), \quad x(t_0) = x_0, \quad x(t_f) = x_f, \quad u_c(t) \in U_c,$$

$$(t, x(t)) \notin \mathcal{S}_x, \qquad (9.1)$$

$$\Delta x(t) = F_d(x(t), u_d(t), t), \quad u_d(t) \in U_d, \quad (t, x(t)) \in \mathcal{S}_x, \qquad (9.2)$$

where $t \geq 0$, $x(t) \in \mathcal{D} \subseteq \mathbb{R}^n$ is the state vector, $\mathcal{D}$ is an open set with $0 \in \mathcal{D}$, $(u_c(t), u_d(t_k)) \in U_c \times U_d \subseteq \mathbb{R}^{m_c} \times \mathbb{R}^{m_d}$, $t \in [t_0, t_f]$, $k \in \mathbb{Z}_{[t_0, t_f)}$, is the hybrid control input, $x(t_0) = x_0$ is given, $x(t_f) = x_f$ is fixed, $F_c : \mathcal{D} \times U_c \times \mathbb{R} \to \mathbb{R}^n$ is Lipschitz continuous and satisfies $F_c(0, 0, t) = 0$ for every $t \in [t_0, t_f]$, $F_d : \mathcal{S}_x \times U_d \to \mathbb{R}^n$ is continuous and satisfies $F_d(0, 0, t) = 0$ for every $t \in [t_0, t_f]$, and $\mathcal{S}_x \subset [0, \infty) \times \mathbb{R}^n$. Then determine the control inputs $(u_c(t), u_d(t_k)) \in U_c \times U_d$, $t \in [t_0, t_f]$, $k \in \mathbb{Z}_{[t_0, t_f)}$, such that the hybrid performance functional

$$J(x_0, u_c(\cdot), u_d(\cdot), t_0) = \int_{t_0}^{t_f} L_c(x(t), u_c(t), t) dt$$

$$+ \sum_{k \in \mathbb{Z}_{[t_0, t_f)}} L_d(x(t_k), u_d(t_k), t_k) \qquad (9.3)$$

is minimized over all admissible control inputs $(u_c(\cdot), u_d(\cdot)) \in \mathcal{U}_c \times \mathcal{U}_d$, where $L_c : \mathcal{D} \times U_c \times \mathbb{R} \to \mathbb{R}$ and $L_d : \mathcal{S}_x \times U_d \to \mathbb{R}$ are given.

Next, we present a hybrid version of Bellman's principle of optimality which provides necessary and sufficient conditions, with a given hybrid control $(u_c(t), u_d(t_k)) \in U_c \times U_d$, $t \geq t_0$, $k \in \mathbb{Z}_{[t, t_0)}$, for minimizing the performance functional (9.3).

Lemma 9.1 *Let $(u_c^*(\cdot), u_d^*(\cdot)) \in \mathcal{U}_c \times \mathcal{U}_d$ be an optimal hybrid control that generates the trajectory $x(t)$, $t \in [t_0, t_f]$, with $x(t_0) = x_0$. Then the trajectory $x(\cdot)$ from (t_0, x_0) to (t_f, x_f) is optimal if and only if for all $t', t'' \in [t_0, t_f]$, the portion of the trajectory $x(\cdot)$ going from $(t', x(t'))$ to $(t'', x(t''))$ optimizes the same cost functional over $[t', t'']$, where $x(t') = x_1$ is a point on the optimal trajectory generated by $(u_c^*(\cdot), u_d^*(\cdot))$.*

Proof. Let $u_c^*(\cdot) \in \mathcal{U}_c$ and $u_d^*(\cdot) \in \mathcal{U}_d$ solve the Impulsive Optimal Control Problem and let $x(t)$, $t \in [t_0, t_f]$, be the solution to (9.1) and (9.2) generated by $u_c^*(\cdot)$ and $u_d^*(\cdot)$. Next, *ad absurdum*, suppose there exist $t' \geq t_0$, $t'' \leq t_f$, and $\hat{u}_c(t)$, $t \in [t', t'']$, $\hat{u}_d(t_k)$, $k \in \mathbb{Z}_{[t', t'')}$, such that

$$\int_{t'}^{t''} L_c(\hat{x}(t), \hat{u}_c(t), t) dt + \sum_{k \in \mathbb{Z}_{[t', t'')}} L_d(\hat{x}(t_k), \hat{u}_d(t_k), t_k)$$

$$< \int_{t'}^{t''} L_c(x(t), u_c^*(t), t)\mathrm{d}t + \sum_{k \in \mathbb{Z}_{[t',t'')}} L_d(x(t_k), u_d^*(t_k), t_k), \quad (9.4)$$

where $\hat{x}(t)$ is a solution of (9.1) and (9.2) for all $t \in [t', t'']$ with $u_c(t) = \hat{u}_c(t)$, $u_d(t_k) = \hat{u}_d(t_k)$, $\hat{x}(t') = x(t')$, and $\hat{x}(t'') = x(t'')$. Now, define

$$u_{co}(t) \triangleq \begin{cases} u_c^*(t), & [t_0, t'], \\ \hat{u}_c(t), & [t', t''], \\ u_c^*(t), & [t'', t_f], \end{cases} \qquad u_{do}(t_k) \triangleq \begin{cases} u_d^*(t_k), & k \in \mathbb{Z}_{[t_0,t')}, \\ \hat{u}_d(t_k), & k \in \mathbb{Z}_{[t',t'')}, \\ u_d^*(t_k), & k \in \mathbb{Z}_{[t'',t_f)}. \end{cases}$$

Then,

$$J(x_0, u_{co}(\cdot), u_{do}(\cdot), t_0) = \int_{t_0}^{t_f} L_c(x(t), u_{co}(t), t)\mathrm{d}t$$

$$+ \sum_{k \in \mathbb{Z}_{[t_0,t_f)}} L_d(x(t_k), u_{do}(t_k), t_k)$$

$$= \int_{t_0}^{t'} L_c(x(t), u_c^*(t), t)\mathrm{d}t$$

$$+ \sum_{k \in \mathbb{Z}_{[t_0,t')}} L_d(x(t_k), u_d^*(t_k), t_k)$$

$$+ \int_{t'}^{t''} L_c(\hat{x}(t), \hat{u}_c(t), t)\mathrm{d}t$$

$$+ \sum_{k \in \mathbb{Z}_{[t',t'')}} L_d(\hat{x}(t_k), \hat{u}_d(t_k), t_k)$$

$$+ \int_{t''}^{t_f} L_c(x(t), u_c^*(t), t)\mathrm{d}t$$

$$+ \sum_{k \in \mathbb{Z}_{[t'',t_f)}} L_d(x(t_k), u_d^*(t_k), t_k)$$

$$< \int_{t_0}^{t'} L_c(x(t), u_c^*(t), t)\mathrm{d}t$$

$$+ \sum_{k \in \mathbb{Z}_{[t_0,t')}} L_d(x(t_k), u_d^*(t_k), t_k)$$

$$+ \int_{t'}^{t''} L_c(x(t), u_c^*(t), t)\mathrm{d}t$$

$$+ \sum_{k \in \mathbb{Z}_{[t',t'')}} L_d(x(t_k), u_d^*(t_k), t_k)$$

$$+ \int_{t''}^{t_f} L_c(x(t), u_c^*(t), t)dt$$

$$+ \sum_{k \in \mathbb{Z}_{[t'', t_f)}} L_d(x(t_k), u_d^*(t_k), t_k)$$

$$= J(x_0, u_c^*(\cdot), u_d^*(\cdot), t_0),$$

which is a contradiction.

Conversely, if $(u_c^*(\cdot), u_d^*(\cdot))$ minimizes $J(\cdot, \cdot, \cdot, \cdot)$ over $[t', t'']$ and $k \in \mathbb{Z}_{[t', t'')}$ for all $t' \geq t_0$ and $t'' \leq t_f$, then $(u_c^*(\cdot), u_d^*(\cdot))$ minimizes $J(\cdot, \cdot, \cdot, \cdot)$ over $[t_0, t_f]$.   □

Next, let $(u_c^*(t), u_d^*(t_k))$, $t \in [t_0, t_f]$, $k \in \mathbb{Z}_{[t_0, t_f)}$, solve the Impulsive Optimal Control Problem and define the optimal cost $J^*(x_0, t_0) \triangleq J(x_0, u_c^*(\cdot), u_d^*(\cdot), t_0)$. Furthermore, define, for $p \in \mathbb{R}^n$ and $q : \mathbb{R}^n \times \mathbb{R} \to \mathbb{R}$, the Hamiltonians

$$H_c(x, u_c, p(x, t), t) \triangleq L_c(x, u_c, t) + p^T(x, t)F_c(x, u_c, t), \quad (9.5)$$

$$H_d(x, u_d, q(x, t_k), t_k) \triangleq L_d(x, u_d, t_k) + q(x + F_d(x, u_d, t_k), t_k)$$
$$- q(x, t_k). \quad (9.6)$$

Theorem 9.1 *Let $J^*(x, t)$ denote the minimal cost for the Impulsive Optimal Control Problem with $x_0 = x$ and $t_0 = t$, and assume that $J^*(\cdot, \cdot)$ is continuously differentiable in x. Then*

$$0 = \frac{\partial J^*(x(t), t)}{\partial t} + \min_{u_c(\cdot) \in \mathcal{U}_c} H_c(x(t), u_c(t), p(x(t), t), t),$$

$$(t, x(t)) \notin \mathcal{S}_x, \quad (9.7)$$

$$0 = \min_{u_d(\cdot) \in \mathcal{U}_d} H_d(x(t), u_d(t), q(x(t), t), t), \quad (t, x(t)) \in \mathcal{S}_x, \quad (9.8)$$

where $p(x(t), t) \triangleq \left(\frac{\partial J^*(x(t), t)}{\partial x} \right)^T$ and $q(x(t), t) \triangleq J^*(x(t), t)$. Furthermore, if $(u_c^*(\cdot), u_d^*(\cdot))$ solves the Impulsive Optimal Control Problem, then

$$0 = \frac{\partial J^*(x(t), t)}{\partial t} + H_c(x(t), u_c^*(t), p(x(t), t), t), \quad (t, x(t)) \notin \mathcal{S}_x, \quad (9.9)$$

$$0 = H_d(x(t), u_d^*(t), q(x(t), t), t), \quad (t, x(t)) \in \mathcal{S}_x. \quad (9.10)$$

Proof. Let $(t, x(t)) \notin \mathcal{S}_x$. It follows from Lemma 9.1 that for small enough $\varepsilon > 0$ and $t' \in [t, t + \varepsilon]$,

$$J^*(x(t), t) = \min_{(u_c(\cdot), u_d(\cdot)) \in \mathcal{U}_c \times \mathcal{U}_d} \left[\int_t^{t_f} L_c(x(s), u_c(s), s)ds \right.$$

$$+ \sum_{k\in\mathbb{Z}_{[t,t_{\mathrm{f}})}} L_{\mathrm{d}}(x(t_k), u_{\mathrm{d}}(t_k), t_k) \Bigg]$$

$$= \min_{u_{\mathrm{c}}(\cdot)\in\mathcal{U}_{\mathrm{c}}} \int_t^{t'} L_{\mathrm{c}}(x(s), u_{\mathrm{c}}(s), s)\mathrm{d}s$$

$$+ \min_{(u_{\mathrm{c}}(\cdot), u_{\mathrm{d}}(\cdot))\in\mathcal{U}_{\mathrm{c}}\times\mathcal{U}_{\mathrm{d}}} \Bigg[\int_{t'}^{t_{\mathrm{f}}} L_{\mathrm{c}}(x(s), u_{\mathrm{c}}(s), s)\mathrm{d}s$$

$$+ \sum_{k\in\mathbb{Z}_{[t',t_{\mathrm{f}})}} L_{\mathrm{d}}(x(t_k), u_{\mathrm{d}}(t_k), t_k) \Bigg]$$

$$= \min_{u_{\mathrm{c}}(\cdot)\in\mathcal{U}_{\mathrm{c}}} \Bigg[\int_t^{t'} L_{\mathrm{c}}(x(s), u_{\mathrm{c}}(s), s)\mathrm{d}s + J^*(x(t'), t') \Bigg],$$

or, equivalently,

$$0 = \min_{u_{\mathrm{c}}(\cdot)\in\mathcal{U}_{\mathrm{c}}} \Bigg[\frac{1}{t'-t} \left[J^*(x(t'), t') - J^*(x(t), t) \right]$$

$$+ \frac{1}{t'-t} \int_t^{t'} L_{\mathrm{c}}(x(s), u_{\mathrm{c}}(s), s)\mathrm{d}s \Bigg].$$

Letting $t' \to t$ yields

$$0 = \min_{u_{\mathrm{c}}(\cdot)\in\mathcal{U}_{\mathrm{c}}} \left[\frac{\mathrm{d}J^*(x(t), t)}{\mathrm{d}t} + L_{\mathrm{c}}(x(t), u_{\mathrm{c}}(t), t) \right].$$

Now, (9.7) and (9.9) follow by noting that

$$\frac{\mathrm{d}J^*(x(t), t)}{\mathrm{d}t} = \frac{\partial J^*(x(t), t)}{\partial t} + \frac{\partial J^*(x(t), t)}{\partial x} F_{\mathrm{c}}(x(t), u_{\mathrm{c}}(t), t).$$

Next, let $(t, x(t)) \in \mathcal{S}_x$. It follows from Lemma 9.1 that

$$J^*(x(t), t) = \min_{(u_{\mathrm{c}}(\cdot), u_{\mathrm{d}}(\cdot))\in\mathcal{U}_{\mathrm{c}}\times\mathcal{U}_{\mathrm{d}}} \Bigg[L_{\mathrm{d}}(x(t), u_{\mathrm{d}}(t), t)$$

$$+ \int_{t+}^{t_{\mathrm{f}}} L_{\mathrm{c}}(x(t), u_{\mathrm{c}}(t), t)\mathrm{d}t + \sum_{k\in\mathbb{Z}_{[t+,t_{\mathrm{f}})}} L_{\mathrm{d}}(x(t_k), u_{\mathrm{d}}(t_k), t_k) \Bigg]$$

$$= \min_{u_{\mathrm{d}}(\cdot)\in\mathcal{U}_{\mathrm{d}}} L_{\mathrm{d}}(x(t), u_{\mathrm{d}}(t), t) + J^*(x(t^+), t^+)$$

$$= \min_{u_{\mathrm{d}}(\cdot) \in \mathcal{U}_{\mathrm{d}}} L_{\mathrm{d}}(x(t), u_{\mathrm{d}}(t), t) + J^*(x(t) + F_{\mathrm{d}}(x(t), u_{\mathrm{d}}^*(t), t), t),$$

which implies (9.8) and (9.10). □

Next, we provide a converse result to Theorem 9.1.

Theorem 9.2 *Suppose there exists a continuously differentiable function* $V : \mathcal{D} \times \mathbb{R} \to \mathbb{R}$ *and an optimal control* $(u_{\mathrm{c}}^*(\cdot), u_{\mathrm{d}}^*(\cdot))$ *such that* $V(x(t_{\mathrm{f}}), t_{\mathrm{f}}) = 0$,

$$0 = \frac{\partial V(x, t)}{\partial t} + H_{\mathrm{c}}\left(x, u_{\mathrm{c}}^*(t), \left(\frac{\partial V(x, t)}{\partial x}\right)^{\mathrm{T}}, t\right), \quad (t, x) \notin \mathcal{S}_x, \quad (9.11)$$

$$0 = H_{\mathrm{d}}(x, u_{\mathrm{d}}^*(t), V(x, t), t), \quad (t, x) \in \mathcal{S}_x, \quad (9.12)$$

$$H_{\mathrm{c}}\left(x, u_{\mathrm{c}}^*(t), \left(\frac{\partial V(x, t)}{\partial x}\right)^{\mathrm{T}}, t\right) \le H_{\mathrm{c}}\left(x, u_{\mathrm{c}}(t), \left(\frac{\partial V(x, t)}{\partial x}\right)^{\mathrm{T}}, t\right),$$
$$(t, x) \notin \mathcal{S}_x, \quad u_{\mathrm{c}}(\cdot) \in \mathcal{U}_{\mathrm{c}}, \quad (9.13)$$

$$H_{\mathrm{d}}(x, u_{\mathrm{d}}^*(t), V(x, t), t) \le H_{\mathrm{d}}(x, u_{\mathrm{d}}(t), V(x, t), t),$$
$$(t, x) \in \mathcal{S}_x, \quad u_{\mathrm{d}}(\cdot) \in \mathcal{U}_{\mathrm{d}}, \quad (9.14)$$

where $H_{\mathrm{c}}(\cdot, \cdot, \cdot, \cdot)$ *and* $H_{\mathrm{d}}(\cdot, \cdot, \cdot, \cdot)$ *are given by (9.5) and (9.6), respectively. Then* $(u_{\mathrm{c}}^*(\cdot), u_{\mathrm{d}}^*(\cdot))$ *solves the Impulsive Optimal Control Problem, that is,*

$$J^*(x_0, t_0) = J(x_0, u_{\mathrm{c}}^*(\cdot), u_{\mathrm{d}}^*(\cdot), t_0) \le J(x_0, u_{\mathrm{c}}(\cdot), u_{\mathrm{d}}(\cdot), t_0),$$
$$(u_{\mathrm{c}}(\cdot), u_{\mathrm{d}}(\cdot)) \in \mathcal{U}_{\mathrm{c}} \times \mathcal{U}_{\mathrm{d}}, \quad (9.15)$$

and

$$J^*(x_0, t_0) = V(x_0, t_0). \quad (9.16)$$

Proof. Let $x(t)$, $t \ge t_0$, satisfy (9.1) and (9.2) and, for all $(t, x(t)) \notin \mathcal{S}_x$, define

$$\dot{V}(x(t), t) \triangleq \frac{\partial V(x(t), t)}{\partial t} + \frac{\partial V(x(t), t)}{\partial x} F_{\mathrm{c}}(x(t), u_{\mathrm{c}}(t), t).$$

Then, with $u_{\mathrm{c}}(t) = u_{\mathrm{c}}^*(t)$, it follows from (9.11) that

$$0 = \dot{V}(x(t), t) + L_{\mathrm{c}}(x(t), u_{\mathrm{c}}^*(t), t), \quad (t, x(t)) \notin \mathcal{S}_x. \quad (9.17)$$

Furthermore, it follows from (9.12) that

$$0 = V(x(t) + F_d(x(t), u_d^*(t), t)) - V(x(t), t) + L_d(x(t), u_d^*(t), t),$$
$$(t, x(t)) \in \mathcal{S}_x. \quad (9.18)$$

Now, noting that $V(x(t_f), t_f) = 0$, it follows from (9.17) and (9.18) that

$$J^*(x_0, t_0) = J(x_0, u_c^*(\cdot), u_d^*(\cdot), t_0)$$
$$= \int_{t_0}^{t_f} L_c(x(t), u_c^*(t), t) dt + \sum_{k \in \mathbb{Z}_{[t_0, t_f)}} L_d(x(t_k), u_d^*(t_k), t_k)$$
$$= V(x_0, t_0).$$

Next, for all $(u_c(\cdot), u_d(\cdot)) \in \mathcal{U}_c \times \mathcal{U}_d$ it follows from (9.11)–(9.14) that

$$J(x_0, u_c(\cdot), u_d(\cdot), t_0) = \int_{t_0}^{t_f} L_c(x(t), u_c(t), t) dt$$
$$+ \sum_{k \in \mathbb{Z}_{[t_0, t_f)}} L_d(x(t_k), u_d(t_k), t_k)$$
$$= \int_{t_0}^{t_f} \left[-\dot{V}(x(t), t) + \frac{\partial V(x(t), t)}{\partial t} \right.$$
$$\left. + H_c\left(x(t), u_c(t), \left(\frac{\partial V(x(t), t)}{\partial x}\right)^{\mathrm{T}}, t\right) \right] dt$$
$$+ \sum_{k \in \mathbb{Z}_{[t_0, t_f)}} [V(x(t_k), t_k) - V(x(t_k)$$
$$+ F_d(x(t), u_d(t_k), t_k))$$
$$+ H_d(x(t_k), u_d(t_k), V(x(t_k), t_k), t_k)]$$
$$\geq \int_{t_0}^{t_f} -\dot{V}(x(t), t) dt + \sum_{k \in \mathbb{Z}_{[t_0, t_f)}} [V(x(t_k), t_k)$$
$$- V(x(t_k) + F_d(x(t), u_d(t_k), t_k))]$$
$$+ \int_{t_0}^{t_f} \left[\frac{\partial V(x(t), t)}{\partial t} \right.$$
$$\left. + H_c\left(x(t), u_c^*(t), \left(\frac{\partial V(x(t), t)}{\partial x}\right)^{\mathrm{T}}, t\right) \right] dt$$
$$+ \sum_{k \in \mathbb{Z}_{[t_0, t_f)}} H_d(x(t_k), u_d^*(t_k), V(x(t_k), t_k), t_k)$$

$$= V(x_0, t_0) - V(x(t_\mathrm{f}), t_\mathrm{f})$$
$$= J^*(x_0, t_0),$$

which completes the proof.   □

Next, we use Theorem 9.2 to characterize optimal hybrid *feedback* controllers for nonlinear impulsive dynamical systems. In order to obtain time-invariant controllers, we restrict our attention to state-dependent impulsive dynamical systems with non-Zeno solutions and optimality notions over the infinite horizon with an infinite number of resetting times. Hence, the Impulsive Optimal Control Problem becomes

$$\dot{x}(t) = F_\mathrm{c}(x(t), u_\mathrm{c}(t)), \quad x(0) = x_0, \quad x(t) \notin \mathcal{Z}_x, \quad (9.19)$$
$$\Delta x(t) = F_\mathrm{d}(x(t), u_\mathrm{d}(t)), \quad x(t) \in \mathcal{Z}_x, \quad\quad\quad\quad (9.20)$$

where $\mathcal{Z}_x \subset \mathcal{D}$, and $u_\mathrm{c}(\cdot)$ and $u_\mathrm{d}(\cdot)$ are restricted to the class of *admissible* hybrid controls consisting of measurable functions such that $(u_\mathrm{c}(t), u_\mathrm{d}(t_k)) \in U_\mathrm{c} \times U_\mathrm{d}$ for all $t \geq 0$ and $k \in \mathbb{Z}_{[0,\infty)}$, where the constraint set $U_\mathrm{c} \times U_\mathrm{d}$ is given with $(0, 0) \in U_\mathrm{c} \times U_\mathrm{d}$. To address the optimal nonlinear feedback control problem let $\phi_\mathrm{c} : \mathcal{D} \to U_\mathrm{c}$ be such that $\phi_\mathrm{c}(0) = 0$ and let $\phi_\mathrm{d} : \mathcal{Z}_x \to U_\mathrm{d}$. If $(u_\mathrm{c}(t), u_\mathrm{d}(t_k)) = (\phi_\mathrm{c}(x(t)), \phi_\mathrm{d}(x(t_k)))$, where $x(t)$, $t \geq 0$, satisfies (9.1) and (9.2), then $(u_\mathrm{c}(\cdot), u_\mathrm{d}(\cdot))$ is a *hybrid feedback control*. Given the hybrid feedback control $(u_\mathrm{c}(t), u_\mathrm{d}(t_k)) = (\phi_\mathrm{c}(x(t)), \phi_\mathrm{d}(x(t_k)))$, the closed-loop state-dependent impulsive dynamical system has the form

$$\dot{x}(t) = F_\mathrm{c}(x(t), \phi_\mathrm{c}(x(t))), \quad x(0) = x_0, \quad x(t) \notin \mathcal{Z}_x, \quad (9.21)$$
$$\Delta x(t) = F_\mathrm{d}(x(t), \phi_\mathrm{d}(x(t))), \quad x(t) \in \mathcal{Z}_x. \quad\quad\quad\quad (9.22)$$

Now, we present the main theorem for characterizing hybrid feedback controllers that guarantee closed-loop stability and minimize a hybrid nonlinear-nonquadratic performance functional over the infinite horizon. For the statement of this result, recall that with $\mathcal{S}_x \triangleq [0, \infty) \times \mathcal{Z}_x$ it follows from Assumptions A1 and A2 of Chapter 2 that the resetting times $t_k(= \tau_k(x_0))$ are well defined and distinct for every trajectory of (9.21) and (9.22). Furthermore, define the set of regulation hybrid controllers by

$$\mathcal{C}(x_0) \triangleq \{(u_\mathrm{c}(\cdot), u_\mathrm{d}(\cdot)) : (u_\mathrm{c}(\cdot), u_\mathrm{d}(\cdot)) \text{ is admissible and } x(\cdot) \text{ given by}$$
$$(9.19) \text{ and } (9.20) \text{ satisfies } x(t) \to 0 \text{ as } t \to \infty\}.$$

Theorem 9.3 *Consider the nonlinear controlled impulsive system (9.21) and (9.22) with hybrid performance functional*

$$J(x_0, u_c(\cdot), u_d(\cdot)) = \int_0^\infty L_c(x(t), u_c(t))dt + \sum_{k \in \mathbb{Z}_{[0,\infty)}} L_d(x(t_k), u_d(t_k)),$$
(9.23)

where $(u_c(\cdot), u_d(\cdot))$ *is an admissible hybrid control. Assume there exist a continuously differentiable function* $V : \mathcal{D} \to \mathbb{R}$ *and a hybrid control law* $\phi_c : \mathcal{D} \to U_c$ *and* $\phi_d : \mathcal{Z}_x \to U_d$ *such that* $V(0) = 0$, $V(x) > 0$, $x \neq 0$, $\phi_c(0) = 0$, *and*

$$V'(x)F_c(x, \phi_c(x)) < 0, \quad x \notin \mathcal{Z}_x, \quad x \neq 0, \tag{9.24}$$
$$V(x + F_d(x, \phi_d(x))) - V(x) \leq 0, \quad x \in \mathcal{Z}_x, \tag{9.25}$$
$$H_c(x, \phi_c(x)) = 0, \quad x \notin \mathcal{Z}_x, \tag{9.26}$$
$$H_d(x, \phi_d(x)) = 0, \quad x \in \mathcal{Z}_x, \tag{9.27}$$
$$H_c(x, u_c) \geq 0, \quad x \notin \mathcal{Z}_x, \quad u_c \in U_c, \tag{9.28}$$
$$H_d(x, u_d) \geq 0, \quad x \in \mathcal{Z}_x, \quad u_d \in U_d, \tag{9.29}$$

where

$$H_c(x, u_c) \triangleq L_c(x, u_c) + V'(x)F_c(x, u_c), \tag{9.30}$$
$$H_d(x, u_d) \triangleq L_d(x, u_d) + V(x + F_d(x, u_d)) - V(x). \tag{9.31}$$

Then, with the hybrid feedback control $(u_c(\cdot), u_d(\cdot)) = (\phi_c(x(\cdot)), \phi_d(x(\cdot)))$, *there exists a neighborhood of the origin* $\mathcal{D}_0 \subseteq \mathcal{D}$ *such that if* $x_0 \in \mathcal{D}_0$, *the zero solution* $x(t) \equiv 0$ *of the closed-loop system (9.21) and (9.22) is locally asymptotically stable. Furthermore,*

$$J(x_0, \phi_c(x(\cdot)), \phi_d(x(\cdot))) = V(x_0), \quad x_0 \in \mathcal{D}_0. \tag{9.32}$$

In addition, if $x_0 \in \mathcal{D}_0$ *then the hybrid feedback control* $(u_c(\cdot), u_d(\cdot)) = (\phi_c(x(\cdot)), \phi_d(x(\cdot)))$ *minimizes* $J(x_0, u_c(\cdot), u_d(\cdot))$ *in the sense that*

$$J(x_0, \phi_c(x(\cdot)), \phi_d(x(\cdot))) = \min_{(u_c(\cdot), u_d(\cdot)) \in \mathcal{C}(x_0)} J(x_0, u_c(\cdot), u_d(\cdot)). \tag{9.33}$$

Finally, if $\mathcal{D} = \mathbb{R}^n$, $U_c = \mathbb{R}^{m_c}$, $U_d = \mathbb{R}^{m_d}$, *and* $V(x) \to \infty$ *as* $\|x\| \to \infty$, *then the zero solution* $x(t) \equiv 0$ *of the closed-loop system (9.21) and (9.22) is globally asymptotically stable.*

Proof. Local and global asymptotic stability is a direct consequence of (9.24) and (9.25) by applying Theorem 2.1 to the closed-loop system (9.21) and (9.22). Conditions (9.32) and (9.33) are a direct consequence of Theorem 9.2, with $V(x,t) = V(x)$, $t_0 = 0$, $t_f \to \infty$, and using the fact that $\lim_{t \to \infty} V(x(t)) = 0$ and $\lim_{k \to \infty} V(x(t_k)) = 0$.

$\square$

Note that (9.26) and (9.27) are the steady-state hybrid Hamilton-Jacobi-Bellman equation for the nonlinear hybrid system (9.19) and (9.20) with the hybrid performance criterion $J(x_0, u_c(\cdot), u_d(\cdot))$ given by (9.23). Furthermore, Theorem 9.3 guarantees optimality with respect to the set of admissible stabilizing hybrid controllers $\mathcal{C}(x_0)$. However, it is important to note that an explicit characterization of $\mathcal{C}(x_0)$ is not required. In addition, the optimal stabilizing hybrid *feedback* control law $(u_c, u_d) = (\phi_c(x), \phi_d(x))$ is independent of the initial condition x_0. Finally, in order to assure asymptotic stability of the hybrid closed-loop system (9.21) and (9.22), Theorem 9.3 requires that $V(\cdot)$ satisfy (9.24) and (9.25) which implies that $V(\cdot)$ is a Lyapunov function for the hybrid closed-loop system (9.21) and (9.22).

Next, we specialize Theorem 9.3 to linear impulsive systems. For the following result let $A_c \in \mathbb{R}^{n \times n}$, $B_c \in \mathbb{R}^{n \times m_c}$, $A_d \in \mathbb{R}^{n \times n}$, $B_d \in \mathbb{R}^{n \times m_d}$, $R_{1c} \in \mathbb{R}^{n \times n}$, $R_{2c} \in \mathbb{R}^{m_c \times m_c}$, $R_{1d} \in \mathbb{R}^{n \times n}$, and $R_{2d} \in \mathbb{R}^{m_d \times m_d}$ be given, where R_{1c}, R_{2c}, R_{1d}, and R_{2d} are positive definite.

Corollary 9.1 *Consider the linear controlled impulsive system*

$$\dot{x}(t) = A_c x(t) + B_c u_c(t), \quad x(0) = x_0, \quad x(t) \notin \mathcal{Z}_x, \quad (9.34)$$
$$\Delta x(t) = (A_d - I_n)x(t) + B_d u_d(t), \quad x(t) \in \mathcal{Z}_x, \quad (9.35)$$

with quadratic hybrid performance functional

$$J(x_0, u_c(\cdot), u_d(\cdot)) = \int_0^\infty [x^T(t)R_{1c}x(t) + u_c^T(t)R_{2c}u_c(t)]dt$$
$$+ \sum_{k \in \mathbb{Z}_{[0,\infty)}} [x^T(t_k)R_{1d}x(t_k) + u_d^T(t_k)R_{2d}u_d(t_k)],$$

$$(9.36)$$

where $(u_c(\cdot), u_d(\cdot))$ is an admissible hybrid control. Furthermore, assume there exists a positive-definite matrix $P \in \mathbb{R}^{n \times n}$ such that

$$0 = x^T(A_c^T P + PA_c + R_{1c} - PB_c R_{2c}^{-1} B_c^T P)x, \quad x \notin \mathcal{Z}_x, \quad (9.37)$$
$$0 = x^T(A_d^T PA_d - P + R_{1d} - A_d^T PB_d(R_{2d} + B_d^T PB_d)^{-1}B_d^T PA_d)x,$$
$$x \in \mathcal{Z}_x. \quad (9.38)$$

Then, the zero solution $x(t) \equiv 0$ to (9.34) and (9.35) is globally asymptotically stable with the hybrid feedback controller

$$u_c = \phi_c(x) = -R_{2c}^{-1}B_c^T Px, \quad x \notin \mathcal{Z}_x, \quad (9.39)$$

$$u_\mathrm{d} = \phi_\mathrm{d}(x) = -(R_{2\mathrm{d}} + B_\mathrm{d}^\mathrm{T} P B_\mathrm{d})^{-1} B_\mathrm{d}^\mathrm{T} P A_\mathrm{d} x, \quad x \in \mathcal{Z}_x, \quad (9.40)$$

and

$$J(x_0, \phi_\mathrm{c}(\cdot), \phi_\mathrm{d}(\cdot)) = x_0^\mathrm{T} P x_0, \quad x_0 \in \mathbb{R}^n. \quad (9.41)$$

Furthermore,

$$J(x_0, \phi_\mathrm{c}(\cdot), \phi_\mathrm{d}(\cdot)) = \min_{(u_\mathrm{c}(\cdot), u_\mathrm{d}(\cdot)) \in \mathcal{C}(x_0)} J(x_0, u_\mathrm{c}(\cdot), u_\mathrm{d}(\cdot)), \quad (9.42)$$

where $\mathcal{C}(x_0)$ is the set of regulation hybrid controllers for (9.34) and (9.35) and $x_0 \in \mathbb{R}^n$.

Proof. The result is a direct consequence of Theorem 9.3 with $F_\mathrm{c}(x, u_\mathrm{c}) = A_\mathrm{c} x + B_\mathrm{c} u_\mathrm{c}$, $L_\mathrm{c}(x, u_\mathrm{c}) = x^\mathrm{T} R_{1\mathrm{c}} x + u_\mathrm{c}^\mathrm{T} R_{2\mathrm{c}} u_\mathrm{c}$, for $x \notin \mathcal{Z}_x$, $F_\mathrm{d}(x, u_\mathrm{d}) = (A_\mathrm{d} - I_n) x + B_\mathrm{d} u_\mathrm{d}$, $L_\mathrm{d}(x, u_\mathrm{d}) = x^\mathrm{T} R_{1\mathrm{d}} x + u_\mathrm{d}^\mathrm{T} R_{2\mathrm{d}} u_\mathrm{d}$, for $x \in \mathcal{Z}_x$, $V(x) = x^\mathrm{T} P x$, $\mathcal{D} = \mathbb{R}^n$, and $U_\mathrm{c} \times U_\mathrm{d} = \mathbb{R}^{m_\mathrm{c}} \times \mathbb{R}^{m_\mathrm{d}}$. Specifically, it follows from (9.37) that $H_\mathrm{c}(x, \phi_\mathrm{c}(x)) = 0$, $x \notin \mathcal{Z}_x$, and hence, $V'(x) F_\mathrm{c}(x, \phi_\mathrm{c}(x)) < 0$ for all $x \neq 0$ and $x \notin \mathcal{Z}_x$. Similarly, it follows from (9.38) that $H_\mathrm{d}(x, \phi_\mathrm{d}(x)) = 0$, $x \in \mathcal{Z}_x$, and hence, $V(x + F_\mathrm{d}(x, \phi_\mathrm{d}(x))) - V(x) < 0$ for all $x \neq 0$ and $x \in \mathcal{Z}_x$. Thus, $H_\mathrm{c}(x, u_\mathrm{c}) = H_\mathrm{c}(x, u_\mathrm{c}) - H_\mathrm{c}(x, \phi_\mathrm{c}(x)) = [u_\mathrm{c} - \phi_\mathrm{c}(x)]^\mathrm{T} R_{2\mathrm{c}} [u_\mathrm{c} - \phi_\mathrm{c}(x)] \geq 0$, $x \notin \mathcal{Z}_x$, and $H_\mathrm{d}(x, u_\mathrm{d}) = H_\mathrm{d}(x, u_\mathrm{d}) - H_\mathrm{d}(x, \phi_\mathrm{d}(x)) = [u_\mathrm{d} - \phi_\mathrm{d}(x)]^\mathrm{T} (R_{2\mathrm{d}} + B_\mathrm{d}^\mathrm{T} P B_\mathrm{d}) [u_\mathrm{d} - \phi_\mathrm{d}(x)] \geq 0$, $x \in \mathcal{Z}_x$, so that all conditions of Theorem 9.3 are satisfied. Finally, since $V(\cdot)$ is radially unbounded, the zero solution $x(t) \equiv 0$ to (9.34) and (9.35) with $u_\mathrm{c}(t) = \phi_\mathrm{c}(x(t)) = -R_{2\mathrm{c}}^{-1} B_\mathrm{c}^\mathrm{T} P x(t)$, $x(t) \notin \mathcal{Z}_x$, and $u_\mathrm{d}(t) = \phi_\mathrm{d}(x(t)) = -(R_{2\mathrm{d}} + B_\mathrm{d}^\mathrm{T} P B_\mathrm{d})^{-1} B_\mathrm{d}^\mathrm{T} P A_\mathrm{d} x(t)$, $x(t) \in \mathcal{Z}_x$, is globally asymptotically stable. $\qquad\square$

The optimal hybrid feedback control law $(\phi_\mathrm{c}(x), \phi_\mathrm{d}(x))$ in Corollary 9.1 is derived using the properties of $H_\mathrm{c}(x, u_\mathrm{c})$ and $H_\mathrm{d}(x, u_\mathrm{d})$ as defined in Theorem 9.3. Specifically, since

$$H_\mathrm{c}(x, u_\mathrm{c}) = x^\mathrm{T} R_{1\mathrm{c}} x + u_\mathrm{c}^\mathrm{T} R_{2\mathrm{c}} u_\mathrm{c} + x^\mathrm{T} (A_\mathrm{c}^\mathrm{T} P + P A_\mathrm{c}) x + 2 x^\mathrm{T} P B_\mathrm{c} u_\mathrm{c},$$
$$x \notin \mathcal{Z}_x, \quad (9.43)$$

and

$$H_\mathrm{d}(x, u_\mathrm{d}) = x^\mathrm{T} R_{1\mathrm{d}} x + u_\mathrm{d}^\mathrm{T} R_{2\mathrm{d}} u_\mathrm{d} + (A_\mathrm{d} x + B_\mathrm{d} u_\mathrm{d})^\mathrm{T} P (A_\mathrm{d} x + B_\mathrm{d} u_\mathrm{d})$$
$$- x^\mathrm{T} P x, \quad x \in \mathcal{Z}_x, \quad (9.44)$$

it follows that $\frac{\partial^2 H_\mathrm{c}}{\partial u_\mathrm{c}^2} = R_{2\mathrm{c}} > 0$ and $\frac{\partial^2 H_\mathrm{d}}{\partial u_\mathrm{d}^2} = R_{2\mathrm{d}} + B_\mathrm{d}^\mathrm{T} P B_\mathrm{d} > 0$. Now, $\frac{\partial H_\mathrm{c}}{\partial u_\mathrm{c}} = 2 R_{2\mathrm{c}} u_\mathrm{c} + 2 B_\mathrm{c}^\mathrm{T} P x = 0$, $x \notin \mathcal{Z}_x$, and $\frac{\partial H_\mathrm{d}}{\partial u_\mathrm{d}} = 2(R_{2\mathrm{d}} + B_\mathrm{d}^\mathrm{T} P B_\mathrm{d}) u_\mathrm{d} +$

$2B_\mathrm{d}^\mathrm{T} P A_\mathrm{d} x = 0$, $x \in \mathcal{Z}_x$, give the unique global minimum of $H_\mathrm{c}(x, u_\mathrm{c})$, $x \notin \mathcal{Z}_x$, and $H_\mathrm{d}(x, u_\mathrm{d})$, $x \in \mathcal{Z}_x$, respectively. Hence, since $\phi_\mathrm{c}(x)$ minimizes $H_\mathrm{c}(x, u_\mathrm{c})$ on $x \notin \mathcal{Z}_x$ and $\phi_\mathrm{d}(x)$ minimizes $H_\mathrm{d}(x, u_\mathrm{d})$ on $x \in \mathcal{Z}_x$, it follows that $\phi_\mathrm{c}(x)$ satisfies $\frac{\partial H_\mathrm{c}}{\partial u_\mathrm{c}} = 0$ and $\phi_\mathrm{d}(x)$ satisfies $\frac{\partial H_\mathrm{d}}{\partial u_\mathrm{d}} = 0$, or, equivalently, $\phi_\mathrm{c}(x) = -R_{2\mathrm{c}}^{-1} B_\mathrm{c}^\mathrm{T} P x$, $x \notin \mathcal{Z}_x$, and $\phi_\mathrm{d}(x) = -(R_{2\mathrm{d}} + B_\mathrm{d}^\mathrm{T} P B_\mathrm{d})^{-1} B_\mathrm{d}^\mathrm{T} P A_\mathrm{d} x$, $x \in \mathcal{Z}_x$.

For given $R_{1\mathrm{c}}$, $R_{2\mathrm{c}}$, $R_{1\mathrm{d}}$, and $R_{2\mathrm{d}}$, (9.37) and (9.38) can be solved using constrained nonlinear programming methods using the structure of $\mathcal{Z}_x$. For example, in the case where $\mathcal{Z}_x$ is characterized by the hyperplane $\mathcal{Z}_x = \{x \in \mathbb{R}^n : H x = 0\}$, where $H \in \mathbb{R}^{m \times n}$, it follows that (9.38) holds when $x \in \mathcal{N}(H)$ and (9.37) holds when $x \in [\mathcal{N}(H)]^\perp = \mathcal{R}(H^\mathrm{T})$, where $\mathcal{N}(H)$ denotes the null space of H, $\mathcal{R}(H^\mathrm{T})$ denotes the range space of H^T, and $[\ ]^\perp$ denotes orthogonal complement. Now, reformulating $\mathcal{Z}_x$ as $\{x \in \mathbb{R}^n : E x = 0\}$, where E is an elementary matrix composed of zeroes and ones such that the columns of E span the null space of H, and using the fact that $P > 0$, (9.37) and (9.38) will hold for $P > 0$ with a specific internal matrix structure. This of course reduces the number of free elements in P satisfying (9.37) and (9.38).

Alternatively, to avoid the complexity in solving (9.37) and (9.38), an inverse optimal control problem can be solved, wherein $R_{1\mathrm{c}}$, $R_{2\mathrm{c}}$, $R_{1\mathrm{d}}$, and $R_{2\mathrm{d}}$ are arbitrary. In this case, (9.37) and (9.38) are implied by

$$0 = A_\mathrm{c}^\mathrm{T} P + P A_\mathrm{c} + R_{1\mathrm{c}} - P B_\mathrm{c} R_{2\mathrm{c}}^{-1} B_\mathrm{c}^\mathrm{T} P, \tag{9.45}$$

$$0 = A_\mathrm{d}^\mathrm{T} P A_\mathrm{d} - P + R_{1\mathrm{d}} - A_\mathrm{d}^\mathrm{T} P B_\mathrm{d}(R_{2\mathrm{d}} + B_\mathrm{d}^\mathrm{T} P B_\mathrm{d})^{-1} B_\mathrm{d}^\mathrm{T} P A_\mathrm{d}. \tag{9.46}$$

Since $R_{1\mathrm{c}}$, $R_{2\mathrm{c}}$, $R_{1\mathrm{d}}$, and $R_{2\mathrm{d}}$ are arbitrary, (9.45) and (9.46) can be cast as a LMI feasibility problem involving

$$P > 0, \quad \begin{bmatrix} A_\mathrm{c}^\mathrm{T} P + P A_\mathrm{c} & P B_\mathrm{c} \\ B_\mathrm{c}^\mathrm{T} P & -R_{2\mathrm{c}} \end{bmatrix} < 0, \tag{9.47}$$

$$\begin{bmatrix} A_\mathrm{d}^\mathrm{T} P A_\mathrm{d} - P & A_\mathrm{d}^\mathrm{T} P B_\mathrm{d} \\ B_\mathrm{d}^\mathrm{T} P A_\mathrm{d} & -(R_{2\mathrm{d}} + B_\mathrm{d}^\mathrm{T} P B_\mathrm{d}) \end{bmatrix} < 0. \tag{9.48}$$

9.3 Inverse Optimal Control for Nonlinear Affine Impulsive Systems

In this section, we use the results of Section 9.2 to obtain controllers that are predicated on an *inverse optimal hybrid control problem*. In particular, to avoid the complexity in solving the steady-state hybrid

Hamilton-Jacobi-Bellman equation we do not attempt to minimize a *given* cost functional, but rather, we parameterize a family of stabilizing hybrid controllers that minimize some *derived* cost functional that provides flexibility in specifying the control law. The performance integrand is shown to explicitly depend on the nonlinear impulsive system dynamics, the Lyapunov function of the closed-loop system, and the stabilizing hybrid feedback control law, wherein the coupling is introduced via the hybrid Hamilton-Jacobi-Bellman equation. Hence, by varying the parameters in the Lyapunov function and the performance integrand, the proposed framework can be used to characterize a class of globally stabilizing hybrid controllers that can meet the closed-loop system response constraints.

Consider the state-dependent affine (in the control) impulsive dynamical system

$$\dot{x}(t) = f_c(x(t)) + G_c(x(t))u_c(t), \quad x(0) = x_0, \quad x(t) \notin \mathcal{Z}_x, \quad (9.49)$$
$$\Delta x(t) = f_d(x(t)) + G_d(x(t))u_d(t), \quad x(t) \in \mathcal{Z}_x. \quad (9.50)$$

Furthermore, we consider performance integrands $L_c(x, u_c)$ and $L_d(x, u_d)$ of the form

$$L_c(x, u_c) = L_{1c}(x) + u_c^T R_{2c}(x)u_c, \quad (9.51)$$
$$L_d(x, u_d) = L_{1d}(x) + u_d^T R_{2d}(x)u_d, \quad (9.52)$$

where $L_{1c} : \mathbb{R}^n \to \mathbb{R}$ and satisfies $L_{1c}(x) \geq 0$, $x \in \mathbb{R}^n$, $R_{2c} : \mathbb{R}^n \to \mathbb{P}^{m_c}$, $L_{1d} : \mathcal{Z}_x \to \mathbb{R}$ and satisfies $L_{1d}(x) \geq 0$, $x \in \mathbb{R}^n$, and $R_{2d} : \mathcal{Z}_x \to \mathbb{P}^{m_d}$ so that (9.3) becomes

$$J(x_0, u_c(\cdot), u_d(\cdot)) = \int_0^\infty [L_{1c}(x(t)) + u_c^T(t)R_{2c}(x(t))u_c(t)]dt$$
$$+ \sum_{k \in \mathbb{Z}_{[0,\infty)}} [L_{1d}(x(t_k)) + u_d^T(t_k)R_{2d}(x(t_k))u_d(t_k)].$$

$$(9.53)$$

Corollary 9.2 *Consider the nonlinear impulsive controlled system (9.49) and (9.50) with performance functional (9.53). Assume there exists a continuously differentiable function $V : \mathbb{R}^n \to \mathbb{R}$, and functions $P_{12} : \mathcal{Z}_x \to \mathbb{R}^{1 \times m_d}$ and $P_2 : \mathcal{Z}_x \to \mathbb{N}^{m_d}$ such that $V(0) = 0$, $V(x) > 0$, $x \in \mathbb{R}^n$, $x \neq 0$,*

$$V'(x)[f_c(x) - \tfrac{1}{2}G_c(x)R_{2c}^{-1}(x)G_c^T(x)V'^T(x)] < 0, \quad x \notin \mathcal{Z}_x, \quad x \neq 0, \quad (9.54)$$

$$V(x + f_{\mathrm{d}}(x) - \tfrac{1}{2}G_{\mathrm{d}}(x)(R_{2\mathrm{d}}(x) + P_2(x))^{-1}P_{12}^{\mathrm{T}}(x)) - V(x) \leq 0,$$
$$x \in \mathcal{Z}_x, \qquad (9.55)$$

$$V(x + f_{\mathrm{d}}(x) + G_{\mathrm{d}}(x)u_{\mathrm{d}}) = V(x + f_{\mathrm{d}}(x)) + P_{12}(x)u_{\mathrm{d}} + u_{\mathrm{d}}^{\mathrm{T}}P_2(x)u_{\mathrm{d}},$$
$$x \in \mathcal{Z}_x, \quad u_{\mathrm{d}} \in \mathbb{R}^{m_{\mathrm{d}}}, \qquad (9.56)$$

where u_{d} is admissible, and

$$V(x) \to \infty \ \ as \ \ \|x\| \to \infty. \qquad (9.57)$$

Then the zero solution $x(t) \equiv 0$ of the closed-loop system

$$\dot{x}(t) = f_{\mathrm{c}}(x(t)) + G_{\mathrm{c}}(x(t))\phi_{\mathrm{c}}(x(t)), \quad x(0) = x_0, \quad x(t) \notin \mathcal{Z}_x, (9.58)$$
$$\Delta x(t) = f_{\mathrm{d}}(x(t)) + G_{\mathrm{d}}(x(t))\phi_{\mathrm{d}}(x(t)), \quad x(t) \in \mathcal{Z}_x, \qquad (9.59)$$

is globally asymptotically stable with the hybrid feedback control law

$$\phi_{\mathrm{c}}(x) = -\tfrac{1}{2}R_{2\mathrm{c}}^{-1}(x)G_{\mathrm{c}}^{\mathrm{T}}(x)V'^{\mathrm{T}}(x), \quad x \notin \mathcal{Z}_x, \qquad (9.60)$$
$$\phi_{\mathrm{d}}(x) = -\tfrac{1}{2}(R_{2\mathrm{d}}(x) + P_2(x))^{-1}P_{12}^{\mathrm{T}}(x), \quad x \in \mathcal{Z}_x, \qquad (9.61)$$

and performance functional (9.53), with

$$L_{1\mathrm{c}}(x) = \phi_{\mathrm{c}}^{\mathrm{T}}(x)R_{2\mathrm{c}}(x)\phi_{\mathrm{c}}(x) - V'(x)f_{\mathrm{c}}(x), \qquad (9.62)$$
$$L_{1\mathrm{d}}(x) = \phi_{\mathrm{d}}^{\mathrm{T}}(x)(R_{2\mathrm{d}}(x) + P_2(x))\phi_{\mathrm{d}}(x) - V(x + f_{\mathrm{d}}(x)) + V(x),$$
$$(9.63)$$

is minimized in the sense that

$$J(x_0, \phi_{\mathrm{c}}(x(\cdot)), \phi_{\mathrm{d}}(x(\cdot))) = \min_{(u_{\mathrm{c}}(\cdot), u_{\mathrm{d}}(\cdot)) \in \mathcal{C}(x_0)} J(x_0, u_{\mathrm{c}}(\cdot), u_{\mathrm{d}}(\cdot)),$$
$$x_0 \in \mathbb{R}^n. \quad (9.64)$$

Finally,

$$J(x_0, \phi_{\mathrm{c}}(x(\cdot)), \phi_{\mathrm{d}}(x(\cdot))) = V(x_0), \quad x_0 \in \mathbb{R}^n. \qquad (9.65)$$

Proof. The result is a direct consequence of Theorem 9.3 with $\mathcal{D} = \mathbb{R}^n$, $U_{\mathrm{c}} = \mathbb{R}^{m_{\mathrm{c}}}$, $U_{\mathrm{d}} = \mathbb{R}^{m_{\mathrm{d}}}$, $F_{\mathrm{c}}(x, u_{\mathrm{c}}) = f_{\mathrm{c}}(x) + G_{\mathrm{c}}(x)u_{\mathrm{c}}$, $F_{\mathrm{d}}(x, u_{\mathrm{d}}) = f_{\mathrm{d}}(x) + G_{\mathrm{d}}(x)u_{\mathrm{d}}$, $L_{\mathrm{c}}(x, u_{\mathrm{c}}) = L_{1\mathrm{c}}(x) + u_{\mathrm{c}}^{\mathrm{T}}R_{2\mathrm{c}}(x)u_{\mathrm{c}}$, and $L_{\mathrm{d}}(x, u_{\mathrm{d}}) = L_{1\mathrm{d}}(x) + u_{\mathrm{d}}^{\mathrm{T}}R_{2\mathrm{d}}(x)u_{\mathrm{d}}$. Specifically, with (9.51) and (9.52) the Hamiltonians have the form

$$H_{\mathrm{c}}(x, u_{\mathrm{c}}) = L_{1\mathrm{c}}(x) + u_{\mathrm{c}}^{\mathrm{T}}R_{2\mathrm{c}}(x)u_{\mathrm{c}} + V'(x)(f_{\mathrm{c}}(x) + G_{\mathrm{c}}(x)u_{\mathrm{c}}),$$
$$x \notin \mathcal{Z}_x, \quad u_{\mathrm{c}} \in \mathbb{R}^{m_{\mathrm{c}}}, \qquad (9.66)$$

$$H_{\mathrm{d}}(x, u_{\mathrm{d}}) = L_{1\mathrm{d}}(x) + V(x + f_{\mathrm{d}}(x)) + P_{12}(x)u_{\mathrm{d}}$$
$$+ u_{\mathrm{d}}^{\mathrm{T}}(R_{2\mathrm{d}}(x) + P_2(x))u_{\mathrm{d}} - V(x), \quad x \in \mathcal{Z}_x, \quad u_{\mathrm{d}} \in \mathbb{R}^{m_{\mathrm{d}}}.$$
$$(9.67)$$

Now, the hybrid feedback control law (9.60) and (9.61) is obtained by setting $\frac{\partial H_{\mathrm{c}}}{\partial u_{\mathrm{c}}} = 0$ and $\frac{\partial H_{\mathrm{d}}}{\partial u_{\mathrm{d}}} = 0$. With (9.60) and (9.61) it follows that (9.54) and (9.55) imply (9.24) and (9.25), respectively. Next, since $V(\cdot)$ is continuously differentiable and $x = 0$ is a local minimum of $V(\cdot)$, it follows that $V'(0) = 0$, and hence, it follows that $\phi_{\mathrm{c}}(0) = 0$.

Next, with $L_{1\mathrm{c}}(x)$ and $L_{1\mathrm{d}}(x)$ given by (9.62) and (9.63), respectively, and $\phi_{\mathrm{c}}(x)$ and $\phi_{\mathrm{d}}(x)$ given by (9.60) and (9.61), (9.26) and (9.27) hold. Finally, since

$$
\begin{aligned}
H_{\mathrm{c}}(x, u_{\mathrm{c}}) &= H_{\mathrm{c}}(x, u_{\mathrm{c}}) - H_{\mathrm{c}}(x, \phi_{\mathrm{c}}(x)) \\
&= [u_{\mathrm{c}} - \phi_{\mathrm{c}}(x)]^{\mathrm{T}} R_{2\mathrm{c}}(x)[u_{\mathrm{c}} - \phi_{\mathrm{c}}(x)], \quad x \notin \mathcal{Z}_x, \quad (9.68)
\end{aligned}
$$

$$
\begin{aligned}
H_{\mathrm{d}}(x, u_{\mathrm{d}}) &= H_{\mathrm{d}}(x, u_{\mathrm{d}}) - H_{\mathrm{d}}(x, \phi_{\mathrm{d}}(x)) \\
&= [u_{\mathrm{d}} - \phi_{\mathrm{d}}(x)]^{\mathrm{T}} (R_{2\mathrm{d}}(x) + P_2(x))[u_{\mathrm{d}} - \phi_{\mathrm{d}}(x)], \quad x \in \mathcal{Z}_x, \\
&\hspace{8cm} (9.69)
\end{aligned}
$$

where $R_{2\mathrm{c}}(x) > 0$, $x \notin \mathcal{Z}_x$, and $R_{2\mathrm{d}}(x) + P_2(x) > 0$, $x \in \mathcal{Z}_x$, conditions (9.28) and (9.29) hold. The result now follows as a direct consequence of Theorem 9.3. $\qquad\square$

Note that (9.54) and (9.55) are equivalent to

$$
\dot{V}(x) \triangleq V'(x)[f_{\mathrm{c}}(x) + G_{\mathrm{c}}(x)\phi_{\mathrm{c}}(x)] < 0, \quad x \notin \mathcal{Z}_x, \quad x \neq 0, \quad (9.70)
$$

$$
\Delta V(x) \triangleq V(x + f_{\mathrm{d}}(x) + G_{\mathrm{d}}(x)\phi_{\mathrm{d}}(x)) - V(x) \leq 0, \quad x \in \mathcal{Z}_x, \quad (9.71)
$$

with $\phi_{\mathrm{c}}(x)$ and $\phi_{\mathrm{d}}(x)$ given by (9.60) and (9.61), respectively. Furthermore, conditions (9.70) and (9.71) with $V(0) = 0$ and $V(x) > 0$, $x \in \mathbb{R}^n$, $x \neq 0$, assure that $V(x)$ is a Lyapunov function for the impulsive closed-loop system (9.58) and (9.59).

9.4 Nonlinear Hybrid Control with Polynomial and Multilinear Performance Functionals

In this section, we specialize the results of Section 9.3 to linear impulsive systems controlled by inverse optimal nonlinear hybrid controllers that minimize a derived polynomial and multilinear cost functional. For the results in this section we assume $u_{\mathrm{d}}(t_k) \equiv 0$. Furthermore, let $R_{1\mathrm{c}} \in \mathbb{P}^n$, $R_{1\mathrm{d}} \in \mathbb{P}^n$, $R_{2\mathrm{c}} \in \mathbb{P}^{m_{\mathrm{c}}}$, $\hat{R}_q, \check{R}_q \in \mathbb{N}^n$, $q = 2, \ldots, r$, be given, where r is a positive integer, and define $S_{\mathrm{c}} \triangleq B_{\mathrm{c}} R_{2\mathrm{c}}^{-1} B_{\mathrm{c}}^{\mathrm{T}}$.

Corollary 9.3 *Consider the linear controlled impulsive system*

$$
\dot{x}(t) = A_{\mathrm{c}} x(t) + B_{\mathrm{c}} u_{\mathrm{c}}(t), \quad x(0) = x_0, \quad x(t) \notin \mathcal{Z}_x, \quad (9.72)
$$

$$\Delta x(t) = (A_{\mathrm{d}} - I_n)x(t), \quad x(t) \in \mathcal{Z}_x, \tag{9.73}$$

where $u_{\mathrm{c}}(\cdot)$ is admissible. Assume there exist $P \in \mathbb{P}^n$ and $M_q \in \mathbb{N}^n$, $q = 2, \ldots, r$, such that

$$0 = x^{\mathrm{T}}(A_{\mathrm{c}}^{\mathrm{T}} P + P A_{\mathrm{c}} + R_{1\mathrm{c}} - P B_{\mathrm{c}} R_{2\mathrm{c}}^{-1} B_{\mathrm{c}}^{\mathrm{T}} P)x, \quad x \notin \mathcal{Z}_x, \tag{9.74}$$

$$0 = x^{\mathrm{T}}[(A_{\mathrm{c}} - S_{\mathrm{c}} P)^{\mathrm{T}} M_q + M_q(A_{\mathrm{c}} - S_{\mathrm{c}} P) + \hat{R}_q]x,$$
$$x \notin \mathcal{Z}_x, \quad q = 2, \ldots, r, \tag{9.75}$$

$$0 = x^{\mathrm{T}}(A_{\mathrm{d}}^{\mathrm{T}} P A_{\mathrm{d}} - P + R_{1\mathrm{d}})x, \quad x \in \mathcal{Z}_x, \tag{9.76}$$

$$0 = x^{\mathrm{T}}(A_{\mathrm{d}}^{\mathrm{T}} M_q A_{\mathrm{d}} - M_q + \hat{\hat{R}}_q)x, \quad x \in \mathcal{Z}_x, \quad q = 2, \ldots, r. \tag{9.77}$$

Then the zero solution $x(t) \equiv 0$ of the closed-loop system

$$\dot{x}(t) = A_{\mathrm{c}} x(t) + B_{\mathrm{c}} \phi_{\mathrm{c}}(x(t)), \quad x(0) = x_0, \quad x(t) \notin \mathcal{Z}_x, \tag{9.78}$$

$$\Delta x(t) = (A_{\mathrm{d}} - I_n)x(t), \quad x(t) \in \mathcal{Z}_x, \tag{9.79}$$

is globally asymptotically stable with the feedback control law

$$\phi_{\mathrm{c}}(x) = -R_{2\mathrm{c}}^{-1} B_{\mathrm{c}}^{\mathrm{T}} \left(P + \sum_{q=2}^{r} (x^{\mathrm{T}} M_q x)^{q-1} M_q \right) x, \quad x \notin \mathcal{Z}_x, \tag{9.80}$$

and the performance functional (9.53), with $R_{2\mathrm{c}}(x) = R_{2\mathrm{c}}$ and

$$L_{1\mathrm{c}}(x) = x^{\mathrm{T}} \left(R_{1\mathrm{c}} + \sum_{q=2}^{r} (x^{\mathrm{T}} M_q x)^{q-1} \hat{R}_q + \left[\sum_{q=2}^{r} (x^{\mathrm{T}} M_q x)^{q-1} M_q \right]^{\mathrm{T}} \right.$$

$$\left. \cdot S_{\mathrm{c}} \left[\sum_{q=2}^{r} (x^{\mathrm{T}} M_q x)^{q-1} M_q \right] \right) x, \tag{9.81}$$

$$L_{1\mathrm{d}}(x) = x^{\mathrm{T}} R_{1\mathrm{d}} x$$

$$+ \sum_{q=2}^{r} \frac{1}{q} \left[(x^{\mathrm{T}} \hat{\hat{R}}_q x) \sum_{j=1}^{q} (x^{\mathrm{T}} M_q x)^{j-1} (x^{\mathrm{T}} A_{\mathrm{d}}^{\mathrm{T}} M_q A_{\mathrm{d}} x)^{q-j} \right], \tag{9.82}$$

is minimized in the sense that

$$J(x_0, \phi_{\mathrm{c}}(x(\cdot))) = \min_{u_{\mathrm{c}}(\cdot) \in \mathcal{C}(x_0)} J(x_0, u_{\mathrm{c}}(\cdot)), \quad x_0 \in \mathbb{R}^n. \tag{9.83}$$

Finally,

$$J(x_0, \phi_{\mathrm{c}}(x(\cdot))) = x_0^{\mathrm{T}} P x_0 + \sum_{q=2}^{r} \frac{1}{q} (x_0^{\mathrm{T}} M_q x_0)^q, \quad x_0 \in \mathbb{R}^n. \tag{9.84}$$

Proof. The result is a direct consequence of Corollary 9.2 with $f_c(x) = A_c x$, $f_d(x) = (A_d - I_n)x$, $G_c(x) = B_c$, $G_d(x) = 0$, $u_d = 0$, $R_{2c}(x) = R_{2c}$, $R_{2d}(x) = I_{m_d}$, and $V(x) = x^T P x + \sum_{q=2}^r \frac{1}{q}(x^T M_q x)^q$. Specifically, for $x \notin \mathcal{Z}_x$ it follows from (9.74), (9.75), and (9.80) that

$$V'(x)[f_c(x) - \tfrac{1}{2}G_c(x)R_{2c}^{-1}G_c^T(x)V'^T(x)] = -x^T R_{1c} x$$
$$- \sum_{q=2}^r (x^T M_q x)^{q-1} x^T \hat{R}_q x - \phi_c^T(x) R_{2c} \phi_c(x)$$
$$- x^T \left[\sum_{q=2}^r (x^T M_q x)^{q-1} M_q\right]^T S_c \left[\sum_{q=2}^r (x^T M_q x)^{q-1} M_q\right] x,$$

which implies (9.54). For $x \in \mathcal{Z}_x$ it follows from (9.76) and (9.77) that

$$\Delta V(x) = V(x + f_d(x)) - V(x)$$
$$= -x^T R_{1d} x$$
$$- \sum_{q=2}^r \frac{1}{q}\left[(x^T \hat{R}_q x) \sum_{j=1}^q (x^T M_q x)^{j-1}(x^T A_d^T M_q A_d x)^{q-j}\right],$$

which implies (9.55) with $G_d(x) = 0$. Finally, with $u_d = 0$, (9.56) is automatically satisfied so that all the conditions of Corollary 9.2 are satisfied. $\square$

As noted in Section 9.2, viewing R_{1c}, R_{2c}, $\hat{R}_q$, and $\hat{R}_q$, $q = 2,\ldots,r$, as arbitrary matrices, it follows that (9.74)–(9.77) are implied by a set of Bilinear Matrix Inequalities (BMIs). This considerably minimizes the numerical complexity for solving (9.74)–(9.77).

Finally, we specialize the results of Section 9.3 to linear impulsive systems controlled by inverse optimal hybrid controllers that minimize a derived multilinear functional. For the next result, recall the definition of S_c, let $R_{1c} \in \mathbb{P}^n$, $R_{1d} \in \mathbb{P}^n$, $R_{2c} \in \mathbb{P}^{m_c}$, $\hat{R}_{2q}, \hat{R}_{2q} \in \mathcal{N}^{(2q,n)}$, $q = 2,\ldots,r$, be given, where $\mathcal{N}^{(2q,n)} \triangleq \{\Psi \in \mathbb{R}^{1 \times n^{2q}} : \Psi x^{[2q]} \geq 0, x \in \mathbb{R}^n\}$, and define $\overset{k}{\oplus} A \triangleq A \oplus \cdots \oplus A$ (k times) and $x^{[k]} \triangleq x \otimes \cdots \otimes x$ (k times).

Corollary 9.4 *Consider the linear controlled impulsive system (9.72) and (9.73). Assume there exist $P \in \mathbb{P}^n$ and $\hat{P}_q \in \mathcal{N}^{(2q,n)}$, $q = 2,\ldots,r$,* *such that*

$$0 = x^{\mathrm{T}}(A_c^{\mathrm{T}}P + PA_c + R_{1c} - PB_cR_{2c}^{-1}B_c^{\mathrm{T}}P)x, \quad x \notin \mathcal{Z}_x, \tag{9.85}$$

$$0 = x^{\mathrm{T}}(\hat{P}_q[\overset{2q}{\oplus}(A_c - S_cP)] + \hat{R}_{2q})x, \quad x \notin \mathcal{Z}_x, \quad q = 2, \ldots, r, \tag{9.86}$$

$$0 = x^{\mathrm{T}}(A_d^{\mathrm{T}}PA_d - P + R_{1d})x, \quad x \in \mathcal{Z}_x, \tag{9.87}$$

$$0 = x^{\mathrm{T}}(\hat{P}_q[A_d^{[2q]} - I_n^{[2q]}] + \hat{\hat{R}}_{2q})x, \quad x \in \mathcal{Z}_x, \quad q = 2, \ldots, r. \tag{9.88}$$

Then the zero solution $x(t) \equiv 0$ of the closed-loop system (9.78) and (9.79) is globally asymptotically stable with the feedback control law

$$\phi_c(x) = -R_{2c}^{-1}B_c^{\mathrm{T}}(Px + \tfrac{1}{2}g'^{\mathrm{T}}(x)), \quad x \notin \mathcal{Z}_x, \tag{9.89}$$

where $g(x) \triangleq \sum_{q=2}^r \hat{P}_q x^{[2q]}$, and the performance functional (9.53), with $R_{2c}(x) = R_{2c}$ and

$$L_{1c}(x) = x^{\mathrm{T}}R_{1c}x + \sum_{q=2}^r \hat{R}_{2q}x^{[2q]} + \tfrac{1}{4}g'(x)S_cg'^{\mathrm{T}}(x), \tag{9.90}$$

$$L_{1d}(x) = x^{\mathrm{T}}R_{1d}x + \sum_{q=2}^r \hat{\hat{R}}_{2q}x^{[2q]}, \tag{9.91}$$

is minimized in the sense that

$$J(x_0, \phi_c(x(\cdot))) = \min_{u_c(\cdot) \in \mathcal{C}(x_0)} J(x_0, u_c(\cdot)), \quad x_0 \in \mathbb{R}^n. \tag{9.92}$$

Finally,

$$J(x_0, \phi_c(x(\cdot))) = x_0^{\mathrm{T}}Px_0 + \sum_{q=2}^r \hat{P}_q x_0^{[2q]}, \quad x_0 \in \mathbb{R}^n. \tag{9.93}$$

Proof. The result is a direct consequence of Theorem 9.2 with $f_c(x) = A_cx$, $f_d(x) = (A_d - I_n)x$, $G_c(x) = B_c$, $G_d(x) = 0$, $u_d = 0$, $R_{2c}(x) = R_{2c}$, $R_{2d}(x) = I_{m_d}$, and $V(x) = x^{\mathrm{T}}Px + \sum_{q=2}^r \hat{P}_q x^{[2q]}$. Specifically, for $x \notin \mathcal{Z}_x$ it follows from (9.85), (9.86), and (9.89) that

$$V'(x)[f_c(x) - \tfrac{1}{2}G_c(x)R_{2c}^{-1}(x)G_c^{\mathrm{T}}(x)V'^{\mathrm{T}}(x)] = -x^{\mathrm{T}}R_{1c}x - \sum_{q=2}^r \hat{R}_{2q}x^{[2q]}$$

$$-\phi_c^{\mathrm{T}}(x)R_{2c}\phi_c(x) - \tfrac{1}{4}g'(x)S_cg'^{\mathrm{T}}(x),$$

which implies (9.54). For $x \in \mathcal{Z}_x$ it follows from (9.87) and (9.88) that

$$\Delta V(x) = V(x + f_d(x)) - V(x) = -x^{\mathrm{T}}R_{1d}x - \sum_{q=2}^r \hat{\hat{R}}_{2q}x^{[2q]},$$

which implies (9.55) with $G_d(x) = 0$. Finally, with $u_d = 0$, (9.56) is automatically satisfied so that all the conditions of Corollary 9.2 are satisfied. $\square$

9.5 Gain, Sector, and Disk Margins for Optimal Hybrid Regulators

For continuous-time nonlinear dynamical systems with continuous flows, the problem of guaranteed stability margins for optimal and inverse optimal regulators is well established [46,82,127,128]. Specifically, nonlinear inverse optimal controllers that minimize a *meaningful* nonlinear-nonquadratic performance criterion involving a nonlinear-nonquadratic, nonnegative-definite function of the state and a quadratic positive-definite function of the control were shown to possess sector margin guarantees to component decoupled input nonlinearities lying in the conic sector $(\frac{1}{2}, \infty)$. These results also hold for disk margin guarantees, where asymptotic stability of the closed-loop system is guaranteed in the face of a dissipative dynamic input operator. In addition, an equivalence between dissipativity with respect to a quadratic supply rate and optimality of a nonlinear regulator also holds.

In this section, we develop sufficient conditions for hybrid gain, sector, and disk margins guarantees for nonlinear impulsive dynamical systems controlled by optimal and inverse optimal hybrid regulators. Furthermore, we develop a hybrid counterpart of the return difference inequality for continuous-time systems to provide connections between dissipativity and optimality of nonlinear hybrid controllers. In particular, we show that unlike the case for continuous-time systems with continuous flows, the equivalence between dissipativity and optimality of hybrid controllers breaks down. However, we do show that optimal hybrid controllers imply dissipativity with respect to a quadratic supply rate. To develop hybrid gain, sector, and disk margins for optimal hybrid regulators, the following corollary to Theorem 6.2 is necessary.

Corollary 9.5 *Consider the closed-loop system consisting of the non-linear impulsive dynamical systems $\mathcal{G}$ given by (6.1)–(6.4) and $\mathcal{G}_c$ given by (6.5)–(6.8), and assume $\mathcal{G}$ and $\mathcal{G}_c$ are zero-state observable. Let $a_c, b_c, a_{cc}, b_{cc}, \delta_c, a_d, b_d, a_{dc}, b_{dc}, \delta_d \in \mathbb{R}$ be such that $b_c, b_d > 0$, $0 < a_c + b_c$, $0 < a_d + b_d$, $0 < 2\delta_c < b_c - a_c$, $0 < 2\delta_d < b_d - a_d$, $a_{cc} = a_c + \delta_c$,*

$a_{dc} = a_d + \delta_d$, $b_{cc} = b_c - \delta_c$, $b_{dc} = b_d - \delta_d$, and let $M_c \in \mathbb{R}^{m_c \times m_c}$ and $M_d \in \mathbb{R}^{m_d \times m_d}$ be positive definite. If $\mathcal{G}$ is dissipative with respect to the quadratic supply rate $(s_c(u_c, y_c), s_d(u_d, y_d)) = (\frac{a_c b_c}{a_c + b_c} y_c^T M_c y_c +$
$u_c^T M_c y_c + \frac{1}{a_c + b_c} u_c^T M_c u_c, \frac{a_d b_d}{a_d + b_d} y_d^T M_d y_d + u_d^T M_d y_d + \frac{1}{a_d + b_d} u_d^T M_d u_d)$
and has a radially unbounded storage function, and $\mathcal{G}_c$ is dissipative with respect to the quadratic supply rate $(s_{cc}(u_{cc}, y_{cc}), s_{dc}(u_{dc}, y_{dc})) = (u_{cc}^T M_c y_{cc} - \frac{1}{a_{cc} + b_{cc}} y_{cc}^T M_c y_{cc} - \frac{a_{cc} b_{cc}}{a_{cc} + b_{cc}} u_{cc}^T M_c u_{cc}, u_{dc}^T M_d y_{dc} - \frac{1}{a_{dc} + b_{dc}} y_{dc}^T$
$\cdot M_d y_{dc} - \frac{a_{dc} b_{dc}}{a_{dc} + b_{dc}} u_{dc}^T M_d u_{dc})$ and has a radially unbounded storage function, then the negative feedback interconnection of $\mathcal{G}$ and $\mathcal{G}_c$ is globally asymptotically stable.

Proof. The proof is a direct consequence of Theorem 6.2 with $Q_c = \frac{a_c b_c}{a_c + b_c} M_c$, $S_c = \frac{1}{2} M_c$, $R_c = \frac{1}{a_c + b_c} M_c$, $Q_{cc} = -\frac{1}{a_{cc} + b_{cc}} M_c$, $S_{cc} = \frac{1}{2} M_c$, $R_{cc} = -\frac{a_{cc} b_{cc}}{a_{cc} + b_{cc}} M_c$, $Q_d = \frac{a_d b_d}{a_d + b_d} M_d$, $S_d = \frac{1}{2} M_d$, $R_d = \frac{1}{a_d + b_d} M_d$, $Q_{dc} = -\frac{1}{a_{dc} + b_{dc}} M_d$, $S_{dc} = \frac{1}{2} M_d$, and $R_{dc} = -\frac{a_{dc} b_{dc}}{a_{dc} + b_{dc}} M_d$. Specifically, let $\sigma > 0$ be such that

$$\sigma \left(\frac{\delta_c^2}{(a_c + b_c)^2} - \frac{1}{4} \right) + \frac{1}{4} > 0, \quad \sigma \left(\frac{\delta_d^2}{(a_d + b_d)^2} - \frac{1}{4} \right) + \frac{1}{4} > 0.$$

In this case, $\hat{Q}_c$ and $\hat{Q}_d$ given by (6.16) and (6.17) satisfy $\hat{Q}_c < 0$ and $\hat{Q}_d < 0$ so that all the conditions of Theorem 6.2 are satisfied. $\square$

Now, we consider impulsive nonlinear systems $\mathcal{G}$ of the form given by (6.1)–(6.4) with $l_c = m_c$, $l_d = m_d$, $J_c(x) \equiv 0$, $J_d(x) \equiv 0$, $h_c(x) = -\phi_c(x)$, and $h_d(x) = -\phi_d(x)$, where $\phi_c : \mathbb{R}^n \to \mathbb{R}^{m_c}$ and $\phi_d : \mathcal{Z}_x \to \mathbb{R}^{m_d}$ are such that $\mathcal{G}$ is asymptotically stable with $(u_c, u_d) = (-y_c, -y_d)$. Furthermore, we assume that the system $\mathcal{G}$ is zero-state observable. In this case, $\mathcal{G}$ becomes

$$\dot{x}(t) = f_c(x(t)) + G_c(x(t))u_c(t), \quad x(0) = x_0, \quad x(t) \notin \mathcal{Z}_x, \quad (9.94)$$
$$\Delta x(t) = f_d(x(t)) + G_d(x(t))u_d(t), \quad x(t) \in \mathcal{Z}_x, \quad (9.95)$$
$$y_c(t) = -\phi_c(x(t)), \quad x(t) \notin \mathcal{Z}_x, \quad (9.96)$$
$$y_d(t) = -\phi_d(x(t)), \quad x(t) \in \mathcal{Z}_x. \quad (9.97)$$

Next, we define the hybrid robustness margins for $\mathcal{G}$ given by (9.94)–(9.97). Specifically, consider the negative feedback interconnection of $\mathcal{G}$ and $\Delta(\cdot, \cdot)$ given in Figure 9.1, where $\Delta : \mathbb{R}^{m_c} \times \mathbb{R}^{m_d} \to \mathbb{R}^{m_c} \times \mathbb{R}^{m_d}$ is either a linear operator $\Delta(y_c, y_d) = (\Delta_c y_c, \Delta_d y_d)$, a nonlinear static operator $\Delta(y_c, y_d) = (\sigma_c(y_c), \sigma_d(y_d))$, or a dynamic operator $\Delta(\cdot, \cdot)$. In the case where $\Delta(\cdot, \cdot)$ is a dynamic operator, $(u_c, u_d) = (-\Delta_c(y_c), -\Delta_d(y_d))$. Furthermore, we assume that in the nominal

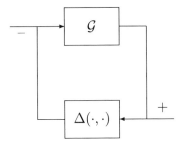

Figure 9.1 Feedback interconnection of $\mathcal{G}$ and $\Delta(\cdot,\cdot)$.

case $\Delta(\cdot,\cdot)$ is such that $(u_{\mathrm{c}}, u_{\mathrm{d}}) = (-\Delta_{\mathrm{c}}(y_{\mathrm{c}}), -\Delta_{\mathrm{d}}(y_{\mathrm{d}})) = (-y_{\mathrm{c}}, -y_{\mathrm{d}})$ so that the nominal closed-loop system is asymptotically stable.

Definition 9.1 *Let $\alpha_{\mathrm{c}}, \beta_{\mathrm{c}}, \alpha_{\mathrm{d}}, \beta_{\mathrm{d}} \in \mathbb{R}$ be such that $0 \leq \alpha_{\mathrm{c}} < 1 < \beta_{\mathrm{c}} < \infty$ and $0 \leq \alpha_{\mathrm{d}} < 1 < \beta_{\mathrm{d}} < \infty$. Then the nonlinear impulsive system $\mathcal{G}$ given by (9.94)–(9.97) is said to have a* hybrid gain margin *$((\alpha_{\mathrm{c}}, \beta_{\mathrm{c}}), (\alpha_{\mathrm{d}}, \beta_{\mathrm{d}}))$ if the negative feedback interconnection of $\mathcal{G}$ and $\Delta(y_{\mathrm{c}}, y_{\mathrm{d}}) = (\Delta_{\mathrm{c}} y_{\mathrm{c}}, \Delta_{\mathrm{d}} y_{\mathrm{d}})$ is globally asymptotically stable for all $\Delta_{\mathrm{c}} = \mathrm{diag}[k_{1\mathrm{c}}, \ldots, k_{m_{\mathrm{c}}\mathrm{c}}]$, where $k_{i\mathrm{c}} \in (\alpha_{\mathrm{c}}, \beta_{\mathrm{c}})$, $i = 1, \ldots, m_{\mathrm{c}}$, and $\Delta_{\mathrm{d}} = \mathrm{diag}[k_{1\mathrm{d}}, \ldots, k_{m_{\mathrm{d}}\mathrm{d}}]$, where $k_{i\mathrm{d}} \in (\alpha_{\mathrm{d}}, \beta_{\mathrm{d}})$, $i = 1, \ldots, m_{\mathrm{d}}$.*

Definition 9.2 *Let $\alpha_{\mathrm{c}}, \beta_{\mathrm{c}}, \alpha_{\mathrm{d}}, \beta_{\mathrm{d}} \in \mathbb{R}$ be such that $0 \leq \alpha_{\mathrm{c}} < 1 < \beta_{\mathrm{c}} < \infty$ and $0 \leq \alpha_{\mathrm{d}} < 1 < \beta_{\mathrm{d}} < \infty$. Then the nonlinear impulsive system $\mathcal{G}$ given by (9.94)–(9.97) is said to have a* hybrid sector margin *$((\alpha_{\mathrm{c}}, \beta_{\mathrm{c}}), (\alpha_{\mathrm{d}}, \beta_{\mathrm{d}}))$ if the negative feedback interconnection of $\mathcal{G}$ and $\Delta(y_{\mathrm{c}}, y_{\mathrm{d}}) = (\sigma_{\mathrm{c}}(y_{\mathrm{c}}), \sigma_{\mathrm{d}}(y_{\mathrm{d}}))$ is globally asymptotically stable for all static nonlinearities $\sigma_{\mathrm{c}} : \mathbb{R}^{m_{\mathrm{c}}} \to \mathbb{R}^{m_{\mathrm{c}}}$ and $\sigma_{\mathrm{d}} : \mathbb{R}^{m_{\mathrm{d}}} \to \mathbb{R}^{m_{\mathrm{d}}}$ such that $\sigma_{\mathrm{c}}(0) = 0$, $\sigma_{\mathrm{d}}(0) = 0$, $\sigma_{\mathrm{c}}(y_{\mathrm{c}}) = [\sigma_{1\mathrm{c}}(y_{1\mathrm{c}}), \ldots, \sigma_{m_{\mathrm{c}}\mathrm{c}}(y_{m_{\mathrm{c}}\mathrm{c}})]$, $\sigma_{\mathrm{d}}(y_{\mathrm{d}}) = [\sigma_{1\mathrm{d}}(y_{1\mathrm{d}}), \ldots, \sigma_{m_{\mathrm{d}}\mathrm{d}}(y_{m_{\mathrm{d}}\mathrm{d}})]$, $\alpha_{\mathrm{c}} y_{ic}^2 < \sigma_{ic}(y_{ic}) y_{ic} < \beta_{\mathrm{c}} y_{ic}^2$, for all $y_{ic} \neq 0$, $i = 1, \ldots, m_{\mathrm{c}}$, and $\alpha_{\mathrm{d}} y_{id}^2 < \sigma_{id}(y_{id}) y_{id} < \beta_{\mathrm{d}} y_{id}^2$, for all $y_{id} \neq 0$, $i = 1, \ldots, m_{\mathrm{d}}$.*

Definition 9.3 *Let $\alpha_{\mathrm{c}}, \beta_{\mathrm{c}}, \alpha_{\mathrm{d}}, \beta_{\mathrm{d}} \in \mathbb{R}$ be such that $0 \leq \alpha_{\mathrm{c}} < 1 < \beta_{\mathrm{c}} < \infty$ and $0 \leq \alpha_{\mathrm{d}} < 1 < \beta_{\mathrm{d}} < \infty$. Then the nonlinear impulsive system $\mathcal{G}$ given by (9.94)–(9.97) is said to have a* hybrid disk margin *$((\alpha_{\mathrm{c}}, \beta_{\mathrm{c}}), (\alpha_{\mathrm{d}}, \beta_{\mathrm{d}}))$ if the negative feedback interconnection of $\mathcal{G}$ and $\Delta(y_{\mathrm{c}}, y_{\mathrm{d}}) = (\Delta_{\mathrm{c}}(y_{\mathrm{c}}), \Delta_{\mathrm{d}}(y_{\mathrm{d}}))$ is globally asymptotically stable for all dynamic operators $\Delta(\cdot,\cdot)$ such that $\Delta(\cdot,\cdot)$ is zero-state observable and there exists $\varepsilon > 0$ such that $\Delta(\cdot,\cdot)$ is dissipative with respect to the hybrid supply rate $(s_{\mathrm{c}}, \varepsilon s_{\mathrm{d}})$, where $s_{\mathrm{c}}(u_{\mathrm{c}}, y_{\mathrm{c}}) = u_{\mathrm{c}}^{\mathrm{T}} y_{\mathrm{c}} - \frac{1}{\hat{\alpha}_{\mathrm{c}} + \hat{\beta}_{\mathrm{c}}} y_{\mathrm{c}}^{\mathrm{T}} y_{\mathrm{c}} -$*

$\frac{\hat{\alpha}_c \hat{\beta}_c}{\hat{\alpha}_c + \hat{\beta}_c} u_c^T u_c$, $s_d(u_d, y_d) = u_d^T y_d - \frac{1}{\hat{\alpha}_d + \hat{\beta}_d} y_d^T y_d - \frac{\hat{\alpha}_d \hat{\beta}_d}{\hat{\alpha}_d + \hat{\beta}_d} u_d^T u_d$, and where $\hat{\alpha}_c = \alpha_c + \delta$, $\hat{\beta}_c = \beta_c - \delta$, and $\delta \in \mathbb{R}$ is such that $0 < 2\delta < \min\{(\beta_c - \alpha_c), (\beta_d - \alpha_d)\}$.

Note that if $\mathcal{G}$ has a hybrid disk margin $((\alpha_c, \beta_c), (\alpha_d, \beta_d))$, then $\mathcal{G}$ has hybrid gain and sector margins $((\alpha_c, \beta_c), (\alpha_d, \beta_d))$. The following key lemma is needed for developing the main result of this section.

Lemma 9.2 *Consider the impulsive nonlinear dynamical system $\mathcal{G}$ given by (9.94)–(9.97) where $(\phi_c(x), \phi_d(x))$ is a stabilizing optimal hybrid control law given by (9.60) and (9.61), and where $V(x)$, $P_{12}(x)$, and $P_2(x)$ are such that $V(0) = 0$, $V(x) > 0$, $x \in \mathbb{R}^n$, $x \neq 0$, and satisfy (9.54)–(9.56). Then for all admissible $u_c(t) \in \mathbb{R}^{m_c}$ and $u_d(t_k) \in \mathbb{R}^{m_d}$, the solution $x(t)$, $t \geq 0$, to (9.94) and (9.95) satisfies*

$$V(x(\hat{t})) - V(x(t)) \leq \int_t^{\hat{t}} \{[u_c(s) + y_c(s)]^T R_{2c}(x(s))[u_c(s) + y_c(s)]$$
$$- u_c^T(s) R_{2c}(x(s)) u_c(s)\} ds,$$
$$t_k < t < \hat{t} \leq t_{k+1}, \quad (9.98)$$
$$V(x(t_k^+)) - V(x(t_k)) \leq \{[u_d(t_k) + y_d(t_k)]^T (R_{2d}(x(t_k)) + P_2(x(t_k)))$$
$$\cdot [u_d(t_k) + y_d(t_k)]$$
$$- u_d^T(t_k) R_{2d}(x(t_k)) u_d(t_k)\}, \quad k \in \overline{\mathbb{Z}}_+. \quad (9.99)$$

Proof. Note that it follows from (9.62) that for all $x \in \mathbb{R}^n \backslash \mathcal{Z}_x$ and $u_c \in \mathbb{R}^{m_c}$,

$$u_c^T R_{2c}(x) u_c \leq L_{1c}(x) + u_c^T R_{2c}(x) u_c$$
$$= \phi_c^T(x) R_{2c}(x) \phi_c(x) - V'(x) f_c(x) + u_c^T R_{2c}(x) u_c$$
$$= y_c^T R_{2c}(x) y_c + 2 y_c^T R_{2c}(x) u_c - V'(x)[f_c(x) + G_c(x) u_c]$$
$$+ u_c^T R_{2c}(x) u_c$$
$$= [u_c + y_c]^T R_{2c}(x)[u_c + y_c] - V'(x)[f_c(x) + G_c(x) u_c],$$

which implies that, for all admissible $u_c(t) \in \mathbb{R}^{m_c}$ and $t \geq 0$, $t \neq t_k$, $k \in \overline{\mathbb{Z}}_+$,

$$u_c^T(t) R_{2c}(x(t)) u_c(t) \leq [u_c(t) + y_c(t)]^T R_{2c}(x(t))[u_c(t) + y_c(t)] - \dot{V}(x(t)).$$

Now, integrating over $[t, \hat{t}]$ yields (9.98).

Next, it follows from (9.63) that for all $x \in \mathcal{Z}_x$ and $u_d \in \mathbb{R}^{m_d}$,

$$u_d^T R_{2d}(x) u_d \leq L_{1d}(x) + u_d^T R_{2d}(x) u_d$$

$$\begin{aligned}
&= \phi_{\mathrm{d}}^{\mathrm{T}}(x)(R_{2\mathrm{d}}(x) + P_2(x))\phi_{\mathrm{d}}(x) - V(x + f_{\mathrm{d}}(x)) + V(x) \\
&\quad + u_{\mathrm{d}}^{\mathrm{T}} R_{2\mathrm{d}}(x) u_{\mathrm{d}} \\
&= y_{\mathrm{d}}^{\mathrm{T}}(R_{2\mathrm{d}}(x) + P_2(x)) y_{\mathrm{d}} + 2y_{\mathrm{d}}^{\mathrm{T}}(R_{2\mathrm{d}}(x) + P_2(x)) u_{\mathrm{d}} \\
&\quad - V(x + f_{\mathrm{d}}(x) + G_{\mathrm{d}}(x) u_{\mathrm{d}}) + V(x) + u_{\mathrm{d}}^{\mathrm{T}} R_{2\mathrm{d}}(x) u_{\mathrm{d}} \\
&\quad + u_{\mathrm{d}}^{\mathrm{T}} P_2(x) u_{\mathrm{d}} \\
&= [u_{\mathrm{d}} + y_{\mathrm{d}}]^{\mathrm{T}}(R_{2\mathrm{d}}(x) + P_2(x))[u_{\mathrm{d}} + y_{\mathrm{d}}] \\
&\quad - V(x + f_{\mathrm{d}}(x) + G_{\mathrm{d}}(x) u_{\mathrm{d}}) + V(x),
\end{aligned}$$

which implies (9.99), for all admissible $u_{\mathrm{d}}(t_k) \in \mathbb{R}^{m_{\mathrm{d}}}$, $k \in \overline{\mathbb{Z}}_+$. □

Note that with $R_{2\mathrm{c}}(x) \equiv I_{m_{\mathrm{c}}}$ and $R_{2\mathrm{d}}(x) \equiv I_{m_{\mathrm{d}}}$, conditions (9.98) and (9.99) are precisely the hybrid counterpart of the return difference condition for continuous-time and discrete-time systems. However, for continuous-time systems with continuous flows an optimal feedback control law $\phi(x)$ satisfying the return difference condition is equivalent to the fact that the continuous-time nonlinear affine system with input u and output $y = -\phi(x)$ is dissipative with respect to the quadratic supply rate $[u + y]^{\mathrm{T}}[u + y] - u^{\mathrm{T}} u$ [128]. Hence, as shown in [128], a feedback control law $\phi(x)$ satisfies the return difference inequality if and only if $\phi(x)$ is optimal with respect to a performance criterion involving a nonnegative-definite weighting function on the state and a quadratic positive-definite function of the control. Alternatively, in the hybrid case, (9.98) and (9.99) are *not* equivalent to the dissipativity of (9.94)–(9.97) due to the presence of $P_2(x)$ in (9.99). However, it follows from Lemma 9.2 and Theorem 3.2 that (9.98) and (9.99) do imply that if $(\phi_{\mathrm{c}}(x), \phi_{\mathrm{d}}(x))$ is a stabilizing optimal hybrid control law, then $\mathcal{G}$ is dissipative with respect to a quadratic hybrid supply rate.

Next, we present our main result which provides hybrid disk margins for the optimal hybrid regulator given by Corollary 9.2. For the following result define

$$\overline{\gamma}_{\mathrm{c}} \triangleq \sup_{x \in \mathbb{R}^n} \sigma_{\max}(R_{2\mathrm{c}}(x)), \quad \underline{\gamma}_{\mathrm{c}} \triangleq \inf_{x \in \mathbb{R}^n} \sigma_{\min}(R_{2\mathrm{c}}(x)), \quad (9.100)$$

$$\overline{\gamma}_{\mathrm{d}} \triangleq \sup_{x \in \mathcal{Z}_x} \sigma_{\max}(R_{2\mathrm{d}}(x) + P_2(x)), \quad \underline{\gamma}_{\mathrm{d}} \triangleq \inf_{x \in \mathcal{Z}_x} \sigma_{\min}(R_{2\mathrm{d}}(x)), \quad (9.101)$$

where $P_2 : \mathcal{Z}_x \to \mathbb{N}^{m_{\mathrm{d}}}$.

Theorem 9.4 *Consider the impulsive nonlinear dynamical system $\mathcal{G}$ given by (9.94)–(9.97) where $(\phi_{\mathrm{c}}(x), \phi_{\mathrm{d}}(x))$ is an optimal stabilizing control law given by (9.60) and (9.61), and where $V(x)$, $P_{12}(x)$,* and $P_2(x)$ are such that $V(0) = 0$, $V(x) > 0$, $x \in \mathbb{R}^n$, $x \neq 0$, and satisfy (9.54)–(9.56). Then $\mathcal{G}$ is dissipative with respect to the hybrid supply rate $(s_c(u_c, y_c), \varepsilon s_d(u_d, y_d)) = (u_c^T y_c + \frac{(1-\theta_c^2)}{2} u_c^T u_c + \frac{1}{2} y_c^T y_c, \varepsilon u_d^T y_d + \frac{\varepsilon(1-\theta_d^2)}{2} u_d^T u_d + \frac{\varepsilon}{2} y_d^T y_d)$ and has a hybrid disk margin $\left((\frac{1}{1+\theta_c}, \frac{1}{1-\theta_c}), (\frac{1}{1+\theta_d}, \frac{1}{1-\theta_d}) \right)$, where $\theta_c \triangleq \sqrt{\underline{\gamma}_c/\overline{\gamma}_c}$, $\theta_d \triangleq \sqrt{\underline{\gamma}_d/\overline{\gamma}_d}$, and $\varepsilon \triangleq \overline{\gamma}_d/\overline{\gamma}_c$.

Proof. Note that for all admissible inputs $u_c(\cdot) \in \mathcal{U}_c$ and $u_d(\cdot) \in \mathcal{U}_d$, it follows from Lemma 9.2 that the solution $x(t)$, $t \geq 0$, to (9.94) and (9.95) satisfies (9.98) and (9.99), which implies that

$$V(x(\hat{t})) - V(x(t)) \leq \int_t^{\hat{t}} \{\overline{\gamma}_c[u_c(s) + y_c(s)]^T[u_c(s) + y_c(s)] \\ - \underline{\gamma}_c u_c^T(s)u_c(s)\}\mathrm{d}s, \quad t_k < t < \hat{t} \leq t_{k+1},$$

and

$$V(x(t_k^+)) - V(x(t_k)) \leq \{\overline{\gamma}_d[u_d(t_k) + y_d(t_k)]^T[u_d(t_k) + y_d(t_k)] \\ - \underline{\gamma}_d u_d^T(t_k)u_d(t_k)\}, \quad k \in \overline{\mathbb{Z}}_+.$$

Hence, with the storage function $V_s(x) = \frac{1}{2\overline{\gamma}_c} V(x)$, it follows from Theorem 3.2 that $\mathcal{G}$ is dissipative with respect to hybrid supply rate $(s_c(u_c, y_c), \varepsilon s_d(u_d, y_d)) = (u_c^T y_c + \frac{(1-\theta_c^2)}{2} u_c^T u_c + \frac{1}{2} y_c^T y_c, \varepsilon u_d^T y_d + \frac{\varepsilon(1-\theta_d^2)}{2} \cdot u_d^T u_d + \frac{\varepsilon}{2} y_d^T y_d)$. Now, the result follows immediately from Corollary 9.5 and Definition 9.3 with $\alpha_c = \frac{1}{1+\theta_c}$, $\beta_c = \frac{1}{1-\theta_c}$, $\alpha_d = \frac{1}{1+\theta_d}$, and $\beta_d = \frac{1}{1-\theta_d}$.   □

Note that in the case where $R_{2c}(x) \equiv I_{m_c}$ it follows that $\theta_c = 1$. Hence, the continuous-time dynamics of $\mathcal{G}$ have a disk margin of $(\frac{1}{2}, \infty)$. This of course does *not* imply that the hybrid optimal nonlinear regulator has hybrid disk margin of $\left((\frac{1}{2}, \infty), (\frac{1}{2}, \infty) \right)$. Next, we provide an alternative result that guarantees hybrid sector and gain margins for the case where $R_{2c}(x)$, $x \in \mathbb{R}^n$, is diagonal.

Theorem 9.5 *Consider the impulsive nonlinear dynamical system $\mathcal{G}$ given by (9.94)–(9.97) where $(\phi_c(x), \phi_d(x))$ is an optimal stabilizing control law given by (9.60) and (9.61), and where $V(x)$, $P_{12}(x)$, and $P_2(x)$ are such that $V(0) = 0$, $V(x) > 0$, $x \in \mathbb{R}^n$, $x \neq 0$, and satisfy (9.54)–(9.56). Furthermore, let $R_{2c}(x) = \mathrm{diag}[r_1(x), \ldots, r_{m_c}(x)]$, where $r_i : \mathbb{R}^n \to (0, \infty)$, $i = 1, \ldots, m_c$. If $\mathcal{G}$ is zero-state observable,* *then $\mathcal{G}$ has a hybrid sector (and hence gain) margin $\left((\frac{1}{2}, \infty), (\frac{1}{1+\theta_{\mathrm{d}}}, \frac{1}{1-\theta_{\mathrm{d}}})\right)$, where $\theta_{\mathrm{d}} \triangleq \sqrt{\underline{\gamma}_{\mathrm{d}}/\overline{\gamma}_{\mathrm{d}}}$.*

Proof. Let $\Delta(y_{\mathrm{c}}, y_{\mathrm{d}}) = (\sigma_{\mathrm{c}}(y_{\mathrm{c}}), \sigma_{\mathrm{d}}(y_{\mathrm{d}}))$, where $\sigma_{\mathrm{c}} : \mathbb{R}^{m_{\mathrm{c}}} \to \mathbb{R}^{m_{\mathrm{c}}}$ and $\sigma_{\mathrm{d}} : \mathbb{R}^{m_{\mathrm{d}}} \to \mathbb{R}^{m_{\mathrm{d}}}$ are static nonlinearities such that $\sigma_{\mathrm{c}}(0) = 0$, $\sigma_{\mathrm{d}}(0) = 0$, $\sigma_{\mathrm{c}}(y_{\mathrm{c}}) = [\sigma_{1\mathrm{c}}(y_{1\mathrm{c}}), \dots, \sigma_{m_{\mathrm{c}}\mathrm{c}}(y_{m_{\mathrm{c}}\mathrm{c}})]^{\mathrm{T}}$, $\sigma_{\mathrm{d}}(y_{\mathrm{d}}) = [\sigma_{1\mathrm{d}}(y_{1\mathrm{d}}),$ $\dots, \sigma_{m_{\mathrm{d}}\mathrm{d}}(y_{m_{\mathrm{d}}\mathrm{d}})]^{\mathrm{T}}$, $\frac{1}{2} y_{ic}^{2} < \sigma_{ic}(y_{ic}) y_{ic} < \infty$, for all $y_{ic} \neq 0$, $i = 1, \dots, m_{\mathrm{c}}$, and $\alpha_{\mathrm{d}} y_{id}^{2} < \sigma_{id}(y_{id}) y_{id} < \beta_{\mathrm{d}} y_{id}^{2}$, for all $y_{id} \neq 0$, $i = 1, \dots, m_{\mathrm{d}}$, where $\alpha_{\mathrm{d}} = \frac{1}{1+\theta_{\mathrm{d}}}$ and $\beta_{\mathrm{d}} = \frac{1}{1-\theta_{\mathrm{d}}}$; or, equivalently, $(\sigma_{id}(y_{id}) - \alpha_{\mathrm{d}} y_{id})(\sigma_{id}(y_{id}) - \beta_{\mathrm{d}} y_{id}) < 0$, for all $y_{id} \neq 0$, $i = 1, \dots, m_{\mathrm{d}}$. In this case, the closed-loop system (9.94)–(9.97) with $(u_{\mathrm{c}}, u_{\mathrm{d}}) = (-\sigma_{\mathrm{c}}(y_{\mathrm{c}}), -\sigma_{\mathrm{d}}(y_{\mathrm{d}}))$ is given by

$$\dot{x}(t) = f_{\mathrm{c}}(x(t)) - G_{\mathrm{c}}(x(t))\sigma_{\mathrm{c}}(-\phi_{\mathrm{c}}(x(t))), \quad x(0) = x_{0}, \quad x(t) \notin \mathcal{Z}_{x},$$
$$(9.102)$$
$$\Delta x(t) = f_{\mathrm{d}}(x(t)) - G_{\mathrm{d}}(x(t))\sigma_{\mathrm{d}}(-\phi_{\mathrm{d}}(x(t))), \quad x(t) \in \mathcal{Z}_{x}. \quad (9.103)$$

Next, consider the Lyapunov function candidate $V(x)$, $x \in \mathbb{R}^{n}$, satisfying (9.54)–(9.56) and let $\dot{V}(x)$ and $\Delta V(x)$ denote the Lyapunov derivative along the closed-loop state trajectories when $x(t) \notin \mathcal{Z}_{x}$ and the Lyapunov difference along the closed-loop state trajectories when $x(t) \in \mathcal{Z}_{x}$, respectively. Now, it follows from (9.54)–(9.56) that, for $x \notin \mathcal{Z}_{x}$,

$$
\begin{aligned}
\dot{V}(x) &= V'(x)f_{\mathrm{c}}(x) - V'(x)G_{\mathrm{c}}(x)\sigma_{\mathrm{c}}(-\phi_{\mathrm{c}}(x)) \\
&\le V'(x)f_{\mathrm{c}}(x) - V'(x)G_{\mathrm{c}}(x)\sigma_{\mathrm{c}}(-\phi_{\mathrm{c}}(x)) + L_{1\mathrm{c}}(x) \\
&= \phi_{\mathrm{c}}^{\mathrm{T}}(x)R_{2\mathrm{c}}(x)\phi_{\mathrm{c}}(x) + 2\phi_{\mathrm{c}}^{\mathrm{T}}(x)R_{2\mathrm{c}}(x)\sigma_{\mathrm{c}}(-\phi_{\mathrm{c}}(x)) \\
&= \sum_{i=1}^{m_{\mathrm{c}}} r_{i}(x)y_{ic}(y_{ic} - 2\sigma_{ic}(y_{ic})) \\
&\le 0,
\end{aligned}
$$

and, for $x \in \mathcal{Z}_{x}$,

$$
\begin{aligned}
\Delta V(x) &= V(x + f_{\mathrm{d}}(x) - G_{\mathrm{d}}(x)\sigma_{\mathrm{d}}(-\phi_{\mathrm{d}}(x))) - V(x) \\
&\le V(x + f_{\mathrm{d}}(x)) - P_{12}(x)\sigma_{\mathrm{d}}(-\phi_{\mathrm{d}}(x)) \\
&\quad + \sigma_{\mathrm{d}}^{\mathrm{T}}(-\phi_{\mathrm{d}}(x))P_{2}(x)\sigma_{\mathrm{d}}(-\phi_{\mathrm{d}}(x)) - V(x) + L_{1\mathrm{d}}(x) \\
&= \phi_{\mathrm{d}}^{\mathrm{T}}(x)(R_{2\mathrm{d}}(x) + P_{2}(x))\phi_{\mathrm{d}}(x) \\
&\quad + 2\phi_{\mathrm{d}}^{\mathrm{T}}(x)(R_{2\mathrm{d}}(x) + P_{2}(x))\sigma_{\mathrm{d}}(-\phi_{\mathrm{d}}(x)) \\
&\quad + \sigma_{\mathrm{d}}^{\mathrm{T}}(-\phi_{\mathrm{d}}(x))P_{2}(x)\sigma_{\mathrm{d}}(-\phi_{\mathrm{d}}(x))
\end{aligned}
$$

$$= [\sigma_\mathrm{d}(y_\mathrm{d}) - y_\mathrm{d}]^\mathrm{T}(R_{2\mathrm{d}}(x) + P_2(x))[\sigma_\mathrm{d}(y_\mathrm{d}) - y_\mathrm{d}]$$
$$- \sigma_\mathrm{d}^\mathrm{T}(y_\mathrm{d})R_{2\mathrm{d}}(x)\sigma_\mathrm{d}(y_\mathrm{d})$$
$$\le \overline{\gamma}_\mathrm{d}[\sigma_\mathrm{d}(y_\mathrm{d}) - y_\mathrm{d}]^\mathrm{T}[\sigma_\mathrm{d}(y_\mathrm{d}) - y_\mathrm{d}] - \underline{\gamma}_\mathrm{d}\sigma_\mathrm{d}^\mathrm{T}(y_\mathrm{d})\sigma_\mathrm{d}(y_\mathrm{d})$$
$$= \overline{\gamma}_\mathrm{d}(1 - \theta_\mathrm{d}^2)\sum_{i=1}^{m_\mathrm{c}}(\sigma_{id}(y_{id}) - \alpha_\mathrm{d}y_{id})(\sigma_{id}(y_{id}) - \beta_\mathrm{d}y_{id})$$
$$\le 0,$$

which, using Theorem 2.1, implies that the closed-loop system (9.102) and (9.103) is Lyapunov stable.

Next, let $\mathcal{R} \triangleq \{x \in \mathbb{R}^n : x \notin \mathcal{Z}_x, \dot{V}(x) = 0\} \cup \{x \in \mathbb{R}^n : x \in \mathcal{Z}_x, \Delta V(x) = 0\}$. Now, note that $\dot{V}(x) = 0$ for all $x \in \mathbb{R}^n \backslash \mathcal{Z}_x$ if and only if $y_\mathrm{c} = 0$ and $\Delta V(x) = 0$ for all $x \in \mathcal{Z}_x$ if and only if $y_\mathrm{d} = 0$. Since $\mathcal{G}$ is zero-state observable it follows that $\mathcal{M} = \{0\}$ is the largest invariant set contained in $\mathcal{R}$. Hence, it follows from Theorem 2.3 that $x(t) \to \mathcal{M} = \{0\}$ as $t \to \infty$. Thus, the closed-loop system (9.102) and (9.103) is globally asymptotically stable for all $\sigma_\mathrm{c}(\cdot)$ and $\sigma_\mathrm{d}(\cdot)$ such that $\frac{1}{2}y_{ic}^2 < \sigma_{ic}(y_{ic})y_{ic} < \infty$, $y_{ic} \ne 0$, $i = 1, \ldots, m_\mathrm{c}$, and $\alpha_\mathrm{d}y_{id}^2 < \sigma_{id}(y_{id})y_{id} < \beta_\mathrm{d}y_{id}^2$, $y_{id} \ne 0$, $i = 1, \ldots, m_\mathrm{d}$, which implies that the nonlinear impulsive system $\mathcal{G}$ given by (9.94)–(9.97) has sector (and hence gain) margins $\left(\left(\frac{1}{2}, \infty\right), \left(\frac{1}{1+\theta_\mathrm{d}}, \frac{1}{1-\theta_\mathrm{d}}\right)\right)$. □

Note that in the case where $R_{2\mathrm{c}}(x)$, $x \in \mathbb{R}^n$, is diagonal, Theorem 9.5 guarantees larger hybrid gain and sector margins than the hybrid gain and sector margin guarantees provided by Theorem 9.4. However, Theorem 9.5 does not provide disk margin guarantees.

Finally, we specialize the results of this section to the case of linear impulsive dynamical systems. For the following results define

$$\overline{\gamma}_\mathrm{c} \triangleq \sigma_{\max}(R_{2\mathrm{c}}), \quad \underline{\gamma}_\mathrm{c} \triangleq \sigma_{\min}(R_{2\mathrm{c}}), \tag{9.104}$$

$$\overline{\gamma}_\mathrm{d} \triangleq \sigma_{\max}(R_{2\mathrm{d}} + B_\mathrm{d}^\mathrm{T}PB_\mathrm{d}), \quad \underline{\gamma}_\mathrm{d} \triangleq \sigma_{\min}(R_{2\mathrm{d}}), \tag{9.105}$$

where $P \in \mathbb{P}^n$ satisfies (9.37) and (9.38).

Corollary 9.6 *Consider the linear impulsive system $\mathcal{G}$ given by (9.34), (9.35), (9.96), and (9.97) where $(\phi_\mathrm{c}(x), \phi_\mathrm{d}(x))$ is an optimal stabilizing control law given by (9.39) and (9.40). Then $\mathcal{G}$ is dissipative with respect to the hybrid supply rate $(s_\mathrm{c}(u_\mathrm{c}, y_\mathrm{c}), \varepsilon s_\mathrm{d}(u_\mathrm{d}, y_\mathrm{d}))$*
$$= (u_\mathrm{c}^\mathrm{T}y_\mathrm{c} + \tfrac{(1-\theta_\mathrm{c}^2)}{2}u_\mathrm{c}^\mathrm{T}u_\mathrm{c} + \tfrac{1}{2}y_\mathrm{c}^\mathrm{T}y_\mathrm{c}, \varepsilon u_\mathrm{d}^\mathrm{T}y_\mathrm{d} + \tfrac{\varepsilon(1-\theta_\mathrm{d}^2)}{2}u_\mathrm{d}^\mathrm{T}u_\mathrm{d} + \tfrac{\varepsilon}{2}y_\mathrm{d}^\mathrm{T}y_\mathrm{d})$$ *and has a hybrid disk margin $\left(\left(\frac{1}{1+\theta_\mathrm{c}}, \frac{1}{1-\theta_\mathrm{c}}\right), \left(\frac{1}{1+\theta_\mathrm{d}}, \frac{1}{1-\theta_\mathrm{d}}\right)\right)$, where $\theta_\mathrm{c} \triangleq \sqrt{\underline{\gamma}_\mathrm{c}/\overline{\gamma}_\mathrm{c}}$,*

$\theta_{\mathrm{d}} \triangleq \sqrt{\underline{\gamma}_{\mathrm{d}}/\overline{\gamma}_{\mathrm{d}}}$, and $\varepsilon \triangleq \overline{\gamma}_{\mathrm{d}}/\overline{\gamma}_{\mathrm{c}}$.

Proof. The result is a direct consequence of Theorem 9.4 with $f_{\mathrm{c}}(x) = A_{\mathrm{c}}x$, $f_{\mathrm{d}}(x) = A_{\mathrm{d}}x$, $G_{\mathrm{c}}(x) = B_{\mathrm{c}}$, $G_{\mathrm{d}}(x) = B_{\mathrm{d}}$, $V(x) = x^{\mathrm{T}}Px$, $L_{1\mathrm{c}} = x^{\mathrm{T}}R_{1\mathrm{c}}x$, $R_{2\mathrm{c}}(x) = R_{2\mathrm{c}}$, $L_{1\mathrm{d}} = x^{\mathrm{T}}R_{1\mathrm{d}}x$, $R_{2\mathrm{d}}(x) = R_{2\mathrm{d}}$, $P_{12}(x) = x^{\mathrm{T}}A_{\mathrm{d}}^{\mathrm{T}}PB_{\mathrm{d}}$, and $P_2(x) = B_{\mathrm{d}}^{\mathrm{T}}PB_{\mathrm{d}}$. $\qquad\square$

Finally, the next result specializes Theorem 9.5 to linear impulsive systems.

Corollary 9.7 *Consider the linear impulsive system $\mathcal{G}$ given by (9.34), (9.35), (9.96), and (9.97) where $(\phi_{\mathrm{c}}(x), \phi_{\mathrm{d}}(x))$ is an optimal stabilizing control law given by (9.39) and (9.40). Furthermore, let $R_{2\mathrm{c}} \in \mathbb{P}^{m_{\mathrm{c}}}$ be diagonal. Then $\mathcal{G}$ has a hybrid sector (and hence gain) margin $\left((\frac{1}{2}, \infty), (\frac{1}{1+\theta_{\mathrm{d}}}, \frac{1}{1-\theta_{\mathrm{d}}}) \right)$, where $\theta_{\mathrm{d}} \triangleq \sqrt{\underline{\gamma}_{\mathrm{d}}/\overline{\gamma}_{\mathrm{d}}}$.*

Proof. The result is a direct consequence of Theorem 9.5 with $f_{\mathrm{c}}(x) = A_{\mathrm{c}}x$, $f_{\mathrm{d}}(x) = A_{\mathrm{d}}x$, $G_{\mathrm{c}}(x) = B_{\mathrm{c}}$, $G_{\mathrm{d}}(x) = B_{\mathrm{d}}$, $V(x) = x^{\mathrm{T}}Px$, $L_{1\mathrm{c}} = x^{\mathrm{T}}R_{1\mathrm{c}}x$, $R_{2\mathrm{c}}(x) = R_{2\mathrm{c}}$, $L_{1\mathrm{d}} = x^{\mathrm{T}}R_{1\mathrm{d}}x$, $R_{2\mathrm{d}}(x) = R_{2\mathrm{d}}$, $P_{12}(x) = x^{\mathrm{T}}A_{\mathrm{d}}^{\mathrm{T}}PB_{\mathrm{d}}$, and $P_2(x) = B_{\mathrm{d}}^{\mathrm{T}}PB_{\mathrm{d}}$. $\qquad\square$

9.6 Inverse Optimal Control for Impulsive Port-Controlled Hamiltonian Systems

In this section, we present results for characterizing hybrid feedback controllers that guarantee closed-loop stability and minimize a hybrid nonlinear-nonquadratic performance functional over the infinite horizon while preserving a closed-loop hybrid Hamiltonian structure. To present these results, consider the general nonlinear impulsive dynamical system

$$\dot{x}(t) = F_{\mathrm{c}}(x(t), u_{\mathrm{c}}(t)), \quad x(0) = x_0, \quad x(t) \notin \mathcal{Z}_x, \qquad (9.106)$$
$$\Delta x(t) = F_{\mathrm{d}}(x(t), u_{\mathrm{d}}(t)), \quad x(t) \in \mathcal{Z}_x, \qquad (9.107)$$

where $F_{\mathrm{c}} : \mathcal{D} \times U_{\mathrm{c}} \to \mathbb{R}^n$ is Lipschitz continuous, $F_{\mathrm{d}} : \mathcal{Z}_x \times U_{\mathrm{d}} \to \mathbb{R}^n$ is continuous, $\mathcal{Z}_x \subset \mathcal{D}$, and $u_{\mathrm{c}}(\cdot)$ and $u_{\mathrm{d}}(\cdot)$ are restricted to the class of admissible hybrid controls such that $(u_{\mathrm{c}}(\cdot), u_{\mathrm{d}}(\cdot)) \in (\mathcal{U}_{\mathrm{c}}, \mathcal{U}_{\mathrm{d}})$. If $(u_{\mathrm{c}}(t), u_{\mathrm{d}}(t)) = (\phi_{\mathrm{c}}(x(t)), \phi_{\mathrm{d}}(x(t_k)))$, where $\phi_{\mathrm{c}} : \mathcal{D} \to U_{\mathrm{c}}$ and $\phi_{\mathrm{d}} : \mathcal{Z}_x \to U_{\mathrm{d}}$, and $x(t)$, $t \geq 0$, and $x(t_k)$, $k \in \overline{\mathbb{Z}}_+$, satisfy (9.106) and (9.107), respectively, then the closed-loop state-dependent impulsive dynamical system has the form

$$\dot{x}(t) = F_{\mathrm{c}}(x(t), \phi_{\mathrm{c}}(x(t))), \quad x(0) = x_0, \quad x(t) \notin \mathcal{Z}_x, \qquad (9.108)$$
$$\Delta x(t) = F_{\mathrm{d}}(x(t), \phi_{\mathrm{d}}(x(t))), \quad x(t) \in \mathcal{Z}_x, \qquad (9.109)$$

where $F_c(x_e, \phi_c(x_e)) = 0$ for some equilibrium point $x_e \in \mathcal{D}$.

The optimal nonlinear hybrid feedback controller $(u_c, u_d) = (\phi_c(x), \phi_d(x))$ that minimizes a nonlinear-nonquadratic hybrid performance criterion is given by Theorem 9.3. The next theorem gives a slightly modified form of Theorem 9.3 as applied to nonzero equilibrium solutions to (9.106) and (9.107). For the statement of this result define the set of regulation hybrid controllers by

$$\mathcal{C}(x_0) \triangleq \{(u_c(\cdot), u_d(\cdot)) : (u_c(\cdot), u_d(\cdot)) \text{ is admissible and } x(\cdot) \text{ given}$$
$$\text{by (9.106) and (9.107) satisfies } x(t) \to x_e \text{ as } t \to \infty\}.$$
$$(9.110)$$

Theorem 9.6 *Consider the nonlinear controlled impulsive dynamical system (9.106) and (9.107) with hybrid performance functional*

$$J(x_0, u_c(\cdot), u_d(\cdot)) = \int_0^\infty L_c(x(t), u_c(t))dt + \sum_{k \in \mathbb{Z}_{[0,\infty)}} L_d(x(t_k), u_d(t_k)),$$
$$(9.111)$$

where $L_c : \mathcal{D} \times U_c \to \mathbb{R}$ and $L_d : \mathcal{Z}_x \times U_d \to \mathbb{R}$ are given. Assume there exist a continuously differentiable function $V : \mathcal{D} \to \mathbb{R}$ and a hybrid control law $\phi_c : \mathcal{D} \to U_c$ and $\phi_d : \mathcal{Z}_x \to U_d$ such that $F_c(x_e, \phi_c(x_e)) = 0$, $V(x_e) = 0$, $V(x) > 0$, $x \in \mathcal{D}$, $x \neq x_e$, and

$$V'(x)F_c(x, \phi_c(x)) \leq 0, \quad x \notin \mathcal{Z}_x, \qquad (9.112)$$
$$V(x + F_d(x, \phi_d(x))) - V(x) \leq 0, \quad x \in \mathcal{Z}_x, \qquad (9.113)$$
$$H_c(x, \phi_c(x)) = 0, \quad x \notin \mathcal{Z}_x, \qquad (9.114)$$
$$H_d(x, \phi_d(x)) = 0, \quad x \in \mathcal{Z}_x, \qquad (9.115)$$
$$H_c(x, u_c) \geq 0, \quad x \notin \mathcal{Z}_x, \quad u_c \in U_c, \qquad (9.116)$$
$$H_d(x, u_d) \geq 0, \quad x \in \mathcal{Z}_x, \quad u_d \in U_d, \qquad (9.117)$$

where

$$H_c(x, u_c) \triangleq L_c(x, u_c) + V'(x)F_c(x, u_c), \qquad (9.118)$$
$$H_d(x, u_d) \triangleq L_d(x, u_d) + V(x + F_d(x, u_d)) - V(x). \qquad (9.119)$$

Then, with the hybrid feedback control $(u_c(\cdot), u_d(\cdot)) = (\phi_c(x(\cdot)), \phi_d(x(\cdot)))$, the equilibrium solution $x(t) \equiv x_e$ of the closed-loop system (9.108) and (9.109) is Lyapunov stable. Furthermore, if $\mathcal{D}_c \subseteq \mathcal{D}$ is a compact positively invariant set and the largest invariant set contained in $\mathcal{R} \triangleq \{x \in \mathcal{D}_c : x \notin \mathcal{Z}_x, V'(x)F_c(x, \phi_c(x)) = 0\} \cup \{x \in \mathcal{D}_c : x \in \mathcal{Z}_x, V(x + F_d(x, \phi_d(x))) = V(x)\}$ is $\mathcal{M} = \{x_e\}$, then the equilibrium *solution $x(t) \equiv x_e$ of the closed-loop system (9.108) and (9.109) is asymptotically stable. Moreover, if $x_0 \in \mathcal{D}_c$, then*

$$J(x_0, \phi_c(x(\cdot)), \phi_d(x(\cdot))) = V(x_0), \quad x_0 \in \mathcal{D}_c, \qquad (9.120)$$

and the hybrid feedback control $(u_c(\cdot), u_d(\cdot)) = (\phi_c(x(\cdot)), \phi_d(x(\cdot)))$ minimizes $J(x_0, u_c(\cdot), u_d(\cdot))$ in the sense that

$$J(x_0, \phi_c(x(\cdot)), \phi_d(x(\cdot))) = \min_{(u_c(\cdot), u_d(\cdot)) \in \mathcal{C}(x_0)} J(x_0, u_c(\cdot), u_d(\cdot)). \ (9.121)$$

Finally, if $\mathcal{D} = \mathbb{R}^n$, $U_c = \mathbb{R}^{m_c}$, $U_d = \mathbb{R}^{m_d}$, and $V(x) \to \infty$ as $\|x\| \to \infty$, then the equilibrium solution $x(t) \equiv x_e$ of the closed-loop system (9.108) and (9.109) is globally asymptotically stable.

Next, we specialize Theorem 9.6 to impulsive port-controlled Hamiltonian systems. In particular, we present an inverse optimal hybrid feedback control problem, wherein we avoid the complexity in solving the steady-state hybrid Hamilton-Jacobi-Bellman equations (9.114) and (9.115) by parameterizing a family of stabilizing hybrid controllers that minimize some derived hybrid cost functional as opposed to minimizing a given hybrid cost functional. The performance integrand provides flexibility in specifying the control law and explicitly depends on the impulsive port-controlled Hamiltonian dynamics, the Hamiltonian function of the closed-loop system, and the stabilizing hybrid feedback control law, wherein the coupling is introduced via the hybrid Hamilton-Jacobi-Bellman equation. Hence, by varying the shaped Hamiltonian, the interconnection and dissipation matrix functions, and the performance integrand, the proposed framework can be used to characterize a class of globally stabilizing hybrid controllers that preserve the hybrid Hamiltonian structure at the closed-loop level. In addition, the inverse optimal hybrid controllers guarantee hybrid disc, sector, and gain margins to multiplicative input uncertainty, and hence, guarantee robustness to unmodeled actuator dynamics.

Consider the impulsive port-controlled Hamiltonian system given by (7.1) and (7.2) with the hybrid performance criterion (9.111). Furthermore, consider performance integrands $L_c(x, u_c)$ and $L_d(x, u_d)$ of the form

$$L_c(x, u_c) = L_{1c}(x) + L_{2c}(x)u_c + u_c^T R_{2c}(x)u_c, \qquad (9.122)$$
$$L_d(x, u_d) = L_{1d}(x) + L_{2d}(x)u_d + u_d^T R_{2d}(x)u_d, \qquad (9.123)$$

where $L_{1c} : \mathcal{D} \to \mathbb{R}$ and satisfies $L_{1c}(x) \geq 0$, $x \in \mathcal{D}$, $L_{2c} : \mathcal{D} \to \mathbb{R}^{1 \times m_c}$, $R_{2c} : \mathcal{D} \to \mathbb{P}^{m_c}$, $L_{1d} : \mathcal{Z}_x \to \mathbb{R}$ and satisfies $L_{1d}(x) \geq 0$, $x \in \mathcal{Z}_x$,

$L_{2d} : \mathcal{Z}_x \to \mathbb{R}^{1 \times m_d}$, $R_{2d} : \mathcal{Z}_x \to \mathbb{P}^{m_d}$ so that (9.111) becomes

$$J(x_0, u_c(\cdot), u_d(\cdot)) = \int_0^\infty [L_{1c}(x(t)) + L_{2c}(x(t))u_c(t)$$
$$+ u_c^T(t) R_{2c}(x(t))u_c(t)]dt$$
$$+ \sum_{k \in \mathbb{Z}_{[0,\infty)}} [L_{1d}(x(t_k)) + L_{2d}(x(t_k))u_d(t_k)$$
$$+ u_d^T(t_k) R_{2d}(x(t_k))u_d(t_k)]. \tag{9.124}$$

Corollary 9.8 *Consider the impulsive port-controlled Hamiltonian system given by (7.1) and (7.2) with performance functional (9.124). Assume there exist functions $\mathcal{H}_s$, $\mathcal{H}_c : \mathcal{D} \to \mathbb{R}$, $\mathcal{J}_{cs}, \mathcal{J}_{ca} : \mathcal{D} \to \mathbb{R}^{n \times n}$, $\mathcal{R}_{cs}, \mathcal{R}_{ca} : \mathcal{D} \to \mathbb{R}^{n \times n}$, $\mathcal{J}_{ds}, \mathcal{J}_{da} : \mathcal{Z}_x \to \mathbb{R}^{n \times n}$, and $\mathcal{R}_{ds}, \mathcal{R}_{da} : \mathcal{Z}_x \to \mathbb{R}^{n \times n}$ such that $\mathcal{H}_s(x) = \mathcal{H}(x) + \mathcal{H}_c(x)$ is continuously differentiable, $\mathcal{J}_{cs}(x) = \mathcal{J}_c(x) + \mathcal{J}_{ca}(x)$, $\mathcal{J}_{cs}(x) = -\mathcal{J}_{cs}^T(x)$, $\mathcal{R}_{cs}(x) = \mathcal{R}_c(x) + \mathcal{R}_{ca}(x)$, $\mathcal{R}_{cs}(x) = \mathcal{R}_{cs}^T(x) \geq 0$, $x \in \mathcal{D}$; $\mathcal{J}_{ds}(x) = \mathcal{J}_d(x) + \mathcal{J}_{da}(x)$, $\mathcal{J}_{ds}(x) = -\mathcal{J}_{ds}^T(x)$, $\mathcal{R}_{ds}(x) = \mathcal{R}_d(x) + \mathcal{R}_{da}(x)$, $\mathcal{R}_{ds}(x) = \mathcal{R}_{ds}^T(x) \geq 0$, $x \in \mathcal{Z}_x$; conditions (7.22), (7.23), and (7.25) are satisfied; and*

$$\frac{\partial^2 \mathcal{H}_s}{\partial x^2}(x) > 0, \quad x \in \mathcal{D}, \tag{9.125}$$

$$[\mathcal{J}_{cs}(x) - \mathcal{R}_{cs}(x) + \frac{1}{2}G_c(x)R_{2c}^{-1}(x)G_c^T(x)]\left(\frac{\partial \mathcal{H}_c}{\partial x}(x)\right)^T$$
$$= [-[\mathcal{J}_{ca}(x) - \mathcal{R}_{ca}(x)]$$
$$- \frac{1}{2}G_c(x)R_{2c}^{-1}(x)G_c^T(x)]\left(\frac{\partial \mathcal{H}}{\partial x}(x)\right)^T - \frac{1}{2}G_c(x)R_{2c}^{-1}(x)L_{2c}^T(x),$$
$$x \notin \mathcal{Z}_x, \tag{9.126}$$

$$[\mathcal{J}_{ds}(x) - \mathcal{R}_{ds}(x)]\left(\frac{\partial \mathcal{H}_c}{\partial x}(x)\right)^T = -[\mathcal{J}_{da}(x) - \mathcal{R}_{da}(x)]\left(\frac{\partial \mathcal{H}}{\partial x}(x)\right)^T$$
$$- \frac{1}{2}G_d(x)\left(R_{2d}(x) + \frac{1}{2}G_d^T(x)\frac{\partial^2 \mathcal{H}_s}{\partial x^2}(x)G_d(x)\right)^{-1}$$
$$\cdot\left(L_{2d}(x) + \frac{\partial \mathcal{H}_s}{\partial x}(x)G_d(x)\right)^T, \quad x \in \mathcal{Z}_x. \tag{9.127}$$

Then the equilibrium solution $x(t) \equiv x_e$ of the closed-loop system given by (7.20) and (7.21) is Lyapunov stable with the hybrid feedback con-

trol law

$$\phi_c(x) = -\frac{1}{2}R_{2c}^{-1}(x)\left(G_c^T(x)\left(\frac{\partial \mathcal{H}_s}{\partial x}(x)\right)^T + L_{2c}^T(x)\right), \quad x \notin \mathcal{Z}_x,$$

(9.128)

$$\phi_d(x) = -\frac{1}{2}\left(R_{2d}(x) + \frac{1}{2}G_d^T(x)\frac{\partial^2 \mathcal{H}_s}{\partial x^2}(x)G_d(x)\right)^{-1}$$

$$\cdot \left(G_d^T(x)\left(\frac{\partial \mathcal{H}_s}{\partial x}(x)\right)^T + L_{2d}^T(x)\right), \quad x \in \mathcal{Z}_x.$$

(9.129)

If, in addition, $\mathcal{D}_c \subseteq \mathcal{D}$ is a compact positively invariant set with respect to (7.20) and (7.21) and the largest invariant set contained in

$$\mathcal{R} \triangleq \left\{ x \in \mathcal{D}_c : x \notin \mathcal{Z}_x, \frac{\partial \mathcal{H}_s}{\partial x}(x)\mathcal{R}_{cs}(x)\left(\frac{\partial \mathcal{H}_s}{\partial x}(x)\right)^T = 0 \right\}$$

$$\cup \left\{ x \in \mathcal{D}_c : x \in \mathcal{Z}_x, \frac{\partial \mathcal{H}_s}{\partial x}(x)[\mathcal{J}_{ds}(x) - \mathcal{R}_{ds}(x) \right.$$

$$\left. + \frac{1}{2}(\mathcal{J}_{ds}(x) - \mathcal{R}_{ds}(x))^T\frac{\partial^2 \mathcal{H}_s}{\partial x^2}(x)(\mathcal{J}_{ds}(x) - \mathcal{R}_{ds}(x))]\left(\frac{\partial \mathcal{H}_s}{\partial x}(x)\right)^T \right.$$

$$\left. = 0 \right\}$$

(9.130)

is $\mathcal{M} = \{x_e\}$, then the equilibrium solution $x(t) \equiv x_e$ of the closed-loop system (7.20) and (7.21) is asymptotically stable. Moreover, the performance functional (9.124), with

$$L_{1c}(x) = \phi_c^T(x)R_{2c}(x)\phi_c(x) - \frac{\partial \mathcal{H}_s}{\partial x}(x)\left[\mathcal{J}_c(x) - \mathcal{R}_c(x)\right]\left(\frac{\partial \mathcal{H}}{\partial x}(x)\right)^T,$$

(9.131)

$$L_{1d}(x) = \phi_d^T(x)\left(R_{2d}(x) + \frac{1}{2}G_d^T(x)\frac{\partial^2 \mathcal{H}_s}{\partial x^2}(x)G_d(x)\right)\phi_d(x)$$

$$- \mathcal{H}_s\left(x + [\mathcal{J}_d(x) - \mathcal{R}_d(x)]\left(\frac{\partial \mathcal{H}}{\partial x}(x)\right)^T\right) + \mathcal{H}_s(x), \quad (9.132)$$

is minimized in the sense that

$$J(x_0, \phi_c(x(\cdot)), \phi_d(x(\cdot))) = \min_{(u_c(\cdot), u_d(\cdot)) \in \mathcal{C}(x_0)} J(x_0, u_c(\cdot), u_d(\cdot)),$$

$$x_0 \in \mathcal{D}_c. \quad (9.133)$$

In addition, $J(x_0, \phi_c(x(\cdot)), \phi_d(x(\cdot))) = \mathcal{H}_s(x_0) - \mathcal{H}_s(x_e)$, $x_0 \in \mathcal{D}_c$. Finally, if $\mathcal{D} = \mathbb{R}^n$, $U_c = \mathbb{R}^{m_c}$, $U_d = \mathbb{R}^{m_d}$, and $\mathcal{H}_s(x) \to \infty$ as $\|x\| \to \infty$, then the above asymptotic stability result is global.

Proof. It follows from (9.126) and (9.127) that the closed-loop system (7.1) and (7.2) has a Hamiltonian structure given by (7.20) and (7.21). Furthermore, (7.22), (7.23), and (9.125) imply that the energy function $\mathcal{H}_s(\cdot)$ has a global minimum at $x = x_e$, and hence, $x = x_e$ is an equilibrium point of the closed-loop system. Now, using the Lyapunov function candidate $V(x) = \mathcal{H}_s(x) - \mathcal{H}_s(x_e)$ with condition (7.25), Lyapunov and asymptotic stability of the closed-loop system follow as in the proof of Theorem 7.1.

Next, with $L_{1c}(x)$ and $L_{1d}(x)$ given by (9.131) and (9.132), respectively, and $\phi_c(x)$ and $\phi_d(x)$ given by (9.128) and (9.129), respectively, (9.114) and (9.115) hold. Finally, since

$$H_c(x, u_c) = H_c(x, u_c) - H_c(x, \phi_c(x))$$
$$= [u_c - \phi_c(x)]^T R_{2c}(x)[u_c - \phi_c(x)], \quad x \notin \mathcal{Z}_x, \qquad (9.134)$$
$$H_d(x, u_d) = H_d(x, u_d) - H_d(x, \phi_d(x))$$
$$= [u_d - \phi_d(x)]^T \left(R_{2d}(x) + \frac{1}{2} G_d^T(x) \frac{\partial^2 \mathcal{H}_s}{\partial x^2}(x) G_d(x) \right)$$
$$\cdot [u_d - \phi_d(x)], \quad x \in \mathcal{Z}_x, \qquad (9.135)$$

where $R_{2c}(x) > 0$, $x \notin \mathcal{Z}_x$, and $R_{2d}(x) + \frac{1}{2} G_d^T(x) \frac{\partial^2 \mathcal{H}_s}{\partial x^2}(x) G_d(x) > 0$, $x \in \mathcal{Z}_x$, conditions (9.116) and (9.117) hold. The result now follows as a direct consequence of Theorem 9.6. $\qquad \square$

In the case where $L_{2c}(x) \equiv 0$, $L_{2d}(x) \equiv 0$, and $R_{2c}(x)$, $x \notin \mathcal{Z}_x$, is a diagonal weighting function, the hybrid controller (9.128) and (9.129) guarantees hybrid sector and gain margins to multiplicative hybrid input uncertainty of $\left((\frac{1}{2}, \infty), (\frac{1}{1+\theta_d}, \frac{1}{1-\theta_d}) \right)$, where $\theta_d = \sqrt{\underline{\gamma}_d / \overline{\gamma}_d}$,

$$\underline{\gamma}_d \triangleq \inf_{x \in \mathcal{Z}_x} \sigma_{\min}(R_{2d}(x)),$$

and

$$\overline{\gamma}_d \triangleq \sup_{x \in \mathcal{Z}_x} \sigma_{\max} \left(R_{2d}(x) + \frac{1}{2} G_d^T(x) \frac{\partial^2 \mathcal{H}_s}{\partial x^2}(x) G_d(x) \right).$$

This follows as a direct consequence of Theorem 9.5.

Chapter Ten

Disturbance Rejection Control for Nonlinear Impulsive Dynamical Systems

10.1 Introduction

In this chapter, we develop an optimality-based theory for disturbance rejection of nonlinear impulsive dynamical systems with bounded energy exogenous disturbances. The key motivation for developing an optimal and inverse optimal nonlinear hybrid control theory that additionally guarantees disturbance rejection is that it provides a class of candidate disturbance rejection hybrid controllers parameterized by the hybrid cost functional that is minimized. In order to address the optimality-based disturbance rejection nonlinear hybrid control problem we extend the nonlinear-nonquadratic, hybrid controller analysis and synthesis framework presented in Chapter 9. Specifically, using nonlinear hybrid dissipativity theory developed in Chapter 3, with appropriate storage functions and supply rates, we transform the nonlinear hybrid disturbance rejection problem into an optimal hybrid control problem. This is accomplished by properly modifying the hybrid cost functional to account for exogenous disturbances so that the solution of the modified optimal nonlinear hybrid control problem serves as the solution to the hybrid disturbance rejection problem.

The main contribution of this chapter is a methodology for designing optimal nonlinear hybrid controllers which guarantee disturbance rejection and minimize a (derived) hybrid performance functional that serves as an upper bound to a nonlinear-nonquadratic hybrid cost functional. In particular, the performance bound can be evaluated in closed form as long as the nonlinear-nonquadratic hybrid cost functional considered is related in a specific way to an underlying Lyapunov function that guarantees stability. This Lyapunov function is shown to be the solution to a steady-state hybrid Hamilton-Jacobi-Isaacs equation for the impulsive controlled system, and plays a key role in constructing the optimal nonlinear disturbance rejection hybrid control law. Furthermore, since our nonlinear-nonquadratic hybrid cost functional is closely related to the structure of the Lyapunov function, the proposed framework provides a class of feedback stabilizing hybrid controllers that minimize a derived hybrid performance functional. Hence, the overall framework provides for a generalization of the Hamilton-Jacobi-Isaacs conditions for addressing the design of optimal and inverse optimal hybrid controllers for nonlinear impulsive systems with exogenous disturbances.

10.2 Nonlinear Impulsive Dynamical Systems with Bounded Disturbances

In this section, we present sufficient conditions for dissipativity for a class of nonlinear impulsive dynamical systems with bounded energy and bounded amplitude disturbances. In addition, we consider the problem of evaluating a performance bound for a nonlinear-nonquadratic hybrid cost functional. The cost bound is evaluated in closed form by relating the hybrid cost functional to an underlying Lyapunov function that guarantees asymptotic stability of the nonlinear impulsive dynamical system.

We begin by considering the nonlinear state-dependent impulsive dynamical system $\mathcal{G}$ given by

$$\dot{x}(t) = f_\mathrm{c}(x(t)) + J_{1\mathrm{c}}(x(t))w_\mathrm{c}(t), \quad x(0) = x_0, \quad x(t) \notin \mathcal{Z}_x,$$
$$w_\mathrm{c}(t) \in \mathcal{W}_\mathrm{c}, \quad (10.1)$$
$$\Delta x(t) = f_\mathrm{d}(x(t)) + J_{1\mathrm{d}}(x(t))w_\mathrm{d}(t), \quad x(t) \in \mathcal{Z}_x, \quad w_\mathrm{d}(t) \in \mathcal{W}_\mathrm{d}, (10.2)$$
$$z_\mathrm{c}(t) = h_\mathrm{c}(x(t)) + J_{2\mathrm{c}}(x(t))w_\mathrm{c}(t), \quad x(t) \notin \mathcal{Z}_x, \quad w_\mathrm{c}(t) \in \mathcal{W}_\mathrm{c}, \quad (10.3)$$
$$z_\mathrm{d}(t) = h_\mathrm{d}(x(t)) + J_{2\mathrm{d}}(x(t))w_\mathrm{d}(t), \quad x(t) \in \mathcal{Z}_x, \quad w_\mathrm{d}(t) \in \mathcal{W}_\mathrm{d}, (10.4)$$

where $t \geq 0$, $x(t) \in \mathcal{D} \subseteq \mathbb{R}^n$, $\mathcal{D}$ is an open set with $0 \in \mathcal{D}$, $w_\mathrm{c}(t) \in \mathcal{W}_\mathrm{c} \subseteq \mathbb{R}^{d_\mathrm{c}}$, $w_\mathrm{d}(t_k) \in \mathcal{W}_\mathrm{d} \subseteq \mathbb{R}^{d_\mathrm{d}}$, $t_k = \tau_k(x_0)$ denotes the kth instant of time at which $x(t)$ intersects $\mathcal{Z}_x$ for a particular trajectory $x(t)$, $z_\mathrm{c}(t) \in \mathbb{R}^{p_\mathrm{c}}$, $z_\mathrm{d}(t_k) \in \mathbb{R}^{p_\mathrm{d}}$, $f_\mathrm{c} : \mathcal{D} \to \mathbb{R}^n$ is Lipschitz continuous and satisfies $f_\mathrm{c}(0) = 0$, $J_{1\mathrm{c}} : \mathcal{D} \to \mathbb{R}^{n \times d_\mathrm{c}}$, $f_\mathrm{d} : \mathcal{D} \to \mathbb{R}^n$ is continuous, $J_{1\mathrm{d}} : \mathcal{D} \to \mathbb{R}^{n \times d_\mathrm{d}}$, $h_\mathrm{c} : \mathcal{D} \to \mathbb{R}^{p_\mathrm{c}}$ and satisfies $h_\mathrm{c}(0) = 0$, $J_{2\mathrm{c}} : \mathcal{D} \to \mathbb{R}^{p_\mathrm{c} \times d_\mathrm{c}}$, $h_\mathrm{d} : \mathcal{D} \to \mathbb{R}^{p_\mathrm{d}}$, $J_{2\mathrm{d}} : \mathcal{D} \to \mathbb{R}^{p_\mathrm{d} \times d_\mathrm{d}}$, and $\mathcal{Z}_x \subset \mathcal{D}$ is the resetting set. Here, we assume that the existence and uniqueness properties of a given state-dependent impulsive dynamical system are satisfied in forward time.

We identify the resetting times by $\tau_k(x_0)$ so that the trajectory of the system (10.1) and (10.2) from the initial condition $x(0) = x_0$ is given by $\psi(t, x_0, w_\mathrm{c})$ for $0 < t \leq \tau_1(x_0)$, where $\psi(\cdot, \cdot, \cdot)$ denotes the solution to the continuous-time dynamics (10.1). If and when the trajectory reaches a state $x_1 \triangleq x(\tau_1(x_0))$ satisfying $x_1 \in \mathcal{Z}_x$, then the state is instantaneously transferred to $x_1^+ \triangleq x_1 + f_d(x_1) + J_{1d}(x_1)w_d$, where $w_d \in \mathcal{W}_d$, according to the resetting law (10.2). The trajectory $x(t)$, $\tau_1(x_0) < t \leq \tau_2(x_0)$, is then given by $\psi(t - \tau_1(x_0), x_1^+, w_c)$, and so on. Note that the solution $x(t)$ of (10.1) and (10.2) is left-continuous, that is, it is continuous everywhere except at the resetting times $\tau_k(x_0)$, and

$$x_k \triangleq x(\tau_k(x_0)) = \lim_{\varepsilon \to 0^+} x(\tau_k(x_0) - \varepsilon), \tag{10.5}$$

$$x_k^+ \triangleq x(\tau_k(x_0)) + f_d(x(\tau_k(x_0))) + J_{1d}(x(\tau_k(x_0)))w_d(\tau_k(x_0)),$$
$$w_d(\tau_k(x_0)) \in \mathcal{W}_d, \tag{10.6}$$

for $k = 1, 2, \ldots$.

We make the following additional assumptions:

A1. If $x \in \overline{\mathcal{Z}_x} \backslash \mathcal{Z}_x$, then there exists $\varepsilon > 0$ such that, for all $0 < \delta < \varepsilon$, $x(\delta, x, w_c) \notin \mathcal{Z}_x$.

A2. If $x \in \mathcal{Z}_x$, then $x + f_d(x) + J_{1d}(x)w_d \notin \mathcal{Z}_x$, $w_d(\tau_k(x_0)) \in \mathcal{W}_d$.

Assumptions A1 and A2 are extensions of Assumptions A1 and A2 of Chapter 2 to impulsive dynamical systems with exogenous disturbances and ensure well posedness of the resetting times. Furthermore, we assume that an infinite number of resettings occur, that is, $\tau_k(x_0) \to \infty$ as $k \to \infty$. As in previous chapters, we denote the resetting times $\tau_k(x_0)$ by t_k and define $\mathbb{Z}_{[0,t)} \triangleq \{k : 0 \leq t_k < t\}$. Finally, for the following result presenting sufficient conditions under which a nonlinear impulsive dynamical system is dissipative with respect to the hybrid supply rate $(s_c(z_c, w_c), s_d(z_d, w_d))$, let $s_c : \mathbb{R}^{p_c} \times \mathbb{R}^{d_c} \to \mathbb{R}$ and $s_d : \mathbb{R}^{p_d} \times \mathbb{R}^{d_d} \to \mathbb{R}$ be given functions.

Proposition 10.1 *Consider the nonlinear impulsive dynamical system $\mathcal{G}$ given by (10.1)–(10.4). Furthermore, assume there exist functions $\Gamma_c : \mathcal{D} \to \mathbb{R}$, $\Gamma_d : \mathcal{D} \to \mathbb{R}$, $P_{1w_d} : \mathcal{D} \to \mathbb{R}^{1 \times d_d}$, $P_{2w_d} : \mathcal{D} \to \mathbb{N}^{d_d}$, and $V : \mathcal{D} \to \mathbb{R}$, where $V(\cdot)$ is a continuously differentiable function, such that*

$$V(0) = 0, \tag{10.7}$$

$$V(x) \geq 0, \quad x \in \mathcal{D}, \tag{10.8}$$

$$V'(x)J_{1c}(x)w_c \leq s_c(z_c, w_c) + \Gamma_c(x), \quad x \notin \mathcal{Z}_x,$$
$$w_c \in \mathcal{W}_c, \quad z_c \in \mathbb{R}^{p_c}, \tag{10.9}$$

$$V'(x)f_{\rm c}(x) + \Gamma_{\rm c}(x) \le 0, \quad x \notin \mathcal{Z}_x, \tag{10.10}$$

$$P_{1w_{\rm d}}(x)w_{\rm d} + w_{\rm d}^{\rm T}P_{2w_{\rm d}}(x)w_{\rm d} \le s_{\rm d}(z_{\rm d}, w_{\rm d}) + \Gamma_{\rm d}(x), \quad x \in \mathcal{Z}_x,$$

$$w_{\rm d} \in \mathcal{W}_{\rm d}, \quad z_{\rm d} \in \mathbb{R}^{p_{\rm d}}, \tag{10.11}$$

$$V(x + f_{\rm d}(x) + J_{1\rm d}(x)w_{\rm d}) = V(x + f_{\rm d}(x)) + P_{1w_{\rm d}}(x)w_{\rm d}$$

$$+ w_{\rm d}^{\rm T}P_{2w_{\rm d}}(x)w_{\rm d}, \quad x \in \mathcal{Z}_x, \quad w_{\rm d} \in \mathcal{W}_{\rm d}, \tag{10.12}$$

$$V(x + f_{\rm d}(x)) - V(x) + \Gamma_{\rm d}(x) \le 0, \quad x \in \mathcal{Z}_x. \tag{10.13}$$

Then the solution $x(t)$, $t \ge 0$, of (10.1) and (10.2) satisfies

$$V(x(T)) \le \int_0^T s_{\rm c}(z_{\rm c}(t), w_{\rm c}(t)){\rm d}t + \sum_{k \in \mathbb{Z}_{[0,T)}} s_{\rm d}(z_{\rm d}(t_k), w_{\rm d}(t_k)) + V(x_0),$$

$$T \ge 0, \quad w_{\rm c}(\cdot) \in \mathcal{L}_2, \quad w_{\rm d}(\cdot) \in \ell_2. \tag{10.14}$$

Proof. Let $x(t)$, $t \ge 0$, satisfy (10.1) and (10.2), and let $w_{\rm c}(\cdot) \in \mathcal{L}_2$ and $w_{\rm d}(\cdot) \in \ell_2$. Then, it follows from (10.9) and (10.10) that

$$\dot{V}(x(t)) \triangleq \frac{{\rm d}V(x(t))}{{\rm d}t}$$

$$= V'(x(t))[f_{\rm c}(x(t)) + J_{1\rm c}(x(t))w_{\rm c}(t)]$$

$$\le V'(x(t))f_{\rm c}(x(t)) + \Gamma_{\rm c}(x(t)) + s_{\rm c}(z_{\rm c}(t), w_{\rm c}(t))$$

$$\le s_{\rm c}(z_{\rm c}(t), w_{\rm c}(t)), \quad x(t) \notin \mathcal{Z}_x, \quad t_k < t \le t_{k+1}. \tag{10.15}$$

Furthermore, it follows from (10.11)–(10.13) that

$$\Delta V(x(t_k)) \triangleq V(x(t_k) + f_{\rm d}(x(t_k)) + J_{1\rm d}(x(t_k))w_{\rm d}(t_k)) - V(x(t_k))$$

$$= V(x(t_k) + f_{\rm d}(x(t_k))) + P_{1w_{\rm d}}(x(t_k))w_{\rm d}(t_k)$$

$$+ w_{\rm d}^{\rm T}(t_k)P_{2w_{\rm d}}(x(t_k))w_{\rm d}(t_k) - V(x(t_k))$$

$$\le s_{\rm d}(z_{\rm d}(t_k), w_{\rm d}(t_k)), \quad x(t_k) \in \mathcal{Z}_x. \tag{10.16}$$

Now, integrating over the interval $[0, T)$ with $\mathbb{Z}_{[0,T)} = \{1, 2, \ldots, i\}$, (10.15) and (10.16) yield

$$\int_0^T s_{\rm c}(z_{\rm c}(s), w_{\rm c}(s)){\rm d}s + \sum_{k \in \mathbb{Z}_{[0,T)}} s_{\rm d}(z_{\rm d}(t_k), w_{\rm d}(t_k))$$

$$= \int_0^{t_1} s_{\rm c}(z_{\rm c}(s), w_{\rm c}(s)){\rm d}s + s_{\rm d}(z_{\rm d}(t_1), w_{\rm d}(t_1))$$

$$+ \int_{t_1^+}^{t_2} s_{\rm c}(z_{\rm c}(s), w_{\rm c}(s)){\rm d}s + s_{\rm d}(z_{\rm d}(t_2), w_{\rm d}(t_2)) + \cdots$$

$$+ \int_{t_{i-1}^+}^{t_i} s_c(z_c(s), w_c(s)) ds + s_d(z_d(t_i), w_d(t_i))$$

$$+ \int_{t_i^+}^{T} s_c(z_c(s), w_c(s)) ds$$

$$\geq V(x(t_1)) - V(x_0) + V(x(t_1) + f_d(x(t_1))) - V(x(t_1))$$
$$+ V(x(t_2)) - V(x(t_1^+)) + V(x(t_2) + f_d(x(t_2)))$$
$$- V(x(t_2)) + \cdots + V(x(t_i)) - V(x(t_{i-1}^+))$$
$$+ V(x(t_i) + f_d(x(t_i))) - V(x(t_i)) + V(x(T)) - V(x(t_i^+))$$
$$= V(x(T)) - V(x_0), \tag{10.17}$$

which proves the result. $\square$

For the next result let $L_c : \mathcal{D} \to \mathbb{R}$ and $L_d : \mathcal{D} \to \mathbb{R}$ be given.

Theorem 10.1 *Consider the nonlinear impulsive dynamical system given by (10.1)–(10.4) with hybrid performance functional*

$$J(x_0) = \int_0^{\infty} L_c(x(t)) dt + \sum_{k \in \mathbb{Z}_{[0,\infty)}} L_d(x(t_k)), \tag{10.18}$$

where $x(t)$, $t \geq 0$, solves (10.1) and (10.2) with $(w_c(t), w_d(t_k)) \equiv (0, 0)$. Assume there exist functions $\Gamma_c : \mathcal{D} \to \mathbb{R}$, $\Gamma_d : \mathcal{D} \to \mathbb{R}$, $P_{1w_d} : \mathcal{D} \to \mathbb{R}^{1 \times d_d}$, $P_{2w_d} : \mathcal{D} \to \mathbb{N}^{d_d}$, and $V : \mathcal{D} \to \mathbb{R}$, where $V(\cdot)$ is continuously differentiable, such that

$$V(0) = 0, \tag{10.19}$$
$$V(x) > 0, \quad x \in \mathcal{D}, \quad x \neq 0, \tag{10.20}$$
$$V'(x) f_c(x) < 0, \quad x \in \mathcal{D}, \quad x \neq 0, \quad x \notin \mathcal{Z}_x, \tag{10.21}$$
$$V'(x) J_{1c}(x) w_c \leq s_c(z_c, w_c) + L_c(x) + \Gamma_c(x),$$
$$\qquad x \notin \mathcal{Z}_x, \quad w_c \in \mathcal{W}_c, \quad z_c \in \mathbb{R}^{p_c}, \tag{10.22}$$
$$V(x + f_d(x)) - V(x) \leq 0, \quad x \in \mathcal{Z}_x, \tag{10.23}$$
$$P_{1w_d}(x) w_d + w_d^T P_{2w_d}(x) w_d \leq s_d(z_d, w_d) + L_d(x) + \Gamma_d(x),$$
$$\qquad x \in \mathcal{Z}_x, \quad w_d \in \mathcal{W}_d, \quad z_d \in \mathbb{R}^{p_d}, \tag{10.24}$$
$$V(x + f_d(x) + J_{1d}(x) w_d) = V(x + f_d(x)) + P_{1w_d}(x) w_d$$
$$\qquad + w_d^T P_{2w_d}(x) w_d, \quad x \in \mathcal{Z}_x, \quad w_d \in \mathcal{W}_d, \tag{10.25}$$
$$L_c(x) + V'(x) f_c(x) + \Gamma_c(x) = 0, \quad x \notin \mathcal{Z}_x, \tag{10.26}$$
$$L_d(x) + V(x + f_d(x)) - V(x) + \Gamma_d(x) = 0, \quad x \in \mathcal{Z}_x. \tag{10.27}$$

Then there exists a neighborhood $\mathcal{D}_0 \subseteq \mathcal{D}$ of the origin such that if $x_0 \in \mathcal{D}_0$, then the zero solution $x(t) \equiv 0$ of the undisturbed (i.e.,

$(w_c(t), w_d(t_k)) \equiv (0,0))$ *system (10.1) and (10.2) is asymptotically stable. If, in addition, $\Gamma_c(x) \geq 0$ and $\Gamma_d(x) \geq 0$, $x \in \mathcal{D}$, then*

$$J(x_0) \leq \mathcal{J}(x_0) = V(x_0), \tag{10.28}$$

where

$$\mathcal{J}(x_0) \triangleq \int_0^\infty [L_c(x(t)) + \Gamma_c(x(t))]dt$$
$$+ \sum_{k \in \mathbb{Z}_{[0,\infty)}} [L_d(x(t_k)) + \Gamma_d(x(t_k))] \tag{10.29}$$

and where $x(t)$, $t \geq 0$, is a solution to (10.1) and (10.2) with $(w_c(t), w_d(t_k)) \equiv (0,0)$. Furthermore, the solution $x(t)$, $t \geq 0$, to (10.1) and (10.2) satisfies the dissipativity constraint

$$\int_0^T s_c(z_c(t), w_c(t))dt + \sum_{k \in \mathbb{Z}_{[0,T)}} s_d(z_d(t_k), w_d(t_k)) + V(x_0) \geq 0,$$
$$T \geq 0, \quad w_c(\cdot) \in \mathcal{L}_2, \quad w_d(\cdot) \in \ell_2. \tag{10.30}$$

Finally, if $\mathcal{D} = \mathbb{R}^n$ and

$$V(x) \to \infty \quad as \quad \|x\| \to \infty, \tag{10.31}$$

then the zero solution $x(t) \equiv 0$ to the undisturbed system (10.1) and (10.2) is globally asymptotically stable.

Proof. Let $x(t)$, $t \geq 0$, satisfy (10.1) and (10.2), and let $w_c(\cdot) \in \mathcal{L}_2$ and $w_d(\cdot) \in \ell_2$. Then

$$\dot{V}(x(t)) \triangleq \frac{d}{dt}V(x(t)) = V'(x(t))[f_c(x(t)) + J_{1c}(x(t))w_c(t)],$$
$$x(t) \notin \mathcal{Z}_x, \quad t_k < t \leq t_{k+1}. \tag{10.32}$$

Hence, with $w_c(t) \equiv 0$, it follows from (10.21) that

$$\dot{V}(x(t)) < 0, \quad x(t) \notin \mathcal{Z}_x, \quad x(t) \neq 0, \quad t_k < t \leq t_{k+1}. \tag{10.33}$$

Furthermore,

$$\Delta V(x(t_k)) \triangleq V(x(t_k) + f_d(x(t_k)) + J_{1d}(x(t_k))w_d(t_k)) - V(x(t_k))$$
$$= V(x(t_k) + f_d(x(t_k))) - V(x(t_k)) + P_{1w_d}(x(t_k))w_d(t_k)$$
$$+ w_d^T(t_k)P_{2w_d}(x(t_k))w_d(t_k), \quad x(t_k) \in \mathcal{Z}_x. \tag{10.34}$$

Hence, with $w_d(t_k) \equiv 0$, it follows from (10.23) that

$$\Delta V(x(t_k)) \le 0, \quad x(t_k) \in \mathcal{Z}_x. \tag{10.35}$$

Thus, using (10.19)–(10.20), (10.33), and (10.35) it follows from Theorem 2.1 that $V(\cdot)$ is a Lyapunov function for (10.1) and (10.2), which proves asymptotic stability of the zero solution $x(t) \equiv 0$ of (10.1) and (10.2) with $(w_c(t), w_d(t_k)) \equiv (0, 0)$. Consequently, $x(t) \to 0$ as $t \to \infty$ for all initial conditions $x_0 \in \mathcal{D}_0$ for some neighborhood $\mathcal{D}_0 \subseteq \mathcal{D}$ of the origin.

Next, for $(w_c(t), w_d(t_k)) \equiv (0, 0)$, (10.33) and (10.35) imply that

$$0 = -\dot{V}(x(t)) + V'(x(t))f_c(x(t)), \quad x(t) \notin \mathcal{Z}_x, \quad t_k < t \le t_{k+1}, \tag{10.36}$$

$$0 = -\Delta V(x(t_k)) + V(x(t_k) + f_d(x(t_k))) - V(x(t_k)), \quad x(t_k) \in \mathcal{Z}_x. \tag{10.37}$$

Hence, if $\Gamma_c(x) \ge 0$, $x \in \mathcal{D}$, (10.26) implies

$$\begin{aligned}
L_c(x(t)) &= -\dot{V}(x(t)) + L_c(x(t)) + V'(x(t))f_c(x(t)) \\
&\le -\dot{V}(x(t)) + L_c(x(t)) + V'(x(t))f_c(x(t)) + \Gamma_c(x(t)) \\
&= -\dot{V}(x(t)), \quad x(t) \notin \mathcal{Z}_x, \quad t_k < t \le t_{k+1}. \tag{10.38}
\end{aligned}$$

Similarly, if $\Gamma_d(x) \ge 0$, $x \in \mathcal{D}$, (10.27) implies

$$\begin{aligned}
L_d(x(t_k)) &= -\Delta V(x(t_k)) + L_d(x(t_k)) + \Delta V(x(t_k)) \\
&\le -\Delta V(x(t_k)) + L_d(x(t_k)) + V(x(t_k) + f_d(x(t_k))) \\
&\quad -V(x(t_k)) + \Gamma_d(x(t_k)) \\
&= -\Delta V(x(t_k)), \quad x(t_k) \in \mathcal{Z}_x. \tag{10.39}
\end{aligned}$$

Now, integrating over the interval $[0, T)$ with $\mathbb{Z}_{[0,T)} = \{1, 2, \ldots, i\}$, (10.38) and (10.39) yield

$$\int_0^T L_c(x(s))ds + \sum_{k \in \mathbb{Z}_{[0,T)}} L_d(x(t_k))$$

$$= \int_0^{t_1} L_c(x(s))ds + L_d(x(t_1))$$

$$+ \int_{t_1^+}^{t_2} L_c(x(s))ds + L_d(x(t_2)) + \cdots$$

$$+ \int_{t_{i-1}^+}^{t_i} L_c(x(s))ds + L_d(x(t_i)) + \int_{t_i^+}^T L_c(x(s))ds$$

$$\leq -V(x(t_1)) + V(x_0) - V(x(t_1) + f_{\mathrm{d}}(x(t_1))) + V(x(t_1))$$
$$-V(x(t_2)) + V(x(t_1^+)) - V(x(t_2) + f_{\mathrm{d}}(x(t_2)))$$
$$+V(x(t_2)) + \cdots - V(x(t_i)) + V(x(t_{i-1}^+))$$
$$-V(x(t_i) + f_{\mathrm{d}}(x(t_i))) + V(x(t_i)) - V(x(T)) + V(x(t_i^+))$$
$$\leq -V(x(t_1)) + V(x_0) - V(x(t_1^+)) + V(x(t_1)) - V(x(t_2))$$
$$+V(x(t_1^+)) - V(x(t_2^+)) + V(x(t_2)) + \cdots - V(x(t_i))$$
$$+V(x(t_{i-1}^+)) - V(x(t_i^+)) + V(x(t_i)) - V(x(T)) + V(x(t_i^+))$$
$$\leq -V(x(T)) + V(x_0). \tag{10.40}$$

Letting $T \to \infty$ and noting that $V(x(T)) \to 0$ for all $x_0 \in \mathcal{D}_0$ yields $J(x_0) \leq V(x_0)$.

Next, let $x(t)$, $t \geq 0$, satisfy (10.1) and (10.2) with $(w_{\mathrm{c}}(t), w_{\mathrm{d}}(t_k)) \equiv (0,0)$. Then, with $L_{\mathrm{c}}(x)$ and $L_{\mathrm{d}}(x)$ replaced by $L_{\mathrm{c}}(x) + \Gamma_{\mathrm{c}}(x)$ and $L_{\mathrm{d}}(x) + \Gamma_{\mathrm{d}}(x)$, respectively, and $J(x_0)$ replaced by $\mathcal{J}(x_0)$, it follows from Theorem 9.2 that $\mathcal{J}(x_0) = V(x_0)$. Finally, since (10.19), (10.20), (10.22), and (10.24)–(10.27) imply (10.7)–(10.13) with $\Gamma_{\mathrm{c}}(x)$ and $\Gamma_{\mathrm{d}}(x)$ replaced by $L_{\mathrm{c}}(x) + \Gamma_{\mathrm{c}}(x)$ and $L_{\mathrm{d}}(x) + \Gamma_{\mathrm{d}}(x)$, respectively, Proposition 10.1 yields

$$V(x(T)) \leq \int_0^T s_{\mathrm{c}}(z_{\mathrm{c}}(t), w_{\mathrm{c}}(t))\mathrm{d}t + \sum_{k \in \mathbb{Z}_{[0,T)}} s_{\mathrm{d}}(z_{\mathrm{d}}(t_k), w_{\mathrm{d}}(t_k)) + V(x_0),$$
$$T \geq 0, \quad w_{\mathrm{c}}(\cdot) \in \mathcal{L}_2, \quad w_{\mathrm{d}}(\cdot) \in \ell_2. \tag{10.41}$$

Now, (10.30) follows by noting that $V(x(T)) \geq 0$, $T \geq 0$.

Finally, for $\mathcal{D} = \mathbb{R}^n$ global asymptotic stability of the zero solution $x(t) \equiv 0$ of the undisturbed system (10.1) and (10.2) is a direct consequence of the radially unbounded condition (10.31) on $V(x)$. $\square$

10.3 Specialization to Dissipative Impulsive Dynamical Systems with Quadratic Supply Rates

In this section, we consider the special case in which the supply rate $(s_{\mathrm{c}}(z_{\mathrm{c}}, w_{\mathrm{c}}), s_{\mathrm{d}}(z_{\mathrm{d}}, w_{\mathrm{d}}))$ is quadratic. Specifically, let $h_{\mathrm{c}} : \mathcal{D} \to \mathbb{R}^{p_{\mathrm{c}}}$, $J_{2\mathrm{c}} : \mathcal{D} \to \mathbb{R}^{p_{\mathrm{c}} \times d_{\mathrm{c}}}$, $h_{\mathrm{d}} : \mathcal{D} \to \mathbb{R}^{p_{\mathrm{d}}}$, $J_{2\mathrm{d}} : \mathcal{D} \to \mathbb{R}^{p_{\mathrm{d}} \times d_{\mathrm{d}}}$, $Q_{\mathrm{c}} \in \mathbb{S}^{p_{\mathrm{c}}}$, $S_{\mathrm{c}} \in \mathbb{R}^{p_{\mathrm{c}} \times d_{\mathrm{c}}}$, $R_{\mathrm{c}} \in \mathbb{S}^{d_{\mathrm{c}}}$, $Q_{\mathrm{d}} \in \mathbb{S}^{p_{\mathrm{d}}}$, $S_{\mathrm{d}} \in \mathbb{R}^{p_{\mathrm{d}} \times d_{\mathrm{d}}}$, $R_{\mathrm{d}} \in \mathbb{S}^{d_{\mathrm{d}}}$, and $(s_{\mathrm{c}}(z_{\mathrm{c}}, w_{\mathrm{c}}), s_{\mathrm{d}}(z_{\mathrm{d}}, w_{\mathrm{d}})) = (z_{\mathrm{c}}^{\mathrm{T}} Q_{\mathrm{c}} z_{\mathrm{c}} + 2z_{\mathrm{c}}^{\mathrm{T}} S_{\mathrm{c}} w_{\mathrm{c}} + w_{\mathrm{c}}^{\mathrm{T}} R_{\mathrm{c}} w_{\mathrm{c}}, z_{\mathrm{d}}^{\mathrm{T}} Q_{\mathrm{d}} z_{\mathrm{d}} + 2z_{\mathrm{d}}^{\mathrm{T}} S_{\mathrm{d}} w_{\mathrm{d}} + w_{\mathrm{d}}^{\mathrm{T}} R_{\mathrm{d}} w_{\mathrm{d}})$, and define

$$\mathcal{N}_{\mathrm{c}}(x) \triangleq R_{\mathrm{c}} + S_{\mathrm{c}}^{\mathrm{T}} J_{2\mathrm{c}}(x) + J_{2\mathrm{c}}^{\mathrm{T}}(x) S_{\mathrm{c}} + J_{2\mathrm{c}}^{\mathrm{T}}(x) Q_{\mathrm{c}} J_{2\mathrm{c}}(x) > 0, \quad x \notin \mathcal{Z}_x,$$

$$(10.42)$$

$$\mathcal{N}_{\mathrm{d}}(x) \triangleq R_{\mathrm{d}} + S_{\mathrm{d}}^{\mathrm{T}} J_{2\mathrm{d}}(x) + J_{2\mathrm{d}}^{\mathrm{T}}(x) S_{\mathrm{d}} + J_{2\mathrm{d}}^{\mathrm{T}}(x) Q_{\mathrm{d}} J_{2\mathrm{d}}(x) > P_{2w_{\mathrm{d}}}(x),$$
$$x \in \mathcal{Z}_x. \quad (10.43)$$

Furthermore, let $L_{\mathrm{c}}(x) \geq 0$, $x \notin \mathcal{Z}_x$, and $L_{\mathrm{d}}(x) \geq 0$, $x \in \mathcal{Z}_x$. Then

$$
\begin{aligned}
\Gamma_{\mathrm{c}}(x) =\ & [\tfrac{1}{2} V'(x) J_{1\mathrm{c}}(x) - h_{\mathrm{c}}^{\mathrm{T}}(x)(Q_{\mathrm{c}} J_{2\mathrm{c}}(x) + S_{\mathrm{c}})] \\
& \cdot \mathcal{N}_{\mathrm{c}}^{-1}(x) [\tfrac{1}{2} V'(x) J_{1\mathrm{c}}(x) - h_{\mathrm{c}}^{\mathrm{T}}(x)(Q_{\mathrm{c}} J_{2\mathrm{c}}(x) + S_{\mathrm{c}})]^{\mathrm{T}} \\
& - h_{\mathrm{c}}^{\mathrm{T}}(x) Q_{\mathrm{c}} h_{\mathrm{c}}(x), \quad x \notin \mathcal{Z}_x, \quad\quad (10.44) \\
\Gamma_{\mathrm{d}}(x) =\ & [\tfrac{1}{2} P_{1w_{\mathrm{d}}}(x) - h_{\mathrm{d}}^{\mathrm{T}}(x)(Q_{\mathrm{d}} J_{2\mathrm{d}}(x) + S_{\mathrm{d}})] \\
& \cdot (\mathcal{N}_{\mathrm{d}}(x) - P_{2w_{\mathrm{d}}}(x))^{-1} [\tfrac{1}{2} P_{1w_{\mathrm{d}}}(x) - h_{\mathrm{d}}^{\mathrm{T}}(x)(Q_{\mathrm{d}} J_{2\mathrm{d}}(x) + S_{\mathrm{d}})]^{\mathrm{T}} \\
& - h_{\mathrm{d}}^{\mathrm{T}}(x) Q_{\mathrm{d}} h_{\mathrm{d}}(x), \quad x \in \mathcal{Z}_x, \quad\quad (10.45)
\end{aligned}
$$

satisfy (10.22) and (10.24), respectively, since in this case

$$
\begin{aligned}
& L_{\mathrm{c}}(x) + \Gamma_{\mathrm{c}}(x) + s_{\mathrm{c}}(z_{\mathrm{c}}, w_{\mathrm{c}}) - V'(x) J_{1\mathrm{c}}(x) w_{\mathrm{c}} \\
& = L_{\mathrm{c}}(x) + [\tfrac{1}{2} V'(x) J_{1\mathrm{c}}(x) - h_{\mathrm{c}}^{\mathrm{T}}(x)(Q_{\mathrm{c}} J_{2\mathrm{c}}(x) + S_{\mathrm{c}}) - \mathcal{N}_{\mathrm{c}}(x) w_{\mathrm{c}}] \\
& \quad \cdot \mathcal{N}_{\mathrm{c}}^{-1}(x) [\tfrac{1}{2} V'(x) J_{1\mathrm{c}}(x) - h_{\mathrm{c}}^{\mathrm{T}}(x)(Q_{\mathrm{c}} J_{2\mathrm{c}}(x) + S_{\mathrm{c}}) - \mathcal{N}_{\mathrm{c}}(x) w_{\mathrm{c}}]^{\mathrm{T}} \\
& \geq 0, \quad x \notin \mathcal{Z}_x, \quad\quad (10.46)
\end{aligned}
$$

and

$$
\begin{aligned}
& L_{\mathrm{d}}(x) + \Gamma_{\mathrm{d}}(x) + s_{\mathrm{d}}(z_{\mathrm{d}}, w_{\mathrm{d}}) - P_{1w_{\mathrm{d}}}(x) w_{\mathrm{d}} - w_{\mathrm{d}}^{\mathrm{T}} P_{2w_{\mathrm{d}}}(x) w_{\mathrm{d}} \\
& = L_{\mathrm{d}}(x) + [\tfrac{1}{2} P_{1w_{\mathrm{d}}}(x) - h_{\mathrm{d}}^{\mathrm{T}}(x)(Q_{\mathrm{d}} J_{2\mathrm{d}}(x) + S_{\mathrm{d}}) \\
& \quad - w_{\mathrm{d}}^{\mathrm{T}}(\mathcal{N}_{\mathrm{d}}^{\mathrm{T}}(x) - P_{2w_{\mathrm{d}}}^{\mathrm{T}}(x))] \\
& \quad \cdot (\mathcal{N}_{\mathrm{d}}(x) - P_{2w_{\mathrm{d}}}(x))^{-1} [\tfrac{1}{2} P_{1w_{\mathrm{d}}}(x) - h_{\mathrm{d}}^{\mathrm{T}}(x)(Q_{\mathrm{d}} J_{2\mathrm{d}}(x) + S_{\mathrm{d}}) \\
& \quad - w_{\mathrm{d}}^{\mathrm{T}}(\mathcal{N}_{\mathrm{d}}^{\mathrm{T}}(x) - P_{2w_{\mathrm{d}}}^{\mathrm{T}}(x))]^{\mathrm{T}} \\
& \geq 0, \quad x \in \mathcal{Z}_x. \quad\quad (10.47)
\end{aligned}
$$

Corollary 10.1 *Let $L_{\mathrm{c}}(x) \geq 0$, $L_{\mathrm{d}}(x) \geq 0$, $x \in \mathcal{D}$, and consider the nonlinear impulsive dynamical system given by (10.1)–(10.4) with hybrid performance functional*

$$J(x_0) \triangleq \int_0^{\infty} L_{\mathrm{c}}(x(t)) \mathrm{d}t + \sum_{k \in \mathbb{Z}_{[0,\infty)}} L_{\mathrm{d}}(x(t_k)), \quad (10.48)$$

where $x(t)$, $t \geq 0$, is a solution to (10.1) and (10.2) with $(w_{\mathrm{c}}(t), w_{\mathrm{d}}(t_k)) \equiv (0,0)$. Assume there exist functions $P_{1w_{\mathrm{d}}} : \mathcal{D} \to \mathbb{R}^{1 \times d_{\mathrm{d}}}$ and

$P_{2w_{\mathrm{d}}} : \mathcal{D} \to \mathbb{N}^{d_{\mathrm{d}}}$, and a continuously differentiable function $V : \mathcal{D} \to \mathbb{R}$ such that

$$V(0) = 0, \tag{10.49}$$

$$V(x) > 0, \quad x \in \mathcal{D}, \quad x \neq 0, \tag{10.50}$$

$$V'(x)f_{\mathrm{c}}(x) < 0, \quad x \notin \mathcal{Z}_x, \quad x \neq 0, \tag{10.51}$$

$$V(x + f_{\mathrm{d}}(x)) - V(x) \leq 0, \quad x \in \mathcal{Z}_x, \tag{10.52}$$

$$V(x + f_{\mathrm{d}}(x) + J_{1\mathrm{d}}(x)w_{\mathrm{d}}) = V(x + f_{\mathrm{d}}(x)) + P_{1w_{\mathrm{d}}}(x)w_{\mathrm{d}}$$
$$+ w_{\mathrm{d}}^{\mathrm{T}} P_{2w_{\mathrm{d}}}(x)w_{\mathrm{d}}, \quad x \in \mathcal{Z}_x, \quad w_{\mathrm{d}} \in \mathcal{W}_{\mathrm{d}}, \tag{10.53}$$

$$\gamma_{\mathrm{d}}^2 I_{d_{\mathrm{d}}} - J_{2\mathrm{d}}^{\mathrm{T}}(x)J_{2\mathrm{d}}(x) - P_{2w_{\mathrm{d}}}(x) > 0, \quad x \in \mathcal{Z}_x, \tag{10.54}$$

$$L_{\mathrm{c}}(x) + V'(x)f_{\mathrm{c}}(x) + \Gamma_{\mathrm{c}}(x) = 0, \quad x \notin \mathcal{Z}_x, \tag{10.55}$$

$$L_{\mathrm{d}}(x) + V(x + f_{\mathrm{d}}(x)) - V(x) + \Gamma_{\mathrm{d}}(x) = 0, \quad x \in \mathcal{Z}_x. \tag{10.56}$$

where

$$\Gamma_{\mathrm{c}}(x) = [\tfrac{1}{2}V'(x)J_{1\mathrm{c}}(x) + h_{\mathrm{c}}^{\mathrm{T}}(x)J_{2\mathrm{c}}(x)][\gamma_{\mathrm{c}}^2 I_{d_{\mathrm{c}}} - J_{2\mathrm{c}}^{\mathrm{T}}(x)J_{2\mathrm{c}}(x)]^{-1}$$
$$\cdot [\tfrac{1}{2}V'(x)J_{1\mathrm{c}}(x) + h_{\mathrm{c}}^{\mathrm{T}}(x)J_{2\mathrm{c}}(x)]^{\mathrm{T}} + h_{\mathrm{c}}^{\mathrm{T}}(x)h_{\mathrm{c}}(x), \quad x \notin \mathcal{Z}_x,$$
$$\tag{10.57}$$

$$\Gamma_{\mathrm{d}}(x) = [\tfrac{1}{2}P_{1w_{\mathrm{d}}}(x) + h_{\mathrm{d}}^{\mathrm{T}}(x)J_{2\mathrm{d}}(x)][\gamma_{\mathrm{d}}^2 I_{d_{\mathrm{d}}} - J_{2\mathrm{d}}^{\mathrm{T}}(x)J_{2\mathrm{d}}(x) - P_{2w_{\mathrm{d}}}(x)]^{-1}$$
$$\cdot [\tfrac{1}{2}P_{1w_{\mathrm{d}}}(x) + h_{\mathrm{d}}^{\mathrm{T}}(x)J_{2\mathrm{d}}(x)]^{\mathrm{T}} + h_{\mathrm{d}}^{\mathrm{T}}(x)h_{\mathrm{d}}(x), \quad x \in \mathcal{Z}_x, \tag{10.58}$$

$\gamma_{\mathrm{c}} > 0$, and $\gamma_{\mathrm{d}} > 0$. Then there exists a neighborhood $\mathcal{D}_0 \subseteq \mathcal{D}$ of the origin such that if $x_0 \in \mathcal{D}_0$, then the zero solution $x(t) \equiv 0$ of the undisturbed (i.e., $(w_{\mathrm{c}}(t), w_{\mathrm{d}}(t_k)) \equiv (0,0)$) system (10.1) and (10.2) is asymptotically stable. Furthermore,

$$J(x_0) \leq \mathcal{J}(x_0) = V(x_0), \tag{10.59}$$

where

$$\mathcal{J}(x_0) \triangleq \int_0^\infty [L_{\mathrm{c}}(x(t)) + \Gamma_{\mathrm{c}}(x(t))]\mathrm{d}t$$
$$+ \sum_{k \in \mathbb{Z}_{[0,\infty)}} [L_{\mathrm{d}}(x(t_k)) + \Gamma_{\mathrm{d}}(x(t_k))], \tag{10.60}$$

and where $x(t)$, $t \geq 0$, is a solution to (10.1) and (10.2) with $(w_{\mathrm{c}}(t), w_{\mathrm{d}}(t_k)) \equiv (0,0)$. Furthermore, the solution $x(t)$, $t \geq 0$, to (10.1) and (10.2) satisfies the nonexpansivity constraint

$$\int_0^T z_{\mathrm{c}}^{\mathrm{T}}(t)z_{\mathrm{c}}(t)\mathrm{d}t + \sum_{k \in \mathbb{Z}_{[0,T)}} z_{\mathrm{d}}^{\mathrm{T}}(t_k)z_{\mathrm{d}}(t_k) \leq \gamma_{\mathrm{c}}^2 \int_0^T w_{\mathrm{c}}^{\mathrm{T}}(t)w_{\mathrm{c}}(t)\mathrm{d}t$$

$$+\gamma_d^2 \sum_{k \in \mathbb{Z}_{[0,T)}} w_d^T(t_k)w_d(t_k) + V(x_0),$$

$$w_c(\cdot) \in \mathcal{L}_2, \quad w_d(\cdot) \in \ell_2. \quad (10.61)$$

Finally, if $\mathcal{D} = \mathbb{R}^n$ and

$$V(x) \to \infty \quad as \quad \|x\| \to \infty,$$

then the zero solution $x(t) \equiv 0$ to the undisturbed system (10.1) and (10.2) is globally asymptotically stable.

Proof. With $Q_c = -I_{p_c}$, $Q_d = -I_{p_d}$, $S_c = 0$, $S_d = 0$, $R_c = \gamma_c^2 I_{d_c}$, and $R_d = \gamma_d^2 I_{d_d}$, it follows from (10.46) and (10.47) that $\Gamma_c(x)$ and $\Gamma_d(x)$ given by (10.57) and (10.58), respectively, satisfy (10.22) and (10.24). The result now follows as a direct consequence of Theorem 10.1. $\square$

Note that if $L_c(x) = h_c^T(x)h_c(x)$ and $L_d(x) = h_d^T(x)h_d(x)$ in Corollary 11.3, then $\Gamma_c(x)$ and $\Gamma_d(x)$ can be chosen as

$$\Gamma_c(x) = [\tfrac{1}{2}V'(x)J_{1c}(x) + h_c^T(x)J_{2c}(x)][\gamma_c^2 I_{d_c} - J_{2c}^T(x)J_{2c}(x)]^{-1}$$
$$\cdot[\tfrac{1}{2}V'(x)J_{1c}(x) + h_c^T(x)J_{2c}(x)]^T, \quad x \notin \mathcal{Z}_x, \quad (10.62)$$

$$\Gamma_d(x) = [\tfrac{1}{2}P_{1w_d}(x) + h_d^T(x)J_{2d}(x)][\gamma_d^2 I_{d_d} - J_{2d}^T(x)J_{2d}(x) - P_{2w_d}(x)]^{-1}$$
$$\cdot[\tfrac{1}{2}P_{1w_d}(x) + h_d^T(x)J_{2d}(x)]^T, \quad x \in \mathcal{Z}_x. \quad (10.63)$$

Next, we specialize the results in Corollary 10.1 to linear impulsive dynamical systems. Specifically, letting $f_c(x) = A_c x$, $f_d(x) = (A_d - I_n)x$, $J_{1c}(x) = D_c$, $J_{1d}(x) = D_d$, $h_c(x) = E_c x$, $h_d(x) = E_d x$, $J_{2c}(x) = 0$, $J_{2d}(x) = 0$, $L_c(x) = x^T R_c x$, $L_d(x) = x^T R_d x$, and $V(x) = x^T P x$, where $A_c \in \mathbb{R}^{n \times n}$, $A_d \in \mathbb{R}^{n \times n}$, $D_c \in \mathbb{R}^{n \times d_c}$, $D_d \in \mathbb{R}^{n \times d_d}$, $E_c \in \mathbb{R}^{p_c \times n}$, $E_d \in \mathbb{R}^{p_d \times n}$, $R_c \triangleq E_c^T E_c > 0$, $R_d \triangleq E_d^T E_d > 0$, and $P \in \mathbb{P}^n$ satisfies

$$0 = x^T(A_c^T P + PA_c + \gamma_c^{-2}PD_cD_c^T P + R_c)x, \quad x \notin \mathcal{Z}_x, \quad (10.64)$$
$$0 = x^T(A_d^T PA_d - P + A_d^T PD_d(\gamma_d^2 I_{d_d} - D_d^T PD_d)^{-1}D_d^T PA_d + R_d)x,$$
$$x \in \mathcal{Z}_x, \quad (10.65)$$

it follows from (10.62) and (10.63) that $\Gamma_c(x)$ and $\Gamma_d(x)$ can be chosen as $\Gamma_c(x) = \gamma_c^{-2}x^T PD_cD_c^T Px$ and $\Gamma_d(x) = x^T A_d^T PD_d(\gamma_d^2 I_{d_d} - D_d^T PD_d)^{-1} D_d^T PA_d x$, where $\gamma_d^2 I_{d_d} - D_d^T PD_d > 0$. Hence, with $x_0 = 0$, Corollary 10.1 implies that

$$\int_0^T x^T(t)R_cx(t)\mathrm{d}t + \sum_{k \in \mathbb{Z}_{[0,T)}} x^T(t_k)R_dx(t_k) \leq \gamma_c^2 \int_0^T w_c^T(t)w_c(t)\mathrm{d}t$$

$$+\gamma_{\mathrm{d}}^2 \sum_{k\in\mathbb{Z}_{[0,T)}} w_{\mathrm{d}}^{\mathrm{T}}(t_k)w_{\mathrm{d}}(t_k),$$

$$w_{\mathrm{c}}(\cdot)\in\mathcal{L}_2, \quad w_{\mathrm{d}}(\cdot)\in\ell_2. \quad (10.66)$$

Corollary 10.2 *Let* $L_{\mathrm{c}}(x)\geq 0$, $L_{\mathrm{d}}(x)\geq 0$, $x\in\mathcal{D}$, $p_{\mathrm{c}}=d_{\mathrm{c}}$, *and* $p_{\mathrm{d}}=d_{\mathrm{d}}$, *and consider the nonlinear impulsive dynamical system given by (10.1)–(10.4) with hybrid performance functional*

$$J(x_0)\triangleq\int_0^\infty L_{\mathrm{c}}(x(t))\mathrm{dt}+\sum_{k\in\mathbb{Z}_{[0,\infty)}} L_{\mathrm{d}}(x(t_k)), \quad (10.67)$$

where $x(t)$, $t\geq 0$, *is a solution to (10.1) and (10.2) with* $(w_{\mathrm{c}}(t), w_{\mathrm{d}}(t_k))\equiv(0,0)$. *Assume there exist functions* $P_{1w_{\mathrm{d}}}:\mathcal{D}\to\mathbb{R}^{1\times d_{\mathrm{d}}}$ *and* $P_{2w_{\mathrm{d}}}:\mathcal{D}\to\mathbb{N}^{d_{\mathrm{d}}}$, *and a continuously differentiable function* $V:\mathcal{D}\to\mathbb{R}$ *such that*

$$V(0)=0, \quad (10.68)$$
$$V(x)>0, \quad x\in\mathcal{D}, \quad x\neq 0, \quad (10.69)$$
$$V'(x)f_{\mathrm{c}}(x)<0, \quad x\notin\mathcal{Z}_x, \quad x\neq 0, \quad (10.70)$$
$$V(x+f_{\mathrm{d}}(x))-V(x)\leq 0, \quad x\in\mathcal{Z}_x, \quad (10.71)$$
$$V(x+f_{\mathrm{d}}(x)+J_{1\mathrm{d}}(x)w_{\mathrm{d}})=V(x+f_{\mathrm{d}}(x))+P_{1w_{\mathrm{d}}}(x)w_{\mathrm{d}}$$
$$+w_{\mathrm{d}}^{\mathrm{T}}P_{2w_{\mathrm{d}}}(x)w_{\mathrm{d}}, \quad x\in\mathcal{Z}_x, \quad w_{\mathrm{d}}\in\mathcal{W}_{\mathrm{d}}, \quad (10.72)$$
$$J_{2\mathrm{d}}(x)+J_{2\mathrm{d}}^{\mathrm{T}}(x)-P_{2w_{\mathrm{d}}}(x)>0, \quad x\in\mathcal{Z}_x, \quad (10.73)$$
$$L_{\mathrm{c}}(x)+V'(x)f_{\mathrm{c}}(x)+\Gamma_{\mathrm{c}}(x)=0, \quad x\notin\mathcal{Z}_x, \quad (10.74)$$
$$L_{\mathrm{d}}(x)+V(f_{\mathrm{d}}(x))-V(x)+\Gamma_{\mathrm{d}}(x)=0, \quad x\in\mathcal{Z}_x, \quad (10.75)$$

where

$$\Gamma_{\mathrm{c}}(x)=[\tfrac{1}{2}V'(x)J_{1\mathrm{c}}(x)-h_{\mathrm{c}}^{\mathrm{T}}(x)][J_{2\mathrm{c}}(x)+J_{2\mathrm{c}}^{\mathrm{T}}(x)]^{-1}$$
$$\cdot[\tfrac{1}{2}V'(x)J_{1\mathrm{c}}(x)-h_{\mathrm{c}}^{\mathrm{T}}(x)]^{\mathrm{T}}, \quad x\notin\mathcal{Z}_x, \quad (10.76)$$
$$\Gamma_{\mathrm{d}}(x)=[\tfrac{1}{2}P_{1w_{\mathrm{d}}}(x)-h_{\mathrm{d}}^{\mathrm{T}}(x)][J_{2\mathrm{d}}(x)+J_{2\mathrm{d}}^{\mathrm{T}}(x)-P_{2w_{\mathrm{d}}}(x)]^{-1}$$
$$\cdot[\tfrac{1}{2}P_{1w_{\mathrm{d}}}(x)-h_{\mathrm{d}}^{\mathrm{T}}(x)]^{\mathrm{T}}, \quad x\in\mathcal{Z}_x. \quad (10.77)$$

Then there exists a neighborhood $\mathcal{D}_0\subseteq\mathcal{D}$ *of the origin such that if* $x_0\in\mathcal{D}_0$, *then the zero solution* $x(t)\equiv 0$ *of the undisturbed (i.e.,* $(w_{\mathrm{c}}(t),w_{\mathrm{d}}(t_k))\equiv(0,0))$ *system (10.1) and (10.2) is asymptotically stable. Furthermore,*

$$J(x_0)\leq\mathcal{J}(x_0)=V(x_0), \quad (10.78)$$

where

$$J(x_0) \triangleq \int_0^\infty [L_c(x(t)) + \Gamma_c(x(t))]dt + \sum_{k \in \mathbb{Z}_{[0,\infty)}} [L_d(x(t_k)) + \Gamma_d(x(t_k))]$$

(10.79)

and where $x(t)$, $t \geq 0$, is a solution to (10.1) and (10.2) with $(w_c(t),$ $w_d(t_k)) \equiv (0,0)$. Furthermore, the solution $x(t)$, $t \geq 0$, to (10.1) and (10.2) satisfies the passivity constraint

$$\int_0^T 2w_c^T(t)z_c(t)dt + \sum_{k \in \mathbb{Z}_{[0,T)}} 2w_d^T(t_k)z_d(t_k) + V(x_0) \geq 0,$$

$$T \geq 0, \quad w_c(\cdot) \in \mathcal{L}_2, \quad w_d(\cdot) \in \ell_2. \quad (10.80)$$

Finally, if $\mathcal{D} = \mathbb{R}^n$ and

$$V(x) \to \infty \quad as \quad \|x\| \to \infty,$$

then the zero solution $x(t) \equiv 0$ to the undisturbed system (10.1) and (10.2) is globally asymptotically stable.

Proof. With $p_c = d_c$, $p_d = d_d$, $Q_c = 0$, $Q_d = 0$, $S_c = I_{d_c}$, $S_d = I_{d_d}$, $R_c = 0$, and $R_d = 0$, it follows from (10.46) and (10.47) that $\Gamma_c(x)$ and $\Gamma_d(x)$ given by (10.76) and (10.77), respectively, satisfy (10.22) and (10.24). The result now follows as a direct consequence of Theorem 10.1. $\square$

To specialize Corollary 10.2 to linear impulsive dynamical systems, let $f_c(x) = A_c x$, $f_d(x) = (A_d - I_n)x$, $J_{1c}(x) = D_c$, $J_{1d}(x) = D_d$, $h_c(x) = E_c x$, $h_d(x) = E_d x$, $J_{2c}(x) = E_{\infty c}$, $J_{2d}(x) = E_{\infty d}$, $L_c(x) = x^T R_c x$, $L_d(x) = x^T R_d x$, $V(x) = x^T P x$, $\Gamma_c(x) = x^T [D_c^T P - E_c]^T (E_{\infty c} + E_{\infty c}^T)^{-1}[D_c^T P - E_c]x$, and $\Gamma_d(x) = x^T (D_d^T P A_d - E_d)^T (E_{\infty d} + E_{\infty d}^T - D_d^T P D_d)^{-1} (D_d^T P A_d - E_d)x$, where $E_{\infty c} + E_{\infty c}^T > 0$, $E_{\infty d} + E_{\infty d}^T - D_d^T P D_d > 0$, and where $A_c \in \mathbb{R}^{n \times n}$, $A_d \in \mathbb{R}^{n \times n}$, $D_c \in \mathbb{R}^{n \times d_c}$, $D_d \in \mathbb{R}^{n \times d_d}$, $E_c \in \mathbb{R}^{p_c \times n}$, $E_d \in \mathbb{R}^{p_d \times n}$, $E_{\infty c} \in \mathbb{R}^{d_c \times d_c}$, $E_{\infty d} \in \mathbb{R}^{d_d \times d_d}$, $R_c \triangleq E_c^T E_c > 0$, $R_d \triangleq E_d^T E_d > 0$, and $P \in \mathbb{P}^n$ satisfies

$$0 = x^T(A_c^T P + P A_c + (D_c^T P - E_c)^T (E_{\infty c} + E_{\infty c}^T)^{-1}$$
$$\cdot (D_c^T P - E_c) + R_c)x, \quad x \notin \mathcal{Z}_x, \quad (10.81)$$
$$0 = x^T(A_d^T P A_d - P + (D_d^T P A_d - E_d)^T (E_{\infty d} + E_{\infty d}^T - D_d^T P D_d)^{-1}$$
$$\cdot (D_d^T P A_d - E_d) + R_d)x, \quad x \in \mathcal{Z}_x. \quad (10.82)$$

Now, it follows from Corollary 10.2, with $x_0 = 0$, that

$$\int_0^T 2w_{\mathrm{c}}^{\mathrm{T}}(t)z_{\mathrm{c}}(t)\mathrm{d}t + \sum_{k\in\mathbb{Z}_{[0,T)}} 2w_{\mathrm{d}}^{\mathrm{T}}(t_k)z_{\mathrm{d}}(t_k) \geq 0,$$

$$w_{\mathrm{c}}(\cdot) \in \mathcal{L}_2, \quad w_{\mathrm{d}}(\cdot) \in \ell_2. \quad (10.83)$$

Next, define the subsets of bounded disturbances

$$\mathcal{W}_{\mathrm{c}\beta_{\mathrm{c}}} \triangleq \left\{ w_{\mathrm{c}}(\cdot) \in \mathcal{L}_2 : \int_0^T w_{\mathrm{c}}^{\mathrm{T}}(t)w_{\mathrm{c}}(t)\mathrm{d}t \leq \beta_{\mathrm{c}}, \quad T \geq 0 \right\}, \quad (10.84)$$

$$\mathcal{W}_{\mathrm{d}\beta_{\mathrm{d}}} \triangleq \left\{ w_{\mathrm{d}}(\cdot) \in \ell_2 : \sum_{k\in\mathbb{Z}_{[0,T)}} w_{\mathrm{d}}^{\mathrm{T}}(t_k)w_{\mathrm{d}}(t_k) \leq \beta_{\mathrm{d}}, \quad T \geq 0 \right\}, \quad (10.85)$$

where $\beta_{\mathrm{c}} > 0$ and $\beta_{\mathrm{d}} > 0$. Furthermore, let $L_{\mathrm{c}} : \mathcal{D} \to \mathbb{R}$ and $L_{\mathrm{d}} : \mathcal{D} \to \mathbb{R}$ be such that $L_{\mathrm{c}}(x) \geq 0$ and $L_{\mathrm{d}}(x) \geq 0$, $x \in \mathcal{D}$.

Theorem 10.2 *Let $\gamma_{\mathrm{c}} > 0$ and $\gamma_{\mathrm{d}} > 0$, and consider the nonlinear impulsive dynamical system (10.1) and (10.2) with performance functional (10.18). Assume there exist functions $P_{1w_{\mathrm{d}}} : \mathcal{D} \to \mathbb{R}^{1\times d_{\mathrm{d}}}$ and $P_{2w_{\mathrm{d}}} : \mathcal{D} \to \mathbb{N}^{d_{\mathrm{d}}}$, and a continuously differentiable function $V : \mathcal{D} \to \mathbb{R}$ such that*

$$V(0) = 0, \quad (10.86)$$

$$V(x) > 0, \quad x \in \mathcal{D}, \quad x \neq 0, \quad (10.87)$$

$$V'(x)f_{\mathrm{c}}(x) < 0, \quad x \notin \mathcal{Z}_x, \quad x \neq 0, \quad (10.88)$$

$$V(x + f_{\mathrm{d}}(x)) - V(x) \leq 0, \quad x \in \mathcal{Z}_x, \quad (10.89)$$

$$V(x + f_{\mathrm{d}}(x) + J_{1\mathrm{d}}(x)w_{\mathrm{d}}) = V(x + f_{\mathrm{d}}(x)) + P_{1w_{\mathrm{d}}}(x)w_{\mathrm{d}}$$
$$+ w_{\mathrm{d}}^{\mathrm{T}}P_{2w_{\mathrm{d}}}(x)w_{\mathrm{d}}, \quad x \in \mathcal{Z}_x, \quad w_{\mathrm{d}} \in \mathcal{W}_{\mathrm{d}}, \quad (10.90)$$

$$\frac{\gamma_{\mathrm{d}}}{\beta_{\mathrm{d}}}I_{d_{\mathrm{d}}} - P_{2w_{\mathrm{d}}}(x) > 0, \quad x \in \mathcal{Z}_x, \quad (10.91)$$

$$L_{\mathrm{c}}(x) + V'(x)f_{\mathrm{c}}(x) + \frac{\beta_{\mathrm{c}}}{4\gamma_{\mathrm{c}}}V'(x)J_{1\mathrm{c}}(x)J_{1\mathrm{c}}^{\mathrm{T}}(x)V'^{\mathrm{T}}(x) = 0, \quad x \notin \mathcal{Z}_x,$$
$$(10.92)$$

$$L_{\mathrm{d}}(x) + V(x + f_{\mathrm{d}}(x)) - V(x)$$
$$+ \frac{1}{4}P_{1w_{\mathrm{d}}}(x)(\frac{\gamma_{\mathrm{d}}}{\beta_{\mathrm{d}}}I_{d_{\mathrm{d}}} - P_{2w_{\mathrm{d}}}(x))^{-1}P_{1w_{\mathrm{d}}}^{\mathrm{T}}(x) = 0, \quad x \in \mathcal{Z}_x. \quad (10.93)$$

Then there exists a neighborhood $\mathcal{D}_0 \subseteq \mathcal{D}$ of the origin such that if $x_0 \in \mathcal{D}_0$, then the zero solution $x(t) \equiv 0$ of the undisturbed (i.e.,

$(w_c(t), w_d(t_k)) \equiv (0,0)))$ *system (10.1) and (10.2) is asymptotically stable. If, in addition, $\Gamma_c(x) \geq 0$, $x \notin \mathcal{Z}_x$ and $\Gamma_d(x) \geq 0$, $x \in \mathcal{Z}_x$, then*

$$J(x_0) \leq \mathcal{J}(x_0) = V(x_0), \tag{10.94}$$

where

$$\mathcal{J}(x_0) \triangleq \int_0^\infty [L_c(x(t)) + \Gamma_c(x(t))]dt + \sum_{k \in \mathbb{Z}_{[0,\infty)}} [L_d(x(t_k)) + \Gamma_d(x(t_k))], \tag{10.95}$$

$$\Gamma_c(x) = \frac{\beta_c}{4\gamma_c} V'(x) J_{1c}(x) J_{1c}^T(x) V'^T(x), \tag{10.96}$$

$$\Gamma_d(x) = \frac{1}{4} P_{1w_d}(x) \left(\frac{\gamma_d}{\beta_d} I_{dd} - P_{2w_d}(x) \right)^{-1} P_{1w_d}^T(x), \tag{10.97}$$

and where $x(t)$, $t \geq 0$, is a solution to (10.1) and (10.2) with $(w_c(t),$ $w_d(t_k)) \equiv (0,0)$. Furthermore, if $x_0 = 0$, then the solution $x(t)$, $t \geq 0$, to (10.1) and (10.2) satisfies

$$V(x(T)) \leq \gamma, \quad \gamma = \gamma_c + \gamma_d, \quad T \geq 0, \quad w_c(\cdot) \in \mathcal{W}_{c\beta_c},$$
$$w_d(\cdot) \in \mathcal{W}_{d\beta_d}. \tag{10.98}$$

Finally, if $\mathcal{D} = \mathbb{R}^n$ and

$$V(x) \to \infty \quad as \quad \|x\| \to \infty, \tag{10.99}$$

then the zero solution $x(t) \equiv 0$ to the undisturbed system (10.1) and (10.2) is globally asymptotically stable.

Proof. The proofs for local and global asymptotic stability, as well as the performance bound (10.94) are identical to the proofs of local and global asymptotic stability given in Theorem 10.1 and performance bound (10.28). Next, with $(s_c(z_c, w_c), s_d(z_d, w_d)) = (\frac{\gamma_c}{\beta_c} w_c^T w_c, \frac{\gamma_d}{\beta_d} w_d^T w_d)$, and $\Gamma_c(x)$ and $\Gamma_d(x)$ given by (10.96) and (10.97), respectively, it follows from Proposition 10.1 that

$$V(x(T)) \leq \frac{\gamma_c}{\beta_c} \int_0^T w_c^T(t) w_c(t) dt + \frac{\gamma_d}{\beta_d} \sum_{k \in \mathbb{Z}_{[0,\infty)}} w_d^T(t_k) w_d(t_k), \quad T \geq 0,$$
$$w_c(\cdot) \in \mathcal{W}_{c\beta_c}, \quad w_d(\cdot) \in \mathcal{W}_{d\beta_d}, \tag{10.100}$$

which yields (10.98). $\qquad\qquad\square$

10.4 Optimal Controllers for Nonlinear Impulsive Dynamical Systems with Bounded Disturbances

In this section, we consider a hybrid control problem involving a notion of optimality with respect to an *auxiliary cost* which guarantees a bound on the worst-case value of a nonlinear-nonquadratic hybrid cost functional over a prescribed set of bounded exogenous disturbances. The optimal hybrid feedback controllers are derived as a direct consequence of Theorem 10.1 and provide a generalization of the hybrid Hamilton-Jacobi-Bellman conditions for time invariant, infinite-horizon problems considered in Chapter 9. In particular, we develop nonlinear hybrid feedback controllers for nonlinear impulsive state-dependent dynamical systems with bounded energy disturbances that additionally minimize a nonlinear-nonquadratic hybrid cost functional. To address the optimal hybrid control problem let $\mathcal{D} \subset \mathbb{R}^n$ be an open set with $0 \in \mathcal{D}$. Furthermore, let $\mathcal{W}_c \subseteq \mathbb{R}^{d_c}$ and $\mathcal{W}_d \subseteq \mathbb{R}^{d_d}$, and let $s_c : \mathbb{R}^{p_c} \times \mathbb{R}^{d_c} \to \mathbb{R}$ and $s_d : \mathbb{R}^{p_d} \times \mathbb{R}^{d_d} \to \mathbb{R}$ be given functions.

Consider the controlled nonlinear impulsive dynamical system

$$\dot{x}(t) = F_c(x(t), u_c(t)) + J_{1c}(x(t))w_c(t), \quad x(0) = x_0, \quad x(t) \notin \mathcal{Z}_x,$$
$$w_c(t) \in \mathcal{W}_c, \qquad (10.101)$$
$$\Delta x(t) = F_d(x(t), u_d(t)) + J_{1d}(x(t))w_d(t), \quad x(t) \in \mathcal{Z}_x, \quad w_d(t) \in \mathcal{W}_d,$$
$$(10.102)$$

with performance variables

$$z_c(t) = h_c(x(t), u_c(t)) + J_{2c}(x(t))w_c(t), \quad x(t) \notin \mathcal{Z}_x, \quad w_c(t) \in \mathcal{W}_c,$$
$$(10.103)$$
$$z_d(t) = h_d(x(t), u_d(t)) + J_{2d}(x(t))w_d(t), \quad x(t) \in \mathcal{Z}_x, \quad w_d(t) \in \mathcal{W}_d,$$
$$(10.104)$$

where $F_c : \mathbb{R}^n \times \mathbb{R}^{m_c} \to \mathbb{R}^n$ satisfies $F_c(0,0) = 0$, $J_{1c} : \mathbb{R}^n \to \mathbb{R}^{n \times d_c}$, $F_d : \mathbb{R}^n \times \mathbb{R}^{m_d} \to \mathbb{R}^n$, $J_{1d} : \mathbb{R}^n \to \mathbb{R}^{n \times d_d}$, $h_c : \mathbb{R}^n \times \mathbb{R}^{m_c} \to \mathbb{R}^{p_c}$ satisfies $h_c(0,0) = 0$, $J_{2c} : \mathbb{R}^n \to \mathbb{R}^{p_c \times d_c}$, $h_d : \mathbb{R}^n \times \mathbb{R}^{m_d} \to \mathbb{R}^{p_d}$, $J_{2d} : \mathbb{R}^n \to \mathbb{R}^{p_d \times d_d}$, and the hybrid control $(u_c(\cdot), u_d(\cdot))$ is restricted to the class of admissible controls consisting of measurable functions such that $(u_c(t), u_d(t_k)) \in U_c \times U_d$ for all $t \geq 0$ and $k \in \mathbb{Z}_{[0,\infty)}$, where the control constraint sets U_c and U_d are given with $(0,0) \in U_c \times U_d$.

Given a hybrid control law $(\phi_c(\cdot), \phi_d(\cdot))$ and a hybrid feedback control law $(u_c(t), u_d(t)) = (\phi_c(x(t)), \phi_d(x(t)))$, the closed-loop system shown in Figure 10.1 has the form

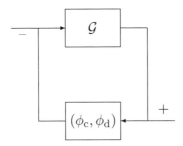

Figure 10.1 Feedback interconnection of $\mathcal{G}$ and $\mathcal{G}_c$.

$$\dot{x}(t) = F_c(x(t), \phi_c(x(t))) + J_{1c}(x(t))w_c(t), \quad x(0) = x_0, \quad x(t) \notin \mathcal{Z}_x,$$
$$w_c(t) \in \mathcal{W}_c, \quad (10.105)$$
$$\Delta x(t) = F_d(x(t), \phi_d(x(t))) + J_{1d}(x(t))w_d(t), \quad x(t) \in \mathcal{Z}_x,$$
$$w_d(t) \in \mathcal{W}_d, \quad (10.106)$$
$$z_c(t) = h_c(x(t), \phi_c(x(t))) + J_{2c}(x(t))w_c(t), \quad x(t) \notin \mathcal{Z}_x,$$
$$w_c(t) \in \mathcal{W}_c, \quad (10.107)$$
$$z_d(t) = h_d(x(t), \phi_d(x(t))) + J_{2d}(x(t))w_d(t), \quad x(t) \in \mathcal{Z}_x,$$
$$w_d(t) \in \mathcal{W}_d. \quad (10.108)$$

We assume that the mappings $\phi_c : \mathcal{D} \to U_c$ and $\phi_d : \mathcal{D} \to U_d$ satisfy sufficient regularity conditions such that the hybrid closed-loop system (10.105) and (10.106) has a unique solution forward in time.

Next, we present an extension of Theorem 9.3 for characterizing hybrid feedback controllers that guarantee stability, minimize an auxiliary hybrid performance functional, and guarantee that the input-output map of the closed-loop system is dissipative, nonexpansive, and passive for bounded input disturbances. For the statement of these results let $L_c : \mathcal{D} \times U_c \to \mathbb{R}$ and $L_d : \mathcal{D} \times U_d \to \mathbb{R}$, and define the set of asymptotically stabilizing hybrid controllers for the nonlinear impulsive dynamical system with $(w_c(t), w_d(t_k)) \equiv (0, 0)$ by

$$\mathcal{C}(x_0) \triangleq \{(u_c(\cdot), u_d(\cdot)) : (u_c(\cdot), u_d(\cdot)) \text{ is admissible and } x(\cdot) \text{ given by}$$
$$(10.101) \text{ and } (10.102) \text{ satisfies } x(t) \to 0 \text{ as } t \to \infty$$
$$\text{with } (w_c(t), w_d(t_k)) \equiv (0, 0)\}. \quad (10.109)$$

Theorem 10.3 *Consider the nonlinear controlled impulsive dynam-*

ical system (10.101)–(10.104) with hybrid performance functional

$$J(x_0, u_c(\cdot), u_d(\cdot)) = \int_0^\infty L_c(x(t), u_c(t))dt + \sum_{k \in \mathbb{Z}_{[0,\infty)}} L_d(x(t_k), u_d(t_k)),$$

$$(10.110)$$

where $(u_c(\cdot), u_d(\cdot))$ is an admissible hybrid control. Assume there exist functions $\Gamma_c : \mathcal{D} \times U_c \to \mathbb{R}$, $\Gamma_d : \mathcal{D} \times U_d \to \mathbb{R}$, $P_{1w_d} : \mathcal{D} \times U_d \to \mathbb{R}^{1 \times d_d}$, $P_{2w_d} : \mathcal{D} \times U_d \to \mathbb{N}^{d_d}$, a continuously differentiable function $V : \mathcal{D} \to \mathbb{R}$, and a hybrid control law $\phi_c : \mathcal{D} \to U_c$ and $\phi_d : \mathcal{D} \to U_d$ such that

$$V(0) = 0, \tag{10.111}$$

$$V(x) > 0, \quad x \in \mathcal{D}, \quad x \neq 0, \tag{10.112}$$

$$\phi_c(0) = 0, \tag{10.113}$$

$$V'(x)F_c(x, \phi_c(x)) < 0, \quad x \notin \mathcal{Z}_x, \quad x \neq 0, \tag{10.114}$$

$$V'(x)J_{1c}(x)w_c \leq s_c(z_c, w_c) + L_c(x, \phi_c(x)) + \Gamma_c(x, \phi_c(x)), \quad x \notin \mathcal{Z}_x,$$
$$w_c \in \mathcal{W}_c, \quad z_c \in \mathbb{R}^{p_c}, \tag{10.115}$$

$$V(x + F_d(x, \phi_d(x))) - V(x) \leq 0, \quad x \in \mathcal{Z}_x, \tag{10.116}$$

$$P_{1w_d}(x, \phi_d(x))w_d + w_d^T P_{2w_d}(x, \phi_d(x))w_d \leq s_d(z_d, w_d)$$
$$+ L_d(x, \phi_d(x)) + \Gamma_d(x, \phi_d(x)), \quad x \in \mathcal{Z}_x, \quad w_d \in \mathcal{W}_d, \quad z_d \in \mathbb{R}^{p_d}$$

$$(10.117)$$

$$V(x + F_d(x, u_d) + J_{1d}(x)w_d) = V(x + F_d(x, u_d))$$
$$+ P_{1w_d}(x, u_d)w_d + w_d^T P_{2w_d}(x, u_d)w_d,$$
$$x \in \mathcal{Z}_x, \quad u_d \in U_d, \quad w_d \in \mathcal{W}_d, \tag{10.118}$$

$$H_c(x, \phi_c(x)) = 0, \quad x \notin \mathcal{Z}_x, \tag{10.119}$$

$$H_d(x, \phi_d(x)) = 0, \quad x \in \mathcal{Z}_x, \tag{10.120}$$

$$H_c(x, u_c) \geq 0, \quad x \notin \mathcal{Z}_x, \quad u_c \in U_c, \tag{10.121}$$

$$H_d(x, u_d) \geq 0, \quad x \in \mathcal{Z}_x, \quad u_d \in U_d, \tag{10.122}$$

where

$$H_c(x, u_c) \triangleq L_c(x, u_c) + \Gamma_c(x, u_c) + V'(x)F_c(x, u_c), \tag{10.123}$$

$$H_d(x, u_d) \triangleq L_d(x, u_d) + \Gamma_d(x, u_d) + V(x + F_d(x, u_d)) - V(x). \tag{10.124}$$

Then, with the hybrid feedback control $(u_c(\cdot), u_d(\cdot)) = (\phi_c(x(\cdot)), \phi_d(x(\cdot)))$, there exists a neighborhood $\mathcal{D}_0 \subseteq \mathcal{D}$ of the origin such that if $x_0 \in \mathcal{D}_0$ and $(w_c(t), w_d(t_k)) \equiv (0, 0)$, the zero solution $x(t) \equiv 0$ of the closed-loop system (10.105) and (10.106) is asymptotically stable. If, *in addition,* $\Gamma_c(x, \phi_c(x)) \geq 0$, $x \notin \mathcal{Z}_x$, *and* $\Gamma_d(x, \phi_d(x)) \geq 0$, $x \in \mathcal{Z}_x$, *then*

$$J(x_0, \phi_c(x(\cdot)), \phi_d(x(\cdot))) \leq \mathcal{J}(x_0, \phi_c(x(\cdot)), \phi_d(x(\cdot))) = V(x_0), \quad (10.125)$$

where

$$\mathcal{J}(x_0, u_c(\cdot), u_d(\cdot)) \triangleq \int_0^\infty [L_c(x(t), u_c(t)) + \Gamma_c(x(t), u_c(t))]dt$$
$$+ \sum_{k \in \mathbb{Z}_{[0,\infty)}} [L_d(x(t_k), u_d(t_k)) + \Gamma_d(x(t_k), u_d(t_k))],$$

$$(10.126)$$

and where $(u_c(\cdot), u_d(\cdot))$ *is admissible and* $x(t)$, $t \geq 0$, *is a solution to (10.101) and (10.102) with* $(w_c(t), w_d(t_k)) \equiv (0,0)$. *In addition, if* $x_0 \in \mathcal{D}_0$ *then the hybrid feedback control* $(u_c(\cdot), u_d(\cdot)) = (\phi_c(x(\cdot)), \phi_d(x(\cdot)))$ *minimizes* $\mathcal{J}(x_0, u_c(\cdot), u_d(\cdot))$ *in the sense that*

$$\mathcal{J}(x_0, \phi_c(x(\cdot)), \phi_d(x(\cdot))) = \min_{(u_c(\cdot), u_d(\cdot)) \in \mathcal{C}(x_0)} \mathcal{J}(x_0, u_c(\cdot), u_d(\cdot)).$$

$$(10.127)$$

Furthermore, the solution $x(t)$, $t \geq 0$, *to (10.105) and (10.106) satisfies the dissipativity constraint*

$$\int_0^T s_c(z_c(t), w_c(t))dt + \sum_{k \in \mathbb{Z}_{[0,T)}} s_d(z_d(t_k), w_d(t_k)) + V(x_0) \geq 0,$$

$$T \geq 0, \quad w_c(\cdot) \in \mathcal{L}_2, \quad w_d(\cdot) \in \ell_2. \quad (10.128)$$

Finally, if $\mathcal{D} = \mathbb{R}^n$, $U_c = \mathbb{R}^{m_c}$, $U_d = \mathbb{R}^{m_d}$, *and*

$$V(x) \to \infty \quad as \quad \|x\| \to \infty, \quad (10.129)$$

then the zero solution $x(t) \equiv 0$ *of the undisturbed closed-loop system (10.105) and (10.106) is globally asymptotically stable.*

Proof. Local and global asymptotic stability is a direct consequence of (10.111)–(10.114) and (10.116) by applying Theorem 10.1 to the closed-loop system (10.105) and (10.106). Furthermore, using (10.119) and (10.120), the performance bound (10.125) is a restatement of (10.28) as applied to the closed-loop system. Next, let $(u_c(\cdot), u_d(\cdot)) \in \mathcal{C}(x_0)$ and let $x(t)$, $t \geq 0$, be the solution to (10.101) and (10.102) with $(w_c(t), w_d(t_k)) \equiv (0,0)$. Then (10.127) follows from Theorem 9.3 with $L_c(x, u_c)$ and $L_d(x, u_d)$ replaced by $L_c(x, u_c) +$

$\Gamma_c(x, u_c)$ and $L_d(x, u_d) + \Gamma_d(x, u_d)$, respectively, and $J(x_0, u_c(\cdot), u_d(\cdot))$ replaced by $\mathcal{J}(x_0, u_c(\cdot), u_d(\cdot))$. Finally, using (10.115), (10.117), and (10.118), condition (10.128) is a restatement of (10.30) as applied to the closed-loop system. $\qquad\square$

Next, we specialize Theorem 10.3 to linear impulsive dynamical systems with bounded energy disturbances. Specifically, we consider the case in which $F_c(x, u_c) = A_c x + B_c u_c$, $J_{1c}(x) = D_c$, $h_c(x, u_c) = E_{1c} x + E_{2c} u_c$, and $J_{2c}(x) = E_{\infty c}$, where $A_c \in \mathbb{R}^{n \times n}$, $B_c \in \mathbb{R}^{n \times m_c}$, $D_c \in \mathbb{R}^{n \times d_c}$, $E_{1c} \in \mathbb{R}^{p_c \times n}$, $E_{2c} \in \mathbb{R}^{p_c \times m_c}$, and $E_{\infty c} \in \mathbb{R}^{p_c \times d_c}$, and $F_d(x, u_d) = A_d x + B_d u_d$, $J_{1d}(x) = D_d$, $h_d(x, u_d) = E_{1d} x + E_{2d} u_d$, and $J_{2d}(x) = E_{\infty d}$, where $A_d \in \mathbb{R}^{n \times n}$, $B_d \in \mathbb{R}^{n \times m_d}$, $D_d \in \mathbb{R}^{n \times d_d}$, $E_{1d} \in \mathbb{R}^{p_d \times n}$, $E_{2d} \in \mathbb{R}^{p_d \times m_d}$, and $E_{\infty d} \in \mathbb{R}^{p_d \times d_d}$.

First, we consider the case where $(s_c(z_c, w_c), s_d(z_d, w_d)) = (\gamma_c^2 w_c^{\mathrm{T}} w_c - z_c^{\mathrm{T}} z_c, \gamma_d^2 w_d^{\mathrm{T}} w_d - z_d^{\mathrm{T}} z_d)$, where $\gamma_c > 0$ and $\gamma_d > 0$ are given. For the following result assume $E_{\infty c} = 0$, $R_{12c} \triangleq E_{1c}^{\mathrm{T}} E_{2c} = 0$, $E_{\infty d} = 0$, and $R_{12d} \triangleq E_{1d}^{\mathrm{T}} E_{2d} = 0$, and define $R_{1c} \triangleq E_{1c}^{\mathrm{T}} E_{1c} > 0$, $R_{2c} \triangleq E_{2c}^{\mathrm{T}} E_{2c} > 0$, $S_c \triangleq B_c R_{2c}^{-1} B_c^{\mathrm{T}}$, $R_{1d} \triangleq E_{1d}^{\mathrm{T}} E_{1d} > 0$, $R_{2d} \triangleq E_{2d}^{\mathrm{T}} E_{2d} > 0$, $R_{2ad} \triangleq R_{2d} + B_d^{\mathrm{T}} P B_d + B_d^{\mathrm{T}} P D_d (\gamma_d^2 I_{d_d} - D_d^{\mathrm{T}} P D_d)^{-1} D_d^{\mathrm{T}} P B_d$, $P_{ad} \triangleq B_d^{\mathrm{T}} P A_d + B_d^{\mathrm{T}} P D_d (\gamma_d^2 I_{d_d} - D_d^{\mathrm{T}} P D_d)^{-1} D_d^{\mathrm{T}} P A_d$, for arbitrary $P \in \mathbb{R}^{n \times n}$ when the indicated inverse exists.

Corollary 10.3 *Consider the linear impulsive controlled dynamical system*

$$\dot{x}(t) = A_c x(t) + B_c u_c(t) + D_c w_c(t), \quad x(t) \notin \mathcal{Z}_x, \quad w_c(\cdot) \in \mathcal{L}_2,$$
$$(10.130)$$
$$\Delta x(t) = A_d x(t) + B_d u_d(t) + D_d w_d(t), \quad x(t) \in \mathcal{Z}_x, \quad w_d(\cdot) \in \ell_2,$$
$$(10.131)$$
$$z_c(t) = E_{1c} x(t) + E_{2c} u_c(t), \quad x(t) \notin \mathcal{Z}_x, \qquad (10.132)$$
$$z_d(t) = E_{1d} x(t) + E_{2d} u_d(t), \quad x(t) \in \mathcal{Z}_x, \qquad (10.133)$$

with hybrid performance functional

$$J(x_0, u_c(\cdot), u_d(\cdot)) = \int_0^\infty [x^{\mathrm{T}}(t) R_{1c} x(t) + u_c^{\mathrm{T}}(t) R_{2c} u_c(t)] dt$$
$$+ \sum_{k \in \mathbb{Z}_{[0,\infty)}} [x^{\mathrm{T}}(t_k) R_{1d} x(t_k) + u_d^{\mathrm{T}}(t_k) R_{2d} u_d(t_k)],$$
$$(10.134)$$

where $(u_c(\cdot), u_d(\cdot))$ is admissible. Assume there exists a positive-definite matrix $P \in \mathbb{R}^{n \times n}$ such that

$$0 = x^T(A_c^T P + PA_c + R_{1c} + \gamma_c^{-2} PD_c D_c^T P - PS_c P)x, \quad x \notin \mathcal{Z}_x,$$

$$\text{(10.135)}$$

$$0 < \gamma_d^2 I_{d_d} - D_d^T PD_d,$$

$$\text{(10.136)}$$

$$0 = x^T(A_d^T PA_d - P + R_{1d} + A_d^T PD_d(\gamma_d^2 I_{d_d} - D_d^T PD_d)^{-1} D_d^T PA_d$$

$$- P_{ad}^T R_{2ad}^{-1} P_{ad})x, \quad x \in \mathcal{Z}_x.$$

$$\text{(10.137)}$$

Then, with the hybrid feedback control law

$$u_c = \phi_c(x) = -R_{2c}^{-1} B_c^T Px, \quad x \notin \mathcal{Z}_x, \qquad \text{(10.138)}$$

$$u_d = \phi_d(x) = -R_{2ad}^{-1} P_{ad}x, \quad x \in \mathcal{Z}_x, \qquad \text{(10.139)}$$

the zero solution $x(t) \equiv 0$ of the undisturbed (i.e., $(w_c(t), w_d(t_k)) \equiv (0,0)$) system (10.130) and (10.131) is globally asymptotically stable for all $x_0 \in \mathbb{R}^n$ and

$$J(x_0, \phi_c(x(\cdot)), \phi_d(x(\cdot))) \leq \mathcal{J}(x_0, \phi_c(x(\cdot)), \phi_d(x(\cdot))) = x_0^T Px_0,$$

$$\text{(10.140)}$$

where

$$\mathcal{J}(x_0, u_c(\cdot), u_d(\cdot)) \triangleq \int_0^\infty [x^T(t)(R_{1c} + \gamma_c^{-2} PD_d D_d^T P)x(t)$$

$$+ u_c^T(t)R_{2c}u_c(t)]dt$$

$$+ \sum_{k \in \mathbb{Z}_{[0,\infty)}} [(D_d^T P(A_d x(t_k) + B_d u_d(t_k)))^T$$

$$\cdot (\gamma_d^2 I_{d_d} - D_d^T PD_d)^{-1}(D_d^T P(A_d x(t_k) + B_d u_d(t_k)))$$

$$+ x^T(t_k)R_{1d}x(t_k) + u_d^T(t_k)R_{2d}(x(t_k))u_d(t_k)],$$

$$\text{(10.141)}$$

and where $(u_c(\cdot), u_d(\cdot))$ is admissible and $x(t)$, $t \geq 0$, is a solution to (10.130) and (10.131) with $(w_c(t), w_d(t_k)) \equiv (0,0)$. Furthermore,

$$\mathcal{J}(x_0, \phi_c(x(\cdot)), \phi_d(x(\cdot))) = \min_{(u_c(\cdot), u_d(\cdot)) \in \mathcal{C}(x_0)} \mathcal{J}(x_0, u_c(\cdot), u_d(\cdot)),$$

$$\text{(10.142)}$$

where $\mathcal{C}(x_0)$ is the set of asymptotically stabilizing hybrid controllers for the system (10.130) and (10.131) with $(w_c(t), w_d(t_k)) \equiv (0,0)$ and $x_0 \in \mathbb{R}^n$. Furthermore, if $x_0 = 0$ then, with $(u_c, u_d) = (\phi_c(x), \phi_d(x))$, *the solution $x(t)$, $t \geq 0$, to (10.130) and (10.131) satisfies the nonexpansivity constraint*

$$\int_0^T z_c(t)^T z_c(t)dt + \sum_{k \in \mathbb{Z}_{[0,T)}} z_d(t_k)^T z_d(t_k)$$

$$\leq \gamma_c^2 \int_0^T w_c(t)^T w_c(t)dt + \gamma_d^2 \sum_{k \in \mathbb{Z}_{[0,T)}} w_d(t_k)^T w_d(t_k),$$

$$T \geq 0, \quad w_c(\cdot) \in \mathcal{L}_2, \quad w_d(\cdot) \in \ell_2. \quad (10.143)$$

Proof. The result is a direct consequence of Theorem 10.3 with $F_c(x, u_c) = A_c x + B_c u_c$, $J_{1c}(x) = D_c$, $L_c(x, u_c) = x^T R_{1c} x + u_c^T R_{2c} u_c$, $\Gamma_c(x, u_c) = \gamma_c^{-2} x^T P D_c D_c^T P x$, $F_d(x, u_d) = A_d x + B_d u_d$, $J_{1d}(x) = D_d$, $L_d(x, u_d) = x^T R_{1d} x + u_d^T R_{2d} u_d$, $\Gamma_d(x, u_d) = \frac{1}{2} P_{1w_d}(x, u_d)(\gamma_d^2 I_{d_d} - P_{2w_d}(x, u_d))^{-1}[\frac{1}{2} P_{1w_d}^T(x, u_d)]$, $P_{1w_d}(x, u_d) = 2[A_d x + B_d u_d]^T P D_d$, $P_{2w_d}(x, u_d) = D_d^T P D_d$, $V(x) = x^T P x$, $\mathcal{D} = \mathbb{R}^n$, $U_c = \mathbb{R}^{m_c}$, and $U_d = \mathbb{R}^{m_d}$.

Specifically, conditions (10.111)–(10.114) and (10.116) are trivially satisfied. Now, forming $x^T(10.135)x$ for $x \notin \mathcal{Z}_x$ it follows that, after some algebraic manipulations, $V'(x)J_{1c}(x)w_c \leq s_c(z_c, w_c) + L_c(x, \phi_c(x), w_c) + \Gamma_c(x, \phi_c(x), w_c)$ for all $x \notin \mathcal{Z}_x$ and $w_c \in \mathcal{W}_c$. Similarly, forming $x^T(10.137)x$ for $x \in \mathcal{Z}_x$ it follows that, after some algebraic manipulations, $P_{1w_d}(x, \phi_d(x))w_d + w_d^T P_{2w_d}(x, \phi_d(x)))w_d \leq s_d(z_d, w_d) + L_d(x, \phi_d(x)) + \Gamma_d(x, \phi_d(x))$ for all $x \in \mathcal{Z}_x$ and $w_d \in \mathcal{W}_d$. Furthermore, it follows from (10.135) that $H_c(x, \phi_c(x)) = 0$ and $H_c(x, u_c) = H_c(x, u_c) - H(x, \phi_c(x)) = [u_c - \phi_c(x)]^T R_{2c}[u_c - \phi_c(x)] \geq 0$, $x \notin \mathcal{Z}_x$. Similarly, it follows from (10.137) that $H_d(x, \phi_d(x)) = 0$ and $H_d(x, u_d) = H_d(x, u_d) - H(x, \phi_d(x)) = [u_d - \phi_d(x)]^T R_{2ad}[u_d - \phi_d(x)] \geq 0$, $x \notin \mathcal{Z}_x$, so that all conditions of Theorem 10.3 are satisfied. Finally, since $V(\cdot)$ is radially unbounded, (10.130) and (10.131) with

$$u_c(t) = \phi_c(x(t)) = -R_{2c}^{-1} B_c^T P x(t), \quad x(t) \notin \mathcal{Z}_x, \quad (10.144)$$
$$u_d(t_k) = \phi_d(x(t_k)) = -R_{2ad}^{-1} P_{ad} x(t_k), \quad x(t_k) \in \mathcal{Z}_x, \quad (10.145)$$

is globally asymptotically stable. $\qquad \square$

The optimal hybrid feedback control law $(\phi_c(x), \phi_d(x))$ in Corollary 10.3 is derived using the properties of $H_c(x, u_c)$ and $H_d(x, u_d)$ as defined in Theorem 10.3. Specifically, since

$$H_c(x, u_c) = x^T(A_c^T P + P A_c + R_{1c} + \gamma_c^{-2} P D_c D_c^T P)x + u_c^T R_{2c} u_c$$
$$+ 2x^T P B_c u_c, \quad x \notin \mathcal{Z}_x, \quad (10.146)$$

and

$$H_d(x, u_d) = x^T(A_d^T P A_d - P + R_{1d} + A_d^T P D_d(\gamma_d^2 I_{d_d} - D_d^T P D_d)^{-1}$$
$$\cdot D_d^T P A_d)x + u_d^T R_{2ad} x + 2u_d^T P_{ad} x, \quad x \in \mathcal{Z}_x, \quad (10.147)$$

it follows that $\frac{\partial^2 H_c}{\partial u_c^2} = R_{2c} > 0$ and $\frac{\partial^2 H_d}{\partial u_d^2} = R_{2ad} > 0$. Now, $\frac{\partial H_c}{\partial u_c} = 2R_{2c}u_c + 2B_c^T P x = 0$, $x \notin \mathcal{Z}_x$, and $\frac{\partial H_d}{\partial u_d} = 2R_{2ad}u_d + 2P_{ad}x = 0$, $x \in \mathcal{Z}_x$, give the unique global minimum of $H_c(x, u_c)$, $x \notin \mathcal{Z}_x$, and $H_d(x, u_d)$, $x \in \mathcal{Z}_x$, respectively. Hence, since $\phi_c(x)$ minimizes $H_c(x, u_c)$ on $x \notin \mathcal{Z}_x$ and $\phi_d(x)$ minimizes $H_d(x, u_d)$ on $x \in \mathcal{Z}_x$, it follows that $\phi_c(x)$ satisfies $\frac{\partial H_c}{\partial u_c} = 0$ and $\phi_d(x)$ satisfies $\frac{\partial H_d}{\partial u_d} = 0$, or, equivalently, $\phi_c(x) = -R_{2c}^{-1} B_c^T P x$, $x \notin \mathcal{Z}_x$, and $\phi_d(x) = -R_{2ad}^{-1} P_{ad} x$, $x \in \mathcal{Z}_x$. Similar remarks hold for the hybrid controllers developed in Corollary 10.4.

Next, we consider the case where $p_c = d_c$, $p_d = d_d$, $s_c(z_c, w_c) = 2w_c^T z_c$, and $s_d(z_d, w_d) = 2w_d^T z_d$. For the following result define $R_{0c} \triangleq (E_{\infty c} + E_{\infty c}^T)^{-1}$, $R_{2sc} \triangleq R_{2c} + E_{2c}^T R_{0c} E_{2c}$, $R_{1sc} \triangleq E_{1c}^T (I_{d_c} + R_{0c}) E_{1c} - E_{1c}^T R_{0c} E_{2c} R_{2sc}^{-1} E_{2c}^T R_{0c} E_{1c}$, $B_{sc} \triangleq B_c - D_c R_{0c} E_{2c}$, $A_{sc} \triangleq A_c - (B_{sc} R_{2sc}^{-1} E_{2c}^T + D_c) R_{0c} E_{1c}$, $S_{sc} \triangleq B_{sc} R_{2sc}^{-1} B_{sc}^T$, $R_{0d} \triangleq E_{\infty d} + E_{\infty d}^T - D_d^T P D_d$, $R_{2sd} \triangleq B_d^T P B_d + R_{2d} + (D_d^T P B_d - E_{2d})^T R_{0d}^{-1}(D_d^T P B_d - E_{2d})$, and $P_{sd} \triangleq B_d^T P A_d + (D_d^T P B_d - E_{2d})^T R_{0d}^{-1}(D_d^T P A_d - E_{1d})$, when the indicated inverse exists. Furthermore, assume that $E_{1c}^T E_{1c} > 0$ and $E_{2c}^T E_{2c} > 0$. Note that using Schur complements it can be shown that $R_{1sc} > 0$.

Corollary 10.4 *Consider the linear impulsive dynamical system given by (10.130) and (10.131) with performance variables*

$$z_c(t) = E_{1c}x(t) + E_{2c}u_c(t) + E_{\infty c}w_c(t), \quad x(t) \notin \mathcal{Z}_x, \quad w_c(\cdot) \in \mathcal{L}_2,$$
$$(10.148)$$
$$z_d(t) = E_{1d}x(t) + E_{2d}u_d(t) + E_{\infty d}w_d(t), \quad x(t) \in \mathcal{Z}_x, \quad w_d(\cdot) \in \ell_2,$$
$$(10.149)$$

and hybrid performance functional (10.134). Assume there exists a positive-definite matrix $P \in \mathbb{R}^{n \times n}$ such that

$$0 = x^T(A_{sc}^T P + P A_{sc} + R_{1sc} + P D_c R_{0c} D_c^T P - P S_{sc} P)x, \quad x \notin \mathcal{Z}_x,$$
$$(10.150)$$
$$0 < E_{\infty d} + E_{\infty d}^T - D_d^T P D_d, \quad (10.151)$$
$$0 = x^T(A_d^T P A_d - P + R_{1d} + (D_d^T P A_d - E_{1d})^T R_{0d}^{-1}(D_d^T P A_d - E_{1d})$$

$$-P_{\mathrm{sd}}^{\mathrm{T}}R_{2\mathrm{sd}}^{-1}P_{\mathrm{sd}})x, \quad x \in \mathcal{Z}_x. \tag{10.152}$$

Then, with the hybrid feedback control law

$$u_{\mathrm{c}} = \phi_{\mathrm{c}}(x) = -R_{2\mathrm{sc}}^{-1}(B_{\mathrm{sc}}^{\mathrm{T}}P + E_{2\mathrm{c}}^{\mathrm{T}}R_{0\mathrm{c}}E_{1\mathrm{c}})x, \quad x \notin \mathcal{Z}_x, \tag{10.153}$$

$$u_{\mathrm{d}} = \phi_{\mathrm{d}}(x) = -R_{2\mathrm{sd}}^{-1}P_{\mathrm{sd}}x, \quad x \in \mathcal{Z}_x, \tag{10.154}$$

the zero solution $x(t) \equiv 0$ of the undisturbed (i.e., $(w_{\mathrm{c}}(t), w_{\mathrm{d}}(t_k)) \equiv (0,0)$) system (10.130) and (10.131) is globally asymptotically stable for all $x_0 \in \mathbb{R}^n$ and

$$J(x_0, \phi_{\mathrm{c}}(x(\cdot)), \phi_{\mathrm{d}}(x(\cdot))) \le \mathcal{J}(x_0, \phi_{\mathrm{c}}(x(\cdot)), \phi_{\mathrm{d}}(x(\cdot))) = x_0^{\mathrm{T}}Px_0, \tag{10.155}$$

where

$$\begin{aligned}
\mathcal{J}(x_0, u_{\mathrm{c}}(\cdot), u_{\mathrm{d}}(\cdot)) \triangleq \int_0^\infty & [x^{\mathrm{T}}(t)(R_{1\mathrm{c}} + (D_{\mathrm{c}}^{\mathrm{T}}P - E_{1\mathrm{c}})^{\mathrm{T}} \\
& \cdot R_{0\mathrm{c}}(D_{\mathrm{c}}^{\mathrm{T}}P - E_{1\mathrm{c}}))x(t) + u_{\mathrm{c}}^{\mathrm{T}}(t)R_{2\mathrm{c}}u_{\mathrm{c}}(t) \\
& -2x^{\mathrm{T}}(t)(D_{\mathrm{c}}^{\mathrm{T}}P - E_{1\mathrm{c}})^{\mathrm{T}}R_{0\mathrm{c}}E_{2\mathrm{c}}u_{\mathrm{c}}(t)]\mathrm{d}t \\
& + \sum_{k \in \mathbb{Z}_{[0,\infty)}} [x^{\mathrm{T}}(t_k)(R_{1\mathrm{d}} + (D_{\mathrm{d}}^{\mathrm{T}}PA_{\mathrm{d}} - E_{1\mathrm{d}})^{\mathrm{T}} \\
& \cdot R_{0\mathrm{d}}^{-1}(D_{\mathrm{d}}^{\mathrm{T}}PA_{\mathrm{d}} - E_{1\mathrm{d}}))x(t_k) \\
& + u_{\mathrm{d}}^{\mathrm{T}}(t_k)(R_{2\mathrm{sd}} - B_{\mathrm{d}}^{\mathrm{T}}PB_{\mathrm{d}})u_{\mathrm{d}}(t_k) \\
& + 2x^{\mathrm{T}}(t_k)(D_{\mathrm{d}}^{\mathrm{T}}PA_{\mathrm{d}} - E_{1\mathrm{d}})^{\mathrm{T}}R_{0\mathrm{d}}^{-1}(D_{\mathrm{d}}^{\mathrm{T}}PB_{\mathrm{d}} \\
& - E_{2\mathrm{d}})u_{\mathrm{d}}(t_k)], \tag{10.156}
\end{aligned}$$

and where $(u_{\mathrm{c}}(\cdot), u_{\mathrm{d}}(\cdot))$ is admissible and $x(t)$, $t \ge 0$, is a solution to (10.130) and (10.131) with $(w_{\mathrm{c}}(t), w_{\mathrm{d}}(t_k)) \equiv (0,0)$. Furthermore,

$$\mathcal{J}(x_0, \phi_{\mathrm{c}}(x(\cdot)), \phi_{\mathrm{d}}(x(\cdot))) = \min_{(u_{\mathrm{c}}(\cdot), u_{\mathrm{d}}(\cdot)) \in \mathcal{C}(x_0)} \mathcal{J}(x_0, u_{\mathrm{c}}(\cdot), u_{\mathrm{d}}(\cdot)), \tag{10.157}$$

where $\mathcal{C}(x_0)$ is the set of asymptotically stabilizing hybrid controllers for the system (10.130) and (10.131) with $(w_{\mathrm{c}}(t), w_{\mathrm{d}}(t_k)) \equiv (0,0)$ and $x_0 \in \mathbb{R}^n$. Furthermore, if $x_0 = 0$ then, with $(u_{\mathrm{c}}, u_{\mathrm{d}}) = (\phi_{\mathrm{c}}(x), \phi_{\mathrm{d}}(x))$, the solution $x(t)$, $t \ge 0$, to (10.130) and (10.131) satisfies the passivity constraint

$$\int_0^T 2w_{\mathrm{c}}^{\mathrm{T}}(t)z_{\mathrm{c}}(t)\mathrm{d}t + \sum_{k \in \mathbb{Z}_{[0,T)}} 2w_{\mathrm{d}}^{\mathrm{T}}(t_k)z_{\mathrm{d}}(t_k) \ge 0, \quad T \ge 0,$$

$$w_{\mathrm{c}}(\cdot) \in \mathcal{L}_2, \quad w_{\mathrm{d}}(\cdot) \in \ell_2. \tag{10.158}$$

Proof. The result is a direct consequence of Theorem 10.3 with
$F_c(x, u_c) = A_c x + B_c u_c$, $J_{1c}(x) = D_c$, $L_c(x, u_c) = x^T R_{1c} x + u_c^T R_{2c} u_c$,
$\Gamma_c(x, u_c) = [(D_c^T P - E_{1c})x - E_{2c} u_c]^T R_{0c}[(D_c^T P - E_{1c})x - E_{2c} u_c]$,
$F_d(x, u_d) = A_d x + B_d u_d$, $J_{1d}(x) = D_d$, $L_d(x, u_d) = x^T R_{1d} x + u_d^T R_{2d} u_d$,
$\Gamma_d(x, u_d) = [(D_d^T P A_d - E_{1d})x + (D_d^T P B_d - E_{2d})u_d]^T R_{0d}^{-1}[(D_d^T P A_d - E_{1d})x + (D_d^T P B_d - E_{2d})u_d]$, $P_{1w_d}(x, u_d) = 2(A_d x + B_d u_d)^T P D_d$,
$P_{2w_d}(x, u_d) = D_d^T P D_d$, $V(x) = x^T P x$, $\mathcal{D} = \mathbb{R}^n$, $U_c = \mathbb{R}^{m_c}$, and
$U_d = \mathbb{R}^{m_d}$.

Specifically, conditions (10.111)–(10.114) and (10.116) are trivially
satisfied. Now, forming $x^T(10.150)x$ for all $x \notin \mathcal{Z}_x$ it follows that, after some algebraic manipulations, $V'(x)J_{1c}(x)w_c \leq s_c(z_c, w_c) + L_c(x,
\phi_c(x), w_c) + \Gamma_c(x, \phi_c(x), w_c)$ for all $x \notin \mathcal{Z}_x$ and $w_c \in \mathcal{W}_c$. Similarly,
forming $x^T(10.152)x$ for all $x \in \mathcal{Z}_x$ it follows that, after some al-
gebraic manipulations, $P_{1w_d}(x, \phi_d(x))w_d + w_d^T P_{2w_d}(x, \phi_d(x)))w_d \leq
s_d(z_d, w_d) + L_d(x, \phi_d(x)) + \Gamma_d(x, \phi_d(x))$ for all $x \in \mathcal{Z}_x$ and $w_d \in
\mathcal{W}_d$. Furthermore, it follows from (10.150) that $H_c(x, \phi_c(x)) = 0$ and
$H_c(x, u_c) = H_c(x, u_c) - H(x, \phi_c(x)) = [u_c - \phi_c(x)]^T R_{2c}[u_c - \phi_c(x)] \geq 0$,
$x \notin \mathcal{Z}_x$. Similarly, it follows from (10.152) that $H_d(x, \phi_d(x)) = 0$ and
$H_d(x, u_d) = H_d(x, u_d) - H(x, \phi_d(x)) = [u_d - \phi_d(x)]^T R_{2sd}[u_d - \phi_d(x)] \geq
0$, $x \notin \mathcal{Z}_x$, so that all conditions of Theorem 10.3 are satisfied. Finally,
since $V(\cdot)$ is radially unbounded, (10.130) and (10.131) with

$$u_c(t) = \phi_c(x(t)) = -R_{2sc}^{-1}(B_{sc}^T P + E_{2c}^T R_{0c} E_{1c})x(t), \quad x(t) \notin \mathcal{Z}_x,$$
$$(10.159)$$

$$u_d(t_k) = \phi_d(x(t_k)) = -R_{2sd}^{-1} P_{sd} x(t_k), \quad x(t_k) \in \mathcal{Z}_x, \qquad (10.160)$$

is globally asymptotically stable. $\square$

10.5 Optimal and Inverse Optimal Nonlinear-Nonquadratic Control for Affine Systems with $\mathcal{L}_2$ Disturbances

In this section, we specialize Theorem 10.3 to affine (in the control)
nonlinear impulsive dynamical systems of the form

$$\dot{x}(t) = f_c(x(t)) + G_c(x(t))u_c(t) + J_{1c}(x(t))w_c(t), \quad x(0) = x_0,$$
$$x(t) \notin \mathcal{Z}_x, \quad w_c(\cdot) \in \mathcal{L}_2, \qquad (10.161)$$
$$\Delta x(t) = f_d(x(t)) + G_d(x(t))u_d(t) + J_{1d}(x(t))w_d(t), \quad x(t) \in \mathcal{Z}_x,$$
$$w_d(\cdot) \in \ell_2, \qquad (10.162)$$
$$z_c(t) = h_c(x(t)) + J_c(x(t))u_c(t), \quad x(t) \notin \mathcal{Z}_x, \qquad (10.163)$$
$$z_d(t) = h_d(x(t)) + J_d(x(t))u_d(t), \quad x(t) \in \mathcal{Z}_x, \qquad (10.164)$$

where $t \geq 0$, $f_c : \mathbb{R}^n \to \mathbb{R}^n$ is Lipschitz continuous and satisfies $f_c(0) = 0$, $G_c : \mathbb{R}^n \to \mathbb{R}^{n \times m_c}$, $J_{1c} : \mathbb{R}^n \to \mathbb{R}^{n \times d_c}$, $f_d : \mathbb{R}^n \to \mathbb{R}^n$ is continuous, $G_d : \mathbb{R}^n \to \mathbb{R}^{n \times m_d}$, $J_{1d} : \mathbb{R}^n \to \mathbb{R}^{n \times d_d}$, $h_c : \mathbb{R}^n \to \mathbb{R}^{p_c}$ and satisfies $h_c(0) = 0$, $J_c : \mathbb{R}^n \to \mathbb{R}^{p_c \times m_c}$, $h_d : \mathbb{R}^n \to \mathbb{R}^{p_d}$, $J_d : \mathbb{R}^n \to \mathbb{R}^{p_d \times m_d}$, $\mathcal{D} = \mathbb{R}^n$, $U_c = \mathbb{R}^{m_c}$, $U_d = \mathbb{R}^{m_d}$, and $\mathcal{Z}_x \subset \mathbb{R}^n$ is the resetting set.

First, we consider the nonexpansivity case so that $(s_c(z_c, w_c), s_d(z_d, w_d)) = (\gamma_c^2 w_c^T w_c - z_c^T z_c, \gamma_d^2 w_d^T w_d - z_d^T z_d)$, where $\gamma_c > 0$ and $\gamma_d > 0$. For the following result, we consider performance integrands $L_c(x, u_c)$ and $L_d(x, u_d)$ of the form

$$L_c(x, u_c) = L_{1c}(x) + L_{2c}(x)u_c + u_c^T R_{2c}(x)u_c, \quad x \notin \mathcal{Z}_x, \quad (10.165)$$
$$L_d(x, u_d) = L_{1d}(x) + L_{2d}(x)u_d + u_d^T R_{2d}(x)u_d, \quad x \in \mathcal{Z}_x, \quad (10.166)$$

where $L_{1c} : \mathbb{R}^n \to \mathbb{R}$, $L_{2c} : \mathbb{R}^n \to \mathbb{R}^{1 \times m_c}$, $R_{2c} : \mathbb{R}^n \to \mathbb{P}^{m_c}$, $L_{1d} : \mathbb{R}^n \to \mathbb{R}$, $L_{2d} : \mathbb{R}^n \to \mathbb{R}^{1 \times m_d}$, $R_{2d} : \mathbb{R}^n \to \mathbb{P}^{m_d}$ so that (10.110) becomes

$$
\begin{aligned}
J(x_0, u_c(\cdot), u_d(\cdot)) = &\int_0^\infty [L_{1c}(x(t)) + L_{2c}(x(t))u_c(t) \\
&+ u_c(t)^T R_{2c}(x(t))u_c(t)]dt \\
&+ \sum_{k \in \mathbb{Z}_{[0,\infty)}} [L_{1d}(x(t_k)) + L_{2d}(x(t_k))u_d(t_k) \\
&+ u_d(t_k)^T R_{2d}(x(t_k))u_d(t_k)].
\end{aligned}
$$
$$(10.167)$$

Corollary 10.5 *Consider the nonlinear impulsive controlled dynamical system (10.161)–(10.164) with performance functional (10.167). Assume there exist functions $P_{1u_d} : \mathbb{R}^n \to \mathbb{R}^{1 \times m_d}$, $P_{2u_d} : \mathbb{R}^n \to \mathbb{N}^{m_d}$, $P_{1w_d} : \mathbb{R}^n \to \mathbb{R}^{1 \times d_d}$, $P_{2w_d} : \mathbb{R}^n \to \mathbb{N}^{d_d}$, $P_{u_d w_d} : \mathbb{R}^n \to \mathbb{R}^{m_d \times d_d}$, $L_{2c} : \mathbb{R}^n \to \mathbb{R}^{1 \times m_c}$, $L_{2d} : \mathbb{R}^n \to \mathbb{R}^{1 \times m_d}$, and a continuously differentiable function $V : \mathbb{R}^n \to \mathbb{R}$ such that*

$$L_{2c}(0) = 0, \quad (10.168)$$
$$V(0) = 0, \quad (10.169)$$
$$V(x) > 0, \quad x \in \mathcal{D}, \quad x \neq 0, \quad (10.170)$$

$$V'(x)[f_c(x) - \tfrac{1}{2}G_c(x)R_{2ac}^{-1}(x)(L_{2c}^T(x) + G_c^T(x)V'^T(x) + 2J_c^T(x)h_c(x))]$$
$$+ \Gamma_c(x, \phi_c(x)) < 0, \quad x \notin \mathcal{Z}_x, \quad x \neq 0, \quad (10.171)$$
$$V(x + f_d(x)) + P_{1u_d}(x)\phi_d(x) + \phi_d^T(x)P_{2u_d}(x)\phi_d(x) - V(x)$$
$$+ \Gamma_d(x, \phi_d(x)) < 0, \quad x \in \mathcal{Z}_x, \quad (10.172)$$
$$V(x + f_d(x) + G_d(x)u_d + J_{1d}(x)w_d) = V(x + f_d(x)) + P_{1u_d}(x)u_d$$
$$+ u_d^T P_{2u_d}(x)u_d + P_{1w_d}(x)w_d + u_d^T P_{u_d w_d}(x)w_d + w_d^T P_{2w_d}(x)w_d,$$

$$x \in \mathcal{Z}_x, \quad u_\mathrm{d} \in \mathbb{R}^{m_\mathrm{d}}, \quad w_\mathrm{d} \in \mathbb{R}^{d_d}, \quad (10.173)$$

$$\gamma_\mathrm{d}^2 I_{d_\mathrm{d}} - P_{2w_\mathrm{d}}(x) > 0, \quad x \in \mathcal{Z}_x, \quad (10.174)$$

and

$$V(x) \to \infty \quad as \quad \|x\| \to \infty, \quad (10.175)$$

where

$$\phi_\mathrm{c}(x) = -\tfrac{1}{2} R_{2ac}^{-1}(x)[L_{2c}^\mathrm{T}(x) + G_\mathrm{c}^\mathrm{T}(x)V'^\mathrm{T}(x) + 2J_\mathrm{c}^\mathrm{T}(x)h_\mathrm{c}(x)], \quad x \notin \mathcal{Z}_x,$$
$$(10.176)$$

$$\phi_\mathrm{d}(x) = -R_{2ad}^{-1}(x)P_a(x), \quad x \in \mathcal{Z}_x, \quad (10.177)$$

and where

$$R_{2ac}(x) \triangleq R_{2c}(x) + J_\mathrm{c}^\mathrm{T}(x)J_\mathrm{c}(x), \quad (10.178)$$

$$R_{2ad}(x) \triangleq R_{2d}(x) + \tfrac{1}{4}P_{u_\mathrm{d}w_\mathrm{d}}(x)(\gamma_\mathrm{d}^2 I_\mathrm{d} - P_{2w_\mathrm{d}}(x))^{-1}P_{u_\mathrm{d}w_\mathrm{d}}^\mathrm{T}(x)$$
$$+ P_{2u_\mathrm{d}}(x) + J_\mathrm{d}^\mathrm{T}(x)J_\mathrm{d}(x), \quad (10.179)$$

$$P_a(x) \triangleq \tfrac{1}{2}[L_{2d}^\mathrm{T}(x) + P_{1u_\mathrm{d}}^\mathrm{T}(x) + \tfrac{1}{2}P_{u_\mathrm{d}w_\mathrm{d}}(x)(\gamma_\mathrm{d}^2 I_{d_\mathrm{d}} - P_{2w_\mathrm{d}}(x))^{-1}$$
$$\cdot P_{1w_\mathrm{d}}^\mathrm{T}(x) + 2J_\mathrm{d}^\mathrm{T}(x)h_\mathrm{d}(x)], \quad (10.180)$$

$$\Gamma_\mathrm{c}(x, u_\mathrm{c}) = \tfrac{1}{4\gamma_\mathrm{c}^2}V'(x)J_{1c}(x)J_{1c}^\mathrm{T}(x)V'(x)$$
$$+ [h_\mathrm{c}(x) + J_\mathrm{c}(x)u_\mathrm{c}]^\mathrm{T}[h_\mathrm{c}(x) + J_\mathrm{c}(x)u_\mathrm{c}], \quad (10.181)$$

$$\Gamma_\mathrm{d}(x, u_\mathrm{d}) = \tfrac{1}{4}[P_{1w_\mathrm{d}}(x) + u_\mathrm{d}^\mathrm{T}P_{u_\mathrm{d}w_\mathrm{d}}(x)](\gamma_\mathrm{d}^2 I_{d_\mathrm{d}} - P_{2w_\mathrm{d}}(x))^{-1}$$
$$\cdot [P_{1w_\mathrm{d}}(x) + u_\mathrm{d}^\mathrm{T}P_{u_\mathrm{d}w_\mathrm{d}}(x)]^\mathrm{T}$$
$$+ (h_\mathrm{d}(x) + J_\mathrm{d}(x)u_\mathrm{d})^\mathrm{T}(h_\mathrm{d}(x) + J_\mathrm{d}(x)u_\mathrm{d}), \quad (10.182)$$

where $\gamma_\mathrm{c} > 0$, $\gamma_\mathrm{d} > 0$, $(u_\mathrm{c}(\cdot), u_\mathrm{d}(\cdot))$ is admissible, and $x(t)$, $t \geq 0$, solves (10.161) and (10.162) with $(w_\mathrm{c}(\cdot), w_\mathrm{d}(\cdot)) \equiv (0, 0)$. Then the zero solution $x(t) \equiv 0$ of the undisturbed (i.e., $(w_\mathrm{c}(\cdot), w_\mathrm{d}(\cdot)) \equiv (0, 0)$) closed-loop system

$$\dot{x}(t) = f_\mathrm{c}(x(t)) + G_\mathrm{c}(x(t))\phi_\mathrm{c}(x(t)), \quad x(0) = x_0, \quad x(t) \notin \mathcal{Z}_x,$$
$$(10.183)$$

$$\Delta x(t) = f_\mathrm{d}(x(t)) + G_\mathrm{d}(x(t))\phi_\mathrm{d}(x(t)), \quad x(t) \in \mathcal{Z}_x, \quad (10.184)$$

is globally asymptotically stable with the hybrid feedback control law

$$\phi_\mathrm{c}(x) = -\tfrac{1}{2} R_{2ac}^{-1}(x)[L_{2c}^\mathrm{T}(x) + G_\mathrm{c}^\mathrm{T}(x)V'^\mathrm{T}(x) + 2J_\mathrm{c}(x)h_\mathrm{c}(x)], \quad x \notin \mathcal{Z}_x,$$
$$(10.185)$$

$$\phi_\mathrm{d}(x) = -\tfrac{1}{2} R_{2ad}^{-1}(x)P_a(x), \quad x \in \mathcal{Z}_x. \quad (10.186)$$

Furthermore, the hybrid performance functional (10.167) satisfies

$$J(x_0, \phi_c(x(\cdot)), \phi_d(x(\cdot))) \le J(x_0, \phi_c(x(\cdot)), \phi_d(x(\cdot))) = V(x_0), (10.187)$$

where

$$J(x_0, u_c(\cdot), u_d(\cdot)) \triangleq \int_0^\infty [L_c(x(t), u_c(t)) + \Gamma_c(x(t), u_c(t))]dt$$

$$+ \sum_{k \in \mathbb{Z}_{[0,\infty)}} [L_d(x(t_k), u_d(t_k)) + \Gamma_d(x(t_k), u_d(t_k))].$$

$$(10.188)$$

In addition, the hybrid performance functional (10.188), with

$$L_{1c}(x) = \phi_c^{\mathrm{T}}(x) R_{2ac} \phi_c(x) - V'(x) f_c(x) - h_c^{\mathrm{T}}(x) h_c(x)$$
$$- \tfrac{1}{4\gamma_c^2} V'(x) J_{1c}(x) J_{1c}^{\mathrm{T}}(x) V'^{\mathrm{T}}(x), \qquad (10.189)$$

$$L_{1d}(x) = P_a(x) R_{2ad}^{-1}(x) P_a(x) - [V(x + f_d(x)) - V(x)$$
$$+ \tfrac{1}{4} P_{1w_d}(x)(\gamma_d^2 I_{dd} - P_{2w_d}(x))^{-1} P_{1w_d}^{\mathrm{T}}(x) + h_d^{\mathrm{T}}(x) h_d(x)],$$

$$(10.190)$$

is minimized in the sense that

$$J(x_0, \phi_c(x(\cdot)), \phi_d(x(\cdot))) = \min_{(u_c(\cdot), u_d(\cdot)) \in \mathcal{C}(x_0)} J(x_0, u_c(\cdot), u_d(\cdot)).$$

$$(10.191)$$

Finally, with $(u_c(\cdot), u_d(\cdot)) = (\phi_c(x(\cdot)), \phi_d(x(\cdot)))$, the solution $x(t)$, $t \ge 0$, of the closed-loop system (10.183) and (10.184) satisfies non-expansivity constraint

$$\int_0^T z_c^{\mathrm{T}}(t) z_c(t) dt + \sum_{k \in \mathbb{Z}_{[0,T)}} z_d^{\mathrm{T}}(t_k) z_d(t_k) \le \gamma_c^2 \int_0^T w_c^{\mathrm{T}}(t) w_c(t) dt$$

$$+ \gamma_d^2 \sum_{k \in \mathbb{Z}_{[0,T)}} w_d^{\mathrm{T}}(t_k) w_d(t_k) + V(x_0), \quad w_c(\cdot) \in \mathcal{L}_2, \quad w_d(\cdot) \in \ell_2.$$

$$(10.192)$$

Proof. The result is a direct consequence of Theorem 10.3 with $F_c(x, u_c) = f_c(x) + G_c(x)u_c$, $z_c = h_c(x) + J_c(x)u_c$, $L_c(x, u_c) = L_{1c}(x) + L_{2c}(x)u_c + u_c^{\mathrm{T}} R_{2c}(x)u_c$, $J_{2c}(x) = 0$, $\Gamma_c(x, u_c)$ given by (10.181), $F_d(x, u_d) = f_d(x) + G_d(x)u_d$, $z_d = h_d(x) + J_d(x)u_d$, $L_d(x, u_d) = L_{1d}(x) + L_{2d}(x)u_d + u_d^{\mathrm{T}} R_{2d}(x)u_d$, $J_{2d}(x) = 0$, $\Gamma_d(x, u_d)$ given by (10.182), $\mathcal{D} = \mathbb{R}^n$, $U_c = \mathbb{R}^{m_c}$, and $U_d = \mathbb{R}^{m_d}$. Specifically, conditions (10.111) and (10.112) and (10.114)–(10.118) are trivially satisfied by replacing $P_{1w_d}(x, u_d)$ by $P_{1w_d}(x) + u_d^T P_{u_d w_d}(x)$. Furthermore, with (10.161), (10.162), (10.165), (10.166), (10.181), and (10.182), the hybrid Hamiltonians have the form

$$H_c(x, u_c) = L_{1c}(x) + L_{2c}(x)u_c + u_c^T R_{2c}(x)u_c$$
$$+ V'(x)(f_c(x) + G_c(x)u_c) + \frac{1}{4\gamma_c^2}V'(x)J_{1c}(x)J_{1c}^T(x)V'(x)$$
$$+ [h_c(x) + J_c(x)u_c]^T[h_c(x) + J_c(x)u_c], \qquad (10.193)$$
$$H_d(x, u_d) = L_{1d}(x) + L_{2d}(x)u_d + u_d^T R_{2d}(x)u_d$$
$$+ V(x + f_d(x) + G_d(x)u_d) - V(x)$$
$$+ \frac{1}{4}[P_{1w_d}(x) + u_d^T P_{u_d w_d}(x)](\gamma_d^2 I_{d_d} - P_{2w_d}(x))^{-1}$$
$$\cdot [P_{1w_d}(x) + u_d^T P_{u_d w_d}(x)]^T$$
$$+ (h_d(x) + J_d(x)u_d)^T(h_d(x) + J_d(x)u_d). \qquad (10.194)$$

Now, the hybrid feedback control law (10.185) and (10.186) is obtained by setting $\frac{\partial H_c}{\partial u_c} = 0$ and $\frac{\partial H_d}{\partial u_d} = 0$.

Next, since $V(\cdot)$ is continuously differentiable and $x = 0$ is a local minimum of $V(\cdot)$, it follows that $V'(0) = 0$, and hence, since by assumption $h_c(0) = 0$ and $L_{2c}(0) = 0$, it follows that $\phi_c(0) = 0$, which proves (10.113). Next, with $L_{1c}(x)$ and $L_{1d}(x)$ given by (10.189) and (10.190), respectively, and $\phi_c(x)$ and $\phi_d(x)$ given by (10.185) and (10.186), (10.119) and (10.120) hold. Finally, since

$$H_c(x, u_c) = H_c(x, u_c) - H_c(x, \phi_c(x))$$
$$= [u_c - \phi_c(x)]^T R_{2ac}(x)[u_c - \phi_c(x)], \quad x \notin \mathcal{Z}_x, \quad (10.195)$$
$$H_d(x, u_d) = H_d(x, u_d) - H_d(x, \phi_d(x))$$
$$= [u_d - \phi_d(x)]^T R_{2ad}(x)[u_d - \phi_d(x)], \quad x \in \mathcal{Z}_x, \quad (10.196)$$

and $R_{2ac}(x) > 0$, $x \notin \mathcal{Z}_x$ and $R_{2ad}(x) > 0$, $x \in \mathcal{Z}_x$, conditions (10.121) and (10.122) hold. The result now follows as a direct consequence of Theorem 10.3. $\qquad \square$

Next, we specialize Theorem 10.3 to the passivity case. Specifically, we consider the case where $p_c = d_c$, $p_d = d_d$, and $(s_c(z_c, w_c), s_d(z_d, w_d)) = (2z_c^T w_c, 2z_d^T w_d)$. For the following result we consider performance variables

$$z_c(t) = h_c(x(t)) + J_c(x(t))u_c(t) + J_{2c}(x(t))w_c(t), \quad x(t) \notin \mathcal{Z}_x,$$
$$w_c(\cdot) \in \mathcal{L}_2, \quad (10.197)$$
$$z_d(t) = h_d(x(t)) + J_d(x(t))u_d(t) + J_{2d}(x(t))w_d(t), \quad x(t) \in \mathcal{Z}_x,$$
$$w_d(\cdot) \in \ell_2, \quad (10.198)$$

where $t \geq 0$, $h_c : \mathbb{R}^n \to \mathbb{R}^{p_c}$ satisfies $h_c(0) = 0$, $J_c : \mathbb{R}^n \to \mathbb{R}^{p_c \times m_c}$, $J_{2c} : \mathbb{R}^n \to \mathbb{R}^{p_c \times p_c}$ and satisfies $J_{2c}(x) + J_{2c}^{\mathrm{T}}(x) > 0$, $x \notin \mathcal{Z}_x$, $h_d : \mathbb{R}^n \to \mathbb{R}^{p_d}$, $J_d : \mathbb{R}^n \to \mathbb{R}^{p_d \times m_d}$, and $J_{2d} : \mathbb{R}^n \to \mathbb{R}^{p_d \times p_d}$ satisfies $J_{2d}(x) + J_{2d}^{\mathrm{T}}(x) > 0$, $x \in \mathcal{Z}_x$. Furthermore, we consider performance integrands $L_c(x, u_c)$ and $L_d(x, u_d)$ of the form given by (10.165) and (10.166).

Corollary 10.6 *Consider the nonlinear impulsive controlled dynamical system (10.161), (10.162), (10.197), and (10.198) with performance functional (10.167). Assume there exist functions $P_{1u_d} : \mathbb{R}^n \to \mathbb{R}^{1 \times m_d}$, $P_{2u_d} : \mathbb{R}^n \to \mathbb{N}^{m_d \times m_d}$, $P_{1w_d} : \mathbb{R}^n \to \mathbb{R}^{1 \times d_d}$, $P_{2w_d} : \mathbb{R}^n \to \mathbb{N}^{d_d}$, $P_{u_d w_d} : \mathbb{R}^n \to \mathbb{R}^{m_d \times d_d}$, $L_{2c} : \mathbb{R}^n \to \mathbb{R}^{1 \times m_c}$, $L_{2d} : \mathbb{R}^n \to \mathbb{R}^{1 \times m_d}$, and a continuously differentiable function $V : \mathbb{R}^n \to \mathbb{R}$ such that*

$$L_{2c}(0) = 0, \tag{10.199}$$

$$V(0) = 0, \tag{10.200}$$

$$V(x) > 0, \quad x \in \mathcal{D}, \quad x \neq 0, \tag{10.201}$$

$$V'(x)[f_c(x) - \tfrac{1}{2}G_c(x)R_{2sc}^{-1}(x)(L_{2c}^{\mathrm{T}}(x) + J_c^{\mathrm{T}}(x)R_{0c}(x)$$
$$\cdot [2h_c(x) - J_{1c}^{\mathrm{T}}(x)V'^{\mathrm{T}}(x)])] + \Gamma_c(x, \phi_c(x)) < 0, \quad x \notin \mathcal{Z}_x, \ x \neq 0, \tag{10.202}$$

$$V(x + f_d(x)) + P_{1u_d}(x)\phi_d(x) + \phi_d^{\mathrm{T}}(x)P_{2u_d}(x)\phi_d(x)$$
$$- V(x) + \Gamma_d(x, \phi_d(x)) < 0, \quad x \in \mathcal{Z}_x, \tag{10.203}$$

$$V(x + f_d(x) + G_d(x)u_d + J_{1d}(x)w_d) = V(x + f_d(x)) + P_{1u_d}(x)u_d$$
$$+ u_d^{\mathrm{T}} P_{2u_d}(x)u_d + P_{1w_d}(x)w_d + u_d^{\mathrm{T}} P_{u_d w_d}(x)w_d + w_d^{\mathrm{T}} P_{2w_d}(x)w_d,$$
$$x \in \mathcal{Z}_x, \quad u_d \in \mathbb{R}^{m_d}, \quad w_d \in \mathbb{R}^{d_d}, \tag{10.204}$$

$$J_{2d}(x) + J_{2d}^{\mathrm{T}}(x) - P_{2w_d}(x) > 0, \quad x \in \mathcal{Z}_x, \quad w_d \in \mathbb{R}^{d_d}, \tag{10.205}$$

and

$$V(x) \to \infty \ as \ \|x\| \to \infty, \tag{10.206}$$

where

$$\phi_c(x) = -\tfrac{1}{2}R_{2sc}^{-1}(x)[L_{2c}^{\mathrm{T}}(x) + G_c^{\mathrm{T}}(x)V'^{\mathrm{T}}(x)$$
$$+ J_c^{\mathrm{T}}(x)R_{0c}(x)(2h_c(x) - J_{1c}^{\mathrm{T}}(x)V'^{\mathrm{T}}(x))], \quad x \notin \mathcal{Z}_x, \tag{10.207}$$

$$\phi_d(x) = -R_{2sd}^{-1}(x)P_{sd}(x), \quad x \in \mathcal{Z}_x, \tag{10.208}$$

and where

$$R_{0c}(x) \triangleq (J_{2c}(x) + J_{2c}^{\mathrm{T}}(x))^{-1}, \tag{10.209}$$

$$R_{2sc}(x) \triangleq R_{2c}(x) + J_c^T(x)R_{0c}(x)J_c(x), \tag{10.210}$$

$$R_{2sd}(x) \triangleq R_{2d}(x) + (\tfrac{1}{2}P_{u_dw_d}(x) - J_d^T(x))(J_{2d}(x) + J_{2d}^T(x)$$
$$-P_{2w_d}(x))^{-1}(\tfrac{1}{2}P_{u_dw_d}(x) - J_d^T(x))^T + P_{2u_d}(x), \tag{10.211}$$

$$P_{sd}(x) \triangleq \tfrac{1}{2}[L_{2d}^T(x) + P_{1u_d}^T(x) + 2(\tfrac{1}{2}P_{u_dw_d}^T(x) - J_d(x))^T(J_{2d}(x)$$
$$+J_{2d}^T(x) - P_{2w_d}(x))^{-1}(\tfrac{1}{2}P_{u_dw_d}^T(x) - J_d(x))],$$

$$\Gamma_c(x, u_c) = [\tfrac{1}{2}J_{1c}^T(x)V'^T(x) - (h_c(x) + J_c(x)u_c)]^T R_{0c}(x)$$
$$\cdot [\tfrac{1}{2}J_{1c}^T(x)V'^T(x) - (h_c(x) + J_c(x)u_c)], \quad x \notin \mathcal{Z}_x, \tag{10.212}$$

$$\Gamma_d(x, u_d) = [\tfrac{1}{2}P_{1w_d}^T(x) - h_d(x) + (\tfrac{1}{2}P_{u_dw_d}^T(x) - J_d(x))u_d]^T$$
$$\cdot (J_{2d}(x) + J_{2d}^T(x) - P_{2w_d}(x))^{-1}$$
$$\cdot [\tfrac{1}{2}P_{1w_d}^T(x) - h_d(x) + (\tfrac{1}{2}P_{u_dw_d}^T(x) - J_d(x))u_d], \quad x \in \mathcal{Z}_x, \tag{10.213}$$

where $\gamma_c > 0$, $\gamma_d > 0$, $(u_c(\cdot), u_d(\cdot))$ is admissible, and $x(t)$, $t \geq 0$, is a solution to (10.161) and (10.162) with $(w_c(\cdot), w_d(\cdot)) \equiv (0, 0)$. Then the zero solution $x(t) \equiv 0$ of the undisturbed (i.e., $((w_c(\cdot), w_d(\cdot)) \equiv (0, 0))$) closed-loop system

$$\dot{x}(t) = f_c(x(t)) + G_c(x(t))\phi_c(x(t)), \quad x(0) = x_0, \quad x(t) \notin \mathcal{Z}_x, \tag{10.214}$$

$$\Delta x(t) = f_d(x(t)) + J_d(x(t))\phi_d(x(t)), \quad x(t) \in \mathcal{Z}_x, \tag{10.215}$$

is globally asymptotically stable with hybrid feedback control law

$$\phi_c(x) = -\tfrac{1}{2}R_{2sc}^{-1}(x)[L_{2c}^T(x) + G_c^T(x)V'^T(x) + J_c^T(x)R_{0c}(x)(2h_c(x)$$
$$-J_{1c}^T(x)V'^T(x))], \quad x \notin \mathcal{Z}_x, \tag{10.216}$$

$$\phi_d(x) = -\tfrac{1}{2}R_{2sd}^{-1}(x)P_{sd}(x), \quad x \in \mathcal{Z}_x. \tag{10.217}$$

Furthermore, the hybrid performance functional (10.167) satisfies

$$J(x_0, \phi_c(x(\cdot)), \phi_d(x(\cdot))) \leq \mathcal{J}(x_0, \phi_c(x(\cdot)), \phi_d(x(\cdot))) = V(x_0), \tag{10.218}$$

where

$$\mathcal{J}(x_0, u_c(\cdot), u_d(\cdot)) \triangleq \int_0^\infty [L_c(x(t)) + \Gamma_c(x(t))]dt$$
$$+ \sum_{k \in \mathbb{Z}_{[0,\infty)}} [L_d(x(t_k)) + \Gamma_d(x(t_k))]. \tag{10.219}$$

In addition, the hybrid performance functional (10.219), with

$$L_{1c}(x) = \phi_c^T(x)R_{2sc}\phi_c(x) - V'(x)f_c(x) - [\tfrac{1}{2}J_{1c}^T(x)V'^T(x) - h_c(x)]^T$$

$$\cdot R_{0c}(x)[\tfrac{1}{2}J_{1c}^{\mathrm{T}}(x)V'^{\mathrm{T}}(x) - h_{c}(x)]h_{c}^{\mathrm{T}}(x)h_{c}(x)$$

$$- \tfrac{1}{4\gamma_c}V'(x)J_{1c}(x)J_{1c}^{\mathrm{T}}(x)V'^{\mathrm{T}}(x), \tag{10.220}$$

$$L_{1d}(x) = P_{sd}(x)R_{2sd}^{-1}(x)P_{sd}(x)$$
$$- [V(x + f_{d}(x)) - V(x) + (\tfrac{1}{2}P_{1w_d}^{\mathrm{T}}(x) - h_{d}(x))^{\mathrm{T}}$$
$$\cdot (J_{2d}(x) + J_{2d}^{\mathrm{T}}(x) - P_{2w_d}(x))^{-1}(\tfrac{1}{2}P_{1w_d}^{\mathrm{T}}(x) - h_{d}(x))], \tag{10.221}$$

is minimized in the sense that

$$\mathcal{J}(x_0, \phi_c(x(\cdot)), \phi_d(x(\cdot))) = \min_{(u_c(\cdot), u_d(\cdot)) \in \mathcal{C}(x_0)} \mathcal{J}(x_0, u_c(\cdot), u_d(\cdot)). \tag{10.222}$$

Finally, with $(u_c(\cdot), u_d(\cdot)) = (\phi_c(x(\cdot)), \phi_d(x(\cdot)))$, *the solution* $x(t)$, $t \geq 0$, *of the closed-loop system (10.214) and (10.215) satisfies the passivity constraint*

$$\int_0^T 2z_c^{\mathrm{T}}(t)w_c(t)dt + \sum_{k \in \mathbb{Z}_{[0,T)}} 2z_d^{\mathrm{T}}(t_k)w_d(t_k) \geq V(x_0),$$

$$w_c(\cdot) \in \mathcal{L}_2, \quad w_d(\cdot) \in \ell_2. \tag{10.223}$$

Proof. The result is a direct consequence of Theorem 10.3 with $F_c(x, u_c) = f_c(x) + G_c(x)u_c$, $z_c = h_c(x) + J_c(x)u_c$, $L_c(x, u_c) = L_{1c}(x) + L_{2c}(x)u_c + u_c^{\mathrm{T}}R_{2c}(x)u_c$, $\Gamma_c(x, u_c)$ given by (10.212), $F_d(x, u_d) = f_d(x) + G_d(x)u_d$, $z_d = h_d(x) + J_d(x)u_d$, $L_d(x, u_d) = L_{1d}(x) + L_{2d}(x)u_d + u_d^{\mathrm{T}}R_{2d}(x)u_d$, $\Gamma_d(x, u_d)$ given by (10.213), $\mathcal{D} = \mathbb{R}^n$, $U_c = \mathbb{R}^{m_c}$, and $U_d = \mathbb{R}^{m_d}$. Specifically, conditions (10.111) and (10.112) and (10.114)–(10.118) are trivially satisfied by replacing $P_{1w_d}(x, u_d)$ by $P_{1w_d}(x) + u_d^{\mathrm{T}}P_{u_d w_d}(x)$. Furthermore, with (10.161), (10.162), (10.165), (10.166), (10.212), and (10.213), the hybrid Hamiltonians have the form

$$H_c(x, u_c) = L_{1c}(x) + L_{2c}(x)u_c + u_c^{\mathrm{T}}R_{2c}(x)u_c$$
$$+ V'(x)(f_c(x) + G_c(x)u_c) + \Gamma_c(x, u_c), \tag{10.224}$$

$$H_d(x, u_d) = L_{1d}(x) + L_{2d}(x)u_d + u_d^{\mathrm{T}}R_{2d}(x)u_d$$
$$+ V(x + f_d(x) + G_d(x)u_d) - V(x)$$
$$+ [\tfrac{1}{2}P_{1w_d}^{\mathrm{T}}(x) - h_d(x) - (\tfrac{1}{2}P_{u_d w_d}^{\mathrm{T}}(x) - J_d(x))u_d]^{\mathrm{T}}$$
$$\cdot (J_{2d}(x) + J_{2d}^{\mathrm{T}}(x) - P_{2w_d}(x))^{-1}$$
$$\cdot [\tfrac{1}{2}P_{1w_d}^{\mathrm{T}}(x) - h_d(x) - (\tfrac{1}{2}P_{u_d w_d}^{\mathrm{T}}(x) - J_d(x))u_d]. \tag{10.225}$$

Now, the hybrid feedback control law (10.216) and (10.217) is obtained by setting $\frac{\partial H_c}{\partial u_c} = 0$ and $\frac{\partial H_d}{\partial u_d} = 0$. The remainder of the proof now follows as in the proof of Corollary 10.5. $\square$

Chapter Eleven

Robust Control for Nonlinear Uncertain Impulsive Dynamical Systems

11.1 Introduction

Although the theory of impulsive dynamical systems has received considerable attention in the literature [12, 14, 39, 61, 62, 79, 93, 148], robust analysis and control design techniques for uncertain nonlinear impulsive dynamical systems remain relatively undeveloped. In this chapter, we extend the analysis and control design framework for nonlinear impulsive dynamical systems developed in Chapters 2 and 9 to address robustness considerations for impulsive dynamical systems. In particular, we build on the results of Chapter 9 to develop an optimality-based framework for addressing the problem of nonlinear-nonquadratic optimal hybrid control for *uncertain* nonlinear impulsive dynamical systems with structured parametric uncertainty. Specifically, using a Lyapunov bounding framework, the robust nonlinear hybrid control problem is transformed into an optimal hybrid control problem by modifying a nonlinear-nonquadratic hybrid cost functional to account for system parametric uncertainty.

The main contribution of this chapter is a methodology for designing nonlinear hybrid controllers which provide robust stability and robust performance over a prescribed range of impulsive system uncertainty. The present framework extends the guaranteed cost control approach [24, 60] to nonlinear impulsive dynamical systems by utilizing a hybrid performance bound to provide robust performance in addition to robust stability. In particular, the performance bound can be evaluated in closed form as long as the nonlinear-nonquadratic hybrid cost functional considered is related in a specific way to an underlying Lyapunov function that guarantees robust stability over a prescribed uncertainty set. This Lyapunov function is shown to be a solution to the steady-state form of the hybrid Hamilton-Jacobi-Bellman equation for the nominal impulsive dynamical system and plays a key role in constructing the optimal nonlinear robust hybrid control law. Hence, the overall framework provides a generalization of the hybrid Hamilton-Jacobi-Bellman conditions developed in Chapter 9 for addressing the design of robust optimal hybrid controllers for nonlinear uncertain impulsive dynamical systems.

A key feature of the present framework is that since the necessary and sufficient hybrid Hamilton-Jacobi-Bellman optimality conditions are obtained for a modified nonlinear-nonquadratic hybrid performance functional rather than the original hybrid performance functional, *globally* optimal controllers are guaranteed to provide both robust stability and performance. Of course, since our approach allows us to minimize a given hybrid Hamiltonian, the resulting robust nonlinear hybrid controllers provide the best worst-case performance over the robust stability range.

11.2 Robust Stability Analysis of Nonlinear Uncertain Impulsive Dynamical Systems

In this section, we present sufficient conditions for robust stability for a class of nonlinear uncertain impulsive dynamical systems. Specifically, we consider the problem of evaluating a performance bound for a nonlinear-nonquadratic hybrid cost functional depending upon a class of nonlinear uncertain impulsive dynamical systems. It turns out that the cost bound can be evaluated in closed form as long as the hybrid cost functional is related in a specific way to an underlying Lyapunov function that guarantees robust stability over a prescribed uncertainty set. Hence, the overall framework guarantees robust stability and performance for nonlinear uncertain impulsive dynamical systems, where robust performance here refers to a guaranteed bound on the worst-case value of a nonlinear-nonquadratic hybrid cost functional over a prescribed uncertainty set.

In this chapter, we restrict our attention to nonlinear state-dependent uncertain impulsive dynamical systems $\mathcal{G}$ given by

$$\dot{x}(t) = f_c(x(t)), \quad x(0) = x_0, \quad x(t) \notin \mathcal{Z}, \qquad (11.1)$$
$$\Delta x(t) = f_d(x(t)), \quad x(t) \in \mathcal{Z}, \qquad (11.2)$$

where $t \geq 0$, $x(t) \in \mathcal{D} \subseteq \mathbb{R}^n$, $\mathcal{D}$ is an open set with $0 \in \mathcal{D}$, $f_c(\cdot) \in \mathcal{F}_c \subset \{f_c : \mathcal{D} \to \mathbb{R}^n : f_c(0) = 0\}$, where $f_c(\cdot)$ is Lipschitz continuous, $f_d(\cdot) \in \mathcal{F}_d \subset \{f_d : \mathcal{D} \to \mathbb{R}^n\}$, where $f_d(\cdot)$ is continuous, and $\mathcal{Z} \subset \mathcal{D}$ is the resetting set. Here, $\mathcal{F}_c$ and $\mathcal{F}_d$ denote the class of nonlinear uncertain impulsive dynamical systems with $f_{c0}(\cdot) \in \mathcal{F}_c$ and $f_{d0}(\cdot) \in \mathcal{F}_d$ defining the nominal nonlinear impulsive dynamical system for the continuous-time and the resetting dynamics, respectively. Furthermore, we assume existence and uniqueness of solutions for the state-dependent impulsive dynamical system (11.1) and (11.2) in forward time.

For a particular trajectory $x(t)$, we let $\tau_k(x_0)$ denote the kth instant of time at which $x(t)$ intersects $\mathcal{Z}$. Thus, the trajectory of the system (11.1) and (11.2) from the initial condition $x(0) = x_0$ is given by $\psi(t, x_0)$ for $0 < t \leq \tau_1(x_0)$, where $\psi(\cdot, x_0)$ is the solution to (11.1) with initial condition $x_0 \in \mathcal{D}$. If and when the trajectory reaches a state $x_1 \triangleq x(\tau_1(x_0))$ satisfying $x_1 \in \mathcal{Z}$, then the state is instantaneously transferred to $x_1^+ \triangleq x_1 + f_\mathrm{d}(x_1)$ according to the resetting law (11.2). The trajectory $x(t)$, $\tau_1(x_0) < t \leq \tau_2(x_0)$, is then given by $\psi(t - \tau_1(x_0), x_1^+)$, and so on. Note that the solution $x(t)$ of (11.1) and (11.2) is left-continuous, that is, it is continuous everywhere except at the resetting times $\tau_k(x_0)$, and

$$x_k \triangleq x(\tau_k(x_0)) = \lim_{\varepsilon \to 0^+} x(\tau_k(x_0) - \varepsilon), \tag{11.3}$$

$$x_k^+ \triangleq x(\tau_k(x_0)) + f_\mathrm{d}(x(\tau_k(x_0))), \tag{11.4}$$

for $k = 1, 2, \ldots$.

The following additional assumptions are similar to Assumptions A1 and A2 of Chapter 2, and ensure well-posedness of the resetting times.

A1. If $x \in \overline{\mathcal{Z}} \backslash \mathcal{Z}$, then there exists $\varepsilon > 0$ such that, for all $0 < \delta < \varepsilon$, $x(\delta, x) \notin \mathcal{Z}$.

A2. If $x \in \mathcal{Z}$, then $x + f_\mathrm{d}(x) \notin \mathcal{Z}$, $f_\mathrm{d}(\cdot) \in \mathcal{F}_\mathrm{d}$.

For the following result let $L_\mathrm{c} : \mathcal{D} \to \mathbb{R}$ and $L_\mathrm{d} : \mathcal{D} \to \mathbb{R}$. Within the context of robustness analysis, it is assumed that the zero solution $x(t) \equiv 0$ of the nominal nonlinear impulsive dynamical system (11.1) and (11.2) is asymptotically stable. Furthermore, we assume that an infinite number of resettings occur. For the following result and the remainder of the chapter we denote the resetting times $\tau_k(x_0)$ by t_k and define $\mathcal{F} \triangleq \mathcal{F}_\mathrm{c} \times \mathcal{F}_\mathrm{d}$ and $\mathbb{Z}_{[0,t)} \triangleq \{k : 0 \leq t_k < t\}$.

Theorem 11.1 *Consider the nonlinear uncertain impulsive dynamical system $\mathcal{G}$ given by (11.1) and (11.2), where $(f_\mathrm{c}(\cdot), f_\mathrm{d}(\cdot)) \in \mathcal{F}$, with the hybrid performance functional*

$$J_{(f_\mathrm{c}, f_\mathrm{d})}(x_0) \triangleq \int_0^\infty L_\mathrm{c}(x(t))\mathrm{d}t + \sum_{k \in \mathbb{Z}_{[0,\infty)}} L_\mathrm{d}(x(t_k)). \tag{11.5}$$

Furthermore, assume there exist functions $\Gamma_c : \mathcal{D} \to \mathbb{R}$, $\Gamma_d : \mathcal{D} \to \mathbb{R}$, and $V : \mathcal{D} \to \mathbb{R}$, where $V(\cdot)$ is a continuously differentiable function, such that

$$V(0) = 0, \tag{11.6}$$

$$V(x) > 0, \quad x \in \mathcal{D}, \quad x \neq 0, \tag{11.7}$$

$$V'(x)f_c(x) \leq V'(x)f_{c0}(x) + \Gamma_c(x), \quad x \notin \mathcal{Z}, \quad f_c(\cdot) \in \mathcal{F}_c, \tag{11.8}$$

$$V'(x)f_{c0}(x) + \Gamma_c(x) < 0, \quad x \notin \mathcal{Z}, \quad x \neq 0, \tag{11.9}$$

$$V(x + f_d(x)) \leq V(x + f_{d0}(x)) + \Gamma_d(x), \quad x \in \mathcal{Z}, \quad f_d(\cdot) \in \mathcal{F}_d, \tag{11.10}$$

$$V(x + f_{d0}(x)) - V(x) + \Gamma_d(x) \leq 0, \quad x \in \mathcal{Z}, \tag{11.11}$$

$$L_c(x) + V'(x)f_{c0}(x) + \Gamma_c(x) = 0, \quad x \notin \mathcal{Z}, \tag{11.12}$$

$$L_d(x) + V(x + f_{d0}(x)) - V(x) + \Gamma_d(x) = 0, \quad x \in \mathcal{Z}, \tag{11.13}$$

where $(f_{c0}(\cdot), f_{d0}(\cdot)) \in \mathcal{F}$ defines the nominal nonlinear impulsive dynamical system. Then there exists a neighborhood $\mathcal{D}_0 \subseteq \mathcal{D}$ of the origin such that if $x_0 \in \mathcal{D}_0$, then the zero solution $x(t) \equiv 0$ to (11.1) and (11.2) is asymptotically stable for all $(f_c(\cdot), f_d(\cdot)) \in \mathcal{F}$, and

$$\sup_{(f_c(\cdot), f_d(\cdot)) \in \mathcal{F}} J_{(f_c, f_d)}(x_0) \leq \mathcal{J}(x_0) = V(x_0), \tag{11.14}$$

where

$$\mathcal{J}(x_0) \triangleq \int_0^\infty [L_c(x(t)) + \Gamma_c(x(t))] \mathrm{d}t$$

$$+ \sum_{k \in \mathbb{Z}_{[0,\infty)}} [L_d(x(t_k)) + \Gamma_d(x(t_k))], \tag{11.15}$$

and where $x(t)$, $t \geq 0$, is a solution to (11.1) and (11.2) with $(f_c(x(\cdot)), f_d(x(\cdot))) = (f_{c0}(x(\cdot)), f_{d0}(x(\cdot)))$. Finally, if $\mathcal{D} = \mathbb{R}^n$ and

$$V(x) \to \infty \quad as \quad \|x\| \to \infty, \tag{11.16}$$

then the zero solution $x(t) \equiv 0$ to (11.1) and (11.2) is globally asymptotically stable for all $(f_c(\cdot), f_d(\cdot)) \in \mathcal{F}$.

Proof. Let $(f_c(\cdot), f_d(\cdot)) \in \mathcal{F}$ and let $x(t)$, $t \geq 0$, satisfy (11.1) and (11.2). Then,

$$\dot{V}(x(t)) \triangleq \frac{\mathrm{d}}{\mathrm{d}t} V(x(t)) = V'(x(t))f_c(x(t)), \quad x(t) \notin \mathcal{Z}. \tag{11.17}$$

Hence, it follows from (11.8) and (11.9) that

$$\dot{V}(x(t)) < 0, \quad x(t) \notin \mathcal{Z}, \quad x(t) \neq 0. \tag{11.18}$$

Furthermore,

$$\Delta V(x(t_k)) \triangleq V(x(t_k) + f_{\mathrm{d}}(x(t_k))) - V(x(t_k)), \quad x(t_k) \in \mathcal{Z}. \tag{11.19}$$

Hence, it follows from (11.10) and (11.11) that

$$\Delta V(x(t_k)) \leq 0, \quad x(t_k) \in \mathcal{Z}. \tag{11.20}$$

Thus, using (11.6), (11.7), (11.18), and (11.20) it follows from Theorem 2.1 that $V(\cdot)$ is a Lyapunov function for (11.1) and (11.2), which proves asymptotic stability of the zero solution $x(t) \equiv 0$ to (11.1) and (11.2) for all $(f_{\mathrm{c}}(\cdot), f_{\mathrm{d}}(\cdot)) \in \mathcal{F}$. Consequently, $x(t) \to 0$ as $t \to \infty$ for all initial conditions $x_0 \in \mathcal{D}_0$ for some neighborhood $\mathcal{D}_0 \subseteq \mathcal{D}$ of the origin.

Now, (11.17) implies that

$$0 = -\dot{V}(x(t)) + V'(x(t))f_{\mathrm{c}}(x(t)), \quad x(t) \notin \mathcal{Z}, \tag{11.21}$$

and hence, using (11.8) and (11.12),

$$\begin{aligned} L_{\mathrm{c}}(x(t)) &= -\dot{V}(x(t)) + L_{\mathrm{c}}(x(t)) + V'(x(t))f_{\mathrm{c}}(x(t)) \\ &\leq -\dot{V}(x(t)) + L_{\mathrm{c}}(x(t)) + V'(x(t))f_{\mathrm{c}0}(x(t)) + \Gamma_{\mathrm{c}}(x(t)) \\ &= -\dot{V}(x(t)), \quad x(t) \notin \mathcal{Z}. \end{aligned} \tag{11.22}$$

Similarly, (11.19) implies that

$$0 = -\Delta V(x(t_k)) + V(x(t_k) + f_{\mathrm{d}}(x(t_k))) - V(x(t_k)), \quad x(t_k) \in \mathcal{Z}, \tag{11.23}$$

and hence, using (11.10) and (11.13),

$$\begin{aligned} L_{\mathrm{d}}(x(t_k)) &= -\Delta V(x(t_k)) + L_{\mathrm{d}}(x(t_k)) + V(x(t_k)) \\ &\quad + f_{\mathrm{d}}(x(t_k))) - V(x(t_k)) \\ &\leq -\Delta V(x(t_k)) + L_{\mathrm{d}}(x(t_k)) + V(x(t_k) + f_{\mathrm{d}0}(x(t_k))) \\ &\quad - V(x(t_k)) + \Gamma_{\mathrm{d}}(x(t_k)) \\ &= -\Delta V(x(t_k)), \quad x(t_k) \in \mathcal{Z}. \end{aligned} \tag{11.24}$$

Next, integrating over the interval $[0, t)$ with $\mathbb{Z}_{[0,t)} = \{1, 2, \dots, i\}$, (11.22) and (11.24) yield

$$\int_0^t L_{\mathrm{c}}(x(s))\mathrm{d}s + \sum_{k \in \mathbb{Z}_{[0,t)}} L_{\mathrm{d}}(x(t_k)) = \int_0^{t_1} L_{\mathrm{c}}(x(s))\mathrm{d}s + L_{\mathrm{d}}(x(t_1))$$

$$+ \int_{t_1^+}^{t_2} L_c(x(s))\mathrm{d}s + L_d(x(t_2))$$

$$+ \cdots + \int_{t_{i-1}^+}^{t_i} L_c(x(s))\mathrm{d}s + L_d(x(t_i)) + \int_{t_{i^+}}^{t} L_c(x(s))\mathrm{d}s$$

$$\leq -V(x(t_1)) + V(x_0) - V(x(t_1) + f_d(x(t_1))) + V(x(t_1))$$
$$-V(x(t_2)) + V(x(t_1^+)) - V(x(t_2) + f_d(x(t_2)))$$
$$+V(x(t_2)) + \cdots - V(x(t_i)) + V(x(t_{i-1}^+))$$
$$-V(x(t_i) + f_d(x(t_i))) + V(x(t_i)) - V(x(t)) + V(x(t_i^+))$$
$$\leq -V(x(t_1)) + V(x_0) - V(x(t_1^+)) + V(x(t_1)) - V(x(t_2))$$
$$+V(x(t_1^+)) - V(x(t_2^+)) + V(x(t_2)) + \cdots - V(x(t_i))$$
$$+V(x(t_{i-1}^+)) - V(x(t_i^+)) + V(x(t_i)) - V(x(t)) + V(x(t_i^+))$$
$$\leq -V(x(t)) + V(x_0). \tag{11.25}$$

Letting $t \to \infty$ and noting that $V(x(t)) \to 0$ for all $x_0 \in \mathcal{D}_0$ yields $J_{(f_c,f_d)}(x_0) \leq V(x_0)$.

Next, let $x(t)$, $t \geq 0$, satisfy (11.1) and (11.2) with $(f_c(x(\cdot)), f_d(x(\cdot))) = (f_{c0}(x(\cdot)), f_{d0}(x(\cdot)))$. Then it follows from (11.12) that

$$L_c(x(t)) + \Gamma_c(x(t)) = -\dot{V}(x(t)) + L_c(x(t))$$
$$+V'(x(t))f_{c0}(x(t)) + \Gamma_c(x(t))$$
$$= -\dot{V}(x(t)), \quad x(t) \notin \mathcal{Z}. \tag{11.26}$$

Similarly, it follows from (11.13) that

$$L_d(x(t_k)) + \Gamma_d(x(t_k)) = -\Delta V(x(t_k)) + L_d(x(t_k)) + V(x(t_k)$$
$$+f_{d0}(x(t_k))) - V(x(t_k)) + \Gamma_d(x(t_k))$$
$$= -\Delta V(x(t_k)), \quad x(t_k) \in \mathcal{Z}. \tag{11.27}$$

Now, integrating over the interval $[0, t)$ with $\mathbb{Z}_{[0,t)} = \{1, 2, \ldots, i\}$, (11.26) and (11.27) yield

$$\int_0^t [L_c(x(t)) + \Gamma_c(x(t))]\mathrm{d}t + \sum_{k \in \mathbb{Z}_{[0,t)}} [L_d(x(t_k)) + \Gamma_d(x(t_k))]$$
$$= -V(x(t)) + V(x_0). \tag{11.28}$$

Letting $t \to \infty$ and noting that $V(x(t)) \to 0$ for all $x_0 \in \mathcal{D}_0$ yields $\mathcal{J}(x_0) = V(x_0)$.

Finally, for $\mathcal{D} = \mathbb{R}^n$ and for all $(f_c(\cdot), f_d(\cdot)) \in \mathcal{F}$, global asymptotic stability of the zero solution $x(t) \equiv 0$ to (11.1) and (11.2) is a direct consequence of Theorem 2.1 using the radially unbounded condition (11.16) on $V(x)$, $x \in \mathbb{R}^n$. □

Theorem 11.1 provides sufficient conditions for robust stability of a class of nonlinear uncertain impulsive dynamical systems given by (11.1) and (11.2). Specifically, (11.6) and (11.7) assume that $V(x)$ is a Lyapunov function candidate for the nonlinear uncertain impulsive dynamical system (11.1) and (11.2). Conditions (11.8)–(11.11) imply $\dot{V}(x(t)) < 0$, $x(t) \notin \mathcal{Z}$, $t > 0$, and $\Delta V(x(t_k)) \leq 0$, $x(t_k) \in \mathcal{Z}$, $k \in \mathbb{Z}_+$, for $x(\cdot)$ satisfying (11.1) and (11.2) for all $(f_c(\cdot), f_d(\cdot)) \in \mathcal{F}$, and hence, $V(\cdot)$ is a Lyapunov function guaranteeing robust stability of the nonlinear uncertain impulsive dynamical system (11.1) and (11.2).

It is important to note that conditions (11.9) and (11.11) are *verifiable* conditions since they are independent of the uncertain system parameters $(f_c(\cdot), f_d(\cdot)) \in \mathcal{F}$. To apply Theorem 11.1 we specify the bounding functions $\Gamma_c(\cdot)$ and $\Gamma_d(\cdot)$ for the uncertainty set $\mathcal{F}_c \times \mathcal{F}_d$ such that $\Gamma_c(\cdot)$ and $\Gamma_d(\cdot)$ bound $\mathcal{F}_c \times \mathcal{F}_d$. In [24, 59, 60, 67] the uncertainty set $\mathcal{F}$ and bounding functions $\Gamma_c(\cdot)$ and $\Gamma_d(\cdot)$ are given concrete forms for continuous-time and discrete-time systems. Since impulsive dynamical systems involve a hybrid formulation of continuous-time and discrete-time dynamics, identical constructions can be developed for uncertain impulsive dynamical systems once the theoretical basis of the approach is established. For further details see [24, 59, 60, 67].

If $\mathcal{F}$ consists of only the nominal nonlinear impulsive dynamical system $(f_{c0}(\cdot), f_{d0}(\cdot))$, then $\Gamma_c(x) = 0$ and $\Gamma_d(x) = 0$ for all $x \in \mathcal{D}$ satisfy (11.8) and (11.10), respectively, and hence, $J_{(f_{c0}, f_{d0})}(x_0) = \mathcal{J}(x_0)$. Finally, a worst-case upper bound to the nonlinear-nonquadratic hybrid performance functional is given in terms of a Lyapunov function which can be interpreted in terms of an auxiliary cost defined for the nominal impulsive dynamical system.

Next, we specialize Theorem 11.1 to nonlinear uncertain impulsive dynamical systems of the form

$$\dot{x}(t) = f_{c0}(x(t)) + \Delta f_c(x(t)), \quad x(0) = x_0, \quad x(t) \notin \mathcal{Z}, \quad (11.29)$$
$$\Delta x(t) = f_{d0}(x(t)) + \Delta f_d(x(t)), \quad x(t) \in \mathcal{Z}, \quad (11.30)$$

where $t \geq 0$, $f_{c0} : \mathcal{D} \to \mathbb{R}^n$ and satisfies $f_{c0}(0) = 0$, $f_{d0} : \mathcal{D} \to \mathbb{R}^n$, and $(f_{c0} + \Delta f_c, f_{d0} + \Delta f_d) \in \mathcal{F} = \mathcal{F}_c \times \mathcal{F}_d$. Here, $\mathcal{F} = \mathcal{F}_c \times \mathcal{F}_d$ is such that

$$\mathcal{F}_c = \{f_{c0} + \Delta f_c : \mathcal{D} \to \mathbb{R}^n : \Delta f_c \in \mathbf{\Delta}_c\}, \quad (11.31)$$
$$\mathcal{F}_d = \{f_{d0} + \Delta f_d : \mathcal{D} \to \mathbb{R}^n : \Delta f_d \in \mathbf{\Delta}_d\}, \quad (11.32)$$

where $(\boldsymbol{\Delta}_{\mathrm{c}}, \boldsymbol{\Delta}_{\mathrm{d}})$ are given nonlinear uncertainty sets of nonlinear per-turbations Δf_{c} and Δf_{d} of the nominal system dynamics $f_{\mathrm{c}0}(\cdot) \in \mathcal{F}_{\mathrm{c}}$ and $f_{\mathrm{d}0}(\cdot) \in \mathcal{F}_{\mathrm{d}}$. Since $\mathcal{F}_{\mathrm{c}} \subset \{f_{\mathrm{c}} : \mathcal{D} \to \mathbb{R}^n : f_{\mathrm{c}}(0) = 0\}$ it follows that $\Delta f_{\mathrm{c}}(0) = 0$ for all $\Delta f_{\mathrm{c}} \in \boldsymbol{\Delta}_{\mathrm{c}}$. For the remainder of this chapter define $\boldsymbol{\Delta} \triangleq \boldsymbol{\Delta}_{\mathrm{c}} \times \boldsymbol{\Delta}_{\mathrm{d}}$.

Corollary 11.1 *Consider the nonlinear uncertain impulsive dynam-ical system (11.29) and (11.30) with the hybrid performance func-tional (11.5). Furthermore, assume there exist functions $\Gamma_{\mathrm{c}} : \mathcal{D} \to \mathbb{R}$, $\Gamma_{\mathrm{d}} : \mathcal{D} \to \mathbb{R}$, $P_{1f_{\mathrm{d}}} : \mathcal{D} \to \mathbb{R}^{1 \times n}$, $P_{2f_{\mathrm{d}}} : \mathcal{D} \to \mathbb{N}^n$, and $V : \mathcal{D} \to \mathbb{R}$, where $V(\cdot)$ is a continuously differentiable function, such that (11.6), (11.7), (11.9), and (11.11)–(11.13) hold, and*

$$V'(x)\Delta f_{\mathrm{c}}(x) \le \Gamma_{\mathrm{c}}(x), \quad x \notin \mathcal{Z}, \quad \Delta f_{\mathrm{c}}(\cdot) \in \boldsymbol{\Delta}_{\mathrm{c}}, \quad (11.33)$$
$$\Delta f_{\mathrm{d}}^{\mathrm{T}}(x)P_{1f_{\mathrm{d}}}^{\mathrm{T}}(x) + P_{1f_{\mathrm{d}}}(x)\Delta f_{\mathrm{d}}(x) + \Delta f_{\mathrm{d}}^{\mathrm{T}}(x)P_{2f_{\mathrm{d}}}(x)\Delta f_{\mathrm{d}}(x) \le \Gamma_{\mathrm{d}}(x),$$
$$x \in \mathcal{Z}, \quad \Delta f_{\mathrm{d}}(\cdot) \in \boldsymbol{\Delta}_{\mathrm{d}}, \ (11.34)$$
$$V(x + f_{\mathrm{d}0}(x) + \Delta f_{\mathrm{d}}(x)) = V(x + f_{\mathrm{d}0}(x)) + \Delta f_{\mathrm{d}}^{\mathrm{T}}(x)P_{1f_{\mathrm{d}}}^{\mathrm{T}}(x)$$
$$+ P_{1f_{\mathrm{d}}}(x)\Delta f_{\mathrm{d}}(x) + \Delta f_{\mathrm{d}}^{\mathrm{T}}(x)P_{2f_{\mathrm{d}}}(x)\Delta f_{\mathrm{d}}(x),$$
$$x \in \mathcal{Z}, \quad \Delta f_{\mathrm{d}}(\cdot) \in \boldsymbol{\Delta}_{\mathrm{d}}. \ (11.35)$$

Then there exists a neighborhood $\mathcal{D}_0 \subseteq \mathcal{D}$ of the origin such that if $x_0 \in \mathcal{D}_0$, then the zero solution $x(t) \equiv 0$ to (11.29) and (11.30) is asymptotically stable for all $(\Delta f_{\mathrm{c}}(\cdot), \Delta f_{\mathrm{d}}(\cdot)) \in \boldsymbol{\Delta}$, and the hybrid performance functional (11.5) satisfies

$$\sup_{(\Delta f_{\mathrm{c}}(\cdot), \Delta f_{\mathrm{d}}(\cdot)) \in \boldsymbol{\Delta}} J_{(\Delta f_{\mathrm{c}}, \Delta f_{\mathrm{d}})}(x_0) \le \mathcal{J}(x_0) = V(x_0), \quad (11.36)$$

where

$$\mathcal{J}(x_0) \triangleq \int_0^\infty [L_{\mathrm{c}}(x(t)) + \Gamma_{\mathrm{c}}(x(t))]\mathrm{d}t$$
$$+ \sum_{k \in \mathbb{Z}_{[0,\infty)}} [L_{\mathrm{d}}(x(t_k)) + \Gamma_{\mathrm{d}}(x(t_k))], \quad (11.37)$$

and where $x(t)$, $t \ge 0$, is a solution to (11.29) and (11.30) with $(\Delta f_{\mathrm{c}}(x(t)), \Delta f_{\mathrm{d}}(x(t_k))) = (0,0)$. Finally, if $\mathcal{D} = \mathbb{R}^n$ and $V(x)$, $x \in \mathbb{R}^n$, satisfies (11.16), then the zero solution $x(t) \equiv 0$ to (11.29) and (11.30) is globally asymptotically stable for all $(\Delta f_{\mathrm{c}}(\cdot), \Delta f_{\mathrm{d}}(\cdot)) \in \boldsymbol{\Delta}$.

Proof. The result is a direct consequence of Theorem 11.1 with $f_{\mathrm{c}}(x) = f_{\mathrm{c}0}(x) + \Delta f_{\mathrm{c}}(x)$, $f_{\mathrm{d}}(x) = f_{\mathrm{d}0}(x) + \Delta f_{\mathrm{d}}(x)$, and $V(x + f_{\mathrm{d}}(x))$

given by (11.35). Specifically, it follows from (11.33) and (11.9) that $V'(x)f_{\rm c}(x) \leq V'(x)f_{\rm c0}(x) + \Gamma_{\rm c}(x) < 0$ for all $x \neq 0$, $x \notin \mathcal{Z}$, and $\Delta f_{\rm c}(\cdot) \in \boldsymbol{\Delta}_{\rm c}$. Furthermore, it follows from (11.34), (11.35), and (11.11) that $V(x + f_{\rm d}(x)) \leq V(x + f_{\rm d0}(x)) + \Gamma_{\rm d}(x)$ for all $x \in \mathcal{Z}$ and $\Delta f_{\rm d}(\cdot) \in \boldsymbol{\Delta}_{\rm d}$. Hence, all the conditions of Theorem 11.1 are satisfied. $\qquad\square$

The following corollary specializes Theorem 11.1 to a class of linear uncertain impulsive dynamical systems. Specifically, we consider $\mathcal{F} = \mathcal{F}_{\rm c} \times \mathcal{F}_{\rm d}$ to be the set of linear uncertain functions

$$\mathcal{F}_{\rm c} = \{(A_{\rm c} + \Delta A_{\rm c})x : x \in \mathbb{R}^n,\ A_{\rm c} \in \mathbb{R}^{n \times n},\ \Delta A_{\rm c} \in \boldsymbol{\Delta}_{A_{\rm c}}\},$$
$$\mathcal{F}_{\rm d} = \{(A_{\rm d} + \Delta A_{\rm d} - I_n)x : x \in \mathbb{R}^n,\ A_{\rm d} \in \mathbb{R}^{n \times n},\ \Delta A_{\rm d} \in \boldsymbol{\Delta}_{A_{\rm d}}\},$$

where $(\boldsymbol{\Delta}_{A_{\rm c}}, \boldsymbol{\Delta}_{A_{\rm d}}) \subset \mathbb{R}^{n \times n} \times \mathbb{R}^{n \times n}$ are given bounded uncertainty sets of uncertain perturbations $\Delta A_{\rm c}$ and $\Delta A_{\rm d}$ of the nominal system matrices $A_{\rm c}$ and $A_{\rm d}$ such that $0_n \in \boldsymbol{\Delta}_{A_{\rm c}}$ and $0_n \in \boldsymbol{\Delta}_{A_{\rm d}}$. In this case, $\boldsymbol{\Delta} = \boldsymbol{\Delta}_{A_{\rm c}} \times \boldsymbol{\Delta}_{A_{\rm d}}$.

Corollary 11.2 *Let $R_{\rm c} \in \mathbb{P}^n$ and $R_{\rm d} \in \mathbb{N}^n$. Consider the linear state-dependent uncertain impulsive dynamical system*

$$\dot{x}(t) = (A_{\rm c} + \Delta A_{\rm c})x(t), \quad x(0) = x_0, \quad x(t) \notin \mathcal{Z}, \quad t \geq 0, \quad (11.38)$$
$$\Delta x(t) = (A_{\rm d} + \Delta A_{\rm d} - I_n)x(t), \quad x(t) \in \mathcal{Z}, \quad\quad\quad (11.39)$$

with the hybrid quadratic performance functional

$$J_{(\Delta A_{\rm c}, \Delta A_{\rm d})}(x_0) \triangleq \int_0^\infty x^{\rm T}(t) R_{\rm c} x(t) {\rm d}t + \sum_{k \in \mathbb{Z}_{[0,\infty)}} x^{\rm T}(t_k) R_{\rm d} x(t_k),$$

$$(11.40)$$

where $(\Delta A_{\rm c}, \Delta A_{\rm d}) \in \boldsymbol{\Delta}$. Let $\Omega_{\rm c} : \mathcal{N}_{\rm P} \subseteq \mathbb{S}^n \to \mathbb{N}^n$ and $\Omega_{\rm d} : \mathcal{N}_{\rm P} \subseteq \mathbb{S}^n \to \mathbb{N}^n$ be such that

$$x^{\rm T}(\Delta A_{\rm c}^{\rm T} P + P \Delta A_{\rm c})x \leq x^{\rm T} \Omega_{\rm c}(P)x, \quad x \notin \mathcal{Z}, \quad \Delta A_{\rm c} \in \boldsymbol{\Delta}_{A_{\rm c}}, (11.41)$$
$$x^{\rm T}(\Delta A_{\rm d}^{\rm T} P A_{\rm d} + A_{\rm d}^{\rm T} P \Delta A_{\rm d} + \Delta A_{\rm d}^{\rm T} P \Delta A_{\rm d})x \leq x^{\rm T} \Omega_{\rm d}(P)x, \quad x \in \mathcal{Z},$$
$$\Delta A_{\rm d} \in \boldsymbol{\Delta}_{A_{\rm d}}, (11.42)$$

where $P \in \mathcal{N}_{\rm P}$. Furthermore, suppose there exists $P \in \mathbb{P}^n$ satisfying

$$0 = x^{\rm T}(A_{\rm c}^{\rm T} P + P A_{\rm c} + \Omega_{\rm c}(P) + R_{\rm c})x, \quad x \notin \mathcal{Z}, \quad (11.43)$$
$$0 = x^{\rm T}(A_{\rm d}^{\rm T} P A_{\rm d} - P + \Omega_{\rm d}(P) + R_{\rm d})x, \quad x \in \mathcal{Z}. \quad (11.44)$$

Then the zero solution $x(t) \equiv 0$ to (11.38) and (11.39) is globally asymptotically stable for all $(\Delta A_c, \Delta A_d) \in \boldsymbol{\Delta}$, and the hybrid quadratic performance functional (11.40) satisfies

$$\sup_{(\Delta A_c, \Delta A_d) \in \boldsymbol{\Delta}} J_{(\Delta A_c, \Delta A_d)}(x_0) \leq J(x_0) = x_0^{\mathrm{T}} P x_0, \quad x_0 \in \mathbb{R}^n, \quad (11.45)$$

where

$$J(x_0) \triangleq \int_0^\infty x^{\mathrm{T}}(t)(\Omega_c(P) + R_c)x(t)\mathrm{d}t$$
$$+ \sum_{k \in \mathbb{Z}_{[0,\infty)}} x^{\mathrm{T}}(t_k)(\Omega_d(P) + R_d)x(t_k), \quad (11.46)$$

and where $x(t)$, $t \geq 0$, is a solution to (11.38) and (11.39) with $(\Delta A_c, \Delta A_d) = (0, 0)$.

Proof. The result is a direct consequence of Theorem 11.1 with $f_c(x) = (A_c + \Delta A_c)x$, $f_{c0}(x) = A_c x$, $L_c(x) = x^{\mathrm{T}} R_c x$, $\Gamma_c(x) = x^{\mathrm{T}} \Omega_c(P)x$, $f_d(x) = (A_d + \Delta A_d - I_n)x$, $f_{d0}(x) = (A_d - I_n)x$, $L_d(x) = x^{\mathrm{T}} R_d x$, $\Gamma_d(x) = x^{\mathrm{T}} \Omega_d(P)x$, $V(x) = x^{\mathrm{T}} P x$, and $\mathcal{D} = \mathbb{R}^n$. Specifically, conditions (11.6) and (11.7) are trivially satisfied. Now, $V'(x)f_c(x) = x^{\mathrm{T}}(A_c^{\mathrm{T}} P + P A_c)x + x^{\mathrm{T}}(\Delta A_c^{\mathrm{T}} P + P \Delta A_c)x$, $x \notin \mathcal{Z}$ and $\Delta A_c \in \boldsymbol{\Delta}_{A_c}$, and hence, it follows from (11.41) that $V'(x)f_c(x) \leq V'(x)f_{c0}(x) + \Gamma_c(x) = x^{\mathrm{T}}(A_c^{\mathrm{T}} P + P A_c + \Omega_c(P))x$, $x \notin \mathcal{Z}$. Similarly, $V(x + f_d(x)) - V(x) = x^{\mathrm{T}}(A_d^{\mathrm{T}} P A_d - P)x + x^{\mathrm{T}}(\Delta A_d^{\mathrm{T}} P A_d + A_d^{\mathrm{T}} P \Delta A_d + \Delta A_d^{\mathrm{T}} P \Delta A_d)x$, $x \in \mathcal{Z}$, and $\Delta A_d \in \boldsymbol{\Delta}_{A_d}$, and hence, it follows from (11.42) that $V(x + f_d(x)) - V(x) \leq V(x + f_{d0}(x)) - V(x) + \Gamma_d(x) = x^{\mathrm{T}}(A_d^{\mathrm{T}} P A_d - P + \Omega_d(P))x$, for all $x \in \mathcal{Z}$.

Furthermore, it follows from (11.43) that $L_c(x) + V'(x)f_{c0}(x) + \Gamma_c(x) = 0$, $x \notin \mathcal{Z}$, and hence, $V'(x)f_{c0}(x) + \Gamma_c(x) < 0$, for all $x \neq 0$, $x \notin \mathcal{Z}$. Similarly, it follows from (11.44) that $L_d(x) + V(x + f_{d0}(x)) - V(x) + \Gamma_d(x) = 0$, $x \in \mathcal{Z}$, and hence, $V(x + f_{d0}(x)) - V(x) + \Gamma_d(x) \leq 0$, $x \in \mathcal{Z}$, so that all the conditions of Theorem 11.1 are satisfied. Finally, since $V(x)$, $x \in \mathbb{R}^n$, is radially unbounded, the zero solution $x(t) \equiv 0$ to (11.38) and (11.39) is globally asymptotically stable for all $(\Delta A_c, \Delta A_d) \in \boldsymbol{\Delta}$. $\square$

Corollary 11.2 generalizes Theorem 4.1 of [24] involving quadratic Lyapunov bounds for addressing robust stability and performance analysis of linear uncertain dynamical systems to linear uncertain impulsive dynamical systems.

11.3 Optimal Robust Control for Nonlinear Uncertain Impulsive Dynamical Systems

In this section, we consider a control problem for nonlinear uncertain impulsive dynamical systems involving a notion of optimality with respect to an *auxiliary hybrid cost* which guarantees a bound on the worst-case value of a nonlinear-nonquadratic hybrid cost criterion over a prescribed uncertainty set. The optimal robust hybrid time-invariant feedback controllers are derived as a direct consequence of Theorem 11.1 and provide a generalization of the hybrid Hamilton-Jacobi-Bellman conditions for state-dependent impulsive dynamical systems with optimality notions over the infinite horizon for addressing robust feedback controllers of nonlinear uncertain impulsive dynamical systems.

To address the robust optimal control problem let $\mathcal{D} \subset \mathbb{R}^n$ be an open set with $0 \in \mathcal{D}$ and let $U_c \subseteq \mathbb{R}^{m_c}$ and $U_d \subseteq \mathbb{R}^{m_d}$, where $0 \in U_c$ and $0 \in U_d$. Furthermore, let $\mathcal{F}_c \subset \{F_c : \mathcal{D} \times U_c \to \mathbb{R}^n : F_c(0,0) = 0\}$ and $\mathcal{F}_d \subset \{F_d : \mathcal{D} \times U_d \to \mathbb{R}^n\}$. Next, consider the nonlinear uncertain impulsive controlled dynamical system

$$\dot{x}(t) = F_c(x(t), u_c(t)), \quad x(0) = x_0, \ x(t) \notin \mathcal{Z}_x, \ u_c(t) \in U_c, \quad (11.47)$$
$$\Delta x(t) = F_d(x(t), u_d(t)), \quad x(t) \in \mathcal{Z}_x, \ u_d(t) \in U_d, \quad (11.48)$$

where $t \geq 0$, $(F_c(\cdot, \cdot), F_d(\cdot, \cdot)) \in \mathcal{F}$, $\mathcal{F} \triangleq \mathcal{F}_c \times \mathcal{F}_d$, $\mathcal{Z}_x \subset \mathcal{D}$, and $(u_c(t), u_d(t_k)) \in U_c \times U_d$, $k \in \overline{\mathbb{Z}}_+$, is the hybrid control input where the control constraint sets U_c and U_d are given. We assume $(0,0) \in U_c \times U_d$, $F_c : \mathcal{D} \times U_c \to \mathbb{R}^n$ is Lipschitz continuous and $F_d : \mathcal{D} \times U_d \to \mathbb{R}^n$ is continuous. To address the robust optimal nonlinear hybrid feedback control problem, let $\phi_c : \mathcal{D} \to U_c$ be such that $\phi_c(0) = 0$ and let $\phi_d : \mathcal{D} \to U_d$. If $(u_c(t), u_d(t_k)) = (\phi_c(x(t)), \phi_d(x(t_k)))$, where $x(t)$, $t \geq 0$, satisfies (11.47) and (11.48), then $(u_c(\cdot), u_d(\cdot))$ is a *hybrid feedback control*. Given the hybrid feedback control $(u_c(t), u_d(t_k)) = (\phi_c(x(t)), \phi_d(x(t_k)))$, the closed-loop state-dependent impulsive dynamical system has the form

$$\dot{x}(t) = F_c(x(t), \phi_c(x(t))), \quad x(0) = x_0, \ x(t) \notin \mathcal{Z}_x, \ t \geq 0, (11.49)$$
$$\Delta x(t) = F_d(x(t), \phi_d(x(t))), \quad x(t) \in \mathcal{Z}_x, \quad (11.50)$$

for all $(F_c(\cdot, \cdot), F_d(\cdot, \cdot)) \in \mathcal{F}$.

Next, we present sufficient conditions for characterizing robust non-linear hybrid feedback controllers that guarantee robust stability over a class of nonlinear uncertain impulsive dynamical systems and minimize an auxiliary hybrid performance functional. For the statement of this result let $L_c : \mathcal{D} \times U_c \to \mathbb{R}$, $L_d : \mathcal{D} \times U_d \to \mathbb{R}$, and define the set of asymptotically stabilizing hybrid controllers for the nominal nonlinear impulsive dynamical system $(F_{c0}(\cdot, \cdot), F_{d0}(\cdot, \cdot))$ by

$$\mathcal{C}(x_0) \triangleq \{(u_c(\cdot), u_d(\cdot)) : (u_c(\cdot), u_d(\cdot)) \text{ is admissible and the zero}$$
$$\text{solution } x(t) \equiv 0 \text{ to (11.47) and (11.48) is asymptotically}$$
$$\text{stable with } (F_c(\cdot, \cdot), F_d(\cdot, \cdot)) = (F_{c0}(\cdot, \cdot), F_{d0}(\cdot, \cdot))\}.$$

Theorem 11.2 *Consider the nonlinear uncertain impulsive dynamical system (11.47) and (11.48) with the hybrid performance functional*

$$J_{(F_c, F_d)}(x_0, u_c(\cdot), u_d(\cdot)) = \int_0^\infty L_c(x(t), u(t)) dt$$
$$+ \sum_{k \in \mathbb{Z}_{[0,\infty)}} L_d(x(t_k), u_d(t_k)), \quad (11.51)$$

where $(F_c(\cdot, \cdot), F_d(\cdot, \cdot)) \in \mathcal{F}$ and $(u_c(\cdot), u_d(\cdot))$ is an admissible hybrid control. Assume there exist functions $V : \mathcal{D} \to \mathbb{R}$, $\Gamma_c : \mathcal{D} \times U_c \to \mathbb{R}$, $\Gamma_d : \mathcal{D} \times U_d \to \mathbb{R}$, and a hybrid control law $\phi_c : \mathcal{D} \to U_c$ and $\phi_d : \mathcal{D} \to U_d$, where $V(\cdot)$ is a continuously differentiable function, such that

$$V(0) = 0, \quad (11.52)$$
$$V(x) > 0, \quad x \in \mathcal{D}, \quad x \neq 0, \quad (11.53)$$
$$\phi_c(0) = 0, \quad (11.54)$$
$$V'(x)F_c(x, \phi_c(x)) \leq V'(x)F_{c0}(x, \phi_c(x)) + \Gamma_c(x, \phi_c(x)), \quad x \notin \mathcal{Z}_x,$$
$$F_c(\cdot, \cdot) \in \mathcal{F}_c, \quad (11.55)$$
$$V'(x)F_{c0}(x, \phi_c(x)) + \Gamma_c(x, \phi_c(x)) < 0, \quad x \notin \mathcal{Z}_x, \quad x \neq 0, \quad (11.56)$$
$$V(x + F_d(x, \phi_d(x))) \leq V(x + F_{d0}(x, \phi_d(x))) + \Gamma_d(x, \phi_d(x)),$$
$$x \in \mathcal{Z}_x, \quad F_d(\cdot, \cdot) \in \mathcal{F}_d, \quad (11.57)$$
$$V(x + F_{d0}(x, \phi_d(x))) - V(x) + \Gamma_d(x, \phi_d(x)) \leq 0, \quad x \in \mathcal{Z}_x, \quad (11.58)$$
$$H_c(x, \phi_c(x)) = 0, \quad x \notin \mathcal{Z}_x, \quad (11.59)$$
$$H_d(x, \phi_d(x)) = 0, \quad x \in \mathcal{Z}_x, \quad (11.60)$$
$$H_c(x, u_c(x)) \geq 0, \quad x \notin \mathcal{Z}_x, \quad u_c \in U_c, \quad (11.61)$$
$$H_d(x, u_d(x)) \geq 0, \quad x \in \mathcal{Z}_x, \quad u_d \in U_d, \quad (11.62)$$

where $(F_{c0}(\cdot, \cdot), F_{d0}(\cdot, \cdot)) \in \mathcal{F}$ defines the nominal impulsive dynamical system and

$$H_c(x, u_c) \triangleq L_c(x, u_c) + V'(x)F_{c0}(x, u_c) + \Gamma_c(x, u_c), \quad (11.63)$$
$$H_d(x, u_d) \triangleq L_d(x, u_d) + V(x + F_{d0}(x, u_d)) - V(x) + \Gamma_d(x, u_d). \quad (11.64)$$

Then, with the hybrid feedback control $(u_c(\cdot), u_d(\cdot)) = (\phi_c(x(\cdot)), \phi_d$
$(x(\cdot)))$, *there exists a neighborhood of the origin* $\mathcal{D}_0 \subseteq \mathcal{D}$ *such that if*
$x_0 \in \mathcal{D}_0$, *the zero solution* $x(t) \equiv 0$ *of the closed-loop system (11.49)*
and (11.50) is asymptotically stable for all $(F_c(\cdot, \cdot), F_d(\cdot, \cdot)) \in \mathcal{F}$. *Fur-*
thermore,

$$\sup_{(F_c(\cdot, \cdot), F_d(\cdot, \cdot)) \in \mathcal{F}} J_{(F_c, F_d)}(x_0, \phi_c(x(\cdot)), \phi_d(x(\cdot))) \leq \mathcal{J}(x_0, \phi_c(\cdot), \phi_d(\cdot))$$

$$= V(x_0), \quad x_0 \in \mathcal{D}_0, \tag{11.65}$$

where

$$\mathcal{J}(x_0, u_c(\cdot), u_d(\cdot)) \triangleq \int_0^\infty [L_c(x(t), u_c(t)) + \Gamma_c(x(t), u_c(t))]\mathrm{d}t$$

$$+ \sum_{k \in \mathbb{Z}_{[0,\infty)}} [L_d(x(t_k), u_d(t_k)) + \Gamma_d(x(t_k), u_d(t_k))],$$

$$\tag{11.66}$$

and where $(u_c(\cdot), u_d(\cdot))$ *is an admissible hybrid control and* $x(t)$, $t \geq$
0, *solves (11.47) and (11.48) with* $(F_c(x(\cdot), u_c(\cdot)), F_d(x(\cdot), u_d(\cdot))) =$
$(F_{c0}(x(\cdot), u_c(\cdot)), F_{d0}(x(\cdot), u_d(\cdot)))$. *In addition, if* $x_0 \in \mathcal{D}_0$ *then the*
hybrid feedback control $(u_c(\cdot), u_d(\cdot)) = (\phi_c(x(\cdot)), \phi_d(x(\cdot)))$ *minimizes*
$\mathcal{J}(x_0, u_c(\cdot), u_d(\cdot))$ *in the sense that*

$$\mathcal{J}(x_0, \phi_c(x(\cdot)), \phi_d(x(\cdot))) = \min_{(u_c(\cdot), u_d(\cdot)) \in \mathcal{C}(x_0)} \mathcal{J}(x_0, u_c(\cdot), u_d(\cdot)). \tag{11.67}$$

Finally, if $\mathcal{D} = \mathbb{R}^n$, $U_c = \mathbb{R}^{m_c}$, $U_d = \mathbb{R}^{m_d}$, *and* $V(x) \to \infty$ *as* $\|x\| \to$
∞, *then the zero solution* $x(t) \equiv 0$ *of the closed-loop system (11.49)*
and (11.50) is globally asymptotically stable for all $(F_c(\cdot, \cdot), F_d(\cdot, \cdot)) \in$
$\mathcal{F}$.

Proof. Local and global asymptotic stability are a direct consequence of (11.52)–(11.58) by applying Theorem 11.1 to the closed-loop
system (11.49) and (11.50). Furthermore, using (11.59) and (11.60),
condition (11.65) is a restatement of (11.14) as applied to the closed-
loop system (11.49) and (11.50). Next, let $(u_c(\cdot), u_d(\cdot)) \in \mathcal{C}(x_0)$ and
let $x(\cdot)$ be the solution of (11.47) and (11.48) with $(F_c(\cdot, \cdot), F_d(\cdot, \cdot)) =$
$(F_{c0}(\cdot, \cdot), F_{d0}(\cdot, \cdot))$. Then it follows that

$$0 = -\dot{V}(x(t)) + V'(x(t))F_{c0}(x(t), u_c(t)), \quad x(t) \notin \mathcal{Z}_x, \tag{11.68}$$

$$0 = -\Delta V(x(t_k)) + V(x(t_k) + F_{d0}(x(t_k), u_d(t_k))) - V(x(t_k)),$$

$$x(t_k) \in \mathcal{Z}_x. \tag{11.69}$$

Hence,

$$
\begin{aligned}
L_{\mathrm{c}}(x(t)&, u_{\mathrm{c}}(t)) + \Gamma_{\mathrm{c}}(x(t), u_{\mathrm{c}}(t)) \\
&= -\dot{V}(x(t)) + L_{\mathrm{c}}(x(t), u_{\mathrm{c}}(t)) + V'(x(t))F_{\mathrm{c}0}(x(t), u_{\mathrm{c}}(t)) \\
&\quad + \Gamma_{\mathrm{c}}(x(t), u_{\mathrm{c}}(t)) \\
&= -\dot{V}(x(t)) + H_{\mathrm{c}}(x(t), u_{\mathrm{c}}(t)), \quad x(t) \notin \mathcal{Z}_x.
\end{aligned} \tag{11.70}
$$

Similarly,

$$
\begin{aligned}
L_{\mathrm{d}}(x(t_k)&, u_{\mathrm{d}}(t_k)) + \Gamma_{\mathrm{d}}(x(t_k), u_{\mathrm{d}}(t_k)) \\
&= -\Delta V(x(t_k)) + L_{\mathrm{d}}(x(t_k), u_{\mathrm{d}}(t_k)) + \Delta V(x(t_k)) + \Gamma_{\mathrm{d}}(x(t_k), u_{\mathrm{d}}(t_k)) \\
&= -\Delta V(x(t_k)) + H_{\mathrm{d}}(x(t_k), u_{\mathrm{d}}(t_k)), \quad x(t_k) \in \mathcal{Z}_x.
\end{aligned} \tag{11.71}
$$

Now, using (11.63), (11.64), and (11.66), and the fact that $(u_{\mathrm{c}}(\cdot), u_{\mathrm{d}}(\cdot))$ $\in \mathcal{C}(x_0)$, it follows that

$$
\begin{aligned}
\mathcal{J}(x_0, u_{\mathrm{c}}(\cdot), u_{\mathrm{d}}(\cdot)) &= \int_0^\infty [-\dot{V}(x(t)) + H_{\mathrm{c}}(x(t), u_{\mathrm{c}}(t))]\mathrm{d}t \\
&\quad + \sum_{k \in \mathbb{Z}_{[0,\infty)}} [-\Delta V(x(t_k)) + H_{\mathrm{d}}(x(t_k), u_{\mathrm{d}}(t_k))] \\
&= -\lim_{t \to \infty} V(x(t)) + V(x_0) + \int_0^\infty H_{\mathrm{c}}(x(t), u_{\mathrm{c}}(t))\mathrm{d}t \\
&\quad + \sum_{k \in \mathbb{Z}_{[0,\infty)}} H_{\mathrm{d}}(x(t_k), u_{\mathrm{d}}(t_k)) \\
&\geq V(x_0) \\
&= \mathcal{J}(x_0, \phi_{\mathrm{c}}(x(\cdot)), \phi_{\mathrm{d}}(x(\cdot))),
\end{aligned} \tag{11.72}
$$

which yields (11.67). Condition (11.65) can be shown similarly as in the proof of Theorem 11.1 by using (11.55)–(11.58). ◻

If $\mathcal{F}$ consists of only the nominal nonlinear impulsive closed-loop system $(F_{\mathrm{c}0}(\cdot, \cdot), F_{\mathrm{d}0}(\cdot, \cdot))$, then $\Gamma_{\mathrm{c}}(x, u_{\mathrm{c}}) = 0$ and $\Gamma_{\mathrm{d}}(x, u_{\mathrm{d}}) = 0$ for all $x \in \mathcal{D}$ and $(u_{\mathrm{c}}, u_{\mathrm{d}}) \in U_{\mathrm{c}} \times U_{\mathrm{d}}$, and hence, $J(x_0, u_{\mathrm{c}}(\cdot), u_{\mathrm{d}}(\cdot)) = \mathcal{J}(x_0, u_{\mathrm{c}}(\cdot), u_{\mathrm{d}}(\cdot))$. In this case, Theorem 11.2 specializes to Theorem 9.3. Theorem 11.2 guarantees optimality with respect to the set of admissible stabilizing controllers $\mathcal{C}(x_0)$. However, it is important to note that an explicit characterization of $\mathcal{C}(x_0)$ is not required. In addition, the optimal robustly stabilizing hybrid feedback control law $(u_{\mathrm{c}}(t), u_{\mathrm{d}}(t_k)) = (\phi_{\mathrm{c}}(x(t)), \phi_{\mathrm{d}}(x(t_k)))$ is independent of the initial condition x_0.

Next, we specialize Theorem 11.2 to linear uncertain impulsive dynamical systems. Specifically, in this case we consider $\mathcal{F} = \mathcal{F}_c \times \mathcal{F}_d$ to be the set of uncertain functions given by

$$\mathcal{F}_c = \{(A_c + \Delta A_c)x + B_c u_c : x \in \mathbb{R}^n, \ u_c \in \mathbb{R}^{m_c}, \ A_c \in \mathbb{R}^{n \times n},$$
$$B_c \in \mathbb{R}^{n \times m_c}, \ \Delta A_c \in \boldsymbol{\Delta}_{A_c}\},$$
$$\mathcal{F}_d = \{(A_d + \Delta A_d - I_n)x + B_d u_d : x \in \mathbb{R}^n, \ u_d \in \mathbb{R}^{m_d}, \ A_d \in \mathbb{R}^{n \times n},$$
$$B_d \in \mathbb{R}^{n \times m_d}, \ \Delta A_d \in \boldsymbol{\Delta}_{A_d}\}.$$

For the following result let $R_{1c} \in \mathbb{P}^n$, $R_{2c} \in \mathbb{P}^{m_c}$, $R_{1d} \in \mathbb{N}^n$, and $R_{2d} \in \mathbb{N}^{m_d}$ be given and define $R_{2ad} \triangleq R_{2d} + B_d^T P B_d + \Omega_{d_{u_d u_d}}(P)$ for arbitrary $P \in \mathbb{P}^n$ and $\Omega_{d_{u_d u_d}} : \mathbb{N}^n \to \mathbb{N}^{m_d}$.

Corollary 11.3 *Consider the linear state-dependent uncertain impulsive controlled dynamical system*

$$\dot{x}(t) = (A_c + \Delta A_c)x(t) + B_c u_c(t), \quad x(0) = x_0, \quad x(t) \notin \mathcal{Z}, \quad t \geq 0,$$
$$\tag{11.73}$$
$$\Delta x(t) = (A_d + \Delta A_d - I_n)x(t) + B_d u_d(t), \quad x(t) \in \mathcal{Z}, \tag{11.74}$$

with the hybrid quadratic performance functional

$$J_{(\Delta A_c, \Delta A_d)}(x_0, u_c(\cdot), u_d(\cdot)) \triangleq \int_0^\infty [x^T(t) R_{1c} x(t) + u_c^T(t) R_{2c} u_c(t)] dt$$
$$+ \sum_{k \in \mathbb{Z}_{[0,\infty)}} [x^T(t_k) R_{1d} x(t_k)$$
$$+ u_d^T(t_k) R_{2d} u_d(t_k)], \tag{11.75}$$

where $(u_c(\cdot), u_d(\cdot))$ is admissible and $(\Delta A_c, \Delta A_d) \in \boldsymbol{\Delta}$. Furthermore, assume there exist $P \in \mathbb{P}^n$, $\Omega_c : \mathbb{P}^n \to \mathbb{N}^n$, $\Omega_{d_{xx}} : \mathbb{P}^n \to \mathbb{N}^n$, $\Omega_{d_{xu_d}} : \mathbb{N}^n \to \mathbb{R}^{n \times m_d}$, and $\Omega_{d_{u_d u_d}} : \mathbb{N}^n \to \mathbb{N}^{m_d}$, such that

$$x^T(\Delta A_c^T P + P \Delta A_c)x \leq x^T \Omega_c(P)x, \quad x \notin \mathcal{Z}, \quad \Delta A_c \in \boldsymbol{\Delta}_{A_c}, \tag{11.76}$$
$$x^T(\Delta A_d^T P A_d + A_d^T P \Delta A_d - \Delta A_d P B_d R_{2ad}^{-1}(B_d^T P A_d + \Omega_{d_{xu_d}}^T(P))$$
$$- (B_d^T P A_d + \Omega_{d_{xu_d}}^T(P))^T R_{2ad}^{-1} B_d^T P \Delta A_d + \Delta A_d^T P \Delta A_d)x$$
$$\leq x^T(\Omega_{d_{xx}}(P) - \Omega_{d_{xu_d}}(P) R_{2ad}^{-1}(B_d^T P A_d + \Omega_{d_{xu_d}}^T(P))$$
$$- (B_d^T P + \Omega_{d_{xu_d}}^T(P))^T R_{2ad}^{-1} \Omega_{d_{xu_d}}^T(P) + (B_d^T P A_d + \Omega_{d_{xu_d}}^T(P))^T$$
$$\cdot R_{2ad}^{-1} \Omega_{d_{u_d u_d}}(P) R_{2ad}^{-1}(B_d^T P A_d + \Omega_{d_{xu_d}}^T(P)))x,$$
$$x \in \mathcal{Z}, \quad \Delta A_d \in \boldsymbol{\Delta}_{A_d}, \tag{11.77}$$

and

$$0 = x^{\mathrm{T}}(A_{\mathrm{c}}^{\mathrm{T}}P + PA_{\mathrm{c}} + R_{1\mathrm{c}} + \Omega_{\mathrm{c}}(P) - PB_{\mathrm{c}}R_{2\mathrm{c}}^{-1}B_{\mathrm{c}}^{\mathrm{T}}P)x, \quad x \notin \mathcal{Z}, \tag{11.78}$$

$$0 < R_{2\mathrm{d}} + B_{\mathrm{d}}^{\mathrm{T}}PB_{\mathrm{d}} + \Omega_{\mathrm{d}_{u_{\mathrm{d}}u_{\mathrm{d}}}}(P), \tag{11.79}$$

$$0 = x^{\mathrm{T}}(A_{\mathrm{d}}^{\mathrm{T}}PA - P + R_{1\mathrm{d}} + \Omega_{\mathrm{d}_{xx}}(P)$$
$$- (B_{\mathrm{d}}^{\mathrm{T}}PA_{\mathrm{d}} + \Omega_{\mathrm{d}_{xu_{\mathrm{d}}}}^{\mathrm{T}}(P))^{\mathrm{T}}R_{2\mathrm{ad}}^{-1}(B_{\mathrm{d}}^{\mathrm{T}}PA_{\mathrm{d}} + \Omega_{\mathrm{d}_{xu_{\mathrm{d}}}}^{\mathrm{T}}(P)))x, \ x \in \mathcal{Z}. \tag{11.80}$$

Then, with the hybrid feedback control law

$$u_{\mathrm{c}} = \phi_{\mathrm{c}}(x) = -R_{2\mathrm{c}}^{-1}B_{\mathrm{c}}^{\mathrm{T}}Px, \quad x \notin \mathcal{Z}_x,$$
$$u_{\mathrm{d}} = \phi_{\mathrm{d}}(x) = -R_{2\mathrm{ad}}^{-1}(B_{\mathrm{d}}^{\mathrm{T}}PA_{\mathrm{d}} + \Omega_{\mathrm{d}_{xu_{\mathrm{d}}}}^{\mathrm{T}}(P))x, \quad x \in \mathcal{Z}_x,$$

the zero solution $x(t) \equiv 0$ to (11.73) and (11.74) is globally asymptotically stable for all $x_0 \in \mathbb{R}^n$ and $(\Delta A_{\mathrm{c}}, \Delta A_{\mathrm{d}}) \in \boldsymbol{\Delta}_{A_{\mathrm{c}}} \times \boldsymbol{\Delta}_{A_{\mathrm{d}}}$, and

$$\sup_{(\Delta A_{\mathrm{c}}, \Delta A_{\mathrm{d}}) \in \boldsymbol{\Delta}} J_{(\Delta A_{\mathrm{c}}, \Delta A_{\mathrm{d}})}(x_0, \phi_{\mathrm{c}}(x(\cdot)), \phi_{\mathrm{d}}(x(\cdot))) \leq \mathcal{J}(x_0, \phi_{\mathrm{c}}(\cdot), \phi_{\mathrm{d}}(\cdot))$$

$$= x_0^{\mathrm{T}}Px_0, \quad x_0 \in \mathbb{R}^n, \tag{11.81}$$

where

$$\mathcal{J}(x_0, u_{\mathrm{c}}(\cdot), u_{\mathrm{d}}(\cdot)) \triangleq \int_0^{\infty} [x^{\mathrm{T}}(t)R_{1\mathrm{c}}x(t) + u_{\mathrm{c}}^{\mathrm{T}}(t)R_{2\mathrm{c}}u_{\mathrm{c}}(t)$$
$$+ x^{\mathrm{T}}(t)\Omega_{\mathrm{c}}(P)x(t)]\mathrm{d}t$$
$$+ \sum_{k \in \mathbb{Z}_{[0,\infty)}} [x^{\mathrm{T}}(t_k)R_{1\mathrm{d}}x(t_k) + u_{\mathrm{d}}^{\mathrm{T}}(t_k)R_{2\mathrm{d}}u_{\mathrm{d}}(t_k)$$
$$+ x^{\mathrm{T}}(t_k)\Omega_{\mathrm{d}_{xx}}(P)x(t_k) + 2x^{\mathrm{T}}(t_k)\Omega_{\mathrm{d}_{xu_{\mathrm{d}}}}(P)u_{\mathrm{d}}(t_k)$$
$$+ u_{\mathrm{d}}^{\mathrm{T}}(t_k)\Omega_{\mathrm{d}_{u_{\mathrm{d}}u_{\mathrm{d}}}}(P)u_{\mathrm{d}}(t_k)], \tag{11.82}$$

and where $(u_{\mathrm{c}}(\cdot), u_{\mathrm{d}}(\cdot))$ is admissible and $x(t)$, $t \geq 0$, is a solution to (11.73) and (11.74) with $(\Delta A_{\mathrm{c}}, \Delta A_{\mathrm{d}}) = (0,0)$. Furthermore,

$$\mathcal{J}(x_0, \phi_{\mathrm{c}}(x(\cdot)), \phi_{\mathrm{d}}(x(\cdot))) = \min_{(u_{\mathrm{c}}(\cdot), u_{\mathrm{d}}(\cdot)) \in \mathcal{C}(x_0)} \mathcal{J}(x_0, u_{\mathrm{c}}(\cdot), u_{\mathrm{d}}(\cdot)), \tag{11.83}$$

where $\mathcal{C}(x_0)$ is the set of asymptotically stabilizing hybrid controllers for the nominal impulsive dynamical system and $x_0 \in \mathbb{R}^n$.

Proof. The result is direct consequence of Theorem 11.2 with $F_c(x, u_c) = (A_c + \Delta A_c)x + B_c u_c$, $F_{c0}(x, u_c) = A_c x + B_c u_c$, $L_c(x, u_c) = x^T R_{1c} x + u_c^T R_{2c} u_c$, $\Gamma_c(x, u_c) = x^T \Omega_c(P)x$, $F_d(x, u_d) = (A_d + \Delta A_d - I_n)x + B_d u_d$, $F_{d0}(x, u_d) = (A_d - I_n)x + B_d u_d$, $L_d(x, u_d) = x^T R_{1d} x + u_d^T R_{2d} u_d$, $\Gamma_d(x, u_d) = x^T \Omega_{d_{xx}}(P)x + 2x^T \Omega_{d_{xu_d}}(P)u_d + u_d^T \Omega_{d_{u_d u_d}}(P)u_d$, $V(x) = x^T P x$, $\mathcal{D} = \mathbb{R}^n$, $U_c = \mathbb{R}^{m_c}$, and $U_d = \mathbb{R}^{m_d}$. Specifically, conditions (11.52) and (11.53) are trivially satisfied. Now, forming $x^T(11.76)x$, for $x \notin \mathcal{Z}_x$, it follows that, after some algebraic manipulation, $V'(x)F_c(x, \phi_c(x)) \leq V'(x)F_{c0}(x, \phi_c(x)) + \Gamma_c(x, \phi_c(x))$, for all $x \notin \mathcal{Z}_x$ and $\Delta A_c \in \mathbf{\Delta}_{A_c}$.

Similarly, forming $x^T(11.77)x$, for $x \in \mathcal{Z}_x$, it follows that, after some algebraic manipulation, $V(x + F_d(x, \phi_d(x))) - V(x) \leq V(x + F_{d0}(x, \phi_d(x))) - V(x) + \Gamma_d(x, \phi_d(x))$, for all $x \in \mathcal{Z}_x$ and $\Delta A_d \in \mathbf{\Delta}_{A_d}$. Furthermore, it follows from (11.78) that $H_c(x, \phi_c(x)) = 0$, $x \notin \mathcal{Z}_x$, and hence, $V'(x)F_{c0}(x, \phi_c(x_0)) + \Gamma_c(x, \phi_c(x)) < 0$, for all $x \neq 0$, $x \notin \mathcal{Z}_x$. Similarly, it follows from (11.80) that $H_d(x, \phi_d(x)) = 0$, $x \in \mathcal{Z}_x$, and hence, $V(x + F_{d0}(x, \phi_d(x))) - V(x) + \Gamma_d(x, \phi_d(x)) \leq 0$ for all $x \in \mathcal{Z}$. Thus, $H_c(x, u_c) = H_c(x, u_c) - H_c(x, \phi_c(x)) = [u_c - \phi_c(x)]^T R_{2c}[u_c - \phi_c(x)] \geq 0$, $x \notin \mathcal{Z}_x$, and $H_d(x, u_d) = H_d(x, u_d) - H_d(x, \phi_d(x)) = [u_d - \phi_d(x)]^T R_{2ad}[u_d - \phi_d(x)] \geq 0$, $x \in \mathcal{Z}$, so that all the conditions of Theorem 11.2 are satisfied.

Finally, since $V(\cdot)$ is radially unbounded, the zero solution $x(t) \equiv 0$ to (11.73) and (11.74), with

$$u_c(t) = \phi_c(x(t)) = -R_{2c}^{-1} B_c^T P x(t), \quad x(t) \notin \mathcal{Z}_x,$$
$$u_d(t_k) = \phi_d(x(t_k)) = -R_{2ad}^{-1}(B_d^T P A_d + \Omega_{d_{xu_d}}^T(P))x(t_k), \quad x(t_k) \in \mathcal{Z}_x,$$

is globally asymptotically stable for all $(\Delta A_c, \Delta A_d) \in \mathbf{\Delta}_{A_c} \times \mathbf{\Delta}_{A_d}$. $\square$

The optimal hybrid feedback control law $(\phi_c(x), \phi_d(x))$ in Corollary 11.3 is derived using the properties of $H_c(x, u_c)$ and $H_d(x, u_d)$ as defined in Theorem 11.2. Specifically, since

$$
\begin{aligned}
H_c(x, u_c) = {} & x^T R_{1c} x + u_c^T R_{2c} u_c + x^T(A_c^T P + P A_c)x + 2x^T P B_c u_c \\
& + x^T \Omega_c(P)x, \quad x \notin \mathcal{Z}_x,
\end{aligned}
\tag{11.84}
$$

and

$$
\begin{aligned}
H_d(x, u_d) = {} & x^T R_{1d} x + u_d^T R_{2d} u_d + (A_d x + B_d u_d)^T P(A_d x + B_d u_d) \\
& - x^T P x + x^T \Omega_{d_{xx}}(P)x + 2x^T \Omega_{d_{xu_d}}(P)u_d \\
& + u_d^T \Omega_{d_{u_d u_d}}(P)u_d, \quad x \in \mathcal{Z}_x,
\end{aligned}
\tag{11.85}
$$

it follows that $\frac{\partial^2 H_c}{\partial u_c^2} = R_{2c} > 0$ and $\frac{\partial^2 H_d}{\partial u_d^2} = R_{2ad} > 0$. Now, $\frac{\partial H_c}{\partial u_c} = 2R_{2c}u_c + 2B_c^{\mathrm{T}}Px = 0$, $x \notin \mathcal{Z}_x$, and $\frac{\partial H_d}{\partial u_d} = 2R_{2ad}u_d + 2(B_d^{\mathrm{T}}PA_d + \Omega_{d_{xu_d}}(P))x = 0$, $x \in \mathcal{Z}_x$, give the unique global minimum of $H_c(x, u_c)$, $x \notin \mathcal{Z}_x$, and $H_d(x, u_d)$, $x \in \mathcal{Z}_x$, respectively. Hence, since $\phi_c(x)$ minimizes $H_c(x, u_c)$ on $x \notin \mathcal{Z}_x$ and $\phi_d(x)$ minimizes $H_d(x, u_d)$ on $x \in \mathcal{Z}_x$, it follows that $\phi_c(x)$ satisfies $\frac{\partial H_c}{\partial u_c} = 0$ and $\phi_d(x)$ satisfies $\frac{\partial H_d}{\partial u_d} = 0$, or, equivalently, $\phi_c(x) = -R_{2c}^{-1}B_c^{\mathrm{T}}Px$, $x \notin \mathcal{Z}_x$, and $\phi_d(x) = -R_{2ad}^{-1}(B_d^{\mathrm{T}}PA_d + \Omega_{d_{xu_d}}^{\mathrm{T}}(P))x$, $x \in \mathcal{Z}_x$.

11.4 Inverse Optimal Robust Control for Nonlinear Affine Uncertain Impulsive Dynamical Systems

In this section, we specialize Theorem 11.2 to affine uncertain systems. The controllers obtained are predicated on a *robust inverse optimal hybrid control problem*. In particular, to avoid the complexity in solving the steady-state robustified hybrid Hamilton-Jacobi-Bellman equation we do not attempt to minimize a given hybrid cost functional, but rather, we parameterize a family of robustly stabilizing hybrid controllers that minimize some derived hybrid cost functional that provides flexibility in specifying the robust hybrid control law. The performance integrand is shown to explicitly depend on the nominal nonlinear impulsive system dynamics, the Lyapunov function of the uncertain impulsive closed-loop system, and the robustly stabilizing hybrid feedback control law, wherein the coupling is introduced via the robustified hybrid Hamilton-Jacobi-Bellman equation. Hence, by varying the parameters in the Lyapunov function and the performance integrand, the proposed framework can be used to characterize a class of globally robustly stabilizing hybrid controllers that can meet the closed-loop system response constraints over a prescribed range of system parametric uncertainty.

Consider the state-dependent affine (in the control) uncertain impulsive dynamical system

$$\dot{x}(t) = f_{c0}(x(t)) + \Delta f_c(x(t)) + G_c(x(t))u_c(t), \quad x(0) = x_0,$$
$$x(t) \notin \mathcal{Z}_x, \quad (11.86)$$
$$\Delta x(t) = f_{d0}(x(t)) + \Delta f_d(x(t)) + G_d(x(t))u_d(t), \quad x(t) \in \mathcal{Z}_x, \quad (11.87)$$

where $t \geq 0$, $f_{c0} : \mathcal{D} \to \mathbb{R}^n$ and satisfies $f_{c0}(0) = 0$, $f_{d0} : \mathcal{D} \to \mathbb{R}^n$, $\mathcal{D} = \mathbb{R}^n$, $U_c = \mathbb{R}^{m_c}$, $U_d = \mathbb{R}^{m_d}$, and $(\Delta f_c, \Delta f_d) \in \boldsymbol{\Delta} = \boldsymbol{\Delta}_c \times \boldsymbol{\Delta}_d$, where $\boldsymbol{\Delta}_c$ and $\boldsymbol{\Delta}_d$ are as in (11.31) and (11.32).

In this section, no explicit structure is assumed for the elements of $\boldsymbol{\Delta}$. To establish the theoretical basis of our approach, consider performance integrands $L_c(x, u_c)$ and $L_d(x, u_d)$ of the form

$$L_c(x, u_c) = L_{1c}(x) + u_c^T R_{2c}(x) u_c, \tag{11.88}$$

$$L_d(x, u_d) = L_{1d}(x) + u_d^T R_{2d}(x) u_d, \tag{11.89}$$

where $L_{1c} : \mathbb{R}^n \to \mathbb{R}$ and satisfies $L_{1c}(x) \geq 0$, $x \in \mathbb{R}^n$, $R_{2c} : \mathbb{R}^n \to \mathbb{P}^{m_c}$, $L_{1d} : \mathbb{R}^n \to \mathbb{R}$ and satisfies $L_{1d}(x) \geq 0$, $x \in \mathbb{R}^n$, and $R_{2d} : \mathbb{R}^n \to \mathbb{P}^{m_d}$ so that (11.51) becomes

$$J_{(\Delta f_c, \Delta f_d)}(x_0, u_c(\cdot), u_d(\cdot)) = \int_0^\infty [L_{1c}(x(t)) + u_c^T(t) R_{2c}(x(t)) u_c(t)] dt$$

$$+ \sum_{k \in \mathbb{Z}_{[0,\infty)}} [L_{1d}(x(t_k)) + u_d^T(t_k) R_{2d}(x(t_k)) u_d(t_k)]. \tag{11.90}$$

Corollary 11.4 *Consider the nonlinear uncertain impulsive dynamical system (11.86) and (11.87) with the hybrid performance functional (11.90). Assume there exist a continuously differentiable function $V : \mathbb{R}^n \to \mathbb{R}$, and functions $P_{12} : \mathbb{R}^n \to \mathbb{R}^{1 \times m_d}$, $P_2 : \mathbb{R}^n \to \mathbb{N}^{m_d}$, $P_{1f_d} : \mathbb{R}^n \to \mathbb{R}^{1 \times n}$, $P_{2f_d} : \mathbb{R}^n \to \mathbb{N}^n$, $P_{u_d f_d} : \mathbb{R}^n \to \mathbb{R}^{m_d \times n}$, $\Gamma_{c_{xx}} : \mathbb{R}^n \to \mathbb{R}$, $\Gamma_{d_{xx}} : \mathbb{R}^n \to \mathbb{R}$, $\Gamma_{d_{xu_d}} : \mathbb{R}^n \to \mathbb{R}^{1 \times m_d}$, and $\Gamma_{d_{u_d u_d}} : \mathbb{R}^n \to \mathbb{N}^{m_d}$ such that*

$$V(0) = 0, \tag{11.91}$$

$$V(x) > 0, \quad x \in \mathbb{R}^n, \quad x \neq 0, \tag{11.92}$$

$$V'(x) \Delta f_c(x) \leq \Gamma_{c_{xx}}(x), \quad x \notin \mathcal{Z}_x, \quad \Delta f_c \in \boldsymbol{\Delta}_c, \tag{11.93}$$

$$V'(x)[f_{c0}(x) + G_c(x)\phi_c(x)] + \Gamma_{c_{xx}}(x) < 0, \quad x \notin \mathcal{Z}_x, \quad x \neq 0, \tag{11.94}$$

$$P_{1f_d}(x)\Delta f_d(x) + \Delta f_d^T(x) P_{1f_d}^T(x) + \Delta f_d^T(x) P_{2f_d}(x)\Delta f_d(x)$$

$$+ \phi_d^T(x) P_{u_d f_d}(x)\Delta f_d(x) + \Delta f_d^T(x) P_{u_d f_d}^T(x)\phi_d(x) \leq \Gamma_{d_{xx}}(x)$$

$$+ \Gamma_{d_{xu_d}}(x)\phi_d(x) + \phi_d^T(x) \Gamma_{d_{u_d u_d}}(x)\phi_d(x),$$

$$x \in \mathcal{Z}_x, \quad \Delta f_d(\cdot) \in \boldsymbol{\Delta}_d, \tag{11.95}$$

$$V(x + f_{d0}(x)) - V(x) + P_{12}(x)\phi_d(x) + \phi_d^T(x) P_2(x)\phi_d(x) + \Gamma_{d_{xx}}(x)$$

$$+ \Gamma_{d_{xu_d}}(x)\phi_d(x) + \phi_d^T(x)\Gamma_{d_{u_d u_d}}(x)\phi_d(x) \leq 0, \quad x \in \mathcal{Z}_x, \tag{11.96}$$

$$V(x + f_{d0}(x) + G_d(x)u_d) = V(x + f_{d0}(x)) + P_{12}(x)u_d$$

$$+ u_d^T P_2(x)u_d, \quad x \in \mathcal{Z}_x, \quad u_d \in \mathbb{R}^{m_d}, \tag{11.97}$$

$$V(x + f_{d0}(x) + \Delta f_d(x) + G_d(x)u_d) = V(x + f_{d0}(x) + G_d(x)u_d)$$

$$+ P_{1f_d}(x)\Delta f_d(x) + \Delta f_d^T(x) P_{1f_d}^T(x) + \Delta f_d^T(x) P_{2f_d}(x)\Delta f_d(x)$$

$$+ u_d^T P_{u_d f_d}(x)\Delta f_d(x) + \Delta f_d^T(x) P_{u_d f_d}^T(x)u_d,$$

$$x \in \mathcal{Z}_x, \quad u_d \in \mathbb{R}^{m_d}, \quad \Delta f_d(\cdot) \in \boldsymbol{\Delta}_d, \tag{11.98}$$

and

$$V(x) \to \infty \ \ as \ \ \|x\| \to \infty, \tag{11.99}$$

where

$$\phi_{\mathrm{c}}(x) = -\tfrac{1}{2} R_{2\mathrm{c}}^{-1}(x) G_{\mathrm{c}}^{\mathrm{T}}(x) V'^{\mathrm{T}}(x), \quad x \notin \mathcal{Z}_x, \tag{11.100}$$

$$\phi_{\mathrm{d}}(x) = -\tfrac{1}{2}(R_{2\mathrm{d}}(x) + P_2(x) + \Gamma_{\mathrm{d}_{u_{\mathrm{d}}u_{\mathrm{d}}}}(x))^{-1}(P_{12}(x) + \Gamma_{\mathrm{d}_{xu_{\mathrm{d}}}}(x))^{\mathrm{T}},$$
$$x \in \mathcal{Z}_x. \tag{11.101}$$

Then the zero solution $x(t) \equiv 0$ to the impulsive closed-loop system

$$\dot{x}(t) = f_{\mathrm{c}}(x(t)) + \Delta f_{\mathrm{c}}(x(t)) + G_{\mathrm{c}}(x(t))\phi_{\mathrm{c}}(x(t)), \quad x(0) = x_0,$$
$$x(t) \notin \mathcal{Z}_x, \tag{11.102}$$

$$\Delta x(t) = f_{\mathrm{d}}(x(t)) + \Delta f_{\mathrm{d}}(x(t)) + G_{\mathrm{d}}(x(t))\phi_{\mathrm{d}}(x(t)), \quad x(t) \in \mathcal{Z}_x, \tag{11.103}$$

is globally asymptotically stable for all $(\Delta f_{\mathrm{c}}, \Delta f_{\mathrm{d}}) \in \boldsymbol{\Delta}$ with the hybrid feedback control law (11.100) and (11.101). Furthermore, the hybrid performance functional (11.90) satisfies

$$\sup_{(\Delta f_{\mathrm{c}}, \Delta f_{\mathrm{d}}) \in \mathcal{F}} J_{(\Delta f_{\mathrm{c}}, \Delta f_{\mathrm{d}})}(x_0, \phi_{\mathrm{c}}(x(\cdot)), \phi_{\mathrm{d}}(x(\cdot)))$$
$$\leq \mathcal{J}(x_0, \phi_{\mathrm{c}}(x(\cdot)), \phi_{\mathrm{d}}(x(\cdot))) = V(x_0), \quad x_0 \in \mathbb{R}^n, \tag{11.104}$$

where

$$\mathcal{J}(x_0, u_{\mathrm{c}}(\cdot), u_{\mathrm{d}}(\cdot)) \triangleq \int_0^\infty [L_{\mathrm{c}}(x(t), u_{\mathrm{c}}(t)) + \Gamma_{\mathrm{c}}(x(t), u_{\mathrm{c}}(t))]\mathrm{d}t$$
$$+ \sum_{k \in \mathbb{Z}_{[0,\infty)}} [L_{\mathrm{d}}(x(t_k), u_{\mathrm{d}}(t_k)) + \Gamma_{\mathrm{d}}(x(t_k), u_{\mathrm{d}}(t_k))],$$
$$\tag{11.105}$$

and

$$\Gamma_{\mathrm{c}}(x, u_{\mathrm{c}}) = \Gamma_{\mathrm{c}_{xx}}(x), \tag{11.106}$$

$$\Gamma_{\mathrm{d}}(x, u_{\mathrm{d}}) = \Gamma_{\mathrm{d}_{xx}}(x) + \Gamma_{\mathrm{d}_{xu_{\mathrm{d}}}}(x)u_{\mathrm{d}} + u_{\mathrm{d}}^{\mathrm{T}}\Gamma_{\mathrm{d}_{u_{\mathrm{d}}u_{\mathrm{d}}}}(x)u_{\mathrm{d}}, \tag{11.107}$$

and where $(u_{\mathrm{c}}(\cdot), u_{\mathrm{d}}(\cdot))$ is an admissible control and $x(t)$, $t \geq 0$, is a solution of (11.86) and (11.87) with $(\Delta f_{\mathrm{c}}, \Delta f_{\mathrm{d}}) = (0, 0)$. In addition, the hybrid performance functional (11.105), with

$$L_{1\mathrm{c}}(x) = \phi_{\mathrm{c}}^{\mathrm{T}}(x) R_{2\mathrm{c}}(x)\phi_{\mathrm{c}}(x) - V'(x)f_{\mathrm{c}0}(x) - \Gamma_{\mathrm{c}_{xx}}(x), \tag{11.108}$$

$$L_{1\mathrm{d}}(x) = \phi_{\mathrm{d}}^{\mathrm{T}}(x)(R_{2\mathrm{d}}(x) + P_2(x) + \Gamma_{\mathrm{d}_{u_{\mathrm{d}}u_{\mathrm{d}}}}(x))\phi_{\mathrm{d}}(x)$$
$$- V(x + f_{\mathrm{d}0}(x)) + V(x) - \Gamma_{\mathrm{d}_{xx}}(x), \tag{11.109}$$

is minimized in the sense that

$$\mathcal{J}(x_0, \phi_{\mathrm{c}}(x(\cdot)), \phi_{\mathrm{d}}(x(\cdot))) = \min_{(u_{\mathrm{c}}(\cdot), u_{\mathrm{d}}(\cdot)) \in \mathcal{C}(x_0)} \mathcal{J}(x_0, u_{\mathrm{c}}(\cdot), u_{\mathrm{d}}(\cdot)).$$

$$(11.110)$$

Proof. The result is a direct consequence of Theorem 11.2 with $\mathcal{D} = \mathbb{R}^n$, $U_{\mathrm{c}} = \mathbb{R}^{m_{\mathrm{c}}}$, $U_{\mathrm{d}} = \mathbb{R}^{m_{\mathrm{d}}}$, $F_{\mathrm{c}}(x, u_{\mathrm{c}}) = f_{\mathrm{c}0}(x) + \Delta f_{\mathrm{c}}(x) + G_{\mathrm{c}}(x)u_{\mathrm{c}}$, $F_{\mathrm{c}0}(x, u_{\mathrm{c}}) = f_{\mathrm{c}0}(x) + G_{\mathrm{c}}(x)u_{\mathrm{c}}$, $L_{\mathrm{c}}(x, u_{\mathrm{c}})$ given by (11.88), $\Gamma_{\mathrm{c}}(x, u_{\mathrm{c}})$ given by (11.106), $F_{\mathrm{d}}(x, u_{\mathrm{d}}) = f_{\mathrm{d}0}(x) + \Delta f_{\mathrm{d}}(x) + G_{\mathrm{d}}(x)u_{\mathrm{d}}$, $F_{\mathrm{d}0}(x, u_{\mathrm{d}}) = f_{\mathrm{d}0}(x) + G_{\mathrm{d}}(x)u_{\mathrm{d}}$, $L_{\mathrm{d}}(x, u_{\mathrm{d}})$ given by (11.89), and $\Gamma_{\mathrm{d}}(x, u_{\mathrm{d}})$ given by (11.107). Specifically, with (11.86)–(11.89), (11.106), and (11.107), the hybrid Hamiltonians have the form

$$\begin{aligned} H_{\mathrm{c}}(x, u_{\mathrm{c}}) &= L_{1\mathrm{c}}(x) + u_{\mathrm{c}}^{\mathrm{T}} R_{2\mathrm{c}}(x) u_{\mathrm{c}} + V'(x)(f_{\mathrm{c}0}(x) + G_{\mathrm{c}}(x)u_{\mathrm{c}}) \\ &\quad + \Gamma_{\mathrm{c}_{xx}}(x), \quad x \notin \mathcal{Z}_x, \quad u_{\mathrm{c}} \in \mathbb{R}^{m_{\mathrm{c}}}, \end{aligned} \tag{11.111}$$

$$\begin{aligned} H_{\mathrm{d}}(x, u_{\mathrm{d}}) &= L_{1\mathrm{d}}(x) + u_{\mathrm{d}}^{\mathrm{T}} R_{2\mathrm{d}}(x) u_{\mathrm{d}} + V(x + f_{\mathrm{d}0}(x) + G_{\mathrm{d}}(x)u_{\mathrm{d}}) \\ &\quad - V(x) + \Gamma_{\mathrm{d}_{xx}}(x) + \Gamma_{\mathrm{d}_{xu_{\mathrm{d}}}}(x)u_{\mathrm{d}} + u_{\mathrm{d}}^{\mathrm{T}} \Gamma_{\mathrm{d}_{u_{\mathrm{d}}u_{\mathrm{d}}}}(x)u_{\mathrm{d}}, \\ &\quad\quad\quad x \in \mathcal{Z}_x, \quad u_{\mathrm{d}} \in \mathbb{R}^{m_{\mathrm{d}}}. \tag{11.112} \end{aligned}$$

Now, the hybrid feedback control law (11.100) and (11.101) is obtained by setting $\frac{\partial H_{\mathrm{c}}}{\partial u_{\mathrm{c}}} = 0$ and $\frac{\partial H_{\mathrm{d}}}{\partial u_{\mathrm{d}}} = 0$. With (11.100) and (11.101) it follows that (11.93)–(11.98) imply (11.55)–(11.58).

Next, since $V(\cdot)$ is continuously differentiable and $x = 0$ is a local minimum of $V(\cdot)$, it follows that $V'(0) = 0$, and hence, it follows that $\phi_{\mathrm{c}}(0) = 0$, which proves (11.54). Now, with $L_{1\mathrm{c}}(x)$ and $L_{1\mathrm{d}}(x)$ given by (11.108) and (11.109), respectively, and $\phi_{\mathrm{c}}(x)$ and $\phi_{\mathrm{d}}(x)$ given by (11.100) and (11.101), (11.59) and (11.60) hold. Finally, since

$$\begin{aligned} \mathcal{H}_{\mathrm{c}}(x, u_{\mathrm{c}}) &= H_{\mathrm{c}}(x, u_{\mathrm{c}}) - H_{\mathrm{c}}(x, \phi_{\mathrm{c}}(x)) \\ &= [u_{\mathrm{c}} - \phi_{\mathrm{c}}(x)]^{\mathrm{T}} R_{2\mathrm{c}}(x)[u_{\mathrm{c}} - \phi_{\mathrm{c}}(x)], \quad x \notin \mathcal{Z}_x, \tag{11.113} \end{aligned}$$

$$\begin{aligned} \mathcal{H}_{\mathrm{d}}(x, u_{\mathrm{d}}) &= H_{\mathrm{d}}(x, u_{\mathrm{d}}) - H_{\mathrm{d}}(x, \phi_{\mathrm{d}}(x)) \\ &= [u_{\mathrm{d}} - \phi_{\mathrm{d}}(x)]^{\mathrm{T}} (R_{2\mathrm{d}}(x) + P_2(x) + \Gamma_{\mathrm{d}_{u_{\mathrm{d}}u_{\mathrm{d}}}}(x))[u_{\mathrm{d}} - \phi_{\mathrm{d}}(x)], \\ &\quad\quad\quad x \in \mathcal{Z}_x, \tag{11.114} \end{aligned}$$

where $R_{2\mathrm{c}}(x) > 0$, $x \notin \mathcal{Z}_x$, and $R_{2\mathrm{d}}(x) + P_2(x) + \Gamma_{\mathrm{d}_{u_{\mathrm{d}}u_{\mathrm{d}}}}(x) > 0$, $x \in \mathcal{Z}_x$, conditions (11.61) and (11.62) hold. The result now follows as a direct consequence of Theorem 11.2. $\qquad\square$

Note that (11.93)–(11.98) imply

$$\dot{V}(x) \triangleq V'(x)[f_{\mathrm{c}0}(x) + \Delta f_{\mathrm{c}}(x) + G_{\mathrm{c}}(x)\phi_{\mathrm{c}}(x)] < 0,$$

$$x \notin \mathcal{Z}_x, \quad x \neq 0, \quad \Delta f_{\mathrm{c}}(\cdot) \in \mathbf{\Delta}_{\mathrm{c}}, \quad (11.115)$$

$$\Delta V(x) \triangleq V(x + f_{\mathrm{d}0}(x) + \Delta f_{\mathrm{d}}(x) + G_{\mathrm{d}}(x)\phi_{\mathrm{d}}(x)) - V(x) \leq 0,$$
$$x \in \mathcal{Z}_x, \quad \Delta f_{\mathrm{d}}(\cdot) \in \mathbf{\Delta}_{\mathrm{d}}, \quad (11.116)$$

with $\phi_{\mathrm{c}}(x)$ and $\phi_{\mathrm{d}}(x)$ given by (11.100) and (11.101), respectively. Furthermore, conditions (11.115) and (11.116) with $V(0) = 0$ and $V(x) > 0$, $x \in \mathbb{R}^n$, $x \neq 0$, assure that $V(x)$ is a Lyapunov function for the impulsive closed-loop system (11.102) and (11.103).

11.5 Robust Nonlinear Hybrid Control with Polynomial Performance Functionals

In this section, we specialize the results of Section 11.4 to linear uncertain impulsive dynamical systems controlled by inverse optimal nonlinear hybrid controllers that minimize a derived polynomial cost functional. Specifically, assume $\mathcal{F} = \mathcal{F}_{\mathrm{c}} \times \mathcal{F}_{\mathrm{d}}$ to be the set of uncertain functions given by

$$\mathcal{F}_{\mathrm{c}} = \{(A_{\mathrm{c}} + \Delta A_{\mathrm{c}})x + B_{\mathrm{c}}u_{\mathrm{c}} : x \in \mathbb{R}^n, \ u_{\mathrm{c}} \in \mathbb{R}^{m_{\mathrm{c}}}, \ A_{\mathrm{c}} \in \mathbb{R}^{n \times n},$$
$$B_{\mathrm{c}} \in \mathbb{R}^{n \times m_{\mathrm{c}}}, \ \Delta A_{\mathrm{c}} \in \mathbf{\Delta}_{A_{\mathrm{c}}}\}, \quad (11.117)$$
$$\mathcal{F}_{\mathrm{d}} = \{(A_{\mathrm{d}} + \Delta A_{\mathrm{d}} - I_n)x : x \in \mathbb{R}^n, \ A_{\mathrm{d}} \in \mathbb{R}^{n \times n}, \ \Delta A_{\mathrm{d}} \in \mathbf{\Delta}_{A_{\mathrm{d}}}\}.$$
$$(11.118)$$

For the results in this section we assume $u_{\mathrm{d}}(t_k) \equiv 0$. Furthermore, let $R_{1\mathrm{c}} \in \mathbb{P}^n$, $R_{1\mathrm{d}} \in \mathbb{N}^n$, $R_{2\mathrm{c}} \in \mathbb{P}^{m_{\mathrm{c}}}$, $\hat{R}_q, \hat{\hat{R}}_q \in \mathbb{N}^n$, $q = 2, \ldots, r$, be given, where r is a positive integer, and define $S_{\mathrm{c}} \triangleq B_{\mathrm{c}} R_{2\mathrm{c}}^{-1} B_{\mathrm{c}}^{\mathrm{T}}$.

Corollary 11.5 *Consider the linear uncertain controlled impulsive dynamical system*

$$\dot{x}(t) = (A_{\mathrm{c}} + \Delta A_{\mathrm{c}})x(t) + B_{\mathrm{c}}u_{\mathrm{c}}(t), \quad x(0) = x_0, \quad x(t) \notin \mathcal{Z}_x,$$
$$(11.119)$$
$$\Delta x(t) = (A_{\mathrm{d}} + \Delta A_{\mathrm{d}} - I_n)x(t), \quad x(t) \in \mathcal{Z}_x, \quad (11.120)$$

where $u_{\mathrm{c}}(\cdot)$ is admissible input and $(\Delta A_{\mathrm{c}}, \Delta A_{\mathrm{d}}) \in \mathbf{\Delta}$. Let $\Omega_{\mathrm{c}} : \mathcal{N}_{\mathrm{P}} \subseteq \mathbb{S}^n \to \mathbb{N}^n$ and $\Omega_{\mathrm{d}} : \mathcal{N}_{\mathrm{P}} \subseteq \mathbb{S}^n \to \mathbb{N}^n$ be such that

$$x^{\mathrm{T}}(\Delta A_{\mathrm{c}}^{\mathrm{T}} P + P \Delta A_{\mathrm{c}})x \leq x^{\mathrm{T}} \Omega_{\mathrm{c}}(P)x, \quad x \notin \mathcal{Z}_x, \quad \Delta A_{\mathrm{c}} \in \mathbf{\Delta}_{A_{\mathrm{c}}},$$
$$(11.121)$$
$$x^{\mathrm{T}}(\Delta A_{\mathrm{d}}^{\mathrm{T}} P A_{\mathrm{d}} + A_{\mathrm{d}}^{\mathrm{T}} P \Delta A_{\mathrm{d}} + \Delta A_{\mathrm{d}}^{\mathrm{T}} P \Delta A_{\mathrm{d}})x \leq x^{\mathrm{T}} \Omega_{\mathrm{d}}(P)x, \quad x \in \mathcal{Z}_x,$$
$$\Delta A_{\mathrm{d}} \in \mathbf{\Delta}_{A_{\mathrm{d}}}, \quad (11.122)$$

where $P \in \mathcal{N}_P$. Assume there exist $P \in \mathbb{P}^n$ and $M_q \in \mathbb{N}^n$, $q = 2, \ldots, r$, such that

$$0 = x^T(A_c^T P + PA_c + R_{1c} + \Omega_c(P) - PS_cP)x, \quad x \notin \mathcal{Z}_x, \quad (11.123)$$

$$0 = x^T[(A_c - S_cP)^T M_q + M_q(A_c - S_cP) + \hat{R}_q]x, \quad x \notin \mathcal{Z}_x,$$
$$q = 2, \ldots, r, \quad (11.124)$$

$$0 = x^T(A_d^T PA_d - P + R_{1d} + \Omega_d(P))x, \quad x \in \mathcal{Z}_x, \quad (11.125)$$

$$0 = x^T(A_d^T M_q A_d - M_q + \hat{R}_q)x, \quad x \in \mathcal{Z}_x, \quad q = 2, \ldots, r. \quad (11.126)$$

Then the zero solution $x(t) \equiv 0$ of the uncertain impulsive closed-loop system

$$\dot{x}(t) = (A_c + \Delta A_c)x(t) + B_c\phi_c(x(t)), \quad x(0) = x_0, \quad x(t) \notin \mathcal{Z}_x, \quad (11.127)$$

$$\Delta x(t) = (A_d + \Delta A_d - I_n)x(t), \quad x(t) \in \mathcal{Z}_x, \quad (11.128)$$

is globally asymptotically stable with the feedback control law

$$\phi_c(x) = -R_{2c}^{-1} B_c^T \left(P + \sum_{q=2}^{r} (x^T M_q x)^{q-1} M_q \right) x, \quad x \notin \mathcal{Z}_x, \quad (11.129)$$

and the hybrid performance functional (11.90) satisfies

$$\sup_{(\Delta A_c, \Delta A_d) \in \Delta} J_{(\Delta A_c, \Delta A_d)}(x_0, \phi_c(x(\cdot))) \leq \mathcal{J}(x_0, \phi_c(x(\cdot)))$$

$$= x_0^T Px_0 + \sum_{q=2}^{r} \frac{1}{q} (x_0^T M_q x_0)^q, \quad x_0 \in \mathbb{R}^n, \quad (11.130)$$

where

$$\mathcal{J}(x_0, u_c(\cdot)) \triangleq \int_0^\infty [L_{1c}(x(t)) + u_c^T(t)R_{2c}(x(t))u_c(t) + \Gamma_c(x(t))]dt$$

$$+ \sum_{k \in \mathbb{Z}_{[0,\infty)}} [L_d(x(t_k)) + \Gamma_d(x(t_k))], \quad (11.131)$$

and where $u_c(\cdot)$ is admissible, $x(t)$, $t \geq 0$, is a solution to (11.119) and (11.120) with $(\Delta A_c, \Delta A_d) = (0, 0)$, and

$$\Gamma_c(x) = x^T \left(\Omega_c(P) + \sum_{q=2}^{r} (x^T M_q x)^{q-1} \Omega_c(M_q) \right) x, \quad (11.132)$$

$$\Gamma_{\mathrm{d}}(x) = x^{\mathrm{T}}\Omega_{\mathrm{d}}(P)x + \sum_{q=2}^{r}\frac{1}{q}\left[(x^{\mathrm{T}}\hat{R}_q x)\sum_{j=1}^{q}(x^{\mathrm{T}}M_q x)^{j-1}((x^{\mathrm{T}}(A_{\mathrm{d}}^{\mathrm{T}}M_q A_{\mathrm{d}}\right.$$

$$\left. +\Omega_{\mathrm{d}}(M_q))x)^{q-j} - (x^{\mathrm{T}}A_{\mathrm{d}}^{\mathrm{T}}M_q A_{\mathrm{d}}x)^{q-j})\right]. \tag{11.133}$$

In addition, the hybrid performance functional (11.90), with $R_{2\mathrm{c}}(x) = R_{2\mathrm{c}}$ and

$$L_{1\mathrm{c}}(x) = x^{\mathrm{T}}\left(R_{1\mathrm{c}} + \sum_{q=2}^{r}(x^{\mathrm{T}}M_q x)^{q-1}\hat{R}_q\right.$$

$$\left. + \left[\sum_{q=2}^{r}(x^{\mathrm{T}}M_q x)^{q-1}M_q\right]^{\mathrm{T}}S_{\mathrm{c}}\left[\sum_{q=2}^{r}(x^{\mathrm{T}}M_q x)^{q-1}M_q\right]\right)x,$$

$$\tag{11.134}$$

$$L_{1\mathrm{d}}(x) = x^{\mathrm{T}}R_{1\mathrm{d}}x$$

$$+ \sum_{q=2}^{r}\frac{1}{q}\left[(x^{\mathrm{T}}\hat{R}_q x)\sum_{j=1}^{q}(x^{\mathrm{T}}M_q x)^{j-1}(x^{\mathrm{T}}A_{\mathrm{d}}^{\mathrm{T}}M_q A_{\mathrm{d}}x)^{q-j}\right],$$

$$\tag{11.135}$$

is minimized in the sense that

$$\mathcal{J}(x_0, \phi_{\mathrm{c}}(x(\cdot))) = \min_{u_{\mathrm{c}}(\cdot)\in\mathcal{C}(x_0)}\mathcal{J}(x_0, u_{\mathrm{c}}(\cdot)), \quad x_0 \in \mathbb{R}^n, \tag{11.136}$$

where $\mathcal{C}(x_0)$ is the set of asymptotically stabilizing controllers for the nominal impulsive dynamical system and $x_0 \in \mathbb{R}^n$.

Proof. The result is a direct consequence of Corollary 11.4 with $f_{\mathrm{c}0}(x) = A_{\mathrm{c}}x$, $\Delta f_{\mathrm{c}}(x) = \Delta A_{\mathrm{c}}x$, $\Delta A_{\mathrm{c}} \in \boldsymbol{\Delta}_{A_{\mathrm{c}}}$, $G_{\mathrm{c}}(x) = B_{\mathrm{c}}$, $L_{1\mathrm{c}}(x)$ given by (11.134), $\Gamma_{\mathrm{c}}(x, u_{\mathrm{c}}) = \Gamma_{\mathrm{c}}(x)$ given by (11.132), $f_{\mathrm{d}0}(x) = (A_{\mathrm{d}} - I_n)x$, $\Delta f_{\mathrm{d}}(x) = \Delta A_{\mathrm{d}}x$, $\Delta A_{\mathrm{d}} \in \boldsymbol{\Delta}_{A_{\mathrm{d}}}$, $G_{\mathrm{d}}(x) = 0$, $u_{\mathrm{d}} = 0$, $L_{1\mathrm{d}}(x)$ given by (11.135), $\Gamma_{\mathrm{d}}(x, u_{\mathrm{d}}) = \Gamma_{\mathrm{d}}(x)$ given by (11.133), $R_{2\mathrm{c}}(x) = R_{2\mathrm{c}}$, $R_{2\mathrm{d}}(x) = I_{m_{\mathrm{d}}}$, and $V(x) = x^{\mathrm{T}}Px + \sum_{q=2}^{r}\frac{1}{q}(x^{\mathrm{T}}M_q x)^q$. Specifically, conditions (11.91)–(11.93), (11.95), and (11.98) are trivially satisfied.

Now, for $x \notin \mathcal{Z}_x$ it follows from (11.123), (11.124), and (11.129) that

$$V'(x)[f_{\mathrm{c}0}(x) - \tfrac{1}{2}G_{\mathrm{c}}(x)R_{2\mathrm{c}}^{-1}G_{\mathrm{c}}^{\mathrm{T}}(x)V'^{\mathrm{T}}(x)] + \Gamma_{\mathrm{c}}(x) = -x^{\mathrm{T}}R_{1\mathrm{c}}x$$

$$-\sum_{q=2}^{r}(x^{\mathrm{T}}M_qx)^{q-1}x^{\mathrm{T}}\hat{R}_qx$$

$$-\phi_{\mathrm{c}}^{\mathrm{T}}(x)R_{2\mathrm{c}}\phi_{\mathrm{c}}(x)$$

$$-x^{\mathrm{T}}\left[\sum_{q=2}^{r}(x^{\mathrm{T}}M_qx)^{q-1}M_q\right]^{\mathrm{T}}S_{\mathrm{c}}\left[\sum_{q=2}^{r}(x^{\mathrm{T}}M_qx)^{q-1}M_q\right]x,$$

$$(11.137)$$

which implies (11.94). Similarly, for $x \in \mathcal{Z}_x$ it follows from (11.125) and (11.126) that

$$V(x+f_{\mathrm{d}0}(x)) - V(x) + \Gamma_{\mathrm{d}}(x) = -x^{\mathrm{T}}R_{1\mathrm{d}}x$$

$$-\sum_{q=2}^{r}\frac{1}{q}\left[(x^{\mathrm{T}}\hat{R}_qx)\sum_{j=1}^{q}(x^{\mathrm{T}}M_qx)^{j-1}(x^{\mathrm{T}}A_{\mathrm{d}}^{\mathrm{T}}M_qA_{\mathrm{d}}x)^{q-j}\right],$$

$$(11.138)$$

which implies (11.96) with $\phi_{\mathrm{d}}(x) = 0$. Finally, with $u_{\mathrm{d}} = 0$, (11.97) is automatically satisfied so that all the conditions of Corollary 11.4 are satisfied. □

Chapter Twelve

Hybrid Dynamical Systems

12.1 Introduction

As discussed in Chapter 1, hybrid dynamical systems involve an interacting countable collection of dynamical systems possessing a mixture of continuous-time dynamics and discrete-time dynamics that include impulsive dynamical systems, hierarchical systems, and switching systems as special cases. Even though numerous results focusing on specific forms of hybrid systems have been developed in the literature (see for example [51] and the numerous references therein), the development of a general model for hybrid dynamical systems has received little attention in the literature. Notable exceptions include [30, 121, 169]. In particular, the authors in [121, 169] introduce a general (undisturbed) hybrid dynamical system model whose flow is defined on an arbitrary metric space evolving over a notion of generalized abstract time. For this class of hybrid dynamical systems the authors in [121, 169] provide a thorough treatment of Lyapunov stability and Lagrange stability results. Alternatively, the authors in [30] give a unified framework for an autonomous hybrid control model, and develop an optimal control framework for synthesizing hybrid controllers.

In this chapter, we present a unified dynamical systems framework for a general class of systems possessing left-continuous flows, that is, *left-continuous dynamical systems*.[1] A left-continuous dynamical system is a precise mathematical object and is defined on the semi-infinite interval as a mapping between vector spaces satisfying an appropriate set of axioms and includes hybrid inputs and hybrid outputs that take their values in appropriate vector spaces. The notion of a left-continuous dynamical system introduced in this chapter generalizes virtually all existing notions of dynamical systems and includes hybrid, impulsive, and switching dynamical systems as special cases. Furthermore, using generalized left-continuous storage func-

---

[1] Right-continuous dynamical systems, that is, systems possessing right-continuous flows, can also be analogously considered.

tions and hybrid supply rates we extend the notions of dissipativity theory and exponential dissipativity theory introduced in Chapter 3 to left-continuous dynamical systems. As in the impulsive dynamical systems case, the overall approach provides an interpretation of a generalized hybrid energy balance of a left-continuous dynamical system in terms of the stored or, accumulated generalized energy, dissipated energy over the continuous-time dynamics, and dissipated energy at the resetting events. Furthermore, we show that the set of all possible storage functions of a left-continuous dynamical system forms a convex set, and is bounded from below by the system's available stored generalized energy which can be recovered from the system, and bounded from above by the system's required generalized energy supply needed to transfer the system from an initial state of minimum generalized energy to a given state. Finally, using the concepts of dissipativity and exponential dissipativity for left-continuous systems, we develop feedback interconnection stability results for left-continuous dynamical systems.

12.2 Left-Continuous Dynamical Systems

In this section, we introduce the notion of a left-continuous dynamical system. The following definition is concerned with left-continuous dynamical systems, or systems with left-continuous flows. For this definition $\mathcal{U} \triangleq \mathcal{U}_c \times \mathcal{U}_d$ is an input space and consists of bounded left-continuous U-valued functions on the semi-infinite interval $[0, \infty)$. The set $U \triangleq U_c \times U_d$, where $U_c \subseteq \mathbb{R}^{m_c}$ and $U_d \subseteq \mathbb{R}^{m_d}$, contains the set of input values, that is, for every $u = (u_c, u_d) \in \mathcal{U}$ and $t \in [0, \infty)$, $u(t) \in U$, $u_c(t) \in U_c$, and $u_d(t) \in U_d$. Furthermore, $\mathcal{Y} \triangleq \mathcal{Y}_c \times \mathcal{Y}_d$ is an output space and consists of bounded left-continuous Y-valued functions on the semi-infinite interval $[0, \infty)$. The set $Y \triangleq Y_c \times Y_d$, where $Y_c \subseteq \mathbb{R}^{l_c}$ and $Y_d \subseteq \mathbb{R}^{l_d}$, contains the set of output values, that is, for every $y = (y_c, y_d) \in \mathcal{Y}$ and $t \in [0, \infty)$, $y(t) \in Y$, $y_c(t) \in Y_c$, and $y_d(t) \in Y_d$. Finally, $\mathcal{D} \subseteq \mathbb{R}^n$ and $\|\cdot\|$ denotes the Euclidean norm.

Definition 12.1 *A left-continuous dynamical system on $\mathcal{D}$ is the octuple $(\mathcal{D}, \mathcal{U}, U, \mathcal{Y}, Y, s, h_c, h_d)$, where $s : [0, \infty) \times [0, \infty) \times \mathcal{D} \times \mathcal{U} \to \mathcal{D}$, $h_c : \mathcal{D} \times U_c \to Y_c$, and $h_d : \mathcal{D} \times U_d \to Y_d$ are such that the following axioms hold:*

   i) (Left-continuity): For every $t_0 \in [0, \infty)$, $x_0 \in \mathcal{D}$, and $u \in \mathcal{U}$, $s(t_0, \cdot, x_0, u)$ is left-continuous, that is, $\lim_{\tau \to t^-} s(\tau, t_0, x_0, u) =$

$s(t, t_0, x_0, u)$ *for all* $t \in (t_0, \infty)$.

ii) *(Consistency): For every* $x_0 \in \mathcal{D}$, $u \in \mathcal{U}$, *and* $t_0 \in [0, \infty)$, $s(t_0, t_0, x_0, u) = x_0$.

iii) *(Determinism): For every* $t_0 \in [0, \infty)$ *and* $x_0 \in \mathcal{D}$, $s(t, t_0, x_0, u_1)$ $= s(t, t_0, x_0, u_2)$ *for all* $t \in [t_0, \infty)$ *and* $u_1, u_2 \in \mathcal{U}$ *satisfying* $u_1(\tau) = u_2(\tau)$, $\tau \in [t_0, t]$.

iv) *(Semi-group property):* $s(t_2, t_0, x_0, u) = s(t_2, t_1, s(t_1, t_0, x_0, u), u)$ *for all* $t_0, t_1, t_2 \in [0, \infty)$, $t_0 \le t_1 \le t_2$, $x_0 \in \mathcal{D}$, *and* $u \in \mathcal{U}$.

v) *(Read-out map):*[2] *There exists* $y \in \mathcal{Y}$ *such that* $y(t) = (h_c(s(t, t_0, x_0, u), u_c(t)), h_d(s(t, t_0, x_0, u), u_d(t)))$ *for all* $x_0 \in \mathcal{D}$, $u \in \mathcal{U}$, $t_0 \in [0, \infty)$, *and* $t \in [t_0, \infty)$.

The notion of a left-continuous dynamical system introduced in Definition 12.1 is defined on the semi-infinite time interval $[0, \infty)$. Instead of the semi-infinite time interval, a completely ordered time set $\mathbb{T}$ having a hybrid topological structure involving isolated (discrete) points and closed sets homeomorphic to intervals on the real line can be considered [169]. In this case, continuous-time and discrete events can be defined on $\mathbb{T} \subset [0, \infty) \times \mathcal{T}$, where $\mathcal{T}$ is a countable subset of $[0, \infty)$. Alternatively, in our formulation continuous-time and discrete events are defined on a single semi-infinite interval $[0, \infty)$, where the graph of the semi-infinite interval corresponding to discrete events is piecewise constant and left-continuous.

At each moment of time $t \in [0, \infty)$, the left-continuous dynamical system $\mathcal{G}$ receives a hybrid input $u(t)$ and generates a hybrid output $y(t)$. The values of the hybrid input are taken from the fixed set $U = U_c \times U_d$. Furthermore, over a time segment the hybrid input function is not arbitrary but belongs to the input class $\mathcal{U}$. In addition, each hybrid output $y(t)$ belongs to the fixed set $Y = Y_c \times Y_d$ with $y(\cdot) \in \mathcal{Y}$ over a given time segment, where $\mathcal{Y}$ denotes an output space. In general, the hybrid output of $\mathcal{G}$ depends on both the present hybrid input of $\mathcal{G}$ and the past history of $\mathcal{G}$. Hence, the hybrid output at some time t depends on the *state* $s(t, t_0, x_0, u)$ of $\mathcal{G}$ which effectively serves as an information storage (memory) of past history. Furthermore, the determinism axiom assures that the state, and hence, the hybrid output before some time t are not influenced by the values of the hybrid output after time t. Hence, future hybrid inputs to $\mathcal{G}$ do not

---

[2]More generally, the read-out maps h_c and h_d can be explicit functions of time, that is, $h_c : [0, \infty) \times \mathcal{D} \times U_c \to Y_c$ and $h_d : [0, \infty) \times \mathcal{D} \times U_d \to Y_d$.

effect past and present hyrbid outputs of $\mathcal{G}$. Finally, we note that the hybrid read-out map is memoryless in the sense that hybrid outputs only depend on the instantaneous (present) values of the state and hybrid input.

Henceforth, we denote the left-continuous dynamical system $(\mathcal{D}, \mathcal{U}, U, \mathcal{Y}, Y, s, h_\mathrm{c}, h_\mathrm{d})$ by $\mathcal{G}$. Furthermore, we refer to $s(t, t_0, x_0, u)$, $t \geq t_0$, as the *trajectory* of $\mathcal{G}$ corresponding to $x_0 \in \mathcal{D}$, $t_0 \in [0, \infty)$, and $u \in \mathcal{U}$, and for a given trajectory $s(t, t_0, x_0, u)$, $t \geq t_0$, we refer to $t_0 \in [0, \infty)$ as an *initial time* of $\mathcal{G}$, $x_0 \in \mathcal{D}$ as an *initial condition* of $\mathcal{G}$, and $u \in \mathcal{U}$ as an *input* to $\mathcal{G}$. The trajectory $s(t, t_0, x_0, u)$, $t \geq t_0$, of $\mathcal{G}$ is *bounded* if there exists $\gamma > 0$ such that $\|s(t, t_0, x_0, u)\| < \gamma$, $t \geq t_0$.

The dynamical system $\mathcal{G}$ is *isolated* if the input space consists of one constant element only, that is, $u(t) \equiv u^*$, and the dynamical system $\mathcal{G}$ is *undisturbed* if $u^* = 0$. Furthermore, an *equilibrium point* of the undisturbed dynamical system $\mathcal{G}$ is a point $x \in \mathcal{D}$ satisfying $s(t, t_0, x, 0) = x$, $t \geq t_0$. An equilibrium point $x \in \mathcal{D}$ of the undisturbed dynamical system $\mathcal{G}$ is *uniformly Lyapunov stable* if, for all $\varepsilon > 0$ and $t_0 \in [0, \infty)$, there exists $\delta = \delta(\varepsilon) > 0$ such that if $\|x - x_0\| < \delta$, then $\|x - s(t, t_0, x_0, 0)\| < \varepsilon$, $t \geq t_0$. An equilibrium point $x \in \mathcal{D}$ of the undisturbed dynamical system $\mathcal{G}$ is *uniformly asymptotically stable* if x is uniformly Lyapunov stable and there exists $\delta > 0$ such that if $\|x - x_0\| < \delta$, then $\lim_{t \to \infty} s(t, t_0, x_0, 0) = x$. An equilibrium point $x \in \mathcal{D}$ of the undisturbed dynamical system $\mathcal{G}$ is *uniformly exponentially stable* if there exist positive constants α, β, and δ such that if $\|x - x_0\| < \delta$, then $\|x - s(t, t_0, x_0, 0)\| \leq \alpha \|x - x_0\| e^{-\beta t}$, $t \geq t_0$. Finally, an equilibrium point $x \in \mathcal{D}$ of the undisturbed dynamical system $\mathcal{G}$ is *globally uniformly asymptotically* (respectively, *uniformly exponentially*) *stable* if uniform asymptotic (respectively, uniform exponential) stability holds for any $x_0 \in \mathbb{R}^n$ and $t_0 \in [0, \infty)$. The next definition provides a specialization of Definition 12.1 to the case of stationary left-continuous dynamical systems, wherein $\mathcal{U}$ is closed under the shift operator.

Definition 12.2 *A stationary left-continuous dynamical system on $\mathcal{D}$ is the octuple $(\mathcal{D}, \mathcal{U}, U, \mathcal{Y}, Y, s, h_\mathrm{c}, h_\mathrm{d})$, where $s : [0, \infty) \times [0, \infty) \times \mathcal{D} \times \mathcal{U} \to \mathcal{D}$, $h_\mathrm{c} : \mathcal{D} \times U_\mathrm{c} \to Y_\mathrm{c}$, and $h_\mathrm{d} : \mathcal{D} \times U_\mathrm{d} \to Y_\mathrm{d}$ are such that Axioms i)–v) hold, and*

    vi) (Stationarity): For every $t_0, t \in [0, \infty)$, $t \geq t_0$, $\tau, T \in \mathbb{R}$, $x_0 \in \mathcal{D}$, $u, u_T \in \mathcal{U}$, such that $u_T(t) = u(t + T)$, $t \in [0, \infty)$, $s(t + \tau, t_0 + \tau, x_0, u_\tau) = s(t, t_0, x_0, u)$.

Note that without loss of generality, for a given stationary left-continuous dynamical system $\mathcal{G}$, we can set $t_0 = 0$ by redefining $u(t) \triangleq u(t - t_0)$. Hence, we will denote the trajectory $s(t, t_0, x_0, u)$, $t \geq t_0$, of a stationary left-continuous dynamical system $\mathcal{G}$ as $s(t, 0, x_0, u)$, $t \geq 0$. In the following, for every $t_0 \in [0, \infty)$, $x_0 \in \mathcal{D}$, and $u \in \mathcal{U}$, let $\mathcal{T}_{t_0, x_0, u} \subseteq [t_0, \infty)$ denote a dense subset of the semi-infinite interval $[t_0, \infty)$ such that $\mathcal{T}^c_{t_0, x_0, u} \triangleq [t_0, \infty) \backslash \mathcal{T}_{t_0, x_0, u}$ is (finitely or infinitely) countable. For notational convenience we write $\mathcal{T}$ and $\mathcal{T}^c$ for $\mathcal{T}_{t_0, x_0, u}$ and $\mathcal{T}^c_{t_0, x_0, u}$, respectively. Furthermore, we refer to $\mathcal{T}^c$ as the set of *resetting times* or *resetting events*.

Definition 12.3 *A strong left-continuous dynamical system on $\mathcal{D}$ is the octuple $(\mathcal{D}, \mathcal{U}, U, \mathcal{Y}, Y, s, h_{\mathrm{c}}, h_{\mathrm{d}})$, where $s : [0, \infty) \times [0, \infty) \times \mathcal{D} \times \mathcal{U} \to \mathcal{D}$, $h_{\mathrm{c}} : \mathcal{D} \times U_{\mathrm{c}} \to Y_{\mathrm{c}}$, and $h_{\mathrm{d}} : \mathcal{D} \times U_{\mathrm{d}} \to Y_{\mathrm{d}}$ are such that Axioms i)–vi) hold, and*

vii) *(Quasi-continuous dependence): For every $t_0 \in [0, \infty)$, $x_0 \in \mathcal{D}$, and $u \in \mathcal{U}$, there exists $\mathcal{T} \subseteq [t_0, \infty)$ such that $[t_0, \infty) \backslash \mathcal{T}$ is countable and for every $\varepsilon > 0$ and $t \in \mathcal{T}$, there exists $\delta(\varepsilon, x_0, u, t) > 0$, such that if $\|x_0 - x\| < \delta(\varepsilon, x_0, u, t)$, $x \in \mathcal{D}$, then $\|s(t, t_0, x_0, u) - s(t, t_0, x, u)\| < \varepsilon$.*

In applying Definition 12.1 it may be convenient to replace Axiom *vii)* with a stronger condition which may be easier to verify in practice. The following proposition provides sufficient conditions for $\mathcal{G}$ to be a strong left-continuous dynamical system.

Proposition 12.1 *Let $\mathcal{G}$ be a stationary left-continuous dynamical system such that the following condition holds:*

vii)′ *For every $t_0 \in [0, \infty)$, $x_0 \in \mathcal{D}$, $u \in \mathcal{U}$, $\varepsilon, \eta > 0$, and $T \in \mathcal{T}$, there exists $\delta(\varepsilon, x_0, u, T) > 0$ such that if $\|x_0 - x\| < \delta(\varepsilon, x_0, u, T)$, $x \in \mathcal{D}$, then, for every $t \in \mathcal{T} \cap [0, T]$ such that $|t - \tau| > \eta$ and for all $\tau \in \mathcal{T}^c \cap [0, T]$, $\|s(t, t_0, x_0, u) - s(t, t_0, x, u)\| < \varepsilon$. Furthermore, if $t \in \mathcal{T}$ is an accumulation point of $\mathcal{T}^c$, then $s(t, t_0, \cdot, u)$ is continuous for all $u \in \mathcal{U}$.*

Then $\mathcal{G}$ is a strong left-continuous dynamical system.

Proof. Let $x_0 \in \mathcal{D}$, $u \in \mathcal{U}$, $\hat{t} \in \mathcal{T}$ be such that $\hat{t}$ is not an accumulation point of $\mathcal{T}^c$, and let $T(\hat{t}) \in \mathcal{T}$ be such that $\hat{t} \leq T(\hat{t})$. Furthermore, let $\eta(\hat{t}) > 0$ be such that $|\hat{t} - \tau| > \eta$, for every $\tau \in \mathcal{T}^c$. Then it follows from *vii)′* that for every $\varepsilon > 0$ there exists $\delta(\varepsilon, x_0, u, T) > 0$, such that if $\|x_0 - y\| < \delta(\varepsilon, x_0, u, T)$, $y \in \mathcal{D}$, then $\|s(\hat{t}, t_0, x_0, u) - s(\hat{t}, t_0, y, u)\| < \varepsilon$, which implies $vii)$ with $t = \hat{t}$. Now, the result is immediate since $\hat{t}$ is arbitrary in the set of all times such that $\hat{t}$ is not an accumulation point of $\mathcal{T}^c$ and, by assumption, if $t \in \mathcal{T}_{x_0}$ is an accumulation point of $\mathcal{T}^c$, then $s(t, t_0, \cdot, u)$ is continuous at x_0. $\square$

The next result considers undisturbed (i.e., $u(t) \equiv 0$) stationary left-continuous dynamical systems. In particular, we show that $\mathcal{G}$ is a stationary left-continuous dynamical system satisfying Axiom $vii)'$ if and only if the trajectory of $\mathcal{G}$ is *jointly continuous between resetting events*, that is, for every $\varepsilon > 0$ and $k \in \overline{\mathbb{Z}}_+$, there exists $\delta = \delta(\varepsilon, k) > 0$ such that if $|t - t'| + \|x_0 - y\| < \delta$, then $\|s(t, 0, x_0, 0) - s(t', 0, y, 0)\| < \varepsilon$, where $x_0, y \in \mathcal{D}$, $t \in (\tau_k(x_0), \tau_{k+1}(x_0)]$, and $t' \in (\tau_k(y), \tau_{k+1}(y)]$. For this result we assume that $\mathcal{T}$ in Definition 12.3 is given by $\mathcal{T}_{x_0} \triangleq \{t \in [0, \infty) : s(t, 0, x, 0) = s(t^+, 0, x_0, 0)\}$ so that $[0, \infty) \backslash \mathcal{T}_{x_0}$ corresponds to the (countable) set of resetting times where the trajectory $s(\cdot, 0, x_0, 0)$ is discontinuous. Furthermore, we let $\tau_i(x_0)$, $i = 1, 2, \ldots$, where $\tau_0(x_0) \triangleq 0$ and $\tau_1(x_0) < \tau_2(x_0) < \cdots$, denote the resetting times, that is, $\{\tau_1(x_0), \tau_2(x_0), \ldots\} = [0, \infty) \backslash \mathcal{T}_{x_0}$. Finally, we assume that for every $i = 1, 2, \ldots$, $\tau_i(\cdot)$ is continuous.

Proposition 12.2 *Consider the undisturbed (i.e., $u(t) \equiv 0$) dynamical system $\mathcal{G}$ satisfying Axioms i), ii), and iv). Then $\mathcal{G}$ is a stationary left-continuous dynamical system satisfying Axiom vii)' if and only if the trajectory $s(t, 0, x_0, 0)$, $t \geq 0$, of $\mathcal{G}$ is jointly continuous between resetting events.*

Proof. Assume $\mathcal{G}$ is a stationary left-continuous dynamical system, let $\varepsilon > 0$, and let $k \in \overline{\mathbb{Z}}_+$. Since, by assumption, $\tau_k(\cdot)$ is continuous, it follows that for sufficiently small $\delta_1 > 0$, $\tau_k(x)$ and $\tau_{k+1}(x)$, $x \in \mathcal{B}_{\delta_1}(x_0)$, where $\mathcal{B}_{\delta_1}(x_0)$ denotes the open ball centered at x_0 with radius δ_1, are well defined and finite. Hence, it follows from Axiom $vii)'$ that $s(t, 0, \cdot, 0)$, $t \in (\tau_k(\cdot), \tau_{k+1}(\cdot)]$, is uniformly bounded on $\mathcal{B}_{\delta_1}(x_0)$. Now, since $\mathcal{G}$ is continuous between resetting events it follows that for $\varepsilon > 0$ and $k \in \overline{\mathbb{Z}}_+$ there exists $\hat{\delta} = \hat{\delta}(\varepsilon, k) > 0$ such that if $|t - t'| < \hat{\delta}$, then

$$\|s(t, 0, x, 0) - s(t', 0, x, 0)\| < \frac{\varepsilon}{3}, \quad x \in \mathcal{B}_{\delta_1}(x_0), \ t, t' \in (\tau_k(x), \tau_{k+1}(x)].$$
$$(12.1)$$

Next, it follows from the continuity of $\tau_k(\cdot)$ that for every sufficiently small $\lambda > 0$ and $k \in \overline{\mathbb{Z}}_+$, $\underline{\tau}_k(\lambda, x_0) \triangleq \inf_{x \in \mathcal{B}_\lambda(x_0)} \tau_k(x)$ and $\overline{\tau}_k(\lambda, x_0) \triangleq$

$\sup_{x \in \mathcal{B}_\lambda(x_0)} \tau_k(x)$ are well defined and $\lim_{\lambda \to 0} \underline{\tau}_k(\lambda, x_0) = \lim_{\lambda \to 0} \overline{\tau}_k(\lambda, x_0) = \tau_k(x_0)$. (Note that $\underline{\tau}_k(\lambda, x_0) \le \tau_k(x) \le \overline{\tau}_k(\lambda, x_0)$, for all $x \in \mathcal{B}_\lambda(x_0)$.) Hence, there exists $\delta' = \delta'(\hat{\delta}) > 0$ such that $\overline{\tau}_k(\delta', x_0) - \underline{\tau}_k(\delta', x_0) < \hat{\delta}$ and $\overline{\tau}_{k+1}(\delta', x_0) - \underline{\tau}_{k+1}(\delta', x_0) < \hat{\delta}$. Next, let $\eta > 0$ be such that

$$\overline{\tau}_k(\delta', x_0) - \underline{\tau}_k(\delta', x_0) < \tau_k(x_0) - \underline{\tau}_k(\delta', x_0) + \eta < \hat{\delta}, \tag{12.2}$$

$$\overline{\tau}_{k+1}(\delta', x_0) - \underline{\tau}_{k+1}(\delta', x_0) < \eta + \overline{\tau}_{k+1}(\delta', x_0) - \tau_{k+1}(x_0) < \hat{\delta}. \tag{12.3}$$

Then, it follows from Axiom $vii)'$ that there exists $\delta'' = \delta''(\varepsilon, \eta, k) > 0$ such that

$$\|s(t, 0, x_0, 0) - s(t, 0, y, 0)\| < \frac{\varepsilon}{3}, \quad y \in \mathcal{B}_{\delta''}(x_0),$$
$$t \in (\tau_k(x_0) + \eta, \tau_{k+1}(x_0) - \eta). \tag{12.4}$$

Now, if $|t - t'| + \|x_0 - y\| < \delta$, where $\delta = \min\{\delta_1, \delta', \delta'', \hat{\delta}\}$, $t \in (\tau_k(x_0) + \eta, \tau_{k+1}(x_0) - \eta)$, and $t' \in (\tau_k(y), \tau_{k+1}(y)]$, then it follows from (12.1), (12.4), and the triangular inequality for vector norms that

$$\|s(t, 0, x_0, 0) - s(t', 0, y, 0)\| \le \|s(t, 0, x_0, 0) - s(t, 0, y, 0)\|$$
$$+ \|s(t, 0, y, 0) - s(t', 0, y, 0)\|$$
$$< \frac{2}{3}\varepsilon$$
$$< \varepsilon. \tag{12.5}$$

Finally, if $|t - t'| + \|x_0 - y\| < \delta$, where $t \in (\tau_k(x_0), \tau_{k+1}(x_0)] \setminus (\tau_k(x_0) + \eta, \tau_{k+1}(x_0) - \eta)$ and $t' \in (\tau_k(y), \tau_{k+1}(y)]$, then conditions (12.2) and (12.3) imply that there exists $t'' \in (\tau_k(x_0) + \eta, \tau_{k+1}(x_0) - \eta)$ such that $|t - t''| < \hat{\delta}$ and $|t' - t''| < \hat{\delta}$. Hence, by (12.1) and (12.4) it follows that

$$\|s(t, 0, x_0, 0) - s(t', 0, y, 0)\| \le \|s(t, 0, x_0, 0) - s(t'', 0, x_0, 0)\|$$
$$+ \|s(t'', 0, x_0, 0) - s(t'', 0, y, 0)\|$$
$$+ \|s(t'', 0, y, 0) - s(t', 0, y, 0)\|$$
$$< \varepsilon,$$

which establishes that $\mathcal{G}$ is jointly continuous between resetting events.

To show that joint continuity of $s(t, 0, x_0, 0)$, $t \ge 0$, between resetting events implies that $\mathcal{G}$ is a stationary left-continuous dynamical system satisfying Axiom $vii)'$, let $\varepsilon, \eta > 0$, $T \in \mathcal{T}_{x_0}$, and suppose $\tau_k(x_0) < T < \tau_{k+1}(x_0)$. Then, it follows from the joint continuity of $\mathcal{G}$ that there exists $\delta' = \delta'(\varepsilon, k) > 0$ such that if $|t - t'| +$

$\|x_0 - y\| < \delta'$, then $\|s(t, 0, x_0, 0) - s(t', 0, y, 0)\| < \varepsilon$, where $x_0, y \in \mathcal{D}$, $t \in (\tau_k(x_0), \tau_{k+1}(x_0)]$, and $t' \in (\tau_k(y), \tau_{k+1}(y)]$. Now, it follows that there exists $\delta'' = \delta''(x_0, \eta, k) > 0$ such that $\overline{\tau}_k(\delta'', x_0) - \tau_k(x_0) < \eta$ and $\tau_{k+1}(x_0) - \underline{\tau}_{k+1}(\delta'', x_0) < \eta$. Note that the above inequalities guarantee that if $t = t' \in (\tau_k(x_0) + \eta, \tau_{k+1}(x_0) - \eta)$, then $t \in (\tau_k(x_0), \tau_{k+1}(x_0)]$ and $t' \in (\tau_k(y), \tau_{k+1}(y)]$, $y \in \mathcal{B}_{\delta''}(x_0)$. Furthermore, letting $\delta_k = \delta_k(\varepsilon, \eta, x_0, k) = \min\{\delta', \delta''\}$, it follows from the joint continuity of $\mathcal{G}$ that for $t = t' \in (\tau_k(x_0) + \eta, \tau_{k+1}(x_0) - \eta)$, $\|s(t, 0, x_0, 0) - s(t, 0, y, 0)\| < \varepsilon$, $y \in \mathcal{B}_{\delta_k}(x_0)$.

Similarly, we can obtain $\delta_{k-1} = \delta_{k-1}(\varepsilon, \eta, x_0, k) > 0$ such that an analogous inequality can be constructed for all $y \in \mathcal{B}_{\delta_{k-1}}(x_0)$ and $t \in (\tau_{k-1}(x_0) + \eta, \tau_k(x_0) - \eta)$. Recursively repeating this procedure for $m = k - 2, \ldots, 1$, and choosing $\delta = \delta(\varepsilon, \eta, x_0, k) = \delta(\varepsilon, \eta, x_0, T) = \min\{\delta_1, \ldots, \delta_k\}$, it follows that $\|s(t, 0, x_0, 0) - s(t, 0, y, 0)\| < \varepsilon$, $y \in \mathcal{B}_\delta(x_0)$, $t \in [0, T]$, and $|t - \tau_l(x_0)| > \eta, l = 1, \ldots, k$, which implies that $\mathcal{G}$ is a stationary left-continuous dynamical system satisfying Axiom $vii)'$. $\square$

12.3 Specialization to Hybrid and Impulsive Dynamical Systems

In this section, we show that hybrid dynamical systems [30, 169] and impulsive dynamical systems [12, 14, 79, 93, 148] are a specialization of left-continuous dynamical systems. We start our presentation by considering a definition of a controlled hybrid dynamical system that includes the definition given in [30] as a special case. For this definition let $\mathcal{Q} \subseteq \overline{\mathbb{Z}}_+$, where $\overline{\mathbb{Z}}_+$ denotes the set of nonnegative integers.

Definition 12.4 *A hybrid dynamical system* $\mathcal{G}_H$ *is the 13-tuple* $(\mathcal{D}, \mathcal{Q}, \mathcal{U}, U, \mathcal{Y}, Y, q, x, s_c, f_d, \mathcal{S}, h_c, h_d)$, *where* $q : [0, \infty) \times [0, \infty) \times \mathcal{D} \times \mathcal{Q} \times \mathcal{U} \to \mathcal{Q}$, $x : [0, \infty) \times [0, \infty) \times \mathcal{D} \times \mathcal{Q} \times \mathcal{U} \to \mathcal{D}$, $s_c = \{s_{cq}\}_{q \in \mathcal{Q}}$, $s_{cq} : [0, \infty) \times [0, \infty) \times \mathcal{D} \times \mathcal{U} \to \mathcal{D}$, $\mathcal{S} = \{\mathcal{S}_q\}_{q \in \mathcal{Q}}$, $\mathcal{S}_q \subset [0, \infty) \times \mathcal{D} \times \mathcal{U}$, $f_d = \{f_{dq}\}_{q \in \mathcal{Q}}$, $f_{dq} : \mathcal{S}_q \to \mathcal{D} \times \mathcal{Q}$, $h_c : \mathcal{D} \times U_c \to Y_c$, *and* $h_d : \mathcal{D} \times U_d \to Y_d$ *are such that the following axioms hold:*

  i) *For every* $q \in \mathcal{Q}$, $t_0 \in [0, \infty)$, *and* $u \in \mathcal{U}$, $s_{cq}(\cdot, t_0, \cdot, u)$ *is jointly continuous on* $[t_0, \infty) \times \mathcal{D}$.

  ii) *For every* $q \in \mathcal{Q}$, $t_0 \in [0, \infty)$, $x_0 \in \mathcal{D}$, *and* $u \in \mathcal{U}$, $s_{cq}(t_0, t_0, x_0, u) = x_0$.

  iii) *For every* $q \in \mathcal{Q}$, $t_0 \in [0, \infty)$, *and* $x_0 \in \mathcal{D}$, $s_{cq}(t, t_0, x_0, u_1) =$

$s_{cq}(t, t_0, x_0, u_2)$ for all $t \in [t_0, \infty)$ and $u_1, u_2 \in \mathcal{U}$ satisfying $u_1(\tau) = u_2(\tau)$, $\tau \in [t_0, t]$.

iv) For every $q \in \mathcal{Q}$, $t_0, t_1, t_2 \in [0, \infty)$, $t_0 \leq t_1 \leq t_2$, $x_0 \in \mathcal{D}$, and $u \in \mathcal{U}$, $s_{cq}(t_2, t_0, x_0, u) = s_{cq}(t_2, t_1, s_{cq}(t_1, t_0, x_0, u), u)$.

v) For every $q_0 \in \mathcal{Q}$, $t_0 \in [0, \infty)$, $x_0 \in \mathcal{D}$, and $u \in \mathcal{U}$, $q(\cdot)$ and $x(\cdot)$ are such that $q(t, t_0, x_0, q_0, u) = q_0$ and $x(t, t_0, x_0, q_0, u) = s_{cq_0}(t, t_0, x_0, u)$, for all $t_0 \leq t \leq t_1$, where $t_1 \triangleq \min\{t \geq t_0 : (t, s_{cq_0}(t, t_0, x_0, u), u(t)) \notin \mathcal{S}_{q_0}\}$ exists. Furthermore, $[x^\mathrm{T}(t_1^+, t_0, x_0, q_0, u), q^\mathrm{T}(t_1^+, t_0, x_0, q_0, u)]^\mathrm{T} = f_{dq_0}(t_1, x(t_1), u(t_1)) + [x^\mathrm{T}(t_1, t_0, x_0, q_0, u), q^\mathrm{T}(t_1, t_0, x_0, q_0, u)]^\mathrm{T}$ and for $(x_1, q_1) \triangleq (x(t_1^+, t_0, x_0, q_0, u), q(t_1^+, t_0, x_0, q_0, u))$, $q(\cdot)$ and $x(\cdot)$ are such that $q(t, t_0, x_0, q_0, u) = q_1$ and $x(t, t_0, x_0, q_0, u) = s_{cq_1}(t, t_1, x_1, u)$, for all $t_1 < t \leq t_2$, where $t_2 \triangleq \min\{t > t_1 : (t, s_{cq_1}(t, t_1, x_1, u), u(t)) \notin \mathcal{S}_{q_1}\}$ exists, and so on.

vi) There exists $y \in \mathcal{Y}$ such that $y(t) = (h_c(x(t, t_0, x_0, q_0, u), u_c(t)), h_d(x(t, t_0, x_0, q_0, u), u_d(t)))$ for all $x_0 \in \mathcal{D}$, $u \in \mathcal{U}$, $t_0 \in [0, \infty)$, and $t \in [t_0, \infty)$.

It follows from Definition 12.4 that hybrid dynamical systems involve switchings between a countable collection of continuous dynamical systems. To ensure that the switchings or resetting times are well defined and distinct we make the following additional assumptions:

A1. If $(t, x(t, t_0, x_0, q_0, u), u(t)) \in \overline{\mathcal{S}}_q \backslash \mathcal{S}_q$, where $\overline{\mathcal{S}}_q$ denotes the closure of the set $\mathcal{S}_q$, then there exists $\varepsilon > 0$ such that, for all $0 < \delta < \varepsilon$,

$$s_{cq}(t + \delta, t, x(t, t_0, x_0, q_0, u), u) \notin \mathcal{S}_q.$$

A2. If $(t_k, x(t_k, t_0, x_0, q_0, u), u(t_k)) \in \partial \mathcal{S}_q \cap \mathcal{S}_q$, where $\partial \mathcal{S}_q$ denotes the boundary of the set $\mathcal{S}_q$, then there exists $\varepsilon > 0$ such that, for all $0 < \delta < \varepsilon$,

$$s_{c\hat{q}}(t_k + \delta, t_k, x(t_k^+, t_0, x_0, q_0, u), u) \notin \mathcal{S}_{\hat{q}}, \quad \hat{q} \in \mathcal{Q}.$$

Assumption A1 ensures that if a trajectory reaches the closure of $\mathcal{S}_q$ at a point that does not belong to $\mathcal{S}_q$, then the trajectory must be directed away from $\mathcal{S}_q$, that is, a trajectory cannot enter $\mathcal{S}_q$ through a point that belongs to the closure of $\mathcal{S}_q$ but not to $\mathcal{S}_q$. Equivalently,

A1 implies that a trajectory can only reach $\mathcal{S}_q$ through a point belonging to both $\mathcal{S}_q$ and its boundary. Furthermore, A2 ensures that when a trajectory intersects the boundary of a resetting set $\mathcal{S}_q$, it instantaneously exits $\mathcal{S}_q$ and the continuous-time dynamics becomes the active element of the hybrid dynamical system. Since a continuous trajectory starting outside $\mathcal{S}_q$ and intersecting the interior of $\mathcal{S}_q$ must first intersect the boundary of $\mathcal{S}_q$, it follows that no trajectory can reach the interior of $\mathcal{S}_q$.

To show that $\mathcal{G}_H$ is a left-continuous dynamical system, let s : $[0,\infty) \times [0,\infty) \times (\mathcal{D} \times \mathcal{Q}) \times \mathcal{U} \rightarrow \mathcal{D} \times \mathcal{Q}$ be such that $s(t_0, t_0, (x_0, q_0), u) = (x_0, q_0)$, and for every $k = 1, 2, \ldots,$

$$s(t, t_0, (x_0, q_0), u) = (s_{cq_{k-1}}(t, t_{k-1}, x_{k-1}, u), q_{k-1}), \quad t_{k-1} < t \leq t_k,$$
(12.6)
$$s(t_k^+, t_0, (x_0, q_0), u) = f_{dq_{k-1}}(t_k, x_k, u(t_k)) + [x_k^T, q_k^T]^T.$$
(12.7)

Note that s satisfies Axioms $i)$–$v)$ of Definition 12.1 so that the controlled hybrid dynamical system $\mathcal{G}_H$ generates a left-continuous dynamical system on $\mathcal{D} \times \mathcal{Q}$ given by the octuple $(\mathcal{D} \times \mathcal{Q}, \mathcal{U}, U, \mathcal{Y}, Y, s, h_c,$ $h_d)$. Since the resetting events $\mathcal{T}^c = \{t_1, t_2, \ldots\}$ can be a function of time t, the system state $x(t, t_0, x_0, q_0, u)$, and the system input u, hybrid dynamical systems can involve system jumps at variable times, and hence, in general are time-varying left-continuous dynamical systems. In the case where the resetting events are defined by a prescribed sequence of times which are independent of the system trajectories and system inputs, that is, $\mathcal{S}_q = \mathcal{T}_q \times \mathcal{D} \times \mathcal{U}$, where $\mathcal{T}_q \subset [0,\infty)$ and $q \in \mathcal{Q}$ is a closed discrete set, we refer to $\mathcal{G}_H$ as a *time-dependent hybrid dynamical system*. Alternatively, in the case where the resetting events are defined by the manifold $\mathcal{S}_q = [0,\infty) \times \mathcal{S}_{xq} \times \mathcal{U}$, where $\mathcal{S}_{xq} \subset \mathcal{D}$, $q \in \mathcal{Q}$, that is, $\mathcal{S}_q$ is independent of time and the inputs, we refer to $\mathcal{G}_H$ as a *state-dependent hybrid dynamical system*. More generally, if the resetting events are defined by the manifold $\mathcal{S}_q = ([0,\infty) \times \mathcal{S}_{xq} \times \mathcal{U}) \cup ([0,\infty) \times \mathcal{D} \times \mathcal{S}_{uq})$, where $\mathcal{S}_{uq} \subset \mathcal{U}$, $q \in \mathcal{Q}$, we refer to $\mathcal{G}_H$ as an *input/state-dependent hybrid dynamical system*. Note that if $\{s_{cq}\}_{q \in \mathcal{Q}}$ are continuous trajectories such that Axiom $vi)$ in Definition 12.2 holds, then state- and input/state-dependent hybrid dynamical systems are stationary left-continuous dynamical systems.

As in the case of impulsive dynamical systems, the analysis of hybrid dynamical systems can be quite involved. In particular, such systems can exhibit Zenoness and beating, as well as confluence. Even though A1 and A2 allow for the possibility of confluence and Zeno solutions, A2 precludes the possibility of beating. In the case of stationary state-dependent hybrid dynamical systems several interesting observations can be made regarding quasi-continuous dependence and Zeno solutions. Specifically, if the first resetting time is continuous with respect to all initial conditions and all system solutions are non-Zeno, then the hybrid dynamical system can be shown to satisfy Axiom vii). Furthermore, if the second resetting time is continuous with respect to all initial conditions on the resetting surfaces $\mathcal{S}_q$, $q \in \mathcal{Q}$, and all convergent solutions starting from $\mathcal{D} \backslash \cup_{q \in \mathcal{Q}} \overline{\mathcal{S}}_q$ are Zeno, then all the trajectories approach the set $\cup_{q \in \mathcal{Q}} \overline{\mathcal{S}}_q \backslash \mathcal{S}_q$ as $t \to \infty$. The proofs of these facts follow as in the proofs of Propositions 2.1 and 2.3.

The notion of a controlled hybrid dynamical system $\mathcal{G}_{\mathrm{H}}$ given by Definition 12.4 generalizes all the existing notions of dynamical systems wherein the state space has a fixed dimension. For example, if $\mathcal{Q} = \{q\}$ and $\mathcal{S} = \varnothing$, then $\mathcal{G}_{\mathrm{H}}$ denotes a continuous-time dynamical system with a continuous flow [165]. Alternatively, if $\mathcal{Q} = \{q\}$, $\mathcal{S} = \mathcal{S}_q$, and s_{cq} denotes the solution to an ordinary differential equation

$$\dot{x}_q(t) = f_{cq}(t, x_q(t), u_c(t)), \quad x_q(t_0) = x_0, \quad t \geq t_0, \tag{12.8}$$

where $x_q(t) \in \mathcal{D}$, $t \geq t_0$, and $f_{cq} : [0, \infty) \times \mathcal{D} \times U_c \to \mathbb{R}^n$, then the hybrid dynamical system $\mathcal{G}_{\mathrm{H}}$ is characterized by the impulsive differential equation

$$\dot{x}(t) = f_{cq}(t, x(t), u_c(t)), \quad x(t_0) = x_0, \quad (t, x(t), u_c(t)) \notin \mathcal{S}_q, \tag{12.9}$$
$$\Delta x(t) = f_{dq}(t, x(t), u_d(t)), \quad (t, x(t), u_c(t)) \in \mathcal{S}_q. \tag{12.10}$$

More generally, if $\mathcal{Q}$ is a (finitely or infinitely) countable set and $\{s_{cq}\}_{q \in \mathcal{Q}}$ denote the solutions to a set of ordinary differential equations, then $\mathcal{G}_{\mathrm{H}}$ can be represented by a set of coupled ordinary differential equations and difference equations or, equivalently, a set of impulsive differential equations with discontinuous vector fields. Specifically, for every $q \in \mathcal{Q}$, let s_{cq} denote the solution to the ordinary differential equation

$$\dot{x}_q(t) = f_{cq}(t, x_q(t), u_c(t)), \quad x_q(t_0) = x_0, \quad t \geq t_0, \tag{12.11}$$

where $x_q(t) \in \mathcal{D}$, $t \geq t_0$, and $f_{cq} : [0, \infty) \times \mathcal{D} \times U_c \to \mathbb{R}^n$. In this case, the hybrid dynamical system $\mathcal{G}_{\mathrm{H}}$ is characterized by the impulsive differential equation

$$\begin{bmatrix} \dot{x}(t) \\ \dot{q}(t) \end{bmatrix} = \begin{bmatrix} f_{cq(t)}(t, x(t), u_c(t)) \\ 0 \end{bmatrix}, \quad \begin{bmatrix} x(t_0) \\ q(t_0) \end{bmatrix} = \begin{bmatrix} x_0 \\ q_0 \end{bmatrix},$$
$$(t, x(t), u_c(t)) \notin \mathcal{S}_{q(t)}, \tag{12.12}$$

$$\left[\begin{array}{c} \Delta x(t) \\ \Delta q(t) \end{array} \right] = f_{\mathrm{d}q(t)}(t, x(t), u_{\mathrm{d}}(t)), \quad (t, x(t), u_{\mathrm{c}}(t)) \in \mathcal{S}_{q(t)}. \quad (12.13)$$

Finally, note that if $\Delta x(t) = 0$ in (12.13), then (12.12) specializes to the case of switched hybrid systems involving continuous flows but discontinuous vector fields [29, 101, 140], that is, a Filippov dynamical system.

We close this section by noting that several of the classical hybrid dynamical system models developed in the literature [7, 11, 31, 130, 157, 167] are a special case of the impulsive dynamical system (12.12) and (12.13). Specifically, the Witsenhausen model [167], the Tavernini model [157], the Nerode-Kohn model [130], and the Antsaklis-Stiver-Lemmon model [7] are special cases of an autonomous version of (12.12) and (12.13) with $\Delta x(t) \equiv 0$, $u_{\mathrm{c}}(t) \equiv 0$, and $u_{\mathrm{d}}(t) \equiv 0$. Hence, these models belong to the class of switched hybrid system models with continuous flows and discontinuous vector fields. Alternatively, the Back-Guckenheimer-Myers model [11] is a special case of an autonomous version of (12.12) and (12.13) with $u_{\mathrm{c}}(t) \equiv 0$ and $u_{\mathrm{d}}(t) \equiv 0$. Finally, the Brockett models [31] are a special case of an autonomous version of (12.12) and (12.13) with $\Delta x(t) \equiv 0$. For a further discussion of these models the interested reader is referred to [30].

12.4 Stability Analysis of Left-Continuous Dynamical Systems

In this section, we present uniform Lyapunov, uniform asymptotic, and uniform exponential stability results for left-continuous dynamical systems. Furthermore, for strong left-continuous dynamical systems we present an invariant set stability theorem that generalizes the Krasovskii-LaSalle invariance principle to systems with left-continuous flows. For the statement of the following result we define

$$\dot{V}(t, s(t, t_0, x_0, u)) \triangleq \lim_{\tau \to t^-} \frac{1}{t - \tau} \left[V(t, s(t, t_0, x_0, u)) \right.$$
$$\left. - V(\tau, s(\tau, t_0, x_0, u)) \right], \quad (12.14)$$

for a given continuous function $V : [t_0, \infty) \times \mathcal{D} \to [0, \infty)$, whenever the limit on the right-hand side exists. Note that $V(t, s(t, t_0, x_0, u))$ is left-continuous on $[t_0, \infty)$, and is continuous everywhere on $[t_0, \infty)$ except on the discrete set $\mathcal{T}^c$. Furthermore, we assume that the origin is an equilibrium point of the undisturbed left-continuous dynamical system $\mathcal{G}$.

Theorem 12.1 *Suppose there exist a continuous function $V : [0, \infty)$ $\times \mathcal{D} \to [0, \infty)$ and class $\mathcal{K}$ functions $\alpha(\cdot)$ and $\beta(\cdot)$ satisfying*

$$\alpha(\|x\|) \leq V(t, x) \leq \beta(\|x\|), \quad x \in \mathcal{D}, \quad t \in [t_0, \infty), \quad (12.15)$$
$$V(t, s(t, t_0, x_0, 0)) \leq V(t, s(\tau, t_0, x_0, 0)), \quad t \geq \tau \geq t_0. \quad (12.16)$$

Then the equilibrium point $x = 0$ of the undisturbed left-continuous dynamical system $\mathcal{G}$ is uniformly Lyapunov stable. If, in addition, for every $x_0 \in \mathcal{D}$, $V(\cdot)$ is such that $\dot{V}(s(t, t_0, x_0, 0))$, $t \in \mathcal{T}$, exists and

$$\dot{V}(t, s(t, t_0, x_0, 0)) \leq -\gamma(\|s(t, t_0, x_0, 0)\|), \quad t \in \mathcal{T}, \quad (12.17)$$

where $\gamma : [0, \infty) \to [0, \infty)$ is a class $\mathcal{K}$ function, then the equilibrium point $x = 0$ of the undisturbed left-continuous dynamical system $\mathcal{G}$ is uniformly asymptotically stable. Alternatively, if there exist scalars $\varepsilon, \hat{\alpha}, \hat{\beta} > 0$, and $p \geq 1$ such that

$$\hat{\alpha}\|x\|^p \leq V(t, x) \leq \hat{\beta}\|x\|^p, \quad x \in \mathcal{D}, \quad t \in [0, \infty), \quad (12.18)$$
$$\dot{V}(t, s(t, t_0, x_0, 0)) \leq -\varepsilon V(t, s(t, t_0, x_0, 0)), \quad t \in \mathcal{T}, \quad (12.19)$$

then the equilibrium point $x = 0$ of the undisturbed left-continuous dynamical system $\mathcal{G}$ is uniformly exponentially stable. Finally, if $\mathcal{D} = \mathbb{R}^n$ and $\alpha(\cdot)$ is a class $\mathcal{K}_\infty$ function, then (12.17) implies (respectively, (12.18) and (12.19) imply) that the equilibrium point $x = 0$ of the undisturbed left-continuous dynamical system $\mathcal{G}$ is globally uniformly asymptotically (respectively, exponentially) stable.

Proof. *i*) It follows from (12.16) that $V(t, s(t, t_0, x_0, 0))$, $t \geq t_0$, is a nonincreasing function of time. Moreover, for all $t \in (t_0, \infty)$,

$$V(t + \sigma, s(t + \sigma, t_0, x_0, 0)) \leq V(t - \sigma, s(t - \sigma, t_0, x_0, 0)), \quad (12.20)$$

for every sufficiently small $\sigma > 0$. Since $V(\cdot, \cdot)$ is continuous on $[0, \infty) \times \mathcal{D}$, letting $\sigma \to 0$ yields

$$V(t^+, s(t^+, t_0, x_0, 0)) \leq V(t, s(t, t_0, x_0, 0)), \quad t \in [t_0, \infty). \quad (12.21)$$

Next, let $\varepsilon > 0$ be such that $\mathcal{B}_\varepsilon(0) \triangleq \{x \in \mathbb{R}^n : \|x\| < \varepsilon\} \subset \mathcal{D}$, define $\eta \triangleq \alpha(\varepsilon)$, and define $\mathcal{D}_\eta \triangleq \{x \in \mathcal{B}_\varepsilon(0) : \text{ there exists } t \in [0, \infty) \text{ such that } V(t, x) < \eta\}$. Now, since $V(t, s(t, t_0, x_0, 0))$ is a nonincreasing function of time, $\mathcal{D}_\eta \times [0, \infty)$ is a positive invariant set with respect to the left-continuous dynamical system $\mathcal{G}$. Next, let $\delta = \delta(\varepsilon) > 0$ be such that $\beta(\delta) = \alpha(\varepsilon)$. Hence, it follows from (12.15) that for all $(x_0, t_0) \in \mathcal{B}_\delta(0) \times [0, \infty)$,

$$\alpha(\|s(t, t_0, x_0, 0)\|) \leq V(t, s(t, t_0, x_0, 0)) \leq V(t_0, x_0) < \beta(\delta) = \alpha(\varepsilon),$$
$$t \geq t_0, \quad (12.22)$$

and hence, $s(t, t_0, x_0, 0) \in \mathcal{B}_\varepsilon(0)$, $t \geq t_0$, establishing uniform Lyapunov stability of the equilibrium point $x = 0$ of $\mathcal{G}$.

ii) Uniform Lyapunov stability follows from *i)*. Next, let $\varepsilon > 0$ and $\delta = \delta(\varepsilon) > 0$ be such that for every $x_0 \in \mathcal{B}_\delta(0)$, $s(t, t_0, x_0, 0) \in \mathcal{B}_\varepsilon(0)$, $t \geq t_0$, for all $t_0 \in [0, \infty)$ (the existence of such a (δ, ε) pair follows from uniform Lyapunov stability), and assume that (12.17) holds. Since by (12.16) $V(t, s(t, t_0, x_0, 0))$ is a nonincreasing function of time and, since $V(\cdot, \cdot)$ is bounded from below, it follows from the Bolzano-Weierstass theorem [146] that there exists $L \geq 0$ such that $\lim_{t \to \infty} V(t, s(t, t_0, x_0, 0)) = L$.

Now, suppose, *ad absurdum*, for some $x_0 \in \mathcal{B}_\delta(0)$ and $t_0 \in [0, \infty)$, $L > 0$. Since $V(\cdot, \cdot)$ is continuous and $V(t_0, 0) = 0$ for all $t_0 \in [0, \infty)$ it follows that $\mathcal{D}_L \triangleq \{x \in \mathcal{B}_\varepsilon(0) : V(t, x) \leq L \text{ for all } t \in [0, \infty)\}$ is nonempty and $s(t, t_0, x_0, 0) \notin \mathcal{D}_L$, $t \geq t_0$. Thus, as in the proof of *i)*, there exists $\hat{\delta} > 0$ such that $\mathcal{B}_{\hat{\delta}}(0) \subset \mathcal{D}_L$. Hence, it follows from (12.21) and (12.17) that for any $x_0 \in \mathcal{B}_\delta(0)$ and $t \geq t_0$,

$$V(t, s(t, t_0, x_0, 0)) = V(t_0, x_0) + \int_{t_0}^{t} \dot{V}(\tau, s(\tau, t_0, x_0, 0)) \mathrm{d}\tau$$
$$+ \sum_{i \in \mathbb{Z}_{[t_0, t)}} [V(t_i^+, s(t_i^+, t_0, x_0, 0))$$
$$- V(t_i, s(t_i, t_0, x_0, 0))]$$
$$\leq V(t_0, x_0) - \int_{t_0}^{t} \gamma(\|s(\tau, t_0, x_0, 0)\|) \mathrm{d}\tau$$
$$\leq V(t_0, x_0) - \gamma(\hat{\delta}) t, \quad (12.23)$$

where $\mathbb{Z}_{[t_0, t)} \triangleq \{k \in \overline{\mathbb{Z}}_+ : t_0 < t_k \leq t\}$ and $[0, \infty) \backslash \mathcal{T} = \{t_1, t_2, \ldots\}$.

Letting $t \geq \frac{V(t_0, x_0) - L}{\gamma(\hat{\delta})}$, it follows that $V(t, x(t)) \leq L$, which is a contradiction. Hence, $L = 0$, and, since $x_0 \in \mathcal{B}_\delta(0)$ and $t_0 \in [0, \infty)$ was chosen arbitrarily, it follows that $V(t, s(t, t_0, x_0, 0)) \to 0$ as $t \to \infty$ for all $x_0 \in \mathcal{B}_\delta(0)$ and $t_0 \in [0, \infty)$. Now, since $V(t, s(t, t_0, x_0, 0)) \geq \alpha(\|s(t, t_0, x_0, 0)\|) \geq 0$ it follows that $\alpha(\|s(t, t_0, x_0, 0)\|) \to 0$ or, equivalently, $s(t, t_0, x_0, 0) \to 0$, $t \to \infty$, establishing uniform asymptotic stability of the equilibrium point $x = 0$ of $\mathcal{G}$.

iii) Let $\varepsilon > 0$ and $\eta \triangleq \alpha(\varepsilon)$ be given as in the proof of *i)*. Now, (12.19) implies that $\dot{V}(t, s(t, t_0, x_0, 0)) \leq 0$, $t \in \mathcal{T}$, $x_0 \in \mathcal{D}$, $t_0 \in$

$[0, \infty)$, and hence, using (12.21), it follows that $V(t, s(t, t_0, x_0, 0))$ is a nonincreasing function of time and $\mathcal{D}_\eta \times [0, \infty) \subset \mathcal{D} \times [0, \infty)$ is a positive invariant set with respect to $\mathcal{G}$. It follows from (12.19) that for all $(x_0, t_0) \in \mathcal{D}_\eta \times [0, \infty)$ with $t_k \in [0, \infty) \backslash \mathcal{T}$, $k \in \overline{\mathbb{Z}}_+$,

$$\dot{V}(t, s(t, t_0, x_0, 0)) \leq -\varepsilon V(t, s(t, t_0, x_0, 0)), \quad t_0 \leq t \leq t_1, \quad (12.24)$$

which implies that

$$V(t, s(t, t_0, x_0, 0)) \leq V(t_0, x_0) e^{-\varepsilon(t-t_0)}, \quad t_0 \leq t \leq t_1. \quad (12.25)$$

Similarly,

$$\dot{V}(t, s(t, t_0, x_0, 0)) \leq -\varepsilon V(t, s(t, t_0, x_0, 0)), \quad t_1 < t \leq t_2, \quad (12.26)$$

which, using (12.21) and (12.25), yields

$$\begin{aligned} V(t, s(t, t_0, x_0, 0)) &\leq V(t_1^+, s(t_1^+, t_0, x_0, 0)) e^{-\varepsilon(t-t_1)} \\ &\leq V(t_1, s(t_1, t_0, x_0, 0)) e^{-\varepsilon(t-t_1)} \\ &\leq V(t_0, x_0) e^{-\varepsilon(t_1-t_0)} e^{-\varepsilon(t-t_1)} \\ &= V(t_0, x_0) e^{-\varepsilon(t-t_0)}, \quad t_1 < t \leq t_2. \end{aligned} \quad (12.27)$$

Recursively repeating the above arguments for $t_k < t \leq t_{k+1}$, $k = 3, 4, \ldots$, it follows that

$$V(t, s(t, t_0, x_0, 0)) \leq V(t_0, x_0) e^{-\varepsilon(t-t_0)}, \quad t \geq t_0. \quad (12.28)$$

Now, it follows from (12.18) and (12.28) that for all $t \geq t_0$

$$\begin{aligned} \hat{\alpha} \| s(t, t_0, x_0, 0) \|^p &\leq V(t, s(t, t_0, x_0, 0)) \\ &\leq V(t_0, x_0) e^{-\varepsilon(t-t_0)} \\ &\leq \hat{\beta} \| x_0 \|^p e^{-\varepsilon(t-t_0)}, \end{aligned} \quad (12.29)$$

and hence,

$$\| s(t, t_0, x_0, 0) \| \leq \left(\frac{\hat{\beta}}{\hat{\alpha}} \right)^{1/p} \| x_0 \| e^{-\frac{\varepsilon}{p}(t-t_0)}, \quad t \geq t_0, \quad (12.30)$$

establishing exponential stability of the equilibrium point $x = 0$ of $\mathcal{G}$. The global results follow using standard arguments. $\qquad \square$

Next, we present a generalization of the Krasovskii-LaSalle invariance principle to strong left-continuous dynamical systems. This result is predicated on the generalized positive limit set theorem (Theorem 2.2) for systems with left-continuous flows satisfying the quasi-continuous dependence property given in Axiom *vii*). For the statement of the next result define the γ-level set

$$V^{-1}(\gamma) \triangleq \{x \in \mathcal{D}_c : V(x) = \gamma\},$$

where $\gamma \in \mathbb{R}$, $\mathcal{D}_c \subseteq \mathcal{D}$, and $V : \mathcal{D}_c \to \mathbb{R}$ is a continuous function, and let $\mathcal{M}_\gamma$ denote the largest invariant set (with respect to the strong left-continuous dynamical system $\mathcal{G}$) contained in $V^{-1}(\gamma)$.

Theorem 12.2 *Let $s(t, 0, x_0, 0)$, $t \geq 0$, denote a trajectory of the undisturbed strong left-continuous dynamical system $\mathcal{G}$ and let $\mathcal{D}_c \subset \mathcal{D}$ be a compact positively invariant set with respect to $\mathcal{G}$. Assume there exists a continuous function $V : \mathcal{D}_c \to \mathbb{R}$ such that $V(s(t, 0, x_0, 0)) \leq V(s(\tau, 0, x_0, 0))$, $0 \leq \tau \leq t$, for all $x_0 \in \mathcal{D}_c$. If $x_0 \in \mathcal{D}_c$, then $s(t, 0, x_0, 0) \to \mathcal{M} \triangleq \cup_{\gamma \in \mathbb{R}} \mathcal{M}_\gamma$ as $t \to \infty$. If, in addition, $0 \in \overset{\circ}{\mathcal{D}}_c$, $V(0) = 0$, $V(x) > 0$, $x \in \mathcal{D}_c$, $x \neq 0$, and for every $x_0 \in \mathcal{D}_c$ there exists an unbounded infinite sequence $\{\tau_n\}_{n=1}^\infty$ such that $V(s(\tau_{n+1}, 0, x_0, 0)) < V(s(\tau_n, 0, x_0, 0))$, $n = 1, 2, \ldots$, then the origin is an asymptotically stable equilibrium point of the undisturbed strong left-continuous dynamical system $\mathcal{G}$.*

Proof. Since $V(\cdot)$ is continuous on the compact set $\mathcal{D}_c$, there exists $\beta \in \mathbb{R}$ such that $V(x) \geq \beta$, $x \in \mathcal{D}_c$. Hence, since $V(s(t, 0, x_0, 0))$, $t \geq 0$, is nonincreasing, $\gamma_{x_0} \triangleq \lim_{t \to \infty} V(s(t, 0, x_0, 0))$, $x_0 \in \mathcal{D}_c$, exists. Now, for every $y \in \omega(x_0)$ there exists an increasing unbounded sequence $\{t_n\}_{n=0}^\infty$ with $t_0 = 0$, such that $s(t_n, 0, x_0, 0) \to y$ as $n \to \infty$, and, since $V(\cdot)$ is continuous, it follows that $V(y) = V(\lim_{n \to \infty} s(t_n, 0, x_0, 0)) = \lim_{n \to \infty} V(s(t_n, 0, x_0, 0)) = \gamma_{x_0}$. Hence, $y \in V^{-1}(\gamma_{x_0})$ for all $y \in \omega(x_0)$, or, equivalently, $\omega(x_0) \subseteq V^{-1}(\gamma_{x_0})$. Now, since $\mathcal{D}_c$ is compact and positively invariant, it follows that $s(t, 0, x_0, 0)$, $t \geq 0$, is bounded for all $x_0 \in \mathcal{D}_c$, and hence, it follows from Theorem 2.2 that $\omega(x_0)$ is a nonempty, compact invariant set. Thus, $\omega(x_0)$ is a subset of the largest invariant set contained in $V^{-1}(\gamma_{x_0})$, that is, $\omega(x_0) \subseteq \mathcal{M}_{\gamma_{x_0}}$. Hence, for all $x_0 \in \mathcal{D}_c$, $\omega(x_0) \subseteq \mathcal{M}$. Since $s(t, 0, x_0, 0) \to \omega(x_0)$ as $t \to \infty$, it follows that $s(t, 0, x_0, 0) \to \mathcal{M}$ as $t \to \infty$.

Finally, if $V(0) = 0$, $V(x) > 0$, $x \in \mathcal{D}_c$, $x \neq 0$, and for every $x_0 \in \mathcal{D}_c$ there exists an unbounded sequence $\{\tau_n\}_{n=1}^\infty$, with $\tau_1 = 0$, such that $V(s(\tau_{n+1}, 0, x_0, 0)) < V(s(\tau_n, 0, x_0, 0))$, $n = 1, 2, \ldots$, then $V(s(t, 0, x_0, 0))$ is a nonincreasing function of time, and hence, there exists $\gamma_{x_0} \geq 0$ such that $\lim_{t \to \infty} V(s(t, 0, x_0, 0)) = \gamma_{x_0}$. Now, suppose *ad absurdum,* $\gamma_{x_0} > 0$. Since $\mathcal{D}_c$ is a compact positively invariant set with respect to $\mathcal{G}$, it follows that $s(t, 0, x_0, 0)$ is bounded for all $x_0 \in \mathcal{D}_c$, and hence, it follows from Theorem 2.2 that $\omega(x_0)$ is a nonempty, compact invariant set. Thus, $\omega(x_0) \subseteq \mathcal{M}_{\gamma_{x_0}}$. Since for every $x_0 \in \mathcal{D}_c$, $V(s(\tau_{n+1}, 0, x_0, 0)) < V(s(\tau_n, 0, x_0, 0))$, $n = 1, 2, \ldots$, it follows that $\mathcal{M}_{\gamma_{x_0}}$ is not an invariant set which is a contradiction. Hence, $\gamma_{x_0} = 0$ and, since $V(\cdot)$ is continuous and $V(0) = 0$, it follows that $\mathcal{M}$ contains no invariant set other than the set $\{0\}$, and hence, the origin is an asymptotically stable equilibrium point of the undisturbed strong left-continuous dynamical system $\mathcal{G}$. $\qquad\qquad\square$

12.5 Dissipative Left-Continuous Dynamical Systems: Input-Output and State Properties

In this section, we extend the notion of dissipative dynamical systems to develop the concept of dissipativity for left-continuous dynamical systems. The presentation here very closely parallels the presentation given in Section 3.2, and hence, the comments are brief and many of the proofs are omitted. We begin by considering the left-continuous dynamical system $\mathcal{G}$ with input $u = (u_c, u_d)$ and output $y = (y_c, y_d)$. Recall that a function $(s_c(u_c, y_c), s_d(u_d, y_d))$, where $s_c : U_c \times Y_c \to \mathbb{R}$ and $s_d : U_d \times Y_d \to \mathbb{R}$ are such that $s_c(0,0) = 0$ and $s_d(0,0) = 0$, is a hybrid supply rate[3] if $s_c(u_c, y_c)$ is locally integrable, that is, for all input-output pairs $u_c(t) \in U_c$, $y_c(t) \in Y_c$, $s_c(\cdot, \cdot)$ satisfies $\int_t^{\hat{t}} |s_c(u_c(s), y_c(s))| \, ds < \infty, t, \hat{t} \geq 0$. Note that since all input-output pairs $u_d(t_k) \in U_d$, $y_d(t_k) \in Y_d$, are defined for the resetting events $t_k \in \mathcal{T}^c$, $s_d(\cdot, \cdot)$ satisfies $\sum_{k \in \mathbb{Z}_{[t,\hat{t})}} |s_d(u_d(t_k), y_d(t_k))| < \infty$, where $k \in \mathbb{Z}_{[t,\hat{t})} \triangleq \{k : t \leq t_k < \hat{t}\}$. For the remainder of this chapter, we use the notation $s(t, t_0, x_0, u)$, $t \geq t_0$, and $x(t)$, $t \geq t_0$, interchangeably to denote the trajectory of $\mathcal{G}$ with initial time t_0, initial condition x_0, and input u.

Definition 12.5 *A left-continuous dynamical system $\mathcal{G}$ is* dissipative *with respect to the hybrid supply rate (s_c, s_d) if the* dissipation in-

---

[3]More generally, the hybrid supply rate (s_c, s_d) can be an explicit function of time, that is, $s_c : [0, \infty) \times U_c \times Y_c \to \mathbb{R}$ and $s_d : [0, \infty) \times U_d \times Y_d \to \mathbb{R}$.

equality

$$0 \leq \int_{t_0}^{T} s_c(u_c(t), y_c(t)) \mathrm{d}t + \sum_{k \in \mathbb{Z}_{[t_0, T)}} s_d(u_d(t_k), y_d(t_k)), \quad T \geq t_0,$$

(12.31)

is satisfied for all $T \geq t_0$ and all $(u_c(\cdot), u_d(\cdot)) \in \mathcal{U}_c \times \mathcal{U}_d$ with $x_0 = 0$. A left-continuous dynamical system $\mathcal{G}$ is exponentially dissipative with respect to the hybrid supply rate (s_c, s_d) if there exists a constant $\varepsilon > 0$, such that the dissipation inequality (12.31) is satisfied, with $s_c(u_c(t), y_c(t))$ replaced by $e^{\varepsilon t} s_c(u_c(t), y_c(t))$ and $s_d(u_d(t_k), y_d(t_k))$ replaced by $e^{\varepsilon t_k} s_d(u_d(t_k), y_d(t_k))$, for all $T \geq t_0$ and $(u_c(\cdot), u_d(\cdot)) \in \mathcal{U}_c \times \mathcal{U}_d$ with $x_0 = 0$. A left-continuous dynamical system is lossless with respect to the hybrid supply rate (s_c, s_d) if $\mathcal{G}$ dissipative with respect to the hybrid supply rate (s_c, s_d) and the dissipation inequality (12.31) is satisfied as an equality for all $T \geq t_0$ and $(u_c(\cdot), u_d(\cdot)) \in \mathcal{U}_c \times \mathcal{U}_d$ with $x_0 = s(T, t_0, x_0, u) = 0$.

Next, define the *available storage* $V_a(t_0, x_0)$ of the left-continuous dynamical system $\mathcal{G}$ by

$$V_a(t_0, x_0) \triangleq - \inf_{(u_c(\cdot), u_d(\cdot)), T \geq t_0} \left[\int_{t_0}^{T} s_c(u_c(t), y_c(t)) \mathrm{d}t \right.$$

$$\left. + \sum_{k \in \mathbb{Z}_{[t_0, T)}} s_d(u_d(t_k), y_d(t_k)) \right].$$

(12.32)

Note that $V_a(t_0, x_0) \geq 0$ for all $(t_0, x_0) \in \mathbb{R} \times \mathcal{D}$ since $V_a(t_0, x_0)$ is the supremum over a set of numbers containing the zero element ($T = t_0$). It follows from (12.32) that the available storage of a left-continuous dynamical system $\mathcal{G}$ is the maximum amount of generalized stored energy which can be extracted from $\mathcal{G}$ at any time T. Furthermore, define the *available exponential storage* of the left-continuous dynamical system $\mathcal{G}$ by

$$V_a(t_0, x_0) \triangleq - \inf_{(u_c(\cdot), u_d(\cdot)), T \geq t_0} \left[\int_{t_0}^{T} e^{\varepsilon t} s_c(u_c(t), y_c(t)) \mathrm{d}t \right.$$

$$\left. + \sum_{k \in \mathbb{Z}_{[t_0, T)}} e^{\varepsilon t_k} s_d(u_d(t_k), y_d(t_k)) \right],$$

(12.33)

where $\varepsilon > 0$.

Definition 12.6 *Consider the left-continuous dynamical system* $\mathcal{G}$ *with input* $u = (u_c, u_d)$, *output* $y = (y_c, y_d)$, *and hybrid supply rate* (s_c, s_d). *A continuous nonnegative-definite function* $V_s : \mathbb{R} \times \mathcal{D} \to \mathbb{R}$ *satisfying* $V_s(t, 0) = 0$, $t \in \mathbb{R}$, *and*

$$V_s(T, x(T)) \le V_s(t_0, x_0) + \int_{t_0}^{T} s_c(u_c(t), y_c(t))\mathrm{d}t$$

$$+ \sum_{k \in \mathbb{Z}_{[t_0, T)}} s_d(u_d(t_k), y_d(t_k)), \qquad (12.34)$$

where $x(T) = s(T, t_0, x_0, u)$, $T \ge t_0$, *and* $(u_c(t), u_d(t_k)) \in U_c \times U_d$, *is called a* storage function *for* $\mathcal{G}$. *A continuous nonnegative-definite function* $V_s : \mathbb{R} \times \mathcal{D} \to \mathbb{R}$ *satisfying* $V_s(t, 0) = 0$, $t \in \mathbb{R}$, *and*

$$e^{\varepsilon T} V_s(T, x(T)) \le e^{\varepsilon t_0} V_s(t_0, x_0) + \int_{t_0}^{T} e^{\varepsilon t} s_c(u_c(t), y_c(t))\mathrm{d}t$$

$$+ \sum_{k \in \mathbb{Z}_{[t_0, T)}} e^{\varepsilon t_k} s_d(u_d(t_k), y_d(t_k)), \qquad (12.35)$$

where $\varepsilon > 0$, *is called an* exponential storage function *for* $\mathcal{G}$.

Note that for every $t_0 \in [0, \infty)$, $x_0 \in \mathcal{D}$, and $u \in \mathcal{U}$, $V_s(t, s(t, t_0, x_0, u))$ is left-continuous on $[t_0, \infty)$, and is continuous everywhere on $[t_0, \infty)$ except on $\mathcal{T}^c$.

Definition 12.7 *A left-continuous dynamical system* $\mathcal{G}$ *is* completely reachable *if for all* $(t_0, x_0) \in \mathbb{R} \times \mathcal{D}$, *there exist a finite time* $t_i \le t_0$, *square integrable input* $u_c(t)$ *defined on* $[t_i, t_0]$, *and input* $u_d(t_k)$ *defined on* $k \in \mathbb{Z}_{[t_i, t_0)}$, *such that* $s(t_0, t_i, 0, u) = x_0$.

Theorem 12.3 *Consider the left-continuous dynamical system* $\mathcal{G}$ *with input* $u = (u_c, u_d)$ *and output* $y = (y_c, y_d)$, *and assume that* $\mathcal{G}$ *is completely reachable. Then* $\mathcal{G}$ *is dissipative (respectively, exponentially dissipative) with respect to the hybrid supply rate* (s_c, s_d) *if and only if the available system storage* $V_a(t_0, x_0)$ *given by (12.32) (respectively, the available exponential system storage* $V_a(t_0, x_0)$ *given by (12.33)) is finite for all* $t_0 \in \mathbb{R}$ *and* $x_0 \in \mathcal{D}$ *and* $V_a(t, 0) = 0$, $t \in \mathbb{R}$. *Moreover, if* $V_a(t, 0) = 0$, $t \in \mathbb{R}$, *and* $V_a(t_0, x_0)$ *is finite for all* $t_0 \in \mathbb{R}$ *and* $x_0 \in \mathcal{D}$, *then* $V_a(t, x)$, $(t, x) \in \mathbb{R} \times \mathcal{D}$, *is a storage function (respectively, exponential storage function) for* $\mathcal{G}$. *Finally, all storage functions (respectively, exponential storage functions)* $V_s(t, x)$, $(t, x) \in \mathbb{R} \times \mathcal{D}$, *for* $\mathcal{G}$ *satisfy*

$$0 \le V_a(t, x) \le V_s(t, x), \quad (t, x) \in \mathbb{R} \times \mathcal{D}. \qquad (12.36)$$

Proof. The proof is identical to that of the impulsive dynamical system case given in Theorem 3.1 and, hence, is omitted. □

The following corollary is immediate from Theorem 12.3.

Corollary 12.1 *Consider the left-continuous dynamical system $\mathcal{G}$ and assume that $\mathcal{G}$ is completely reachable. Then $\mathcal{G}$ is dissipative (respectively, exponentially dissipative) with respect to the hybrid supply rate (s_c, s_d) if and only if there exists a continuous storage function (respectively, exponential storage function) $V_s(t, x)$, $(t, x) \in \mathbb{R} \times \mathcal{D}$, satisfying (12.34) (respectively, (12.35)).*

The next result gives necessary and sufficient conditions for dissipativity and exponential dissipativity over an interval $t \in (t_k, t_{k+1}]$ involving the consecutive resetting times $t_k, t_{k+1} \in \mathcal{T}^c$.

Theorem 12.4 *$\mathcal{G}$ is dissipative with respect to the hybrid supply rate (s_c, s_d) if and only if there exists a continuous, nonnegative-definite function $V_s : \mathbb{R} \times \mathcal{D} \to \mathbb{R}$ such that, for all $k \in \overline{\mathbb{Z}}_+$,*

$$V_s(\hat{t}, s(\hat{t}, t_0, x_0, u)) - V_s(t, s(t, t_0, x_0, u)) \leq \int_t^{\hat{t}} s_c(u_c(s), y_c(s)) ds,$$

$$t_k < t \leq \hat{t} \leq t_{k+1}, \quad (12.37)$$

$$V_s(t_k^+, s(t_k^+, t_0, x_0, u)) - V_s(t_k, s(t_k, t_0, x_0, u)) \leq s_d(u_d(t_k), y_d(t_k)). \quad (12.38)$$

Furthermore, $\mathcal{G}$ is exponentially dissipative with respect to the hybrid supply rate (s_c, s_d) if and only if there exist a continuous, nonnegative-definite function $V_s : \mathbb{R} \times \mathcal{D} \to \mathbb{R}$ and a scalar $\varepsilon > 0$ such that

$$e^{\varepsilon \hat{t}} V_s(\hat{t}, s(\hat{t}, t_0, x_0, u)) - e^{\varepsilon t} V_s(t, s(t, t_0, x_0, u))$$

$$\leq \int_t^{\hat{t}} e^{\varepsilon s} s_c(u_c(s), y_c(s)) ds, \quad t_k < t \leq \hat{t} \leq t_{k+1}, \quad (12.39)$$

$$V_s(t_k^+, s(t_k^+, t_0, x_0, u)) - V_s(t_k, s(t_k, t_0, x_0, u)) \leq s_d(u_d(t_k), y_d(t_k)). \quad (12.40)$$

Finally, $\mathcal{G}$ is lossless with respect to the hybrid supply rate (s_c, s_d) if and only if there exists a continuous, nonnegative-definite function $V_s : \mathbb{R} \times \mathcal{D} \to \mathbb{R}$ such that (12.37) and (12.38) are satisfied as equalities.

Proof. Let $k \in \overline{\mathbb{Z}}_+$ and suppose $\mathcal{G}$ is dissipative with respect to the hybrid supply rate (s_c, s_d). Then, there exists a continuous nonnegative-definite function $V_s : \mathbb{R} \times \mathcal{D} \to \mathbb{R}$ such that (12.34) holds. Now, since for $t_k < t \le \hat{t} \le t_{k+1}$, $\mathbb{Z}_{[t,\hat{t})} = \emptyset$, (12.37) is immediate. Next, note that

$$V_s(t_k^+, s(t_k^+, t_0, x_0, u)) - V_s(t_k, s(t_k, t_0, x_0, u))$$

$$\le \int_{t_k}^{t_k^+} s_c(u_c(s), y_c(s))\mathrm{d}s + s_d(u_d(t_k), y_d(t_k)), \qquad (12.41)$$

which, since $\mathbb{Z}_{[t_k, t_k^+)} = \{k\}$, implies (12.38).

Conversely, suppose (12.37) and (12.38) hold, let $\hat{t} \ge t \ge 0$, and let $\mathbb{Z}_{[t,\hat{t})} = \{i, i+1, \ldots, j\}$. (Note that if $\mathbb{Z}_{[t,\hat{t})} = \emptyset$ the converse is a direct consequence of (12.34).) In this case, it follows from (12.37) and (12.38) that

$$
\begin{aligned}
V_s(\hat{t}, x(\hat{t})) &- V_s(t, x(t)) \\
&= V_s(\hat{t}, x(\hat{t})) \\
&\quad - V_s(t_j^+, x(t_j^+)) + V_s(t_j^+, x(t_j^+)) - V_s(t_{j-1}^+, x(t_{j-1}^+)) \\
&\quad + V_s(t_{j-1}^+, x(t_{j-1}^+)) - \cdots - V_s(t_i^+, x(t_i^+)) + V_s(t_i^+, x(t_i^+)) \\
&\quad - V_s(t, x(t)) \\
&\le \int_t^{\hat{t}} s_c(u_c(s), y_c(s))\mathrm{d}s + \sum_{k \in \mathbb{Z}_{[t,\hat{t})}} s_d(u_d(t_k), y_d(t_k)),
\end{aligned}
$$

which implies that $\mathcal{G}$ is dissipative with respect to the hybrid supply rate (s_c, s_d).

Finally, similar constructions show that $\mathcal{G}$ is exponentially dissipative (respectively, lossless) with respect to the hybrid supply rate (s_c, s_d) if and only if (12.39) and (12.40) are satisfied (respectively, (12.37) and (12.38) are satisfied as equalities). $\square$

If in Theorem 12.4 $\dot{V}_s(\cdot, s(\cdot, t_0, x_0, u))$ exists almost everywhere on $[t_0, \infty)$ except the discrete set $\mathcal{T}^c$, then an equivalent statement for dissipativeness of the left-continuous dynamical system $\mathcal{G}$ with respect to the hybrid supply rate (s_c, s_d) is

$$\dot{V}_s(t, s(t, t_0, x_0, u)) \le s_c(u_c(t), y_c(t)), \quad t_k < t \le t_{k+1}, \quad (12.42)$$

$$\Delta V_s(t_k, s(t_k, t_0, x_0, u)) \le s_d(u_d(t_k), y_d(t_k)), \quad k \in \overline{\mathbb{Z}}_+, \qquad (12.43)$$

where $\Delta V_s(t_k, s(t_k, t_0, x_0, u)) \triangleq V_s(t_k^+, s(t_k^+, t_0, x_0, u)) - V_s(t_k, s(t_k, t_0, x_0, u))$, $k \in \overline{\mathbb{Z}}_+$, denotes the difference of the storage function $V_s(t, x)$ at the times t_k, $k \in \overline{\mathbb{Z}}_+$, of the left-continuous dynamical system $\mathcal{G}$. Furthermore, an equivalent statement for exponential dissipativeness of the left-continuous dynamical system $\mathcal{G}$ with respect to the hybrid supply rate (s_c, s_d) is given by

$$\dot{V}_s(t, s(t, t_0, x_0, u)) + \varepsilon V_s(t, s(t, t_0, x_0, u)) \leq s_c(u_c(t), y_c(t)),$$
$$t_k < t \leq t_{k+1}, \qquad (12.44)$$

where $\varepsilon > 0$, and (12.43).

The following theorem provides sufficient conditions for guaranteeing that all storage functions (respectively, exponential storage functions) of a given dissipative (respectively, exponentially dissipative) left-continuous dynamical system are positive definite. For this result we need the following definition.

Definition 12.8 *A left-continuous dynamical system $\mathcal{G}$ with input $u = (u_c, u_d)$ and output $y = (y_c, y_d)$ is* zero-state observable *if $(u_c(t), u_d(t_k)) \equiv (0, 0)$ and $(y_c(t), y_d(t_k)) \equiv (0, 0)$ implies $s(t, t_0, x_0, u) \equiv 0$. Furthermore, a left-continuous dynamical system $\mathcal{G}$ is* minimal *if it is zero-state observable and completely reachable.*

Theorem 12.5 *Consider the left-continuous dynamical system $\mathcal{G}$ and assume that $\mathcal{G}$ is completely reachable and zero-state observable. Furthermore, assume that $\mathcal{G}$ is dissipative (respectively, exponentially dissipative) with respect to the hybrid supply rate (s_c, s_d) and there exist functions $\kappa_c : Y_c \to U_c$ and $\kappa_d : Y_d \to U_d$ such that $\kappa_c(0) = 0$, $\kappa_d(0) = 0$, $s_c(\kappa_c(y_c), y_c) < 0$, $y_c \neq 0$, and $s_d(\kappa_d(y_d), y_d) < 0$, $y_d \neq 0$. Then all the storage functions (respectively, exponential storage functions) $V_s(t, x)$, $(t, x) \in \mathbb{R} \times \mathcal{D}$, for $\mathcal{G}$ are positive definite, that is, $V_s(\cdot, 0) = 0$ and $V_s(t, x) > 0$, $(t, x) \in \mathbb{R} \times \mathcal{D}$, $x \neq 0$.*

Proof. The proof is identical to that of the impulsive dynamical system case given in Theorem 3.3 and, hence, is omitted. $\square$

Next, we introduce the concept of required supply of a left-continuous dynamical system $\mathcal{G}$. Specifically, define the *required supply* $V_r(t_0, x_0)$ of the left-continuous dynamical system $\mathcal{G}$ by

$$V_r(t_0, x_0) \triangleq \inf_{(u_c(\cdot), u_d(\cdot)),\, T \leq t_0} \left[\int_T^{t_0} s_c(u_c(t), y_c(t)) dt \right.$$
$$\left. + \sum_{k \in \mathbb{Z}_{[T, t_0)}} s_d(u_d(t_k), y_d(t_k)) \right], \qquad (12.45)$$

where $T \leq t_0$ and $u \in \mathcal{U}$ are such that $s(t_0, T, 0, u) = x_0$. Note that $V_r(t_0, 0) = 0$, $t_0 \in \mathbb{R}$. It follows from (12.45) that the required supply of a left-continuous dynamical system is the minimum amount of generalized energy which can be delivered to the left-continuous dynamical system in order to transfer it from a zero initial state to a given state x_0. Similarly, define the *required exponential supply* of the left-continuous dynamical system $\mathcal{G}$ by

$$V_r(t_0, x_0) \triangleq \inf_{(u_c(\cdot), u_d(\cdot)), \, T \leq t_0} \left[\int_T^{t_0} e^{\varepsilon t} s_c(u_c(t), y_c(t)) dt \right.$$

$$\left. + \sum_{k \in \mathbb{Z}_{[T, t_0)}} e^{\varepsilon t_k} s_d(u_d(t_k), y_d(t_k)) \right], \qquad (12.46)$$

where $\varepsilon > 0$, and $T \leq t_0$ and $u \in \mathcal{U}$ are such that $s(t_0, T, 0, u) = x_0$.

Next, using the notion of required supply, we show that all storage functions for a left-continuous dynamical system are bounded from above by the required supply and bounded from below by the available storage.

Theorem 12.6 *Consider the left-continuous dynamical system $\mathcal{G}$ and assume that $\mathcal{G}$ is completely reachable. Then $\mathcal{G}$ is dissipative (respectively, exponentially dissipative) with respect to the hybrid supply rate (s_c, s_d) if and only if $0 \leq V_r(t, x) < \infty$, $t \in \mathbb{R}$, $x \in \mathcal{D}$. Moreover, if $V_r(t, x)$ is finite and nonnegative for all $(t_0, x_0) \in \mathbb{R} \times \mathcal{D}$, then $V_r(t, x)$, $(t, x) \in \mathbb{R} \times \mathcal{D}$, is a storage function (respectively, exponential storage function) for $\mathcal{G}$. Finally, all storage functions (respectively, exponential storage functions) $V_s(t, x)$, $(t, x) \in \mathbb{R} \times \mathcal{D}$, for $\mathcal{G}$ satisfy*

$$0 \leq V_a(t, x) \leq V_s(t, x) \leq V_r(t, x) < \infty, \quad (t, x) \in \mathbb{R} \times \mathcal{D}. \quad (12.47)$$

Proof. The proof is identical to that of the impulsive dynamical system case given in Theorem 3.4 and, hence, is omitted. $\square$

Theorem 12.7 *Consider the left-continuous dynamical system $\mathcal{G}$ and assume $\mathcal{G}$ is completely reachable to and from the origin. Then $\mathcal{G}$ is lossless with respect to the hybrid supply rate (s_c, s_d) if and only if there exists a continuous storage function $V_s(t, x)$, $(t, x) \in \mathbb{R} \times \mathcal{D}$, satisfying (12.34) as an equality. Furthermore, if $\mathcal{G}$ is lossless with respect to the hybrid supply rate (s_c, s_d), then $V_a(t, x) = V_r(t, x)$, and hence, the storage function $V_s(t, x)$, $(t, x) \in \mathbb{R} \times \mathcal{D}$, is unique and is* *given by*

$$V_s(t_0, x_0) = -\int_{t_0}^{T_+} s_c(u_c(t), y_c(t)) dt - \sum_{k \in \mathbb{Z}_{[t_0, T_+)}} s_d(u_d(t_k), y_d(t_k))$$

$$= \int_{-T_-}^{t_0} s_c(u_c(t), y_c(t)) dt + \sum_{k \in \mathbb{Z}_{[-T_-, t_0)}} s_d(u_d(t_k), y_d(t_k))$$

$$(12.48)$$

with $x(t_0) = x_0$, $x_0 \in \mathcal{D}$, *and* $(u_c(\cdot), u_d(\cdot)) \in \mathcal{U}_c \times \mathcal{U}_d$, *for any* $T_+ > t_0$
and $T_- > -t_0$ *such that* $s(-T_-, t_0, x_0, u) = 0$ *and* $s(T_+, t_0, x_0, u) = 0$.

Proof. The proof is identical to that of the impulsive dynamical
system case given in Theorem 3.5 and, hence, is omitted. $\qquad\square$

Finally, we provide two definitions of left-continuous dynamical systems which are dissipative (respectively, exponentially dissipative)
with respect to hybrid supply rates of a specific form.

Definition 12.9 *A left-continuous dynamical system* $\mathcal{G}$ *with input*
$u = (u_c, u_d)$, *output* $y = (y_c, y_d)$, $m_c = l_c$, *and* $m_d = l_d$ *is pas-*
sive (respectively, exponentially passive) if $\mathcal{G}$ *is dissipative (respec-*
tively, exponentially dissipative) with respect to the hybrid supply rate
$(s_c(u_c, y_c), s_d(u_d, y_d)) = (2u_c^{\mathrm{T}} y_c, 2u_d^{\mathrm{T}} y_d)$.

Definition 12.10 *A left-continuous dynamical system* $\mathcal{G}$ *with input*
$u = (u_c, u_d)$ *and output* $y = (y_c, y_d)$ *is nonexpansive (respectively,*
exponentially nonexpansive) if $\mathcal{G}$ *is dissipative (respectively, exponen-*
tially dissipative) with respect to the hybrid supply rate $(s_c(u_c, y_c),$
$s_d(u_d, y_d)) = (\gamma_c^2 u_c^{\mathrm{T}} u_c - y_c^{\mathrm{T}} y_c, \gamma_d^2 u_d^{\mathrm{T}} u_d - y_d^{\mathrm{T}} y_d)$, *where* γ_c *and* $\gamma_d > 0$
are given.

In light of the above definitions, the following result is immediate.

Proposition 12.3 *Consider the left-continuous dynamical system* $\mathcal{G}$
with input $u = (u_c, u_d)$, *output* $y = (y_c, y_d)$, *storage function* $V_s(\cdot, \cdot)$,
and hybrid supply rate (s_c, s_d). *Then the following statements hold:*

   i) If $\mathcal{G}$ *is dissipative,* $s_c(0, y_c) \leq 0$, $y_c \in Y_c$, $s_d(0, y_d) \leq 0$, $y_d \in Y_d$,
   and $V_s(\cdot, \cdot)$ *satisfies (12.15), then the equilibrium point* $x = 0$ *of*
   the undisturbed system $\mathcal{G}$ *is Lyapunov stable.*

ii) *If $\mathcal{G}$ is exponentially dissipative, $s_c(0, y_c) \leq 0$, $y_c \in Y_c$, $s_d(0, y_d)$ ≤ 0, $y_d \in Y_d$, and $V_s(\cdot, \cdot)$ satisfies (12.15), then the equilibrium point $x = 0$ of the undisturbed system $\mathcal{G}$ is asymptotically stable. If, in addition, $V_s(\cdot, \cdot)$ satisfies (12.18), then the equilibrium point $x = 0$ of the undisturbed system $\mathcal{G}$ is exponentially stable.*

iii) *If $\mathcal{G}$ is passive (respectively, nonexpansive), then the equilibrium point $x = 0$ of the undisturbed system $\mathcal{G}$ is Lyapunov stable.*

iv) *If $\mathcal{G}$ is exponentially passive (respectively, exponentially nonexpansive) and $V_s(\cdot, \cdot)$ satisfies (12.15), then the equilibrium point $x = 0$ of the undisturbed system $\mathcal{G}$ is asymptotically stable. If, in addition, $V_s(\cdot, \cdot)$ satisfies (12.18), then the equilibrium point $x = 0$ of the undisturbed system $\mathcal{G}$ is exponentially stable.*

v) *If $\mathcal{G}$ is a strong left-continuous dynamical system, zero-state observable, and nonexpansive, then the equilibrium point $x = 0$ of the undisturbed system $\mathcal{G}$ is asymptotically stable.*

Proof. The result is a direct consequence of Theorems 12.4, 12.1, and 12.2 using standard arguments.   □

12.6 Interconnections of Dissipative Left-Continuous Dynamical Systems

In this section, we consider interconnections of dissipative left-continuous dynamical systems. Specifically, consider a finite collection of left-continuous dynamical systems $\mathcal{G}_\alpha = (\mathcal{D}_\alpha, \mathcal{U}_\alpha, U_\alpha, \mathcal{Y}_\alpha, Y_\alpha, s_\alpha, h_{c\alpha}, h_{d\alpha})$, where α spans over a finite index set $\mathcal{A}$,[4] and consider the spaces $\tilde{\mathcal{U}}$, $\tilde{U}$, $\tilde{\mathcal{Y}}$, and $\tilde{Y}$. Here, the elements of $\mathcal{U}_\alpha$ and $\mathcal{Y}_\alpha$ are internal inputs and outputs, respectively, while the elements of $\tilde{\mathcal{U}}$ and $\tilde{\mathcal{Y}}$ are external inputs and outputs, respectively. Next, we introduce an interconnection function $\mathcal{I} : \tilde{U} \times \Pi_{\alpha \in \mathcal{A}} Y_\alpha \to \tilde{Y} \times \Pi_{\alpha \in \mathcal{A}} U_\alpha$, where $\Pi_{\alpha \in \mathcal{A}}$ denotes the Cartesian set product. Figure 12.1 illustrates the concept of a finite collection of left-continuous dynamical subsystems $\mathcal{G}_\alpha$ interconnected through the interconnection constraint $\mathcal{I}$ yielding an interconnected system $\tilde{\mathcal{G}} = \Pi_{\alpha \in \mathcal{A}} \mathcal{G}_\alpha / \mathcal{I}$. The following definition provides well-posedness conditions for the interconnected system $\tilde{\mathcal{G}}$ to qualify as a left-continuous dynamical system.

---

[4]More generally, countably infinite sets with an appropriate measure on $\mathcal{A}$ can also be considered.

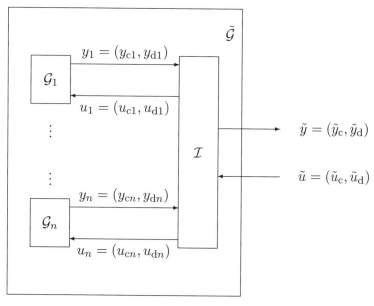

Figure 12.1 System interconnection $\tilde{\mathcal{G}} = \Pi_{\alpha \in \mathcal{A}} \mathcal{G}_\alpha \backslash \mathcal{I}$.

Definition 12.11 *The left-continuous dynamical system* $\tilde{\mathcal{G}} = (\Pi_{\alpha \in \mathcal{A}}$
$\mathcal{D}_\alpha, \tilde{\mathcal{U}}, \tilde{U}, \tilde{\mathcal{Y}}, \tilde{Y}, \Pi_{\alpha \in \mathcal{A}} s_\alpha, \tilde{h}_c, \tilde{h}_d)$ *is an* interconnection *of the left-conti-*
nuous dynamical systems $\mathcal{G}_\alpha = (\mathcal{D}_\alpha, \mathcal{U}_\alpha, U_\alpha, \mathcal{Y}_\alpha, Y_\alpha, s_\alpha, h_{c\alpha}, h_{d\alpha}), \alpha \in$
$\mathcal{A}$, *through the interconnection constraint* $\mathcal{I}$ *if for every* $x_\alpha \in \mathcal{D}_\alpha, \tilde{u} \in$
$\tilde{\mathcal{U}}$, *and* $t \in [t_0, \infty)$, *there exist unique maps* $\psi_\alpha : [0, \infty) \times \Pi_{\alpha \in \mathcal{A}} \mathcal{D}_\alpha \times$
$\tilde{U} \rightarrow U_\alpha$, $(\tilde{h}_c, \tilde{h}_d) : [0, \infty) \times \Pi_{\alpha \in \mathcal{A}} \mathcal{D}_\alpha \times \tilde{U} \rightarrow \tilde{Y}$, *and* $s_\alpha : [0, \infty) \times$
$[0, \infty) \times \Pi_{\alpha \in \mathcal{A}} \mathcal{D}_\alpha \times \tilde{\mathcal{U}} \rightarrow \mathcal{D}_\alpha$, *such that* $u_\alpha(t) = \psi_\alpha(t, s_\alpha(t, t_0, x_\alpha, \tilde{u}), \tilde{u})$
and $\Pi_{\alpha \in \mathcal{A}} s_\alpha$ *satisfies Axioms i)–iv).*

A straightforward but key property of a left-continuous interconnected dynamical system is that if the component subsystems are
dissipative and the interconnection constraint does not introduce any
new supply or dissipation, then the interconnected system is dissipative. Hence, the following result is immediate. For this result let
$\tilde{s}_c : \tilde{U}_c \times \tilde{Y}_c \rightarrow \mathbb{R}$ and $\tilde{s}_d : \tilde{U}_d \times \tilde{Y}_d \rightarrow \mathbb{R}$ be given.

Proposition 12.4 *Let* $\mathcal{G}_\alpha$, $\alpha \in \mathcal{A}$, *be a finite collection of left-conti-*
nuous dissipative dynamical systems with hybrid supply rates $s_\alpha =$
$(s_{c\alpha}(u_{c\alpha}, y_{c\alpha}), s_{d\alpha}(u_{d\alpha}, y_{d\alpha}))$, *where* $s_{c\alpha} : U_{c\alpha} \times Y_{c\alpha} \rightarrow \mathbb{R}$ *and* $s_{d\alpha} :$
$U_{d\alpha} \times Y_{d\alpha} \rightarrow \mathbb{R}$, *and storage functions* $V_{s\alpha}(\cdot, \cdot)$. *Furthermore, let the*
interconnection constraint $\mathcal{I} : \tilde{U} \times \Pi_{\alpha \in \mathcal{A}} Y_\alpha \rightarrow \tilde{Y} \times \Pi_{\alpha \in \mathcal{A}} U_\alpha$ *be such that*
$\tilde{s}_c = \sum_{\alpha \in \mathcal{A}} s_{c\alpha}$ *and* $\tilde{s}_d = \sum_{\alpha \in \mathcal{A}} s_{d\alpha}$. *Then the interconnected system*
$\tilde{\mathcal{G}} = \Pi_{\alpha \in \mathcal{A}} \mathcal{G}_\alpha / \mathcal{I}$ *is dissipative with respect to the hybrid supply rate*

$(\tilde{s}_c, \tilde{s}_d) = (\sum_{\alpha \in \mathcal{A}} s_{c\alpha}, \sum_{\alpha \in \mathcal{A}} s_{d\alpha})$ *and has a storage function* $V_s(\cdot, \cdot) = \sum_{\alpha \in \mathcal{A}} V_{s\alpha}(\cdot, \cdot)$.

Proof. The result is a direct consequence of Theorem 12.4 by summing both sides of the inequalities

$$V_{s\alpha}(\hat{t}, x_\alpha(\hat{t})) - V_{s\alpha}(t, x_\alpha(t)) \leq \int_t^{\hat{t}} s_{c\alpha}(u_{c\alpha}(s), y_{c\alpha}(s)) ds,$$

$$t_k < t \leq \hat{t} \leq t_{k+1}, \quad (12.49)$$

$$V_{s\alpha}(t_k^+, x_\alpha(t_k^+)) - V_{s\alpha}(t_k, x_\alpha(t_k)) \leq s_{d\alpha}(u_d(t_k), y_d(t_k)), \quad (12.50)$$

for $k \in \overline{\mathbb{Z}}_+$, over $\alpha \in \mathcal{A}$, and using the assumptions $s_c = \sum_{\alpha \in \mathcal{A}} s_{c\alpha}$ and $s_d = \sum_{\alpha \in \mathcal{A}} s_{d\alpha}$. $\qquad \square$

The following corollary is a direct consequence of Proposition 12.4.

Corollary 12.2 *Consider the left-continuous dynamical systems* $\mathcal{G}_1$ *and* $\mathcal{G}_2$ *with input-output pairs* $(u_{c1}, u_{d1}; y_{c1}, y_{d1})$ *and* $(u_{c2}, u_{d2}, ; y_{c2}, y_{d2})$, *respectively. Then the following statements hold:*

i) If $\mathcal{G}_1$ *and* $\mathcal{G}_2$ *are passive, then the parallel interconnection of* $\mathcal{G}_1$ *and* $\mathcal{G}_2$ *is passive.*

ii) If $\mathcal{G}_1$ *and* $\mathcal{G}_2$ *are passive, then the negative feedback interconnection of* $\mathcal{G}_1$ *and* $\mathcal{G}_2$ *is passive.*

iii) If $\mathcal{G}_1$ *and* $\mathcal{G}_2$ *are nonexpansive with gains* $(\gamma_{c1}, \gamma_{d1})$ *and* $(\gamma_{c2}, \gamma_{d2})$, *respectively, then the cascade interconnection of* $\mathcal{G}_1$ *and* $\mathcal{G}_2$ *is nonexpansive with gain* $(\gamma_{c1}\gamma_{c2}, \gamma_{d1}\gamma_{d2})$.

Proof. The result is a direct consequence of Proposition 12.4 by noting the interconnection constraints for cascade, parallel, and feedback interconnections are given by $(\tilde{u}_c, \tilde{u}_d) = (u_{c1}, u_{d1})$, $(u_{c2}, u_{d2}) = (y_{c1}, y_{d1})$, $(\tilde{y}_c, \tilde{y}_d) = (y_{c2}, y_{d2})$; $(\tilde{u}_c, \tilde{u}_d) = (u_{c1}, u_{d1}) = (u_{c2}, u_{d2})$, $(\tilde{y}_c, \tilde{y}_d) = (y_{c1} + y_{c2}, y_{d1} + y_{d2})$; and $(\tilde{u}_c, \tilde{u}_d) = (u_{c1} + y_{c2}, u_{d1} + y_{d2})$, $(\tilde{y}_c, \tilde{y}_d) = (y_{c1}, y_{d1}) = (u_{c2}, u_{d2})$, respectively. Now, the result is immediate by noting that the above interconnection constraints satisfy the required constraints on $\tilde{s}_c$ and $\tilde{s}_d$ in Proposition 12.4. $\qquad \square$

Next, we consider stability of feedback interconnections of dissipative left-continuous dynamical systems. Specifically, using the notion of dissipative and exponentially dissipative left-continuous dynamical systems, with appropriate storage functions and hybrid supply rates,

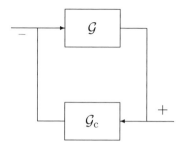

Figure 12.2 Feedback interconnection of $\mathcal{G}$ and $\mathcal{G}_c$.

we construct Lyapunov functions for interconnected left-continuous dynamical systems by appropriately combining storage functions for each subsystem. Here, for simplicity of exposition, we restrict our attention to stationary left-continuous dynamical systems. Furthermore, we assume that for the dynamical system $\mathcal{G}$, $\mathcal{D} = \mathbb{R}^n$, $U_c = \mathbb{R}^{m_c}$, $U_d = \mathbb{R}^{m_d}$, $Y_c = \mathbb{R}^{l_c}$, and $Y_d = \mathbb{R}^{l_d}$.

We begin by considering the negative feedback interconnection of the stationary left-continuous dynamical system $\mathcal{G}$ with the stationary left-continuous feedback system $\mathcal{G}_c$ given by the octuple $(\mathbb{R}^{n_c}, \hat{\mathcal{U}}, \mathbb{R}^{m_{cc}} \times \mathbb{R}^{m_{dc}}, \hat{\mathcal{Y}}, \mathbb{R}^{l_{cc}} \times \mathbb{R}^{l_{dc}}, s_c, h_{cc}, h_{dc})$, where $\hat{\mathcal{U}} \triangleq \mathcal{U}_{cc} \times \mathcal{U}_{dc}$, $\hat{\mathcal{Y}} \triangleq \mathcal{Y}_{cc} \times \mathcal{Y}_{dc}$, $s_c : [0, \infty) \times [0, \infty) \times \mathbb{R}^{n_c} \times \hat{\mathcal{U}} \to \mathbb{R}^{n_c}$, $h_{cc} : \mathbb{R}^{n_c} \times \mathbb{R}^{m_{cc}} \to \mathbb{R}^{l_{cc}}$, and $h_{dc} : \mathbb{R}^{n_c} \times \mathbb{R}^{m_{dc}} \to \mathbb{R}^{l_{dc}}$. We refer to $s_c(t, 0, x_{c0}, \hat{u})$, $t \geq 0$, as the trajectory of $\mathcal{G}_c$ corresponding to an initial condition $x_{c0} \in \mathbb{R}^{n_c}$ and input $\hat{u} = (u_{cc}, u_{dc}) \in \hat{\mathcal{U}}$, where $u_{cc} \in \mathcal{U}_{cc}$ and $u_{dc} \in \mathcal{U}_{dc}$. Furthermore, for $\mathcal{G}_c$, let $\mathcal{T}_c^c$ denote the set of resetting times and let $\mathcal{T}_c$ denote the complement of $\mathcal{T}_c^c$, that is, $[t_0, \infty) \backslash \mathcal{T}_c^c$. Note that with the feedback interconnection given by Figure 12.2, $(u_{cc}, u_{dc}) = (y_c, y_d)$ and $(y_{cc}, y_{dc}) = (-u_c, -u_d)$. For the ensuing results, we assume that the negative feedback interconnection of $\mathcal{G}$ and $\mathcal{G}_c$ is well posed, that is, the feedback interconnection generates an undisturbed stationary left-continuous dynamical system on $\mathbb{R}^n \times \mathbb{R}^{n_c}$ with trajectory $\tilde{s}(t, 0, (x_0, x_{c0}), 0) \triangleq (s(t, 0, x_0, u), s_c(t, 0, x_{c0}, y))$ and initial condition $(x_0, x_{c0}) \in \mathbb{R}^n \times \mathbb{R}^{n_c}$.

Theorem 12.8 *Consider the feedback system consisting of the stationary left-continuous dynamical systems $\mathcal{G}$ and $\mathcal{G}_c$ with input-output pairs $(u_c, u_d; y_c, y_d)$ and $(u_{cc}, u_{dc}; y_{cc}, y_{dc})$, respectively, and with $(u_{cc}, u_{dc}) = (y_c, y_d)$ and $(y_{cc}, y_{dc}) = (-u_c, -u_d)$. Assume $\mathcal{G}$ and $\mathcal{G}_c$ are zero-state observable and dissipative with respect to the hybrid supply* *rates* $(s_{\rm c}(u_{\rm c}, y_{\rm c}), s_{\rm d}(u_{\rm d}, y_{\rm d}))$ *and* $(s_{\rm cc}(u_{\rm cc}, y_{\rm cc}), s_{\rm dc}(u_{\rm dc}, y_{\rm dc}))$, *and with continuous positive definite, radially unbounded storage functions* $V_{\rm s}(\cdot)$ *and* $V_{\rm sc}(\cdot)$, *respectively, such that* $V_{\rm s}(0) = 0$ *and* $V_{\rm sc}(0) = 0$. *Furthermore, assume there exists a scalar* $\sigma > 0$ *such that* $s_{\rm c}(u_{\rm c}, y_{\rm c}) + \sigma s_{\rm cc}(u_{\rm cc}, y_{\rm cc}) \le 0$ *and* $s_{\rm d}(u_{\rm d}, y_{\rm d}) + \sigma s_{\rm dc}(u_{\rm dc}, y_{\rm dc}) \le 0$. *Then the following statements hold:*

i) *The negative feedback interconnection of* $\mathcal{G}$ *and* $\mathcal{G}_{\rm c}$ *is Lyapunov stable.*

ii) *If* $\mathcal{G}$ *and* $\mathcal{G}_{\rm c}$ *are exponentially dissipative with respect to the hybrid supply rates* $(s_{\rm c}(u_{\rm c}, y_{\rm c}), s_{\rm d}(u_{\rm d}, y_{\rm d}))$ *and* $(s_{\rm cc}(u_{\rm cc}, y_{\rm cc}), s_{\rm dc}(u_{\rm dc}, y_{\rm dc}))$, *respectively, and* $V_{\rm s}(\cdot)$ *and* $V_{\rm sc}(\cdot)$ *are such that there exist constants* α, $\alpha_{\rm c}$, β, $\beta_{\rm c} > 0$ *such that*

$$\alpha\|x\|^2 \le V_{\rm s}(x) \le \beta\|x\|^2, \quad x \in \mathbb{R}^n, \tag{12.51}$$
$$\alpha_{\rm c}\|x_{\rm c}\|^2 \le V_{\rm sc}(x_{\rm c}) \le \beta_{\rm c}\|x_{\rm c}\|^2, \quad x_{\rm c} \in \mathbb{R}^{n_{\rm c}}, \tag{12.52}$$

then the negative feedback interconnection of $\mathcal{G}$ *and* $\mathcal{G}_{\rm c}$ *is globally exponentially stable.*

Proof. Let $\tilde{\mathcal{T}}^{\rm c} \triangleq \mathcal{T}^{\rm c} \cup \mathcal{T}_{\rm c}^{\rm c}$, $\tilde{\mathcal{T}} \triangleq [t_0, \infty)\backslash\tilde{\mathcal{T}}^{\rm c}$, and $t_k \in \tilde{\mathcal{T}}^{\rm c}$, $k = 1, 2, \ldots$.

i) Consider the Lyapunov function candidate $V(x, x_{\rm c}) = V_{\rm s}(x) + \sigma V_{\rm sc}(x_{\rm c})$. Now, the corresponding Lyapunov left derivative of $V(x, x_{\rm c})$ along the state trajectories $(x(t), x_{\rm c}(t)) = (s(t, 0, x_0, u), s_{\rm c}(t, 0, x_{\rm c0}, y))$, $t \in (t_k, t_{k+1}]$, is given by

$$\begin{aligned}
\dot{V}(x(t), x_{\rm c}(t)) &= \dot{V}_{\rm s}(x(t)) + \sigma\dot{V}_{\rm sc}(x_{\rm c}(t)) \\
&\le s_{\rm c}(u_{\rm c}(t), y_{\rm c}(t)) + \sigma s_{\rm cc}(u_{\rm cc}(t), y_{\rm cc}(t)) \\
&\le 0, \quad t \in \tilde{\mathcal{T}},
\end{aligned} \tag{12.53}$$

and the Lyapunov difference of $V(x, x_{\rm c})$ at the resetting times $t_k \in \tilde{\mathcal{T}}^{\rm c}$, $k \in \overline{\mathbb{Z}}_+$, is given by

$$\begin{aligned}
\Delta V(x(t_k), x_{\rm c}(t_k)) &= \Delta V_{\rm s}(x(t_k)) + \sigma\Delta V_{\rm sc}(x_{\rm c}(t_k)) \\
&\le s_{\rm d}(u_{\rm d}(t_k), y_{\rm d}(t_k)) + \sigma s_{\rm dc}(u_{\rm dc}(t_k), y_{\rm dc}(t_k)) \\
&\le 0.
\end{aligned} \tag{12.54}$$

Now, Lyapunov stability of the negative feedback interconnection of $\mathcal{G}$ and $\mathcal{G}_{\rm c}$ follows as a direct consequence of Theorem 12.1.

ii) If $\mathcal{G}$ and $\mathcal{G}_c$ are exponentially dissipative and (12.51) and (12.52) hold, then it follows that

$$\dot{V}(x(t), x_c(t)) = \dot{V}_s(x(t)) + \sigma \dot{V}_{sc}(x_c(t))$$
$$\leq -\varepsilon_c V_s(x(t)) - \sigma \varepsilon_{cc} V_{sc}(x_c(t)) + s_c(u_c(t), y_c(t))$$
$$+ \sigma s_{cc}(u_{cc}(t), y_{cc}(t))$$
$$\leq -\min\{\varepsilon_c, \varepsilon_{cc}\} V(x(t), x_c(t)), \quad t_k < t \leq t_{k+1}, \quad (12.55)$$

and $\Delta V(x(t_k), x_c(t_k))$, $t_k \in \tilde{\mathcal{T}}^c$, $k \in \overline{\mathbb{Z}}_+$, satisfies (12.54). Now, Theorem 12.1 implies that the negative feedback interconnection of $\mathcal{G}$ and $\mathcal{G}_c$ is globally exponentially stable. $\square$

The next result presents Lyapunov, asymptotic, and exponential stability of dissipative feedback systems with quadratic hybrid supply rates.

Theorem 12.9 *Let* $Q_c \in \mathbb{S}^{l_c}$, $S_c \in \mathbb{R}^{l_c \times m_c}$, $R_c \in \mathbb{S}^{m_c}$, $Q_d \in \mathbb{S}^{l_d}$, $S_d \in \mathbb{R}^{l_d \times m_d}$, $R_d \in \mathbb{S}^{m_d}$, $Q_{cc} \in \mathbb{S}^{l_{cc}}$, $S_{cc} \in \mathbb{R}^{l_{cc} \times m_{cc}}$, $R_{cc} \in \mathbb{S}^{m_{cc}}$, $Q_{dc} \in \mathbb{S}^{l_{dc}}$, $S_{dc} \in \mathbb{R}^{l_{dc} \times m_{dc}}$, *and* $R_{dc} \in \mathbb{S}^{m_{dc}}$. *Consider the closed-loop system consisting of the left-continuous dynamical systems* $\mathcal{G}$ *and* $\mathcal{G}_c$, *and assume* $\mathcal{G}$ *and* $\mathcal{G}_c$ *are zero-state observable. Furthermore, assume* $\mathcal{G}$ *is dissipative with respect to the quadratic hybrid supply rate* $(s_c(u_c, y_c), s_d(u_d, y_d)) = (y_c^T Q_c y_c + 2y_c^T S_c u_c + u_c^T R_c u_c, y_d^T Q_d y_d + 2y_d^T S_d u_d + u_d^T R_d u_d)$ *and has a radially unbounded storage function* $V_s(\cdot)$, *and* $\mathcal{G}_c$ *is dissipative with respect to the quadratic hybrid supply rate* $(s_{cc}(u_{cc}, y_{cc}), s_{dc}(u_{dc}, y_{dc})) = (y_{cc}^T Q_{cc} y_{cc} + 2y_{cc}^T S_{cc} u_{cc} + u_{cc}^T R_{cc} u_{cc}, y_{dc}^T Q_{dc} y_{dc} + 2y_{dc}^T S_{dc} u_{dc} + u_{dc}^T R_{dc} u_{dc})$ *and has a radially unbounded storage function* $V_{sc}(\cdot)$. *Finally, assume there exists a scalar* $\sigma > 0$ *such that*

$$\hat{Q}_c \triangleq \begin{bmatrix} Q_c + \sigma R_{cc} & -S_c + \sigma S_{cc}^T \\ -S_c^T + \sigma S_{cc} & R_c + \sigma Q_{cc} \end{bmatrix} \leq 0, \qquad (12.56)$$

$$\hat{Q}_d \triangleq \begin{bmatrix} Q_d + \sigma R_{dc} & -S_d + \sigma S_{dc}^T \\ -S_d^T + \sigma S_{dc} & R_d + \sigma Q_{dc} \end{bmatrix} \leq 0. \qquad (12.57)$$

Then the following statements hold:

i) *The negative feedback interconnection of* $\mathcal{G}$ *and* $\mathcal{G}_c$ *is Lyapunov stable.*

ii) *If* $\mathcal{G}$ *and* $\mathcal{G}_c$ *are exponentially dissipative with respect to the hybrid supply rates* $(s_c(u_c, y_c), s_d(u_d, y_d))$ *and* $(s_{cc}(u_{cc}, y_{cc}), s_{dc}(u_{dc}, y_{dc}))$, *respectively, and there exist constants* α, α_c, β, $\beta_c > 0$

such that (12.51) and (12.52) hold, then the negative feedback interconnection of $\mathcal{G}$ and $\mathcal{G}_c$ is globally exponentially stable.

iii) If $\hat{Q}_c < 0$, $\hat{Q}_d < 0$, and $\mathcal{G}$ and $\mathcal{G}_c$ are strong left-continuous dynamical systems, then the negative feedback interconnection of $\mathcal{G}$ and $\mathcal{G}_c$ is globally asymptotically stable.

Proof. Statements *i)* and *ii)* are a direct consequence of Theorem 12.8 by noting

$$s_c(u_c, y_c) + \sigma s_{cc}(u_{cc}, y_{cc}) = \begin{bmatrix} y_c \\ y_{cc} \end{bmatrix}^T \hat{Q}_c \begin{bmatrix} y_c \\ y_{cc} \end{bmatrix}, \qquad (12.58)$$

$$s_d(u_d, y_d) + \sigma s_{dc}(u_{dc}, y_{dc}) = \begin{bmatrix} y_d \\ y_{dc} \end{bmatrix}^T \hat{Q}_d \begin{bmatrix} y_d \\ y_{dc} \end{bmatrix}, \qquad (12.59)$$

and hence, $s_c(u_c, y_c) + \sigma s_{cc}(u_{cc}, y_{cc}) \leq 0$ and $s_d(u_d, y_d) + \sigma s_{dc}(u_{dc}, y_{dc}) \leq 0$.

To show *iii)* consider the Lyapunov function candidate $V(x, x_c) = V_s(x) + \sigma V_{sc}(x_c)$. Noting that $u_{cc} = y_c$ and $y_{cc} = -u_c$ it follows that the corresponding Lyapunov left derivative of $V(x, x_c)$ along the trajectories $(x(t), x_c(t)) = (s(t, 0, x_0, u), s_c(t, 0, x_{c0}, y))$ satisfies

$$
\begin{aligned}
\dot{V}(x(t), x_c(t)) &= \dot{V}_s(x(t)) + \sigma \dot{V}_{sc}(x_c(t)) \\
&\leq s_c(u_c(t), y_c(t)) + \sigma s_{cc}(u_{cc}(t), y_{cc}(t)) \\
&= y_c^T(t) Q_c y_c(t) + 2 y_c^T(t) S_c u_c(t) + u_c^T(t) R_c u_c(t) \\
&\quad + \sigma [y_{cc}^T(t) Q_{cc} y_{cc}(t) + 2 y_{cc}^T(t) S_{cc} u_{cc}(t) + u_{cc}^T(t) R_{cc} u_{cc}(t)] \\
&= \begin{bmatrix} y_c(t) \\ y_{cc}(t) \end{bmatrix}^T \hat{Q}_c \begin{bmatrix} y_c(t) \\ y_{cc}(t) \end{bmatrix} \\
&\leq 0, \quad t_k < t \leq t_{k+1}, \qquad (12.60)
\end{aligned}
$$

and, similarly, the Lyapunov difference satisfies

$$\Delta V(x(t_k), x_c(t_k)) = \begin{bmatrix} y_d(t_k) \\ y_{dc}(t_k) \end{bmatrix}^T \hat{Q}_d \begin{bmatrix} y_d(t_k) \\ y_{dc}(t_k) \end{bmatrix}$$

$$\leq 0, \quad k \in \overline{\mathbb{Z}}_+, \qquad (12.61)$$

which implies that $V(s(\cdot, 0, x_0, u), s_c(\cdot, 0, x_{c0}, y))$ is a nonincreasing function of time. Hence, it follows from Theorem 12.1 that the negative feedback interconnection of $\mathcal{G}$ and $\mathcal{G}_c$ is Lyapunov stable.

Next, since $\mathcal{G}$ and $\mathcal{G}_c$ are zero-state observable it follows that $V(s(t, 0, x_0, 0), s_c(t, 0, x_{c0}, 0)) = V(s(0, 0, x_0, 0, s_c(0, 0, x_{c0}, 0)), t \geq 0$, if and only if $(s(t, 0, x_0, 0), s_c(t, 0, x_{c0}, 0)) = (0, 0), t \geq 0$. Hence, the largest invariant set $\mathcal{M} \triangleq \cup_{\gamma \in \mathbb{R}} \mathcal{M}_\gamma$ contained in $\mathcal{R}$ is the set $\{(0,0)\}$. Now, it follows from Theorem 12.2 that the negative feedback interconnection of $\mathcal{G}$ and $\mathcal{G}_c$ is globally asymptotically stable. $\qquad \square$

The following result is a direct consequence of Theorem 12.9. For this result note that if a left-continuous dynamical system $\mathcal{G}$ is dissipative (respectively, exponentially dissipative) with respect to a hybrid supply rate $(s_c(u_c, y_c), s_d(u_d, y_d)) = (2u_c^T y_c, 2u_d^T y_d)$, then, with $(\kappa_c(y_c), \kappa_d(y_d)) = (-k_c y_c, -k_d y_d)$, where $k_c, k_d > 0$, it follows that $(s_c(u_c, y_c), s_d(u_d, y_d)) = (-2k_c y_c^T y_c, -2k_d y_d^T y_d) < (0,0)$, $(y_c, y_d) \neq (0,0)$. Alternatively, if a left-continuous dynamical system $\mathcal{G}$ is dissipative (respectively, exponentially dissipative) with respect to a hybrid supply rate $(s_c(u_c, y_c), s_d(u_d, y_d)) = (\gamma_c^2 u_c^T u_c - y_c^T y_c, \gamma_d^2 u_d^T u_d - y_d^T y_d)$, where $\gamma_c, \gamma_d > 0$, then, with $(\kappa_c(y_c), \kappa_d(y_d)) = (0,0)$, it follows that $(s_c(u_c, y_c), s_d(u_d, y_d)) = (-y_c^T y_c, -y_d^T y_d) < (0,0)$, $(y_c, y_d) \neq (0,0)$. Hence, if $\mathcal{G}$ is zero-state observable it follows from Theorem 12.5 that all storage functions of $\mathcal{G}$ are positive definite.

Corollary 12.3 *Consider the closed-loop system consisting of the stationary left-continuous dynamical systems $\mathcal{G}$ and $\mathcal{G}_c$, and assume $\mathcal{G}$ and $\mathcal{G}_c$ are zero-state observable. Then the following statements hold:*

i) *If $\mathcal{G}$ and $\mathcal{G}_c$ are exponentially passive with storage functions $V_s(\cdot)$ and $V_{sc}(\cdot)$, respectively, such that (12.51) and (12.52) hold, then the negative feedback interconnection of $\mathcal{G}$ and $\mathcal{G}_c$ is exponentially stable.*

ii) *If $\mathcal{G}$ and $\mathcal{G}_c$ are exponentially nonexpansive with gains $\gamma_c, \gamma_d > 0$ and $\gamma_{cc}, \gamma_{dc} > 0$, and storage functions $V_s(\cdot)$ and $V_{sc}(\cdot)$, respectively, such that (12.51) and (12.52) hold and $\gamma_c \gamma_{cc} \leq 1$ and $\gamma_d \gamma_{dc} \leq 1$, then the negative feedback interconnection of $\mathcal{G}$ and $\mathcal{G}_c$ is exponentially stable.*

Proof. The proof is a direct consequence of Theorem 12.9. Specifically, statement *i)* follows from Theorem 12.9 with $Q_c = 0$, $Q_d = 0$, $Q_{cc} = 0$, $Q_{dc} = 0$, $S_c = I_{m_c}$, $S_d = I_{m_d}$, $S_{cc} = I_{m_{cc}}$, $S_{dc} = I_{m_{dc}}$, $R_c = 0$, $R_d = 0$, $R_{cc} = 0$, and $R_{dc} = 0$. Statement *ii)* follows from Theorem 12.9 with $Q_c = -I_{l_c}$, $Q_d = -I_{l_d}$, $Q_{cc} = -I_{l_{cc}}$, $Q_{dc} = -I_{l_{dc}}$, $S_c = 0$, $S_d = 0$, $S_{cc} = 0$, $S_{dc} = 0$, and $R_c = \gamma_c^2 I_{m_c}$, $R_d = \gamma_d^2 I_{m_d}$, $R_{cc} = \gamma_{cc}^2 I_{m_{cc}}$, and $R_{dc} = \gamma_{dc}^2 I_{m_{dc}}$. $\qquad \square$

Chapter Thirteen

Poincaré Maps and Stability of Periodic Orbits for Hybrid Dynamical Systems

13.1 Introduction

In Chapter 12 a unified dynamical systems framework for a general class of systems possessing left-continuous flows, that is, left-continuous dynamical systems, was developed. Stability results of left-continuous dynamical systems are also considered in Chapter 12. The extension of the Krasovskii-LaSalle invariant set theorem to hybrid and impulsive dynamical systems presented in Chapter 12 provides a powerful tool in analyzing the stability properties of periodic orbits and limit cycles of dynamical systems with impulse effects. However, the periodic orbit of a left-continuous dynamical system is a disconnected set in the n-dimensional state space making the construction of a Lyapunov-like function satisfying the invariance principle a difficult task for high-order nonlinear systems. In such cases, it becomes necessary to seek alternative tools to study the stability of periodic orbits of hybrid and impulsive dynamical systems, especially if the trajectory of the system can be relatively easily integrated.

In this chapter, we generalize Poincaré's theorem to left-continuous dynamical systems, and hence, to hybrid and impulsive dynamical systems. Specifically, we develop necessary and sufficient conditions for stability of periodic orbits based on the stability properties of a fixed point of a discrete-time dynamical system constructed from a Poincaré return map. As opposed to dynamical systems possessing continuous flows requiring the construction of a hyperplane that is transversal to a candidate periodic trajectory necessary for defining the return map, the resetting set, which provides a criterion for determining when the states of the left-continuous dynamical system are to be reset, provides a natural candidate for the transversal surface on which the Poincaré map of a left-continuous dynamical system can be defined. Hence, the Poincaré return map is defined by a subset of the resetting set that induces a discrete-time mapping from this subset onto the resetting set. This mapping traces the left-continuous trajectory of the left-continuous dynamical system from a point on the resetting set to its next corresponding intersection with the resetting set. In the case of impulsive dynamical systems possessing sufficiently smooth resetting manifolds, we show the Poincaré return map can be used to establish a relationship between the stability properties of an impulsive dynamical system with periodic solutions and the stability properties of an equilibrium point of an $(n-1)$th-order discrete-time system. These results are used to analyze the periodic orbits for a verge and folio clock escapement mechanism [145] which exhibits impulsive dynamics.

13.2 Left-Continuous Dynamical Systems with Periodic Solutions

In this section, we generalize Poincaré's theorem to left-continuous dynamical systems. We begin by specializing Definition 12.3 to undisturbed systems (i.e., $u(t) \equiv 0$). For this definition $\mathcal{D} \subseteq \mathbb{R}^n$ and $\mathcal{T}_{x_0} \subseteq [0, \infty)$, $x_0 \in \mathcal{D}$, is a dense subset of the semi-infinite interval $[0, \infty)$ such that $[0, \infty) \backslash \mathcal{T}_{x_0}$ is (finitely or infinitely) countable.

Definition 13.1 *A strong left-continuous dynamical system* on $\mathcal{D}$ is the triple $(\mathcal{D}, [0, \infty), s)$, where $s : [0, \infty) \times \mathcal{D} \to \mathcal{D}$ is such that the following axioms hold:

   i) *(Left-continuity):* $s(\cdot, x_0)$ is left-continuous in t, that is, $\lim_{\tau \to t^-} s(\tau, x_0) = s(t, x_0)$ for all $x_0 \in \mathcal{D}$ and $t \in (0, \infty)$.

   ii) *(Consistency):* $s(0, x_0) = x_0$ for all $x_0 \in \mathcal{D}$.

   iii) *(Semi-group property):* $s(\tau, s(t, x_0)) = s(t+\tau, x_0)$ for all $x_0 \in \mathcal{D}$ and $t, \tau \in [0, \infty)$.

   iv) *(Quasi-continuous dependence):* For every $x_0 \in \mathcal{D}$, there exists $\mathcal{T}_{x_0} \subseteq [0, \infty)$ such that $[0, \infty) \backslash \mathcal{T}_{x_0}$ is countable and for every $\varepsilon > 0$ and $t \in \mathcal{T}_{x_0}$, there exists $\delta(\varepsilon, x_0, t) > 0$ such that if $\|x_0 - y\| < \delta(\varepsilon, x_0, t)$, $y \in \mathcal{D}$, then $\|s(t, x_0) - s(t, y)\| < \varepsilon$.

As in Chapter 12, we denote the strong left-continuous dynamical system $(\mathcal{D}, [0, \infty), s)$ by $\mathcal{G}$. Furthermore, we refer to $s(t, x_0)$, $t \geq 0$, as the *trajectory* of $\mathcal{G}$ corresponding to $x_0 \in \mathcal{D}$, and for a given trajectory $s(t, x_0)$, $t \geq 0$, we refer to $x_0 \in \mathcal{D}$ as an *initial condition* of $\mathcal{G}$. The trajectory $s(t, x_0)$, $t \geq 0$, of $\mathcal{G}$ is *bounded* if there exists $\gamma > 0$ such that $\|s(t, x_0)\| < \gamma$, $t \geq 0$. The next proposition is a specialization of Proposition 12.1 to the case where $\mathcal{G}$ is undisturbed (i.e., $u(t) \equiv 0$).

Proposition 13.1 *Let the triple* $(\mathcal{D}, [0, \infty), s)$, *where* $s : [0, \infty) \times \mathcal{D} \to \mathcal{D}$, *be such that Axioms i)–iii) hold and*

$iv)'$ *For every* $x_0 \in \mathcal{D}$, $\varepsilon, \eta > 0$, *and* $T \in \mathcal{T}_{x_0}$, *there exists* $\delta(\varepsilon, \eta, x_0, T)$ > 0 *such that if* $\|x_0 - y\| < \delta(\varepsilon, \eta, x_0, T)$, $y \in \mathcal{D}$, *then for every* $t \in \mathcal{T}_{x_0} \cap [0, T]$ *such that* $|t - \tau| > \eta$, *for all* $\tau \in [0, T] \backslash \mathcal{T}_{x_0}$, $\|s(t, x_0) - s(t, y)\| < \varepsilon$. *Furthermore, if* $t \in \mathcal{T}_{x_0}$ *is an accumulation point of* $[0, \infty) \backslash \mathcal{T}_{x_0}$, *then* $s(t, \cdot)$ *is continuous at* x_0.

If Axioms $i) - iv)'$ *hold, then* $\mathcal{G}$ *is a strong left-continuous dynamical system.*

Henceforth, we refer to a strong left-continuous dynamical system as a left-continuous dynamical system wherein Axiom $iv)'$ holds in place of Axiom iv). This minor abuse in terminology considerably simplifies the ensuing presentation. Furthermore, we assume that $\mathcal{T}_{x_0}$ in Definition 13.1 is given by $\mathcal{T}_{x_0} \triangleq \{t \in [0, \infty) : s(t, x_0) = s(t^+, x_0)\}$ so that $[0, \infty) \backslash \mathcal{T}_{x_0}$ corresponds to the (countable) set of resetting times, where the trajectory $s(\cdot, x_0)$ is discontinuous. Next, define

$$\mathcal{Z}_{x_0} \triangleq \{x \in \mathbb{R}^n : \text{there exists } t \in [0, \infty) \backslash \mathcal{T}_{x_0} \text{ such that } x = s(t, x_0)\}$$
$$= s([0, \infty) \backslash \mathcal{T}_{x_0}, x_0), \tag{13.1}$$

and $\mathcal{Z} \triangleq \cup_{x_0 \in \mathcal{D}} \mathcal{Z}_{x_0}$. Furthermore, let $\tau_i(x_0)$, $i \in \overline{\mathbb{Z}}_+$, where $\tau_0(x_0) \triangleq 0$ and $\tau_1(x_0) < \tau_2(x_0) < \cdots$, denote the resetting times, that is, $\{\tau_1(x_0), \tau_2(x_0), \ldots\} = [0, \infty) \backslash \mathcal{T}_{x_0}$. Next, we present a key assumption on the resetting times $\tau_i(\cdot)$, $i \in \overline{\mathbb{Z}}_+$.

Assumption 13.1 *For every* $i \in \overline{\mathbb{Z}}_+$, $\tau_i(\cdot)$ *is continuous and for every* $x_0 \in \mathcal{D}$, *there exists* $\varepsilon(x_0) > 0$ *such that* $\tau_{i+1}(x_0) - \tau_i(x_0) \geq \varepsilon(x_0)$, $i \in \overline{\mathbb{Z}}_+$.

The next result is a restatement of Proposition 12.2 and shows that $\mathcal{G}$ is a strong left-continuous dynamical system if and only if the trajectory of $\mathcal{G}$ is jointly continuous between resetting events, that is, for every $\varepsilon > 0$ and $k \in \overline{\mathbb{Z}}_+$ there exists $\delta = \delta(\varepsilon, k) > 0$ such that if $|t - t'| + \|x_0 - y\| < \delta$, then $\|s(t, x_0) - s(t', y)\| < \varepsilon$, where $x_0, y \in \mathcal{D}$, $t \in (\tau_k(x_0), \tau_{k+1}(x_0)]$, and $t' \in (\tau_k(y), \tau_{k+1}(y)]$.

Proposition 13.2 *Consider the dynamical system* $\mathcal{G}$ *satisfying Axioms* $i) - iii)$ *and Assumption 13.1. Then* $\mathcal{G}$ *is a strong left-continuous dynamical system if and only if the trajectory* $s(t, x_0)$, $t \geq 0$, *of* $\mathcal{G}$ *is jointly continuous between resetting events.*

Definition 13.2 *A solution* $s(t, x_0)$ *of* $\mathcal{G}$ *is* periodic *if there exists a finite time* $T > 0$, *known as the* period, *such that* $s(t+T, x_0) = s(t, x_0)$ *for all* $t \geq 0$. *A set* $\mathcal{O} \subset \mathcal{D}$ *is a* periodic orbit *of* $\mathcal{G}$ *if* $\mathcal{O} = \{x \in \mathcal{D} : x = s(t, x_0), 0 \leq t \leq T\}$ *for some periodic solution* $s(t, x_0)$ *of* $\mathcal{G}$.

Note that the set $\mathcal{Z}_{x_0}$ is identical for all $x_0 \in \mathcal{O}$. Furthermore, if for every $x_0 \in \mathbb{R}^n$ there exists $\varepsilon(x_0)$ such that $\tau_{i+1}(x_0) - \tau_i(x_0) \geq \varepsilon(x_0)$, $i \in \overline{\mathbb{Z}}_+$, then it follows that $\mathcal{Z}_{x_0}$ contains a finite number of (isolated) points. Finally, for every $x_0 \in \mathcal{O}$ it follows that $\tau_{i+N}(x_0) = \tau_i(x_0) + T$, $i = 2, 3, \ldots$, where N denotes the number of points in $\mathcal{Z}_{x_0}$.

Next, to extend Poincaré's theorem to hybrid dynamical systems let $\hat{\mathcal{Z}} \subset \mathcal{Z}$ be such that $\mathcal{O} \cap \hat{\mathcal{Z}}$ is a singleton. Note that the existence of such a $\hat{\mathcal{Z}}$ is guaranteed since all the points in $\mathcal{O} \cap \mathcal{Z}$ are isolated. Now, we define the Poincaré return map $P : \hat{\mathcal{Z}} \to \mathcal{Z}$ by

$$P(x) \triangleq s(\tau_{N+1}(x), x), \quad x \in \hat{\mathcal{Z}}. \tag{13.2}$$

Note that if $p \in \mathcal{O} \cap \hat{\mathcal{Z}}$, then $s(\tau_{N+1}(p), p) = p$. Furthermore, if Assumption 13.1 holds then $\tau_{N+1}(\cdot)$ is continuous, and hence it follows that $P(\cdot)$ is well defined. Next, define the discrete-time system given by

$$z(k+1) = P(z(k)), \quad k \in \overline{\mathbb{Z}}_+, \quad z(0) \in \hat{\mathcal{Z}}. \tag{13.3}$$

It is easy to see that p is a fixed point of (13.3). For notational convenience define the set $\Theta_{x_0, \eta} \triangleq \{t \in \mathcal{T}_{x_0} : |t - \tau| > \eta, \ \tau \in [0, \infty) \backslash \mathcal{T}_{x_0}\}$ denoting the set of all nonresetting times that are at least a distance η away from the resetting times.

Next, we introduce the notions of Lyapunov and asymptotic stability of a periodic orbit for the left-continuous dynamical system $\mathcal{G}$.

Definition 13.3 *A periodic orbit* $\mathcal{O}$ *of* $\mathcal{G}$ *is* Lyapunov stable *if for all* $\varepsilon > 0$ *there exists* $\delta = \delta(\varepsilon) > 0$ *such that if* $\mathrm{dist}(x_0, \mathcal{O}) < \delta$, *then* $\mathrm{dist}(s(t, x_0), \mathcal{O}) < \varepsilon$, $t \geq 0$. *A periodic orbit* $\mathcal{O}$ *of* $\mathcal{G}$ *is* asymptotically stable *if it is Lyapunov stable and there exists* $\delta > 0$ *such that if* $\mathrm{dist}(x_0, \mathcal{O}) < \delta$, *then* $\mathrm{dist}(s(t, x_0), \mathcal{O}) \to 0$ *as* $t \to \infty$.

The following key lemma is needed for the main stability result of this section.

Lemma 13.1 *Consider the strong left-continuous dynamical system* $\mathcal{G}$. *Assume the point* $p \in \hat{\mathcal{Z}}$ *generates the periodic orbit* $\mathcal{O} \triangleq \{x \in \mathcal{D} : x = s(t, p), 0 \leq t \leq T\}$, *where* $s(t, p), t \geq 0$, *is the periodic solution with the period* $T = \tau_{N+1}(p)$. *Then the following statements hold:*

i) The periodic orbit $\mathcal{O}$ is Lyapunov stable if and only if for every $\varepsilon > 0$ and for every $p_{\mathcal{O}} \in \overline{\mathcal{O}}$ there exists $\delta' = \delta'(\varepsilon, p_{\mathcal{O}}) > 0$ such that if $x_0 \in \mathcal{B}_{\delta'}(p_{\mathcal{O}})$, then $\mathrm{dist}(s(t, x_0), \mathcal{O}) < \varepsilon,\ t \geq 0$.

ii) The periodic orbit $\mathcal{O}$ is asymptotically stable if and only if it is Lyapunov stable and for every $p_{\mathcal{O}} \in \overline{\mathcal{O}}$ there exists $\delta' = \delta'(p_{\mathcal{O}}) > 0$ such that if $x_0 \in \mathcal{B}_{\delta'}(p_{\mathcal{O}})$, then $\mathrm{dist}(s(t, x_0), \mathcal{O}) \to 0$ as $t \to \infty$.

Proof. *i)* Necessity is immediate. To show sufficiency, assume that for every $\varepsilon > 0$ and for every $p_{\mathcal{O}} \in \overline{\mathcal{O}}$ there exists $\delta' = \delta'(\varepsilon, p_{\mathcal{O}}) > 0$ such that if $x_0 \in \mathcal{B}_{\delta'}(p_{\mathcal{O}})$, then $\mathrm{dist}(s(t, x_0), \mathcal{O}) < \varepsilon,\ t \geq 0$. Here, we assume that $\delta' = \delta'(\varepsilon, p_{\mathcal{O}}) > 0$ is the largest value such that the above distance inequality holds. Next, let $\delta = \delta(\varepsilon) = \inf_{p_{\mathcal{O}} \in \overline{\mathcal{O}}} \delta'(\varepsilon, p_{\mathcal{O}})$ and suppose, *ad absurdum*, that $\delta = 0$. In this case, there exists a sequence $\{p_{\mathcal{O}k}\}_{k=1}^{\infty} \in \overline{\mathcal{O}}$ such that $\lim_{k \to \infty} \delta'(\varepsilon, p_{\mathcal{O}k}) = 0$. Since $\{p_{\mathcal{O}k}\}_{k=1}^{\infty}$ is a bounded sequence, it follows from the Bolzano-Weierstrass theorem [146] that there exists a convergent subsequence $\{q_{\mathcal{O}k}\}_{k=1}^{\infty} \in \{p_{\mathcal{O}k}\}_{k=1}^{\infty}$ such that $\lim_{k \to \infty} q_{\mathcal{O}k} = q$ and $\lim_{k \to \infty} \delta'(\varepsilon, q_{\mathcal{O}k}) = 0$. Note, that since $\overline{\mathcal{O}}$ is closed and $\{q_{\mathcal{O}k}\}_{k=1}^{\infty} \in \overline{\mathcal{O}}$, it follows that $q \in \overline{\mathcal{O}}$, and hence, $\delta'(\varepsilon, q) > 0$. Thus, it follows that there exists $\tilde{q} \in \{q_{\mathcal{O}k}\}_{k=1}^{\infty}$ such that, for sufficiently small $\mu > 0$, $\mathcal{B}_{\delta'(\varepsilon, \tilde{q}) + \mu}(\tilde{q}) \subset \mathcal{B}_{\delta'(\varepsilon, q)}(q)$.

Now, since, for every $p_{\mathcal{O}} \in \overline{\mathcal{O}}$, $\delta' = \delta'(\varepsilon, p_{\mathcal{O}}) > 0$ is assumed to be the largest value such that $\mathrm{dist}(s(t, x_0), \mathcal{O}) < \varepsilon, t \geq 0$, for all $x_0 \in \mathcal{B}_{\delta'}(p_{\mathcal{O}})$, it follows that there exists $x_0' \in \mathcal{B}_{\delta'(\varepsilon, \tilde{q}) + \mu}(\tilde{q})$ and $t' \geq 0$ such that $\mathrm{dist}(s(t', x_0'), \mathcal{O}) > \varepsilon$. However, since $\mathcal{B}_{\delta'(\varepsilon, \tilde{q}) + \mu}(\tilde{q}) \subset \mathcal{B}_{\delta'(\varepsilon, q)}(q)$, then for $x_0' \in \mathcal{B}_{\delta'(\varepsilon, \tilde{q}) + \mu}(\tilde{q})$ it follows that $\mathrm{dist}(s(t, x_0'), \mathcal{O}) < \varepsilon$ for all $t \geq 0$, which is a contradiction. Hence, for every $\varepsilon > 0$ there exists $\delta = \delta(\varepsilon) > 0$ such that for every $p_{\mathcal{O}} \in \overline{\mathcal{O}}$ and $x_0 \in \mathcal{B}_{\delta}(p_{\mathcal{O}})$ it follows that $\mathrm{dist}(s(t, x_0), \mathcal{O}) < \varepsilon, t \geq 0$. Next, given $x_0 \in \mathcal{D}$ such that $\mathrm{dist}(x_0, \mathcal{O}) = \inf_{p_{\mathcal{O}} \in \mathcal{O}} \|x_0 - p_{\mathcal{O}}\| < \delta$, it follows that there exists a point $p^* \in \mathcal{O}$ such that $\mathrm{dist}(x_0, \mathcal{O}) \leq \|x_0 - p^*\| < \delta$, which implies that $x_0 \in \mathcal{B}_{\delta}(p^*)$, and hence, $\mathrm{dist}(s(t, x_0), \mathcal{O}) < \varepsilon, t \geq 0$, establishing Lyapunov stability.

ii) The proof is analogous to *i)* and, hence, is omitted. $\square$

The following theorem generalizes Poincaré's theorem to strong left-continuous dynamical systems by establishing a relationship between the stability properties of the periodic orbit $\mathcal{O}$ and the stability properties of an equilibrium point of the discrete-time system (13.3).

Theorem 13.1 *Consider the strong left-continuous dynamical system $\mathcal{G}$ with the Poincaré return map defined by (13.2). Assume that Assumption 13.1 holds and the point $p \in \hat{\mathcal{Z}}$ generates the periodic orbit*

$\mathcal{O} \triangleq \{x \in \mathcal{D} : x = s(t,p), 0 \le t \le T\}$, *where* $s(t,p), t \ge 0$, *is the periodic solution with the period* $T = \tau_{N+1}(p)$ *such that* $s(\tau_{N+1}(p), p) = p$. *Then the following statements hold:*

    i) $p \in \mathcal{O} \cap \hat{\mathcal{Z}}$ *is a Lyapunov stable fixed point of (13.3) if and only if the periodic orbit* $\mathcal{O}$ *of* $\mathcal{G}$ *generated by* p *is Lyapunov stable.*

    ii) $p \in \mathcal{O} \cap \hat{\mathcal{Z}}$ *is an asymptotically stable fixed point of (13.3) if and only if the periodic orbit* $\mathcal{O}$ *of* $\mathcal{G}$ *generated by* p *is asymptotically stable.*

Proof. *i)* To show necessity, let $\varepsilon > 0$ and note that the set $\mathcal{Z}_p = \{x \in \mathcal{D} : x = s(\tau_l(p), p) = p_l, l = 1, \dots, N\}$ contains N points, where $p \triangleq p_1$. Furthermore, let $p^+ = \lim_{\tau \to 0} s(\tau, p)$ and let $\hat{\varepsilon} > 0$. It follows from joint continuity of solutions of $\mathcal{G}$ that there exists $\hat{\delta} = \hat{\delta}(p, \hat{\varepsilon})$ such that if $\|x_0' - p^+\| + |t - t'| < \hat{\delta}$, then $\|s(t, x_0') - s(t', p^+)\| < \hat{\varepsilon}$, where $t \in (0, \tau_1(x_0')]$ and $t' \in (0, \tau_1(p^+)]$. Next, as shown in the proof of Proposition 12.2, it follows that $\lim_{\lambda \to 0} \underline{\tau}_1(\lambda, p^+) = \lim_{\lambda \to 0} \overline{\tau}_1(\lambda, p^+) = \tau_1(p^+)$. Hence, choosing $\delta' = \delta'(p, \hat{\varepsilon}) > 0$ such that $\delta' < \frac{\hat{\delta}}{2}$ and $\overline{\tau}_1(\delta', p^+) - \underline{\tau}_1(\delta', p^+) + \mu < \frac{\hat{\delta}}{2}$, where μ is a sufficiently small constant, it follows from the joint continuity property that, since $\|x_0' - p^+\| + |t - \tau_1(p^+)| < \hat{\delta}$,

$$\|s(t, x_0') - s(\tau_1(p^+), p^+)\| < \hat{\varepsilon}, \quad x_0' \in \mathcal{B}_{\delta'}(p^+),$$
$$t \in [\underline{\tau}_1(\delta', p^+) - \mu, \tau_1(x_0')]. \quad (13.4)$$

Next, let $\hat{\eta} > 0$ be such that $\hat{\eta} < \tau_1(p^+) - \underline{\tau}_1(\delta', p^+) + \mu$. Now, it follows from the strong quasi-continuous dependence property $iv)'$ that there exists $\delta'' = \delta''(p, \hat{\varepsilon}) > 0$ such that

$$\|s(t, p^+) - s(t, x_0')\| < \hat{\varepsilon}, \quad x_0' \in \mathcal{B}_{\delta''}(p^+), \quad t \in [0, \tau_1(p^+) - \hat{\eta}). \quad (13.5)$$

Now, let $\tilde{\delta} = \min\{\delta', \delta''\}$ and note that $\underline{\tau}_1(\delta', p^+) \le \underline{\tau}_1(\tilde{\delta}, p^+)$. Since $\hat{\eta}$ is such that $\underline{\tau}_1(\delta', p^+) - \mu \in [0, \tau_1(p^+) - \hat{\eta})$, it follows from (13.4) and (13.5) that

$$\text{dist}(s(t, x_0'), \mathcal{O}) < \hat{\varepsilon}, \quad x_0' \in \mathcal{B}_{\tilde{\delta}}(p^+), \quad t \in [0, \tau_1(x_0')]. \quad (13.6)$$

Using similar arguments as above, it can be shown that the resetting event is continuous with respect to the state, that is, for $\tilde{\delta} > 0$ there exists $\delta_1 = \delta_1(\tilde{\delta}) = \delta_1(\hat{\varepsilon})$ such that $\|x^+ - p^+\| < \tilde{\delta}$ for all $x \in \mathcal{B}_{\delta_1}(p) \cap \mathcal{Z}$, where $x^+ = \lim_{\tau \to 0} s(\tau, x)$. Hence, it follows that for $\hat{\varepsilon} > 0$, there exists $\delta_1 = \delta_1(\hat{\varepsilon})$ such that

$$\text{dist}(s(t, x_0'), \mathcal{O}) < \hat{\varepsilon}, \quad x_0' \in \mathcal{B}_{\delta_1}(p) \cap \mathcal{Z}, \quad t \in [0, \tau_2(x_0')]. \quad (13.7)$$

Similarly, for every point in $\mathcal{Z}_p$ there exists a neighborhood such that an analogous condition to (13.7) holds. Specifically, for $\varepsilon > 0$ and $p_N \in \mathcal{Z}_p$ there exists $\delta_N = \delta_N(\varepsilon) < \varepsilon$ such that $\mathrm{dist}(s(t, x_0'), \mathcal{O}) < \varepsilon$ for all $x_0' \in \mathcal{B}_{\delta_N}(p_N) \cap \mathcal{Z}$, $t \in [0, \tau_2(x_0')]$. Analogously, for $p_{N-1} \in \mathcal{Z}_p$, there exists $\delta_{N-1} = \delta_{N-1}(\delta_N) = \delta_{N-1}(\varepsilon)$ such that $\mathrm{dist}(s(t, x_0'), \mathcal{O}) < \delta_N < \varepsilon$ for all $x_0' \in \mathcal{B}_{\delta_{N-1}}(p_{N-1}) \cap \mathcal{Z}$, $t \in [0, \tau_2(x_0')]$. Recursively repeating this procedure and using the semi-group property $iii)$, it follows that for $\varepsilon > 0$ there exists $\delta_1 = \delta_1(\varepsilon) > 0$ such that

$$\mathrm{dist}(s(t, x_0'), \mathcal{O}) < \varepsilon, \quad x_0' \in \mathcal{B}_{\delta_1}(p) \cap \mathcal{Z}, \quad t \in [0, \tau_{N+1}(x_0')]. \quad (13.8)$$

Next, it follows from Lyapunov stability of the fixed point $p \in \mathcal{Z}_p$ of the discrete-time dynamical system (13.3) that, for $\delta_1 > 0$, there exists $\delta_1' = \delta_1'(\delta_1) > 0$ such that $\|z(k+1) - p\| = \|P(z(k)) - p\| < \delta_1$ for all $z(0) \in \mathcal{B}_{\delta_1'}(p) \cap \mathcal{Z}$. Hence, using (13.8) and the semi-group property $iii)$, it follows that

$$\mathrm{dist}(s(t, x_0'), \mathcal{O}) < \varepsilon, \quad x_0' \in \mathcal{B}_{\delta_1'}(p) \cap \mathcal{Z}, \quad t \geq 0. \quad (13.9)$$

Using similar arguments as above, for every $p_{\mathcal{O}} \in \overline{\mathcal{O}}$ there exists $\delta = \delta(\varepsilon, p_{\mathcal{O}})$ such that

$$\mathrm{dist}(s(t, x_0), \mathcal{O}) < \varepsilon, \quad x_0 \in \mathcal{B}_\delta(p_{\mathcal{O}}), \quad t \in [0, \tau_m(p_{\mathcal{O}})], \quad (13.10)$$

where m is the number of resettings required for $s(t, x_0), t \geq 0$, to reach $\mathcal{B}_{\delta_1'}(p) \cap \mathcal{Z}$. Finally, it follows from (13.10), (13.9), and the semi-group property $iii)$ that

$$\mathrm{dist}(s(t, x_0), \mathcal{O}) < \varepsilon, \quad x_0 \in \mathcal{B}_\delta(p_{\mathcal{O}}), \quad t \geq 0, \quad (13.11)$$

which, using Lemma 13.1, proves Lyapunov stability of the periodic orbit $\mathcal{O}$.

Next, we show sufficiency. Assume that $\mathcal{O}$ is a Lyapunov stable periodic orbit. Furthermore, choose $\varepsilon > 0$ and let $\hat{\varepsilon} \in (0, \varepsilon]$ be such that there does not exist a point of $\mathcal{Z}_p$ in $\mathcal{B}_{\hat{\varepsilon}}(p)$ other than $p \in \mathcal{Z}_p$. Note that $\hat{\varepsilon} > 0$ exists since $\mathcal{Z}_p$ is a finite set. Now, using the fact that $\mathcal{G}$ is left-continuous, it follows that for sufficiently small $\hat{\delta} > 0$ there exists $\tilde{\delta} = \tilde{\delta}(\hat{\delta})$ such that $\hat{\delta} \leq \tilde{\delta} < \hat{\varepsilon}$ and

$$\mathrm{dist}(x, \mathcal{O}) > \hat{\delta}, \quad x \in \mathcal{B}_{\hat{\varepsilon}}(p) \backslash \mathcal{B}_{\tilde{\delta}}(p) \cap \mathcal{Z}. \quad (13.12)$$

Here, we let $\tilde{\delta} > 0$ be the smallest value such that (13.12) holds. Note that in this case

$$\lim_{\hat{\delta} \to 0} \tilde{\delta}(\hat{\delta}) = 0. \quad (13.13)$$

Now, it follows from Assumption 13.1 and the joint continuity of solutions of $\mathcal{G}$ that for $\hat{\varepsilon} > 0$ there exists $\delta'(\hat{\varepsilon}) > 0$ such that

$$\|s(\tau_{N+1}(x_0'), x_0') - p\| < \hat{\varepsilon}, \quad x_0' \in \mathcal{B}_{\delta'}(p) \cap \hat{\mathcal{Z}}. \tag{13.14}$$

Hence, using (13.13), we can choose $\hat{\delta} = \hat{\delta}(\hat{\varepsilon}) > 0$ such that $\tilde{\delta}(\hat{\delta}) \leq \delta'(\hat{\varepsilon})$. Next, it follows from the Lyapunov stability of $\mathcal{O}$ that for $\hat{\delta} = \hat{\delta}(\hat{\varepsilon}) > 0$ there exists $\delta = \delta(\hat{\delta}) = \delta(\hat{\varepsilon}) > 0$ such that if $x_0 \equiv z(0) \in \mathcal{B}_\delta(p) \cap \hat{\mathcal{Z}}$, then $\mathrm{dist}(s(t, x_0), \mathcal{O}) < \hat{\delta}, t \geq 0$. Now, using (13.12) and (13.14), it follows that

$$\|z(k+1) - p\| = \|P(z(k)) - p\| = \|s(\tau_{N+1}(z(k)), z(k)) - p\|$$
$$< \tilde{\delta}$$
$$< \hat{\varepsilon}$$
$$\leq \varepsilon, \quad z(0) \in \mathcal{B}_\delta(p) \cap \hat{\mathcal{Z}}, \quad k \in \overline{\mathbb{Z}}_+, \tag{13.15}$$

where $z(k), k \in \overline{\mathbb{Z}}_+$, satisfies (13.3). Thus, (13.15) establishes that $p \in \mathcal{Z}_p$ is a Lyapunov stable fixed point of (13.3).

ii) To show necessity, assume that $p \in \mathcal{O} \cap \hat{\mathcal{Z}}$ is an asymptotically stable fixed point of (13.3). Hence, it follows from i) that the periodic orbit $\mathcal{O}$ is Lyapunov stable and there exists $\delta' > 0$ such that if $z(0) \equiv x_0' \in \mathcal{B}_{\delta'}(p) \cap \hat{\mathcal{Z}}$, then $z(k+1) = P(z(k)) = s(\tau_{(N+1)\cdot(k+1)}(x_0'), x_0') \to p$ as $k \to \infty$. Now, Definition 13.3 implies that a periodic orbit $\mathcal{O}$ of $\mathcal{G}$ is asymptotically stable if it is Lyapunov stable and there exists $\delta > 0$ such that if $\mathrm{dist}(x_0', \mathcal{O}) < \delta$, then for every $\varepsilon > 0$ there exists $T = T(\varepsilon, x_0') > 0$ such that $\mathrm{dist}(s(t, x_0'), \mathcal{O}) < \varepsilon$ for all $t > T$. Next, using similar arguments as in i), for any $\varepsilon > 0$ there exists $\hat{\delta} = \hat{\delta}(\varepsilon) > 0$ such that

$$\mathrm{dist}(s(t, x_0'), \mathcal{O}) < \varepsilon, \quad x_0' \in \mathcal{B}_{\hat{\delta}}(p) \cap \hat{\mathcal{Z}}, \quad t \in [0, \tau_{N+1}(x_0')]. \tag{13.16}$$

Now, it follows from the asymptotic stability of p that for every $x_0' \in \mathcal{B}_{\delta'}(p) \cap \hat{\mathcal{Z}}$ there exists $K = K(\hat{\delta}, x_0') = K(\varepsilon, x_0') \in \overline{\mathbb{Z}}_+$ such that

$$\|s(\tau_{(N+1)\cdot k}(x_0'), x_0') - p\| < \hat{\delta}, \quad k > K. \tag{13.17}$$

Choose $l > K$ and let $T = T(\varepsilon, x_0') = \tau_{(N+1)\cdot l}(x_0')$. Then, it follows from (13.16) and (13.17) that for a given $\varepsilon > 0$ there exists $T = T(\varepsilon, x_0') > 0$ such that

$$\mathrm{dist}(s(t, x_0'), \mathcal{O}) < \varepsilon, \quad x_0' \in \mathcal{B}_{\delta'}(p) \cap \hat{\mathcal{Z}}, \quad t > T, \tag{13.18}$$

and hence, $\mathrm{dist}(s(t, x_0'), \mathcal{O}) \to 0$ as $t \to \infty$ for all $x_0' \in \mathcal{B}_{\delta'}(p) \cap \hat{\mathcal{Z}}$.

Finally, using similar arguments as in i) it can be shown that for every $p_\mathcal{O} \in \overline{\mathcal{O}}$ there exists $\delta = \delta(p_\mathcal{O}) > 0$ such that $\|s(\tau_m(x_0), x_0) - p\| < \delta'$ for all $x_0 \in \mathcal{B}_\delta(p_\mathcal{O})$, where m is the number of resettings required for $s(t, x_0), t \geq 0$, to reach $\mathcal{B}_{\delta'}(p) \cap \hat{\mathcal{Z}}$. This argument along with (13.18), the semi-group property iii), and ii) of Lemma 13.1 implies asymptotic stability of the periodic orbit $\mathcal{O}$.

Finally, we show sufficiency. Assume that $\mathcal{O}$ is an asymptotically stable periodic orbit of $\mathcal{G}$. Hence, $p \in \hat{\mathcal{Z}}$ is a Lyapunov stable fixed point of (13.3) and there exists $\delta > 0$ such that if $x_0 \in \mathcal{B}_p(\delta)$, then for every sequence $\{t_k\}_{k=0}^\infty$ such that $t_k \to \infty$ as $k \to \infty$,

$$\mathrm{dist}(s(t_k, x_0), \mathcal{O}) \to 0, \quad k \to \infty. \tag{13.19}$$

Next, choose $\hat{\delta} \in (0, \delta]$ such that there are no points of $\mathcal{Z}_p$ in $\mathcal{B}_{\hat{\delta}}(p)$ other than $p \in \hat{\mathcal{Z}}$. Once again, $\hat{\delta} > 0$ exists since $\mathcal{Z}_p$ is a finite set. Since $p \in \hat{\mathcal{Z}}$ is a Lyapunov stable fixed point of (13.3) it follows that for $\hat{\delta} > 0$ there exists $\tilde{\delta} = \tilde{\delta}(\hat{\delta}) > 0$ such that if $z(0) \equiv x_0 \in \mathcal{B}_{\tilde{\delta}}(p) \cap \hat{\mathcal{Z}}$, then $z(k+1) = P(z(k)) = s(\tau_{(N+1)\cdot(k+1)}(x_0), x_0) \in \mathcal{B}_{\hat{\delta}}(p) \cap \hat{\mathcal{Z}}, k \in \overline{\mathbb{Z}}_+$.

Next, choose a sequence $\{t_k\}_{k=0}^\infty = \{\tau_{(N+1)\cdot k}(x_0)\}_{k=0}^\infty, x_0 \in \mathcal{B}_{\tilde{\delta}}(p) \cap \hat{\mathcal{Z}}$, and note that $\tau_{(N+1)\cdot k}(x_0) \to \infty$ as $k \to \infty$. Hence, it follows from (13.19) that

$$\begin{aligned} \mathrm{dist}(z(k+1), \mathcal{O}) &= \mathrm{dist}(P(z(k)), \mathcal{O}) \\ &= \mathrm{dist}(s(\tau_{(N+1)\cdot(k+1)}(x_0), x_0), \mathcal{O}) \\ &\to 0, \quad k \to \infty, \quad x_0 \in \mathcal{B}_{\tilde{\delta}}(p) \cap \hat{\mathcal{Z}}. \end{aligned} \tag{13.20}$$

Since $p \in \mathcal{Z}_p$ is the only point of $\mathcal{O}$ in $\mathcal{B}_{\hat{\delta}}(p) \cap \hat{\mathcal{Z}}$, (13.20) implies that $\mathrm{dist}(z(k+1), p) \to 0$ as $k \to \infty$ for all $z(0) \equiv x_0 \in \mathcal{B}_{\tilde{\delta}}(p) \cap \hat{\mathcal{Z}}$, which establishes asymptotic stability of the fixed point $p \in \mathcal{Z}_p$ of (13.3). $\square$

13.3 Specialization to Impulsive Dynamical Systems

In this section, we specialize Poincaré's theorem for strong left-continuous dynamical systems to state-dependent impulsive dynamical systems. Recall that a state-dependent impulsive dynamical system $\mathcal{G}$ has the form

$$\dot{x}(t) = f_\mathrm{c}(x(t)), \quad x(0) = x_0, \quad x(t) \notin \mathcal{Z}_x, \tag{13.21}$$
$$\Delta x(t) = f_\mathrm{d}(x(t)), \quad x(t) \in \mathcal{Z}_x, \tag{13.22}$$

where $t \geq 0$, $x(t) \in \mathcal{D} \subseteq \mathbb{R}^n$, $\mathcal{D}$ is an open set, $f_c : \mathcal{D} \to \mathbb{R}^n$, $f_d : \mathcal{Z}_x \to \mathbb{R}^n$ is continuous, and $\mathcal{Z}_x \subset \mathcal{D}$ is the resetting set. Here, we assume that the continuous-time dynamics $f_c(\cdot)$ are such that the solution to (13.21) is jointly continuous in t and x_0 between resetting events. A sufficient condition ensuring this is Lipschitz continuity of $f_c(\cdot)$.

As in Section 13.2, for a particular trajectory $x(t)$ we denote the resetting times of (13.21) and (13.22) by $\tau_k(x_0)$, that is, the kth instant of time at which $x(t)$ intersects $\mathcal{Z}_x$. Thus, the trajectory of the system (13.21) and (13.22) from the initial condition $x(0) = x_0 \in \mathcal{D}$ is given by $\psi(t, x_0)$ for $0 < t \leq \tau_1(x_0)$, where $\psi(t, x_0)$ denotes the solution to the continuous-time dynamics (13.21). If and when the trajectory reaches a state $x_1 \triangleq x(\tau_1(x_0))$ satisfying $x_1 \in \mathcal{Z}_x$, then the state is instantaneously transferred to $x_1^+ \triangleq x_1 + f_d(x_1)$ according to the resetting law (13.22). The solution $x(t)$, $\tau_1(x_0) < t \leq \tau_2(x_0)$, is then given by $\psi(t - \tau_1(x_0), x_1^+)$, and so on for all $x_0 \in \mathcal{D}$. Note that the solution $x(t)$ of (13.21) and (13.22) is left-continuous, that is, it is continuous everywhere except at the resetting time $\tau_k(x_0)$, and

$$x_k \triangleq x(\tau_k(x_0)) = \lim_{\varepsilon \to 0^+} x(\tau_k(x_0) - \varepsilon), \qquad (13.23)$$

$$x_k^+ \triangleq x(\tau_k(x_0)) + f_d(x(\tau_k(x_0))), \qquad (13.24)$$

for $k = 1, 2, \ldots$.

Here, we assume that assumptions A1 and A2 of Chapter 2 hold, that is:

A1. If $x(t) \in \overline{\mathcal{Z}}_x \backslash \mathcal{Z}_x$, then there exists $\varepsilon > 0$ such that, for all $0 < \delta < \varepsilon$, $x(t + \delta) \notin \mathcal{Z}_x$.

A2. If $x \in \mathcal{Z}_x$, then $x + f_d(x) \notin \mathcal{Z}_x$.

Note that it follows from the definition of $\tau_k(\cdot)$ that $\tau_1(x) > 0$, $x \notin \mathcal{Z}_x$, and $\tau_1(x) = 0$, $x \in \mathcal{Z}_x$. Furthermore, since for every $x \in \mathcal{Z}_x$, $x + f_d(x) \notin \mathcal{Z}_x$, it follows that $\tau_2(x) = \tau_1(x) + \tau_1(x + f_d(x)) > 0$. Finally, note that it follows from A1 and A2 that the resetting times $\tau_k(x_0)$ are well defined and distinct.

Recall that since *not* every bounded solution of an impulsive dynamical system over a forward time interval can be extended to infinity due to Zeno solutions, we assume that $f_c(\cdot)$ and $f_d(\cdot)$ are such that existence and uniqueness of solutions for (13.21) and (13.22) are satisfied in forward time. In this section we assume that $f_c(\cdot)$ and $f_d(\cdot)$ are such that $\tau_k(x_0) \to \infty$ as $k \to \infty$ for all $x_0 \in \mathcal{D}$. In light of the above, note that the solution to (13.21) and (13.22) with initial condition $x_0 \in \mathcal{D}$ denoted by $s(t, x_0)$, $t \geq 0$, is $i)$ left-continuous, that is, $\lim_{\tau \to t^-} s(\tau, x_0) = s(t, x_0)$ for all $x_0 \in \mathcal{D}$ and $t \in (0, \infty)$; *ii*) consistent, that is, $s(0, x_0) = x_0$, for all $x_0 \in \mathcal{D}$; and *iii*) satisfies the semi-group property, that is, $s(\tau, s(t, x_0)) = s(t+\tau, x_0)$ for all $x_0 \in \mathcal{D}$ and $t, \tau \in [0, \infty)$. To see this, note that $s(0, x_0) = x_0$ for all $x_0 \in \mathcal{D}$ and

$$
s(t, x_0) = \begin{cases} \psi(t, x_0), & 0 \le t \le \tau_1(x_0), \\ \psi(t - \tau_k(x_0), s(\tau_k(x_0), x_0) + f_{\mathrm{d}}(s(\tau_k(x_0), x_0))), \\ & \tau_k(x_0) < t \le \tau_{k+1}(x_0), \\ \psi(t - \tau(x_0), s(\tau(x_0), x_0)), & t \ge \tau(x_0), \end{cases}
$$

$$(13.25)$$

where $\tau(x_0) \triangleq \sup_{k \ge 0} \tau_k(x_0)$, which implies that $s(\cdot, x_0)$ is left-continuous. Furthermore, uniqueness of solutions implies that $s(t, x_0)$ satisfies the semi-group property $s(\tau, s(t, x_0)) = s(t+\tau, x_0)$ for all $x_0 \in \mathcal{D}$ and $t, \tau \in [0, \infty)$.

Next, we present two key assumptions on the structure of the resetting set $\mathcal{Z}_x$. Specifically, we assume that the resetting set $\mathcal{Z}_x$ is such that the following assumptions hold:

A3. There exists a continuously differentiable function $\mathcal{X} : \mathcal{D} \to \mathbb{R}$ such that the resetting set $\mathcal{Z}_x = \{x \in \mathcal{D} : \mathcal{X}(x) = 0\}$; moreover, $\mathcal{X}'(x) \ne 0$, $x \in \mathcal{Z}_x$.

A4. $\frac{\partial \mathcal{X}(x)}{\partial x} f_{\mathrm{c}}(x) \ne 0$, $x \in \mathcal{Z}_x$.

It follows from A3 that the resetting set $\mathcal{Z}_x$ is an embedded submanifold [81], while A4 assures that the solution of $\mathcal{G}$ is not tangent to the resetting set $\mathcal{Z}_x$. The following proposition shows that under Assumptions A3 and A4, the resetting times $\tau_k(\cdot)$ are continuous at $x_0 \in \mathcal{D}$ for all $k \in \overline{\mathbb{Z}}_+$.

Proposition 13.3 *Consider the nonlinear state-dependent impulsive dynamical system $\mathcal{G}$ given by (13.21) and (13.22). Assume that A3 and A4 hold. Then $\tau_k(\cdot)$ is continuous at $x_0 \in \mathcal{D}$, where $0 < \tau_k(x_0) < \infty$, for all $k \in \overline{\mathbb{Z}}_+$.*

Proof. First, it follows from Proposition 2.2 that $\tau_1(\cdot)$ is continuous at $x_0 \in \mathcal{D}$, where $0 < \tau_1(x_0) < \infty$. Since $f_{\mathrm{c}}(\cdot)$ is such that the solutions to (13.21) are jointly continuous in t and x_0, it follows that $\psi(\cdot, \cdot)$ is continuous in both its arguments. Furthermore, note that

$\psi(\tau_1(x), x) = s(\tau_1(x), x), x \in \mathcal{D}$. Next, it follows from the definition of $\tau_k(x)$ that for every $x \in \mathcal{D}$ and $k \in \{1, 2, \ldots\}$,

$$\tau_k(x) = \tau_{k-j}(x) + \tau_j[s(\tau_{k-j}(x), x) + f_d(s(\tau_{k-j}(x), x))], \quad j = 1, \ldots, k, \tag{13.26}$$

where $\tau_0(x) \triangleq 0$. Hence, since $f_d(\cdot)$ is continuous, it follows from (13.26) that $\tau_2(x) = \tau_1(x) + \tau_1[s(\tau_1(x), x) + f_d(s(\tau_1(x), x))]$ is also continuous on $\mathcal{D}$. By recursively repeating this procedure for $k = 3, 4, \ldots$, it follows that $\tau_k(x)$ is a continuous function on $\mathcal{D}$ for all $k \in \overline{\mathbb{Z}}_+$. □

Since $f_c(\cdot)$ and $f_d(\cdot)$ are such that the Axioms i)–iii) hold for the state-dependent impulsive dynamical system $\mathcal{G}$, and $\mathcal{G}$ is jointly continuous between resetting events, then, with Assumptions A3 and A4 satisfied, it follows from Propositions 13.2 and 13.3 that the state-dependent impulsive dynamical system $\mathcal{G}$ is a strong left-continuous dynamical system. Hence, the following corollary to Theorem 13.1 is immediate.

Corollary 13.1 *Consider the impulsive dynamical system $\mathcal{G}$ given by (13.21) and (13.22) with the Poincaré return map defined by (13.2). Assume that A3 and A4 hold, and the point $p \in \hat{\mathcal{Z}}_x$ generates the periodic orbit $\mathcal{O} \triangleq \{x \in \mathcal{D} : x = s(t, p), 0 \leq t \leq T\}$, where $s(t, p)$, $t \geq 0$, is the periodic solution with the period $T = \tau_{N+1}(p)$ such that $s(\tau_{N+1}(p), p) = p$. Then the following statements hold:*

i) *$p \in \mathcal{O} \cap \hat{\mathcal{Z}}_x$ is a Lyapunov stable fixed point of (13.3) if and only if the periodic orbit $\mathcal{O}$ of $\mathcal{G}$ generated by p is Lyapunov stable.*

ii) *$p \in \mathcal{O} \cap \hat{\mathcal{Z}}_x$ is an asymptotically stable fixed point of (13.3) if and only if the periodic orbit $\mathcal{O}$ of $\mathcal{G}$ generated by p is asymptotically stable.*

Corollary 13.1 gives necessary and sufficient conditions for Lyapunov and asymptotic stability of a periodic orbit of the state-dependent impulsive dynamical system $\mathcal{G}$ based on the stability properties of a fixed point of the n-dimensional discrete-time dynamical system involving the Poincaré map (13.2). Next, as is the case of the classical Poincaré theorem, we present a specialization of Corollary 13.1 that allows us to analyze the stability of periodic orbits by replacing the nth-order impulsive dynamical system by an $(n-1)$th-order discrete-time dynamical system.

To present this result assume, without loss of generality, that $\frac{\partial \mathcal{X}(x)}{\partial x_n}$ $\neq 0, x \in \mathcal{Z}_x$, where $x = [x_1, \ldots, x_n]^{\mathrm{T}}$. Then, it follows from the implicit function theorem [89] that $x_n = g(x_1, \ldots, x_{n-1})$, where $g(\cdot)$ is a continuously differentiable function at $x_r \triangleq [x_1, \ldots, x_{n-1}]^{\mathrm{T}}$ such that $[x_r^{\mathrm{T}}, g(x_r)]^{\mathrm{T}} \in \mathcal{Z}_x$. Note that in this case $P : \hat{\mathcal{Z}}_x \to \mathcal{Z}_x$ in (13.3) is given by $P(x) \triangleq [P_1(x), \ldots, P_n(x)]^{\mathrm{T}}$, where

$$P_n(x_r, g(x_r)) = g(P_1(x_r, g(x_r)), \ldots, P_{n-1}(x_r, g(x_r))). \qquad (13.27)$$

Hence, we can reduce the n-dimensional discrete-time system (13.3) to the $(n-1)$-dimensional discrete-time system given by

$$z_r(k+1) = \mathcal{P}_r(z_r(k)), \quad k \in \overline{\mathbb{Z}}_+, \qquad (13.28)$$

where $z_r \in \mathbb{R}^{n-1}$, $[z_r^{\mathrm{T}}(\cdot), g(z_r(\cdot))]^{\mathrm{T}} \in \mathcal{Z}_x$, and

$$\mathcal{P}_r(x_r) \triangleq \begin{bmatrix} P_1(x_r, g(x_r)) \\ \vdots \\ P_{n-1}(x_r, g(x_r)) \end{bmatrix}. \qquad (13.29)$$

Note that it follows from (13.27) and (13.29) that $p \triangleq [p_r^{\mathrm{T}}, g(p_r)]^{\mathrm{T}} \in \hat{\mathcal{Z}}_x$ is a fixed point of (13.3) if and only if p_r is a fixed point of (13.28).

Corollary 13.2 *Consider the impulsive dynamical system $\mathcal{G}$ given by (13.21) and (13.22) with the Poincaré return map defined by (13.2). Assume that A3 and A4 hold, $\frac{\partial \mathcal{X}(x)}{\partial x_n} \neq 0$, $x \in \mathcal{Z}_x$, and the point $p \in \hat{\mathcal{Z}}_x$ generates the periodic orbit $\mathcal{O} \triangleq \{x \in \mathcal{D} : x = s(t, p), 0 \le t \le T\}$, where $s(t, p)$, $t \ge 0$, is the periodic solution with the period $T = \tau_{N+1}(p)$ such that $s(\tau_{N+1}(p), p) = p$. Then the following statements hold:*

i) *For $p = [p_r^{\mathrm{T}}, g(p_r)]^{\mathrm{T}} \in \mathcal{O} \cap \hat{\mathcal{Z}}_x$, p_r is a Lyapunov stable fixed point of (13.28) if and only if the periodic orbit $\mathcal{O}$ is Lyapunov stable.*

ii) *For $p = [p_r^{\mathrm{T}}, g(p_r)]^{\mathrm{T}} \in \mathcal{O} \cap \hat{\mathcal{Z}}_x$, p_r is an asymptotically stable fixed point of (13.28) if and only if the periodic orbit $\mathcal{O}$ is asymptotically stable.*

Proof. i) To show necessity, assume that p_r is a Lyapunov stable fixed point of (13.28) and let $\varepsilon > 0$. Then it follows from the continuity of $g(\cdot)$ that there exists $\delta' = \delta'(\varepsilon) > 0$ such that

$$\|g(x_r) - g(p_r)\| < \frac{\varepsilon}{2}, \quad x_r \in \mathcal{B}_{\delta'}(p_r). \qquad (13.30)$$

Choosing $\delta' < \frac{\varepsilon}{2}$, it follows from the Lyapunov stability of p_r that for $\delta' > 0$ there exists $\delta = \delta(\varepsilon) < \delta'$ such that

$$\|z_r(k+1) - p_r\| = \|\mathcal{P}_r(z_r(k)) - p_r\| < \delta' < \frac{\varepsilon}{2}, \quad z_r(0) \in \mathcal{B}_\delta(p_r), \quad (13.31)$$

where $z(k) = [z_r(k)^T, g(z_r(k))]^T \in \mathcal{Z}_x$, $k \in \overline{\mathbb{Z}}_+$, satisfies (13.3).
   If $z(0) \in \mathcal{B}_\delta(p) \cap \hat{\mathcal{Z}}_x$, that is,

$$\begin{aligned}
\|z(0) - p\| &= \|[z_r(0)^T, g(z_r(0))]^T - [p_r^T, g(p_r)]^T\| \\
&\leq \|[z_r(0)^T, 0]^T - [p_r^T, 0]^T\| \\
&\quad + \|[0, \ldots, 0, g(z_r(0))]^T - [0, \ldots, 0, g(p_r)]^T\| \\
&< \delta, \qquad\qquad\qquad\qquad\qquad\qquad\qquad\qquad\qquad (13.32)
\end{aligned}$$

then it follows from (13.32) that $z_r(0) \in \mathcal{B}_\delta(p_r)$. Hence, using (13.27), (13.30), and (13.31), it follows that

$$\begin{aligned}
\|z(k+1) - p\| &= \|P(z(k)) - p\| \\
&= \|[\mathcal{P}_r^T(z_r(k)), P_n(z_r(k), g(z_r(k)))]^T - [p_r^T, g(p_r)]^T\| \\
&\leq \|[\mathcal{P}_r^T(z_r(k)), 0]^T - [p_r^T, 0]^T\| \\
&\quad + \|[0, \ldots, 0, P_n(z_r(k), g(z_r(k)))]^T - [0, \ldots, 0, g(p_r)]^T\| \\
&< \delta' + \frac{\varepsilon}{2} \\
&< \varepsilon. \qquad\qquad\qquad\qquad\qquad\qquad\qquad\qquad\qquad (13.33)
\end{aligned}$$

Thus, for a given $\varepsilon > 0$ there exists $\delta = \delta(\varepsilon) > 0$ such that if $z(0) \in \mathcal{B}_\delta(p) \cap \hat{\mathcal{Z}}_x$, then $\|z(k+1) - p\| < \varepsilon$, which establishes Lyapunov stability of p for the discrete-time system (13.3). Now, Lyapunov stability of the periodic orbit $\mathcal{O}$ follows as a direct consequence of Theorem 13.1.
   Next, to show sufficiency, assume that the periodic orbit $\mathcal{O}$ is Lyapunov stable. In this case, it follows from Theorem 13.1 that $p = [p_r^T, g(p_r)]^T \in \mathcal{O} \cap \hat{\mathcal{Z}}_x$ is a Lyapunov stable fixed point of (13.3). Hence, for every $\varepsilon > 0$ there exists $\delta' = \delta'(\varepsilon) > 0$ such that

$$\|z(k+1) - p\| = \|P(z(k)) - p\| < \varepsilon, \quad k \in \overline{\mathbb{Z}}_+, \quad z(0) \in \mathcal{B}_{\delta'}(p) \cap \hat{\mathcal{Z}}_x. \qquad (13.34)$$

Now, it follows from the continuity of $g(\cdot)$ that for $\delta' > 0$ there exists $\delta = \delta(\varepsilon) > 0$ such that

$$\|g(x_r) - g(p_r)\| < \frac{\delta'}{2}, \quad x_r \in \mathcal{B}_\delta(p_r). \qquad (13.35)$$

Choosing $\delta < \frac{\delta'}{2}$ and $z_r(0) \in \mathcal{B}_\delta(p_r)$, it follows from (13.35) that

$$\|z(0) - p\| = \|[z_r(0)^T, g(z_r(0))]^T - [p_r^T, g(p_r)]^T\| < \frac{\delta'}{2} + \frac{\delta'}{2} = \delta',$$
(13.36)

that is, $z(0) \in \mathcal{B}_{\delta'}(p) \cap \hat{\mathcal{Z}}_x$. Hence, it follows from (13.34), (13.36), and (13.33) that for a given $\varepsilon > 0$, there exists $\delta = \delta(\varepsilon) > 0$ such that $\|z_r(k+1) - p_r\| = \|\mathcal{P}_r(z_r(k)) - p_r\| < \varepsilon$ for all $z_r(0) \in \mathcal{B}_\delta(p_r)$, which establishes Lyapunov stability of p_r for (13.28).

$ii)$ To show necessity, assume that p_r is an asymptotically stable fixed point of (13.28). Now, it follows from $i)$ that $p = [p_r^T, g(p_r)]^T \in \mathcal{O} \cap \hat{\mathcal{Z}}_x$ is a Lyapunov stable fixed point of (13.3) and there exists $\delta > 0$ such that

$$\|z_r(k+1) - p_r\| = \|\mathcal{P}_r(z_r(k)) - p_r\| \to 0, \quad k \to \infty, \quad z_r(0) \in \mathcal{B}_\delta(p_r).$$
(13.37)

If $z(0) \in \mathcal{B}_\delta(p) \cap \hat{\mathcal{Z}}_x$, then, as shown in $i)$, $z_r(0) \in \mathcal{B}_\delta(p_r)$. Hence, it follows from (13.37), the continuity of $g(\cdot)$, and the representation given in (13.33) and (13.27) that $\|z(k+1) - p\| \to 0$ as $k \to \infty$ for all $z(0) \in \mathcal{B}_\delta(p) \cap \hat{\mathcal{Z}}_x$. This establishes asymptotic stability of p for (13.3). Now, asymptotic stability of the periodic orbit $\mathcal{O}$ of $\mathcal{G}$ follows as a direct consequence of Theorem 13.1.

Finally, to show sufficiency, assume that the periodic orbit $\mathcal{O}$ is asymptotically stable. In this case, it follows from Theorem 13.1 that $p = [p_r^T, g(p_r)]^T \in \mathcal{O} \cap \hat{\mathcal{Z}}_x$ is an asymptotically stable fixed point of (13.3) and, by $i)$, p_r is a Lyapunov stable fixed point of (13.28). Hence, there exists $\delta' > 0$ such that

$$\begin{aligned}\|z(k+1) - p\| &= \|P(z(k)) - p\| \\ &= \|[\mathcal{P}_r^T(z_r(k)), P_n(z_r(k), g(z_r(k)))]^T - [p_r^T, g(p_r)]^T\| \\ &\to 0, \quad k \to \infty, \quad z(0) \in \mathcal{B}_{\delta'}(p) \cap \hat{\mathcal{Z}}_x.\end{aligned}$$
(13.38)

Using similar arguments as in $i)$, there exists $\delta > 0$ such that if $z_r(0) \in \mathcal{B}_\delta(p_r)$, then $z(0) \in \mathcal{B}_{\delta'}(p) \cap \hat{\mathcal{Z}}_x$. Thus, it follows from (13.38) that

$$\|z_r(k+1) - p_r\| = \|\mathcal{P}_r(z_r(k)) - p_r\| \to 0, \quad k \to \infty, \quad z_r(0) \in \mathcal{B}_\delta(p_r),$$
(13.39)

which establishes asymptotic stability of p_r for (13.28). $\qquad\square$

13.4 Limit Cycle Analysis of a Verge and Foliot Clock Escapement

In the remainder of this chapter, we use impulsive differential equations and Poincaré maps to model the dynamics of a verge and foliot clock escapement mechanism, determine conditions under which the dynamical system possesses a stable limit cycle, and analyze the period and amplitude of the oscillations of this limit cycle. Although clocks are one of the most important instruments in science and technology, it is not widely appreciated that feedback control has been essential to the development of accurate timekeeping. As described in [120], feedback control played a role in the operation of ancient water clocks in the form of regulated valves. Alternative timekeeping devices, such as sundials, hourglasses, and burning candles, were developed as well, although each of these had disadvantages.

Mechanical clocks were developed in the 12th century to keep both time and the calendar, including the prediction of astronomical events [49, 97]. Although early mechanical clocks were expensive, large, and not especially accurate (they were often set using sundials), this technology for timekeeping had inherent advantages of accuracy and reliability as mechanical technology improved.

The crucial component of a mechanical clock is the *escapement*, which is a device for producing precisely regulated motion. The earliest escapement is the weight-driven *verge and foliot* escapement, which dates from the late 13th century. The feedback nature of the verge and foliot escapement is discussed in [102], which points out that this mechanism is a work of "pure genius."[1] The authors of [102] have performed an important service in identifying this device as a contribution of automatic control technology.

It is interesting to note that the verge and foliot escapement was

---

[1]It is important to note here that far more complex and ingenious differential geared mechanisms were inspired, developed, and built by the ancient Greeks over twelve centuries before the development of the first mechanical clock. These technological marvels included Heron's automata and, arguably the greatest fundamental mechanical invention of all time, the Antikythera mechanism. The Antikythera mechanism, most likely inspired by Archimedes, was built around 76 B.C. and was a device for calculating the motions of the stars and planets, as well as for keeping time and calendar. This first analog computer involving a complex array of meshing gears was a quintessential hybrid system that unequivocally shows the singular sophistication, capabilities, and imagination of the ancient Greeks. And as in the case of the origins of much of modern science and mathematics, it shows that modern engineering can also be traced back to the great cosmic theorists of ancient Greece.

the only mechanical escapement known from the time of its inception until the middle of the 17th century. In 1657 Huygens modified the verge and foliot escapement by replacing the foliot with a pendulum swinging in a vertical plane and the crown gear mounted horizontally. However, the basic paddle/gear teeth interaction remained the same. The next escapement innovation was the invention of the anchor or recoil escapement by Hooke in 1651 in which a pendulum-driven lever arm alternately engages gear teeth in the same plane. Subsequent developments invoking additional refinements include the deadbeat escapement of Graham and the grasshopper escapement of Harrison. The latter device played a crucial role when the British government sought novel technologies for determining longitude at sea [156]. For details on these and other escapements, see [22, 48, 72, 141]. Since escapements produce oscillations from stored energy, they can be analyzed as self-oscillating dynamical systems. For details, see [5].

The present development considers only the verge and foliot escapement, which consists of a pair of rotating rigid bodies which interact through collisions. These collisions constitute feedback action which give rise to a limit cycle. This limit cycle provides the crown gear with a constant average angular velocity that determines the clock speed for accurate timekeeping.

The verge and foliot is analyzed in [102] under elastic and inelastic conditions. For the latter case expressions were obtained for the period of the limit cycle and for the crown gear angular velocity at certain points in time. Because of the presence of collisions, a hybrid continuous-discrete model was used to account for instantaneous changes in velocity.

13.5 Modeling

The verge and foliot escapement mechanism shown in Figure 13.1 consists of two rigid bodies rotating on bearings. For simplicity we assume that these bearings are frictionless. The *crown gear* has teeth spaced equally around its perimeter. The *verge and foliot*, which henceforth will be referred to as the *verge*, has two paddles that engage the teeth of the crown gear through alternating collisions. We ignore sliding of the paddles along the crown gear teeth, which may occur in practice. For the orientation shown in Figure 13.1, there is an upper paddle and a lower paddle.

Collisions involving the upper paddle impart a positive torque impulse to the verge, while those involving the lower paddle impart a

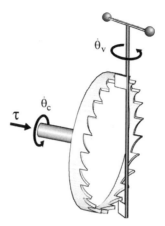

Figure 13.1 Verge and foliot escapement mechanism. The angular velocities
of the crown gear and verge are $\dot{\theta}_c$ and $\dot{\theta}_v$, respectively, with the
sign convention shown. There is a constant torque τ applied to
the crown gear with positive direction shown.

negative torque impulse to the verge. Each collision imparts a negative torque impulse which acts to retard the motion of the crown gear.
The mechanism is driven by a constant torque applied to the crown
gear. This torque is usually provided by a mass hanging from a rope
which is wound around the shaft. The verge spins freely at all times
except at the instant a collision takes place. Energy is assumed to
leave the system only through the collisions. The amount of energy
lost during each collision is a function of the system geometry as well
as the coefficient of restitution e realized in the collision.

The crown gear and verge have inertias I_c and I_v, contact radii r_c
and r_v, and angular velocities $\dot{\theta}_c$ and $\dot{\theta}_v$, respectively. The velocities
immediately before and after a collision are denoted by the subscripts
0 and 1, respectively, as in $\dot{\theta}_{c_0}$ and $\dot{\theta}_{c_1}$. The motion of the crown gear
and verge is governed by the differential equations

$$\ddot{\theta}_c(t) = \frac{1}{I_c}\tau - \frac{r_c}{I_c}F(\theta_c(t), \theta_v(t), \dot{\theta}_c(t), \dot{\theta}_v(t)), \qquad (13.40)$$

$$\ddot{\theta}_v(t) = \begin{cases} +\frac{r_v}{I_v}F(\theta_c(t), \theta_v(t), \dot{\theta}_c(t), \dot{\theta}_v(t)), & \text{upper}, \\[2mm] -\frac{r_v}{I_v}F(\theta_c(t), \theta_v(t), \dot{\theta}_c(t), \dot{\theta}_v(t)), & \text{lower}, \end{cases} \qquad (13.41)$$

where the first expression in (13.41) applies to collisions between the
crown gear and the upper paddle, and the second expression applies to
collisions between the crown gear and the lower paddle. The function

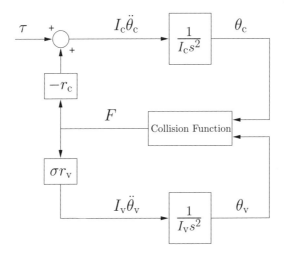

Figure 13.2 System block diagram showing the interconnection of the crown
gear and verge rigid bodies through the collision block.

$F(\theta_c(t), \theta_v(t), \dot\theta_c(t), \dot\theta_v(t))$ is the collision force, which is zero when the
crown gear and verge are not in contact and is impulsive at the instant
of impact. The collision force function F acts equally and oppositely
on the crown gear and verge. Defining

$$\sigma \triangleq \begin{cases} +1, & \text{upper,} \\ -1, & \text{lower,} \end{cases} \tag{13.42}$$

(13.41) can be written in the form

$$\ddot\theta_v(t) = \sigma \frac{r_v}{I_v} F(\theta_c(t), \theta_v(t), \dot\theta_c(t), \dot\theta_v(t)). \tag{13.43}$$

A system diagram is shown in Figure 13.2.

To determine the collision force function, we integrate (13.40) and
(13.43) across a collision event to obtain

$$\dot\theta_{c_1} - \dot\theta_{c_0} = \lim_{\Delta t \to 0} \left(\frac{1}{I_c} \int_{t-\Delta t}^{t+\Delta t} \tau \, ds - \frac{r_c}{I_c} \int_{t-\Delta t}^{t+\Delta t} F(s) \, ds \right), \tag{13.44}$$

$$\dot\theta_{v_1} - \dot\theta_{v_0} = \lim_{\Delta t \to 0} \left(\sigma \frac{r_v}{I_v} \int_{t-\Delta t}^{t+\Delta t} F(s) \, ds \right). \tag{13.45}$$

Eliminating the integrated collision force from (13.44) and (13.45)
yields

$$\frac{\sigma I_v}{r_v} \dot\theta_{v_0} + \frac{I_c}{r_c} \dot\theta_{c_0} = \frac{\sigma I_v}{r_v} \dot\theta_{v_1} + \frac{I_c}{r_c} \dot\theta_{c_1}, \tag{13.46}$$

which is an expression of conservation of linear momentum at the instant of a collision. Expression (13.46) can be rewritten as

$$M_v V_{v0} + M_c V_{c0} = M_v V_{v1} + M_c V_{c1}, \qquad (13.47)$$

where $M_c \triangleq \frac{I_c}{r_c^2}$ and $M_v \triangleq \frac{I_v}{r_v^2}$ are the effective crown gear mass and effective verge mass, respectively, and $V_c \triangleq r_c \dot{\theta}_c$ and $V_v \triangleq \sigma r_v \dot{\theta}_v$ are the tangential velocities of the crown gear and the verge, respectively.

The coefficient of restitution e relates the linear velocities of the crown gear and the verge before and after the collision according to

$$V_{c1} - V_{v1} = -e(V_{c0} - V_{v0}), \qquad (13.48)$$

which accounts for the loss of kinetic energy in a collision. Solving (13.46) and (13.48) yields

$$\Delta \dot{\theta}_c = -\frac{M_v(1+e)}{r_c(M_v + M_c)} V_{c0} + \sigma \frac{M_v(1+e)}{r_c(M_v + M_c)} V_{v0},$$

$$\Delta \dot{\theta}_v = \sigma \frac{M_c(1+e)}{r_v(M_v + M_c)} V_{c0} - \frac{M_c(1+e)}{r_v(M_v + M_c)} V_{v0}, \qquad (13.49)$$

where

$$\Delta \dot{\theta}_c \triangleq \dot{\theta}_{c_1} - \dot{\theta}_{c_0}, \qquad \Delta \dot{\theta}_v \triangleq \dot{\theta}_{v_1} - \dot{\theta}_{v_0}, \qquad (13.50)$$

are the impulsive changes in angular velocity when a collision occurs. These quantities depend on the geometry as well as the velocities immediately before the collision. The integral of the impulsive force function over a collision event is

$$\int_{t_0}^{t_1} F(s)\, ds = \frac{M_c M_v (1+e)}{M_v + M_c} (V_{c0} - V_{v0}), \qquad (13.51)$$

where t_0 is a time slightly before the collision and t_1 is a time slightly after the collision.

13.6 Impulsive Differential Equation Model

In this section, we rewrite the equations of motion of the escapement mechanism in the form of an impulsive differential equation. To describe the dynamics of the verge and foliot escapement mechanism as an impulsive differential equation, define the state

$$x = [\, x_1 \ x_2 \ x_3 \ x_4 \,]^{\mathrm{T}} \triangleq [\, \theta_c \ \theta_v \ \dot{\theta}_c \ \dot{\theta}_v \,]^{\mathrm{T}}, \qquad (13.52)$$

where x_1 is the position of the crown gear, that is, the counterclock-wise angle swept by the line connecting the center of the crown gear and the zeroth tooth from the 12 o'clock position; x_2 is the position of the verge, that is, the deviation of the mean line of the angular offset between two paddles from the vertical plane perpendicular to the plane of the crown gear; x_3 is the angular velocity of the crown gear; and x_4 is the angular velocity of the verge.

Between collisions the state satisfies

$$\dot{x}(t) = \begin{bmatrix} 0 & 0 & 1 & 0 \\ 0 & 0 & 0 & 1 \\ 0 & 0 & 0 & 0 \\ 0 & 0 & 0 & 0 \end{bmatrix} x(t) + \begin{bmatrix} 0 \\ 0 \\ 1/I_c \\ 0 \end{bmatrix} \tau, \tag{13.53}$$

while the resetting function is given by

$$f_d(x) = \begin{bmatrix} 0 & 0 & 0 & 0 \\ 0 & 0 & 0 & 0 \\ 0 & 0 & -r_c\, G_c & \sigma r_v\, G_c \\ 0 & 0 & \sigma r_c\, G_v & -r_v\, G_v \end{bmatrix} x, \tag{13.54}$$

where

$$G_c \triangleq \frac{\frac{I_v}{r_v^2}(1+e)}{r_c \left(\frac{I_v}{r_v^2} + \frac{I_c}{r_c^2} \right)}, \qquad G_v \triangleq \frac{\frac{I_c}{r_c^2}(1+e)}{r_v \left(\frac{I_v}{r_v^2} + \frac{I_c}{r_c^2} \right)}. \tag{13.55}$$

The resetting set is

$$\mathcal{Z}_x = \left\{ \bigcup_{m=0}^{n} \mathcal{Z}_{x\,m}^{\text{upper}} \right\} \bigcup \left\{ \bigcup_{m=0}^{n} \mathcal{Z}_{x\,m}^{\text{lower}} \right\}, \tag{13.56}$$

where, for $m = 0, \ldots, n$,

$$\mathcal{Z}_{x\,m}^{\text{upper}} = \{ x : r_c \sin(x_1 - m\,\alpha_c) = r_v \tan(x_2 + \alpha_v/2),$$
$$r_c x_3 - r_v x_4 > 0,$$
$$(m - 1/2)\alpha_c + 2p\pi \leq x_1 \leq (m + 1/2)\alpha_c + 2p\pi,$$
$$p \in \{0, 1, 2, \ldots\}\}, \tag{13.57}$$
$$\mathcal{Z}_{x\,m}^{\text{lower}} = \{ x : r_c \sin(m\alpha_c - x_1) = r_v \tan(-x_2 + \alpha_v/2),$$
$$r_c x_3 + r_v x_4 > 0,$$
$$(m - 1/2)\alpha_c + (2p - 1)\pi \leq x_1 \leq (m + 1/2)\alpha_c + (2p - 1)\pi,$$
$$p \in \{0, 1, 2, \ldots\}\}, \tag{13.58}$$

where α_c is the angle between neighboring teeth on the crown gear, α_v is the angular offset of the paddles about the vertical axis, m is the index of the crown gear tooth involved in the collision, and p is the number of full rotations of the crown gear. The crown gear teeth are numbered from 0 to n clockwise, or opposite the direction of increasing θ_c, beginning at $\theta_c = 0$. There must be an odd number of crown gear teeth for the mechanism to function correctly, and thus n is even.

13.7 Characterization of Periodic Orbits

In this section we characterize periodic orbits of the clock escapement mechanism, which henceforth we denote by $\mathcal{G}$. First we integrate the continuous-time dynamics (13.53) to obtain

$$\theta_{c_2} = \theta_{c_0} + \dot{\theta}_{c_1} \Delta t + \frac{\tau}{2I_c} \Delta t^2, \tag{13.59}$$

$$\theta_{v_2} = \theta_{v_0} + \dot{\theta}_{v_1} \Delta t, \tag{13.60}$$

where θ_{c_2} and θ_{v_2} are evaluated immediately before the next collision and Δt is the elapsed time between two successive collisions. For an initial collision involving the upper paddle we have

$$r_c \sin(\theta_{c_0} - m\alpha_c) = r_v \tan(\theta_{v_0} + \alpha_v/2). \tag{13.61}$$

The index m' of the crown gear tooth involved in the subsequent lower collision is given by

$$m' = m + \pi/\alpha_c + 1/2, \tag{13.62}$$

so that the condition

$$r_c \sin(m\alpha_c + \pi + \alpha_c/2 - \theta_{c_2}) = r_v \tan(-\theta_{v_2} + \alpha_v/2) \tag{13.63}$$

must be satisfied at the lower collision. Substituting (13.59) and (13.60) into (13.63) yields

$$r_c \sin\left(\theta_{c_0} + \dot{\theta}_{c_1}\Delta t + \frac{\tau}{2I_c}\Delta t^2 - \left(m + \frac{1}{2}\right)\alpha_c\right)$$
$$= r_v \tan\left(-\theta_{v_0} - \dot{\theta}_{v_1}\Delta t + \frac{\alpha_v}{2}\right). \tag{13.64}$$

A small-angle approximation of (13.61) and (13.64) implies

$$r_c(\theta_{c_0} - m\alpha_c) = r_v\left(\theta_{v_0} + \frac{\alpha_v}{2}\right), \tag{13.65}$$

$$r_{\mathrm{c}} \left(\theta_{\mathrm{c}0} + \dot{\theta}_{\mathrm{c}1} \Delta t + \frac{\tau}{2I_{\mathrm{c}}} \Delta t^2 - \left(m + \frac{1}{2} \right) \alpha_{\mathrm{c}} \right)$$

$$= r_{\mathrm{v}} \left(-\theta_{\mathrm{v}0} - \dot{\theta}_{\mathrm{v}1} \Delta t + \frac{\alpha_{\mathrm{v}}}{2} \right). \tag{13.66}$$

Subtracting (13.65) from (13.66) yields

$$r_{\mathrm{c}}\dot{\theta}_{\mathrm{c}1} \Delta t + \frac{r_{\mathrm{c}}\tau}{2I_{\mathrm{c}}} \Delta t^2 - \frac{1}{2} r_{\mathrm{c}}\alpha_{\mathrm{c}} = -r_{\mathrm{v}}\dot{\theta}_{\mathrm{v}1} \Delta t - 2r_{\mathrm{v}}\theta_{\mathrm{v}0}. \tag{13.67}$$

Analogous expressions hold for collisions involving the lower paddle. Solving for Δt yields

$$\Delta t = \frac{-d + \sqrt{d^2 + (r_{\mathrm{c}}\,\alpha_{\mathrm{c}} - 4\sigma r_{\mathrm{v}}\theta_{\mathrm{v}0})r_{\mathrm{c}}\tau/I_{\mathrm{c}}}}{r_{\mathrm{c}}\tau/I_{\mathrm{c}}}, \tag{13.68}$$

where

$$d = \frac{[(2+e)M_{\mathrm{c}} - eM_{\mathrm{v}}]\,r_{\mathrm{c}}\dot{\theta}_{\mathrm{c}0}}{M_{\mathrm{v}} + M_{\mathrm{c}}} + \sigma \frac{[(2+e)M_{\mathrm{v}} - eM_{\mathrm{c}}]\,r_{\mathrm{v}}\dot{\theta}_{\mathrm{v}0}}{M_{\mathrm{v}} + M_{\mathrm{c}}}. \tag{13.69}$$

Furthermore, the kinetic energy ΔT lost in a collision is given by

$$\Delta T = \frac{M_{\mathrm{v}}\,M_{\mathrm{c}}}{2\,(M_{\mathrm{v}} + M_{\mathrm{c}})} \,(e^2 - 1)\,(r_{\mathrm{v}}\,\dot{\theta}_{\mathrm{v}0} - \sigma r_{\mathrm{c}}\,\dot{\theta}_{\mathrm{c}0})^2. \tag{13.70}$$

Next, we specify conditions that characterize a periodic orbit in the $(\dot{\theta}_{\mathrm{c}}, \dot{\theta}_{\mathrm{v}})$ plane. The first condition

$$\dot{\theta}_{\mathrm{v}1} = -\dot{\theta}_{\mathrm{v}0} \tag{13.71}$$

requires the verge to reverse direction at every collision. This condition ensures that the absolute value of the verge speed is constant with time. On the other hand, the crown gear will lose speed with every collision and then gain speed between collisions. Thus, the second condition

$$\dot{\theta}_{\mathrm{c}1} = \dot{\theta}_{\mathrm{c}0} - \frac{\tau\,\Delta t}{I_{\mathrm{c}}} \tag{13.72}$$

requires the crown gear speed to be the same before each collision. The third condition

$$\theta_{\mathrm{v}0} = -\theta_{\mathrm{v}2} \tag{13.73}$$

requires the range of motion of the verge between collisions to be centered at $\theta_{\mathrm{v}0} = 0$. This condition keeps the motion of the verge from wandering out of the range of angles within which the mechanism

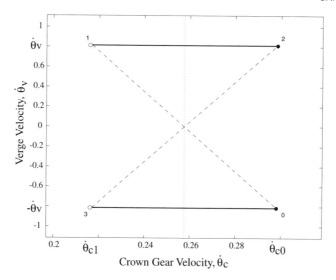

Figure 13.3 Velocity phase portrait of a representative periodic orbit. Instant 0 is prior to a collision with the upper paddle, 1 is after the collision, 2 is prior to the successive lower paddle collision, and 3 is after the second collision. The continuous-time trajectories are shown with solid lines and the impulsive jumps are shown with dashed lines. The average crown gear velocity is shown with a vertical dotted line.

will work properly. A representative periodic orbit satisfying (13.71), (13.72), and (13.73) is shown in Figure 13.3.

Next, we derive some properties of periodic orbits satisfying (13.71), (13.72), and (13.73). The average crown gear velocity $\overline{\dot{\theta}_c}$ of a periodic orbit is given by

$$\overline{\dot{\theta}_c} = \frac{\Delta T}{\tau \, \Delta t}. \tag{13.74}$$

To obtain an expression for $\overline{\dot{\theta}_c}$ as a function of the applied torque and geometric parameters it follows from (13.46), (13.48), and (13.68)–(13.74) that

$$\overline{\dot{\theta}_c} = \frac{\sqrt{\tau}}{2r_c} \sqrt{\left(\frac{1-e}{1+e}\right) \frac{(M_c + M_v)\alpha_c}{M_c M_v}}. \tag{13.75}$$

Figure 13.4 shows the sensitivity of $\overline{\dot{\theta}_c}$ to changes in the parameters e and M_v.

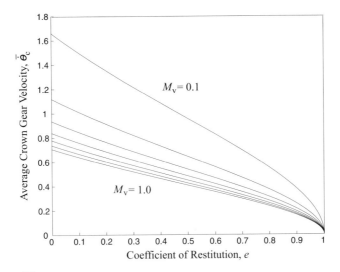

Figure 13.4 $\overline{\dot{\theta}}_c$ versus e for several values of M_v. The system parameters are $M_v = 0.10, 0.25, 0.40, 0.55, 0.70, 0.85$, and 1.00, increasing from the top curve to the bottom curve, $M_c = 1$, $r_c = 1$, $\tau = 1$, and $\alpha_c = 1$.

The crown gear velocity before a collision is given by

$$\dot{\theta}_{c0} = \frac{(1-e)\,M_c + 2\,M_v}{(1-e)(M_c + M_v)}\,\overline{\dot{\theta}}_c, \qquad (13.76)$$

and the crown gear velocity after a collision is given by

$$\dot{\theta}_{c1} = \frac{(1-e)\,M_c - 2\,e\,M_v}{(1-e)(M_c + M_v)}\,\overline{\dot{\theta}}_c. \qquad (13.77)$$

The verge velocity is given by

$$\dot{\theta}_v = \pm\frac{1}{2\,r_v}\sqrt{\left(\frac{1+e}{1-e}\right)\frac{M_c\,\alpha_c\,\tau}{(M_c + M_v)\,M_v}}, \qquad (13.78)$$

where the verge velocity is positive following a collision involving the upper paddle and negative following a collision involving the lower paddle. Finally, the period of this periodic orbit is

$$2\,\Delta t = \frac{\alpha_c}{\overline{\dot{\theta}}_c}. \qquad (13.79)$$

13.8 Limit Cycle Analysis of the Clock Escapement Mechanism

In this section, we use the results of Section 13.3 to show that the periodic orbit generated by the escapement mechanism is asymptotically stable. For convenience, we denote the periodic orbit values of $\dot{\theta}_{c1}$, $\dot{\theta}_{c0}$, and $\dot{\theta}_{v}$ given by (13.77), (13.76), and (13.78) by a, b, and $\pm c$, respectively. The following assumption is needed.

Assumption 13.2 $\alpha_{v} < \frac{\pi}{2}$.

It follows from (13.73) that between consecutive collisions *on the periodic orbit*, the mean line of the angular offset between two paddles sweeps an angle of α_{v}, that is, $\alpha_{v} = c\Delta t$. This assures the existence of a fixed point of (13.28) for the escapement mechanism. Furthermore, Assumption 13.2 assures that $\mathcal{X} : \mathcal{D} \to \mathbb{R}$ is continuously differentiable. To see this, note that $\mathcal{X}(\cdot)$ is determined by (13.57) and (13.58). Now, in order for $\mathcal{X}(\cdot)$ to be continuously differentiable we need to avoid $x_2 + \frac{\alpha_{v}}{2} = \pm\frac{\pi}{2}$ and $-x_2 + \frac{\alpha_{v}}{2} = \pm\frac{\pi}{2}$. Since the position of the verge is always within the range $(-\frac{\alpha_{v}}{2} - \varepsilon, \frac{\alpha_{v}}{2} + \varepsilon)$, where $\varepsilon > 0$ is small, it follows that in order to avoid the singularity $\pm\frac{\pi}{2}$ we need to make sure that $\alpha_{v} + \varepsilon \neq \frac{\pi}{2}$ which can be achieved by assuming $\alpha_{v} < \frac{\pi}{2}$.

Without loss of generality, suppose that the trajectory $s(t, x_0)$, $t \geq 0$, of $\mathcal{G}$ starts from a point in the four-dimensional state space associated with the upper paddle collision such that its projection onto the three-dimensional subspace lies in a sufficiently small neighborhood of the point $(x_2, x_3, x_4) = (-\frac{\alpha_{v}}{2}, b, -c)$. Then, we can construct a three-dimensional, discrete-time system that identifies the next point (x_2, x_3, x_4) on the trajectory right before the next upper paddle collision. This iterative procedure can be captured by the nonlinear difference equation

$$
\begin{bmatrix} x_2(k+1) \\ x_3(k+1) \\ x_4(k+1) \end{bmatrix} = \begin{bmatrix} f_2(x_2(k), x_3(k), x_4(k)) \\ f_3(x_2(k), x_3(k), x_4(k)) \\ f_4(x_2(k), x_3(k), x_4(k)) \end{bmatrix}, \tag{13.80}
$$

where $f_2(\cdot, \cdot, \cdot)$, $f_3(\cdot, \cdot, \cdot)$, and $f_4(\cdot, \cdot, \cdot)$ are given in the Appendix A. It follows from (13.71)–(13.73) that the point $(-\frac{\alpha_{v}}{2}, b, -c)$ is a fixed point of (13.80). Next, it follows from standard discrete-time stability theory that if $\rho(J(-\frac{\alpha_{v}}{2}, b, -c)) < 1$, where

$$
J\left(-\frac{\alpha_{v}}{2}, b, -c\right)
$$

$$
\triangleq
\left.
\begin{bmatrix}
\frac{\partial f_2(x_2,x_3,x_4)}{\partial x_2} & \frac{\partial f_2(x_2,x_3,x_4)}{\partial x_3} & \frac{\partial f_2(x_2,x_3,x_4)}{\partial x_4} \\[4pt]
\frac{\partial f_3(x_2,x_3,x_4)}{\partial x_2} & \frac{\partial f_3(x_2,x_3,x_4)}{\partial x_3} & \frac{\partial f_3(x_2,x_3,x_4)}{\partial x_4} \\[4pt]
\frac{\partial f_4(x_2,x_3,x_4)}{\partial x_2} & \frac{\partial f_4(x_2,x_3,x_4)}{\partial x_3} & \frac{\partial f_4(x_2,x_3,x_4)}{\partial x_4}
\end{bmatrix}
\right|_{(x_2,x_3,x_4)=(-\frac{\alpha_v}{2},b,-c)}
\tag{13.81}
$$

and $\rho(\cdot)$ denotes the spectral radius, then the point $(x_2, x_3, x_4) = (-\frac{\alpha_v}{2}, b, -c)$ is a locally asymptotically stable fixed point of (13.80).

Next, we introduce the discrete-time dynamical system

$$
\begin{bmatrix}
\hat{x}_2(k+1) \\
\hat{x}_3(k+1) \\
\hat{x}_4(k+1)
\end{bmatrix}
=
\begin{bmatrix}
f_2^{(n)}(\hat{x}_2(k), \hat{x}_3(k), \hat{x}_4(k)) \\
f_3^{(n)}(\hat{x}_2(k), \hat{x}_3(k), \hat{x}_4(k)) \\
f_4^{(n)}(\hat{x}_2(k), \hat{x}_3(k), \hat{x}_4(k))
\end{bmatrix},
\tag{13.82}
$$

where $f_i^{(n)}(\hat{x}_2(k), \hat{x}_3(k), \hat{x}_4(k))$, $i = 2, 3, 4$, denotes the n-time composition operator of $f_i(\cdot, \cdot, \cdot)$, $i = 2, 3, 4$, with $f_2(\cdot, \cdot, \cdot)$, $f_3(\cdot, \cdot, \cdot)$, and $f_4(\cdot, \cdot, \cdot)$ and n is the number of crown gear teeth. Note that $(\hat{x}_2, \hat{x}_3, \hat{x}_4) = (-\frac{\alpha_v}{2}, b, -c)$ is a fixed point of (13.82).

Proposition 13.4 *Consider the impulsive dynamical system $\mathcal{G}$. If $\rho(J(-\frac{\alpha_v}{2}, b, -c)) < 1$, then the point $(\hat{x}_2, \hat{x}_3, \hat{x}_4) = (-\frac{\alpha_v}{2}, b, -c)$ is a locally asymptotically stable fixed point of (13.82). Alternatively, if $\rho(J(-\frac{\alpha_v}{2}, b, -c)) > 1$, then the fixed point $(-\frac{\alpha_v}{2}, b, -c)$ of (13.82) is unstable.*

Proof. Given a continuously differentiable function $f : \mathbb{R}^n \to \mathbb{R}^n$, consider the N-time composition operator of $f(\cdot)$ with itself, that is, $h(x) \triangleq f^{(N)}(x)$, $x \in \mathbb{R}^n$, $N \in \overline{\mathbb{Z}}_+$. Now, using the chain rule for vector valued functions it follows that

$$
\frac{\partial h(x)}{\partial x} = \left.\frac{\partial f(s_1)}{\partial s_1}\right|_{s_1=f^{(N-1)}(x)} \cdot \left.\frac{\partial f(s_2)}{\partial s_2}\right|_{s_2=f^{(N-2)}(x)}
$$
$$
\cdots \left.\frac{\partial f(s_N)}{\partial s_N}\right|_{s_N=f^0(x)}, \quad x \in \mathbb{R}^n,
\tag{13.83}
$$

where $f^0(x) \triangleq x$. Next, since $(-\frac{\alpha_v}{2}, b, -c)$ is a fixed point of the system (13.80), it follows that the Jacobian matrix $\hat{J}(\hat{x}_2(k), \hat{x}_3(k), \hat{x}_4(k))$ of the discrete-time system (13.82) evaluated at the fixed point $(-\frac{\alpha_v}{2}, b, -c)$ is given by the N-time product of $J(-\frac{\alpha_v}{2}, b, -c)$, that is,

$$
\hat{J}\left(-\frac{\alpha_v}{2}, b, -c\right) = J^N\left(-\frac{\alpha_v}{2}, b, -c\right).
\tag{13.84}
$$

Hence,

$$
\rho\left(\hat{J}\left(-\frac{\alpha_v}{2}, b, -c\right)\right) = \rho^N\left(J\left(-\frac{\alpha_v}{2}, b, -c\right)\right).
\tag{13.85}
$$

Now, it follows that if $\rho(J(-\frac{\alpha_v}{2}, b, -c)) < 1$, then $\rho(\hat{J}(-\frac{\alpha_v}{2}, b, -c)) < 1$, which implies that $(-\frac{\alpha_v}{2}, b, -c)$ is a locally asymptotically stable fixed point of (13.82). Alternatively, if $\rho(J(-\frac{\alpha_v}{2}, b, -c)) > 1$, then $\rho(\hat{J}(-\frac{\alpha_v}{2}, b, -c)) > 1$, which implies that the fixed point $(-\frac{\alpha_v}{2}, b, -c)$ of (13.82) is unstable.   □

Next, it follows from the uniqueness of solutions of $\mathcal{G}$ and the fact that the initial conditions (x_1', x_2, x_3, x_4) and (x_1, x_2, x_3, x_4), where $x_1 = x_1' + 2\pi$, give rise to identical solutions for $\mathcal{G}$, that the point $x_0 = (0, -\frac{\alpha_v}{2}, b, -c)$ is a fixed point for the discrete-time system capturing the state of $\mathcal{G}$ immediately before every $(np+1)$th upper paddle collision for $p = 0, 1, 2, \ldots$. Note that whenever an upper paddle collision occurs, the position of the crown gear is completely defined by the position of the verge, and the relation between them results from the collision condition, that is, $x_1 = f_1(x_2)$, where $f_1 : \mathbb{R} \to \mathbb{R}$ is defined by (13.57). Thus, the aforementioned four-dimensional system has the form

$$
\begin{bmatrix}
\tilde{x}_1(k+1) \\
\tilde{x}_2(k+1) \\
\tilde{x}_3(k+1) \\
\tilde{x}_4(k+1)
\end{bmatrix}
=
\begin{bmatrix}
f_1(f_2^{(n)}(\tilde{x}_2(k), \tilde{x}_3(k), \tilde{x}_4(k))) \\
f_2^{(n)}(\tilde{x}_2(k), \tilde{x}_3(k), \tilde{x}_4(k)) \\
f_3^{(n)}(\tilde{x}_2(k), \tilde{x}_3(k), \tilde{x}_4(k)) \\
f_4^{(n)}(\tilde{x}_2(k), \tilde{x}_3(k), \tilde{x}_4(k))
\end{bmatrix},
\qquad (13.86)
$$

where $f_1(\cdot)$ is given by

$$
f_1(x_2) = \arcsin\left(\frac{r_v}{r_c} \tan\left(x_2 + \frac{\alpha_v}{2}\right)\right).
\qquad (13.87)
$$

Next, we identify the periodic orbit generated by the point $x_0 = (0, -\frac{\alpha_v}{2}, b, -c)$. For any point on this orbit with $(x_3, x_4) = (z, c)$, $z \in (a, b]$, it follows that $z = a + \frac{\tau}{I_c} t_z$, where t_z is the time spanned for the crown gear to restore its velocity from the value of a to z. Thus, this point can be characterized by

$$
x_z =
\begin{bmatrix}
x_{10} + \frac{a(z-a)}{\tau} I_c + \frac{(z-a)^2}{2\tau} I_c \\
-\frac{\alpha_v}{2} + \frac{c(z-a)}{\tau} I_c \\
z \\
c
\end{bmatrix},
\qquad (13.88)
$$

where $x_{10} = l\alpha_c, l = 0, 1, 2, \ldots, n - 1$. Similarly, every point on the orbit with $(x_3, x_4) = (z, -c), z \in (a, b]$, is characterized by

$$
x'_z = \begin{bmatrix} x'_{10} + \frac{a(z-a)}{\tau}I_c + \frac{(z-a)^2}{2\tau}I_c \\ \frac{\alpha_v}{2} - \frac{c(z-a)}{\tau}I_c \\ z \\ -c \end{bmatrix}, \tag{13.89}
$$

where $x'_{10} = \frac{2l+1}{2}\alpha_c, l = 0, 1, 2, \ldots, n - 1$. Since the initial conditions (x'_1, x_2, x_3, x_4) and (x_1, x_2, x_3, x_4), where $x_1 = x'_1 + 2\pi$, give rise to identical solutions for $\mathcal{G}$, it follows that $\mathcal{O} \triangleq \{y \in \mathbb{R}^4 : y = x_z\} \cup \{y \in \mathbb{R}^4 : y = x'_z\}$ is the periodic orbit of $\mathcal{G}$. The expressions given by (13.88) and (13.89) imply that points $x_0 = (x_{10}, -\frac{\alpha_v}{2}, b, -c) \in \mathcal{Z}_x$ or $x_0 = (x'_{10}, \frac{\alpha_v}{2}, b, c) \in \mathcal{Z}_x$ generate $\mathcal{O}$. Next, we show that $\mathcal{O}$ is asymptotically stable. For this result let $\mathcal{D}$ be a sufficiently small neighborhood of $\mathcal{O}$ for which the state of $\mathcal{G}$ is defined.

Theorem 13.2 *Consider the impulsive dynamical system $\mathcal{G}$. Then the following statements hold:*

i) *If $\rho(J(-\frac{\alpha_v}{2}, b, -c)) < 1$, then the periodic orbit $\mathcal{O}$ of $\mathcal{G}$ generated by $x_0 = (x_{10}, -\frac{\alpha_v}{2}, b, -c) \in \mathcal{Z}_x$ or $x_0 = (x'_{10}, \frac{\alpha_v}{2}, b, c) \in \mathcal{Z}_x$ is asymptotically stable.*

ii) *If $\rho(J(-\frac{\alpha_v}{2}, b, -c)) > 1$, then the periodic orbit $\mathcal{O}$ of $\mathcal{G}$ generated by $x_0 = (x_{10}, -\frac{\alpha_v}{2}, b, -c) \in \mathcal{Z}_x$ or $x_0 = (x'_{10}, \frac{\alpha_v}{2}, b, c) \in \mathcal{Z}_x$ is unstable.*

Proof. First, we show that the assumptions of Corollary 13.2 hold for $\mathcal{G}$. To see that Assumption A3 holds, note that for $\mathcal{Z}_{xm}^{\text{upper}}$ given by (13.57) with a small-angle approximation, $\mathcal{X}'(x) = [r_c, -r_v, 0, 0] \neq 0, x \in \mathcal{Z}_{xm}^{\text{upper}}$, and $\frac{\partial \mathcal{X}(x)}{\partial x_1} \neq 0, x \in \mathcal{Z}_{xm}^{\text{upper}}$. Furthermore, for $\mathcal{Z}_{xm}^{\text{lower}}$ given by (13.58) with a small-angle approximation, $\mathcal{X}'(x) = [r_c, r_v, 0, 0] \neq 0, x \in \mathcal{Z}_{xm}^{\text{lower}}$, and $\frac{\partial \mathcal{X}(x)}{\partial x_1} \neq 0, x \in \mathcal{Z}_{xm}^{\text{lower}}$. Note, that in both cases $\mathcal{X}(\cdot)$ is a continuously differentiable function by Assumption 13.2. To see that Assumption A4 holds, note that $\frac{\partial \mathcal{X}(x)}{\partial x}f_c(x) = r_c x_3 - r_v x_4 > 0, x \in \mathcal{Z}_{xm}^{\text{upper}}$, and $\frac{\partial \mathcal{X}(x)}{\partial x}f_c(x) = r_c x_3 + r_v x_4 > 0, x \in \mathcal{Z}_{xm}^{\text{lower}}$. Next, to show i) assume that $\rho(J(-\frac{\alpha_v}{2}, b, -c)) < 1$. Then it follows from Proposition 13.4 that the fixed point $(-\frac{\alpha_v}{2}, b, -c)$ of (13.82) is asymptotically stable, and hence, by Corollary 13.2, the periodic orbit $\mathcal{O}$ of $\mathcal{G}$ is asymptotically stable. Finally, the proof to ii) follows analogously. $\square$

The condition $\rho(J(-\frac{\alpha_v}{2}, b, -c)) < 1$ guarantees (local) asymptotic stability of the escapement mechanism. Alternatively, it follows from physical considerations that for each choice of clock parameters, if the value of the coefficient of restitution e is sufficiently close to 1, then the escapement mechanism dissipates less energy during a collision event than it gains from the rotational torque between collisions. This leads to instability of the mechanism. However, the Jacobian matrix J is sufficiently complex that we have been unable to show analytically the explicit dependence of the spectral radius of J on the parameter e.

13.9 Numerical Simulation of an Escapement Mechanism

In this section we numerically integrate the equations of motion (13.53)–(13.58) to illustrate convergence of the trajectories to a stable limit cycle. We choose the parameters $\tau = 1$ N·m, $e = 0.05$, $I_c = 10$ kg·m^2, $I_v = 0.15$ kg·m^2, $r_c = 1$ m, $r_v = 0.3$ m, and $\alpha_c = 24$ deg. For these parameters it follows from (13.75)–(13.79) that the periodic orbit has an average crown gear velocity of 0.257 rad/sec, a crown gear velocity of 0.297 rad/sec prior to collisions, a crown gear velocity of 0.216 rad/sec after collisions, a verge speed of 0.813 rad/sec, and a period of 1.63 sec. Furthermore, the eigenvalues of the Jacobian matrix (13.81) are $\lambda_1 = -0.7191$, $\lambda_2 = 0.2072$, and $\lambda_3 = -0.0149$, which implies that the fixed point $(-\frac{\alpha_v}{2}, b, -c)$ of (13.82) is asymptotically stable, and hence, by Theorem 13.2 the periodic orbit of the escapement mechanism is asymptotically stable. An initial verge position of $\theta_{v0} = 0$ is chosen for all simulations. We assume that the verge and the crown gear are in contact at the start of the simulation, which determines the crown gear's initial position.

For a collection of four initial conditions, Figure 13.5 shows the trajectories of the system in terms of the verge and crown gear velocities $\dot{\theta}_v$ and $\dot{\theta}_c$. For each choice of initial conditions it can be seen that the trajectory approaches a periodic orbit, which is discontinuous due to the impulsive nature of the collisions. Numerical computation of the amplitude and period of this orbit from the simulation data yields 0.257 rad/sec and 1.63 sec, respectively, which agrees with the values given by (13.75) and (13.79).

The kinetic energy time histories of the verge, crown gear, and total system are shown in Figure 13.6 for the system considered in Figure 13.5. It can be seen that the verge kinetic energy converges, whereas the crown gear and total system kinetic energies converge to periodic

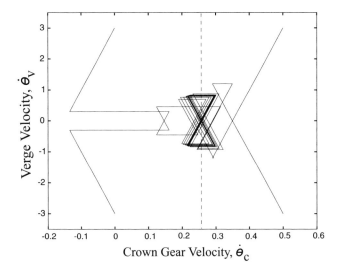

Figure 13.5 Escapement phase portrait from four initial conditions showing
convergence to a periodic orbit. Initial conditions are $(\dot{\theta}_c, \dot{\theta}_v) =$
$(0.5, 3)$, $(0.5, -3)$, $(0, 3)$, and $(0, -3)$. The average crown gear
velocity from (13.75) is plotted with a dotted line.

signals.

For two values of the coefficient of restitution, Figure 13.7 shows
the time history of the crown gear velocity $\dot{\theta}_c$ as it approaches the
periodic orbit given by (13.75) and (13.79). The average velocity
and orbit period are 0.2449 rad/sec and 1.7104 sec, respectively, for
$e = 0.1$, and 0.1354 rad/sec and 3.0942 sec, respectively, for $e = 0.6$.
A full orbit cycle appears as two consecutive saw-tooth patterns in
Figure 13.7.

Finally, instability of the escapement mechanism implies that the
escapement mechanism gains more energy from the rotational torque
between collisions than it loses during collisions. To illustrate that
$\rho(J(-\frac{\alpha_v}{2}, b, -c)) > 1$ leads to an unstable limit cycle, let $\tau = 10$ N·m,
$e = 0.05$, $I_c = 7$ kg·m^2, $I_v = 0.15$ kg·m^2, $r_c = 3$ m, $r_v = 0.3$ m,
and $\alpha_c = 24$ deg, so that the eigenvalues of the Jacobian matrix
(13.81) are $\lambda_1 = 1.8559$, $\lambda_2 = 0.1775$, and $\lambda_3 = 0.0046$. Since the
fixed point $(-\frac{\alpha_v}{2}, b, -c)$ of the discrete-time system (13.82) is unstable, it follows from Theorem 13.2 that the periodic orbit of the
escapement mechanism is also unstable. Figure 13.8 shows the nonconverging velocity phase portrait of the system. Finally, Figure 13.9
shows $\rho(J(-\frac{\alpha_v}{2}, b, -c))$ versus the coefficient of restitution e for several values of the torque τ.

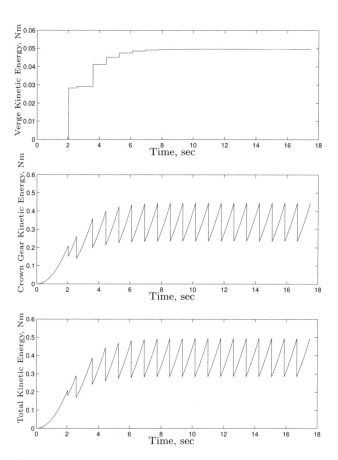

Figure 13.6 Verge, crown gear, and total kinetic energy time histories starting from rest.

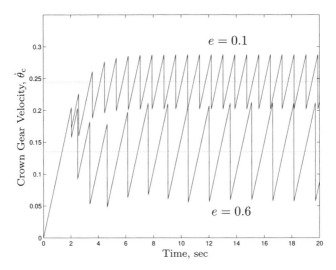

Figure 13.7 Time histories of the crown gear velocity $\dot{\theta}_c$ starting from rest with coefficients of restitution of 0.1 and 0.6. The values of $\overline{\dot{\theta}_c}$ from (13.75) are plotted as dotted lines.

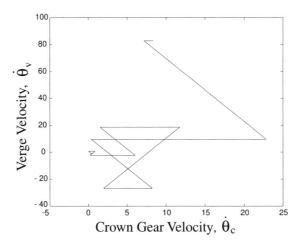

Figure 13.8 Velocity phase portrait for the unstable escapement.

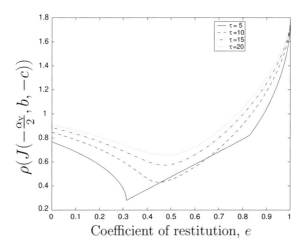

Figure 13.9 $\rho(J(-\frac{\alpha_v}{2}, b, -c))$ versus coefficient of restitution for different values of torque τ.

Appendix A

System Functions for the Clock Escapement Mechanism

The purpose of this appendix is to characterize the functions $f_2(x_2(k), x_3(k), x_4(k))$, $f_3(x_2(k), x_3(k), x_4(k))$, and $f_4(x_2(k), x_3(k), x_4(k))$ given in (13.80). For this analysis we denote the intermediate states of $\mathcal{G}$ between two consecutive upper paddle collisions as follows: (x_1', x_2', x_3', x_4') denotes the state of $\mathcal{G}$ immediately after the upper paddle collision, $(x_1'', x_2'', x_3'', x_4'')$ denotes the state of $\mathcal{G}$ immediately before the lower paddle collision, and $(x_1''', x_2''', x_3''', x_4''')$ denotes the state of $\mathcal{G}$ immediately after the lower paddle collision. Hence, using (13.54), the next point immediately after the initial upper paddle collision is given by

$$
\begin{bmatrix} x_2'(k) \\ x_3'(k) \\ x_4'(k) \end{bmatrix} = \begin{bmatrix} 0 & 0 & 0 \\ 0 & -r_c G_c & r_v G_c \\ 0 & r_c G_v & -r_v G_v \end{bmatrix} \begin{bmatrix} x_2(k) \\ x_3(k) \\ x_4(k) \end{bmatrix} + \begin{bmatrix} x_2(k) \\ x_3(k) \\ x_4(k) \end{bmatrix}. \quad \text{(A.1)}
$$

Since the verge moves with constant velocity and the crown gear moves with acceleration $\frac{\tau}{I_c}$ on the continuous part of the trajectory, the next intermediate point immediately before the lower paddle collision is given by

$$
\begin{bmatrix} x_2''(k) \\ x_3''(k) \\ x_4''(k) \end{bmatrix} = \begin{bmatrix} x_2'(k) \\ x_3'(k) \\ x_4'(k) \end{bmatrix} + \begin{bmatrix} x_4'(k)\Delta t_1(x_2(k), x_3(k), x_4(k)) \\ \frac{\tau}{I_c}\Delta t_1(x_2(k), x_3(k), x_4(k)) \\ 0 \end{bmatrix}, \quad \text{(A.2)}
$$

where $\Delta t_1(x_2(k), x_3(k), x_4(k))$ is the time between successive collisions of the upper and lower paddles, respectively.

Similarly, using (13.54), the next intermediate point immediately after the lower paddle collision is given by

$$
\begin{bmatrix} x_2'''(k) \\ x_3'''(k) \\ x_4'''(k) \end{bmatrix} = \begin{bmatrix} 0 & 0 & 0 \\ 0 & -r_c G_c & -r_v G_c \\ 0 & -r_c G_v & -r_v G_v \end{bmatrix} \begin{bmatrix} x_2''(k) \\ x_3''(k) \\ x_4''(k) \end{bmatrix} + \begin{bmatrix} x_2''(k) \\ x_3''(k) \\ x_4''(k) \end{bmatrix}.
$$
$$
\text{(A.3)}
$$

Hence,

$$
\begin{bmatrix} x_2(k+1) \\ x_3(k+1) \\ x_4(k+1) \end{bmatrix} = \begin{bmatrix} x_2'''(k) \\ x_3'''(k) \\ x_4'''(k) \end{bmatrix} + \begin{bmatrix} x_4'''(k)\Delta t_2((x_2(k), x_3(k), x_4(k)) \\ \frac{\tau}{I_c}\Delta t_2((x_2(k), x_3(k), x_4(k)) \\ 0 \end{bmatrix}
$$

(A.4)

is the state of (13.80) associated with the instant immediately before the next upper paddle collision, where $\Delta t_2(x_2(k), x_3(k), x_4(k))$ is the time between successive collisions of the lower and upper paddles, respectively. Now, using (A.1)–(A.4) we obtain

$$
\begin{bmatrix} x_2(k+1) \\ x_3(k+1) \\ x_4(k+1) \end{bmatrix} = A \begin{bmatrix} x_2(k) \\ x_3(k) \\ x_4(k) \end{bmatrix} + \begin{bmatrix} 0 & 0 & 0 \\ 0 & -r_cG_c & -r_vG_c \\ 0 & -r_cG_v & -r_vG_v \end{bmatrix}
$$
$$
\cdot \begin{bmatrix} x_4'(k)\Delta t_1(x_2(k), x_3(k), x_4(k)) \\ \frac{\tau}{I_c}\Delta t_1(x_2(k), x_3(k), x_4(k)) \\ 0 \end{bmatrix}
$$
$$
+ \begin{bmatrix} 1 & 0 & 0 \\ 0 & 1 & 0 \\ 0 & 0 & 1 \end{bmatrix} \begin{bmatrix} x_4'(k)\Delta t_1(x_2(k), x_3(k), x_4(k)) \\ \frac{\tau}{I_c}\Delta t_1(x_2(k), x_3(k), x_4(k)) \\ 0 \end{bmatrix}
$$
$$
+ \begin{bmatrix} x_4'''(k)\Delta t_2(x_2(k), x_3(k), x_4(k)) \\ \frac{\tau}{I_c}\Delta t_2(x_2(k), x_3(k), x_4(k)) \\ 0 \end{bmatrix},
$$

(A.5)

where

$$
A = \begin{bmatrix} 0 & 0 & 0 \\ 0 & -r_cG_c & -r_vG_c \\ 0 & -r_cG_v & -r_vG_v \end{bmatrix} \begin{bmatrix} 0 & 0 & 0 \\ 0 & -r_cG_c & r_vG_c \\ 0 & r_cG_v & -r_vG_v \end{bmatrix}
$$
$$
+ \begin{bmatrix} 0 & 0 & 0 \\ 0 & -r_cG_c & -r_vG_c \\ 0 & -r_cG_v & -r_vG_v \end{bmatrix}
$$
$$
+ \begin{bmatrix} 0 & 0 & 0 \\ 0 & -r_cG_c & r_vG_c \\ 0 & r_cG_v & -r_vG_v \end{bmatrix} + \begin{bmatrix} 1 & 0 & 0 \\ 0 & 1 & 0 \\ 0 & 0 & 1 \end{bmatrix}.
$$

(A.6)

Next, we compute $\Delta t_1(x_2(k), x_3(k), x_4(k))$ and $\Delta t_2(x_2(k), x_3(k), x_4(k))$ as in Section 13.7. Specifically, integrating the continuous-time dynamics after the upper paddle collision we obtain

$$
x_1''(k) = x_1(k) + x_3'(k)\Delta t_1 + \frac{\tau}{2I_c}\Delta t_1^2,
$$

(A.7)

$$x_2''(k) = x_2(k) + x_4'(k)\Delta t_1, \tag{A.8}$$

where $x_1(k)$ is the position of the crown gear immediately before the initial upper paddle collision. Using (13.57) the condition for the initial collision involving the upper paddle is

$$r_c \sin(x_1(k) - m\alpha_c) = r_v \tan\left(x_2(k) + \frac{\alpha_v}{2}\right), \tag{A.9}$$

where m is the index of the crown gear tooth involved in the collision. Furthermore, using (13.62), (13.63), (A.7), and (A.8) yields

$$r_c \sin\left(-x_1(k) - x_3'(k)\Delta t_1 - \frac{\tau}{2I_c}\Delta t_1^2 + m\alpha_c + \pi + \frac{\alpha_c}{2}\right)$$
$$= r_v \tan\left(-x_2(k) - x_4'(k)\Delta t_1 + \frac{\alpha_v}{2}\right). \tag{A.10}$$

Since $\sin(\pi - \alpha) = \sin\alpha$ and, for small angles, $\sin\alpha \approx \alpha$, we can rewrite (A.9) and (A.10) as

$$r_c(x_1(k) - m\alpha_c) = r_v\left(x_2(k) + \frac{\alpha_v}{2}\right), \tag{A.11}$$

$$r_c\left(x_1(k) + x_3'(k)\Delta t_1 + \frac{\tau}{2I_c}\Delta t_1^2 - m\alpha_c - \frac{\alpha_c}{2}\right)$$
$$= r_v\left(-x_2(k) - x_4'(k)\Delta t_1 + \frac{\alpha_v}{2}\right). \tag{A.12}$$

Subtracting (A.12) from (A.11) yields

$$\frac{r_c\tau}{2I_c}\Delta t_1^2 + (r_c x_3'(k) + r_v x_4'(k))\Delta t_1 + 2r_v x_2(k) - \frac{r_c\alpha_c}{2} = 0, \tag{A.13}$$

which further implies

$$\Delta t_1 = \frac{-(r_c x_3'(k) + r_v x_4'(k))}{\frac{r_c\tau}{I_c}}$$
$$+ \frac{\sqrt{(r_c x_3'(k) + r_v x_4'(k))^2 + \frac{r_c\tau}{I_c}(r_c\alpha_c - 4r_v x_2(k))}}{\frac{r_c\tau}{I_c}}. \tag{A.14}$$

Using (A.1) it follows that

$$r_c x_3'(k) + r_v x_4'(k) = r_c(-r_c G_c x_3(k) + r_v G_c x_4(k) + x_3(k))$$
$$+ r_v(r_c G_v x_3(k) - r_v G_v x_4(k) + x_4(k))$$
$$= (-r_c^2 G_c + r_c + r_v r_c G_v)x_3(k)$$
$$+ (-r_v^2 G_v + r_v + r_c r_v G_c)x_4(k)$$
$$= -\alpha x_3(k) - \beta x_4(k), \tag{A.15}$$

where

$$\alpha = -(-r_{\rm c}^2 G_{\rm c} + r_{\rm c} + r_{\rm v} r_{\rm c} G_{\rm v}), \quad \beta = -(-r_{\rm v}^2 G_{\rm v} + r_{\rm v} + r_{\rm c} r_{\rm v} G_{\rm c}). \quad (A.16)$$

Thus, we obtain

$$\Delta t_1(x_2(k), x_3(k), x_4(k)) = \frac{\alpha x_3(k) + \beta x_4(k)}{\frac{r_{\rm c}\tau}{I_{\rm c}}}$$
$$+ \frac{\sqrt{(\alpha x_3(k) + \beta x_4(k))^2 + \lambda}}{\frac{r_{\rm c}\tau}{I_{\rm c}}}, \quad (A.17)$$

where $\lambda \triangleq \lambda(x_2(k)) = \frac{r_{\rm c}\tau}{I_{\rm c}}(r_{\rm c}\alpha_{\rm c} - 4r_{\rm v}x_2(k))$.

Next, it follows from (A.3) that $x_3'''(k)$ and $x_4'''(k)$ are the velocities of the crown gear and the verge, respectively, immediately after the lower paddle collision, and hence, the positions of the crown gear and the verge before the successive upper paddle collision are given by

$$x_1(k+1) = x_1'''(k) + x_3'''(k)\Delta t_2 + \frac{\tau}{2I_{\rm c}}\Delta t_2^2,$$
$$x_2(k+1) = x_2'''(k) + x_4'''(k)\Delta t_2,$$

where $x_1'''(k)$ and $x_2'''(k)$ are positions of the crown gear and the verge, respectively, immediately after the lower paddle collision. Using a similar procedure as outlined above, the conditions for the lower and upper paddle collisions, respectively, are given by

$$r_{\rm c}\sin\left(m\alpha_{\rm c} + \pi + \frac{\alpha_{\rm c}}{2} - x_1''(k)\right) = r_{\rm v}\tan\left(-x_2'''(k) + \frac{\alpha_{\rm v}}{2}\right), \quad (A.18)$$

$$r_{\rm c}\sin\left(x_1'''(k) + x_3'''(k)\Delta t_2 + \frac{\tau}{2I_{\rm c}}\Delta t_2^2 - (m+1)\alpha_{\rm c}\right)$$
$$= r_{\rm v}\tan\left(x_2'''(k) + x_4'''(k)\Delta t_2 + \frac{\alpha_{\rm v}}{2}\right), \quad (A.19)$$

which can be approximated by

$$r_{\rm c}\left(x_1'''(k) - m\alpha_{\rm c} - \frac{\alpha_{\rm c}}{2}\right) = r_{\rm v}\left(-x_2'''(k) + \frac{\alpha_{\rm v}}{2}\right), \quad (A.20)$$

$$r_{\rm c}\left(x_1'''(k) + x_3'''(k)\Delta t_2 + \frac{\tau}{2I_{\rm c}}\Delta t_2^2 - (m+1)\alpha_{\rm c}\right)$$
$$= r_{\rm v}\left(x_2'''(k) + x_4'''(k)\Delta t_2 + \frac{\alpha_{\rm v}}{2}\right). \quad (A.21)$$

Subtracting (A.21) from (A.20) gives

$$\frac{r_c\tau}{2I_c}\Delta t_2^2 + (r_c x_3'''(k) - r_v x_4'''(k))\Delta t_2 - \frac{r_c\alpha_c}{2} - 2r_v x_2'''(k) = 0,$$

so that

$$\Delta t_2 = \frac{-(r_c x_3'''(k) - r_v x_4'''(k))}{\frac{r_c\tau}{I_c}}$$

$$+ \frac{\sqrt{(r_c x_3'''(k) - r_v x_4'''(k))^2 + \frac{r_c\tau}{I_c}(r_c\alpha_c + 4r_v x_2'''(k))}}{\frac{r_c\tau}{I_c}}. \quad (A.22)$$

From (A.1)–(A.3) and (A.17) it follows that

$$x_3''''(k) = \left((-r_c G_c + 1)^2 - r_v G_c r_c G_v + \alpha\left(-G_c + \frac{1}{r_c}\right)\right) x_3(k)$$

$$+ \left(r_v G_c(-r_c G_c + 1) + r_v^2 G_v G_c - r_v G_c + \beta\left(-G_c + \frac{1}{r_c}\right)\right) x_4(k)$$

$$+ (-G_c + \frac{1}{r_c})\sqrt{(\alpha x_3(k) + \beta x_4(k))^2 + \lambda}. \quad (A.23)$$

Similarly, from (A.1)–(A.3) and (A.17) it follows that

$$x_4''''(k) = (r_c^2 G_c G_v - r_c G_v + (-r_v G_v + 1)r_c G_v - \alpha G_v)x_3(k)$$

$$+ (-r_c G_v r_v G_c + (-r_v G_v + 1)^2 - \beta G_v)x_4(k)$$

$$- G_v\sqrt{(\alpha x_3(k) + \beta x_4(k))^2 + \lambda}. \quad (A.24)$$

Now, using (A.23) and (A.24) we obtain

$$r_c x_3''''(k) - r_v x_4''''(k) = -\gamma x_3(k) - \delta x_4(k) - \nu\sqrt{(\alpha x_3(k) + \beta x_4(k))^2 + \lambda},$$
$$(A.25)$$

where

$$\gamma = -(r_c(-r_c G_c + 1)^2 - 2r_c^2 r_v G_c G_v + r_c G_v^2 r_v^2$$
$$+ \alpha(1 + r_v G_v - r_c G_c)), \quad (A.26)$$

$$\delta = -(-r_v r_c^2 G_c^2 + 2r_v^2 r_c G_v G_c - r_v(-r_v G_v + 1)^2$$
$$+ \beta(1 + r_v G_v - r_c G_c)), \quad (A.27)$$

$$\nu = -(1 + r_v G_v - r_c G_c). \quad (A.28)$$

Next, using $x_2'''(k) = x_2(k) + x_4'(k)\Delta t_1$, (A.1), and (A.17), it follows that

$$x_2'''(k) = x_2(k) + \frac{1}{\frac{r_c\tau}{I_c}}(r_c G_v x_3(k) - r_v G_v x_4(k) + x_4(k))$$

$$\cdot (\alpha x_3(k) + \beta x_4(k) + \sqrt{(\alpha x_3(k) + \beta x_4(k))^2 + \lambda}). \quad (A.29)$$

Thus, (A.22) can be rewritten as

$$\Delta t_2(x_2(k), x_3(k), x_4(k)) = \frac{\gamma x_3(k) + \delta x_4(k)}{\frac{r_c \tau}{I_c}}$$

$$+ \frac{\nu \sqrt{(\alpha x_3(k) + \beta x_4(k))^2 + \lambda}}{\frac{r_c \tau}{I_c}}$$

$$+ \frac{1}{\frac{r_c \tau}{I_c}} \Big((\gamma x_3(k) + \delta x_4(k) + \nu \sqrt{(\alpha x_3(k) + \beta x_4(k))^2 + \lambda})^2$$

$$+ 4 r_v (r_c G_v x_3(k) - r_v G_v x_4(k) + x_4(k))$$

$$\cdot (\alpha x_3(k) + \beta x_4(k) + \sqrt{(\alpha x_3(k) + \beta x_4(k))^2 + \lambda}) + \mu \Big)^{\frac{1}{2}},$$

$$(A.30)$$

where $\mu \triangleq \mu(x_2(k)) = \frac{r_c \tau}{I_c}(r_c \alpha_c + 4 r_v x_2(k))$. Finally, it follows from (A.5) using (A.17) and (A.30) that

$$\begin{bmatrix} x_2(k+1) \\ x_3(k+1) \\ x_4(k+1) \end{bmatrix} = A \begin{bmatrix} x_2(k) \\ x_3(k) \\ x_4(k) \end{bmatrix}$$

$$+ \begin{bmatrix} \frac{1}{\frac{r_c \tau}{I_c}} \tilde{g}(x_3(k), x_4(k))\bar{g}(x_2(k), x_3(k), x_4(k)) \\ (-G_c + \frac{1}{r_c})(\alpha x_3(k) + \beta x_4(k) + \sqrt{(\alpha x_3(k) + \beta x_4(k))^2 + \lambda}) \\ -G_v(\alpha x_3(k) + \beta x_4(k) + \sqrt{(\alpha x_3(k) + \beta x_4(k))^2 + \lambda}) \end{bmatrix}$$

$$+ \begin{bmatrix} \frac{1}{\frac{r_c \tau}{I_c}} \tilde{f}(x_2(k), x_3(k), x_4(k))\bar{f}(x_2(k), x_3(k), x_4(k)) \\ \frac{1}{r_c}(\gamma x_3(k) + \delta x_4(k) + \nu \sqrt{(\alpha x_3(k) + \beta x_4(k))^2 + \lambda}) \\ 0 \end{bmatrix}$$

$$+ \begin{bmatrix} 0 \\ \frac{1}{r_c} \hat{f}(x_2(k), x_3(k), x_4(k)) \\ 0 \end{bmatrix}, \qquad (A.31)$$

where

$$\tilde{g}(x_3(k), x_4(k)) = r_c G_v x_3(k) - r_v G_v x_4(k) + x_4(k),$$

$$\bar{g}(x_2(k), x_3(k), x_4(k)) = \alpha x_3(k) + \beta x_4(k)$$

$$+ \sqrt{(\alpha x_3(k) + \beta x_4(k))^2 + \lambda},$$

$$\tilde{f}(x_2(k), x_3(k), x_4(k)) = \xi x_3(k) + \zeta x_4(k)$$

$$- G_v \sqrt{(\alpha x_3(k) + \beta x_4(k))^2 + \lambda},$$

$$\bar{f}(x_2(k), x_3(k), x_4(k)) = \gamma x_3(k) + \delta x_4(k),$$

$$+\nu\sqrt{(\alpha x_3(k) + \beta x_4(k))^2 + \lambda}$$
$$+\hat{f}(x_2(k), x_3(k), x_4(k)),$$
$$\hat{f}(x_2(k), x_3(k), x_4(k)) = \Big((\gamma x_3(k) + \delta x_4(k)$$
$$+\nu\sqrt{(\alpha x_3(k) + \beta x_4(k))^2 + \lambda})^2$$
$$+4r_{\mathrm{v}}\tilde{g}(x_3(k), x_4(k))\bar{g}(x_2(k), x_3(k), x_4(k))$$
$$+\mu\Big)^{\frac{1}{2}},$$
$$\xi = r_{\mathrm{c}}^2 G_{\mathrm{v}}G_{\mathrm{c}} - r_{\mathrm{c}}G_{\mathrm{v}} - \alpha G_{\mathrm{v}} + (-r_{\mathrm{v}}G_{\mathrm{v}} + 1)r_{\mathrm{c}}G_{\mathrm{v}},$$
$$\zeta = -r_{\mathrm{c}}r_{\mathrm{v}}G_{\mathrm{c}}G_{\mathrm{v}} - \beta G_{\mathrm{v}} + (-r_{\mathrm{v}}G_{\mathrm{v}} + 1)^2.$$

Now, using (A.31) we can characterize the functions $f_2(x_2(k), x_3(k),$ $x_4(k))$, $f_3(x_2(k), x_3(k), x_4(k))$, and $f_4(x_2(k), x_3(k), x_4(k))$ appearing in (13.80); namely,

$$f_2(x_2(k), x_3(k), x_4(k)) = \hat{a}_{11}x_2(k) + \hat{a}_{12}x_3(k) + \hat{a}_{13}x_4(k)$$
$$+\frac{1}{\frac{r_{\mathrm{c}}\tau}{I_{\mathrm{c}}}}\tilde{g}(x_3(k), x_4(k))\bar{g}(x_2(k), x_3(k), x_4(k))$$
$$+\frac{1}{\frac{r_{\mathrm{c}}\tau}{I_{\mathrm{c}}}}\tilde{f}(x_2(k), x_3(k), x_4(k))$$
$$\cdot\bar{f}(x_2(k), x_3(k), x_4(k)), \tag{A.32}$$
$$f_3(x_2(k), x_3(k), x_4(k)) = \hat{a}_{21}x_2(k) + \hat{a}_{22}x_3(k) + \hat{a}_{23}x_4(k)$$
$$+\omega\sqrt{(\alpha x_3(k) + \beta x_4(k))^2 + \lambda}$$
$$+\frac{1}{r_{\mathrm{c}}}\hat{f}(x_2(k), x_3(k), x_4(k)), \tag{A.33}$$
$$f_4(x_2(k), x_3(k), x_4(k)) = \hat{a}_{31}x_2(k) + \hat{a}_{32}x_3(k) + \hat{a}_{33}x_4(k)$$
$$-G_{\mathrm{v}}\sqrt{(\alpha x_3(k) + \beta x_4(k))^2 + \lambda}, \tag{A.34}$$

where $\omega = \frac{\nu}{r_{\mathrm{c}}} - G_{\mathrm{c}} + \frac{1}{r_{\mathrm{c}}}$ and $\hat{a}_{ij}$ denotes the (i, j) component of the matrix $\hat{A}$ given by

$$\hat{A} = A + \begin{bmatrix} 0 & 0 & 0 \\ 0 & \alpha(-G_{\mathrm{c}} + \frac{1}{r_{\mathrm{c}}}) & \beta(-G_{\mathrm{c}} + \frac{1}{r_{\mathrm{c}}}) \\ 0 & -\alpha G_{\mathrm{v}} & -\beta G_{\mathrm{v}} \end{bmatrix} + \begin{bmatrix} 0 & 0 & 0 \\ 0 & \frac{\gamma}{r_{\mathrm{c}}} & \frac{\delta}{r_{\mathrm{c}}} \\ 0 & 0 & 0 \end{bmatrix}.$$
$$\tag{A.35}$$

Bibliography

[1] M. A. Aizerman and F. R. Gantmacher, *Absolute Stability of Regulator Systems*. San Francisco, CA: Holden-Day, 1964.

[2] B. D. O. Anderson, "A system theory criterion for positive real matrices," *SIAM J. Control Optim.*, vol. 5, pp. 171–182, 1967.

[3] B. D. O. Anderson and S. Vongpanitlerd, *Network Analysis and Synthesis: A Modern Systems Theory Approach*. Englewood Cliffs, NJ: Prentice-Hall, 1973.

[4] D. H. Anderson, *Compartmental Modeling and Tracer Kinetics*. New York, NY: Springer-Verlag, 1983.

[5] A. A. Andronov, A. A. Vitt, and S. E. Khaikin, *Theory of Oscillators*. New York, NY: Dover Publications, 1966.

[6] P. J. Antsaklis and A. Nerode, eds., "Special issue on hybrid control systems," *IEEE Trans. Autom. Control*, vol. 43, no. 4, 1998.

[7] P. J. Antsaklis, J. A. Stiver, and M. D. Lemmon, "Hybrid system modeling and autonomous control systems," in *Hybrid Systems* (R. L. Grossman, A. Nerode, A. P. Ravn, and H. Rischel, eds.), pp. 366–392, New York, NY: Springer-Verlag, 1993.

[8] T. M. Apostol, *Mathematical Analysis*. Reading, MA: Addison-Wesley, 1957.

[9] E. Awad and F. E. C. Culick, "On the existence and stability of limit cycles for longitudinal acoustic modes in a combustion chamber," *Combust. Sci. Technol.*, vol. 9, pp. 195–222, 1986.

[10] E. Awad and F. E. C. Culick, "The two-mode approximation to nonlinear acoustics in combustion chambers I. Exact solution for second order acoustics," *Combust. Sci. Technol.*, vol. 65, pp. 39–65, 1989.

[11] A. Back, J. Guckenheimer, and M. Myers, "A dynamical simulation facility for hybrid systems," in *Hybrid Systems* (R. L. Grossman, A. Nerode, A. P. Ravn, and H. Rischel, eds.), pp. 255–267, New York, NY: Springer, 1993.

[12] D. D. Bainov and P. S. Simeonov, *Systems with Impulse Effect: Stability, Theory and Applications.* Chichester, U.K.: Ellis Horwood, 1989.

[13] D. D. Bainov and P. S. Simeonov, *Impulsive Differential Equations: Periodic Solutions and Applications.* Essex, U.K.: Longman Scientific & Technical, 1993.

[14] D. D. Bainov and P. S. Simeonov, *Impulsive Differential Equations: Asymptotic Properties of the Solutions.* Singapore: World Scientific, 1995.

[15] E. A. Barbashin and N. N. Krasovskii, "On the stability of motion in the large," *Dokl. Akad. Nauk.*, vol. 86, pp. 453–456, 1952.

[16] M. Bardi and I. C. Dolcetta, *Optimal Control and Viscosity Solutions of Hamilton-Jacobi-Bellman Equations.* Boston, MA: Birkhauser, 1997.

[17] G. Barles, "Deterministic impulse control problems," *SIAM J. Control Optim.*, vol. 23, pp. 419–432, 1985.

[18] G. Barles, "Quasi-variational inequalities and first-order Hamilton-Jacobi equations," *Nonlinear Analysis*, vol. 9, pp. 131–148, 1985.

[19] R. Bellman, "Vector Lyapunov functions," *SIAM J. Control*, vol. 1, pp. 32–34, 1962.

[20] A. Berman, M. Neumann, and R. J. Stern, *Nonnegative Matrices in Dynamic Systems.* New York, NY: Wiley and Sons, 1989.

[21] A. Berman and R. J. Plemmons, *Nonnegative Matrices in the Mathematical Sciences.* New York, NY: Academic Press, 1979.

[22] D. S. Bernstein, "Feedback control: An invisible thread in the history of technology," *IEEE Control Syst. Mag.*, vol. 22, no. 2, pp. 53–68, 2002.

[23] D. S. Bernstein and S. P. Bhat, "Nonnegativity, reducibility, and semistability of mass action kinetics," in *Proc. IEEE Conf. Dec. Control* (Phoenix, AZ), pp. 2206–2211, 1999.

[24] D. S. Bernstein and W. M. Haddad, "Robust stability and performance analysis for state space systems via quadratic Lyapunov bounds," *SIAM J. Matrix Anal. Appl.*, vol. 11, pp. 236–271, 1990.

[25] D. S. Bernstein and D. C. Hyland, "Compartmental modeling and second-moment analysis of state space systems," *SIAM J. Matrix Anal. Appl.*, vol. 14, pp. 880–901, 1993.

[26] A. M. Bloch and J. E. Mardsen, "Stabilization of rigid body dynamics by the energy-Casimir method," *Syst. Control Lett.*, vol. 14, pp. 341–346, 1990.

[27] S. Boyd, L. E. Ghaoui, E. Feron, and V. Balakrishnan, *Linear Matrix Inequalities in System and Control Theory*. Philadelphia, PA: SIAM studies in applied mathematics, 1994.

[28] R. M. Brach, *Mechanical Impact Dynamics*. New York, NY: Wiley, 1991.

[29] M. S. Branicky, "Multiple-Lyapunov functions and other analysis tools for switched and hybrid systems," *IEEE Trans. Autom. Control*, vol. 43, pp. 475–482, 1998.

[30] M. S. Branicky, V. S. Borkar, and S. K. Mitter, "A unified framework for hybrid control: Model and optimal control theory," *IEEE Trans. Autom. Control*, vol. 43, pp. 31–45, 1998.

[31] R. W. Brockett, "Hybrid models for motion control systems," in *Essays in Control: Perspectives in the Theory and its Applications* (H. L. Trentleman and J. C. Willems, eds.), pp. 29–53, Boston, MA: Birkhäuser, 1993.

[32] B. Brogliato, *Nonsmooth Impact Mechanics: Models, Dynamics, and Control*. London, U.K.: Springer-Verlag, 1996.

[33] B. Brogliato, *Nonsmooth Mechanics*. London, U.K.: Springer-Verlag, 1999.

[34] B. Brogliato, *Impacts in Mechanical Systems*. Berlin: Springer-Verlag, 2000.

[35] R. T. Bupp, D. S. Bernstein, V. Chellaboina, and W. M. Haddad, "Resetting virtual absorbers for vibration control," *J. Vibr. Control*, vol. 6, pp. 61–83, 2000.

[36] R. T. Bupp, D. S. Bernstein, and V. T. Coppola, "A benchmark problem for nonlinear control design," *Int. J. Robust Nonlinear Control*, vol. 8, pp. 307–310, 1998.

[37] C. Byrnes, W. Lin, and B. K. Ghosh, "Stabilization of discrete-time nonlinear systems by smooth state feedback," *Syst. Control Lett.*, vol. 21, pp. 255–263, 1993.

[38] C. I. Byrnes and W. Lin, "Losslessness, feedback equivalence and the global stabilization of discrete-time nonlinear systems," *IEEE Trans. Autom. Control*, vol. 39, pp. 83–98, 1994.

[39] V. Chellaboina, S. P. Bhat, and W. M. Haddad, "An invariance principle for nonlinear hybrid and impulsive dynamical systems," *Nonlinear Analysis*, vol. 53, pp. 527–550, 2003.

[40] V. Chellaboina and W. M. Haddad, "Stability margins of discrete-time nonlinear-nonquadratic optimal regulators," *Int. J. Syst. Sci.*, vol. 33, pp. 577–584, 2002.

[41] E. A. Coddington and N. Levinson, *Theory of Ordinary Differential Equations*. New York, NY: McGraw-Hill, 1955.

[42] W. A. Coppel, *Stability and Asymptotic Behavior of Differential Equations*. Boston, MA: D. C. Heath and Co., 1965.

[43] F. E. C. Culick, "Nonlinear behavior of acoustic waves in combustion chambers I," *Acta Astronautica*, vol. 3, pp. 715–734, 1976.

[44] A. F. Filippov, *Differential Equations with Discontinuous Right-Hand Sides*. Mathematics and its applications (Soviet series), Dordrecht, The Netherlands: Kluwer Academic Publishers, 1988.

[45] D. M. Foster and M. R. Garzia, "Nonhierarchical communications networks: An application of compartmental modeling," *IEEE Trans. Communications*, vol. 37, pp. 555–564, 1989.

[46] R. A. Freeman and P. V. Kokotović, "Inverse optimality in robust stabilization," *SIAM J. Control Optim.*, vol. 34, pp. 1365–1391, 1996.

[47] R. E. Funderlic and J. B. Mankin, "Solution of homogeneous systems of linear equations arising from compartmental models," *SIAM J. Sci. Statist. Comput.*, vol. 2, pp. 375–383, 1981.

[48] W. J. Gazely, *Clock and Watch Escapements*. London, U.K.: Heywood, 1956.

[49] J. Gimpel, *The Medieval Machine: The Industrial Revolution of the Middle Ages*. New York, NY: Penguin Books, 1976.

[50] K. Godfrey, *Compartmental Models and their Applications*. New York, NY: Academic Press, 1983.

[51] R. Grossman, A. Nerode, A. Ravn, and H. Rischel, eds., *Hybrid Systems*. New York, NY: Springer-Verlag, 1993.

[52] L. T. Grujić, A. A. Martynyuk, and M. Ribbens-Pavella, *Large Scale Systems Stability Under Structural and Singular Perturbations*. Berlin: Springer-Verlag, 1987.

[53] W. M. Haddad and D. S. Bernstein, "Explicit construction of quadratic Lyapunov functions for the small gain, positivity, circle, and Popov theorems and their application to robust stability. Part I: Continuous-time theory," *Int. J. Robust. Nonlin. Control*, vol. 3, pp. 313–339, 1993.

[54] W. M. Haddad and D. S. Bernstein, "Explicit construction of quadratic Lyapunov functions for the small gain, positivity, circle, and Popov theorems and their application to robust stability. Part II: Discrete-time theory," *Int. J. Robust. Nonlin. Control*, vol. 4, pp. 249–365, 1994.

[55] W. M. Haddad and D. S. Bernstein, "Parameter-dependent Lyapunov functions and the Popov criterion in robust analysis and synthesis," *IEEE Trans. Autom. Control*, vol. 40, pp. 536–543, 1995.

[56] W. M. Haddad and V. Chellaboina, "Dissipativity theory and stability of feedback interconnections for hybrid dynamical systems," *J. Math. Prob. Engin.*, vol. 7, pp. 299–355, 2001.

[57] W. M. Haddad and V. Chellaboina, "Stability and dissipativity theory for nonnegative dynamical systems: A unified analysis framework for biological and physiological systems," *Nonlinear Analysis: Real World Applications*, vol. 6, pp. 35–65, 2005.

[58] W. M. Haddad, V. Chellaboina, and E. August, "Stability and dissipativity theory for discrete-time nonnegative and compartmental dynamical systems," *Int. J. Control*, vol. 76, pp. 1845–1861, 2003.

[59] W. M. Haddad, V. Chellaboina, and J. L. Fausz, "Robust nonlinear feedback control for uncertain linear systems with nonquadratic performance criteria," *Syst. Control Lett.*, vol. 33, pp. 327–338, 1998.

[60] W. M. Haddad, V. Chellaboina, J. L. Fausz, and A. Leonessa, "Optimal nonlinear robust control for nonlinear uncertain systems," *Int. J. Control*, vol. 73, pp. 329–342, 2000.

[61] W. M. Haddad, V. Chellaboina, and N. A. Kablar, "Nonlinear impulsive dynamical systems. Part I: Stability and dissipativity," *Int. J. Control*, vol. 74, pp. 1631–1658, 2001.

[62] W. M. Haddad, V. Chellaboina, and N. A. Kablar, "Nonlinear impulsive dynamical systems. Part II: Stability of feedback interconnections and optimality," *Int. J. Control*, vol. 74, pp. 1659–1677, 2001.

[63] W. M. Haddad, V. Chellaboina, and S. G. Nersesov, "Hybrid nonnegative and compartmental dynamical systems," *J. Math. Prob. Engin.*, vol. 8, pp. 493–515, 2002.

[64] W. M. Haddad, V. Chellaboina, and S. G. Nersesov, "A system-theoretic foundation for thermodynamics: Energy flow, energy balance, energy equipartition, entropy, and ectropy," in *Proc. Amer. Control Conf.* (Boston, MA), pp. 396–417, 2004.

[65] W. M. Haddad, V. Chellaboina, and S. G. Nersesov, *Thermodynamics: A Dynamical Systems Approach*. Princeton, NJ: Princeton University Press, 2005.

[66] W. M. Haddad, J. P. How, S. R. Hall, and D. S. Bernstein, "Extensions of mixed-μ bounds to monotonic and odd monotonic nonlinearities using absolute stability theory," *Int. J. Control*, vol. 60, pp. 905–951, 1994.

[67] W. M. Haddad, H.-H. Huang, and D. S. Bernstein, "Robust stability and performance via fixed-order dynamic compensation: The discrete-time case," *IEEE Trans. Autom. Control*, vol. 38, pp. 776–782, 1993.

[68] W. M. Haddad, Q. Hui, S. G. Nersesov, and V. Chellaboina, "Thermodynamic modeling, energy equipartition, and nonconservation of entropy for discrete-time dynamical systems," in *Proc. Amer. Control Conf.* (Portland, OR), pp. 4832–4837, 2005.

[69] W. M. Haddad, Q. Hui, S. G. Nersesov, and V. Chellaboina, "Thermodynamic modeling, energy equipartition, and nonconservation of entropy for discrete-time dynamical systems," *Adv. Diff. Eqs.*, vol. 2005, pp. 275–318, 2005.

[70] W. M. Haddad, S. G. Nersesov, and V. Chellaboina, "Energy-based control for hybrid port-controlled Hamiltonian systems," *Automatica*, vol. 39, pp. 1425–1435, 2003.

[71] T. Hagiwara and M. Araki, "Design of a stable feedback controller based on the multirate sampling of the plant output," *IEEE Trans. Autom. Control*, vol. 33, pp. 812–819, 1988.

[72] M. V. Headrick, "Origin and evolution of the anchor clock escapement," *IEEE Control Syst. Mag.*, vol. 22, no. 2, pp. 41–52, 2002.

[73] D. J. Hill and P. J. Moylan, "The stability of nonlinear dissipative systems," *IEEE Trans. Autom. Control*, vol. 21, pp. 708–711, 1976.

[74] D. J. Hill and P. J. Moylan, "Stability results for nonlinear feedback systems," *Automatica*, vol. 13, pp. 377–382, 1977.

[75] D. J. Hill and P. J. Moylan, "Dissipative dynamical systems: Basic input-output and state properties," *J. Franklin Inst.*, vol. 309, pp. 327–357, 1980.

[76] L. Hitz and B. D. O. Anderson, "Discrete positive-real functions and their application to system stability," *Proc. IEE*, vol. 116, pp. 153–155, 1969.

[77] R. A. Horn and R. C. Johnson, *Matrix Analysis*. Cambridge, U.K.: Cambridge University Press, 1985.

[78] R. A. Horn and R. C. Johnson, *Topics in Matrix Analysis*. Cambridge, U.K.: Cambridge University Press, 1995.

[79] S. Hu, V. Lakshmikantham, and S. Leela, "Impulsive differential systems and the pulse phenomena," *J. Math. Anal. Appl.*, vol. 137, pp. 605–612, 1989.

[80] A. Isidori, *Nonlinear Control Systems: An Introduction.* New York, NY: Springer-Verlag, 1989.

[81] A. Isidori, *Nonlinear Control Systems.* Berlin: Springer-Verlag, 1995.

[82] D. H. Jacobson, *Extensions of Linear-Quadratic Control Optimization and Matrix Theory.* New York, NY: Academic Press, 1977.

[83] J. A. Jacquez, *Compartmental Analysis in Biology and Medicine,* 2nd ed. Ann Arbor, MI: University of Michigan Press, 1985.

[84] J. A. Jacquez and C. P. Simon, "Qualitative theory of compartmental systems," *SIAM Rev.*, vol. 35, pp. 43–79, 1993.

[85] J. A. Jacquez, C. P. Simon, J. Koopman, L. Sattenspiel, and T. Perry, "Modeling and analyzing HIV transmission: The effect of contact patterns," *Math. Biosci.*, vol. 92, pp. 119–199, 1988.

[86] C. C. Jahnke and F. E. C. Culick, "Application of dynamical systems theory to nonlinear combustion instabilities," *J. Propulsion Power*, vol. 10, pp. 508–517, 1994.

[87] M. Jamshidi, *Large-Scale Systems.* Amsterdam: North-Holland, 1983.

[88] E. Kamke, "Zur Theorie der Systeme gewöhnlicher Differential-Gleichungen. II," *Acta Mathematica*, vol. 58, pp. 57–85, 1931.

[89] H. K. Khalil, *Nonlinear Systems.* Upper Saddle River, NJ: Prentice Hall, 1996.

[90] Y. Kishimoto and D. S. Bernstein, "Thermodynamic modeling of interconnected systems I: Conservative coupling," *J. Sound Vibr.*, vol. 182, pp. 23–58, 1995.

[91] N. N. Krasovskii, *Problems of the Theory of Stability of Motion.* Stanford, CA: Stanford University Press, 1959.

[92] G. K. Kulev and D. D. Bainov, "Stability of sets for systems with impulses," *Bull. Inst. Math. Academia Sinica*, vol. 17, pp. 313–326, 1989.

[93] V. Lakshmikantham, D. D. Bainov, and P. S. Simeonov, *Theory of Impulsive Differential Equations*. Singapore: World Scientific, 1989.

[94] V. Lakshmikantham, S. Leela, and S. Kaul, "Comparison principle for impulsive differential equations with variable times and stability theory," *Nonlinear Analysis*, vol. 22, pp. 499–503, 1994.

[95] V. Lakshmikantham and X. Liu, "On quasi stability for impulsive differential systems," *Nonlinear Analysis*, vol. 13, pp. 819–828, 1989.

[96] V. Lakshmikantham, V. M. Matrosov, and S. Sivasundaram, *Vector Lyapunov Functions and Stability Analysis of Nonlinear Systems*. Dordrecht, Netherlands: Kluwer Academic Publishers, 1991.

[97] D. S. Landes, *Revolution in Time: Clocks and the Making of the Modern World*. Cambridge, MA: Harvard University Press, 2000.

[98] J. P. LaSalle, "Some extensions of Liapunov's second method," *IRE Trans. Circ. Thy.*, vol. 7, pp. 520–527, 1960.

[99] J. P. LaSalle and S. Lefschetz, *Stability by Lyapunov's Direct Method*. New York, NY: Academic Press, 1961.

[100] S. Lefschetz, *Stability of Nonlinear Control Systems*. New York, NY: Academic Press, 1965.

[101] A. Leonessa, W. M. Haddad, and V. Chellaboina, "Nonlinear system stabilization via hierarchical switching control," *IEEE Trans. Autom. Control*, vol. 46, pp. 17–28, 2001.

[102] A. M. Lepschy, G. A. Mian, and U. Viaro, "Feedback control in ancient water and mechanical clocks," *IEEE Trans. Education*, vol. 35, pp. 3–10, 1992.

[103] W. Lin and C. I. Byrnes, "KYP lemma, state feedback and dynamic output feedback in discrete-time bilinear systems," *Syst. Control Lett.*, vol. 23, pp. 127–136, 1994.

[104] W. Lin and C. I. Byrnes, "Passivity and absolute stabilization of a class of discrete-time nonlinear systems," *Automatica*, vol. 31, pp. 263–267, 1995.

[105] X. Liu, "Quasi stability via Lyapunov functions for impulsive differential systems," *Applicable Analysis*, vol. 31, pp. 201–213, 1988.

[106] X. Liu, "Stability results for impulsive differential systems with applications to population growth models," *Dyn. Stab. Syst.*, vol. 9, pp. 163–174, 1994.

[107] R. Lozano, B. Brogliato, O. Egeland, and B. Maschke, *Dissipative Systems Analysis and Control*. London, U.K.: Springer-Verlag, 2000.

[108] K.-Y. Lum, D. S. Bernstein, and V. T. Coppola, "Global stabilization of the spinning top with mass imbalance," *Dyn. Stab. Syst.*, vol. 10, pp. 339–365, 1995.

[109] J. Lunze, "Stability analysis of large-scale systems composed of strongly coupled similar subsystems," *Automatica*, vol. 25, pp. 561–570, 1989.

[110] A. M. Lyapunov, *The General Problem of the Stability of Motion*. Kharkov, Russia: Kharkov Mathematical Society, 1892.

[111] A. M. Lyapunov, "Problème generale de la stabilité du mouvement," in *Annales de la Faculté des Sciences de e'Université de Toulouse* (É. Davaux, ed.), vol. 9, pp. 203–474, 1907. Reprinted by Princeton University Press, Princeton, NJ, 1949.

[112] A. M. Lyapunov, *The General Problem of Stability of Motion* (A. T. Fuller, trans. and ed.). Washington, DC: Taylor and Francis, 1992.

[113] J. Lygeros, D. N. Godbole, and S. Sastry, "Verified hybrid controllers for automated vehicles," *IEEE Trans. Autom. Control*, vol. 43, pp. 522–539, 1998.

[114] H. Maeda, S. Kodama, and F. Kajiya, "Compartmental system analysis: Realization of a class of linear systems with physical constraints," *IEEE Trans. Circuits Syst.*, vol. 24, pp. 8–14, 1977.

[115] H. Maeda, S. Kodama, and Y. Ohta, "Asymptotic behavior of nonlinear compartmental systems: Nonoscillation and stability," *IEEE Trans. Circuits Syst.*, vol. 25, pp. 372–378, 1978.

[116] A. A. Martynyuk, *Stability of Motion of Complex Systems.* Kiev: Naukova Dumka, 1975.

[117] A. A. Martynyuk, *Stability by Liapunov's Matrix Function Method with Applications.* New York, NY: Marcel Dekker, 1998.

[118] A. A. Martynyuk, *Qualitative Methods in Nonlinear Dynamics. Novel Approaches to Liapunov's Matrix Functions.* New York, NY: Marcel Dekker, 2002.

[119] V. M. Matrosov, "Method of vector Liapunov functions of interconnected systems with distributed parameters (Survey)," *Avtomatika Telemekhanika*, vol. 33, pp. 63–75, 1972 (in Russian).

[120] O. Mayr, *The Origins of Feedback Control.* Cambridge, MA: MIT Press, 1970.

[121] A. N. Michel and B. Hu, "Towards a stability theory of general hybrid systems," *Automatica*, vol. 35, pp. 371–384, 1999.

[122] A. N. Michel and R. K. Miller, *Qualitative Analysis of Large Scale Dynamical Systems.* New York, NY: Academic Press, 1977.

[123] V. D. Mil'man and A. D. Myshkis, "On the stability of motion in the presence of impulses," *Sib. Math. J.*, vol. 1, pp. 233–237, 1960.

[124] V. D. Mil'man and A. D. Myshkis, "Approximate methods of solutions of differential equations," in *Random Impulses in Linear Dyanmcial Systems*, pp. 64–81. Kiev: Publ. House. Acad. Sci. Ukr. SSR, 1963.

[125] R. R. Mohler, "Biological modeling with variable compartmental structure," *IEEE Trans. Autom. Control*, vol. 1974, pp. 922–926, 1974.

[126] A. S. Morse, C. C. Pantelides, S. Sastry, and J. M. Schumacher, eds., "Special issue on hybrid control systems," *Automatica*, vol. 35, no. 3, 1999.

[127] P. J. Moylan, "Implications of passivity in a class of nonlinear systems," *IEEE Trans. Autom. Control*, vol. 19, pp. 373–381, 1974.

[128] P. J. Moylan and B. D. O. Anderson, "Nonlinear regulator theory and an inverse optimal control problem," *IEEE Trans. Autom. Control*, vol. 18, pp. 460–465, 1973.

[129] R. J. Mulholland and M. S. Keener, "Analysis of linear compartmental models for ecosystems," *J. Theoret. Biol.*, vol. 44, pp. 105–116, 1974.

[130] A. Nerode and W. Kohn, "Models for hybrid systems: Automata, topologies, stability," in *Hybrid Systems* (R. L. Grossman, A. Nerode, A. P. Ravan, and H. Rischel, eds.), pp. 317–356, New York, NY: Springer-Verlag, 1993.

[131] H. Nijmeijer and A. J. van der Schaft, *Nonlinear Dynamical Control Systems*. New York, NY: Springer, 1990.

[132] Y. Ohta, H. Maeda, and S. Kodama, "Reachability, observability and realizability of continuous-time positive systems," *SIAM J. Control Optim.*, vol. 22, pp. 171–180, 1984.

[133] R. Ortega, A. Loria, R. Kelly, and L. Praly, "On output feedback global stabilization of Euler-Lagrange systems," *Int. J. Control*, vol. 5, pp. 313–324, 1995.

[134] R. Ortega, A. Loria, P. J. Nicklasson, and H. Sira-Ramírez, *Passivity-Based Control of Euler-Lagrange Systems*. London, U.K.: Springer-Verlag, 1998.

[135] R. Ortega and M. W. Spong, "Adaptive motion control of rigid robots: A tutorial," *Automatica*, vol. 25, pp. 877–888, 1989.

[136] R. Ortega, A. van der Schaft, and B. Maschke, "Stabilization of port-controlled Hamiltonian systems via energy balancing," in *Stability and Stabilization of Nonlinear Systems* (D. Aeyels, F. Lamnabhi-Lagarrigue, and A. van der Schaft, eds.), London, U.K.: Springer-Verlag, 1999.

[137] R. Ortega, A. van der Schaft, B. Maschke, and G. Escobar, "Interconnection and damping assignment passivity-based control of port-controlled Hamiltonian systems," *Automatica*, vol. 38, pp. 585–596, 2002.

[138] R. Ortega, A. van der Schaft, B. Maschke, and G. Escobar, "Energy-shaping of port-controlled Hamiltonian systems by interconnection," in *Proc. IEEE Conf. Dec. Control* (Phoenix, AZ), pp. 1646–1651, December 1999.

[139] K. M. Passino, A. N. Michel, and P. J. Antsaklis, "Lyapunov stability of a class of discrete event systems," *IEEE Trans. Autom. Control*, vol. 39, pp. 269–279, 1994.

[140] P. Peleties and R. DeCarlo, "Asymptotic stability of m-switched systems using Lyapunov-like functions," in *Proc. Amer. Control Conf.* (Boston, MA), pp. 1679–1684, 1991.

[141] L. Penman, *Practical Clock Escapements*. Shingle Springs, CA: Clockworks Press, 1998.

[142] V. M. Popov, *Hyperstability of Control Systems*. New York, NY: Springer, 1973.

[143] S. Prajna, A. J. van der Schaft, and G. Meinsma, "An LMI approach to stabilization of linear port-controlled Hamiltonian systems," *Syst. Control Lett.*, vol. 45, pp. 371–385, 2002.

[144] N. Rouche, P. Habets, and M. Laloy, *Stability Theory by Liapunov's Direct Method*. New York, NY: Springer, 1977.

[145] A. V. Roup, D. S. Bernstein, S. G. Nersesov, W. M. Haddad, and V. Chellaboina, "Limit cycle analysis of the verge and foliot clock escapement using impulsive differential equations and Poincaré maps," *Int. J. Control*, vol. 76, pp. 1685–1698, 2003.

[146] H. Royden, *Real Analysis*. Englewood Cliffs, NJ: Prentice Hall, 1988.

[147] M. G. Safonov, *Stability and Robustness of Multivariable Feedback Systems*. Cambridge, MA: MIT Press, 1980.

[148] A. M. Samoilenko and N. A. Perestyuk, *Impulsive Differential Equations*. Singapore: World Scientific, 1995.

[149] W. Sandberg, "On the mathematical foundations of compartmental analysis in biology, medicine and ecology," *IEEE Trans. Circuits Syst.*, vol. 25, pp. 273–279, 1978.

[150] S. Shishkin, R. Ortega, D. Hill, and A. Loria, "On output feedback stabilization of Euler-Lagrange systems with nondissipative forces," *Syst. Control Lett.*, vol. 27, pp. 315–324, 1996.

[151] D. D. Šiljak, *Large-Scale Dynamic Systems: Stability and Structure.* New York, NY: Elsevier/North-Holland, 1978.

[152] D. D. Šiljak, "Complex dynamical systems: Dimensionality, structure and uncertainty," *Large Scale Systems*, vol. 4, pp. 279–294, 1983.

[153] P. S. Simeonov and D. D. Bainov, "The second method of Lyapunov for systems with an impulse effect," *Tamkang J. Math.*, vol. 16, pp. 19–40, 1985.

[154] P. S. Simeonov and D. D. Bainov, "Stability with respect to part of the variables in systems with impulse effect," *J. Math. Anal. Appl.*, vol. 124, pp. 547–560, 1987.

[155] V. A. Sinitsyn, "On stability of solution in inertial navigation problem," *Certain Problems on Dynamics of Mechanical Systems* (Moscow), pp. 46–50, 1991.

[156] D. Sobel and W. J. H. Andrewes, *The Illustrated Longitude.* New York, NY: Walker and Co., 1998.

[157] L. Tavernini, "Differential automata and their discrete simulators," *Nonlinear Analysis*, vol. 11, pp. 665–683, 1987.

[158] C. Tomlin, G. J. Pappas, and S. Sastry, "Conflict resolution for air traffic management: A study in multiagent hybrid systems," *IEEE Trans. Autom. Control*, vol. 43, pp. 509–521, 1998.

[159] A. van der Schaft, "Stabilization of Hamiltonian systems," *Nonlinear Analysis*, vol. 10, pp. 1021–1035, 1986.

[160] A. van der Schaft, L_2-*Gain and Passivity Techniques in Nonlinear Control.* London, U.K.: Springer-Verlag, 2000.

[161] A. van der Schaft and H. Schumacher, *An Introduction to Hybrid Dynamical Systems.* London, U.K.: Springer-Verlag, 2000.

[162] M. Vidyasagar, *Nonlinear Systems Analysis.* Englewood Cliffs, NJ: Prentice-Hall, 1993.

[163] V. I. Vorotnikov, *Partial Stability and Control.* Boston, MA: Birkhäuser, 1998.

[164] T. Ważewski, "Systèmes des équations et des inégalités différentielles ordinaires aux deuxiémes membres monotones et leurs applications," *Ann. Soc. Pol. Math.*, vol. 23, pp. 112–166, 1950.

[165] J. C. Willems, "Dissipative dynamical systems. Part I: General theory," *Arch. Rational Mech. Anal.*, vol. 45, pp. 321–351, 1972.

[166] J. C. Willems, "Dissipative dynamical systems. Part II: Linear systems with quadratic supply rates," *Arch. Rational Mech. Anal.*, vol. 45, pp. 352–393, 1972.

[167] H. S. Witsenhausen, "A class of hybrid-state continuous-time dynamic systems," *IEEE Trans. Autom. Control*, vol. 11, pp. 161–167, 1966.

[168] T. Yang, *Impulsive Control Theory*. Berlin: Springer-Verlag, 2001.

[169] H. Ye, A. N. Michel, and L. Hou, "Stability theory of hybrid dynamical systems," *IEEE Trans. Autom. Control*, vol. 43, pp. 461–474, 1998.

[170] H. Ye, A. N. Michel, and L. Hou, "Stability analysis of systems with impulsive effects," *IEEE Trans. Autom. Control*, vol. 43, pp. 1719–1723, 1998.

[171] G. Zames, "On the input-output stability of time-varying nonlinear feedback systems, part I: Conditions derived using concepts of loop gain, conicity, and positivity," *IEEE Trans. Autom. Control*, vol. 11, pp. 228–238, 1966.

[172] G. Zames, "On the input-output stability of time-varying nonlinear feedback systems, part II: Conditions involving circles in the frequency plane and sector nonlinearities," *IEEE Trans. Autom. Control*, vol. 11, pp. 465–476, 1966.

[173] V. I. Zubov, *The Dynamics of Controlled Systems*. Moscow: Vysshaya Shkola, 1982.

Index

adiabatically isolated system, 276
admissible stabilizing controllers, 398
asymptotically stable, 21, 56
asymptotically stable equilibrium
    point, 9
asymptotically stable matrix, 151
asymptotically stable periodic orbit,
    446
asymptotically stable with respect to
    x_1, 46
asymptotically stable with respect to
    x_1 uniformly in x_{20}, 46
available exponential storage, 89
available exponential storage of a
    left-continuous dynamical system,
    428
available storage, 88
available storage of a left-continuous
    dynamical system, 428

beating, 2, 14, 420
Bellman's principle of optimality, 320
bilinear matrix inequalities, 335
bounded, 70
bounded trajectory, 29
bounded with respect to x_1 uniformly
    in x_2, 64
boundedness, 63

cascade interconnection, 437
class $\mathcal{K}$ function, 48
class $\mathcal{K}_\infty$ function, 48
closed system, 134
combustion control, 199
combustion systems, 199
compartmental systems, 125
completely reachable left-continuous
    dynamical system, 429
completely reachable system, 90
confluence, 2, 14, 420
connected system, 272
connective stability, 148
conservation of momentum, 224
continuous-time dynamics, 12

controller kinetic energy, 267
controller Lagrangian, 267
controller potential energy, 267

damping injection, 250
deadlock, 2
derived cost functional, 331
determinism axiom, 413
disconnected system, 152, 272
disk margin guarantees, 337
dissipation matrix, 152
dissipation rate function, 259
dissipative dynamical system, 81
dissipative dynamical systems, 250
dissipative left-continuous dynamical
    system, 427
dissipative nonnegative system, 136
dissipative system, 88
dynamical invariants, 236

emulated energy, 251, 260, 268
energy shaping, 249
energy-based hybrid control, 251, 267
energy-Casimir functions, 236
entropy function, 273
equilibrium point, 12, 414
essentially nonnegative function, 127
essentially nonnegative matrix, 72, 127
Euler-Lagrange equations, 266
Euler-Lagrange system, 265
exponential storage function, 90
exponential storage function of a
    left-continuous dynamical system,
    429
exponentially dissipative
    left-continuous dynamical system,
    428
exponentially dissipative nonnegative
    system, 136
exponentially dissipative system, 88
exponentially nonaccumulative system,
    142
exponentially nonexpansive
    left-continuous dynamical system,